from: Wynne
to: Kim + Mike
X-mas "94"

The New York Public Library
BOOK OF CHRONOLOGIES

Bruce Wetterau

A STONESONG PRESS BOOK

New York London Toronto Sydney Tokyo Singapore

Publisher Anne Zeman
Senior Editor Kate Kelly
Managing Editor William Hamill
Associate Managing Editor Jill Schoenhaut
Production Manager Robin B. Besofsky
Production Editor Lisa Wolff
Senior Editorial Assistant Susan Lauzau
Indexer Kathy Hohenstein
Designer MRP Design

First Edition

Prentice Hall Press

Simon & Schuster, Inc.
15 Columbus Circle
New York, NY 10023

Manufactured in the United States of America

1 2 3 4 5 6 7 8 9 10

Library of Congress Cataloging-in-Publication Data

Wetterau, Bruce.
The New York Public Library book of chronologies / Bruce Wetterau.
—1st ed.
p. cm.
"A Stonesong Press book."
Includes bibliographical references and index.
ISBN 0-13-620451-1
1. Chronology. Historical. I. New York Public Library.
II. Title.
D11.W47 1990
902'.02—dc20 90-46768
CIP

Introduction

Julius Caesar. William Shakespeare. Galileo. These are among the greatest people in history who have, to borrow the words of Longfellow, left their "footprints in the sands of time." As important as these men are, though, there are multitudes of other people from the past who concern us too. Nor can we forget the fascinating swirl of events, deeds, and works of centuries past. That is the rich diversity, and the great burden, of history. How do you organize such a vast accumulation of factual information so that one can learn from it?

Time itself is the key. Historians long ago discovered the secret of compiling historical material into a simple record of events arranged by date. This chronological presentation not only organizes the material logically one item after another, it also lets historians see how events and trends develop over time.

The *New York Public Library Book of Chronologies* uses this chronological scheme to present in clearest possible terms a vast array of reference information on history and culture. Unlike other chronologies of important dates and events, however, this book features chronologies written for specific subjects. In its 14 chapters, you will find over 250 separate chronologies on topics as diverse as literature, Antarctic exploration, the development of airplanes, football, and famous movies.

The book's distinctive arrangement has some important advantages. As with other chronologies, you can always find information about a person or event by looking up the date under the appropriate subject heading. But because material is arranged by specific topics, you can also easily peruse the whole history of the topic, such as ancient Rome, if you want a review of important events related to it. The detailed Table of Contents lists all the chronologies, section by section. In addition, the extensive index gives you another way to find information—by name. With it you can quickly locate the chronology or chronologies that contain such information as the date when the Battle of Bunker Hill was fought, when Charles Dickens wrote *A Christmas Carol,* or when Thomas Edison invented his light bulb.

The *Book of Chronologies* has been written primarily for the general reader, though historians and other scholars will find it a useful addition to their reference shelves. Because it is so easy to use and contains such a wide range of dates-and-events information, the book is invaluable for fact-checking and quick review. Readers, students, writers, editors, and even puzzle players and trivia buffs will find it especially helpful in this regard.

Even the entries within a specific chronology have been written in as clear and concise terms as possible for this kind of quick reference use. For example, the entry below is from the chronology for "Manned Space Exploration." Each entry, of course, begins with a year (1) and a brief sentence or two of text (2) on the event, person, or other subject matter relating to manned space exploration. Where appropriate, months and dates (3) have been included for selected, well-known events. Birth and death dates (4) have also been included selectively.

①

1967 Flash fire in *Apollo I* space capsule causes first deaths of American astronauts—Virgil I. Grissom (1926–67), Edward H. White II (1930–67), and Roger Chaffee (1935–67)—during routine test at Cape Kennedy, Florida (January 27). Less combustible mixture of oxygen and nitrogen is used for capsule atmosphere thereafter.

④ ③ ②

To supplement the major chronologies, feature items have been displayed in boxes in the text. These contain additional useful reference information or interesting sidelights. For example, in the section on sports, you will find chronological listings of World Series winners and of Baseball Hall of Fame inductees alongside the chronology of key events in the history of baseball.

Every effort was made to produce a book that is a joy to browse, as well as to provide useful reference information. Here and there among the significant dates of history, you will find some unusual items to pique your interest or make you smile: perhaps it will be the first and only flight of Howard Hughes's mammoth "Spruce Goose," or possibly Columbus's obvious (and understandable) confusion about the new lands he discovered—on his final voyage to the New World he declared he was near China, when in fact he was discovering Panama.

And whatever your interests, certain chronologies are sure to catch your eye. Sports fans will enjoy the concise histories of football, baseball, basketball, ice hockey, and more. Science buffs will find an entire section of chronologies covering major advances in each branch of science. Literary lions, artists, musicians, philosophers, and educators are covered, as well as the media of today's world—complete with extensive chronologies of memorable movies and TV shows.

A book of this size and complexity would have required years to research and write single-handedly. Ann Marr, Bruce Beasley, Katherine Somervell, Lilian Brady, Cathy Gedvilaz, Leila Finn, Margaret and George Klosko, and Christine Brant deserve special thanks for their work on researching and writing original chronologies. Laurel Adams, Susan Carter, and Joanna Pounds keyboarded much of the manuscript on computer disks. Everyone on the project shouldered some of the burden of research at one time or another, but Leila Finn, Ronda Wiley, Cynthia Remington, Lilian Brady, Mary St. John, and Laurel Adams took care of a number of special assignments.

Susan Lauzau, Lisa Wolff, and Bob Gilbert did additional fact-checking and made helpful recommendations for additional entries, while copyeditor Bill Drennan gave the manuscript a final polishing. I would also like to thank the New York Public Library for giving the manuscript a careful reading.

Paul Fargis of the Stonesong Press came up with the original idea for this book and has been involved in various aspects of the project since. Lastly, I would like to thank both my editors at Prentice Hall, Ken Wright and Kate Kelly, for their part in bringing this project along from promising idea to bound book.

Every effort has been made to ensure the accuracy and appropriateness of material contained in this book. To that end, before the compiling and cross-checking of chronologies were completed, well over 150 different sources were consulted, including numerous reference works, news indexes, periodicals, and government publications. But in a work of this scope, errors and omissions do sometimes occur. Please let us know if you happen to find them, so that the text of future printings can be corrected.

Bruce Wetterau

Table of Contents

1. **Explorers and Exploration / 1**
   - Exploring the World / 3
     - Voyages of Columbus / 4
     - Voyages of Magellan / 6
     - Discovering the Natural Wonders of North America / 7
     - The Lewis and Clark Expedition / 9
   - Exploring Africa / 13
   - Arctic Exploration / 15
   - Antarctic Exploration / 17
   - The Underwater Frontier / 19
   - Manned Space Exploration / 21
     - Getting Space Exploration Off the Ground / 22
2. **Nations and Empires / 27**
   - North America / 29
     - U.S. Presidents and Vice Presidents / 30
     - Canadian Prime Ministers / 32
     - Mexico's Presidents and Heads of State / 35
   - U.S. Territorial Expansion / 40
   - Central and South America / 42
   - Europe / 45
     - Holy Roman Emperors / 46
     - The British Monarchs / 47
     - French Rulers and Heads of State / 48
     - Spanish Rulers and Heads of State / 52
     - British Prime Ministers / 55
     - German Rulers and Heads of State / 58
     - Russian and Soviet Rulers and Heads of State / 59
     - Italian Rulers and Heads of State / 60
   - Middle East / 70
   - Africa / 72
   - Asia and Pacific / 76
     - Chinese Emperors and Heads of State / 76
     - Japanese Rulers and Heads of State / 78
     - Indian Rulers and Heads of State / 81
     - Australian Heads of State / 85
   - Ancient Mesopotamia / 91
     - Ancient Assyrian and Babylonian Rulers / 93
   - Ancient Greece / 94
   - Ancient Egypt / 95
     - Rulers of Ancient Egypt / 97
   - Roman Empire / 98
     - Roman Emperors / 101
   - Byzantine Empire / 103
   - French Colonial Empire / 108
   - British Empire / 110
3. **Politics and Law / 113**
   - Documents That Shaped World History / 115
   - Historic Treaties / 116
   - U.S. Constitution and Amendments / 119
   - U.S. Presidential Elections / 120
   - Composition of U.S. Congresses / 123
   - The United Nations / 126
     - Nobel Peace Prize Winners / 127
   - The Abolition of Slavery / 130
   - Civil Rights in the United States / 131
   - Women's Rights / 133
   - Major U.S. Supreme Court Decisions / 135
   - U.S. Supreme Court Justices / 136
   - U.S. Criminal Justice System / 138
   - Political Scandals Since 1900 / 139
   - Historic Assassinations / 141
   - Uprisings and Revolts / 142
   - Socialism and Communism / 144
4. **U.S. Business, Commerce, and Economics / 147**
   - Origins of Major U.S. Businesses / 149
   - U.S. Business Magnates and Tycoons / 152
   - The Developing U.S. Economy / 154
     - Winners of the Nobel Prize in Economics / 157
   - New York Stock Exchange / 157
   - U.S. Labor Movement / 159
   - U.S. Chemical Industry / 161
   - U.S. Railroad Industry / 163
   - U.S. Steel Industry / 165
   - U.S. Oil Industry / 168
   - U.S. Automobile Industry / 170
   - U.S. Airline Industry / 171
5. **Technology / 175**
   - Basic Technology / 177
   - Prehistoric Tools and Implements / 190
   - Advances in Agriculture / 192
   - Development of Ships / 195
   - Development of Submarines / 198
   - Clocks and Other Time-Telling Devices / 199
   - Printing and Copying / 201
   - Age of Steam / 203
   - Railroad Development / 204
   - Key Developments in the Industrial Revolution / 205
   - Plastics and Synthetics / 207
   - Photography / 209
   - Electricity / 211
   - Telegraph and Telephone / 214
   - Radio and Television / 216
   - Development of Automobiles / 218
     - Internal Combustion Engine / 220
   - Development of Airplanes / 221
   - Rockets / 224
   - Nuclear Technology / 226
   - Computers / 227
   - The Quest for Speed / 229
6. **The Arts / 231**
   - Novelists, Poets, and Essayists / 233
     - Pulitzer Prize for Poetry / 234
     - Pulitzer Prize for Fiction / 236
     - Pulitzer Prize for History / 239
     - Pulitzer Prize for Biography or Autobiography / 244
     - Nobel Prize for Literature / 254
   - Painters and Sculptors / 263
     - Art of Ancient Greece and Rome / 264
   - Playwrights / 278
     - The Greek and Roman Dramatists / 279
     - Development of the Theater from Elizabethan Times / 281
     - Shakespeare's Plays / 284
     - Pulitzer Prize for Drama / 285
   - Composers / 289
     - Development of Western Music / 290
     - Development of Western Musical Instruments / 292
     - History of European and American Opera / 297
7. **Religion, Philosophy, and Education / 303**
   - The Bible / 305
   - Roman Catholicism / 306
     - Popes / 310
   - Eastern Orthodoxy / 313
     - Ecumenical Councils / 314
   - Protestantism / 315

Anglicanism/Episcopalianism / 317
- Archbishops of Canterbury / 319

Presbyterian Church / 320
Baptist Church / 322
Methodist Church / 323
Judaism / 325
Buddhism / 328
Hinduism / 330
Islam / 332
Confucianism / 334
Philosophy / 335
Education / 339
- U.S. and Canadian Colleges and Universities / 341

8. Architecture and Engineering Feats / 345
- Mounds and Megaliths / 348
- Temples, Palaces, and Monuments of the Ancient World / 349
- Castles and Cathedrals / 351
- Soaring Spans: Bridges Since 1800 / 356
- Taller than Ever Before—Developing the Skyscrapers / 358

9. Science / 361
Medicine / 363
- The Fight Against Disease / 364
- Medications / 365
- Modern Dentistry / 366
- Psychology / 368
- Winners of the Nobel Prize for Physiology or Medicine / 370

Astronomy / 375
- Discovering Our Solar System / 378

Biology / 385
- Unraveling the Mysteries of the Cell / 386
- Evolving Plant and Animal Life / 388
- Timetable of Mammal Evolution / 391

Anthropology and Archaeology / 392
- The Development of Man / 396

Chemistry / 396
- Discovering the Elements / 397
- Winners of the Nobel Prize for Chemistry / 401

Physics / 403
- Winners of the Nobel Prize for Physics / 405

Mathematics / 411
Earth Sciences / 414
- Major Geologic Epochs / 415
- Meteorology / 417
- Exploring Earth's Upper Atmosphere / 420

Environmentalism / 422
- Endangered Species / 423

10. Necessities to Notoriety: An Omnium Gatherum / 425
Foods and Beverages / 427
Clothing: From Egyptian Robes to French Bikinis / 429
Holidays and Festivals / 434
Fads and Crazes Since 1900 / 435
Basic Household Items / 437
Furniture and Furniture Styles / 438
Customs and Traditions / 440
Villains: Crimes and Criminals Since 1900 / 441

11. War and Military History / 445
Wars, from Ancient Greece to the Modern World / 447
English Civil War / 453
American Revolution / 454
French Revolution and Revolutionary Wars / 456
Napoleonic Wars / 458
American Civil War / 460
World War I / 462
Russian Revolution and Civil War / 465
Chinese Civil War / 466
World War II / 468
Korean War / 471
Vietnam War / 472
Battles from Ancient to Modern Times / 475
Military Techniques and Technology / 489
- Military Small Arms Development / 494
- Tanks and Troop Carriers in the 20th Century / 496
- Warships / 497
- Warplanes / 500

12. Accidents and Disasters / 503
Air Disasters / 505
Train Wrecks and Auto Accidents / 507
Fires / 511
Ship Sinkings and Other Accidents at Sea Since 1900 / 513
Natural Disasters / 515
Plagues and Epidemics / 527

13. Media, Entertainment, and Contemporary Music / 531
Newspapers and Magazines / 533
- Pulitzer Prize for Reporting / 535

Radio and Television Broadcasting / 536
Book Publishing / 545
- Bestsellers 1945–1988 / 547

Motion Pictures Industry / 552
- Academy Awards / 553

Recording Industry / 567
- Grammy Awards / 568
- Jazz Greats / 572

14. Sports / 575
Football / 577
- National College Football Champions / 577
- Heisman Trophy Winners / 578
- NFL and AFL Championships / 579
- Super Bowl Game / 581
- Pro Football Hall of Fame Members / 583

Basketball / 586
- NCAA Championships / 587
- NBA Championships / 589

Baseball / 590
- World Series / 591
- National Baseball Hall of Fame Members / 595

Ice Hockey / 598
- Stanley Cup Champions / 599

Soccer / 602
- World Cup Soccer Championships / 602
- NASL Championships / 603

Golf / 605
- United States Open Championships / 605
- Masters Tournament / 607
- Professional Golfer's Association Champions / 608
- British Open / 609

Tennis / 610
- Davis Cup Tournament Championships / 611
- Wimbledon Champions / 613
- United States Open Champions / 615
- French Open Champions / 617

Heavyweight Boxing Championships / 619
America's Cup Races / 619
Olympic Games / 620

Selected Bibliography / 622

Index / 625

1 Explorers and Exploration

Exploring the World

B.C.

c. 2600 First recorded seagoing voyage. Egyptian sailors travel to Byblos in Phoenicia in search of cedarwood.

c. 2000 Mesopotamian traders journey as far east as India. Egyptians trade with Nubia, Ethiopia, and Crete.

c. 1493 Egyptian voyagers, at behest of Queen Hatshepsut (d. 1481 B.C.), journey down Red Sea to Punt (probably present-day Somaliland) in search of myrrh trees.

c. 800 Phoenician traders have probably established trade routes to Gadir, on Atlantic coast of Spain, by this time. Phoenicians, whose colony of Carthage for centuries controls access to western Mediterranean, refuse to allow ships other than their own to sail through Strait of Gibraltar to Atlantic coasts of Europe or Africa.

c. 650 Colaeus, Greek merchant, discovers Strait of Gibraltar for Greece.

c. 700–500 Greek city-states actively engage in setting up colonies in Mediterranean, notably those of Syracuse and Tarentum in modern Italy.

c. 600 Greeks from Ionian city of Phocaea found ancient city of Massilia (modern Marseilles) on distant Atlantic coast of France.

c. 600 Phoenicians, sent by Egyptian pharaoh Necho II (d. 593), sail around Africa about this time, according to contemporary accounts. Sailors take two years to complete voyage from Red Sea, around Africa, and back to Mediterranean by way of Strait of Gibraltar.

510–507 Persian king Darius (550–486 B.C.), having conquered northwestern India, sends Greek Scylax of Caryanda on exploratory voyage down Indus. Scylax sails to its mouth and then follows coast to Red Sea.

c. 500 Dorian and Ionian Greeks colonize Crimean region of Black Sea, far to east of their homelands.

c. 500? Carthaginian Hanno sails with 60 ships on exploratory voyage down western coast of Africa, bringing 30,000 men and women to establish new colonies. Hanno probably ranges as far south as modern Gambia, and he founds six Carthaginian cities on the way.

c. 500 Greek Herodotus (484?–c. 420?) writes and maps his *Histories,* offering complete account of his Mediterranean journeys.

485–465 Persian prince attempts voyage around Africa, beginning at Strait of Gibraltar. He may have sailed as far south along Africa's western coast as the equator before turning back.

334–323 Conquests by Macedonian Greek Alexander the Great (356–323 B.C.) foster greater awareness of eastern lands—notably India, at eastern edge of Alexander's new (and short-lived) empire.

302–290 Greek traveler Megasthenes (c. 350–c. 390 B.C.) journeys to India at various times during this period, going as far east as Ganges River. He provides first written mention of island of Ceylon but does not travel there.

c. 300 Greek navigator Pytheas explores Atlantic coast of Europe and British Isles. His epic voyage also takes him north of modern-day Great Britain to land he calls Thule, which may be Norway or Iceland. On his voyage into Arctic waters, Pytheas sees what he calls the "ever-shining sun" (possibly Arctic summer's midnight sun) and seas as thick as jelly (probably clogged with ice).

285–246 Egyptian ruler Ptolemy II Philadelphus (c. 308–246?) reigns. He orders exploration of Red Sea's western coast to establish anchorages for trading vessels.

c. 146 Polybius (c. 200–118 B.C.), Greek historian, begins his voyage to explore

▶

The Voyages of Columbus

1484 Portuguese court refuses proposal by Christopher Columbus (1451–1506) for voyage seeking westward route to China and Japan.

1486 Columbus presents his proposal to King Ferdinand (1452–1516) and Queen Isabella (1451–1504) of Spain. They establish commission to study the idea.

1490 Commission recommends against funding Columbus's voyage.

1492 Columbus is recalled to Spanish court. He demands to be given titles of knight, grand admiral, and viceroy, as well as share of profits from his discoveries. Though stunned by Columbus's terms, the king and queen eventually meet them.

1492 Columbus's first voyage is launched (August 3) from Palcs with crew of 90 aboard three ships (*Pinta, Nina,* and flagship, *Santa Maria*). Columbus carries royal letters to present to Great Khan and other Oriental sovereigns.

1492 Expedition reaches Canary Islands (August 12). It remains there until September 6.

1492 Columbus writes, "This day we completely lost sight of land, and many men sighed and wept for fear they would not see it again for a long time" (September 9).

1492 Crew becomes restless with fear there will be no wind for return voyage (September 24). "They have said it is insanity and suicidal on my part to risk their lives," Columbus writes.

1492 Weeks pass with no sight of land. Martin Alonso Pinzon (d. 1493), captain of *Pinta,* talks Columbus into altering course toward southwest (October 7). Crew is encouraged by flocks of birds heading in that direction, but days pass with only false sightings of land.

1492 Land is finally sighted amid shouts of "*Tierra! Tierra!*" (October 12). Columbus lands on island he names San Salvador (in modern Bahamas), believing he has found islands lying just off the Orient. He ignores gold jewelry worn by natives and sails on in search of Japan.

1492 Columbus discovers Cuba (October 28). He searches unsuccessfully for Great Khan of Orient and for gold. Discovers tobacco.

1492 *Pinta* becomes separated from other ships (November 21). Columbus believes Pinzon has decided to search for gold on his own.

1492 Columbus discovers Haiti, naming it Española.

1492 *Santa Maria* is wrecked on coast of Española (Christmas day). Columbus leaves 38 men there with enough food and ammunition for one year.

1493 Columbus sails for Spain in *Nina* (January 4).

1493 *Nina* catches up with *Pinta* (January 6) on return voyage. Pinzon apologizes; Columbus writes that Pinzon "acted with greed and arrogance" and was "disloyal and unworthy."

1493 Columbus fears death during terrible storms while recrossing the Atlantic. He seals record of his discoveries in a cask and casts it into sea in case he is drowned (February 14).

1493 Columbus reaches Portugal (March 9) and Spain (March 15).

1493 Pinzon dies (March 20).

1493 Ferdinand and Isabella receive Columbus and bestow wealth and honors on him (April).

1493 Second voyage is launched (September 25) with 17 ships and over 1,000 passengers. Expedition is launched to establish colony at Española and base for exploring China and Japan.

1493 Land is sighted at Marigalante (November 3). Expedition later discovers Puerto Rico and Guadeloupe (November).

1493 Columbus reaches Española and finds that all Spaniards have been killed by natives. He founds settlement called Isabella, first European city in the New World, in a new location. Leaves his brother Diego (1450–1515?) in charge.

1494 Columbus sails west to seek the mainland (April 24). He discovers Jamaica (May 5).

1494 Columbus sails toward Cuba to discover whether it is the mainland. He turns back (June 13), forcing his entire crew to swear an oath that it is the mainland.

1494 Columbus returns to new colony of Isabella (September 29).

1495 Columbus sets out to "pacify" natives (March 24) and takes 500 prisoners, whom he sends to Spain.

1495 Ferdinand and Isabella send Juan de Aguado to new colony of Isabella to investigate the situation (October). Power struggle ensues between Columbus and Aguado.

1496 Columbus returns to Spain (June 11), navigating first two ships built in New World to arrive in Europe.

The Voyages of Columbus

1498	Third voyage is launched (May 30), with six ships and over 200 passengers.
1498	Columbus changes course (July 13) and thus narrowly misses discovery of Amazon.
1498	Columbus discovers Trinidad and briefly lands on mainland of Venezuela (July), mistaking it for an island. At Orinoco River mouth he believes he has discovered one of the four rivers of Paradise and now declares Earth to be pear-shaped rather than round.
1498	Columbus returns to Española and discovers rebellion by Francisco Roldan against leadership by his brothers Diego (1450–1515?) and Bartolomew (1445–1514?). He appeases Roldan and then recants his agreement.
1499	Ferdinand and Isabella appoint Francisco de Bobadilla (d. 1502) to investigate the rebellion (March).
1499	Ferdinand and Isabella, disturbed by continuing reports of Columbus's administrative incompetence, appoint Bobadilla governor and chief magistrate of Española (May 21).
1500	Bobadilla reaches Española (August 23) and hears news of numerous executions of Spaniards ordered by Columbus and his brothers.
1500	Bobadilla arrests Columbus and Diego and ships them in chains back to Spain (September 15).
1500	Columbus arrives in Spain (November) and is ordered freed by Ferdinand and Isabella (November). He weeps with gratitude before the royal court.
1502	Nicolas de Ovando, appointed the new governor of Española, sails from Spain (February 13) with 32 ships and 2,500 passengers.
1502	Columbus launches his fourth and last voyage (May 9) with four ships and 146 men. He is forbidden to go to Española.
1502	Columbus discovers Martinique (June 15) following his quickest transatlantic crossing (21 days).
1502	Columbus attempts to enter Santo Domingo against royal orders, and Ovando turns him away (June).
1502	Columbus explores coast of Honduras while subduing repeated attempted mutinies (September). He writes to Ferdinand and Isabella that he is within 10 days' journey of Ganges River.
1503	Columbus writes (Easter Sunday) that his ships are "rotten, worm-eaten, all in holes."
1503	Columbus explores coast of Panama (May), believing he is "next to Cathay" (China).
1503–04	Columbus is stranded in Jamaica (June). He sends messenger to Ovando to seek help.
1504	Ovando sends two boats to rescue Columbus (late spring).
1504	Columbus, almost paralyzed by arthritis, sets sail for Spain (September 12).
1504	Columbus arrives in Spain (November 7).
1504	Queen Isabella dies (November 26).
1505	King Ferdinand receives Columbus (May) but refuses to restore his governorship of the Indies.
1506	Columbus dies (May 21).

Atlantic coast of Africa, following final destruction of Carthage by Rome.

117 Greek Eudoxus of Cyzicus successfully completes voyages between Egypt and India. He later makes two unsuccessful attempts at circumnavigating Africa, disappearing without a trace on the second.

25 Caesar Augustus sends Aelius Gallus as head of military expedition to explore Arabia.

A.D.

c. 50 Diogenes, traveling inland from the eastern coast of Africa, discovers two lakes thought to be the sources of the Nile.

c. 50 Roman tax collector becomes first (and possibly only) Roman to voyage to Ceylon, when his ship is blown off course from mouth of Red Sea.

▶

The Voyages of Magellan

1513 Portuguese nobleman Ferdinand Magellan (1480?–1521) is wounded while serving in Morocco, causing him to limp for rest of his life.

1517 Magellan goes to Seville, Spain, after falling out of royal favor in Portugal. He renounces his Portuguese citizenship.

1518 King Charles II of Spain accepts Magellan's proposal (March 22) for voyage to discover a navigable strait through South America to Pacific. This would allow Spanish to exploit spice trade while avoiding trip around Portuguese-controlled Africa.

1519 Magellan's fleet of five small but heavily armed ships leaves Sanlúcar de Barrameda with 270 men (September 20). Magellan commands *Trinidad,* while subordinate captains command *San Antonio, Concepción, Victoria,* and *Santiago.* Ships are laden with trading goods.

1519 Fleet sails south, reaching Tenerife in Canary Islands off Africa by September 26. It turns westward toward South America (October 3).

1519 Fleet enters Bay of Rio de Janeiro, off Brazilian coast (December 13), then sails southwestward in search of strait to carry it to Pacific. Magellan fails to find strait as he repeatedly explores coastal estuaries on voyage south.

1520 Fleet reaches Port San Julián on Argentina's southern coast (March 31). With Southern Hemisphere winter closing in, Magellan decides to stop here, ignoring crews' demands to pass the winter farther north, where it is warmer.

1520 Three subordinate captains mutiny (April 2). Outnumbered but determined to keep his expedition together, Magellan sends boatload of men to the mutinying *Victoria,* ostensibly to talk. These men kill *Victoria*'s captain and talk her crew out of mutiny. Magellan then forces surrender of the other two captains.

1520 *Santiago* sinks in storm near mouth of Santa Cruz (May 22). At about this time, Magellan's party finds and befriends a Patagonian native, a giant so tall the Europeans only "came up to the level of his waistbelt."

1520 Magellan sails fleet farther south to Santa Cruz River (August 24) and remains there until start of southern spring (October).

1520 Fleet, under way again, sights and begins to investigate yet another bay (October 21). This time, however, it is the long-sought-after strait, later called Strait of Magellan.

1520 Tortuous passage of the strait (October 21–November 28). Finding a way through demands tireless exploration of bays and narrow channels. *San Antonio* mutinies and sails back to Spain. Remaining crewmen endure dangerous and stormy passage. Magellan bursts into tears of joy as at last he finds strait's Pacific end.

1520 Pacific crossing (November 28–March 6). Though entire crossing is without storm (thus they name ocean the Pacific), it is much longer than anticipated. Plagued by hunger and thirst, the men eat worm-infested provisions, rats, and even leather off yardarms.

1521 Magellan's three remaining ships make first landfall in Pacific at Guam (March 6). They take on desperately needed supplies, though throngs of welcoming indigenous people steal much of the ships' gear.

1521 Fleet reaches Philippines (March 16).

1521 Magellan lands at island of Cebu in Philippines (April 7). He makes alliance for Spain with king of island, who converts to Christianity.

1521 Magellan, aiding his new native ally in war against neighboring island, is killed covering his comrades' retreat during fight on Mactan Island (April 27).

1521 Survivors of battle burn *Concepción* as unseaworthy (May).

1521 *Victoria* and *Trinidad* reach Moluccas (November 8) and load up with spices.

1521 *Trinidad* is wrecked (December 18). Only four of her crew ever reach Spain.

1522 *Victoria,* commanded by Juan Sebastian El Cano, a mutineer in 1520, rounds Cape of Good Hope (February 13). Hostile Portuguese imprison half of *Victoria*'s crew when ship later puts in at Cape Verde Islands.

1522 *Victoria* anchors at Seville, completing first circumnavigation of Earth (September 8). Only 18 of original crew are left after the journey of nearly three years.

Discovering the Natural Wonders of North America

Including mountains, rivers, lakes, and other natural wonders, arranged by the date of discovery by European explorers.

c. 1500 Colima, 13,933-foot volcano in Mexico.

1519 Popocatépetl, 17,887-foot volcano in Mexico, sighted by Spanish conquistadors.

1524 Hudson River sighted by Giovanni da Verrazano (1485?–1528?); explored by Henry Hudson (d. 1611) in 1609.

1534 St. Lawrence, 750-mile river (U.S./Canada), discovered by Jacques Cartier (1491–1557).

1539 Colorado, 1,450-mile river, discovered by Francisco de Vlloa (d. c. 1540).

1540 Grand Canyon discovered by Spanish explorer García López de Cárdenas (fl. 1540).

1540 Rio Grande, 1,880-mile river (U.S./Mexico), discovered by Spanish explorer Francisco de Coronado (1510–44).

1541 Arkansas, 1,450-mile river, probably first sighted by Coronado.

1541 Mississippi, 2,348-mile river, discovered by Hernando de Soto (c. 1500–42).

1550s Orizaba, 18,700-foot-high volcano in Mexico and third-highest peak in North America. First Europeans to see it were Spanish conquistadors.

1612 Huron (U.S./Canada), world's fourth-largest lake, found by Étienne Brulé (1592?–1633).

1615 Ontario (U.S./Canada), lake discovered by Étienne Brulé and Samuel de Champlain (c. 1567–1635).

1622 Superior (U.S./Canada), world's largest freshwater lake, discovered by Étienne Brulé.

1634 Michigan, world's fifth-largest lake, found by Jean Nicolet (1598–1642).

1669 Erie (U.S./Canada), lake discovered by Louis Jolliet (1645–1700).

1669 Ohio, 975-mile river, discovered by Robert Cavalier de La Salle (1643–87).

1673 Missouri, 2,315-mile river, first sighted by Jacques Marquette (1637–75) and Jolliet.

1678 Niagara, 167-foot waterfall (U.S./Canada), found by Father Louis Hennepin (1626–1701).

1690 Saskatchewan, 1,205-mile river (Canada), first explored by Henry Kelsey (d. 1729).

1730s Winnipeg, lake (Canada), discovered by French trapper Jean Baptiste de la Verendrye (fl. 1730).

1769 La Brea tar pits (Los Angeles).

1771 Great Slave Lake, second-largest entirely in Canada, discovered by fur trader Samuel Hearne (1745–92).

1789 Mackenzie, 1,060-mile Canadian river, explored by Alexander Mackenzie (1764–1820).

1792 Columbia, 1,200-mile river (U.S./Canada), discovered by U.S. navigator Robert Gray (1755–1806).

1792 Rainier, 14,408-foot mountain, first sighted by English explorer George Vancouver (1757–98).

1794 McKinley, 20,320-foot Alaskan mountain, first sighted by Vancouver.

1799 Mammoth Cave, Kentucky, found by settlers.

c. 1800 Great Bear Lake, largest lake entirely in Canada, found by fur traders.

1806 Pike's Peak, 14,110-foot mountain in Colorado, discovered by Zebulon M. Pike (1779–1813).

1807 Old Faithful geyser in Yellowstone. John Colter (c. 1775–1813) probably was first white man to visit Yellowstone.

1821 Lassen, volcano first sighted by a Spaniard, Luis Argüello (fl. 1820).

1824 Great Salt Lake, discovered by James Bridger (1804–81).

1831 Yukon, 1,587-mile river (U.S./Canada), first sighted by Russian explorers.

1851 Bridal Veil, 620-foot waterfall in Yosemite.

1851 Ribbon, 1,612-foot waterfall in Yosemite.

1851 Upper Yosemite, 1,300-foot waterfall in Yosemite.

1855 Crater Lake (Oregon), discovered by prospectors.

1864 Whitney, 14,494-foot mountain, highest in U.S. lower 48 states, discovered by Josiah D. Whitney (1819–96).

1901 Carlsbad Caverns, New Mexico, world's largest natural underground caverns.

c. 50 Roman sailors by this time have gained thorough knowledge of eastern coast of Africa south to Zanzibar.

54–68 Roman emperor Nero reigns. During this time he dispatches an expedition in search of Nile headwaters. The explorers push southward as far as White Nile marshlands.

c. 150 Roman Julius Maternus crosses Sahara, probably reaching modern-day Chad on his four-month journey.

c. 200 Ptolemy compiles his *Guide to Geography* at Alexandria.

c. 700 Irish monks sail westward to Faeroe Islands in North Atlantic.

700 Swedish Vikings establish settlements along Baltic and journey east to Black Sea and to Constantinople.

c. 800 Irish voyagers reach Iceland. Some believe that the accounts of St. Brendan "The Navigator" describe landfalls as far west as the New World.

c. 800–950 Age of the Vikings. In pursuit of land and wealth, Vikings set out from their Scandinavian homes south into continental Europe and west to British Isles, Iceland, Greenland, and finally to the New World.

862 Vikings sail Strait of Gibraltar to Balearics and Italian coast.

c. 930 Norwegian Viking Gunnbjörn is blown west by storms and discovers Greenland.

982–86 Eric the Red explores Gunnbjörn's land, returning to Iceland in 985 and departing in 986 for Greenland as head of settlement expedition.

986 Bjarni Herjulfsson, blown off course en route from Iceland to Greenland, sights east coast of North America.

1000 Leif Eriksson journeys to New World, making three landfalls on North American continent, which he names Helluland, Markland, and Vinland.

1265 Nicolò and Maffeo Polo, first European travelers to reach China, visit court of Kublai Khan at Cambaluc (Beijing).

1271–95 Marco Polo (1254–1324) accompanies his father, Nicolò, and uncle on overland journey to China. He spends next 24 years traveling through Asia, including Sumatra, Iran, and India, as Kublai Khan's official representative. The travelers return by sea, visiting Malay and Indian subcontinents.

1298 Marco Polo is imprisoned in Genoa, where he writes his celebrated account of his travels in Asia.

c. 1420–60 Portuguese prince "Henry the Navigator" (1394–1460) finances and directs some 40 years of exploration along western coast of Africa, attempting to expand Portuguese influence and to convert indigenous Africans to Christianity. Trading posts and forts are established, and slave trade becomes profitable.

1492 Italian-born mariner Christopher Columbus (1451–1506) sails westward in search of new route to spice-rich "Indies." With three ships—*Nina, Pinta,* and *Santa Maria*—he discovers America instead, landing in Bahamas and later reaching Cuba, Hispaniola, Venezuela, and other places. (*See The Voyages of Columbus.*)

1497–99 Portuguese navigator Vasco da Gama (c. 1469–1524) sails from Portugal around Africa and becomes first European to travel to India by sea, arriving there in 1499. He establishes eastward sea route for profitable spice trade, thus assuring existence of Portuguese Empire and some 400 years of European presence in Far East (*see Exploring Africa*).

1513 Vasco Nuñez de Balboa (1475?–1519) leads expedition west across Isthmus of Panama and first views Pacific Ocean.

1513–21 Spanish explorer Juan Ponce de León (c. 1460–1521) sails from

▶

The Lewis and Clark Expedition

1803	U.S. President Thomas Jefferson (1743–1826) requests $2,500 from Congress to launch expedition to explore region between Mississippi River and Pacific Ocean (January 18). Congress approves.
1803	Louisiana Purchase gives United States control of vast unexplored area west of Mississippi formerly under French control (April 30).
1804	Meriwether Lewis (1774–1809) and William Clark (1770–1838) launch 30-man expedition from St. Louis (May 14). Commissioned by Jefferson, they are to explore the land, study plant and animal life, and promote friendship with Indians.
1804	Expedition passes La Charette, home of Daniel Boone and last frontier settlement (May 25).
1804	First council with Indians is held at site of Council Bluffs, Iowa (August 25). Lewis and Clark inform Indians of change in government resulting from Louisiana Purchase and promise them U.S. protection.
1804	Teton Sioux Indians are encountered near site of Pierre, South Dakota (September 25). Confrontation is avoided by giving them "great Father's milk"—whiskey.
1804	Expedition reaches Mandan Indian villages near site of Bismarck, North Dakota (October 26). At this point expedition has traveled 1,000 miles up Missouri River at rate of 9 miles per day.
1804	Fort Mandan is established (October, in present-day South Dakota), and expedition spends winter there. French-Canadian trader Toussaint Charbonneau and his Indian wife, Sacajawea (1787–1812), join expedition. With Lewis as midwife, Sacajawea gives birth to son, whom she carries for remainder of trip.
1805	Journey continues upriver (April 5). Party is sent back to St. Louis with letters and trinkets for President.
1805	Mouth of Yellowstone River is reached (April 26).
1805	Expedition reaches Great Falls of Missouri (June 15). It takes one month to transport canoes 18 miles around the falls.
1805	Expedition reaches three forks that combine to form Missouri River (July 25). They are named Jefferson, Madison, and Gallatin rivers.
1805	Source of Missouri River is reached (August 12). Expedition follows a tributary of Columbia River. Shoshone Indians are encountered. Sacajawea, a Shoshone, serves as interpreter and proves invaluable for gaining Indians' trust.
1805	Overland journey of 300 miles begins (August). Expedition follows snowy mountain passes through what becomes Idaho and Montana.
1805	Clearwater Fork of Columbia River is reached (October 7). In canoes built from hollow tree trunks, expedition proceeds toward Pacific. Explorers play violin for Indians and show them telescope and compass in exchange for food.
1805	Clark writes, "Great joy in camp. We are in view of the Ocian [sic]" (November 7).
1805–06	Construction of Fort Clatsop on Pacific (November–March). Lewis and Clark map entire expedition. During winter, explorers kill 150 elk and make 338 pairs of elkskin moccasins for return trip.
1806	Return journey begins (March 23).
1806	Lewis and Clark separate briefly. Lewis explores Missouri River, and Clark follows Yellowstone River south.
1806	Lewis is accidentally shot in leg in Montana after being mistaken for elk by member of his party.
1806	Separate parties under Lewis and Clark meet at mouth of Yellowstone (August 12).
1806	Expedition reaches St. Louis (September 23), ending 6,000-mile exploration.
1806	Lewis is named governor of Louisiana Territory, and Clark is named general of militia and Indian agent.
1809	Lewis dies under questionable circumstances (is murdered or commits suicide).
1814	*History of the Expedition* is published.
1904–05	Complete journals of the expedition are published in eight volumes.

Puerto Rico in search of "fountain of youth." Seeking Bimini, he discovers Florida. (He returns in 1521 with shipload of settlers but is injured by Indians and dies soon after.)

▶

1519–21 Portuguese navigator Ferdinand Magellan (c. 1480–1521) seeks westward passage to the Indies. Sailing from Spain with five ships, he discovers Strait of Magellan at southern tip of South America and crosses Pacific, only to be killed by indigenous peoples in Philippines; 18 crewmen (of 250) survive to complete first circumnavigation of globe, in 1522, proving earth is round (*see The Voyages of Magellan*).

1519–21 Hernando Cortés (1485–1547) arrives in Mexico with 600 men, 15 horses, and some cannons. He leads a quick and successful conquest of the Aztecs and seizes capital city of Tenochtitlán for Spain.

1524 Spanish conqueror Francisco Pizarro (1475?–1541) sails from Panama on his first expedition down the west coast of South America. He discovers and names Peru and first learns of the great Inca Empire. However, he is forced to turn back. He returns after collecting only small amounts of gold (1527).

1528–36 Juan Cabeza de Vaca (1490–1556) explores the American Gulf Plains from Texas to Mexico.

1530s Jacques Cartier discovers St. Lawrence River (1534), prompting further French exploration on the river and west to Great Lakes.

1531–35 Pizarro returns to Peru. He attacks Inca city of Caxamalca on November 15, 1532, and sacks Cuzco, 600 miles farther south, in November 1533. Having successfully conquered the Incas, Pizarro founds city of Lima in 1535.

1539–42 Spanish conqueror Gonzalo Pizarro (1502?–48), accompanied by large force of Spaniards and Indians, explores lands to east of Quito. Entourage crosses Andes and makes its way through rain forest before discovering Napo River, a headwater of the Amazon.

1539–42 Hernando de Soto lands in Tampa Bay, Florida, with 600 men and 200 horses and travels in search of treasure through Georgia, the Carolinas, Tennessee, Alabama, and Oklahoma. He and his companions are first white men to see Mississippi River (1541).

1540–42 Francisco de Coronado (1510–54), searching for legendary Seven Cities of Cibola, travels through American Southwest; he is first European to explore Arizona and New Mexico.

c. 1541 Spanish conqueror Francisco de Orellana (c. 1490–1546), at behest of Gonzalo Pizarro, launches sailing party on Napo River, travels to the Amazon, and continues downstream to the Atlantic (1542).

1565 Spanish sailor Alonso de Arellano sails *San Lucas* from East Indies to Mexico, first such west–east crossing of Pacific.

1567 Historian and sailor Pedro Sarmiento de Gamboa mounts expedition to locate Terra Australis. Sailing west from Peru, he lands in Solomon Islands, not in Australia.

1570 Abraham Ortelius (1527–98) publishes *Theatre of the World*, collection of 70 maps regarded by historians as first modern atlas.

1576 Sir Martin Frobisher (1535–94) searches for Northwest Passage by way of Arctic. (*See Arctic Exploration.*)

1577 British captain Francis Drake (c. 1540–96) leads first voyage around world by the English. His fleet of four ships, including flagship *Golden Hind*, successfully navigates Strait of Magellan and rounds Cape Horn. Explores Pacific coasts of South and North America, visiting native villages and traveling as far north as California and Oregon. Plunders Spanish treasure ships and port settlements

▶

during voyage, returning to England with enormous riches.

1602–07 Bento de Goes journeys from Agra, India, on overland route to China in effort to prove Marco Polo's Cathay and Matteo Ricci's China are same place. He arrives ill at Suchow but verifies Cathay as China.

1603–11 Samuel de Champlain (1567–1635) begins his exploration of eastern coast of North America. Following St. Lawrence River west, he founds Quebec (1608), crosses Lake Huron, and discovers what is now Lake Champlain (1611).

1609–10 English explorer Henry Hudson (d. 1611) sails from England in search of Northwest Passage for Dutch East India Company. He explores Hudson River, establishing Dutch claims to area. In 1610 reaches Hudson Bay. He is abandoned at sea by mutinous crew.

1616 Dutch merchant sailor Dirk Hartog overshoots Spice Islands and lands (October 25) on western coast of Australia.

1626 Jesuit missionaries Estêvão Cacella and João Cabral become first Europeans to reach Bhutan.

1642 Abel Tasman (1603–59) discovers Tasmania (which he names Van Diëmen's Land).

1644 Tasman explores southern coast of New Guinea and northern coast of Australia.

1673 Louis Jolliet (1645–1700) and Jacques Marquette (1637–75) discover upper Mississippi River. They travel as far south as Arkansas River before returning north to Canada (then New France).

1678–82 French explorer-missionary Robert La Salle (1643–87) sails length of Mississippi River, claiming the valley and naming it Louisiana (for Louis XIV of France). He travels through North America building forts, establishing trade posts, and establishing good relations with Indians.

1714 Jesuits Ippolito Desideri and Emmanuel Freyre travel with a Tartar caravan from Lahore into lands skirting Mount Kailas—terrain in the Himalayas no Europeans had traveled before nor would venture through again for 200 years.

1728 Danish navigator Vitus Bering (1680–1741), in Russian service, sails through what is later called Bering Strait into Arctic Ocean, proving that Asia and North America are not joined by land.

1731–43 French-Canadian fur trader Sieur de la Verendrye (1685–1749) sets out on foot to explore western Canada and search for western sea. He and his sons are first Europeans to reach Lake Winnipeg (1733). They end their 1,500-mile overland journey at Fort St. Pierre (1743).

1733–43 Great Northern Expedition. Vitus Bering leads Russian expedition that charts much of Siberian coastline in Arctic. In 1741 Bering discovers Alaska and sails along part of coastline. He is shipwrecked and dies soon after.

1743 Charles Marie de la Condamine (1701–74) explores Amazon to its mouth, providing the first scientific observations of geography and peoples of Amazon Valley.

1768 British navigator James Cook (1728–79) sails around the world, exploring countries and meeting natives of Australia and New Zealand while searching for legendary "southern continent." In 1776 he explores western North American coast for an Atlantic passage. He develops maps and charts that are used for generations.

1789–93 Scottish fur trader Sir Alexander Mackenzie (1764–1820) explores western Canada, following what is now

▶

Mackenzie River to Arctic Ocean (1789), and discovers Fraser River (1793) while blazing trail through Canadian Rockies to the Pacific.

1797–1807 Englishman David Thompson (1770–1857), surveyor for Northwest Company, explores western Canada. He is first European to cross Howse Pass to source of the Columbia River (1807).

1801–03 British navigator Matthew Flinders (1774–1814), who explores Australia's southern coast in detail and charts much of the rest of its coastline, successfully circumnavigates continent.

1804 Meriwether Lewis (1774–1809) and William Clark (1770–1838) explore Louisiana Purchase territory and land beyond it to Pacific Ocean. Their expedition takes them from St. Louis, across Rockies, and along Columbia River to ocean. They return to St. Louis in November 1806 (*see Lewis and Clark Expedition*).

1805–7 American explorer Zebulon M. Pike (1779–1813) leads expedition into Spanish New Mexico and the Rocky Mountains, discovering in Colorado the peak named after him.

1822–26 Scottish explorer Alexander Gordon Laing (1793–1826) explores Niger River Basin of Africa, becoming first white man to reach Timbuktu (1826).

1822–27 Scottish explorer Hugh Clapperton (1788–1827) leads two expeditions, both to northern Nigeria and Lake Chad regions of Africa.

1824–29 American mountain man Jedediah Strong Smith (1799–1831) blazes trails through South Pass, Wyoming (1824) and Mojave Desert. He becomes one of first men to cross Sierra Nevada and Great Salt Desert from west to east (1829).

1826–28 Frenchman René Caillié (1799–1838) leads expedition through North Africa, becoming first white man to visit Timbuktu and live to tell about it, as well as first to cross Sahara.

1831 English scientist Charles Darwin (1809–82) sets sail as naturalist on board *HMS Beagle* for world voyage to accumulate scientific data. His observations of animal life along the way help provide basis for his theory of evolution.

1840–41 English explorer Edward John Eyre (1815–1901) is first European to make major overland journey across Australia. He crosses more than 1,000 miles of waterless Nullarbor Plain before reaching Albany.

1842–46 U.S. Army captain John Charles Frémont (1813–90) explores Wind River Chain of Rocky Mountains and various parts of Colorado and Oregon. He leads expedition from Great Salt Lake to Sutter's Fort in California.

1849–71 Scottish explorer and medical missionary to Botswana David Livingstone (1813–73) sets off north across Kalahari Desert and eventually discovers Zambezi River (1851) as well as Victoria Falls (1855). He is lost sight of after leaving with expedition to discover Nile source (1866) and is "found" five years later by H. M. Stanley (1841–1904) east of Lake Tanganyika (1871).

1858–62 Scottish explorer John McDouall Stuart (1815–66) makes six trips to Australia's interior, reaching continent's geographical center April 22, 1860.

1860–61 Irish explorer Robert O'Hara Burke (1821–61) and English companion William John Wills (1833–61) set out to cross Australia from south to north. Expedition is poorly managed and inadequately provisioned, and although the two men successfully reach Gulf of Carpentaria (December 16), both die of starvation on the return journey (June 1861).

▶

1871–88 Russian Nikolay Przhevalsky (1839–88) explores central Asia, accurately locating and describing major geological features and collecting animals and plants.

1873 Indian Army officer Peter Egerton Warburton is first man to cross Australia from central Alice Springs to western coast, a journey of more than 2,000 miles.

1876–78 Englishman Charles Montagu Doughty (1843–1926) spends two years exploring Arabia, keenly observing the landscape and people and providing an account in his *Travels in Arabia Deserts.*

1878 Finnish-born Swedish explorer Nils Nordenskjöld (1832–1901) becomes first man to sail through Northeast Passage from Norway to the Bering Strait, although he spends one winter frozen in ice near the strait before reaching it. (He is also first to sail all the way around Europe and Asia.)

1893–1938 Swede Sven Hedin (1865–1952) explores deserts of central Asia, Tibetan Plateau, and Himalayas. Unlike earlier explorers in the region, Hedin provided Europe with complete accounts of his journeys.

1906 Norwegian polar explorer Roald Amundsen (1872–1928) becomes first European successfully to navigate ship through Northwest Passage.

1909 Robert E. Peary (1856–1920) leads expedition to reach North Pole (April 6). American Frederick Cook (1865–1940) claims to have reached Pole in 1908, however.

1911 Amundsen is first man to reach South Pole (December 14). British explorer Robert Falcon Scott (1868–1912) reaches it just 35 days after Amundsen (January 18, 1912). Scott and his four companions die of starvation and cold on return trip.

1932 Englishman Harry St. John Philby (1885–1960) makes his celebrated journey into "empty quarter" of Arabia, Rugi al-Khali Desert, a land so bleak the Bedouin scarcely travel there. He discovers remains of legendary Wabar, five craters probably formed by meteorites and thought by Arabs to be site of heavenly fires.

Exploring Africa

1482 Portuguese explorer Diego Cão (fl. c. 1480) discovers mouth of Congo River and explores Africa's western coast.

1488 Portuguese navigator Bartolomeu Dias (c. 1450–1500) rounds Cape of Good Hope after being blown off course and is first European to travel up eastern coast of Africa.

1497 Portuguese navigator Vasco da Gama (c. 1469–1524) sails around African Cape of Good Hope to reach India by sea in 1499. African ports thereafter become important landfalls on sea routes to Orient.

1553 English merchants travel to Morocco to trade cloth for ivory, pepper, and gold.

1652 Cape Town is founded by Dutch East India Company to supply sea traffic to India and Orient.

1684 Dutch settlers arrive in Cape Town, followed by religious refugees from France. They begin South African movement of "trekking," in which colonists explore and settle progressively farther inland.

1768 Scottish explorer James Bruce (1730–94) searches for Nile River source; discovers

▶

instead source of its tributary the Blue Nile, in Abyssinia. His exploration arouses popular interest in African exploration.

1779 Dutch Colonel Robert Jacob Gordon discovers mouth of Orange River and later maps its 500-mile course from South Atlantic to Ural River confluence.

1795 British explorer Mungo Park (1771–1806) searches for Niger River. He succeeds in finding it, but suffers exhaustion and becomes ill on the way. He finally is killed on Niger by attacking natives.

1798 German traveler Frederick Hornemann (1772–1801) becomes first European in modern times to cross Sahara. He does so by joining caravan of African merchants.

1822 British government-sponsored expedition leaves for Tripoli to search for Niger River source. Explorers cross Sahara and reach Lake Chad in 1823 after traveling 1,000 miles overland. There they discover that the lake is not part of the Niger.

1827 French explorer René Caillié (1799–1838), after several unsuccessful attempts, travels with Muslims to reach Timbuktu. He then crosses unexplored western part of Sahara to Tangier, covering 2,800 miles in 538 days.

1830 British explorers John Lander (1807–39) and Richard Lander (1804–34) successfully lead expedition to trace Niger River's course in West Africa. Canoeing downriver to its delta, they are captured and ransomed by natives.

1841 Scottish missionary David Livingstone (1813–73) travels to Africa and there becomes renowned explorer and European spokesman for native peoples. He discovers Zambesi River, Victoria Falls (1855), and Lake Nyasa.

1850 German geographer and historian Heinrich Barth (1821–65) joins British expedition that crosses Sahara to Lake Chad and explores territories to south and southeast. Barth continues on to Timbuktu (1853), traveling over 10,000 miles in five years. His five-volume book on African exploration becomes classic.

c. 1850 German missionaries explore African interior, drawing up careful charts of traveled areas. They return to Europe with tales of snow-capped mountains in tropical East Africa that later prove to be Mount Kenya and Mount Kilimanjaro.

1856–59 British explorers Richard Burton (1821–90) and John Speke (1827–64) search for White Nile's source. They discover Lake Tanganyika, which Burton mistakenly believes is the source. Speke goes on to discover and name Lake Victoria, correctly concluding it is a major Nile source. Speke and Burton, both ill and exhausted, reach coast in 1859. Speke returns to England and announces finding.

1860 Speke returns to Lake Victoria. He visits three kingdoms (in modern Uganda) and explores Victoria Nile River. Discovers and names Ripon Falls, which he believes to be important Nile source.

1863 English friends of Speke, Samuel Baker (1821–93) and wife, further explore upper Nile region. They discover and name Lake Albert and magnificent Murchison Falls at the lake.

1866 David Livingstone sets out to explore perimeter of Lake Albert, believing that a river entering the lake's southern end would be the true Nile source. He disappears for five years.

1871 Welsh journalist Henry Stanley (1841–1904), on assignment for a New York newspaper, searches for and finds Livingstone, greeting him with famous "Dr. Livingstone, I presume?" He explores north end of Lake Tanganyika with Livingstone. Stanley returns from Africa (1873) after Livingstone's death and is met with both acclaim and accusations of being more exploiter than explorer.

1872 English explorer Verney Cameron (1844–94) commands Royal Geographic Society expedition to locate Livingstone. He belatedly encounters servants bearing Livingstone's body (1873). Continues to Lake Tanganyika, travels Lualaba River, and charts Congo-Zambezi watershed. He is

▶

first European to cross equatorial Africa from coast to coast.

1873 Stanley returns to Africa to prove himself an explorer. He later proves Lake Victoria is Nile source, showing that from there it is joined by Blue Nile and then flows south through Sudan and Egypt. Also discovers that the Lualaba joins Congo River and flows to South Atlantic.

1877 Four great African rivers—Zambezi, Niger, Nile, and Congo—have been explored and their courses determined by this date.

1878 Scottish explorer Joseph Thomson (1858–95) journeys to eastern-central Africa with Royal Geographic Society expedition. He travels to Lake Nyasa and Lake Tanganyika. Discovers Lake Rukwa. In 1882 he explores shortest route from sea to Nile headwaters, previously said to be impassable due to hostility of natives, whose numerous attacks he somehow survives.

1888 Stanley discovers Ruwenzori Mountain Range, which proves to be Mountains of the Moon mentioned by Ptolemy in second century.

Arctic Exploration

B.C.

c. 300 Greek trader Pythias lands in a place he calls Thule, variously identified as Iceland or Norway. He is generally considered first Arctic explorer.

A.D.

795 Irish monks begin exploration of Iceland.

874 Norsemen found first colony in Iceland, near Reykjavik.

986 Eric the Red founds Norse colonies in Greenland.

1553 Three English ships set out to discover Northeast Passage from Europe to Far East by way of Arctic waters. Ships become icebound, and only one crew survives overland journey to Moscow.

1565 Dutch trading post is established at what later becomes Russian Arctic city of Archangel.

1576 English navigator Sir Martin Frobisher (1535–94) takes three ships in unsuccessful search for Northwest Passage from North Atlantic to Pacific by way of Arctic.

1578 Frobisher sets out with 15 ships to found settlement of 100 men on Baffin Island. Storms scatter his ships and end the venture.

1587 English navigator John Davis (1550?–1605) completes three years of Arctic exploration, during which he partially maps coasts of Baffin Island and Labrador.

1594–97 Dutch navigator Willem Barents (c. 1550–97) explores Arctic, searching for Northeast Passage. He discovers Novaya Zemlya islands, Bear Island, and Spitsbergen and, after being trapped in the Arctic for a winter, escapes just days before his death.

1610 English navigator Henry Hudson (d. 1611) sets out aboard *Discovery* to find Northwest Passage. He discovers Hudson Bay, but after wintering there his crew mutinies and sets Hudson, his son, and others adrift in a boat. They suffer an unknown fate. The mutiny's leaders are later killed by Eskimos, while the rest are imprisoned upon reaching England.

1728 Danish navigator Vitus Bering (1680–1741), in Russian service, sails from Siberian coast through Bering Strait into Arctic Ocean. Though bad weather obscures view of land bordering strait, he concludes

▶

that Asia and North America are not joined by land, as had been thought.

1733–43 Bering leads Russia's Great Northern Expedition. He maps much of Arctic coastline of Siberia and discovers Alaska. Shipwrecked on return voyage, he dies with most of his crew on Bering Island.

1770 New Siberian Islands are discovered, yielding a fortune in ivory tusks remaining from mammoths that died there.

1773 British sea captain Constantine J. Phipps (1744–92) fails in attempt to sail across North Pole. He gets as far as Spitsbergen.

1776–79 Famed British navigator James Cook (1728–79) sets out on unsuccessful voyage to discover northern Pacific entrance to either Northwest or Northeast Passage. He is killed in disagreement with Hawaiians.

1824 Mapping of northeastern coast of Siberia is completed by Russian Arctic explorer Baron Ferdinand von Wrangel (1794–1870).

1827 British explorer Sir William Parry (1790–1855) is first to use sleighs in attempt to reach North Pole. He reaches 82°45′N.

1831 British explorer James Clark Ross (1800–62) determines position of North Magnetic Pole (June 1) while on Arctic voyage with his uncle Captain John Ross (1777–1862).

1845 British admiral Sir John Franklin (1786–1847) sails into Lancaster Sound looking for Northwest Passage. He and his party die after becoming trapped in remote Arctic wastes. Their remains are finally located in 1859.

1854 British naval commander Robert McClure (1807–73), searching for Franklin's party, discovers Northwest Passage, though he does not sail through it.

1878–79 Swedish scientist and Arctic explorer Adolf Nordenskjöld (1832–1901) successfully navigates Northeast Passage for first time. His steamship *Vega* is icebound for winter of 1878–79 but arrives safely in Alaska (July 22, 1879).

1879 American naval officer George W. DeLong (1844–81) attempts to reach North Pole aboard *Jeanette.* His ship becomes icebound, however, and later sinks.

1882–83 First International Polar Year. Scientists and explorers from 11 countries make scientific observations throughout winter.

1893 Norwegian explorer Fridtjof Nansen (1861–1930), aboard *Fram*, ship designed to withstand being icebound, drifts in Arctic icepack from 1893 to 1895. *Fram* later breaks free and is returned to Norway.

1897 First attempt to fly to North Pole is made in a balloon by Swedish scientist S. A. Andrée (1854–97). Remains of the party are not found until 1930.

1900 Umberto Cagni (1863–1932), of duke of Abruzzi's Italian expedition, reaches northern record of 86°34′N.

1906 Norwegian explorer Roald Amundsen (1872–1928), sailing aboard converted herring boat *Göja,* successfully navigates Northwest Passage for first time. Voyage takes three years.

1909 First men reach the North Pole (April 6). American Arctic explorer Robert E. Peary (1856–1920), Matthew Henson, and 50 Eskimos travel to pole by dogsled from Ellesmere Island.

1909 Long-standing controversy begins when American explorer Dr. Frederick Cook (1865–1940) announces (September 1) that he reached North Pole April 21, 1908, a year before Peary. Peary's claim is generally supported, however.

1926 First flight to North Pole and back (May 9). American polar explorer Rear Admiral Richard E. Byrd (1888–1957) and Floyd Bennett (1890–1928) make the flight aboard Fokker trimotor airplane. Byrd's claim is disputed in 1970s, however.

1928 First flight across Arctic Ocean is made by Australian explorer G. H. Wilkins (1888–1958) and C. B. Eielson.

1932–33 Second International Polar Year.

▶

1937 USSR sets up first floating scientific station, North Pole 1, which drifts south for nine months before being taken off its melting ice floe in Greenland Sea.

1944 First single-season voyage through Northwest Passage is completed aboard schooner by Canadian Mountie Henry A. Larsen.

1950 Soviet station North Pole 2 is set up and maintained for a year.

1952 United States establishes weather station on Ice Island T-3.

1954 First passage of North Pole by deep-draught vessel is made by Canadian icebreaker *HMCS Labrador.*

1958 First underwater crossing of North Pole is made by U.S. nuclear submarine *Nautilus* (August 5).

1959 U.S. nuclear submarine *Skate* is first to surface at North Pole by breaking through the ice.

1961 American researchers set up the drifting sea ice station ARLIS II (Arctic Research Lab Ice Station II) and keep it running until 1965.

1969 Commercial feasibility of passage through Arctic Circle is tested as U.S. icebreaking tanker *Manhattan* smashes through 650 miles of ice.

1977 Soviet nuclear icebreaker *Arktika* becomes first surface ship to reach North Pole.

1978 Southern passage along coast of Siberia is opened by Soviet nuclear icebreaker *Sibir.*

1988 *National Geographic* article uses Peary's scientific data to show that he never reached North Pole, claiming he was off course by as much as 70 miles.

1989 National Geographic Society issues new study, conducted by Navigation Foundation and based on new analytical methods, to verify Peary's account. They support veracity of Peary's claim, saying he "was not a fraud or a fake."

Antarctic Exploration

1738–39 French explorer J.B.C. Bouvet de Lozier navigates through Antarctic ice packs for 1,000 miles without finding land.

1772–75 Englishman James Cook (1728–79) discovers ice packs between 60°S and 70°S but does not discover land. He postulates existence of undiscovered continent beyond the ice packs.

1772 French explorer Yves-Joseph de Kerguelen-Tremarec discovers Antarctic islands.

1820 Russian Fabian Gottlieb von Bellingshausen (1778–1852), Englishman Edward Bransfield (1795–1852), and American Nathaniel B. Palmer (1799–1877) each claim first sighting of Antarctica.

1821 American sea captain John Davis becomes first person to land on Antarctica (February 7).

1821 Von Bellingshausen discovers Alexander Island and Peter Island.

1821 Total of 100 British and American sealers hunt in Antarctic waters.

1838–42 U.S. Navy Lieutenant Charles Wilkes (1798–1877) explores Antarctica and establishes that it is a continent.

1839 British explorer James C. Ross (1800–62) discovers iceless Ross Sea. He discovers volcanoes Mount Terror and Mount Erebus and sails 450 miles along an ice ridge.

▶

1840 French explorer Jules D'Urville leads expedition close to South Magnetic Pole (January 20).

1895 Sixth International Geophysical Conference in London declares Antarctic exploration to be most important exploration yet undone. It encourages new expeditions by scientists worldwide.

1898–99 Belgian ship *Belgica* penetrates ice pack and becomes caught in ice. Crew spends winter in Antarctic, surviving by eating seal meat.

1899 C. E. Borchgrevink and Southern Cross expedition land on Ross ice barrier.

1902–04 Captain Robert F. Scott (1868–1912) and other British National Expedition members discover Peninsula Edward VII; they carry out sledge probes (1901–4).

1902 Scott goes up in balloon for aerial reconnaissance over Antarctica.

1909 French explorer Jean Charcot (1867–1936) investigates Palmer Peninsula coast and discovers Charcot Island.

1907 British explorer Ernest Henry Shackleton (1874–1922) launches expedition to discover South Pole and South Magnetic Pole.

1908 Shackleton expedition climbs 12,280-foot Mount Erebus.

1909 Shackleton expedition comes within 100 miles of South Pole, surpassing by 366 miles any previous Antarctic inland exploration. Dwindling food supply forces end to expedition.

1909 South Magnetic Pole is reached by three members of Shackleton's expedition (January 16).

1909 North Pole is discovered (April 6), intensifying quest for South Pole.

1911 Race for South Pole. Norwegian, German, Australian, British, and Japanese expeditions are launched.

1911 Norwegian explorer Roald Amundsen (1872–1928) launches expedition (October 20) with four crew members, four sleds, 52 dogs, and four-month supply of food.

1911 British explorer Robert F. Scott (1868–1912) departs from McMurdo Sound (November 1) with experimental motor sleds and ponies to haul supplies, both of which prove unsuccessful in difficult Antarctic conditions. Five-man team for final polar assault man-hauls supply sleds toward the pole.

1911 Amundsen expedition becomes first to reach South Pole (December 14) and plants Norwegian flag.

1912 Scott expedition arrives at South Pole 35 days after Amundsen (January 18) and finds Norwegian camp already established. Their supplies exhausted on the return trek, Scott and his entire party die during a blizzard (April). Their remains are found (November 1912) just 11 miles from the supply depot.

1914 Shackleton organizes transantarctic expedition. His ship is crushed in ice pack before completion of the mission.

1928 Development of airplanes enables Sir G. H. Wilkins (1888–1958) to study Antarctic (Palmer) Peninsula via plane.

1929 Expedition under U.S. admiral Richard E. Byrd (1888–1957) establishes Little America base on Ross Ice Shelf.

1929 Byrd makes first flight over South Pole (November 28–29). Expedition discovers and names Rockefeller Plateau and ranges east of Edward VII Peninsula.

1933 Byrd expedition names Edsel Ford Range while carrying out extensive research in area.

1933 Lincoln Ellsworth (1880–1951) unsuccessfully attempts to explore Antarctica by air.

1935 Canadian Herbert Hollick-Kenyon and Ellsworth fly across previously unseen regions. They determine that Antarctica is one continent, not divided by a channel, as previously thought. Eternity Range at base of peninsula is named by Ellsworth.

1946 Byrd expedition, comprised of 13 ships, carries out aerial mapping of region. It discovers new bays, islands, glaciers, and

▶

nine mountain ranges and confirms existence of coal seams.

1949–52 Norwegian-British-Swedish expedition explores Maudheim Base on Queen Maud Land.

1953 France establishes bases in Kerguelen and Crozet islands.

1954 Australia builds Mawson Station on mainland coast of MacRobertson Land.

1955 Argentina establishes General Belgrano Station on Filchner Ice Shelf.

1955 First Antarctic Conference is held, in France.

1957 Twelve International Geophysical Year (IGY) participants build and man Byrd and Amundsen-Scott winter bases. They carry out extensive research of world's last unexplored areas.

1958 British Commonwealth Transantarctic Expedition reaches New Zealand Scott Base on Ross Island by traversing South Pole from Filchner Ice Shelf.

1959 Argentine Naval Transport Command begins commercial tours of Antarctic.

1966 Total of 90 percent of Antarctic has been viewed or photographed by this date.

1967 U.S. Antarctic Mountaineering Expedition scales Vinsear Massif, 16,864-foot peak in Sentinel Range of Ellsworth Highland.

1973 Discovery by United States of petroleum resources in Ross Sea.

1975–76 Italy launches its first expedition of continent.

1989 International Transantarctic Expedition is launched. It is seven-month, 5,000-mile exploration by dogsled, covering areas never before crossed on foot.

The Underwater Frontier

B.C.

300s Alexander the Great (356–323 B.C.) uses glass barrel to descend to ocean bottom, according to legend.

c. 100 Japanese women dive to retrieve food on sea floor, according to account from this time. They hold their breath for up to three minutes.

A.D.

1535 Italians use underwater glass bell to enable divers to explore sunken ships in Italy's Lake Nemi.

1620 Dutch inventor Cornelius van Drebbel (1572–1634) tests early submarine, powered by oarsmen inside enclosed hull in England's Thames River.

1663 Salvagers use diving bell to recover cannon from warship *Vasa,* submerged 100 feet in Stockholm Harbor.

1690 English astronomer Edmund Halley (1656–1742) develops wooden diving bell with air supplied by lead-lined casks lowered from surface. Divers use it to descend more than 50 feet.

1715 Englishman John Lethbridge develops forerunner of diving suit, large wooden cylinder with armholes and glass panel for viewing. It is used to dive more than 60 feet.

1797 German K. H. Klingert develops helmeted diving suit supplied with pumped air.

1819 Augustus Siebe (1788–1872) invents an open-dress diving suit with pumped air similar to Klingert's design, but much more practical.

1837 Siebe develops airtight diving suit with valve for ventilation.

▶

1844 French zoologist Henri Milne-Edwards (1800–1885) conducts first underwater studies of marine life off coast of Sicily.

1863 First submarine run by compressed air, *Le Plongeur,* is invented in France.

1865 Precursor to Aqua-lung is developed in France. It is an air-filled canister worn on back with automatic demand valve.

1879 First self-contained oxygen-rebreather lung is developed in England by Henry Fleuss.

1887 One of first battery-powered submarines is developed in France.

1892 Frenchman Louis Bouton takes first underwater photographs off coast of France.

1898 *Holland VI,* first practical submarine, is built by American inventor John P. Holland (1840–1914). It dives to 75 feet.

1899 Bouton takes first successful underwater remote-control pictures.

1906 John Haldane (1860–1936) of Scotland establishes decompression tables to eliminate hazards of compressed-air diving to 200 feet.

1912 First pressurized, submersible decompression chamber is developed in England by Sir Robert Davis.

1913 Armored diving suit with ball-and-socket joints in arms and legs is invented by German manufacturing company Neufeldt & Kuhnke.

1915 Salvagers set depth record for underwater recovery as they raise submarine *F-4* from 306 feet off coast of Honolulu, Hawaii.

1926 First underwater color photographs using artificial lighting are taken off Dry Tortugas, Florida by *National Geographic* photographer.

1930 Decompression tables for depths up to 300 feet are established by British Admiralty Deep Diving Committee.

1930 Naturalist William Beebe (1877–1962) and designer Otis Barton descend in their bathysphere to record 1,428 feet off Bermuda. They observe many strange deep-sea life forms never seen before.

1934 Beebe-Barton bathysphere dives to new record of 3,028 feet off Bermuda.

1935 Jim Jarratt dives 330 feet off coast of Ireland in "Iron Man" armored suit and locates wreck of *Lusitania.*

1935 Rubber foot fins are invented in France.

1943 French marine biologist Jacques-Yves Cousteau (b. 1911) and Emile Gaguan refine their fully automatic compressed-air Aqua-lung. It enables them to dive to 210 feet in Mediterranean.

1948 Wilfred Bollard of British Royal Navy dives to 540 feet in Lock Fyne, Scotland, breathing mixture of helium and oxygen.

1954 French researchers take bathyscaph *F.N.R.S. 3* to 13,287-foot depth off coast of Dakar.

1960 Swiss oceanographer Jacques Piccard (b. 1922) and Lieutenant Don Walsh take bathyscaph *Trieste* down to record 35,800 feet in Mariana Trench, deepest known place in ocean.

1964 *Sealab 1.* Four U.S. Navy divers remain underwater (193 feet down off Bermuda) for 11 days.

1965 *Conshelf 3.* Six men live underwater for 22 days (328 feet down) in Mediterranean.

1966 Submersibles *Alvin, Aluminaut,* and *Cubmarine* descend to 2,500 feet to retrieve hydrogen bomb lost near Palomares, Spain.

1970 Six divers remain submerged for six days (520 feet down) off coast of Hawaii.

1976 Deepest dive in armored "Jim Suit" reaches depth of 1,440 feet off coast of Spain.

1978 Research ship *Glomar Challenger* collects core samples with remote drill on seafloor 23,104 feet below surface.

1985 Wreckage of *Titanic* is located in North Atlantic. Remote-control equipment finds and photographs it.

Manned Space Exploration

1957 Successful orbit of world's first artificial satellite, *Sputnik 1*, by Soviet Union (October 4), begins "space race" between United States and USSR, stimulating manned space exploration.

1958 United States publicly announces start of its manned space program, Project Mercury (October 7).

1961 Soviet cosmonaut Yuri A. Gagarin (1934–68) makes history when he becomes first man in outer space (April 12). His flight, aboard 10,417-pound *Vostok 1* spacecraft, lasts 1 hour, 48 minutes (one Earth orbit) and reaches maximum altitude of 187 miles.

1961 U.S. astronaut Alan B. Shepard, Jr. (b. 1923) becomes first American in space during 15-minute suborbital flight (May 5). Shepard rides atop *Mercury 3* rocket in *Freedom 7* space capsule, reaching maximum altitude of 116.5 miles.

1961 President John F. Kennedy (1917–63) publicly commits United States to sending a man to moon and back by end of the decade (May 21).

1961 Virgil I. Grissom (1926–67) becomes second American in space during suborbital flight lasting 15 minutes (July 21). Grissom's capsule, *Liberty Bell 7*, sinks during recovery operations after reentry.

1961 Second Soviet cosmonaut launched into space becomes first man to spend more than a day in space (August 7). Gherman S. Titov (b. 1935) completes 17 Earth orbits during 25½-hour flight aboard *Vostok 2*.

1962 John Glenn (b. 1921) becomes first American to orbit Earth (three times) during his 5-hour *Mercury 6* flight (February 20).

1963 U.S. astronaut L. Gordon Cooper (b. 1927) completes 22 Earth orbits aboard *Mercury 9*, last Mercury series flight (May 15–16). Cooper becomes first American to stay in space over one day.

1963 First woman in space, Soviet cosmonaut Valentina V. Tereshkova, stays up for almost 71 hours (48 Earth orbits) aboard *Vostok 6* (June 16).

1964 Three cosmonauts are launched simultaneously for first time in a single capsule, *Voskhod 1* (October 12). Flight lasts just over 24 hours (16 orbits).

1965 First "walk" in space (March 18) by cosmonaut Alexei A. Leonov (b. 1934). Leonov spends 10 minutes outside craft, *Voskhod 2*.

1965 *Gemini 3* carries first American two-man crew into space (March 23). Astronauts Virgil I. Grissom (1926–67) and John W. Young (b. 1930) conduct first in-orbit maneuvers of manned spacecraft during the flight.

1965 *Gemini 4* flight (June 3–7). First space walk (36 minutes) by American and first use of personal propulsion pack during a space walk, both by astronaut Edward H. White II (1930–67).

1965 First space rendezvous of two manned craft (December 15). *Gemini 6A* comes within 1 foot of *Gemini 7*, launched December 4.

1966 First docking of craft, *Gemini 8* with target vehicle, is completed in space (March 16–17) by astronauts Neil A. Armstrong (b. 1930) and David R. Scott (b. 1932).

1967 Flash fire in *Apollo 1* space capsule causes first deaths of American astronauts—Virgil I. Grissom (1926–67), Edward H. White II (1930–67), and Roger Chaffee (1935–67)—during routine test at Cape Kennedy, Florida (January 27). Less combustible mixture of oxygen and nitrogen is used for capsule atmosphere thereafter.

1967 Soviet cosmonaut Vladimir M. Komarov (1927–67) is killed during reentry when his *Soyuz 1* spacecraft crashes because of

▶

Getting Space Exploration Off the Ground

c. 180 Greek writer Lucian (c. 125–85) is earliest known author to write about space travel. In *True History* he describes ship blown to moon by strong wind. Crew encounters sun and moon creatures.

1619 German astronomer Johannes Kepler (1571–1630) publishes last of his three laws on planetary motion. (He also writes fictitious *Somnium or the Astronomy of the Moon*, in which space travelers are propelled to moon by spirits.)

1657 *States and Empires of the Moon*, by French author and poet Savinien Cyrano de Bergerac (1619–55), is published. (In 1662 his *Comic History of the States of the Sun* also is in print.) De Bergerac describes largely fantastic means of space travel, including rising with the dew, magnetic propulsion, and goose power. He also mentions rockets, which must have seemed fantastic at the time.

1783 Two Frenchmen fly over Paris in earliest recorded balloon flight. Balloonists soon discover dangers of low temperatures and lack of oxygen at high altitudes.

1865 French author Jules Verne (1828–1905) writes *From the Earth to the Moon*, in which he describes imaginary lunar voyage. He accurately predicts high speed necessary to leave Earth's atmosphere.

1870 Boston clergyman Edward Hale (1822–1909) publishes serial in *Atlantic Monthly* describing artificial satellite, "Brick Moon," launched by huge, fast-spinning flywheel. It orbits in space to provide information for navigators.

1898 Russian schoolteacher Konstantin Tsiolkovsky (1857–1933) derives mathematical laws that lay foundations for field of astronautics. He foresees space exploration as inevitable part of man's future.

1903 Tsiolkovsky publishes studies of rocket motors and fuels, suggesting that liquid hydrogen and liquid oxygen be used as propellants.

1919 New England physics professor Robert Goddard (1882–1945) publishes mathematical analysis of rocket propulsion as means of overcoming altitude limitations of scientific balloons. His added mention that same principles could be used to reach moon attracts widespread attention.

1923 Rumanian teacher Hermann Oberth (b. 1894) publishes *The Rocket into Interplanetary Space*, describing high-altitude research rockets and manned spaceships as well as large manned space stations. It inspires widespread interest, especially in Germany.

1926 Goddard launches first liquid-propelled rocket. (It flies 184 feet).

1927 First rocket society is organized for actual rocket experimentation, Verein für Raumschiffahrt (Society for Space Travel), in Germany. Oberth later becomes its president. One of society's teenage volunteers, German Wernher von Braun (1912–77), later is hired by German army.

1930 Founding of American Interplanetary Society (later American Institute of Aeronautics and Astronautics.)

1932 German Ordnance Corps begins research on rocket technology under scientists Walter Dornberger (b. 1895) and Wernher von Braun.

1934 First Aggregate-2 rockets, forerunners of German V-2, are successfully fired by von Braun and Dornberger, in Germany.

1935 Goddard has by this date flown rockets of 100 pounds and more as high as 7,500 feet.

1936 Guggenheim Aeronautical Laboratory (California Institute of Technology) begins testing sounding rockets under direction of Hungarian-American physicist Theodore von Karman (1881–1963).

1937 Germans establish rocket research center on Baltic island of Peenemünde. Thousands of engineers and scientists work to develop rockets for German effort.

1942 German V-2 makes its first successful flight (March 5), setting new records for velocity and altitude (53 miles). V-2 rockets are fired against Paris and London in 1944.

1944 Caltech begins American high-altitude rocket research and fires its first rocket, Private A.

1945 First attempt at manned rocket flight takes place, in Germany. Pilot is killed.

1945 U.S. government establishes Operation Paperclip to lure German rocket scientists to work for United States (March).

1945 United States seizes large quantities of rockets and parts from defeated Germans after World War II. More than 120 German scientists, including Dornberger and von Braun, surrender to American troops and move to United States. Soviets capture other German rocket scientists and rocket equipment as well.

1945 White Sands Proving Ground in New Mexico is established for U.S. rocket research. It launches first captured V-2 in 1946.

Getting Space Exploration Off the Ground

1946 Earliest biological experiments dealing with manned space travel begin at Holloman Air Force Base. Scientists send fungus spores up in balloons to be exposed to cosmic radiation.

1948 White Sands researchers launch 9-pound monkey named Albert in V-2 nose cone.

1949 V-2 WAC Corporal sounding rocket reaches outer space, 244 miles in altitude, after launch from White Sands. White Sands also launches first Viking rocket.

1950 First International Astronautical Congress meets, in Paris. During 1951 meeting in London, International Astronautical Federation is formed.

1951 U.S. Aerobee rocket safely carries monkey and several mice to altitude of 236,000 feet.

1953 U.S. physicist proposes "Minimum Orbital Unmanned Satellite of Earth" (MOUSE), hoping to spur development of U.S. space program. Project attracts interest but is never developed.

1955 U.S. rocket scientist suggests using balloons to carry satellites to thinner atmospheres for launching. Satellites would be called "rockoons."

1955 President Dwight D. Eisenhower (1890–1969) announces U.S. plans to launch scientific satellite during International Geophysical Year (1957–58), thereby initiating U.S. satellite program. Soviet Union plans similar launch.

1956 U.S. Viking rocket is sent into space to test telemetry equipment. First test shot of U.S. satellite program.

1957 U.S. attempt to launch Atlas ICBM is unsuccessful (June).

1957 USSR successfully launches 184-pound *Sputnik I*, earth's first artificial satellite (October 4). United States, now "losing" space race, begins crash program in attempt to recover.

1957 Soviets launch *Sputnik II* with dog named Laika aboard (November 3).

1958 U.S. government establishes National Aeronautics and Space Administration (NASA) in attempt to catch up and compete with Soviet Union in space exploration.

1958 United States launches its first satellite, *Explorer I* (January 31). Discovers Van Allen radiation belt. Launches four other satellites and three lunar probes.

1959 United States launches two monkeys into space and retrieves them safely (May).

1959 Soviet-launched *Luna 2* makes first-ever landing on moon (September 12).

1959 Soviet *Luna 3* takes first-ever pictures of moon's dark side (November).

1959 Chimpanzee Sam is launched into space aboard U.S. rocket in experiment to study possible dangers to humans in space (December).

1961 Soviets put first man in space, cosmonaut Yuri Gagarin (April 12).

---

a malfunctioning parachute (April 24). Soviets subsequently completely redesign Soyuz spacecraft.

1968 Astronauts Frank Borman (b. 1928), James A. Lovell, Jr. (b. 1928), and William A. Anders (b. 1933) become first men to orbit moon (10 times) and to see its dark side, during *Apollo 8* moon orbit mission (December 21–27).

1969 Soviets achieve first successful docking of two manned spacecraft. *Soyuz 4* docks with *Soyuz 5* (January 14–15).

1969 *Apollo 10* moon mission (May 18–26) tests lunar lander (LM). Lander descends to within 50,000 feet of moon's surface (May 22).

1969 Historic *Apollo 11* mission lands first men on Moon (July 16–24). Astronauts Neil A. Armstrong, Edwin E. Aldrin, Jr. (b. 1930), and Michael Collins (b. 1930) are aboard 96,698-pound Apollo spacecraft (command and lunar modules). Apollo achieves lunar orbit July 20. After 109 hours, the lunar module (code-named *Eagle*) descends to moon's surface with Armstrong and Aldrin aboard, touching down at 4:17 P.M. (EDT) at Tranquillity Base. Before live television audience on Earth, Armstrong becomes first man to walk on moon (10:56 P.M., EDT). *Apollo 11* leaves

▶

moon July 21 (after 21 hours on surface) and lands on Earth July 24.

1969 Soviets orchestrate first triple launch of manned spacecraft, orbiting *Soyuz 6*, *7*, and *8* simultaneously (October 11–13).

1969 *Apollo 12*, second manned lunar landing (November 14–24). Charles Conrad, Jr. (b. 1930) and Alan L. Bean (b. 1932) land for 31 hours in Ocean of Storms (November 19–20).

1970–72 *Apollo 13* astronauts narrowly escape tragedy when oxygen tank malfunctions during a moon mission (April 11–17). United States launches four other moon missions without incident (*Apollo 14*, January 31–February 9, 1971; *Apollo 15*, July 26–August 7, 1971; *Apollo 16*, April 16–27, 1972; and *Apollo 17*, December 7–19, 1972).

1971 Cosmonauts aboard *Soyuz 9* set a space endurance record of over 17½ days (June 2–19).

1971 Soviets man world's first space station, *Salyut 1* (June 7). The 66-foot-long space station (put into orbit April 18) weighs 40,000 pounds. Cosmonauts Georgi T. Dobrovolski (1928–71), Vladislav N. Volkov (1935–71), and Viktor I. Patsayev (1933–71) remain aboard *Salyut 1* for 23 days. They suffocate during reentry, however (June 29).

1973 U.S. *Skylab*, 120-foot-long orbiting work station, is launched into Earth orbit (May 14). Astronauts Gerald P. Carr (b. 1932), Edward G. Gibson (b. 1936), and William R. Pogue (b. 1930) later set a new space endurance record staying aboard *Skylab* for 84 days (1973–74).

1975 First linkup (July 17) of Soviet and American spacecraft in space (*Soyuz 19* and U.S. *Apollo ASTP* spacecraft).

1977 Soviet cosmonauts begin mission that sets new 96-day space endurance record aboard *Salyut 6* (December 10).

1978 "Guest" cosmonaut, Vladimir Remek of Czechoslovakia, goes into space aboard *Soyuz 28*. Remek is first person in space from a country other than United States or Soviet Union.

1978 Soviet cosmonauts set new space endurance record of 139 days aboard *Salyut 6*.

1979 Soviets begin new record-setting endurance mission of 175 days aboard *Salyut 6* space station.

1980 Cosmonauts begin record-setting 184-day mission aboard *Salyut 6* space station (April 9).

1981 First flight of reusable spacecraft, U.S. Space Shuttle *Columbia* (STS-1) (April 12–14). Astronauts John W. Young (b. 1930) and Robert L. Crippen (b. 1937) pilot 4.58-million-pound *Columbia*, which glides back to Earth April 14, landing (1:21 P.M., EDT) at Edwards Air Force Base after 54½-hour flight. Three other test flights of *Columbia* are flown: November 12–14, 1981; March 22–30, 1982; and June 27–July 4, 1982.

1982 Soviet cosmonauts Anatoliy N. Berezovoy and Velentin Lebedev set new space endurance record of 211 days aboard *Salyut 7*. They return to Earth December 10.

1982 First operational flight of reusable Space Shuttle *Columbia* (November 11–16). *Columbia* carries aloft first four-man crew: Vance D. Brand (b. 1931), Robert F. Overmyer (b. 1936), Joseph P. Allen (b. 1937), and William Lenoir (b. 1939). Five-day mission includes first launch of satellites from the Shuttle.

1983 First American female astronaut is launched into space aboard Shuttle *Challenger* (June 18–24). Crew member Sally K. Ride (b. 1951) goes aloft for just over six days with Robert L. Crippen (b. 1937), Frederick H. Hauck (b. 1941), and John M. Fabian (b. 1939).

1984 Fourth *Challenger* flight, including first-ever untethered space walks using Buck Rogers-type rocket packs (February 3–7).

1984 New space endurance record of 237 days is set by Soviet cosmonauts Leonid Kizim, Vladimir Solovyov, and Oleg Atkov.

▶

1986 Tragic explosion of Shuttle *Challenger* during liftoff at Cape Canaveral, Florida (January 28). It kills all seven astronauts aboard: Francis R. Scobee, Michael J. Smith, Judith A. Resnik, Ronald E. McNair, Ellison S. Onizuka, Gregory B. Jarvis, and Christa McAuliffe. United States halts shuttle flights pending safety studies and redesign of booster rocket.

1986 Soviets launch new space station, called *Mir*, into Earth orbit (February 20).

1987 Soviet cosmonaut Yuri V. Romanenko sets new space endurance record of 326 days in space station *Mir*.

1988 Space Shuttle *Discovery* is launched successfully (September 29), first since 1986 *Challenger* explosion.

1988 Two Soviets, Vladimir Titov and Musa Manarov, set new space endurance record of one year, spent aboard space station *Mir*.

1990 Hubble Space Telescope is deployed in Earth orbit by American astronauts aboard Space Shuttle *Discovery*.

2 Nations and Empires

North America

See also Politics and Law.

A.D.

c. 500 Mayan civilization reaches its height.

c. 750 Toltecs migrate southward into Mexico. They establish flourishing civilization but are driven out by arrival of yet another migrating autocthonous people (c. 1100).

c. 1000 Norseman Leif Ericson explores Canadian coastline. He builds settlement (called Vinland) at Newfoundland.

1325 Aztecs, one of new group of tribes migrating from North, found their capital city of Tenochtitlán. It becomes nucleus of flourishing Aztec culture.

1492 Christopher Columbus (1451–1506), sailing under Spanish flag, discovers Americas and ushers in era North American exploration. (*See Explorers and Exploration.*)

1502–20 Emperor Montezuma (1480?–1520), last Aztec ruler, is in power in Mexico.

1519–21 Spanish conquistador Hernando Cortes (1485–1547) conquers Mexico. Leading small force of about 500 soldiers, he conquers Mayans and Aztecs by guile and by force.

1520 Aztecs unsuccessfully attempt to rebel against Spanish conquistadors. Cortes destroys Tenochtitlán, builds new city (modern Mexico City) in its place, and establishes Spanish control throughout region.

1534–35 French navigator Jacques Cartier (1491–1557) explores region around Gulf of St. Lawrence and returns to Canada to establish Quebec.

1535 Viceroyalty of New Spain is erected in South America. Native population is forcibly Christianized. System of forced labor is set up for work on large plantations owned by Spanish colonists.

1535–50 First Spanish viceroy of Mexico, Antonio de Mendoza (1490?–1552), is in office. Mexico City becomes his capital.

1565 St. Augustine, Florida is founded by Spanish.

1584 Virginia is named for virgin queen, Elizabeth, by Sir Walter Raleigh (1552–1618).

1586 Lost colony of Roanoke is settled. Colonists disappear under unknown circumstances.

1605 French explorer Samuel de Champlain (1567–1635) and others found Port Royal as first permanent settlement in Canada. French subsequently develop fur trade in Canada.

1607 Jamestown colony is established by English settlers in Virginia.

1608 Champlain discovers Lake Champlain (1609) and becomes commandant of New France, French colony in Canada.

1609 English explorer Henry Hudson (fl. 1607–11) explores Hudson River, establishing Dutch claims to area.

1614 Dutch found colony of New Amsterdam in area of modern-day New York City.

1620 Pilgrims land at Plymouth Rock. They establish colony in Massachusetts.

1621 Pilgrims at Plymouth colony celebrate first Thanksgiving.

1623 British establish their first settlement at Nova Scotia.

1636 Harvard College is founded.

1663 Government of France takes over administration of its North American colonies from chartered company.

1664 British capture New Amsterdam from Dutch and rename it New York.

1670 British establish Hudson's Bay Company for development of their territories in Canada.

1677–79 In Culpepper's Rebellion, Carolina colony rebels against British taxation.

1678 French explorer Robert La Salle (1643–87) explores Mississippi River, claiming

▶

U.S. Presidents and Vice Presidents

1789–97 George Washington (1732–99), first President; Vice President 1789–97, John Adams (1735–1826).

1797–1801 John Adams (1735–1826), second President; Vice President 1797–1801, Thomas Jefferson (1743–1826).

1801–9 Thomas Jefferson (1743–1826), third President; Vice President 1801–5, Aaron Burr (1756–1836); 1805–9, George Clinton (1739–1812).

1809–17 James Madison (1751–1836), fourth President; Vice President 1809–12, George Clinton (1739–1812); 1813–14, Elbridge Gerry (1744–1814), who dies in office; 1814–17, none.

1817–25 James Monroe (1758–1831), fifth President; Vice President 1817–25, Daniel Tompkins (1774–1825).

1825–29 John Quincy Adams (1767–1848), sixth President; Vice President 1825–29, John Calhoun (1782–1850).

1829–37 Andrew Jackson (1767–1845), seventh President; Vice President 1829–32, John Calhoun (1782–1850); 1833–37, Martin Van Buren (1782–1862).

1837–41 Martin Van Buren (1782–1862), eighth President; Vice President 1837–41, Richard M. Johnson (1780–1850).

1841 (March–April) William Henry Harrison (1773–1841), ninth President and first to die in office; Vice President 1841 (March–April), John Tyler (1790–1862).

1841–45 John Tyler (1790–1862), 10th President; who completes Harrison's term; Vice President 1841–45, none.

1845–49 James Polk (1795–1849), 11th President; Vice President 1845–49, George Dallas (1792–1864).

1849–50 Zachary Taylor (1784–1850), 12th President, who dies in office; Vice President 1849–50, Millard Fillmore (1800–1874).

1850–53 Millard Fillmore (1800–1874), 13th President; Vice President 1850–53, none.

1853–57 Franklin Pierce (1804–69), 14th President; Vice President 1853, William King (1786–1853), who dies before taking office; 1853–57, none.

1857–61 James Buchanan (1791–1868), 15th President; Vice President 1857–61, John Breckinridge (1821–75).

1861–65 Abraham Lincoln (1809–65), 16th President, assassinated early in his second term; Vice President 1861–65, Hannibal Hamlin (1809–91); 1865, Andrew Johnson (1808–75).

1865–69 Andrew Johnson (1808–75), 17th President; Vice President 1865–69, none.

1869–77 Ulysses S. Grant (1822–85), 18th President; Vice President 1869–73, Schuyler Colfax (1823–85); 1873–75, Henry Wilson (1812–75).

1877–81 Rutherford B. Hayes (1822–93), 19th President; Vice President 1877–81, William Wheeler (1819–87).

1881 (March–September) James A. Garfield (1831–81), 20th President, who is assassinated; Vice President 1881 (March–September), Chester A. Arthur (1829–86).

1881–85 Chester A. Arthur (1829–86), 21st President, who serves out Garfield's term; Vice President 1881–85, none.

1885–89 Grover Cleveland (1837–1908), 22nd President; Vice President 1885, Thomas Hendricks (1819–85), who dies in office; 1885–89, none.

1889–93 Benjamin Harrison (1833–1901), 23rd President; Vice President 1889–93, Levi Morton (1824–1920).

1893–97 Grover Cleveland (1837–1908), 24th President; Vice President 1893–97, Adlai Stevenson (1835–1914).

1897–1901 William McKinley (1843–1901), 25th President, who is assassinated by anarchist; Vice President 1897–99, Garret Hobart (1844–99), who dies in office; 1901 (March–September), Theodore Roosevelt (1858–1919).

1901–9 Theodore Roosevelt (1858–1919), 26th President, elected after completing McKinley's term; Vice President 1901–5, none; 1905–9, Charles Fairbanks (1852–1918.)

1909–13 William H. Taft (1857–1930), 27th President. Vice President 1909–12, James Sherman (1855–1912), who dies in office.

1913–21 Woodrow Wilson (1856–1924), 28th President. Vice President 1913–21, Thomas Marshall (1854–1925).

U.S. Presidents and Vice Presidents

1921–23 Warren G. Harding (1865–1924), 29th President, who dies before completing term; Vice President 1921–23, Calvin Coolidge (1872–1933).

1923–29 Calvin Coolidge (1872–1933), 30th President, elected after completing Harding's term; Vice President 1923–25, none; 1925–29, Charles Dawes (1865–1951).

1929–33 Herbert Hoover (1874–1964), 31st President; Vice President 1929–33, Charles Curtis (1860–1936).

1933–45 Franklin D. Roosevelt (1882–1945), 32nd President, who dies soon after being elected to unprecedented fourth term; Vice President 1933–41, John N. Garner (1868–1967); 1941–45, Henry A. Wallace (1888–1965); 1945 (January–April), Harry S. Truman (1884–1972).

1945–53 Harry S. Truman (1884–1972), 33rd President, who is elected after serving out Roosevelt's term; Vice President 1945–49, none; 1949–53, Alben W. Barkley (1877–1956).

1953–61 Dwight D. Eisenhower (1890–1969), 34th President; Vice President 1953–61, Richard M. Nixon (b. 1913).

1961–63 John F. Kennedy (1917–1963), 35th President, who is assassinated; Vice President 1961–63, Lyndon B. Johnson (1908–73).

1963–69 Lyndon B. Johnson (1908–73), 36th President, who is elected after serving out Kennedy's term; Vice President 1963–65, none; 1965–69, Hubert H. Humphrey (1911–1978).

1969–74 Richard M. Nixon (b. 1913), 37th President, who resigns early in second term; Vice President 1969–73, Spiro T. Agnew (b. 1918), who resigns during term; 1973–74, Gerald R. Ford (b. 1913), who is appointed to finish Agnew's term.

1974–77 Gerald R. Ford (b. 1913), 38th President, who serves out Nixon's term; Vice President 1974–77, Nelson A. Rockefeller (1908–79).

1977–81 Jimmy Carter (b. 1924), 39th President; Vice President 1977–81, Walter F. Mondale (b. 1928).

1981–89 Ronald Reagan (b. 1911), 40th President; Vice President 1981–89, George Bush (b. 1924).

1989– George Bush (b. 1924), 41st President; Vice President 1989– , H. Danforth Quayle (b. 1947).

valley and naming it Louisiana (for Louis XIV of France).

1689–97 King William's War marks beginning of sporadic fighting between France and Great Britain for control of North American territories.

1691 First postal service in colonial America begins.

1702–13 In Queen Anne's War, part of larger War of Spanish Succession, British win important victories over French in Canada and by subsequent treaty gain much of French colonial territories.

1734–35 Freedom of the press in American colonies is established by acquittal of New York newspaperman Peter Zenger (1697–1746) on libel charges.

1744–48 In King George's War, part of larger War of Austrian Succession, fighting in Canada centers on Nova Scotia. Captured territories are returned after the war, however.

1754–63 In French and Indian War, part of larger Seven Years War, English and Indian allies capture Quebec (1759) and soundly defeat French. French lose all territory in Canada and in continental United States between Alleghenies and Mississippi River.

1755 First Canadian post office is opened.

1763–66 Pontiac's Rebellion occurs, bloody Indian revolt against British control in Great Lakes region.

1765 First book is printed in Canada.

1765 British impose Stamp Act, tax on publications, in colonial America. Opposition forces its repeal.

▶

Canadian Prime Ministers			
1867–73	Sir John Macdonald (1815–91)	**1926–30**	W. L. Mackenzie King (1874–1950)
1873–78	Alexander Mackenzie (1822–92)	**1930–35**	Richard Bedford Bennett (Viscount Bennett) (1870–1947)
1878–91	John Alexander Macdonald (1815–91)	**1935–48**	W. L. Mackenzie King (1874–1950)
1891–92	Sir J. C. Abbot	**1948–57**	Louis Stephen Saint Laurent (1882–1973)
1892–94	Sir John S. D. Thompson (1844–94)	**1957–63**	John George Diefenbaker (1895–1979)
1894–96	Sir Mackenzie Bowell (1823–1917)	**1963–68**	Lester B. Pearson (1897–1972)
1896	Sir Charles Tupper (1821–1915)	**1968–79**	Pierre Elliott Trudeau (b. 1919)
1896–1911	Sir Wilfrid Laurier (1841–1919)	**1979–80**	Charles Joseph Clark (b. 1939)
1911–20	Sir Robert Laird Borden (1854–1937)	**1980–84**	Pierre Elliott Trudeau (b. 1919)
1920–21	Arthur Meighen (1874–1960)	**1984**	John N. Turner (b. 1929)
1921–26	W. L. Mackenzie King (1874–1950)	**1984-**	Brian Mulroney (b. 1939)
1926	Arthur Meighen (1874–1960)		

1767 Townshend Acts are passed by British Parliament. British impose import tax on American colonies and contribute to sentiment for revolt against British rule.

1773 Boston Tea Party occurs.

1774 Quebec Act sets up colonial government of Canada, with special provisions for large French-speaking population.

1775–83 In American Revolutionary War Canada remains loyal to British crown and becomes haven for tens of thousands of American Tories after war. English-speaking population increases in relation to French-speaking population as result. (*See Wars and Military History.*)

1776 Continental Congress adopts Declaration of Independence.

1777 Articles of Confederation are approved by Continental Congress. They become effective in 1781.

1786 Corrupt colonial government in Mexico is reformed.

1786–87 Shay's rebellion occurs in Massachusetts.

1787 U.S. Constitutional Congress drafts new Constitution after weak central government under Articles of Confederation proves ineffective.

1789 Constitution is ratified by 11 of 13 states.

1790 District of Columbia is created to provide site for new U.S. national capital.

1791 In Constitutional Act, Britain divides most important Canadian province, Quebec, into Upper (English-speaking) and Lower (French-speaking) Canada. Colonies are administered separately.

1791 Vermont becomes first new U.S. state created after original 13.

1791 U.S. Bill of Rights is ratified.

1794 In Whiskey Rebellion, farmers in Pennsylvania rise against liquor tax.

1800 Thomas Jefferson (1743–1826) is elected U.S. President by vote in the House of Representatives, to break tie in electoral votes.

1801–5 In Tripolitan War, United States wars against Barbary pirates.

1803 *In Marbury* v. *Madison*, U.S. Supreme Court finds act of Congress (Judiciary Act of 1801) unconstitutional. It is first decision voiding an act of Congress.

1803 In Louisiana Purchase, United States buys territory west of Mississippi River from France.

1804 Meriwether Lewis (1774–1809) and William Clark (1770–1838) explore Louisiana Purchase territory and land

▶

beyond it to Pacific Ocean. (*See Explorers and Exploration for information on exploration of American West.*)

1804 Twelfth amendment to U.S. Constitution is ratified, providing for election of President by Electoral College.

1807 *Clermont*, built by American inventor Robert Fulton (1765–1815), becomes first commercially successful steamboat.

1808 Napoleon's French army invades and occupies Spanish-ruled Mexico.

1810 Revolt against Spain's colonial rule begins in Mexico under Miguel Hidalgo y Costilla (1753–1811) as turmoil in Europe resulting from Napoleonic Wars weakens Spain's hold on its New World colonies. Rebels are defeated at Battle of Calderón Bridge (1811).

1812–15 In War of 1812, British forces in Canada repulse several American invasions, though Toronto is burned (1813). British fail in invasion of United States but burn Washington, D.C. British bombardment of Baltimore's Fort McHenry is recounted in "The Star-Spangled Banner" by Francis Scott Key (1779–1843).

1813 Rebels proclaim Mexico constitutional republic. They carry on guerrilla warfare against Spanish rule.

1817 Rush-Bagot Agreement between United States and British Canada establishes nonmilitarized border with Canada.

1818 Convention of 1818 sets border between United States and Canada.

1819 In Transcontinental Treaty, Spain cedes Florida to United States.

1820–21 Missouri Compromise is passed by Congress. It provides for admission of Missouri as slave state and Maine as free state, maintaining balance between free and slave states.

1821 In Treaty of Aquala, Mexico's independence is formally recognized by Spain.

1822–23 Rebel leader Augustín de Iturbide (1783–1824) rules as emperor of short-lived Mexican empire.

1823 U.S. President Monroe articulates Monroe Doctrine.

1823 Mexico is established as federal republic. Constitution is enacted in 1824.

1824 Election of John Quincy Adams (1767–1848) for President is decided in House of Representatives due to lack of majority winner in electoral votes.

1828 U.S. South is hurt by high Tariff of Abominations, which favors America's developing industry by imposing tax on imports.

1829 Spanish military force is driven out of Mexico by General Antonio López de Santa Anna (1794–1876).

1830 South Carolina Canal and Rail Road offers first regular steam railway service in United States.

1831 Southampton Insurrection, unsuccessful slave revolt, occurs in Virginia.

1836 U.S. Senate decides election of Vice President Richard M. Johnson, Democrat of Kentucky, only Vice President so elected.

1836 In Texas War of Independence, Mexican province of Texas, largely populated by Americans, secedes and successfully turns back Santa Anna's armies attempting to retake region.

1837–38 Unsuccessful revolt is sparked by William L. Mackenzie (1795–1861) in Upper and Lower Canada. He escapes to United States.

1839 Lord Durham's Report recommends unification of colonial government in Canada under single administration.

1840 Under Act of Union, British unite Upper and Lower Canada.

1842 Webster-Ashburton Treaty between Great Britain and United States grants latter over half disputed Maine–New Brunswick frontier.

▶

1842–46 U.S. Army captain John Charles Frémont (1813–90) explores West. He leads expedition from Great Salt Lake to Sutter's Fort in California.

1844 Samuel F. B. Morse (1791–1872) begins first telegraph service.

1846 Boundary between Canada and United States-held Oregon Territory is set.

1846–48 Mexican War between United States and Mexico is fought over admission of Texas to United States and other grievances. United States successfully invades Mexico and takes Mexico City (1848). Treaty of Guadalupe Hidalgo (1848), which ends war, cedes all of (modern) American Southwest and California to United States for $15 million.

1848 California gold rush begins.

1850 Compromise of 1850 is effected between pro- and antislavery factions in United States.

1855 Santa Anna's dictatorship in Mexico is overthrown. Liberals attempt to institute reforms.

1857 U.S. Supreme Court in *Dred Scott* v. *Sandford* legalizes slavery in U.S. territories and finds Missouri Compromise of 1820–21 unconstitutional.

1857–61 Period of revolt and political unrest begins in Mexico after establishment of new, liberal constitution at Veracruz. Conservatives, including supporters of church, oppose liberals, who nationalize church lands and impose other reforms.

1859 Harper's Ferry Raid, unsuccessful attempt by abolitionist John Brown (1880–59) to start slave uprising, takes place.

1861 Liberal leader Benito Juárez (1806–72) is elected President amid continuing political turmoil.

1861–62 Joint military expedition by Spain, Great Britain, and France invades Mexico to restore order. Spanish and British withdraw, however, after continued unrest (1862). French continue warfare with Mexican revolutionaries.

1861–65 American Civil War occurs. (*See Wars and Military History.*)

1863 Emancipation Proclamation is issued by President Lincoln. It technically frees slaves in Confederate-held territories.

1863 In draft riots, workers in New York City riot for four days over Civil War draft law with paid deferment provision; over 1,200 die.

1864 Maximilian (1832–67), Austrian archduke, becomes emperor of newly established French-dominated Empire of Mexico.

1865 President Lincoln is shot and killed by John Wilkes Booth at Ford's Theater, Washington, D.C.

1865 Thirteenth amendment to U.S. Constitution is ratified, prohibiting slavery.

1865 New Mexican constitution is established. Rebels continue to fight French occupying forces.

1867 British North America Act creates Dominion of Canada composed of provinces of Quebec, Ontario, New Brunswick, and Nova Scotia. Act serves as Canada's written Constitution.

1867 United States buys Alaska from Russia for $7.2 million.

1867 French Army leaves Mexico. Soon after, empire is overthrown and Maximilian is captured and shot.

1867 Mexican republic is reestablished. Congress meets.

1867–72 Liberal leader Benito Juárez is in power again as President. Rebellion and civil strife begin anew.

1867–73 First Canadian prime minister, Sir John Macdonald (1815–91), takes office.

1868 Impeachment and acquittal of U.S. President Andrew Johnson (1808–75) occur. Process is sparked by differences between Congress and President over Reconstruction in South.

1868 Burlingame Treaty signed between United States and China.

▶

Mexico's Presidents and Heads of State

1824–29	Guadalupe Victoria (Juan Felix Fernandez) (1789–1843)
1829	Vicente Guerrero (1783–1831)
1829–32	Anastasio Bustamante (1780–1853)
1832–33	Manuel Gómez Pedraza (1788–1851)
1833–35	Antonio López de Santa Anna (1794–1876)
1835–36	Miguel Barragán (1789–1836)
1836–37	José Justo Corro (1794–1864)
1837–41	Anastasio Bustamante (1780–1853)
1841–42	Antonio López de Santa Anna (1794–1876)
1842–43	Nicolás Bravo (1787–1854)
1843–44	Valentin Canalizo (1797–1847)
1844	Antonio López de Santa Anna (1794–1876)
1844–45	José Joaquín Herrera (1792–1854)
1846	Mariano Paredes y Arrillaga
1846	Nicholás Bravo (1787–1854)
1847	Antonio López de Santa Anna (1794–1876)
1847	Pedro Maria Anaya (1794–1854)
1847	Antonio López de Santa Anna (1794–1876)
1848	Manuel de la Peña y Peña (1789–1850)
1848–51	José Joaquín de Herrera (1792–1854)
1851–53	Mariano Arista (1802–55)
1853–55	Antonio López de Santa Anna (1794–1876), dictator
1855–57	Ignacio Comonfort (1812–63)
1858–63	Benito Juárez (1806–72), provisional President to 1861
1864–67	Maximilian (1832–67), emperor during French occupation
1867–72	Benito Juárez (1806–72)
1872–76	Sebastián Lerdo de Tejada (1827–89)
1877–80	Porfirio Díaz (1830–1915)
1880–84	Manuel González (1833–93)
1884–1911	Porfirio Díaz (1830–1915)
1911–13	Francisco Indalecio Madero (1873–1913)
1913–14	Provisional presidency
1914–15	Eulalio Gutiérrez (1832–1939)
1915–20	Venustiano Carranza (1859–1920), provisional to 1917
1920–24	Alvaro Obregón (1880–1928)
1924–28	Plutarco Elias Calles (1877–1945)
1928–30	Provisional presidency
1930–32	Pascual Ortiz Rubio (1877–1963)
1932–34	Provisional presidency
1934–40	Lázaro Cárdenas (1895–1970)
1940–46	Manuel Avila Camacho (1897–1955)
1946–52	Miguel Aleman Valdés (1903–83)
1952–58	Adolfo Ruiz Cortines (1891–1973)
1958–64	Adolfo López Mateos (1910–69)
1964–70	Gustavo Diaz Ordaz (1911–79)
1970–76	Luís Echeverria Alvarez (b. 1922)
1976–82	José López Portillo (b. 1920)
1982–88	Miguel de la Madrid Hurtado (b. 1934)
1988–	Carlos Salinas de Cortari (b. 1948)

1869 U.S. transcontinental railroad is completed with driving of golden spike at Promontory, Utah.

1869–70 Riel's rebellion in Canadian Northwest is suppressed. New revolt by Riel (1885) also is unsuccessful.

1870 Red River settlement enters Dominion of Canada as province of Manitoba.

1871 In *Legal Tender Case* (Second), U.S. Supreme Court decides Legal Tender acts of 1862 and 1863, which fall within federal government's powers to meet emergencies.

1871 Great Chicago Fire, one of worst fires in U.S. history, destroys much of city.

1871 British Columbia becomes Canadian province.

1873 Railroad is completed between Veracruz and Mexico City.

1873–96 Canada suffers through prolonged economic depression.

►

1874 Mexican government suppresses religious orders.

1876 Porfirio Díaz (1830–1915), rebel leader in earlier period of unrest, takes up new rebellion against Mexican government.

1877 U.S. Congress declares Rutherford B. Hayes President following disputed election results in four states.

1880s U.S. railroad building boom peaks. Over 70,000 miles of track are laid in this decade.

1881 U.S. President James A. Garfield (1831–81) is shot and killed by disgruntled office-seeker.

1884–1911 Díaz again is in office as Mexican President.

1885 First trans-Canadian railway, Canadian Pacific Railway, is completed.

1886 In Haymarket Square riot, anarchist bomb kills 11 during Chicago labor demonstration.

1886 Statue of Liberty, gift from France, is unveiled in New York.

1890 Sherman Antitrust Act is passed.

1896 In *Plessy* v. *Ferguson*, U.S. Supreme Court upholds separate-but-equal public-school facilities for Negroes.

1896–1911 First French-Canadian prime minister, Sir Wilfrid Laurier (1841–1919), is in office.

1897 Gold rush in Klondike region begins.

1898 In Spanish-American War, United States gains control of Philippines and Cuba.

1901 President McKinley is shot and killed by anarchist.

1903 Era of "muckrakers" begins. U.S. magazine and tabloid reporters expose corruption in politics and business.

1903 In *Champion* v. *Ames*, U.S. Supreme Court approves federal powers to prohibit as well as regulate commerce, thereby establishing so-called federal police power.

1903 Wright Brothers achieve first successful powered flight of heavier-than-air craft, at Kitty Hawk, North Carolina.

1903 Hay-Bunau-Varilla Treaty grants United States rights to Panama Canal Zone.

1903 Alaska boundary is set by joint Canadian–U.S. commission.

1905 Canadian Northwest Territories region is reorganized; provinces of Alberta and Saskatchewan are formed.

1906 San Francisco earthquake destroys 4 square miles of downtown district.

1907 Democrats in Congress criticize government spending, which has doubled in past 10 year's to $1 billion per year.

1908 In *Danbury Hatters Case* (*Loewe* v. *Lawler*), U.S. Supreme Court rules against secondary boycott as restraining trade under Sherman Antitrust Act.

1908 Canadian Civil Service Commission is established.

1909 Government establishes Canadian Navy.

1910 International arbitration court settles Canadian–U.S. differences over Atlantic fishing rights.

1911 Díaz is overthrown by Francisco Indalecio Madero (1873–1913). Mexican Revolution, long period of political unrest, follows.

1913 Sixteenth amendment to U.S. Constitution is ratified, giving Congress power to impose income taxes.

1914 Panama Canal, built by United States, is opened.

1914 U.S. Marines briefly occupy Veracruz after Mexicans seize U.S. ship in harbor there.

1914–18 In World War I Germany's unrestricted submarine warfare against merchant shipping helps bring United States into war on Allied side (1917). (*See Wars and Military History*.)

▶

1914–18 Canada joins Allies in World War I, providing substantial troop support and war materiel.

1916 Mexican rebel leader Pancho Villa (1878–1923) raids Columbus, New Mexico. He eludes U.S. military expedition sent after him.

1917 New Mexican Constitution is enacted. It provides for agrarian reforms and separation of church and state.

1917 Unrest erupts in Canada over newly enacted conscription for World War I, particularly among French-Canadians.

1918 President Wilson presents his 14 Points, peace proposals advanced at Paris Peace Conference after World War I. One proposal calls for establishment of League of Nations, which United States later refuses to join.

1920 Nineteenth amendment to U.S. Constitution is ratified, giving women right to vote.

1920s Roaring Twenties period, postwar era in United States, is marked by economic boom, enactment of Prohibition (1920), and rapid social change.

1921 Four-Power Pacific Treaty among United States, France, Great Britain, and Japan recognizes their respective spheres of influence in Pacific.

1922 In Five-Power Naval Limitation Treaty, United States, Italy, France, Great Britain, and Japan agree to limit size of their navies.

1922 Mexican government seizes 1.9 million acres to effect land reforms.

1923 Teapot Dome scandal, involving corruption in Harding administration, begins to break.

1923–29 Calvin Coolidge (1872–1933) is in office as 30th U.S. President after Harding dies in office.

1926 Canada appoints minister to Washington, D.C. United States reciprocates with minister to Canada.

1926 Church property is nationalized in Mexico.

1927 American aviator Charles Lindbergh (1904–74) makes first solo transatlantic flight, from New York to Paris.

1927 Canada is admitted to League of Nations Council.

1928 United States and nearly every other nation sign Kellogg-Briand Pact, renouncing war.

1929 Stock-market crash marks beginning of Great Depression.

1930 London Naval Treaty establishes permanent parity among U.S., British, and Japanese navies.

1930 Conservative government is voted in Canada for first time in almost decade. Responding to economic crisis of Great Depression, Canada establishes various government relief programs, restricts immigration, and sets high, protective tariff on trade.

1931 Statute of Westminster is enacted by Great Britain, establishing Canada as independent, self-governing member of British Commonwealth of Nations.

1933 Roosevelt convenes special congressional session to halt wave of bank failures resulting from Depression. He subsequently introduces many innovative "New Deal" measures to help ease ill effects of the Depression in United States.

1933 Canada officially abandons gold standard.

1933 Twentieth amendment to U.S. Constitution is ratified, changing terms of office of Congress, President, and Vice President to prevent "lame-duck" sessions of Congress.

1933 Twenty-first amendment to U.S. Constitution is ratified repealing prohibition.

1933–39 Severe drought in U.S. Midwest farming region results in huge dust storms and migration of thousands of farmers from area.

1934–40 Lázaro Cárdenas (1895–1970) is in office as Mexican President. He institutes

▶

land reforms, nationalizes railroads, and seizes foreign-owned oil-producing facilities and mines (1937–38).

1935 In *Sick Chicken Case* (*Schechter Poultry* v. *United States*), U.S. Supreme Court voids National Industrial Recovery Act (1933) as unconstitutional. It is one of several rulings against New Deal legislation.

1935 Bank of Canada is established. It is nationalized in 1938.

1935 First U.S.–Canadian reciprocal trade agreement is established.

1939–45 In World War II, United States remains nominally neutral until Japan's sneak attack on Pearl Harbor (1941). United States emerges from this war as world's leading power. (*See Wars and Military History.*)

1939–45 In World War II, Canada again supplies men and materiel for Allied side. War stimulates growth of Canadian industry as demand for war materiel increases.

1940 Leon Trotsky (1879–1940), Communist refugee from Stalinist Soviet Union, is assassinated in Mexico City.

1941 Canadian economy is regulated by government from this time forward during war years.

1944 Socialist party wins first Canadian victory, in Saskatchewan.

1945 United States drops atomic bomb on Hiroshima (and later Nagasaki) after Japanese refuse to surrender at end of World War II, marking beginning of nuclear age.

1945 Canada, Mexico, and United States become charter members of United Nations.

1946 U.S. Central Intelligence Agency is organized.

1946 U.S. G.I. Bill is passed.

1947 U.S. military branches are united under Department of Defense.

1947 French ship *Grandcamp* explodes in Texas City, Texas harbor, destroying most of city. Over 550 die.

1947 Marshall Plan is proposed to provide massive economic aid to war-torn Europe and to counter Communist expansion there.

1948 In Berlin Blockade, United States supplies West Berlin by airlift until Soviets end blockade in 1949.

1949 North Atlantic Treaty Organization (NATO) is formed.

1949 Newfoundland is admitted as Canadian province.

1950 State Department official Alger Hiss (b. 1904) is accused of being Communist spy but actually is convicted only of perjury charge.

1950 U.S. senator Joseph R. McCarthy (1909–57) begins notorious McCarthy hearings. Senate finally censures McCarthy in 1954.

1950–53 In Korean War, first use of direct military intervention to halt Communist expansion proves costly, but Communists are turned back. (*See Wars and Military History.*)

1951 Japanese Peace Treaty between Japan and 48 nations formally ends World War II hostilities.

1951 Twenty-second amendment to U.S. Constitution is ratified, limiting President's service to two terms in office.

1954 In *Brown* v. *Board of Education of Topeka, Kansas*, U.S. Supreme Court ends long-standing practice of separate-but-equal public-school facilities for blacks. (*See Politics and Law, "Civil Rights in the United States."*)

1957 Eisenhower Doctrine is announced. United States offers economic and military aid to any country resisting Communist takeover.

1958 NORAD, early-warning radar system, is established jointly by United States and Canada.

1959 In "kitchen debate," Vice President Richard M. Nixon (b. 1913) debates

▶

with Soviet leader Khrushchev at U.S. trade show in Moscow.

1959 St. Lawrence Seaway, joint U.S.–Canadian waterway linking Atlantic Ocean and Great Lakes, is completed.

1960 U.S. flag with 50 stars is introduced.

1960s Period of political unrest and social change occurs in United States. Decade is marked by civil-rights protest and race riots by blacks in major cities across country and by mass protests (frequently violent) against war in Vietnam.

1960–75 Vietnam War occurs. (*See Vietnam War in Wars and Military History.*)

1961 U.S. astronaut Alan Shepard (b. 1923) becomes first American in space during 15-minute suborbital flight.

1961 United States suffers major diplomatic setback with botched Bay of Pigs invasion.

1962–63 In Cuban missile crisis, President Kennedy risks nuclear war to force withdrawal of Soviet missiles in Cuba, regaining much world prestige for United States.

1963 Hot-line communications link between Washington, D.C. and Moscow is installed.

1963 United States signs Nuclear Test-Ban Treaty, limiting nuclear testing to underground explosions.

1963 President Kennedy is shot and killed in Dallas by Lee Harvey Oswald. He is succeeded by Lyndon B. Johnson.

1964 Civil Rights Act is passed, ending racial discrimination in public places and establishing equal-opportunity employment program. (*See Politics and Law, "Civil Rights in the United States."*)

1966 In *Miranda* v. *Arizona*, U.S. Supreme Court overturns conviction of confessed rapist and establishes requirement for so-called Miranda warnings.

1968 In separate incidents, civil-rights leader Martin Luther King, Jr. (1929–68) and Senator Robert F. Kennedy (1925–68) are shot and killed.

1969 Nixon begins Strategic Arms Limitation Talks (SALT) with Soviets, seeking to establish détente with them.

1970 Voting age is lowered from 21 to 18 in United States.

1970s Mexico eases its economic problems by increasing oil production. By 1980s, however, oil glut leaves Mexico in serious economic trouble.

1972 President Nixon makes historic visit to China.

1972 President Nixon makes historic first visit by U.S. President to USSR. He signs SALT I antiballistic missile treaty.

1972 In *Furman* v. *Georgia*, U.S. Supreme Court rules against current state laws on capital punishment.

1972–74 In Watergate scandal, President Nixon becomes involved in attempt to cover up burglary of Democratic party headquarters and is forced to resign (1974).

1973 In *Roe* v. *Wade*, U.S. Supreme Court invalidates state laws against abortion for women up to six months pregnant.

1974 President Ford pardons Nixon for Watergate crimes.

1974 U.S. government grants amnesty to those who evaded draft during Vietnam War years.

1976 Eskimos file claim asserting rights to over 750,000 square miles of Canadian territory.

1977 Panama Canal Treaty returns canal to Panama.

1979 Three-Mile Island nuclear-power-plant accident slows further growth of nuclear-power industry due to safety concerns.

1979 Second Strategic Arms Limitation Treaty (SALT II) is signed (United States does not ratify it, however).

1979–81 Fifty-two Americans at U.S. embassy in Iran are held hostage by Iranian revolutionaries.

▶

1980 Canadians in Quebec vote down French separatism in this predominantly French-speaking province.

1980 Canadian government proclaims national anthem in effort to promote unity.

1980s Mexico balances on brink of financial disaster as high interest rates on massive foreign loans and worsening economic problems threaten to drive nation into bankruptcy.

1981 President Reagan is shot and wounded in assassination attempt.

1981 First flight of a reusable spacecraft, U.S. space shuttle *Columbia*.

1982 In Constitution Act, Great Britain reaffirms Canadian autonomy within British Commonwealth and Canada's continued recognition of British monarch. Canada gains right to amend its Constitution, British North America Act of 1867.

1982 Mexican government nationalizes Mexican-owned banks. It devalues peso by 30 percent.

1986 Iran-contra scandal erupts over diversion of funds from sale of arms to Iran to contra rebels in Nicaragua. Some $3 million is reportedly diverted.

1987 President Reagan signs Intermediate Nuclear Forces (INF) Treaty, providing for dismantling of all U.S. and Soviet intermediate-range nuclear weapons.

1987 Quebec provincial government signs Canadian Constitution.

1987 Canada and United States agree to free-trade arrangement.

1987 Mexico's inflation rate hits record 159 percent for this year. Wage and price controls cut inflation to about 52 percent in 1988.

1989 United States successfully pressures banks to new agreement to ease Mexico's massive foreign debt.

1989 Bush administration wins legislative battle for federal bailout of nation's ailing savings-and-loan banks. It will eventually cost $159 billion.

1989 Limits on foreign investment are eased as part of economic reform program in Mexico.

1989–90s Pro-life groups challenge abortion laws. Several states, under a new Supreme Court interpretation of *Roe* v. *Wade*, introduce bills to restrict or proscribe abortion.

1990 United States organizes international blockade in Persian Gulf, Gulf of Oman, and Red Sea in response to Iraqi invasion of Kuwait and threatened Middle Eastern crisis.

U.S. Territorial Expansion

(*See also Explorers and Exploration.*)

States are listed by the date they entered the Union, territories by the date of acquisition. Dates given after the state name are for first exploration by Europeans and for the start of lasting settlement of the territory.

The original 13 states are grouped below by order of admission:

1787 Delaware. Explored 1609; settled 1638.

1787 Pennsylvania. Explored c. 1615; settled 1643.

1787 New Jersey. Explored 1609; settled 1664.

1788 Georgia. Explored 1540; settled 1733 at Savannah.

1788 Connecticut. Explored 1614; settled 1633.

1788 Massachusetts. Possibly explored by Norsemen 11th century; settled 1620 at Plymouth colony by English Puritans.

▶

1788 Maryland. Explored 1498; settled 1634.

1788 South Carolina. Explored 1526; settled 1670.

1788 New Hampshire. Explored 1603; settled c. 1625.

1788 Virginia. Settled 1607 at Jamestown, as first permanent English colony in America.

1788 New York. Explored 1524; settled 1613.

1789 North Carolina. Explored 1524; settled 1600s.

1780 Rhode Island. Explored 1524; settled 1636.

1790 District of Columbia. Created from lands ceded by surrounding states.

1791 Vermont. Explored 1609; settlement 1724. First new state after original 13.

1792 Kentucky. Explored c. 1750; settled 1775.

1796 Tennessee. Explored 1540; settled 1769.

1803 Ohio. Explored 1600s; settled 1788.

1812 Louisiana. Explored 1528; settled c. 1715.

1816 Indiana. Explored 1679; settled c. 1700. Land developers devised state name (c. 1765) by adding *a* to *Indian*.

1817 Mississippi. Explored 1540; settled 1699.

1818 Illinois. Explored 1673; settled c. 1680.

1819 Alabama. Explored 1528; settled 1702.

1820 Maine. Possibly explored by Norsemen 11th century; settled 1624.

1821 Missouri. Explored 1683; settled c. 1735.

1836 Arkansas. Explored 1514; settled 1686.

1837 Michigan. Explored 1618; settled 1668.

1845 Florida. Explored 1513; settled 1565 by Spanish at St. Augustine, (now oldest city in United States).

1845 Texas. Explored c. 1530; settled 1682.

1846 Iowa. Explored 1673; settled c. 1790.

1848 Wisconsin. Explored 1634; settled c. 1670.

1850 California. Explored 1541; settled 1769.

1858 Minnesota. Explored c. 1660; settled c. 1815.

1859 Oregon. Explored 1579; settled 1811.

1861 Kansas. Explored 1541; settled 1727.

1863 West Virginia. Explored 1670s; settled 1726.

1864 Nevada. Explored c. 1770; settled c. 1850.

1867 Midway Islands (territory). Acquired by annexation.

1867 Nebraska. Explored 1541; settled 1847.

1876 Colorado. Explored 1500s; settled, 1858. Admitted during U.S. independence centennial year.

1889 North Dakota. Explored 1738; settled 1812.

1889 South Dakota. Explored c. 1742; settled 1817.

1889 Montana. Explored 1742; settled 1807.

1889 Washington. Explored c. 1540s; settled 1845.

1890 Idaho. Explored 1805; settled 1860.

1890 Wyoming. Explored 1700s; settled 1834.

1896 Utah. Explored 1540; settled 1847 by Mormons at Salt Lake City.

1898 Guam (territory). Acquired from Spain.

1898 Puerto Rico. Acquired from Spain; became self-governing 1952.

1898 Wake Island (territory). Acquired by annexation.

1899 American Samoa (territory). Acquired by treaty.

1903 Canal Zone (territory). Leased from Panama; control largely returned in 1978.

1907 Oklahoma. Explored 1541; settled 1796.

1912 New Mexico. Explored c. 1530; settled 1598.

1912 Arizona. Explored c. 1535; settled 1600s. Admitted February 14, Valentine's Day.

1917 Virgin Islands (territory). Acquired from Denmark.

1959 Alaska. Explored 1741; settled 1784. Gained territorial status 1912.

1959 Hawaii. Explored 1778; settled 1800s. Annexed by United States 1898 with Hawaiian indigenous government consent.

Central and South America

Since 1800

1803–15 Napoleonic Wars in Europe. Spain is occupied by French forces, and Napoleon installs Joseph Bonaparte as king of Spain. Unrest in Europe becomes opportunity for revolt in Spanish New World colonies.

1808 Portuguese monarchy flees to Brazil during troubles in Europe. Rio de Janeiro becomes their capital until their return to Europe in 1821.

1810 Revolutionaries in Argentina oust Spanish colonial viceroy and rule by junta. Region of modern Bolivia, Uruguay, and Paraguay remains in royalist hands, however.

1810 Revolt for independence begins in Venezuela and New Granada (modern Colombia and other territories).

1811 Paraguay gains independence following bloodless revolt against Spanish.

1816 Argentina declares complete independence from Spain.

1816–20 Brazil invades and conquers Uruguay.

1817 Battle of Chacabuco. Rebels under Argentine independence fighter José de San Martín (1778–1850) defeat colonial forces in Chile, beginning successful drive for Chilean independence.

1818 Battle of Maipu. Independence fighters defeat Spanish forces. Rebel Bernardo O'Higgins (1778–1842) rules Chile 1818–23.

1819 Battle of Boyaca. Rebel forces led by Simon Bolívar (1783–1830) defeat Spanish forces, gaining independence for New Granada. Greater Colombia is created with Bolívar as President, 1819–30, and eventually includes modern Colombia, Panama, Venezuela, and Ecuador.

1821 Battle of Carabobo. Venezuela gains independence from Spanish and joins Greater Colombia.

1821 Peru proclaims independence.

1821 Brazil declares independence from Portugal under rule of Pedro I (1798–1834), son of Portuguese king.

1822 Battle of Pichincha. Ecuador gains independence after Spanish forces are defeated by rebels under Antonio Sucre (1795–1830). It joins Greater Colombia.

1824 Battle of Ayachucho. Peru and Bolivia are freed of Spanish rule by Bolívar's victory.

1827 Battle of Ituzaingo. Uruguay gains independence by defeating Brazilian troops.

1831 Republic of New Granada is formed after Ecuador and Venezuela withdraw from Greater Colombia.

1832 Border dispute between Colombia and Ecuador results in war.

1835–52 Argentine dictator Juan Rosas (1793–1877) establishes firm control over Argentine provinces.

1836 Half century of unrest in Uruguay begins.

1836–39 Confederation of Peru and Bolivia is formed following successful invasion of Bolivia by Peru. Chile intercedes militarily and breaks up confederation (1839).

1852 French establish first of notorious penal colonies in French Guiana. Devil's Island is operated until 1951.

1853 Argentina establishes republican government after ousting dictator Juan Rosas. Buenos Aires province secedes (1854), but is defeated and returned to republic 1859.

1864–66 Spain launches unsuccessful invasion of Peru following dispute over financial matters.

1865–70 War of the Triple Alliance, bloody war between Paraguay and allied Brazil, Argentina, and Uruguay. Paraguay initially invades Brazil, but is ultimately invaded

▶

and conquered by Brazil's allies. Paraguay loses over half its prewar population of some 525,000 to various causes.

1867–80 Prolonged civil war in Colombia.

1879–84 War of Pacific is fought over mineral-rich Atcama Desert between Chile and Bolivia, with Peru declaring war on both nations. Chile successfully invades both nations and by subsequent treaties gains all of Bolivia's coastal territory and mineral-rich Peruvian lands as well.

1886 Colombia establishes republican government.

1889 Brazil is established as republic after overthrow of monarchy.

1895 Revolution of 1895 in Ecuador. Liberal government is installed following this revolt against conservative faction, which results in many reforms.

1899 Venezuela boundary dispute over border territory between Venezuela and British Guiana is settled by arbitration.

1903 Colombia rejects treaty sought by United States to build canal across Isthmus of Panama, then part of Colombia. Soon after, revolt establishes independence of Panama from Colombia.

1903 New government of Panama concludes treaty with United States for building Panama Canal.

1912–25 U.S. Marines occupy Nicaragua to quell unrest; they stage second occupation in 1927–33.

1914 Panama Canal is opened.

1914–18 In World War I Argentina remains neutral; Brazil joins Allies.

1930 Long rule of Getulio Vargas in Brazil begins. He remains in power until 1945, is ousted, and then is elected to office 1950–54.

1936 Paraguay installs South America's first fascist regime.

1937 Somoza family begins long period of control over government in Nicaragua.

1939–45 World War II. Argentina remains neutral. Brazil, Colombia, and Peru (1943) join Allies.

1941 Border war is fought between Peru and Ecuador; it erupts again in 1981.

1946–55 Argentine dictator Juan Perón (1895–1974) is in power. He proves popular with Argentines, though his regime is repressive.

1948–58 Violent unrest occurs in Colombia during this period; some 200,000 are killed.

1952–67 In Uruguay, rule by executive commission is established after office of presidency is abolished. Presidency is restored in 1967.

1955 Military coup in Argentina ousts Perón, who is exiled.

1959–61 Rebel leader Fidel Castro (b. 1926) seizes power in Cuba and installs Communist regime, first in West.

1960 Brasília, newly constructed city, becomes capital of Brazil.

1961 U.S.-backed Bay of Pigs invasion of Cuba fails.

1962 Cuban missile crisis occurs.

1967 Guerrilla leader Che Guevara (1928–67) is killed by government troops in Bolivia.

1969 "Soccer War," brief conflict between Honduras and El Salvador, is fought.

1970 Bolivian government nationalizes Gulf Oil Corporation holdings. It pays compensation.

1970–73 Marxist leader Salvador Allende (1908–73) is in office as Chile's President. He institutes program of nationalization and socialization before being ousted in coup (CIA involvement is reported).

1973–74 In Argentina, Juan Perón is back in office as President. He dies in 1974 and is succeeded by his wife, Isabel (b. 1931), who is ousted in military coup (1974).

1978 Mass suicide is committed by followers of American Rev. Jim Jones at religious colony in Guyana.

▶

1978 United States agrees to return full control of Panama Canal to Panama in 1999.

1979 Sandinista guerrillas war against Samoza regime, finally forcing dictator to leave Nicaragua. They subsequently institute pro-communist regime. United States backs contra rebels in war against Sandinistas.

1979 In war in El Salvador, government troops battle leftist rebels (supported by Cuba and Sandinista regime in Nicaragua), and rightist death squads (formed to attack leftists).

1980 Gold rush occurs in Amazon jungle in Brazil.

1982 In Falkland Islands War, Argentina seizes these British-held islands off Argentine coast (April), but British invasion forces Argentine garrison to surrender (June).

1983 United States invades Grenada in Caribbean after Communist-backed coup there. Invasion forces surrender and withdrawal of Cuban troops.

1984 In El Salvador, moderates win in elections despite efforts by leftist guerrillas to disrupt voting.

1984 Argentina reaches agreement with U.S. banks on interest payments on massive foreign debt. It later institutes austerity plan.

1985 In Argentina, top government leaders in office after 1976 coup are convicted of murder and human-rights abuses stemming from activities of their repressive regime.

1985 Colombian leftist terrorist group M-19 seizes Palace of Justice and executes 100 people, 11 of them judges.

1985 Drug trade increases sharply in Colombia, Peru, Bolivia, and Ecuador.

1987 Brazil suspends interest payments on $68 billion of its massive foreign debt.

1988 United States sends troops to Honduras following border incursion by Nicaraguan troops.

1989 Caracas, Venezuela is rocked by riots over price increases, and some 300 people are killed. Country's $33 billion foreign debt is blamed for economic troubles that caused the increases.

1989 Leftist guerrillas in El Salvador fail to block free presidential election. Rightist Republican Nationalist faction wins.

1989 Economic crisis in Argentina sparks riots in Buenos Aires and other cities. Unrest results in election of Peronist Carlos Menem (b. 1930) as President.

1989 Central American peace plan, hammered out by leaders of Central American countries, is adopted to end warfare in Nicaragua. Contras are to disband. Leftist rebels in El Salvador are also asked to begin talks.

1989 Drug lords in Colombia assassinate a leading presidential candidate campaigning against drug traffickers. Murder results in ongoing Colombian government campaign to break up powerful drug cartels.

1989 Chilean voters end military rule under Gen. Augusto Pinochet (b. 1915) by electing opposition leader Patricio Aylwin (b. c. 1918). Democratic government is thus restored to last country in South America still ruled by military.

1989 U.S. troops invade Panama (December 20) in effort to capture Gen. Manuel Noriega (b. 1934), wanted in United States on charges of drug trafficking. Pro-Noriega forces are defeated and Noriega is brought to United States.

1990 Free elections are held in Nicaragua. Pro-communist Sandinista President Daniel Ortega (b. 1945) is ousted by pro-U.S. leader Violeta Barrios de Chamorro (b. 1929). Nicaraguan National Assembly later pardons both contra rebels and pro-Sandinista government workers charged with wrongdoing.

Europe

An Overview

B.C.

c. 1000 Iberians invade region of modern Spain, giving their name to Iberian Peninsula.

c. 600 Greek colonists found trading colony on site of modern Marseilles.

500s Celts cross Pyrenees from France and invade Iberian Peninsula.

c. 300 Parisi tribe founds Paris as Lutèce, fishing village on small island in Seine.

c. 250 Carthage begins to expand commercial interests in Iberian Peninsula.

218–201 Second Punic War. Romans conquer Iberian Peninsula during war with Carthage. They subsequently occupy territory and divide it into provinces. Territory becomes prosperous farming and mining region.

58–51 Romans, led by Julius Caesar (c. 102–44 B.C.), conquer Gaul, which is roughly region of modern France. Gaul becomes thriving province under Roman rule.

A.D.

c. 43 Romans begin conquest of British Isles (complete by 84). They establish thriving trading center called Londinium, site of modern London.

60 Iceni tribal queen Boadicea leads revolt against Roman rule, burning Londinium.

c. 100 First London Bridge (of wood) across Thames is built by Romans.

c. 400 Invasions by Germanic peoples—Vandals, Suevi, and Alans—end long period of stability in Iberian provinces.

419–711 Visigoths establish kingdom in Iberia after subduing invading Germanic tribes. Roman Empire, meanwhile, collapses (476) with Battle of Pavia.

c. 500 In Merovingian kingdom, Franks led by Clovis (481-511) briefly unite much of Gaul. By 511 kingdom is divided among Clovis's heirs; it remains divided for centuries.

500s Anglo-Saxon kingdoms are founded in area of modern England by Germanic peoples from Continent who have invaded British Isles.

c. 505 St. Benedict of Nursia (d. c. 547) founds monastery at Subiaco. He founds another at Monte Cassino in 529. This marks beginning of monasticism in the West.

569–72 Lombards invade and take control of much of northern Italy, though they do not attack Rome at this time. Byzantines retain control only at Ravenna and other coastal cities.

711 Muslims from North Africa conquer Iberian Peninsula. They expand northward until Frankish ruler Charles Martel (688–741) finally stops them at Battle of Poitiers (732), in modern France.

751–54 Pepin the Short (714–68) unites Frankish domains under his rule, thus founding Carolingian kingdom. At request of pope, Pepin invades Italy and defeats Lombards. Pepin donates conquered territory around Rome to pope (756), providing nucleus for what becomes Papal States.

768–814 Charlemagne (c. 742–814), great Frankish conqueror and Holy Roman Emporer, expands Carolingian empire to include area of modern France, Germany, northern Italy, Belgium, Luxembourg, and Netherlands.

787–c. 1000 Danes (Norsemen from Denmark) begin attacks on England in 787, and by

▶

Holy Roman Emperors

Years	Emperor	Years	Emperor
800–814	Charlemagne (742–814)	**1298–1308**	Albert I (Hapsburg) (1250–1308)
814–840	Louis I (the Pious) (778–840)	**1308–13**	Henry VII (Luxembourg) (1269–1313)
840–855	Lothair I (795–855)	**1314–47**	Louis IV (Wittelsbach) (1287–1347)
855–875	Louis II (825–875)	**1314–25**	Frederick III of Germany (1286–1330), rival claimant
875–877	Charles II (the Bald) (823–877)	**1347–78**	Charles IV (Luxembourg) (1316–78)
877–881	Interregnum	**1378–1400**	Wenceslaus (1361–1419)
881–887	Charles III (the Fat) (832–888)	**1400–10**	Rupert of Palatinate (1352–1410)
887–899	Arnulf (850–899)	**1410–11**	Jossus of Moravia
899–911	Louis III (893–911)	**1411–37**	Sigismund of Hungary (1368–1437)
911–918	Conrad I (d. 918)	**1438–39**	Albert II (1397–1439); first Hapsburg emperor
919–936	Henry I (the Fowler) (876–936), first Saxon Emperor	**1440–93**	Frederick III (1415–93)
936–973	Otto I (the Great) (912–973)	**1493–1519**	Maximilian I (1459–1519)
973–983	Otto II (955–983)	**1519–56**	Charles V (1500–58)
983–1002	Otto III (980–1002)	**1556–64**	Ferdinand I (1503–64)
1002–24	Henry II (973–1024)	**1564–76**	Maximilian II (1527–76)
1024–39	Conrad II (990–1039), first Franconian emperor	**1576–1612**	Rudolf II (1552–1612)
1039–56	Henry III (1017–56)	**1612–19**	Matthias (1557–1619)
1056–1106	Henry IV (1050–1106)	**1619–37**	Ferdinand II (1578–1637)
1106–25	Henry V (1081–1125)	**1637–58**	Ferdinand III (1608–58)
1125–37	Lothair II (1070–1137)	**1658–1705**	Leopold I (1640–1705)
1138–52	Conrad III (1093–1152), first Hohenstaufen emperor	**1705–11**	Joseph I (1678–1711)
1152–90	Frederick I (Barbarossa) (1123–90)	**1711–40**	Charles VI (1685–1740)
1190–97	Henry VI (1165–97)	**1742–45**	Charles VII (1697–1745), Bavarian emperor
1212–50	Frederick II (1194–1250)	**1745–65**	Francis I (1708–65), first Hapsburg-Lorraine emperor
1250–54	Conrad IV (1228–54), last Hohenstaufen emperor	**1765–90**	Joseph II (1741–90)
1254–73	Great Interregnum; noblemen rule in absence of imperial authority	**1790–92**	Leopold II (1747–92)
1273–91	Rudolf I (Hapsburg) (1218–91)	**1792–1806**	Francis II (1768–1835), last emperor of Holy Roman Empire; he becomes Francis I of Austria in 1806
1292–98	Adolf of Nassau (1255–98)		

800s control much territory in British Isles.

800s Bohemia, Moravia, and Slovakia (modern Czechoslovakia) are united under aegis of Moravian Empire. Varangians (Norsemen) invade Russia in the 800s, establishing ruling Rurik dynasty of early Russia at Novgorod. Early Russian rulers maintain commercial and cultural ties with Byzantine Empire.

842 Muslims begin raiding Italy. They sack Rome (846) and eventually gain control of Sicily (925).

▶

The British Monarchs

828–839	Egbert (775–839), first of Saxon kings
839–858	Ethelwulf (d. 858)
858–860	Ethelbald (d. 860)
860–866	Ethelbert (d. 866)
866–871	Ethelred I (d. 871)
871–899	Alfred (849–99)
899–924	Edward the Elder (870–924)
924–940	Athelstan (895–940)
940–946	Edmund I (922–946)
946–955	Edred (d. 955)
955–958	Edwy (d. 958)
959–975	Edgar (944–975)
975–978	Edward the Martyr (963–978)
978–1016	Ethelred II (968–1016)
1016	Edmund II (980–1016)
1016–35	Canute (c. 995–1035); first of Danish kings
1035–40	Harold I (d. 1040)
1040–42	Hardecanute (1019–42)
1042–66	Edward the Confessor (1004?–66); Saxon kings are restored
1066	Harold II (1022–66)
1066–87	William I the Conqueror (1027–87); first of Norman kings
1087–1100	William II Rufus (1059–1100)
1100–35	Henry I (Beauclerc) (1068–1135)
1135–54	Stephen (1097–1154)
1154–89	Henry II (1133–89); first of Plantagenet kings
1189–99	Richard I Coeur de Lion (1157–99)
1199–1216	John (Lackland) (1167–1216)
1216–72	Henry III (1207–72)
1272–1307	Edward I (1239–1307)
1307–27	Edward II (1284–1327)
1327–77	Edward III (1312–77)
1377–99	Richard II (1367–1400)
1399–1413	Henry IV (1367–1413); House of Lancaster is in power
1413–22	Henry V (1387–1422)
1422–61	Henry VI (1421–71)
1461–70	Edward IV (1442–83); House of York is in power
1470–71	Henry VI (1421–71)
1471–83	Edward IV (1442–83)
1483	Edward V (1470–83)
1483–85	Richard III (1452–85)
1485–1509	Henry VII (1457–1509); first of Tudor kings
1509–47	Henry VIII (1491–1547)
1547–53	Edward VI (1537–53)
1553	Lady Jane Grey (1537–54); reigns for nine days
1553–58	Mary I (1516–58)
1558–1603	Elizabeth (1553–1603)
1603–25	James I (1566–1625); House of Stuart is in power
1625–49	Charles I (1600–49)
1653–58	Oliver Cromwell (1599–1658), lord protector
1658–59	Richard Cromwell (1626–1712), lord protector
1660–85	Charles II (1630–85); Stuart kings are restored
1685–88	James II (1633–1701)
1689–1702	William III (1650–1702) and Mary II (1662–94)
1702–14	Anne (1665–1714)
1714–27	George I (1660–1727); House of Hanover is in power
1727–60	George II (1683–1760)
1760–1820	George III (1738–1820)
1820–30	George IV (1762–1830)
1830–37	William IV (1765–1837)
1837–1901	Victoria (1819–1901)
1901–10	Edward VII (1841–1910); House of Saxe-Coburg-Gotha is in power
1910–36	George V (1865–1936); House of Windsor is in power
1936	Edward VIII (1894–1972)
1936–52	George VI (1895–1952)
1952–	Elizabeth II (b. 1926)

French Rulers and Heads of State

Beginning with breakup of Carolingian Empire in A.D. 840.

840–77	Charles II the Bald (823–77)
877–79	Louis II (846–79)
879–82	Louis II (c. 863–82); ruled with his brother, Carloman
879–84	Carloman (d. 884)
885–87	Charles III the Fat (839–88)
888–98	Eudes (c. 860–98)
893–923	Charles III the Simple (879–929)
922–923	Robert I (c. 865–923)
923–36	Rudolf, duke of Burgundy (923–36)
936–54	Louis IV (921–54)
954–86	Lothair (941–86)
986–87	Louis V (967–87)
987–996	Hugh Capet (c. 938–96), first of Capetian kings
996–1031	Robert II (c. 970–1031)
1031–60	Henry I (1008–60)
1060–1108	Philip I (1052–1108)
1108–37	Louis VI (1081–1137)
1137–80	Louis VII (c. 1120–80)
1180–1223	Philip II (1165–1223)
1223–26	Louis VIII (1187–1226)
1226–70	St. Louis IX (1214–70)
1270–85	Philip III the Bold (1245–85)
1285–1314	Philip IV the Fair (1268–1314)
1314–16	Louis X (1289–1316)
1316	John I (1316)
1317–22	Philip V (c. 1294–1322)
1322–28	Charles IV the Fair (1294–1328)
1328–50	Philip VI (1293–1350), first of Valois kings
1350–64	John II (1310–64)
1364–80	Charles V the Wise (1337–80)
1380–1422	Charles VI the Well-Beloved (1368–1422)
1422–61	Charles VII the Victorious (1403–61)
1461–83	Louis XI (1423–83)
1483–98	Charles VIII (1470–98)
1498–1515	Louis XII (1462–1515)
1515–47	Francis I (1494–1547)
1547–59	Henry II (1519–59)
1559–60	Francis II (1544–60)
1560–74	Charles IX (1550–74)
1574–89	Henry III (1551–89)
1589–1610	Henry IV (1553–1610), first of Bourbon kings
1610–43	Louis XIII (1601–43)
1643–1715	Louis XIV (1638–1715)
1715–74	Louis XV (1710–74)
1774–92	Louis XVI (1754–93), deposed in French Revolution; Louis XVII (1785–95?), never ruled
1792–1804	Governments of First Republic (National Convention, Directory, and Consulate)
1804–14	Napoleon I (1769–1821), ruler of First Empire
1814–24	Louis XVIII (1755–1824), first king of Restoration monarchy
1824–30	Charles X (1757–1836)
1830–48	Louis Philippe (1773–1850)
1848–51	Louis Napoleon (1808–73), President of Second Republic.
1851–70	Louis Napoleon (as Napoleon III) proclaims himself emperor of Second Empire
1871–73	Louis A. Thiers (1797–1877), first President of Third Republic
1873–79	Marie Edmé Patrice Maurice de MacMahon (1808–93)
1879–87	Jules Grévy (1807–91)
1887–94	Sadi Carnot (1837–94)
1894–95	Jean-Paul Pierre Casimir-Périer (1847–1907)
1895–99	François Félix Faure (1841–99)
1899–1906	Émile Loubet (1838–1929)
1906–13	Armand Fallières (1841–1931)
1913–20	Raymond Poincaré (1860–1934)
1920	Paul Eugène Louis Deschanel (1856–1922)
1920–24	Alexandre Millerand (1859–1943)
1924–31	Gaston Doumergue (1863–1937)
1931–32	Paul Doumer (1857–1932)
1932–40	Albert Lebrun (1871–1950)

French Rulers and Heads of State

1940–44	Philippe Pétain (1856–1951), Vichy government head of state
1942–44	Pierre Laval (1883–1945), Vichy administrative chief
1944–46	Charles de Gaulle (1890–1970), President of provisional government
1946	Félix Gouin (1884-?)
1946	Georges Bidault (1899-?)
1946	Léon Blum (1872–1950)
1947–54	Vincent Auriol (1884–1966), President, Fourth Republic.
1954–59	René Coty (1882–1962)
1959–69	Charles de Gaulle (1890–1970), President, Fifth Republic.
1969–74	Georges Pompidou (1911–74)
1974–81	Valéry Giscard d'Estaing (b. 1926)
1981–	François Mitterrand (b. 1916)

843 Treaty of Verdun divides Charlemagne's great empire among his heirs into western part (roughly modern France), eastern part (roughly Germany), and narrow middle section. Middle kingdom (except Italy, which eventually comes under control of Holy Roman Empire), is soon absorbed by Louis the German (804–76) in eastern (German) part and Charles II the Bald (823–77) in France (western Franks).

c. 845 Norsemen invade France. They eventually settle what becomes duchy of Normandy.

871–99 English king Alfred the Great (849–99) recaptures London from Danes. London by this time has become England's largest city.

930 Althing, oldest body of representative government in Europe, is established by Vikings in Iceland.

962 Pope crowns Otto I (912–73) emperor, constituting Holy Roman Empire (what later becomes Germany) as successor state to Western Roman Empire. Relationship between church and monarchies of Europe is thus further strengthened. Otto and pope reach formal agreement on relationship between church and the Holy Roman Empire, but Otto soon asserts he has right to name pope.

987–96 Hugh Capet (c. 938–96) reigns as first of Capetian kings of Western Franks. He has only nominal rule over other kingdoms in France that are beyond his own domains around Paris. Capet makes Paris his capital.

c. 1000 Port cities including Venice, Naples, and Genoa rise as maritime powers. They are successful in naval battles against Muslims and enjoy economic prosperity as result of increased trade.

1014 Brian Boru defeats Norse in Ireland at Battle of Clontarf.

1016 Danes led by Canute (c. 995–1035) conquer England. Canute rules until his death. Anglo-Saxon rule is restored in 1042.

c. 1031 Central authority breaks down in Muslim Spain as result of civil war. Petty kingdoms form.

c. 1050 First university (for medicine) is established at Salerno. Bologna becomes center for study of law in next century.

1066 Norman Conquest. William I the Conqueror (1027–87), claiming English throne, invades England and defeats his rival at Battle of Hastings.

1075 In investiture controversy between Pope Gregory VII (1020–85) and Holy Roman emperor Henry IV (1050–1106), Gregory denies emperor's right to appoint bishops. He excommunicates and deposes Henry (1077) but is finally driven into exile following Henry's invasion of Italy (1081–83). Reforms instituted by Gregory create power struggle between popes and

▶

	Holy Roman emperors that lasts until 1122.
c. 1080	Pisa and other powerful cities declare themselves city-republics as central authority in Italy breaks down.
1094	El Cid (c. 1040–99), Spanish soldier of fortune who joins Muslims, conquers Valencia and becomes its ruler.
1096–1291	In era of Crusades, French and English kings participate in these unsuccessful military expeditions to recapture Holy Land. *(See Wars and Military History.)*
1122	Diet of Worms ends investiture controversy with compromise that maintains authority of church over Holy Roman Empire.
1127	Kingdom of Two Sicilies is created out of territory conquered by Norman invaders in southern Italy and Sicily.
1147	Village on site of the modern-day Moscow is first mentioned in historic records.
1152–90	Frederick I (Barbarossa) (1123–90) reigns. His reign is marked by attempts to reassert control over rebellious German princes and by unsuccessful invasions of Lombardy.
1169	Norman troops invade Ireland, establishing English dominion on island.
1170	Thomas à Becket, archbishop of Canterbury, is killed for opposing king's efforts to limit church's powers.
1198–1216	Power of papacy is at its height. Pope Innocent III (1161–1216) excommunicates Holy Roman emperor Otto IV (1182–1218) in 1210.
1200	Invasion and conquest of Russian kingdoms. Mongols (of Golden Horde) dominate Russia for next centuries.
1200	University of Paris is chartered, marking rise of Paris as cultural center. University later becomes center of medieval scholasticism.
1200s	Hapsburg King Rudolf I establishes Austrian dynasty that will rule for 700 years.
1208	Albigensian Crusade is launched by Pope Innocent III against Cathar and Waldensian sects in southern France. Total of 20,000 people are massacred by Crusaders at Bezier in 1209.
1212	In battle of Navas de Tolosa, Muslim power in Spain is effectively broken. Muslims now control only Granada. Region of modern Spain is controlled by Aragon and independent Christian kingdoms of Castile and Leon.
1212	Most of London, England burns after fire starts at London Bridge.
1215	Magna Carta is signed (June 19) by English king John (c. 1167–1216). It guarantees such basic liberties as habeas corpus and trial by jury and has major effect on development of English law.
1226	Imperial decree grants Teutonic Knights control of Prussia.
1251	Medicis, one of several families that become powerful in medieval times in various Italian cities, rise in Florence.
1254–73	Great Interregnum. Nobles become dominant power in Holy Roman Empire during period of continued unrest.
1264–67	Barons' War. English barons rebel against inept rule of Henry III (1207–72) and force reforms.
1282	Sicilian Vespers. Sicilians revolt against French rule; French are massacred.
1284	Statute of Wales formally annexes Wales to England.
c. 1300	Bands of criminals, forerunners of modern-day Mafia, begin operating in Sicily.
c. 1300	Mercantile class and such business-oriented institutions as banks and insurance firms rise in Italian cities.
1300s	Prague emerges as a leading European university city.
1305	Seventy-year "Avignon Captivity" of papacy begins when Pope Clement V (1264–1314) moves papacy to Avignon in France.

▶

1314 Battle of Bannockburn. Scottish rebel Robert the Bruce (1274–1329) defeats English forces, ending for time English rule over Scotland.

1326 Moscow becomes seat of Russian Orthodox Church. By this time it has become thriving commercial center.

1337–1453 Hundred Years War between England and France ends with France gaining all English kingdoms on Continent (in northern France). (*See Wars and Military History.*)

1345 Notre Dame, famed Gothic cathedral in Paris, is completed.

1348 Black Death plague begins to spread throughout Europe. It wipes out about quarter or more of Europe's population during next decades.

c. 1350 Renaissance begins in Italy. Sparked by renewed interest in Greek and Roman cultures, this great cultural movement spreads throughout Europe.

1356 Golden Bull of 1356 reorganizes Holy Roman Empire as federation of independent kingdoms. From this time forward, emperor controls only his personal domains.

1367 Timber walls around Kremlin at Moscow are replaced with white stone walls and towers. Palaces and churches subsequently are built inside Kremlin.

1380 Battle of Kulikovo. Russian victory over Mongols marks turning point in Russian struggle against their overlords. By 1382, Moscow becomes capital of expanding Grand Duchy of Moscow.

1381 Peasants' Revolt. Wat Tyler (d. 1381) leads first English popular uprising against laws regulating wages and poll tax.

1386 Commonwealth of Poland-Lithuania is formed. Kracow emerges among wealthiest cities of Europe.

1397 Kalmar Union, designed by Danish queen Margaret, unites Sweden, Norway, and Denmark under a single monarchy.

1419–34 In Hussite Wars, royal armies quash rebellion against Catholic Church in Germany.

1422–61 Charles VII the Victorious (1403–61) reigns as French king during the French victory in Hundred Years' War.

1431 Joan of Arc, French visionary who rallied French armies during Hundred Years War, is captured by English and burned at stake as heretic.

c. 1438 Johann Gutenburg (c. 1400–1468) invents printing press and at Mainz pioneers printing from movable type in Europe.

1438–39 Albert II (1397–1439) reigns as first of Hapsburg emperors. Imperial crown becomes hereditary title of Hapsburgs.

1453 Ottoman Turks capture Constantinople and conquer Greece, establishing rule for 350 years.

1455–85 Wars of the Roses, dynastic wars fought in England by House of York to gain throne from ruling House of Lancaster.

1461–83 Louis XI (1423–83) reigns as French king. He expands royal domains (adding Burgundy, Provence, Anjou, Maine, and Bar) and establishes firm foundation for later absolute monarchy in France.

1462–1505 Ivan III the Great (1440–1505) reigns in Russia. He greatly expands his domains by conquest, and he rebuilds brick Kremlin (fortress) in Moscow with 18 towers and mile-long walls.

1469 Ferdinand II (1452–1516), king of Aragon, and Isabella I (1451–1504), queen of Castile and Leon, are married. Though kingdoms of Aragon and Castile and Leon still are nominally separate, they are administered jointly.

1477 First book is printed in England.

1478 Spanish Inquisition is established.

1478 Albania is conquered by Ottoman Turks and is ruled until 1912 as an Ottoman province.

▶

Spanish Rulers and Heads of State			
1479–1504	Ferdinand V (1452–1516) and Isabella I (1451–1504)	**1808–13**	Joseph Bonaparte (1768–1844)
1504–6	Juana (1479–1555) and Philip I (1478–1506)	**1814–33**	Ferdinand VII (1784–1833)
1506–16	Ferdinand V (1452–1516)	**1833–68**	Isabella II (1830–1904)
1516–56	Charles I (1500–58), first Hapsburg king	**1869–70**	Francisco Serrano y Dominguez (1810–85), regent
1556–98	Philip II (1527–98)	**1870–73**	Amadeo (1845–90), king, House of Savoy
1598–1621	Philip III (1578–1621)	**1873–74**	Emilio Castelar y Ripoll (1832–99), President of First Republic
1621–65	Philip IV (1605–65)	**1874–85**	Alfonso XII (1857–85), Bourbon king
1665–1700	Charles II (1661–1700)	**1886–1931**	Alfonso XIII (1886–1941), Bourbon king
1700–1724	Philip V (1683–1746), first Bourbon king	**1931–36**	Niceto Alcala Zamora y Torres (1877–1949), President of Second Republic
1724	Louis I	**1936–39**	Manuel Azaña (1880–1940), President of Republic.
1724–46	Philip V (restored) (1683–1746)	**1939–75**	Francisco Franco (1892–1975), dictator
1746–59	Ferdinand VI (1713–59)	**1975–**	Juan Carlos (b. 1938), Bourbon king
1759–88	Charles III (1716–88)		
1788–1808	Charles IV (1748–1819)		
1808	Ferdinand VII (1784–1833)		

1482 Hapsburgs consolidate power in "low countries."

1483 English king Edward V (1473–83) is deposed and is believed killed in Tower of London.

1492 Jews are expelled from Spain. Their property is confiscated by Spanish crown.

1492 Christopher Columbus (1451–1506), Genoese navigator in Spanish service, discovers New World for Spain.

1494–1559 Italian Wars. Spain, France, and Holy Roman Empire vie for control of Italy. Spanish Hapsburgs ultimately win control of Italian Peninsula, though Papal States and petty kingdoms (Genoa, Tuscany, and others) are nominally independent.

1500s Copenhagen emerges as a center of European commerce and culture.

1516 King Charles I (1500–1558), heir to both Aragón and Castile and Leon, takes the throne and thereby becomes first king of united Spain. He is later crowned Charles V, reigning over Spain and Holy Roman Empire. Treasure stripped from civilizations of New World is by this time flowing to Spain aboard Spanish galleons.

1517 German theologian Martin Luther (1483–1546) sparks Protestant Reformation with protest against church corruption. (*See Religion, Philosophy, and Education*, "*Protestantism.*")

1520 Anabaptist movement develops in Switzerland and Germany. Anabaptists are widely persecuted by both Catholics and Lutherans.

1523 Sweden breaks with Danish monarchy and installs Gustavus I on a new Swedish throne.

1524–25 Short-lived Peasant's Revolt breaks out in Austria and southern Germany, led by radical Protestant Thomas Muntzer (c. 1490–1525).

1526 Bohemia falls under Hapsburg control. An unsuccessful revolt in 1618 initiates the Thirty Years War, but Hapsburg

▶

control is reinstated and lasts until World War I.

1526–29 Ottoman Turks invade Holy Roman Empire from east and push as far as Vienna before being turned back. Turks retain control over much of Hungary, however.

1533–84 Ivan IV the Terrible (1530–84) reigns as first Russian czar. He continues rapid expansion of domains and begins pushing eastward into Mongol territories.

1534 Church of England is created by Act of Supremacy, separating Church of England from Roman Catholic Church following dispute between King Henry VIII and pope, who refused to annul Henry's marriage.

1544–47 Schmalkaldic War. Schmalkaldic League, formed (1531) as alliance of Lutheran states, is defeated by Holy Roman emperor Charles V.

1545–63 Catholic Counter-Reformation begins at Council of Trent.

1546 King Francis I (reigned 1515–47) begins building Louvre as royal art museum. Many other French rulers subsequently add to structure.

1547 Brittany becomes part of French royal domains.

c. 1550 Forerunners of railroads, wagons run on wooden tracks in mines are used in Holy Roman Empire.

1558–1603 Elizabeth I (1533–1603) reigns as queen of England and restores Protestantism. England enjoys cultural flowering and begins establishing far-flung colonial empire. Puritan movement develops in reaction to church reforms.

1562–98 Wars of religion are fought between French Catholics and Protestants. (*See Wars and Military History.*)

1570 Moscow burns in great fire. Some 200,000 persons are killed.

1571 Holy League, joined by forces sent by King Philip, defeats Muslims at Battle of Lepanto.

1588 King Philip launches Spanish Armada against England to end Protestantism there. Great fleet is destroyed by English navy and by storms, establishing England as naval power.

1597 Serfdom is established in Russia.

1598 Edict of Nantes is issued by King Henry IV of France. It marks end of bloody wars of religion and grants religious freedom to Protestants there.

1600s–1700s Paris reigns as the cultural center of Europe. In early 1700s it gives birth to Enlightenment.

1609–10 Christianized Muslims, called Moriscos, are forced to leave Spain. Much of Spanish learned class is lost as result.

1613–45 Michael reigns as first czar of Romanov line, which rules Russia until 1917.

1618–48 Thirty Years War leaves Holy Roman Empire devastated. Spain ultimately loses control over Netherlands, and France emerges as Europe's leading power. (*See Wars and Military History.*)

1620 English Puritans known as Pilgrims establish American colony at Plymouth Rock to escape religious persecution in England.

1640 Portugal revolts against Spanish rule and regains independence.

1642 English Parliament issues Nineteen Propositions in attempt to limit king's powers. Charles I (1600–49) resists, and war is declared. (*See War and Military History, "English Civil War."*)

1643–1715 Louis XIV (1638–1715) reigns. Most powerful of absolutist monarchs of France, he destroys French finances by lavish spending and costly wars.

1648 Peace of Westphalia ends Thirty Years War. Germany is divided between Catholic and Protestant states. Belgium and Netherlands gain independence.

1649 King Charles I is tried and executed, ending English monarchy for a time. Oliver Cromwell (1599–1658) becomes

▶

leader of the newly proclaimed Commonwealth (republican government).

c. 1650 Dutch create a world empire, including holdings in Indonesia, Ceylon, South Africa, West Indies, and North America (New Amsterdam), and enjoy a trade monopoly with Japan.

1652–78 Dutch Wars, two trade wars between England and United Provinces (Netherlands), in 1652–54 and 1664–67, and general European war resulting from French expansionism, in 1672–78.

1660–85 Charles II (1630–85) reigns as king of England after monarchy is restored by Parliament.

1663 New France, in modern Canada, is established as French colony.

1664 Great plague in London, England begins. Some 100,000 eventually perish.

1666 Great Fire of London destroys nearly whole central city. City is rebuilt of brick and stone.

1667–68 War of Devolution. French claims to Spanish Netherlands (by right of marriage) result in invasion that Netherlands, Sweden, and England successfully oppose.

1676–1878 Russo-Turkish Wars, long series of wars between Russia and Ottoman Empire in which Russia greatly expands its borders at expense of weakening Ottoman Empire.

1680s Hapsburgs drive Ottoman Turks from Hungary and establish rule there.

1682 Lavish Palace of Versailles is completed. It becomes home of French court.

1685 Edict of Nantes is renounced by French king Louis XIV (1638–1714). Mass exodus of Protestants from France follows.

1688 Glorious Revolution. King James II is deposed and replaced by King William III (1650–1702) and Queen Mary II (1662–94), who agree to abide by new English Bill of Rights.

1689–97 War of the Grand Alliance. France invades Holy Roman Empire. (*See Wars and Military History.*)

1690 James II of England is defeated by William of Orange at Battle of the Boyne in Ireland.

1694 Bank of England is formed in London to finance war with France.

1700–21 In second Northern War, with Sweden, Russia gains Baltic territories and emerges as a leading power in region.

1700s French thinkers spawn Enlightenment, widespread intellectual movement promoting rationalism.

1701–14 War of Spanish Succession. After King Charles II (1661–1700) dies, leaving no Hapsburg heir, war erupts over succession by Philip V (1683–1746), member of French House of Bourbon, but he is finally recognized as king. Spanish Bourbon line of kings begins. Austria invades Italy during war and retains control of Milan afterward. (*See Wars and Military History.*)

1702–13 Queen Anne's War. Great Britain gains much of France's Canadian territory. (*See Wars and Military History.*)

1703 Construction of Buckingham Palace, modern-day London residence of British sovereigns, begins.

1707 Act of Union formally unites England and Scotland.

1707–48 France and England compete for control of India. Great Britain is victorious by War of Austrian Succession.

1712 Czar Peter moves Russian capital to St. Petersburg, which he founded in 1703. It remains capital until 1918, though Moscow continues as important commercial and cultural center.

1713 Pragmatic Sanction of 1713 is issued by Emperor Charles VI (1685–1740) to ensure that his daughter, Maria Theresa, will inherit Hapsburg domains. Question of succession nevertheless ends in war.

▶

British Prime Ministers

1715–17	Robert Walpole, first earl of Orford (1676–1745)
1717–18	James Stanhope, first Earl Stanhope (1673–1721)
1718–21	Charles Spencer, third earl of Sunderland (1674–1722)
1721–42	Robert Walpole, first earl of Orford (1676–1745)
1742–43	Spencer Compton, earl of Wilmington (1673–1743)
1743–54	Henry Pelham (1695–1754)
1754–56	Thomas Pelham-Holles, first duke of Newcastle (1693–1768)
1756–57	William Cavendish, fourth duke of Devonshire (1720–64)
1757–62	Thomas Pelham-Holles, first duke of Newcastle (1693–1768)
1762–63	John Stuart, third earl of Bute (1713–92)
1763–65	George Grenville (1712–70)
1765–66	Charles Watson-Wentworth, second marquis of Rockingham (1730–82)
1766–68	William Pitt, first earl of Chatham (1708–78)
1768–70	Augustus Henry Fitzroy, third duke of Grafton (1735–1811)
1770–82	Frederick North, second earl of Guilford (1732–92)
1782	Charles Watson-Wentworth, second duke of Rockingham (1730–82)
1782–83	William Petty, earl of Shelburne (1737–1805)
1783	William Henry Bentinck, third duke of Portland (1783–1809)
1783–1801	William Pitt (1759–1806)
1801–4	Henry Addington, first Viscount Sidmouth (1757–1844)
1804–6	William Pitt (1759–1806)
1806–7	William Wyndham Grenville, Lord Grenville (1759–1834)
1807–9	William Henry Bentinck, third duke of Portland (1738–1809)
1809–12	Spencer Perceval (1762–1812)
1812–27	Robert Banks Jenkinson, second earl of Liverpool (1770–1828)
1827	George Canning (1770–1827)
1827–28	Frederick John Robinson, Viscount Goderich (1782–1859)
1828–30	Arthur Wellesley, first duke of Wellington (1769–1852)
1830–34	Charles Grey, second Earl Grey (1764–1845)
1834	William Lamb, second Viscount Melbourne (1779–1848)
1834–35	Robert Peel (1788–1850)
1835–41	William Lamb, second Viscount Melbourne (1779–1848)
1841–46	Robert Peel (1788–1850)
1846–52	John Russell, first Earl Russell of Kingston-Russell (1792–1878)
1852	Edward George Geoffrey Smith Stanley, 14th earl of Derby (1799–1869)
1852–55	George Hamilton Gordon, fourth earl of Aberdeen (1784–1860)
1855–58	Henry John Temple, third Viscount Palmerston (1784–1865)
1858	Edward George Geoffrey Smith Stanley, 14th earl of Derby (1799–1869)
1858–65	Henry John Temple, third Viscount Palmerston (1784–1865)
1865–66	John Russell, first Earl Russell of Kingston-Russell (1792–1878)
1866–68	Edward George Geoffrey Smith Stanley, 14th earl of Derby (1799–1869)
1868	Benjamin Disraeli, first earl of Beaconsfield (1804–81)
1868–74	William Ewart Gladstone (1809–98)
1874–80	Benjamin Disraeli, first earl of Beaconsfield (1804–81)
1880–85	William Ewart Gladstone (1809–98)
1885–86	Robert Arthur Talbot Gascoyne-Cecil, third marquis of Salisbury (1830–1903)
1886	William Ewart Gladstone (1809–98)
1886–92	Robert Arthur Talbot Gascoyne-Cecil, third marquis of Salisbury (1830–1903)
1892–94	William Ewart Gladstone (1809–98)
1894–95	Archibald Philip Primrose, fifth earl of Rosebery (1847–1929)
1895–1902	Robert Gascoyne-Cecil, third marquis of Salisbury (1830–1903)

British Prime Ministers	
1902–5	Arthur James Balfour, first earl of Balfour (1848–1930)
1905–8	Sir Henry Campbell-Bannerman (1836–1908)
1908–16	Herbert Henry Asquith, first earl of Oxford and Asquith (1852–1928)
1916–22	David Lloyd George, first earl of Dwyfor (1863–1945)
1922–23	Andrew Bonar Law (1858–1923)
1923–24	Stanley Baldwin, first Earl Baldwin of Bewdley (1867–1947)
1924	James Ramsay MacDonald (1866–1937)
1924–29	Stanley Baldwin, first Earl Baldwin of Bewdley (1867–1947)
1929–35	James Ramsay MacDonald (1866–1937)
1935–37	Stanley Baldwin, first Earl Baldwin of Bewdley (1867–1947)
1937–40	Arthur Neville Chamberlain (1869–1940)
1940–45	Sir Winston Leonard Spencer Churchill (1874–1965)
1945–51	Clement Richard Attlee, first Earl Attlee (1883–1967)
1951–55	Sir Winston Leonard Spencer Churchill (1874–1965)
1955–57	Anthony Eden, first earl of Avon (1897–1977)
1957–63	Harold Macmillan (1894–1986)
1963–64	Sir Alec Douglas-Home (b. 1903)
1964–70	Harold Wilson (b. 1916)
1970–74	Edward Heath (b. 1916)
1974–76	Harold Wilson (b. 1916)
1976–79	James Callaghan (b. 1912)
1979–	Margaret Thatcher (b. 1925)

1720 House of Savoy establishes Kingdom of Sardinia by uniting territories of Sardinia, Piedmont, and Savoy. Kingdom later becomes nucleus of modern Italian state.

1740–48 War of Austrian Succession. In war for control of Hapsburg domains, Charles VI's daughter, Maria Theresa, ultimately retains family domains. (*See Wars and Military History.*)

1752 Moscow again burns in great fire. Some 18,000 buildings are destroyed.

1753 British Museum is founded in London.

1754–63 French and Indian War, between Great Britain and France over colonial territories in North America. (*See Wars and Military History.*)

1755 Moscow State University, Russia's first higher education institution, is founded.

1756–63 Seven Years War. France is stripped of many of its colonial possessions. Great Britain and Prussia emerge as important European powers. (*See Wars and Military History.*)

1772–95 Russia, Prussia, and Austria take control of Poland and divide up kingdom, completing dismemberment in 1795.

1775–83 American Revolution. Great Britain loses one of its largest colonies, but soon replaces loss by gaining control of India. (*See Wars and Military History.*)

1783 Spain regains Florida from Great Britain after siding with the American colonies in American Revolution.

1788 Government suspends French *parlements* (law courts) to end resistence to taxes needed to avert bankruptcy of government. French nobles retaliate by instigating popular revolt against government.

1789 French Estates-General meets and, after reconstituting itself as National Assembly, announces it has constitutional powers. King Louis XVI bars National Assembly from its assembly hall. Assembly vows to write constitution.

1789 Bastille is stormed by Paris mob seeking arms stored there, and Declaration of Rights of Man and of Citizen is published, declaring individual's right to

▶

representation and equality and freedom of press, speech, and religion.

1792 French Revolutionary Wars break out in Europe. Russia joins in Second Coalition against French. (*See Wars and Military History.*)

1793 French king Louis XVI is executed for treason.

1793–94 Reign of Terror occurs in France. Radical revolutionists execute some 40,000 royalists and others.

1793–95 Spain wars against revolutionary France, joining Great Britain and other nations in attempt to restore French monarchy, but is defeated and forced into alliance with France.

1796 French armies invade Italy during French Revolutionary Wars. French military leader Napoleon Bonaparte (1769–1821) deposes pope in 1798 and annexes Papal States in 1809.

1798 Wolf Tone leads an unsuccessful Irish uprising against English dominion.

1799 Coup of 18 Brumaire. Napoleon seizes power, becoming first consul of Consulate in France.

1800 Napoleon conquers Spain and regains Louisiana Territory.

1800 Act of Union joins England and Ireland and provides for Irish representation in British Parliament.

1801 War of Oranges. France and Spain conquer and divide Portugal.

1801 Russia annexes Black Sea region of Georgia.

1803 Napoleon agrees to purchase of Louisiana Territory by United States.

1803–15 Napoleonic Wars. French ruler Napoleon unsuccessfully invades Russia, but for time France rules much of Continental Europe. (*See Wars and Military History.*)

1804 Napoleon creates French Empire and has himself proclaimed emperor.

1804 Code Napoleon is issued as comprehensive compilation of French civil laws. It forms basis of modern civil law in France.

1805 Napoleon crowns himself king of Italy.

1806 Holy Roman Empire is dissolved. Francis becomes Francis I of Austria, and Napoleon creates confederation of German states to govern other independent German kingdoms.

1808 Revolt is staged against pro-French policies in Spain. Napoleon invades Spain and names his brother Joseph Bonaparte (1768–1844) king of Spain. Resistance to French rule continues.

1808–14 Peninsular War. British expeditionary force lands on Iberian Peninsula and begins protracted campaign to aid Spanish rebels against French in Spain.

1810–26 Liberal constitution for Spanish government is enacted by first national Cortes in Cadiz, where rebel forces are in control (1812). Spanish colonies in South America take advantage of turmoil in Europe and rebel. Spain loses control of all its major colonies there by 1826.

1812–15 War of 1812, between United States and Great Britain. (*See Wars and Military History.*)

1814 Spanish Monarchy is reestablished under Ferdinand VII (1784–1833) after French are driven out. Constitution is abolished and Inquisition is reinstituted.

1814–24 Louis XVIII (1755–1824) reigns as first king of Restoration monarchy in France.

1815 Napoleon again seizes power as emperor (Hundred Days period). He is defeated at Waterloo and is exiled. New Treaty of Paris costs France additional territory and indemnity.

1815 Austrian and Spanish control over Italy is restored by Congress of Vienna after Napoleon's defeat at Waterloo. Papal States also are restored. Norway is granted independence from Sweden and Kingdom of the Netherlands is formed. Austria emerges as the leading

▶

German Rulers and Heads of State

See also Holy Roman Emperors.

German Kings

1871–88	William I (1797–1888)
1888	Frederick III (of Prussia) (1831–88)
1888–1918	William II (1859–1941)

Chancellors of the Empire

1871–90	Otto von Bismarck (1815–98)
1890–94	Leo von Caprivi (1831–99)
1894–1900	Chlodwig Karl Viktor of Hohenlohe-Schillingsfurst (1819–1901)
1900–19	Bernhard von Bülow (1849–1929)
1909–17	Theobald von Bethmann-Hollweg (1856–1921)
1917	Georg Michaelis (1857–1936)
1917–18	Georg von Hertling (1843–1919)
1918	Maximilian (Prince Max of Baden) (1867–1929)
1918–19	Friedrich Ebert (1871–1925)

Weimar Republic and Nazi Regime, Heads of State

1919–25	Friedrich Ebert (1871–1925), Weimar Republic President
1925–34	Paul von Hindenburg (1847–1934), Weimar Republic President
1934–45	Adolf Hitler (1889–1945), Nazi *Führer*
1945	Karl Dönitz (1891–1980) *Führer* for eight days

Weimar Republic and Nazi Regime, Chancellors

1919	Philipp Scheidemann (1865–1939)
1919–20	Gustav Bauer (1870–1944)
1920	Hermann Muller (1876–1931)
1920–21	Konstantin Fehrenbach (1852–1926)
1921–22	Karl Joseph Wirth (1879–1956)
1922–23	Wilhelm Cuno (1876–1933)
1923	Gustav Stresemann (1878–1929)
1923–25	Wilhelm Marx (1863–1946)
1925–26	Hans Luther (1879–1962)
1926–28	Wilhelm Marx (1863–1946)
1928–30	Hermann Muller (1876–1931)
1930–32	Heinrich Bruning (1885–1970)
1932	Franz von Papen (1879–1969)
1932–33	Kurt von Schleicher (1882–1934)
1933–34	Adolf Hitler (1889–1945)

Federal Republic of Germany (West Germany), Heads of State

1949–59	Theodor Heuss (1884–1963), President
1949–63	Konrad Adenauer (1876–1967), chancellor
1959–69	Heinrich Lubke (1894–1972), President
1963–66	Ludwig Erhard (1897–1977), chancellor
1966–69	Kurt Kiesinger (b. 1904), chancellor
1969–74	Gustav Heinemann (1899–1976), President
1969–74	Willy Brandt (b. 1913), chancellor
1974–79	Walter Scheel (b. 1919), President
1974–82	Helmut Schmidt (b. 1918), chancellor
1979–84	Karl Carstens (b. 1914), President
1982–	Helmut Kohl (b. 1930), chancellor
1984–	Richard von Weizsacker (b. 1920), President

German Democratic Republic (East Germany), Heads of State

1949–60	Wilhelm Pieck (1876–1960), prime minister
1949–64	Otto Grotewohl (1894–1964), chancellor
1960–73	Walter Ulbricht (1893–1973), prime minister
1964–73	Willi Stoph (1914–), chancellor
1973–76	Horst Sinderman (1915–), chancellor
1973–76	Willi Stoph (1914–), prime minister
1976–	Erich Honecker (1912–), prime minister
1976–	Willi Stoph (1914–), chancellor

Russian and Soviet Rulers and Heads of State

Listed by date of reign or term in office.

Czars of Russia

1533–84	Ivan IV the Terrible (1530–84)
1584–98	Feodor I (1551–98)
1598–1605	Boris Godunov (c. 1551–1605)
1605	Fëdor II (1598–1605)
1606–10	Vasily (1552–1612)
1613–45	Michael (1596–1645)
1645–76	Alexis (1629–76)
1676–82	Feodor III (1656–82)
1682–89	Ivan V (1666–96), with Peter I
1689–1725	Peter I the Great (1672–1725)
1725–27	Catherine I (1684?–1727)
1727–30	Peter II (1715–30)
1730–40	Anna Ivanovna (1693–1740)
1740–41	Ivan VI (1740–64)
1741–62	Elizabeth Petrovna (1709–62)
1762	Peter III (1728–62)
1762–96	Catherine II the Great (1729–96)
1796–1801	Paul I (1754–1801)
1801–25	Alexander I (1777–1825)
1825–55	Nicholas I (1796–1855)
1855–81	Alexander II (1818–81)
1881–94	Alexander III (1881–94)
1894–1917	Nicholas II (1868–1918)

Premiers (Chairmen of the Council of Ministers)

1917–24	Vladimir Ilyich Lenin (1870–1924)
1924–30	Alexey Rykov (1881–1930)
1930–41	Vyacheslav Molotov (1890–1986)
1941–53	Joseph Stalin (1879–1953)
1953–55	Georgi Malenkov (b. 1902)
1955–58	Nikolai Bulganin (1895–1975)
1958–64	Nikita Khrushchev (1894–1971)
1964–80	Aleksei Kosygin (1904–80)
1980–85	Nikolai Tikonov (b. 1905)
1985–	Nikolai Ryzhkov (b. 1929)

Presidents (Chairmen of the Presidium)

1923–46	Mikhail Kalinin (1875–1946)
1946–53	Nikolai Shvernik (1888–1970)
1953–60	Kliment Voroshilov (1881–1969)
1960–64	Leonid Brezhnev (1906–82)
1964–65	Anastas Mikoyan (1895–1978)
1965–77	Nikolai Podgorny (1903–83)
1977–82	Leonid Brezhnev (1906–82)
1983–84	Yuri Andropov (1914–84)
1984–85	Konstantin Chernenko (1912–85)
1985–88	Andrei Gromyko (1909–89)
1988–	Mikhail Gorbachev (b. 1931)

power among German states after Napoleonic Wars, leading German Confederation of 40 different states. Kingdom of Poland is formed under Russian czar.

1819 Spain cedes Florida and Oregon Territory to United States in return for U.S. recognition of Spanish claim to Texas.

1821–27 Greeks rebel against Ottoman Turks, culminating in Greek independence.

1827 Treaty of London among France, Great Britain, and Russia provides for direct intervention in Greek war for independence from Ottoman Empire.

1829 Daniel O'Connell spearheads the Irish Catholic Emancipation Act in British Parliament.

1830 French King Charles X (1757–1836) is forced to abdicate after liberal revolt against his reign.

1831 Italian nationalist revolutionary Giuseppe Mazzini (1805–72) establishes Young Italy Movement to promote unified Italy under republican government.

▶

Italian Rulers and Heads of State

Kings of Italy

1861–78	Victor Emmanuel II (1820–78)
1878–1900	Humbert I (1844–1900)
1900–1946	Victor Emmanuel III (1869–1947)
1946	Humbert II (1904–83)

Presidents of the Republic

1946–48	Enrico De Nicola (1877–1959), provisional
1948–55	Luigi Einaudi (1874–1961)
1955–62	Giovanni Gronchi (1887–1978)
1962–64	Antonio Segni (1891–1972)
1964–71	Giuseppe Saragat (1898–)
1971–78	Giovanni Leone (1908–)
1978–85	Alessandro Pertini (b. 1896)
1985–	Francesco Cossiga (b. 1928)

Premiers

1876–78	Agostino Depretis (1813–87)
1878	Benedetto Cairoli (1825–89)
1878–79	Agostino Depretis (1813–87)
1879–81	Benedetto Cairoli (1825–89)
1881–87	Agostino Depretis (1813–87)
1887–91	Francesco Crispi (1819–1901)
1891–92	Antonio Rudini (1839–1908)
1892–93	Giovanni Giolitti (1842–1928)
1893–96	Francesco Crispi (1819–1901)
1896–98	Antonio Rudini (1839–1908)
1898–1900	Luigi Pelloux (1839–1924)
1900–1901	Giuseppe Saracco (1821–1907)
1901–3	Giuseppe Zanardelli (1829–1903)
1903–5	Giovanni Giolitti (1842–1928)
1906	Sidney Sonnino (1847–1921)
1906–09	Giovanni Giolitti (1842–1928)
1909–10	Sidney Sonnino (1847–1921)
1910–11	Luigi Luzzatti (1841–1927)
1911–14	Giovanni Giolitti (1842–1928)
1914–16	Antonio Salandra (1853–1931)
1916–17	Paolo Boselli (1838–1932)
1917–19	Vittorio Emanuele Orlando (1860–1952)
1919–20	Francesco Saverio Nitti (1868–1953)
1920–21	Giovanni Giolitti (1842–1928)
1921–22	Ivanoe Bonomi (1873–1951)
1922	Luigi Facta (1861–1930)
1922–43	Benito Mussolini (1883–1945), dictator
1943–44	Pietro Badoglio (1871–1956)
1944–45	Ivanoe Bonomi (1873–1951)
1945	Ferruccio Parri (1890–1981)
1945–53	Alcide De Gasperi (1881–1954)
1953–54	Giuseppe Pella (1902–81)
1954	Amintore Fanfani (b. 1908)
1954–55	Mario Scelba (b. 1901)
1955–57	Antonio Segni (b. 1895)
1957–58	Adone Zoli (1887–1960)
1958–59	Amintore Fanfani (b. 1908)
1959–60	Antonio Segni (b. 1895)
1960	Fernando Tambroni (1901–63)
1960–63	Amintore Fanfani (b. 1908)
1963	Giovanni Leone (b. 1908)
1963–68	Aldo Moro (1916–78)
1968	Giovanni Leone (b. 1908)
1968–70	Mariano Rumor (b. 1915)
1970–72	Emilio Colombo (b. 1920)
1972–73	Giulio Andreotti (b. 1919)
1973–74	Mariano Rumor (b. 1915)
1974–76	Aldo Moro (1916–78)
1976–79	Giulio Andreotti (b. 1919)
1979–80	Francesco Cossiga (b. 1928)
1980–81	Arnaldo Forlani (b. 1925)
1981–83	Giovanni Spadolini (b. 1925)
1983–87	Bettino Craxi (b. 1934)
1987–	Giovanni Goria (b. 1943)

1833 British end slavery throughout empire.

1833–39 First Carlist War. Supporters of Don Carlos (1788-1855) as rightful heir to Spanish throne mount first unsuccessful campaign after Ferdinand's death (1833). They rebel again in 1873–76.

1840s Potato famine ravages Ireland. Between 1846 and 1851, more than a million people perish from starvation and another million and a half emigrate, mostly to United States.

1844–46 Great Britain's railway building boom reaches its height. Parliament approves construction of over 400 new rail lines.

1848 Revolutions of 1848, liberal revolts in European countries against conservative policies of monarchies. Prussia and other states grant new constitutions. In France, revolt ends with establishment of Second French Republic.

1848 German philosopher Karl Marx (1818–83) and German businessman Friedrich Engels (1820–95) publish *Communist Manifesto*, predicting collapse of capitalism.

1848–50 New Parliament replaces legislature of the German Confederation. Legislators attempt to form central government for all Germany and offer crown to Prussian king Frederick William IV (1795–1861). Austria blocks the move, however (1850).

1848–51 Louis Napoleon (1808–73) is in office as President of Second Republic in France.

1849 Great Palace in Kremlin is completed.

1850 Great Britain grants Australia limited self-government.

1853–56 Crimean War. Great Britain joins with France and other powers to block Russian invasion of declining Ottoman Empire. (*See Wars and Military History.*)

1858 Treaty of Aigun settles border dispute between Russia and China.

1858 Government of India Act. British government takes direct control of India.

1859 Big Ben, famed tower clock with 13-ton bell, is installed in London.

1859–60 Kingdom of Sardinia, under King Victor Emmanuel II (1820–78), wars against Austria. With French aid, Sardinia captures Lombardy.

1860 Italian nationalist Giuseppe Garibaldi (1807–82) leads rebel force that conquers Kingdom of Two Sicilies in southern Italy. He turns this territory over to King Victor Emmanuel to create united Italy.

1861 Moldavia and Walachia, provinces controlled by Ottoman Turks, unite to form Romania, an autonomous Ottoman state.

1861 King Victor Emmanuel creates Kingdom of Italy. Parliament is organized.

1861 Edict of Emancipation is declared by Russian czar Alexander II (1818–81), freeing millions of serfs in Russia.

1861 French send military expedition to Mexico. They set up French-dominated Mexican Empire under Maximilian I.

1861–71 William I (1797–1888) rules in Prussia. Otto von Bismarck (1815–98) becomes his prime minister.

1863 First workers' party is founded, in Germany, by German socialist Ferdinand Lassalle (1825–64).

1866 Seven Weeks War. War between Prussia and Austria is provoked by Prussian prime minister Bismarck. By subsequent treaty, victorious Prussia excludes Austria from confederation of northern German states, precursor of modern Germany.

1867 Russia sells Alaska to United States.

1867 British North America Act creates Dominion of Canada.

1867 The dual monarchy of Austria-Hungary is formed.

1868 In Spain, military revolts against renewed absolutism of monarchy. It establishes constitutional monarchy in 1869.

▶

1870 Rome, controlled until now by pope, is seized by King Victor Emmanuel. This marks essentially last step in creating modern Italian state, although pope refuses to recognize state control of Rome until 1929 (Lateran Treaty).

1870–71 Franco-Prussian War. Prussia successfully invades France and by its victory unifies remaining German states (in South) that were not under its leadership. It gains Alsace-Lorraine. French monarchy is toppled.

1871 Rome becomes capital of Italy.

1871 Commune of Paris. Radical workers revolt against new government and humiliating peace terms it accepted in Franco-Prussian War. Some 17,000 rebels are killed before revolt is quashed.

1871–88 After defeat of Bavarian king Ludwig II, Prussian William I reigns as first emperor of united Germany. Bismarck is his chancellor (1871–90).

1873–74 Short-lived Spanish republic is established. Onset of Second Carlist War (1873–76) leads to reestablishment of monarchy.

1880s Rapid industrialization of Germany and more active policy of colonization, notably in Africa, begins.

1880s France expands its colonial empire in Africa and Indochina.

1881–94 Alexander III (1881–94) reigns as Russian czar. He institutes harsh rule, including forced russification of minorities and extensive persecution of Jews, and helps spread revolutionary sentiment among workers and peasants.

c. 1885 Belgium becomes a major colonial power in Africa.

1889 Gen. Georges Boulanger unsuccessfully attempts to seize power and establish dictatorship in France.

1889 Eiffel Tower is built in downtown Paris for Universal Exposition of 1889.

1892 Bribery of French government officials is revealed by Panama Scandal, which breaks some years after French company fails to complete canal-building project and goes bankrupt.

1894 In Dreyfus Affair, French army officer Alfred Dreyfus (1859–1935) is wrongly accused of treason charges. Court finally overturns his conviction for offering top-secret military information to Germans in 1906.

1895–96 Italy mounts unsuccessful invasion of Ethiopia.

1897 China leases Hong Kong to Britain for 99 years.

1898 Spanish-American War. United States, reacting to Spain's harsh measures against Cuban rebels, wars against Spain, which loses Cuba, Puerto Rico, Philippines, and Guam to United States.

1899–1902 Boer War. British defeat rebels in bloody conflict in South Africa. (*See Wars and Military History.*)

1900 Italian King Humbert I (1844–1900) is killed by anarchist assassin.

1903 Russian socialists split into revolutionary Bolsheviks, led by Vladimir Lenin (1870–1924), and gradualist Mensheviks. Bolsheviks begin modern Communist movement.

1904 Entente Cordiale resolves differences between Britain and France over colonial territories in Africa and Far East.

1904–5 Russo-Japanese War. Japanese deliver Russians humiliating defeat. (*See Wars and Military History.*)

1905 Revolution of 1905. Shooting of demonstrators in St. Petersburg begins period of strikes and political unrest. Czar Nicholas finally issues decree promising liberal government and creation of legislative Duma, with prime minister.

1905 Norway gains independence from Sweden.

1906 Peter Stolypin (1863–1911), new Russian prime minister, begins reform program to give land to peasants.

▶

1906 British Labour Party, federation of trade unions and socialist societies, is formed.

1910 Union of South Africa is created by uniting Natal, Cape of Good Hope, Transvaal, and Orange Free State.

1911 In further colonial expansion, Italians occupy and annex Tripoli and Germany gains Congo from France.

1914–18 World War I. Germany is defeated and its economy wrecked, though war is fought beyond its borders. Italy enters war on the Allied side in return for secret territorial concessions to be awarded after war (including Trieste and other Austrian territories and territory in Africa). Huge Russian losses in war strengthen position of radical revolutionaries and finally topple monarchy. France becomes major battleground and is ravaged during this war. Spain declares neutrality and prospers as producer of war materiel. (*See Wars and Military History.*)

1916 Completion of 5,787-mile Trans-Siberian Railroad linking Moscow and Vladivostok, Russia.

1916 Easter Rebellion in Ireland protests lack of home rule and is followed by period of guerrilla warfare until 1922, when Ireland is established as an independent dominion within the British Commonwealth.

1916 Rasputin (1872–1916), mystic believed to be corrupting Russian royal court, is assassinated by noblemen.

1916 Secret Sykes-Picot Agreement provides for dividing up crumbling Ottoman Empire after World War I.

1917 Finland becomes an independent country.

1917–22 Russian Revolution and Civil War. Bolsheviks seize power after Czar Nicholas abdicates. Fighting between Communists and counterrevolutionaries continues until 1922. (*See Wars and Military History.*)

1917–24 Vladimir Ilyich Lenin (1870–1924) is in power as Soviet Communist Party head. He begins nationalization and collectivization and moves Soviet capital to Moscow.

1918 Former czar Nicholas is executed by Bolsheviks.

1918 In Red Terror, Communists murder tens of thousands of opponents after attempt to assassinate Lenin.

1918 Woman suffrage is enacted in Great Britain.

1918 Mutiny by German sailors becomes full-scale revolt against the monarchy. Republic is proclaimed (November). Communists rebel against socialists, now in control of government. Communist uprisings continue sporadically into 1920.

1918 Czechoslovakia emerges as a new nation in Europe under terms of Treaty of Versailles.

1918 Iceland gains independence from Denmark.

1918 Hungary cedes three-quarters of its former territories to Romania, Yugoslavia, and Czechoslovakia as war reparations. Over half the Hungarian pupulation is dead.

1918–19 Yugoslavia is created from the former Austro-Hungarian lands of Serbia, Slovenia, Croatia, Bosnia, Herzegovina, Montenegro, Macedonia, and Kosovo.

1918–20 Baltic War of Liberation. Estonia, Latvia, and Lithuania gain independence from Russia.

1919 Lenin creates Third Communist International (Comintern), formalizing split between Communists and socialists. Lenin uses Comintern to dominate communism worldwide.

1919 Government of Weimar Republic is in power in Germany. Friedrich Ebert (1871–1925) is first President.

1919 Women get right to vote in Italy.

1919–20 Russo-Polish War. Poland joins Ukrainian nationalists in attempting to gain Ukraine's independence from Communist Russia.

▶

1920 Finland gains independence from USSR.

1920 Ireland is granted home rule (it becomes Irish Free State in 1921). Northern Ireland is created.

1921 Benito Mussolini (1883–1945) organizes Fascist Party in Italy.

1921 Four-Power Pacific Treaty among United States, France, Great Britain, and Japan recognizes their respective spheres of influence in Pacific.

1921 Treaty of Riga establishes boundaries of new Poland.

1921 Nazi Party is organized in Germany by Adolf Hitler (1889–1945).

1921–23 Widespread drought strikes USSR. United States sends aid, but millions still die in famine.

1922 Mussolini's Fascists stage march on Rome in show of force that topples legitimate government. Mussolini is made head of state.

1922 First Congress of Soviets, held at the Bolshoi Theater, formally approves creation of USSR.

1922 Five-Power Naval Limitation Treaty is signed limiting size of navies of major powers.

1923 In Munich Beer Hall Putsch, Hitler fails to start right-wing revolution in Bavaria. While serving eight-month jail term resulting from this incident he writes *Mein Kampf*. In 1925, Hitler is successful and Nazi Party is organized; SS is formed to provide "protection."

1923–25 Spain is governed by military dictatorship under Gen. Miguel Primo de Rivera (1879–1930), following military coup. King remains titular head of Spain.

1924 Mussolini takes advantage of walkout by non-Fascist members of Chamber of Deputies (Matteoti crisis) to seize dictatorial powers.

1924 Joseph Stalin (1879–1953) becomes Soviet Communist leader following Lenin's death. He institutes harsh programs of collectivization and repression and pledges spread of communism worldwide.

1926 Stalin's chief opponent, Leon Trotsky (1879–1940), is ousted. Stalin's position as Soviet dictator is assured.

1928 Kellogg-Briand Pact is signed; multilateral treaty renounces war.

1930 London Naval Treaty is signed. Designed to end naval arms race, it limits number of warships for major powers.

1930s Great Britain, France, and Germany are hit hard by Great Depression. Nazi Party gains strength in German Reichstag in 1930 and in 1932 elections.

1931 Statute of Westminster grants autonomous government to Great Britain's former colonial possessions and creates British Commonwealth.

1931–39 Second Spanish republic is established. King Alfonso is deposed. New constitution (1931) grants universal suffrage, national Cortes, and reduced role of church. In first election held for Cortes (1933), rightist factions, including Carlists and Fascists, gain over 40 percent of seats.

1933 Hindenburg appoints Hitler as chancellor, bowing to Nazi tactics that virtually paralyze German government.

1933 Operation of Nazi concentration camps begins for political opponents, Jews, and others. Mass exterminations of Jews (Holocaust) begin after outbreak of World War II.

1934 Communist leader Sergei Kirov (1888–1934) is shot. Assassination becomes pretext for Stalin's bloody purge of his opponents in party. Millions are believed killed, many by mass executions. Stalin subsequently establishes "cult of personality" to promote his totalitarian rule over USSR and Communist parties worldwide.

1935 Germany reoccupies Rhineland.

▶

1935 Swastika is made part of German national flag.

1935–36 Italy again invades Ethiopia, this time successfully. It annexes country.

1936 Edward VIII (1894–1972) reigns briefly as British king. Threatened with constitutional crisis over his intended marriage to American divorcée Wallis Warfield Simpson, he abdicates; he marries her in 1937.

1936–39 Spanish Civil War is fought between conservative right and leftist factions, including republicans, socialists, and Communists. War ends with fall of Madrid to rightists.

1937–40 Arthur Neville Chamberlain (1869–1940) is British prime minister. He attempts policy of appeasement toward Nazi Germany.

1938 Nazis invade and take control of Austria.

1938 Munich Pact between Nazi Germany and other European powers provides for gradual German occupation of Czechoslovakia's Sudetenland. Hitler nullifies pact by invading Czechoslovakia and Poland in 1939.

1939 Pact of Steel between Italy and Nazi Germany further cements alliance between these Fascist states.

1939 Italy occupies Albania.

1939 Denmark signs nonagression pact with Germany but is invaded by Nazi forces in April 1940.

1939 German battleship *Bismarck* is launched.

1939–45 World War II. Nazi Germany invades Poland in 1939, starting war. Stalin brings USSR into war as ally of Nazi Germany, until Germans launch massive invasion of Soviet Union. (*See Wars and Military History.*)

1939–45 Italy joins Nazi Germany in the war only after the fall of France (1940). After failures in North Africa campaigns, Italy is itself invaded by Allied forces (in 1943, by way of Sicily).

1939–75 Gen. Francisco Franco (1892–1975), leader of the rightist forces, is dictator of Spain. He organizes totalitarian government and restores most privileges of landed aristocracy and church. Spain remains neutral in World War II.

1940 Former Soviet leader Trotsky is assassinated in Mexico.

1944–49 Greek Civil War. British and U.S. aid helps prevent post-World War II Communist takeover in Greece.

1945 Mussolini is executed by Partisans as German resistance to Allied invasion of Italy collapses.

1945 Hitler commits suicide in Berlin as Soviet forces invade city.

1945 Marshal Tito (Josip Broz, 1892–1980) rises to power in newly created Federal People's Republic of Yugoslavia.

1945 Allies occupy Germany as World War II ends in Europe. Plans for reunification soon become stalled over administration of occupation zones.

1945 Italy begins rebuilding and in postwar years undergoes "economic miracle," becoming one of world's leading industrialized nations.

1945–54 Indochina War. Communist nationalists successfully oppose France's attempt to reestablish its colonial rule. French defeat at Dien Bien Phu results in complete French withdrawal from Indochina.

1946 Spain is denied entry to United Nations because of its totalitarian government.

1946 Republican government is organized in Italy after abolition of monarchy. New constitution is enacted in 1947.

1946 Albanian government banishes all foreigners, seizes and nationalizes all domestic assets, closes all churches, and makes agriculture and industry collective.

1946 France enacts new constitution, forming Fourth Republic. French Empire is reorganized as French Union.

▶

1947 Spanish government is reorganized as monarchy, with Franco designated as chief of state for life.

1947 Meeting of Big Four (United States, USSR, Great Britain, and France) on divided Germany's economic future is marked by sharp disagreements with Soviet representatives.

1947 India is granted independence as Great Britain moves to end its involvement in colonialism.

1948 Marshal Tito, who maintained Yugoslavian autonomy as a Communist state without Soviet aid, expels Soviet military advisors. Tito is then expelled from Comintern.

1948 Soviet satellite nations are created. Communist takeovers in seven countries by this date form a buffer between free Europe and USSR.

1948–49 Increasing tensions with United States over divided Germany result in Soviet blockade of West Berlin. United States supplies city by airlift until blockade is lifted in 1949.

1949 United States finds evidence Soviet Union has exploded its first atomic bomb. Nuclear arms race with USSR begins.

1949 Ireland becomes a fully independent republic, although six counties of Ulster, Northern Ireland remain under British control.

1949 Federal Republic of Germany (West Germany) is created from U.S., British, and French occupation zones.

1949 NATO (North Atlantic Treaty Organization) is founded. Headquarters are in Paris 1950–67.

1949 Soviets create German Democratic Republic (East Germany). Tens of thousands of East Germans begin leaving for the West (in addition to millions who migrated westward immediately after the war).

1950 Spanish Government gets badly needed economic aid from United States. It reciprocates by granting United States rights to military bases in Spain (1953).

1954 Italy regains Trieste and part of adjoining territory, with remainder awarded to Yugoslavia.

1954 Rebellion breaks out against French control in Algeria. Eight-year Algerian War begins, with France committing 500,000 troops to struggle.

1955 USSR and its East European satellites form Warsaw Pact as regional defense organization.

1956 Khrushchev announces policy of de-Stalinization. Brief period of liberalization in satellite nations ends when uprising in Hungary, led by Imre Nagy, who spurs declaration of neutrality, end of censorship, open borders, and withdrawal from Warsaw Pact, culminates in Soviet invasion. Nagy is executed.

1956 France and Great Britain send troops to Egypt during Suez crisis.

1957 Soviets launch world's first artificial satellite, *Sputnik I*. "Space race" with United States begins.

1957 Treaty of Rome establishes European Economic Community.

1957 Soviets launch world's first artificial satellite, *Sputnik I*. "Space race" with United States begins.

1957 Treaty of Rome establishes European Economic Community.

1958 So-called cod wars, disputes over Iceland's territorial waters, commence between England, Norway, Denmark, and Iceland.

1958 French settlers in Algiers riot, bringing about election of Charles de Gaulle as French premier and approval of new constitution. Referendum gives French territories right to vote for independence or for self-government within newly formed French Community.

1959 "Kitchen debate" occurs between U.S. Vice President Richard M. Nixon (b. 1913) and Khrushchev at U.S. trade

▶

show in Moscow. Khrushchev visits United States in September as Cold War tensions appear to ease.

1960 Sino-Soviet split begins. It develops into widening rift in 1960s. United States uses the opportunity to improve relations with China.

1960s Rise of new troubles in Ireland with intensified guerrilla fighting in effort by Irish Republican Army to reunite six counties of Northern Ireland with the Republic. Troubles persist into 1990s.

1961 Long a Soviet ally, Albanian leader Hoxha breaks with USSR and aligns with China until 1977, when he completely isolates the country.

1961 Soviets construct Berlin Wall to stop defections to West Berlin, major source of embarrassment to them.

1962 Algeria is granted independence, ending Algerian War.

1962–63 Cuban missile crisis. Soviets lose prestige when U.S. President Kennedy risks nuclear war to force withdrawal of Soviet missiles in Cuba.

1963 USSR signs the Nuclear Test-Ban Treaty with Great Britain, limiting nuclear testing to underground explosions.

1966 Franco grants new, more liberal constitution in Spain in face of growing unrest.

1967 Soviets sign Outer Space Treaty, limiting military uses of outer space and providing for sharing of scientific information.

1967 Military coup led by Col. George Papadopoulos topples Greek government and leads to seven-year military dictatorship.

1968 "Brezhnev Doctrine" is proclaimed, prohibiting any acts of independence in Soviet satellite countries.

1968 USSR signs treaty on the nonproliferation of nuclear weapons.

1968 Soviets invade Czechoslovakia to quash liberalization of government there ("Prague Spring").

1969 Franco names Juan Carlos (b. 1938) as successor in Spain. Carlos will rule as king.

1969 Soviet Union begins Strategic Arms Limitation Talks (SALT) with United States.

1969 Socialist Olaf Palme succeeds more liberal Tage Erlander as prime minister of Sweden. Palme is assassinated in 1986 in a still unsolved crime.

1969 Conference of Communist parties is held in Moscow to discuss Soviet/Chinese differences. China later begins liberalization policy and normalization of relations with West.

1970 Great Britain is allowed to join Common Market.

1970 Heads of West and East Germany meet for first time.

1972 President Nixon makes historic first visit by U.S. President to USSR. He signs SALT I antiballistic missile treaty.

1972 Soviets sign biological warfare ban.

1972 Arab terrorists murder 11 Olympic athletes from Israel at Munich Olympic Games.

1972–74 Unrest in Northern Ireland leads to direct British rule there.

1973 Treaty establishes normal diplomatic relations between West and East Germany.

1974 Nobel Prize-winning author Aleksandr Solzhenitsyn (b. 1918) is exiled from USSR for publishing *The Gulag Archipelago*, about horrors of Soviet prison camps.

1974 British Parliament building is hit by IRA terrorist bomb.

1974 West German chancellor Brandt resigns following revelation that high-ranking member of his staff is East German spy.

1974 Greek military dictatorship is overturned after seven years and parliamentary elections are held.

▶

1976	Communist becomes speaker of Italian Chamber of Deputies, culminating alliances between Christian Democrats and leftists begun in early 1960s.
1977	Spain holds first free elections since outbreak of Civil War.
1977	Italy ends status of Roman Catholicism as state religion.
1978	Constitutional monarchy is established by popular vote in Spain.
1978	Abortion is legalized in Italy.
1978	Former Italian President Aldo Moro (1916–78) is kidnapped and murdered by terrorists.
1978	Italy's first socialist premier takes office.
1979	Terrorist bomb explodes in Italian Senate chambers.
1979	Soviets sign SALT II agreement with United States.
1979	Margaret Thatcher (b. 1925) becomes Britain's first woman prime minister.
1979	British World War II hero Lord Mountbatten (1900–79) is killed by IRA bomb.
1979–89	Afghan Civil War. Soviet troops invade Afghanistan as part of Soviet-backed coup. They continue fighting there until their withdrawal in 1989.
1980	Kremlin is rocked by series of deaths of top party leaders in early 1980s, beginning with Premier Kosygin.
1980	Marshal Tito dies, raising questions about Yugoslavian ability to survive as autonomous Communist state.
1980s	Election of former UN Secretary General Kurt Waldheim as President of Austria strikes controversy after campaign reveals Waldheim's service in German army in World War II.
1981	Andreas Papandreou wins parliamentary majority in Greece. Scandals force his resignation in 1989.
1981	Attempted coup by Spanish civil guardsmen fails.
1981	Soviet-backed government imposes martial law in Poland. It disbands Solidarity in 1982, but reverses policy and legalizes it in 1989.
1981	East German prime minister Erich Honecker (b. 1912) publicly discusses possibility of reunification of Germany for first time.
1981	Socialist François Mitterrand (b. 1916) is elected President of France. He names Communists to cabinet posts and institutes socialization program.
1981	Wedding of British Prince Charles and Lady Diana occurs at London's St. Paul's Cathedral. TV coverage of ceremony is broadcast live worldwide by satellite.
1982	In Falkland Islands War, British force Argentine invaders to surrender.
1983	Mass demonstrations are held in West Germany against U.S. deployment of medium-range nuclear missiles there.
1984	Prime Minister Thatcher escapes IRA bombing unhurt; five others are killed, however.
1985	Soviet party leader Mikhail Gorbachev (b. 1931) takes control of government. He develops policy of *glasnost* (openness) and *perestroika* (restructuring).
1985	French government secret agents reportedly are responsible for sinking Greenpeace ship *Rainbow Warrior*.
1986	Novice German pilot Matthias Rust lands his single-engine plane in Moscow's Red Square after flying unhindered from Finland through Soviet air defenses.
1986	Reagan–Gorbachev summit in Iceland produces no new accords.

▶

1986 Project to build tunnel under English Channel, to speed travel between London and Paris, is approved.

1987 Unprecedented Intermediate Nuclear Forces (INF) Treaty is signed, providing for dismantling of all U.S. and Soviet intermediate-range nuclear weapons.

1987 Soviet leader Gorbachev offers to reduce Soviet forces in Europe unilaterally. This becomes basis for further significant talks on arms reduction between United States and USSR.

1989 Hungary adopts multiparty system, first in Communist state.

1989 In radical policy change to end chronic food shortages, Soviet government gives farmers control of land and selection of crops they grow. New family farm concept reverses long-standing Communist policy of collectivization of farms.

1989 Soviets continue program of liberalizing and decentralizing government in face of severe economic problems. National debt reportedly has reached about $500 billion and is growing faster than that of United States.

1989 Government in Poland is formed from leaders outside Communist Party for first time since Soviets began domination of Eastern Europe.

1989 Prime Minister Thatcher successfully completes unbroken term of 10 years in office, longest for British prime minister in this century.

1989 The Berlin Wall, which since August 31, 1961 has divided East and West Berlin and hindered the political freedom of East Germans, is torn down, a first step on the path to German reunification.

1990 Albania opens border to foreign nationals.

1990 National Salvation Front (provisional government) lifts restrictions on travel of Romanians abroad. Under new rules, most Romanians will qualify for ten-year passports. Romania becomes first Soviet-bloc country to outlaw its Communist Party. The move is forced on National Salvation Front by popular demand.

1990 Czechoslovakian Communist Party turns over more than 100 seats in Federal Assembly (Parliament). A total of 120 new deputies are sworn in, including only nine Communists.

1990 Soviet troops, based in Czechoslovakia since crushing of 1968 "Prague Spring" movement, begin phased withdrawal, to be completed by July 1, 1991.

1990 Supreme Soviet (parliament) of Lithuania formally declares restoration of the republic's independence from Soviet Union. (It had been independent from 1918–40, when it was annexed by USSR along with neighboring Baltic states of Estonia and Latvia.) Lithuania is first Soviet Republic to attempt to secede, increasing tension between USSR and the now-independent republic.

1990 An alliance of conservative parties backed by West Germany Chancellor Helmut Kohl scores surprising triumph in East Germany's general elections, just 18 weeks after East Germany opened its borders and installed a reformist Communist government in November, 1989.

1990 Landmark agreement between Soviet President Mikhail S. Gorbachev and West German Chancellor Helmut Kohl, in which Gorbachev drops his opposition to membership of a united Germany in NATO.

Middle East

See Muslim Empires for early history.

1914–18 World War I brings on final collapse of Ottoman Empire, a German ally, and dismemberment of its extensive domains in Middle East.

1917 British occupy Ottoman territory of Palestine and issue Balfour Declaration supporting Jewish national homeland there.

1918 British occupy Ottoman territory that becomes modern Iraq. Iraq becomes British League of Nations mandate after war.

1920 Modern Turkey is established from part of dismembered Ottoman Empire. Sultan rules until nationalist rebels depose him (1922) and establish republic (1923).

1920 Syria becomes French League of Nations mandate (then including Lebanon, which becomes separate state in 1926).

1921 Iraq is established as kingdom under its first king, Faisal I. British maintain military bases.

1921 British rule over Iran, then British protectorate, is ended by military coup. Iranian Reza Pahlavi (1877–1944) seizes power.

1922 Palestine becomes British League of Nations protectorate.

1922 British grant Egypt independence after nationalist revolts (1919–21), though military forces are to remain. Monarchy is established under Fuad I (1868–1936).

1923 Kemal Atatürk (1881–1938), nationalist rebel leader, becomes first President of Turkey.

1923 Transjordan (modern Jordan) is created by partitioning Palestine.

1925 In Iran, Reza Pahlavi becomes hereditary shah. He remains in power to 1941.

1930s Rise of Nazism and persecution of Jews in Europe increase flow of Jewish immigrants to Palestine. Friction and violence between Arabs and Jews in Palestine increase.

1931 U.S., British, Dutch, and other oil interests form Iraqi Petroleum Company to exploit extensive Iraqi oil fields.

1932 Saudi Arabia is created. First king is Ibn Saud (c. 1880–1953), whose conquests in Arabian Peninsula make up kingdom.

1936 Oil is found in Saudi Arabia.

1939–45 World War II. Great Britain establishes Allied control in Mideast early in war to secure valuable oil fields, though Egypt, Libya, and other parts of North Africa become major battlegrounds. *(See Wars and Military History.)*

1941 British liberate Lebanon from Vichy French. Syria is proclaimed republic following Vichy French defeat. Syria becomes independent in 1944.

1941–79 Muhammad Reza Shah Pahlavi rules Iran. He institutes program of rapid modernization that eventually stirs up revolution among Muslim fundamentalists.

1945 Arab League is founded.

1946 Jordan becomes independent kingdom.

1948 Israel is created in Palestine as national homeland for Jews.

1948–49 First Arab–Israeli War. Israel repulses invasions by Egypt, Iraq, Jordan, Syria, and Lebanon, all opposing creation of Israeli state. Separate armistices end conflict. Israel loses no territory, but about 400,000 Palestinian refugees flee to neighboring Arab countries. Jordan occupies (and annexes in 1950) West Bank.

1950 Israel enacts law of return. Immigrant Jews are granted automatic citizenship.

1951 Iran fails in attempt to nationalize oil industry. British blockade forces Iran to put nationalized industry in foreign hands.

▶

1951 Libya gains independence.

1953 Republic is established in Egypt following nationalist and anti-British unrest in 1952. British agree to withdraw troops (1954), and President Gamal Abdel Nasser (1918–70) becomes President (1954–70).

1956 Suez Crisis. Nasser nationalizes canal and precipitates invasion by Israel, Great Britain, and France. (*See Wars and Military History.*)

1958 Iraq becomes republic. Relations with Soviet Union are established. Military rule is established by 1963.

1958 Oil is discovered in Libya.

1958–61 United Arab Republic—Egypt, Syria, and Yemen—unite, with Nasser as President.

1960 OPEC is founded to control oil production.

1964 Palestinian Arab refugees organize Palestinan Liberation Organization to retake Israeli lands. Group is headed by Yasir Arafat (b. 1929).

1967 Third Arab–Israeli War (Six-Day War). Israelis quickly establish air superiority and emerge victorious from ground engagements on various fronts. They capture Sinai Peninsula, Old City of Jerusalem, West Bank, and Golan Heights.

1969 Libyan dictator Muammar al-Qaddafi (b. 1942) seizes power. He later nationalizes foreign oil facilities and establishes relations with Soviet Union.

1970 Anwar Sadat (1918–81) becomes President of Egypt. He soon ousts Soviet advisers (1972).

1970 PLO chief Yasir Arafat agrees to halt PLO attempt to take over Jordan. PLO subsequently moves headquarters to Lebanon.

1972 Jordanian king Hussein (b. 1935) is wounded in assassination attempt by Palestinian.

1973 Fourth Arab–Israeli War (Yom Kippur War). Egypt, Syria, and Iraq attack Israel on Jewish holy day, but Israelis beat back Arab invaders before fighting ends. Israelis agree to limited withdrawal in Sinai as part of peace accords.

1973 OPEC forces massive price hike for oil following Arab oil embargo month earlier.

1974 Turkey invades Cyprus after pro-Greek faction stages coup there.

1974 Saudis get controlling interest in Aramco oil-production consortium.

1975 Saudi king Faisal (c. 1906–75) is shot by nephew.

1975 Revolt by Kurds in northern Iraq is crushed.

1975– Lebanese Civil War. Muslim factions rebel against the Christian government, prompting invasion by Syria; this temporarily halts fighting.

1978–79 In Muslim fundamentalist revolution in Iran, shah is ousted by supporters of Ayatollah Khomeini (1901–89), who opposes modernization of Iran. Khomeini sets up fundamentalist regime.

1979 Egyptian President Anwar Sadat signs peace treaty with Israel in return for phased Israeli pullout from Sinai.

1979–81 American embassy personnel in Iran are held hostage by Khomeini supporters.

1980 Libyan troops intervene in Chad civil war.

1980 U.S. attempt to rescue American hostages in Iran ends in failure.

1980–89 Iran–Iraq War. Border clashes erupt into full-scale conflict after Muslim fundamentalist revolution in Iran. Bloody war destabilizes Persian Gulf region and leads to U.S. naval presence in area until ceasefire.

1981 High-level officials in Iran's Islamic government are assassinated by bomb explosion. Weeks later new President, prime minister, and other officials are killed by bomb.

1981 President Sadat is assassinated in Egypt by Muslim fundamentalists.

1982 Israel invades Lebanon and forces evacuation of PLO militiamen.

1983–84 U.S. peace-keeping force is stationed in Beirut. Troops are withdrawn in 1984 after truck bomb kills 239 of them in 1983.

▶

1986 United States imposes economic sanctions against Libya in response to Libyan backing for Arab terrorists. United States launches air strike against Tripoli following continued terrorist activities.

1986 United States admits to secret arms sales to Iran.

1987 U.S. Navy frigate *Stark*, operating in Persian Gulf, is hit and badly damaged by Iraqi missile; 37 sailors are killed.

1987 Libyan troops are driven out of Chad by government troops.

1987 Demonstration by pro-Iranian Muslims visiting Muslim holy city of Mecca ends with 400 dead after police open fire on demonstrators.

1987 Intifadah in Israel begins. Riots continue sporadically through 1990.

1988 U.S. Navy warship, already exchanging fire with Iranian gunboats, accidentally shoots down oncoming Iranian jetliner, killing all 290 passengers.

1988 Cease-fire sponsored by UN takes hold in Iran–Iraq war.

1988 PLO issues statement recognizing Israel's right to exist and renouncing terrorism.

1989 Iranian leader Ayatollah Khomeini dies.

1990 Iraqi leader Saddam Hussein invades Kuwait and threatens invasion of Saudi Arabia. American forces are sent in response.

Africa

For more information on Egypt and Libya, see Middle East.

c. 800 B.C. Phoenicians establish colonies in North Africa.

c. 600 B.C.–A.D. 400 The Egyptian-influenced Kingdom of Kush (modern Sudan) flourishes.

A.D.

c. 500–1000 Rise of the Sudanic kingships (from Senegal to Red Sea, Nile sources to Zimbabwe).

600s Arabs conquer North Africa. They spread Islam throughout the region.

700–c. 1500 Mombasa (in modern Kenya) is established as African center for Arab trade in ivory and slaves.

c. 1500s–1600s Maize and cassava are introduced to Africa and quickly become staple crops in Congo region.

1652 Dutch East India company settles Cape of Good Hope.

1788 Sierra Leone is established for colonization of blacks who either had escaped slavery or had served in British armed forces.

1815 British gain Cape Colony in southern Africa from Dutch.

1820–22 Egypt conquers Sudan.

1821 Liberian capital of Monrovia is founded for recolonization of freed slaves. Liberia becomes republic 1847.

1830–37 French invade Algeria. It is declared French possession in 1834.

1833 Boers (Dutch settlers) migrate northward out of British-controlled territory in southern Africa, in Great Trek. Gold and diamonds later are discovered in new lands settled by Boers (Transvaal and Orange Free State).

1841 Scottish missionary David Livingstone (1813–73) travels to Africa and there becomes European spokesman for indigenous peoples of Africa. He discovers Zambesi River, Victoria Falls (1855), and Lake Nyasa. *(See also Exploring Africa)*.

1843 Gambia and Natal in southern Africa are made British colonies.

▶

c. 1850 German missionaries explore African interior, drawing up careful charts of traveled areas. They discover Mount Kenya and Mount Kilimanjaro.

1860 Spanish invade Morocco. They establish influence in region.

1860s French establish presence in what becomes Mauritania.

1861 British establish colonial presence in what becomes modern Nigeria. They begin expanding sphere of influence in region.

1866 David Livingstone sets out to explore Lake Albert in search of Nile's source. He disappears for three years, only to be "found" in 1871 by journalist Henry Stanley (1841–1904). Stanley later proves Lake Victoria is Nile source (1873).

1866 Sultan of Zanzibar builds summer palace on site of Dar-es-Salaam.

1874 Ghana (Gold Coast) in West Africa becomes British colony after Danish and Dutch settlements are purchased.

1875–83 Italian-born French explorer Savorgnan de Brazza (1852–1905) explores central Africa (Congo) and stakes French claims to region.

1877 Four great African rivers—Zambezi, Niger, Nile, and Congo—have been explored and their courses determined by this date.

1885 Madagascar is established as French protectorate. Borders of Portuguese Angola are fixed.

1885 British forces under Sir Charles Gordon (1833–85) are massacred at Khartoum during revolt in Sudan. Sudan is reconquered by British in 1898.

1885 Belgian Congo (modern Zaire) becomes personal domain of Belgian king Leopold II (1835–1909).

1885 German East Africa colony, consisting of modern Tanganyika, Rwanda, and Burundi, is formed.

1886 British establish presence in region of modern Kenya. They soon establish themselves as colonial rulers there.

1889 Rhodesia (modern Zambia and Zimbabwe) is colonized by British. Indigenous uprisings are quashed in 1896–97.

1891 French Congo is established as colony.

1895 French West Africa is created as federation of territories including Dahomey, French Guinea, French Sudan, Ivory Coast, Mauritania, Niger, Senegal, and Upper Volta.

1895–96 Italians launch unsuccessful invasion of Ethiopia.

1899–1902 Boer War, bitter conflict between British and Dutch colonists (Boers) in Transvaal and Orange Free State in South Africa. After hard fighting, Boers are forced to accept British rule. (*See Wars and Military History.*

1904 French Sudan (modern Mali) becomes a French colony. It joins French Union in 1946.

1908 Belgian government takes over rule of Belgian Congo from King Leopold following scandal over brutal exploitation of indigenous people there.

1910 Union of South Africa is created by uniting British possessions of Natal, Cape of Good Hope, Transvaal, and Orange Free State.

1910 French Equatorial Africa is created as a federation of colonies Gabon, French Congo, and Ubangi-Shari-Chad. Chad becomes separate colony within federation in 1920.

1912 Morocco is divided into protectorates under French and Spanish.

1914 Nigeria becomes British protectorate.

1915 German South-west Africa (modern Namibia) is conquered during World War I and made protectorate under South Africa.

1923 Rhodesia is made autonomous colony within British empire.

1934–35 Italy again invades Ethiopia, this time successfully. It creates Italian East Africa in 1936.

▶

1939–45 During World War II, northern Africa becomes important battleground. (*See also Wars and Military History.*)

1941 Ethiopian independence is restored following British invasion and defeat of Italian colonial forces.

1945 Muslim nationalist revolt in Algeria is dealt with harshly by French.

1947 Revolt against French rule occurs in Madagascar.

c. 1950 White-ruled government in South Africa establishes apartheid.

1951 Angola and Mozambique become provinces of Portugal.

1952 Ethiopia is awarded control of eastern region of Eritrea.

1952–60 Mau Mau, a native terrorist movement, erupts against British control in Kenya. It helps bring about Kenyan independence in 1963.

1954–62 Algerian war. In revolt against French control, France commits 500,000 troops to struggle.

1955–72 Sudanese civil war. Christians in South rebel against Muslims in North.

1956 Sudan, Tunisia, and Morocco gain independence.

1957 Ghana gains independence.

1958 Riots by French settlers in Algiers lead to election of Charles De Gaulle (1890–1970) as new French premier. Referendum gives French territories right to vote for independence or for self-government within newly formed French Community.

1958 Madagascar votes to become independent republic within French Community.

1958–59 Colonies of French West Africa and French Equatorial Africa are dissolved when their territories elect self-government within French Community.

1960 Nigeria becomes independent. It establishes republican government in 1963.

1960 Former French colonies of Niger, Mauritania, Mali, French Congo, Chad, and Madagascar establish complete independence.

1960 In Egypt, Soviets provide aid to build Aswan High Dam (completed in 1971).

1960 Belgian Congo gains independence as Republic of the Congo (Modern Zaire).

1960 Independent Republic of Somalia is created from both British and Italian territories.

1961 Sierra Leone gains independence from England.

1960–63 Civil war in Congo over secession of Katanga. UN secretary-general Dag Hammarskjöld (1905–61) dies in plane crash en route to negotiate end to this war.

1961 South Africa becomes independent of British Commonwealth.

1961–75 War for independence of Angola.

1962 Algeria votes for independence and is recognized by France. Some 1 million Europeans in Algeria leave country.

1962 Uganda gains independence from Great Britain.

1962 Burundi gains independence from Belgium.

1963 Coup in former French Congo. Communist form of government is adopted.

1963 Kenya declares independence. It becomes republic in 1964.

1964 Rhodesia is divided. Northern Rhodesia, ruled by blacks, declares independence as Zambia. White-ruled Southern Rhodesia remains autonomous colony and institutes complete racial segregration (1970). Rhodesian civil war (1964–75) between ruling whites and black nationalists begins.

1964 Tanganyika and Zanzibar merge to form Tanzania.

1964–75 War of independence in Mozambique. Country gains independence from Portugal in 1975.

1965 Gambia gains independence from Great Britain.

▶

1966 Botswana is granted independence from Great Britain.

1966 Lesotho, formerly the British crown colony of Basutoland, gains independence.

1966–70 Nigerian civil war. Government conquers secessionist Biafra.

1968–74 Drought-prone Sahel region of Africa is hit; 500,000 perish.

1968–84 Civil war in Chad. Fighting begins with Muslim revolt.

1969– Ethiopian civil war, over secession of Eritrea.

1970s South Africa comes under increasing international pressure to end apartheid government. Pressure continues into 1980s.

1971 Zaire becomes official name of Republic of the Congo. Economy is nationalized in 1975.

1971 Algeria nationalizes French oil facilities in the country. It later becomes major oil supplier to United States.

1971 Idi Amin seizes power in Uganda and establishes harsh rule. He is ousted in 1979.

1973–74 Severe drought occurs in Africa; hundreds of thousands perish in resulting famine.

1974 Guinea-Bissau is granted independence by Portugal.

1974 South Africa unsuccessfully resists UN takeover of protectorate of South-West Africa (Namibia). Meanwhile, civil war between Communist and non-Communist factions stalls efforts to establish Namibian independence.

1974 Long-time Ethiopian emperor Haile Selassie (1892–1975) is deposed in bloodless coup. Military establishes socialist government and nationalizes economy. Soviets back Ethiopia by 1977, and Cuban troops aid fighting in civil war.

1975 Angola gains independence. Soviet-backed faction gains control of Angola, though rebel faction continues fighting.

1975 Botswana joins European Common Market as an associate member.

1977–78 Zaire turns back invasions by Angola. France and United States provide aid for Zaire.

1979 Southern Rhodesia becomes Zimbabwe as new constitution transfers control of country to black majority. Zimbabwe becomes fully independent (of Great Britain) in 1980.

1980 Libyan troops intervene in Chad civil war.

1981 South Africa launches invasions of Angola, Mozambique, and Namibia to combat terrorists operating from these border countries.

1983 Military coup in Nigeria ends brief return to democratic government there.

1983–85 In drought throughout much of Africa, tens of thousands die.

1984 Upper Volta is renamed Burkina Faso.

1985 Julius K. Nyerere, Tanzania's political leader since independence in 1961, steps down. In open elections rather than through political appointment, Ali Hassan Mwinyi is chosen President.

1986 South Africa raids Zimbabwe, Zambia, and Botswana to attack terrorists in these border countries.

1987 Libyan troops are driven out of Chad by government troops.

1987 Strike by black workers in South Africa turns violent. Unrest among South African blacks continues into the 1990s.

1988 United States successfully negotiates peace plan for Namibia. Black majority rule is to be established following withdrawal of Cuban troops from neighboring Angola, which Communist rebels in Namibia have been using as base.

1989 Two-year locust plague in African desert regions eases, ending food shortages in affected areas.

1989 Angolan civil war is halted by cease-fire.

▶

1990 Nelson Mandela (b. 1918), President of African National Congress, is freed by South African state President F. W. de Klerk after 27 years' imprisonment in South African jail.

1990 Liberian government is threatened in coup d'état.

Asia and Pacific

B.C.

4000 Rice is grown as food crop in China.

c. **2600** Chinese begin cultivation of silkworms.

c. **2500** Advanced civilization develops in Indus Valley, in modern Pakistan.

c. **2200–c. 1766** Legendary Hsia dynasty period. First emperors of China are traditionally believed to have ruled during this time.

Chinese Emperors and Heads of State

Dynasties

B.C.	
c. **2200–c. 1766**	Hsia (legendary)
c. **1766–1122**	Shang
1122–221	Chou
202 B.C.– A.D. **220**	Han
A.D.	
220–589	Six dynasties period
589–618	Sui
618–907	T'ang
907–960	Five dynasties period
960–1279	Sung
1279–1368	Yüan (Mongol)
1368–1644	Ming
1644–1912	Ch'ing (Manchu)

Emperors

(Beginning with Ming dynasty)

1368–98	Hung-wu (1328–98), First Ming dynasty emperor
1398–1402	Chien-wen (1377–1440?)
1402–24	Yung-lo (1360–1424)
1424–25	Hung-hsi (1378–1425?)
1425–35	Hsüan-te (1398–1435?)
1435–49	T'ien-shun (1427–64)
1449–57	Ching-t'ai (1428–57)
1457–64	T'ien-shun (1427–64), restored
1464–87	Ch'eng-hua (1447–87)
1487–1505	Hung-chih (1470–1505)
1505–21	Cheng-te (1491–1521)
1521–66	Chia-ching (1507–66)
1566–72	Lung-ch'ing (1537–72)
1572–1620	Wan-li (1563–1620)
1620	T'ai-ch'ang (1582–1620?)
1620–27	T'ien-ch'i (1605–27)
1627–44	Ch'ung-chen (1611–44)
1644–61	Shun Chih (1638–61), first Ch'ing (Manchu) dynasty emperor
1661–1722	K'ang-hsi (1654–1722)

Chinese Emperors and Heads of State

1723–35	Yung Cheng (1678–1735)
1735–96	Ch'ien Lung (1711–99)
1796–1820	Chia Ch'ing (1760–1820)
1821–50	Tao Kuang (1782–1850)
1851–61	Hsien Feng (1831–61)
1862–74	T'ung Chih (1856–75)
1875–1908	Kwang-hsu (1871–1908)
1908–12	Hsuan-t'ung (1906–67)

Heads of State, the Republic

1911–12	Sun Yat-sen (1866–1925), first president (provisional)
1912–16	Yuan Shi-kai (1859–1916) (provisional to 1913)
1916–17	Li Yuan-hung (1864–1928)
1917–18	Feng Kuo-chang (1858–1920)
1918–22	Hsu Shih-chang (1858–1939)
1922–23	Li Yuan-hung (1864–1928)
1923–24	Ts'ao K'un (1862–1938)
1928–32	Chiang Kai-shek (1887–1975)
1932–43	Lin Sen (1867–1943)
1943–49	Chiang Kai-shek (1887–1975)

Communist People's Republic

1949–76	Mao Tse-tung (1893–1976), party chairman
1976–81	Hua Gofeng, party chairman
1981–82	Hu Yaobang (1915–89), party chairman
1982–87	Hu Yaobang, party general secretary
1987–89	Zhao Ziyang (b. 1919), party general secretary
1989–	Jiang Zemin (b. 1926), party general secretary

c. 1500 Invasion of India by Aryans, people of uncertain origin. Aryans first settle in northwestern India (Punjab) and at some point overrun Indus Valley civilization. They introduce their religion, Vedism, in India.

c. 1450 Soybean becomes farm crop in China.

1000 Chinese establish rule in Tonkin (Vietnam) that lasts for next 2,000 years.

723 First city on site of present-day Peking is established by this time or earlier.

551–479 Confucius (551–479 B.C.) lives. Five classics of Confucianism (*Wu Ching*) are traditionally attributed to Confucius.

550–250 Classical age or "age of a hundred philosophers," in China. Confucianism, Taoism, and other religions emerge.

c. 528 Buddha (c. 563–c. 483 B.C.) founds Buddhism amid rising opposition to then traditional Indian religion of Brahmanism, which evolved from Vedism.

509 Persian king Darius I (549–486 B.C.) completes his conquest of Indus Valley in northwestern India.

c. 500 Work begins on Grand Canal to link China's major river systems. It eventually extends 1,000 miles from northern China to Yangtze River Valley in South.

c. 500 Chinese invent crossbow.

c. 400 Chinese develop plow made of iron.

371–289 Mencius (Meng-tzu) lives. He is known as the "second sage," after Confucius.

c. 326 B.C. Greek conqueror Alexander the Great (356–323 B.C.) completes conquest of Persian territory in Indus Valley by Battle of Hydaspes River.

c. 325 Earliest known book on strategy and war, *The Art of War*, is written by Suntzu.

c. 322 Chandragupta Maurya (c. 321–296 B.C.) conquers northern India and creates India's first great native empire, Maurya empire. Chandragupta is said to have relinquished his crown in 287 to become Jainist monk.

273–232 Mauryan king Asoka (d. 232 B.C.) is in power. He expands empire to include

▶

	most of modern India and establishes Buddhism as state religion.
221	During Ch'in dynasty, China is unified under repressive regime.
213	Burning of Confucian classics is ordered by Ch'in emperor Shi Huang-ti (255–210 B.C.).
c. 185	Maurya empire collapses. Hinduism, outgrowth of earlier Brahmanism, begins to spread among petty kingdoms of India. Kingdoms in South trade with Romans and Southeast Asian peoples.
179–104	Confucian thinker Tung Chung-shu lives. He categorizes five cardinal virtues of love, righteousness, decorum, wisdom, and trustworthiness.
141–87	Emperor Wu Ti (156–87 B.C.) makes Confucianism basis of imperial government's political ideology and creates state cult of Confucianism (136 B.C.).
124	First imperial university is founded in China as center for study of the Five Confucian Classics. Its enrollment reaches 30,000 by A.D. 220.
c. 124	Chinese examination system is instituted by Wu Ti. It is administered to all Chinese civil servants until its abolition in 1905.

A.D.

61	Mahayana Buddhism spreads to China.
c. 100	Confucian temples are built in each of China's 2,000 counties.
105	First paper is invented in China. Made mainly of mulberry bark and hemp fibers, it is beaten and spread out to dry.
220	Han dynasty falls. Subsequent period of six dynasties (220–589) is one of war and divisiveness. Taoism and Buddhism spread; Confucianism declines.
304	First recorded use of insects to control other insects that damage crops, in China.
320–30	Chandragupta I (d. 330) rules kingdom in northeastern India. He founds Gupta dynasty and expands his kingdom by conquest, creating nucleus of empire that eventually includes much of northern and central India.
c. 500	Invasion by White Huns ends long rule by Gupta kings. Empire breaks up into smaller kingdoms.
515–97	Chih-i, founder of Buddhist T'ien-t'ai sect in China, lives.
c. 520	Indian scholar Bodhidharma arrives in China. Zen school of Buddhism eventually evolves from his teachings.

▶

Japanese Rulers and Heads of State

356–80	Jingo, regent-empress
592–623	Shotoku, regent
592–628	Suiko Tenno, empress
662?–71	Tenchi, emperor
672–86	Temmu, emperor
686–97	Jito, empress
697–703	Mommu, emperor
703–24	Gemmyo, empress
724–56	Shomu, emperor
758–64	Junnin, emperor
764–70	Shotoku, empress
770–81	Konin, emperor
781–806	Kammu, emperor
1068–72	Sanjo II (1034–73), emperor
1072–86	Shirakawa (1053–1129), emperor
1129–56	Toba, emperor
1155–58	Shirakawa II (1127–92), emperor
1183–1198	Toba II (1180–1239), emperor
1192–99	Minamoto Yoritomo (1147–99), first of shōguns
1199–1203	Minamoto Yorii
1203–1219	Minamoto Sanetomo

Japanese Rulers and Heads of State

1368–94	Ashikaga Yoshimitsu (1358–1408), first Muromachi period shōgun
1443–74	Yoshimasa (1435–90)
1568–82	Oda Nobunaga (1534–82), first Oda period shōgun
1585–98	Toyotomi Hideyoshi (1537–98)
1603–5	Tokugawa Ieyasu (1542–1616), first Tokugawa period shōgun
1605–23	Tokugawa Hidetada (1579–1632)
1623–51	Tokugawa Iemitsu (1604–51)
1651–80	Tokugawa Ietsuna (1641–80)
1680–1709	Tokugawa Tsunayoshi (1646–1709)
1709–12	Tokugawa Ineobu (1662–1712)
1712	Tokugawa Ietsugu (1709–16)
1716–45	Yoshimune (1684–1751)
1745–60	Tokugawa Ieshige (1711–61)
1760–86	Tokugawa Ieharu (1737–86)
1786–1837	Tokugawa Ienari (1773–1841)
1837–53	Tokugawa Ieyoshi (1793–1853)
1853–58	Tokugawa Iesada (1824–58)
1858–66	Tokugawa Iemochi (1846–66)
1867	Tokugawa Keiki (1837–1913)
1867–1912	Mutsuhito (1852–1912), first emperor of Meiji Restoration
1912–26	Taisho (1879–1926)
1926–89	Hirohito (1901–89)
1989–	Akihito (b. 1933)

Japanese Prime Ministers

1885–88	Hirobumi Ito (1841–1909)
1888–89	Kiyotaka Kuroda (1840–1900)
1889–91	Aritomo Yamagata (1838–1922)
1891–92	Masayoshi Matsukata (1840–1924)
1892–96	Hirobumi Ito (1841–1909)
1896–98	Masayoshi Matsukata (1840–1924)
1898	Hirobumi Ito (1841–1909)
1898	Shigenobu Okuma (1838–1922)
1898–1900	Aritomo Yamagata (1838–1922)
1900–1901	Hirobumi Ito (1841–1909)
1901	Kimmochi Saionji (1849–1940)
1901–6	Taro Katsura (1847–1913)
1906–8	Kimmochi Saionji (1849–1940)
1908–11	Taro Katsura (1847–1913)
1911–12	Kimmochi Saionji (1849–1940)
1912–13	Taro Katsura (1847–1913)
1913–14	Gombei Yamamoto (1852–1933)
1914–16	Shigenobu Okuma (1838–1922)
1916–18	Masatake Terauchi (1849–1919)
1918–21	Takashi Hara (1865–1921)
1921–22	Korekiyo Takahashi (1854–1936)
1922–23	Tomosaburo Kato (1859–1923)
1923–24	Gombei Yamamoto (1852–1933)
1924	Keigo Kiyoura (1850–1942)
1924–26	Takaakira Kato (1860–1926)
1926–27	Reijiro Wakatsuki (1866–1949)
1927–29	Giichi Tanaka (1863–1929)
1929–31	Osachi Hamaguchi (1870–1931)
1931	Reijiro Wakatsuki (1866–1949)
1931–32	Ki Tsuyoshi Inukai (1855–1932)
1932	Korekiyo Takahashi (1854–1936)
1932–34	Makoto Saito (1858–1936)
1934–46	Keisuki Okada (1862–1952)
1936–37	Koki Hirota (1878–1948)
1937	Senjuro Hayashi (1870–1943)
1937–39	Fumimaro Konoye (1891–1945)
1939	Kiichiro Hiranuma (1867–1952)
1939–40	Nobuyuki Abe (1875–1953)
1940	Mitsumasa Yonai (1880–1948)
1940–41	Fumimaro Konoye (1891–1945)
1941–44	Hideki Tojo (d. 1948)
1944–45	Kuniaki Koiso (1879–1950)
1945	Kantaro Suzuki (1867–1948)
1945	Haruhiko Higashikuma
1945–46	Kijuro Shidehira (1872–1951)
1946–47	Shigeru Yoshida (1878–1967)
1947–48	Tetsu Katayama (1887–?)
1948	Hitoshi Ashida (1887–1959)
1948–54	Shigeru Yoshida (1878–1967)
1954–56	Ichiro Hatoyama (1883–1959)

Japanese Rulers and Heads of State			
1956–57	Tanzan Ishibashi (1884–1973)	1978–80	Masayoshi Ohira (1910–80)
1957–60	Nobusuke Kishi (1896–1987)	1980–82	Zenko Suzuki (b. 1911)
1960–64	Hayato Ikeda (1899–1965)	1982–87	Yasuhiro Nakasone (b. 1918)
1964–72	Eisaku Sato (1901–75)	1987–89	Noboru Takeshita (b. 1924)
1972–74	Kakuei Tanaka (b. 1918)	1989	Sousuke Uno (b. 1922)
1974–76	Takeo Miki (b. 1907)	1989–	Toshiki Kaifu (b. 1931)
1976–78	Takeo Fukuda (b. 1905)		

581–618 Sui dynasty reigns and restores Chinese unity.

604 Japan's first constitution is enacted. It establishes council of nobles, though emperor retains control of kingdom.

613–81 Shan-tao lives. He founds Pure Land sect (Ching-t'u), which postulates Pure Land, or paradise to which Buddha will lead faithful followers, in China.

618–907 T'ang dynasty is in power. Civil-service examinations are expanded and new edition of Confucian classics, *Wu Ching,* is completed.

621 Buddhism becomes state religion of Japan.

c. 700–1300 Khmer empire (present-day Cambodia, Thailand, southern Laos, and central and southern Vietnam) flourishes. Celebrated temple city Angkor Wat is built.

751 Chinese prisoner taken in battle near Samarkand reveals secret of making paper to Arabs. Paper is not finally introduced in Europe until 12th century.

768–824 Han Yu lives. His polemical attacks on Buddhism are highly influential.

794 Japanese capital is moved to Kyoto.

794–1185 Fujiwara family of noblemen control Japan's imperial court during this time, effectively ruling kingdom as regents of emperor.

800s Ungainly-looking but extremely seaworthy Chinese junks are on regular trading voyages to India and back.

c. 850 Chinese probably have developed crude form of gunpowder.

948 Vietnam becomes independent of, although still greatly influenced by, China.

960–1279 Sung dynasty rulers are in power. They bring about cultural flowering and Neo-Confucian revival. Dynasty eventually falls before invading Mongols.

972 Buddhist canon *Tipitaka* is printed in China.

995–1027 Michinaga (966–1027), most powerful ruler of Fujiwara dynasty, rules Japan as virtual dictator. Kyoto reaches its height of splendor during his reign, and Japanese literature flourishes.

1000 Thai people migrate south from China and come under influence of Khmer empire.

1130–1200 Chu Hsi lives. He codifies Confucianism into comprehensive system of thought.

1156 Period of feudal warfare begins, marking decline of Fujiwara family's power.

1185 Minamoto family becomes Japan's ruling family following Gampei War (1180–85).

1192 Muslims establish their first kingdom in India, Delhi Sultanate, in North. They expand kingdom throughout much of India and spread Islam in subcontinent.

1192–1333 Japanese kingdom is ruled by military dictatorship, Kamakura shōgunate. It is established by great Minamoto family ruler and first shōgun, Yoritomo (1147–99). Kamakura becomes capital.

▶

c. 1200 Chinese develop first rocket-propelled weapons, arrows propelled by black-powder rockets.

c. 1250 Zen Buddhism spreads in Japan.

1259–94 Kublai Khan (1216–94) rules China as first of Yüan (Mongol) dynasty emperors.

1260–90 Kublai Khan builds his fabulous new city on site of Peking and makes it capital of Mongol empire.

1275 Italian explorer Marco Polo (1254?–1324?) reaches city and remains until about 1292.

1281 Unexpected typhoon thwarts attempted invasion of Japan by Mongols led by Kublai Khan (c. 1215–94). *Kamikaze* (divine wind) of World War II in modern times refers to this event.

1300s Maoris migrate from Polynesian islands to New Zealand.

1338–1578 In Muromachi period, Japan is ruled by shōguns of Ashikaga family, which comes to power after Emperor Daigo II (1287–1339) fails to restore imperial power over shōguns. Kyoto is again capital.

c. 1350 Kingdom of Siam is established, with Ayutthaya its capital city.

1392–1910 Korea's Yi dynasty (Kingdom of Chosen) flourishes; it is staunch alley of both Ne Ming and Qing dynasties of China.

1399 Mongol conqueror Tamerlane (c. 1366–1405) sacks city of Delhi and massacres its people. Delhi sultanate collapses before Mongol invasion.

c. 1420 Peking becomes Ming dynasty capital under rule of Emperor Yung-lo (1360–1424). It remains capital of China until 1928.

c. 1420 Emperor Yung-lo builds magnificent walled imperial palace, Forbidden City, at Peking's center.

1498 Portuguese under Vasco da Gama (c. 1469–1524) become first Europeans to reach India, marking beginning of seaborne trade between Europe and India.

1521 Magellan reaches Philippines, leading to Spanish conquest of islands by 1564.

1526 Mongol king Baber (1483–1530) invades India from Afghanistan and conquers northern India, thus founding Mogul empire. Subsequent Mogul rulers unite most of India under their rule.

1527–1602 Li Chih lives. Famous Confucian thinker and heretic, he is part of ongoing reaction to rigid interpretation of Confucianism that has developed during Ming dynasty.

▶

Indian Rulers and Heads of State

Mogul Emperors

1526–30	Baber (1483–1530), first of Mogul emperors
1530–56	Humayun (1507–56)
1556–1605	Akbar the Great (1542–1605)
1605–27	Jahangir (1569–1627)
1627–28	Davar
1628–58	Shah Jahan (1592–1666)
1658–1707	Aurangzeb (1618–1707)
1707–12	Bahadur Shah I (d. 1712)
1712–13	Jahandar Shah
1713–19	Farruk-Siar
1719–48	Muhammad Shah (1700–1806)
1748–54	Ahmed
1754–59	Alamgir
1759–1806	Shah Alam (1728–1806)
1806–37	Akbar II
1837–57	Bahadur Shah II (1768–1862), last Mogul emperor

British Viceroys of India

1856–62	Charles John Canning (Earl Canning) (1812–62)

Indian Rulers and Heads of State

1862–63	James Bruce, eighth earl of Elgin (1811–63)
1864–69	John L. M. Lawrence, first Baron Lawrence (1811–79)
1869–72	Richard Southwell Bourke, earl of Mayo (1822–72)
1872–76	Thomas George Baring, first earl of Northbrook (1826–1904)
1876–80	Edward R. L. Bulwer-Lytton, first earl of Lytton (1831–91)
1880–84	George F. S. Robinson, first marquis of Ripon (1827–1909)
1884–88	Frederick T. H.-T. Blackwood, first marquis of Dufferin and Ava (1826–1902)
1888–93	Henry C. K. Petty-Fitzmaurice, fifth marquis of Lansdowne (1845–1927)
1894–99	Victor Alexander Bruce, ninth earl of Elgin (1849–1917)
1899–1905	George Nathaniel Curzon, first marquis Curzon of Kedleston (1859–1925)
1905–10	Gilbert John Elliot-Murray Kynynmound, fourth earl of Minto (1845–1914)
1910–16	Charles Hardinge, first Baron Hardinge of Penshurst (1858–1944)
1916–21	Frederic J. N. Thesiger, first Viscount Chelmsford (1868–1933)
1921–26	Rufus Daniel Isaacs, first marquis of Reading (1860–1935)
1926–31	Edward F. L. Wood, first earl of Halifax (1881–1959)
1931–36	Freeman Freeman-Thomas, marquis of Willingdon (1866–1941)
1936–43	Victor A. J. Hope, marquis of Linlithgow (1887–1951)
1943–47	Archibald Percival Wavell, first Earl Wavell (1883–1950)
1947–48	Louis Mountbatten, first Earl Mountbatten of Burma (1900–1979)
1948–50	Chakravarti Rajagopalacharia, 1879–1972, governor-general.

Presidents of the Republic

1950–62	Rajendra Prasad (1884–1963)
1962–67	Sarvepalli Radhakrishnan (1888–1975)
1967–69	Zakir Husain (1897–1969)
1969–74	Varahagiri Venkata Giri (1894–1980)
1974–77	Fakhruddin Ali Ahmen (b. 1905)
1977–82	Neelan Sanjiva Reddy (b. 1913)
1982–	Zail Singh (b. 1916)
1987–	Ramaswamy Venkataraman (b. 1910)

Prime Ministers of the Republic

1950–64	Jawaharlal Nehru (1889–1964)
1964–66	Lal Bahadur Shastri (1904–66)
1966–77	Indira Gandhi (1917–84)
1977–79	Morarji R. Desai (b. 1896)
1979–80	Charan Singh (1902–87)
1980–84	Indira Gandhi (1917–84)
1984–	Rajiv Gandhi (b. 1944)

1542 First Europeans, Portuguese, visit Japan. Spanish Jesuit missionary St. Francis Xavier (1506–52) reaches Japan in 1549.

1556 Shaanxi (Shensi) is hit by one of deadliest of all earthquakes; estimated 830,000 persons perish.

1571 Manila is established as Spanish colonial capital.

1592 Japan invades Korea, but Koreans and allied Chinese turn back invasion; they fight off second attack in 1596.

1600s Portuguese in East Indies are for most part ousted by Dutch East India Company.

1603–1868 In Tokugawa period, shōguns of Tokugawa family rule Japan. Strict feudal system gives rise to samurai warriors and promotes feudal warrior ethic. Confucianism spreads to Japan during this time. Tokyo is capital.

1606 Dutch land at Gulf of Carpenteria, Australia and name it New Holland.

1611 Dutch begin trade with Japan.

▶

1612 British East India Company gets trading rights with Mogul empire after defeating Portuguese in India.

1635 Japan's imperial government forbids overseas contacts, thus beginning long period of isolation. Only Dutch are permitted to continue trading with Japan (at Nagasaki).

1637 Japanese Christians unsuccessfully rebel against abuses (Simbara Rebellion). Thereafter government requires all Japanese subjects to register at Buddhist temples.

1641 Dutch seize Malacca on Malay Peninsula from Portuguese.

1642 Dutch explorer Abel J. Tasman sights New Zealand. (*See also Voyages of Discovery.*)

1644–1912 Ch'ing (Manchu) dynasty is in power; it is last dynasty to rule China. Shun Chih (1638–61) is first ruler of this dynasty (1644–61).

1647 Sivaji (1627–80), Hindu rebel against dominant Muslim Mogul empire, begins his conquests in India. He founds Maratha empire in opposition to Moguls.

1653 Spectacular mausoleum, Taj Mahal, is built at Agra, India by Mogul emperor Shah Jahan (1592–1666) to honor his dead wife, Mumtaz Mahal (d. 1631). Some 20,000 workers take part in construction.

1691 British found Calcutta.

1700 Episode of 47 ronin (wandering samurai who have no masters). This samurai feud at Tokyo later figures in Japanese literature.

c. 1700 Chinese gain control of Taiwan.

1707 Mount Fuji erupts for last time.

1707 Mogul empire disintegrates on death of its last great emperor, Aurangzeb (1618–1707). Meanwhile, Maratha empire dominates southern India.

1735–95 Ch'en-lung (1711–99) rules China. Empire reaches its greatest extent during his reign.

1739 Delhi is sacked by invading Persians.

1756 Calcutta is captured from British by indigenous Bengali forces. British prisoners are crowded into and held overnight in infamous Black Hole of Calcutta; many reportedly die.

1756–63 European Seven Years War spills over to colonial territories. In India, British soldier Robert Clive (1725–74) soundly defeats French in Bengal and establishes Great Britain as dominant colonial power in India.

1775–1818 In three Maratha wars (1775–82, 1803–5, and 1817–18), British reduce and conquer Maratha kingdom.

1769 Capt. James Cook claims New Zealand for Great Britain.

1770 Cook claims Australia for Great Britain.

1782 Bangkok is established as the seat of Thai monarchy.

1788 British settlement of Australia begins with landing of convict ships.

1790 In "Skull Famine," severe drought leads to famine and cannibalism in Bombay and other cities.

1796–1804 Buddhist revolt (White Lotus Rebellion) against Manchu rule in China.

1798 Ceylon becomes British crown colony. Island was previously held by Portuguese and Dutch.

1799 Dutch East India Company bankruptcy prompts Holland to establish direct colonial rule in East Indies (present-day Indonesia).

1819 Sir Thomas Stamford Raffles establishes Singapore for British East India Company.

1824 Dutch cede Malacca to British.

1824–86 Burma Wars, three wars by which British colonial forces from India reduce and conquer neighboring kingdom of Burma.

1838–1919 Anglo-Afghan Wars (1838–42, 1878–80, and 1919). British in India establish

▶

temporary control over neighboring Afghanistan. British finally recognize Afghan independence after 1919 war.

1839–42 First Opium War with Great Britain reveals growing influence of Europeans in China. China is forced to make trade concessions.

1841 New Zealand becomes British colony.

1845–49 Sikh Wars (1845–46 and 1848–49). British defeat Sikhs in northwestern India and gain control of Kashmir and Punjab.

1850–64 Taipin Rebellion. Manchu rule is seriously weakened.

1851 Gold is discovered in Victoria, Australia, prompting international gold rush.

1854 United States persuades Japan to end period of isolationism. Japan signs trade treaty with the United States after latter sends fleet of warships under Commodore Matthew Perry (1794–1858) to Japan. Trade and diplomatic relations are opened with European nations in following years.

1856–60 Second Opium War. Great Britain and France defeat Manchu armies and force China to allow continued opium trade and to make other concessions.

1857–59 Sepoys, native Indians serving in British India's colonial army, unsuccessfully revolt against British.

1858 British government takes over colonial rule of India from British East India Company.

1858 China is forced to accept Treaty of Aigun, settling dispute over its northern border with Russia.

1860 Japan establishes its first embassy in the United States.

1867–68 Civil war rages for control of Japanese government. It ends with restoration of emperor's authority and ouster of the shōguns.

1867–1912 Japanese Emperor Mutsuhito (1852–1912) is in power. His reign (Meiji Restoration) brings about modernization of Japan and its rapid rise to world-power status. Mutsuhito designates Tokyo his capital (1869) and issues Charter Oath, abolishing feudalism of past and setting policy of Westernization of Japan.

1870 Tientsin Massacre. Chinese mob kills 20 Frenchmen.

1871 Mint at Osaka, Japan is opened.

1872 Japan's first railroad, between Tokyo and Yokohama, begins service.

1872 Conscription for military service is instituted as part of reorganization of Japanese army.

1873 Post office is established in Japan.

1873 Japan adopts Gregorian calendar.

1873 Japanese emperor Mutsuhito ends ban on Christianity.

1874 Japanese impose Treaty of Kwanghwa, ending Korean isolationism and allowing Japanese into Korean marketplaces.

1874 Fiji becomes British possession.

1875 Japanese emperor formally abolishes feudalism. Council of Elder Statesmen (*genro*) is created.

1877 Satsuma Rebellion. Samurai warriors on Kyushu unsuccessfully rebel against Japan's modernization.

1877 Queen Victoria (1819–1901) becomes empress of India.

1880 China relinquishes claims to Ryukyu Islands, which become part of Japan.

1880 Japan's first legal code is promulgated.

1883–85 French take control of Vietnam and other Chinese domains in Indochina.

1885 Indigenous Indian political party, Indian National Congress, is formed. It promotes Indian nationalism.

1886 Imperial university is established at Tokyo.

1887 Khmer territories not already annexed by Kingdom of Siam (Thailand) become part of French Union of Indochina. Vietnam also is absorbed.

▶

1889 New constitution drafted by Japanese statesman Prince Hirobumi Ito (1841–1909) establishes constitutional monarchy with legislative assembly.

1892 Gilbert Islands in Micronesia become British protectorate. By 1915, Gilbert Islands join Ellice Islands in Polynesia as British administrative unit.

1893 France claims Laos as part of Union of Indochina.

1894–95 First Sino-Japanese War, for control of Korea, long a vassal state of China. Japan invades Manchuria and forces peace terms on China. Japan forces China to give up Taiwan and other territories and to recognize Korea's independence.

1897 Japan adopts gold standard, which it maintains until 1931.

1897 By new agreement China leases Hong Kong to Great Britain for 99 years.

1898 Philippines is ceded to United States from Spain for $20 million.

1898 Peking University is founded.

1900–01 Boxer Rebellion. Members of anti-foreign Chinese secret society lay two-month siege to foreign embassy quarters in Peking. Revolt is reaction to long-standing foreign domination of China.

1901 Australian commonwealth is created by passage of constitution in Parliament.

1902 Japan strikes its first treaty alliance with Western power, Great Britain.

▶

Australian Heads of State

Governors-General

Both Governors-General and those officers who administered the government of the Commonwealth between governor-generalships are listed.

1901–3	John Adrian Louis Hope, Earl of Hopetoun
1901–3	Hallam Tennyson, Lord Tennyson (Acting)
1903–4	Hallam Tennyson, Lord Tennyson
1904–8	Henry Stafford Northcote, Lord Northcote
1908–11	William Humble Ward, Earl of Dudley
1911–14	Thomas Denman, Lord Denman
1914–20	Sir Ronald Crauford Munro-Ferguson
1920–25	Henry William Forster, Lord Forster
1925–30	Sir John Lawrence Baird, Lord Stonehaven
1930–31	Arthur Herbert Tennyson Sommers Cocks, Lord Somers
1931–36	Sir Isaac Alfred Isaacs
1936–45	Alexander Gore Arkwright Hore-Ruthven, Lord Gowrie
1944–45	Sir Winston Dugan (officer administering the government)
1945–47	Prince Henry William Frederick Albert, Duke of Gloucester, Earl of Ulster and Baron Culloden
1947	Sir Winston Dugan (officer administering the government)
1947–53	Sir William John McKell
1953–60	Sir William Joseph Slim
1960–61	William Shepherd Morrison, Viscount Dunrossil
1961	Sir Reginald Alexander Dallas Brooks (officer administering the government)
1961–65	William Philip Sidney, Viscount De L'Isle
1965	Sir Henry Abel Smith (officer administering the government)
1965–69	Richard Gardiner Casey, Lord Casey
1969–74	Sir Paul Memaa Caedwalla Hasluck
1974–77	Sir John Robert Kerr
1977–82	Sir Zelman Cowen
1982–89	Sir Ninian Martin Stephen
1989–	William G. Hayden

Prime Ministers

1901–3	Edmund Barton
1903–4	Alfred Deakin

Australian Heads of State			
1904	J.C. Watson	**1946–49**	J. B. Chifley
1904–5	G. H. Reid	**1949–51**	R. G. Menzies
1905–8	Andrew Fisher	**1951–56**	R. G. Menzies
1909–10	Alfred Deakin	**1956–58**	R. G. Menzies
1910–13	Andrew Fisher	**1958–63**	R. G. Menzies
1913–14	Joseph Cook	**1963–66**	Sir Robert Menzies
1914–15	W. M. Hughs	**1966**	H. E. Holt
1915–16	W. M. Hughs	**1966–67**	H. E. Holt
1916–17	W. M. Hughs	**1967–68**	J. McEwen
1917–18	W. M. Hughs	**1968**	J. G. Gorton
1918–23	W. M. Hughs	**1968–69**	J. G. Gorton
1923–29	S. M. Bruce	**1969–71**	J. G. Gorton
1929–32	J. H. Scullin	**1971–72**	W. McMahon
1932–38	J. A. Lyons	**1972**	E. G. Whitlam
1938–39	J. A. Lyons	**1972–74**	E. G. Whitlam
1939	Sir Earl Page	**1974–75**	E. G. Whitlam
1939–40	R. G. Menzies	**1975**	J. M. Fraser
1940	R. G. Menzies	**1975–77**	J. M. Fraser
1940–41	R. G. Menzies	**1977–80**	J. M. Fraser
1941	A. W. Fadden	**1980–83**	J. M. Fraser
1941–43	J. Curtin	**1983–84**	R. J. Hawke
1943–45	J. Curtin	**1984–87**	R. J. Hawke
1945	F. M. Forde	**1987–90**	R. J. Hawke
1945–46	J. B. Chifley	**1990–**	R. J. Hawke

1904–5 Russo-Japanese War, sparked by Russian designs on Korea. Modernized Japanese navy helps deliver humiliating defeat on Russia. Victory is confirms Japan's rise as major world power.

1905 Ancient Chinese examination system is abolished. Confucian classics are gradually eliminated from school curricula.

1906 Japanese Navy nearly doubles overall tonnage of warships in two years as building program continues.

1906 Muslims in India form Muslim League.

1907 New Zealand becomes independent British dominion.

1908 Japan agrees to limit emigration of laborers to United States by so-called gentlemen's agreement.

1908 U.S. Navy's "Great White Fleet" is sent overseas to show U.S. flag; it arrives in Tokyo Bay.

1908–12 Hsuan-t'ung (1906–67) reigns as last emperor of China.

1910 Korea is annexed as Japanese colony and called province of Chosen.

1910 Constitutional assembly convenes as part of movement toward legislative government promised by China's imperial government.

▶

1911 House in Wuhan, China where bombs are manufactured by secret Combined League Society explodes. Lists of society members and of local army officers are found in debris. Officers on list foment rebellion against central-government (Manchu) troops. Rebellion quickly spreads throughout central and southern China under leadership of Sun Yat-sen (1866–1925).

1911–12 Sun Yat-sen becomes provisional President of rebel Chinese republic.

1912 Boy emperor Hsuan-t'ung abdicates, ending long rule of Manchu emperors. China is declared republic, and Sun Yat-sen steps down as provisional President.

1912 Kuomintang (Nationalist) Party is formed by uniting Combined League Society and other groups. By 1913 it is China's leading political party.

1912–16 Yuan Shi-kai (1859–1916) is in power as President of China. In following years he establishes himself as dictator, sparking renewed unrest.

1913 Sun Yat-sen flees from China after dispute with President Yuan.

1914 Yuan dissolves Chinese Parliament and arranges his election as President for life.

1914–18 World War I. Japan joins Allies in war against Germany. (*See Wars and Military History.*)

1915 British enact Defense of India Act, granting colonial administrators special powers to quell unrest in India during World War I.

1915 Yuan moves to create new constitutional monarchy in China with himself as monarch, but provincial rulers succeed in blocking plan.

1916 Yuan dies. Sun Yat-sen returns to China. Parliament is reconvened briefly until provincial leaders (warlords) take Peking (1917) and dissolve Parliament. Warlords hold real power in China in following years.

1919 Versailles Peace Conference awards Japan Shantung, province in northern China formerly held by Germans. Japan announces it will not accept province after bitter protests in China.

1919 British extend unpopular wartime powers in India after World War I ends, sparking bitter protests led by Indian nationalist Mohandas Gandhi (1869–1948).

1920–22 Gandhi leads his first noncooperation movement against British rule.

1921 Total of 77 million Japanese are counted in country's first census.

1921 Korean assassin stabs and kills Japan's premier Takashi Hara (1856–1921).

1921 Japan signs Four-Power Treaty following Washington Naval Disarmament Conference.

1921 India's first national Parliament is convened.

1921 Peking Union Hospital, first modern hospital in China, is founded.

1922 Indonesia is made part of Kingdom of the Netherlands.

1925 Property restriction for voting rights in Japan is ended.

1926 Women gain right to run for elective offices in India.

1927 Sukarno founds Indonesian Nationalist Party in opposition to Dutch rule; he is jailed.

1927–49 Chinese civil war occurs (*See Wars and Military History.*)

1928 Chinese Nationalists under Chiang Kai-shek (1887–1975) move China's capital to Nanking. Peking is renamed Peiping.

1930 Salt March, led by Gandhi, is organized to defy British colonial rule. Protesters will march to sea and there manufacture salt, in defiance of colonial government monopoly on its production.

1930–31 Gandhi leads new noncooperation movement aimed at full independence

▶

for India.

1931 Baseball game is subject of Japan's first television broadcast.

1931 Japan abandons gold standard. Move leads to boom in exports and makes Japan's economy first to recover from worldwide Great Depression.

1932 Japanese Army leaders unilaterally decide to occupy Manchuria.

1932 Military extremists kill Premier Ki Inukai (1855–1932). Parties no longer participate in national politics.

1932 In Poona Pact, Hindu political leaders agree to grant members of untouchable (lowest) caste increased representation.

1933 Japan withdraws from League of Nations over criticism of its actions in Manchuria.

1936 Revolt by military in Tokyo fails; 17 officers are later sentenced to death in incident.

1936 Indian leader Jawaharlal Nehru (1889–1964) is elected President of Indian National Congress.

1937 Burma becomes self-governing British protectorate.

1937 Japanese exporters reportedly have been flooding overseas markets with shoddy products falsely labeled "Made in USA," to circumvent limits on Japan's exports to these countries.

1937 Constitution and federal government are formally established in India, though Great Britain withholds full independence.

1937–45 Second Sino-Japanese War. Japanese military precipitates war and quickly takes Peking and other major cities. Chinese Nationalists and Communists continue to resist, however, and this war eventually becomes part of larger World War II (*See Wars and Military History.*)

1938 Japanese government calls up 1 million recruits for military duty.

1939 Ho Chi Minh creates Vietminh Party in opposition to French colonialism in Indochina.

1939 Japan joins Germany and Italy as member of the Axis in World War II. Japan invades French Indochina.

1939–45 World War II. (*See Wars and Military History.*)

1940 Gandhi begins yet another noncooperation movement to demand complete independence of India as World War II begins. Indian nationalists continue demands for independence even as Japanese armies advance toward India.

1941 Vichy France accedes to Japanese takeover in Indochina.

1941 Sneak Japanese attack on U.S. naval base at Pearl Harbor marks beginning of broad offensive that nearly makes Japan master of Pacific theater.

1942 Japanese occupy Burma. In response, Allied troops build the Burma Road, linking India with southwestern China and creating principal supply line for Nationalist Chinese forces. Japanese occupation forces also move into Dutch East Indies, Singapore, and Philippines this year.

1943 Tide of war turns against Japan as Allied offensive in Pacific begins.

1945 Sukarno declares independence for Indonesia. Independence is recognized by Dutch in 1949.

1945 United States drops atomic bomb on Hiroshima, then drops second bomb on Nagasaki when Japanese still refuse to surrender. Japan then surrenders and is occupied by U.S. troops.

1945 Korea is divided along 38th parallel in Soviet and American occupation of former Japanese lands.

1946 Emperor Hirohito breaks centuries-long tradition by declaring publicly that divinity of Japanese emperor is myth.

1947 Pakistan becomes British dominion and, in 1956, declares independence from Great Britain.

1947 Louis Mountbatten (1900–79), Earl Mountbatten of Burma, serves as last

▶

British viceroy of India. He oversees transfer of power to indigenous Indian government.

1947 India becomes fully independent. East and West Pakistan are created as separate Muslim states.

1948 Burma gains independence.

1948 Ceylon becomes independent British dominion.

1948 Fanatical Hindu assassinates Gandhi.

1948 New Indian government bans long-standing practice of discrimination against untouchable class.

1949 Prolonged Chinese Civil War ends. Chinese Communists capture Peking and make it capital of Communist China (People's Republic of China).

1949 Chiang Kai-shek establishes Republic of China on Taiwan following retreat of Chinese Nationalists from mainland.

1949 Americans withdraw occupation forces from South Korea.

1950 India and Pakistan establish common bill of rights for religious minorities in effort to end violence between Hindus and Muslims.

1950 China occupies Tibet.

1950 Attempts to reunify North and South Korea fail, and Republic of South Korea is established, with Dr. Syngman Rhee as President. North Korea then invades South Korea, answered by UN counterattack, with landing at Inchon of UN troops under Gen. Douglas MacArthur. (*See also Wars and Military History, "Korean War."*)

1953 China begins five-year plan for agricultural and industrial development, with Soviet aid.

1953 Cambodia gains independence from France and organizes government under Prince Norodom Sihanouk.

1954 Japanese sign defense pact with United States.

1954 Indian government outlaws bigamy.

1954 French defeat at Dien Bien Phu, Vietnam prompts division of the country into North Vietnam and South Vietnam.

1955 Liberal Democratic Party is formed in Japan, beginning its long period of unbroken control of national politics.

1956 Chinese government promises liberalization in "Speech of the 100 Flowers."

1958–60 China's Great Leap Forward, program for industrial development, fails.

1959 Indian troops fight with Chinese along border with Tibet; fighting recurs in 1962.

1960 Split between Chinese and Soviet Communists widens. Chinese hold to inevitability of revolution and resent Soviet "revisionist" ideology.

1961 Park Chung Hee takes control of South Korean government after military coup. He is later elected President (1972) and assassinated (1979).

1963 Malaysia is created by federations of Malaya, Singapore, North Borneo, and Sarawak.

1964 China explodes its first atomic bomb.

1965 Singapore withdraws from Malaysian federation.

1966 Ferdinand Marcos is elected President of Philippines.

1966–77 Indira Gandhi (1917–84), Nehru's daughter, is in office as Indian prime minister.

1969 Conference of Communist parties is held in Moscow to discuss Soviet/Chinese differences. China later begins liberalization policy and normalization of relations with West.

1969 Ho Chi Minh dies.

1970s Japanese exports (notably autos and electronic equipment) to United States and other nations increase dramatically, spurring growth of Japan's "economic miracle."

▶

1971 In Pakistani War, India invades West Pakistan to aid Bangladesh (East Pakistan) in its war for independence.

1971 China launches its first space satellite.

1971 Communist China joins United Nations, at expense of Taiwan.

1971 East Pakistan declares independence from West Pakistan, forming nation of Bangladesh.

1972 Japan regains control of island of Okinawa from United States.

1972 U.S. President Nixon makes historic visit to China, major step in normalizing U.S.–Chinese relations.

1972 Ceylon is declared Republic of Sri Lanka.

1974 India explodes its first nuclear device.

1975 India launches its first satellite.

1975 Indira Gandhi, her popularity waning, is convicted of election fraud. She is permitted to retain prime ministership, however.

1975 Vietnam-dominated Lao People's Democratic Republic takes control of Laotian government.

1975 Pol Pot leads forces of the Khmer Rouge into Cambodian capital of Phnom Penh and establishes new government, Kampuchean People's Republic.

1975 Papua New Guinea declares independence from Australian administration.

1976 In Lockheed bribery scandal, former Japanese prime minister Kakuei Tanaka (b. 1918) is arrested for having taken money in scandal.

1976 Mao's death is marked by state funeral in China.

1977 Indira Gandhi loses vote to retain prime ministership. She is arrested on charges of corruption in office and later serves short prison sentence.

1978 Japan agrees to new trade arrangement in response to U.S. pressure to ease restrictions on imports from United States and other trading partners.

1978 Ellice Islands declare independence from Great Britain as new nation of Tuvalu.

1978–90 Vietnamese invade Cambodia and set up satellite government. Kampuchean People's Republic, however, continues to be recognized by United Nations until Vietnamese withdrawal.

1979 Chinese troops briefly invade Vietnam as reprisal for Vietnam's invasion of Cambodia.

1979 Gilbert Islands declare independence from Great Britain as new nation of Kiribati.

1980s Continuing trade surpluses (world's largest) fuel Japan's economic miracle. Japan becomes world financial center by this decade, with eight of world's 10 largest banks. It is well established as world leader in manufacturing.

1980–84 Indira Gandhi is again in office as prime minister.

1980 Highly publicized trial of Gang of Four, including Mao's widow, begins in Peking. Mao's widow gets suspended death sentence (1983).

1983 Population of India reaches 731 million; of China, 1.05 million.

1983 Filipino opposition leader Benigno Aquino is assassinated, allegedly by Ferdinand Marcos's hit men.

1984 Sikh revolt in Punjab for autonomous state is put down by government troops. Some 400 Sikh rebels die.

1984 Indira Gandhi is assassinated by two Sikhs among her security forces. She is succeeeded by her son, Rajiv Gandhi (b. 1944).

1984 Leak at Union Carbide plant kills over 2,000 in Bhopal, India.

1986 Mass demonstrations by Chinese students against restrictions on freedoms.

1986 Ferdinand Marcos flees Philippines and seeks asylum in Hawaii. Corazon Aquino takes office as Philippine President.

▶

1986 Benazir Bhutto, daughter of former Pakistani president Zulfikar Ali Bhutto, returns to Pakistan from her exile in Europe and tries to relaunch Pakistan People's Party, sparking antigovernment riots. She is ousted in 1990.

1987 Long-standing separatist movement by Tamil rebels in Sri Lanka is ended by accord with Indian government.

1988 Summer Olympics are held in Seoul, South Korea.

1989 Akihito (b. 1933) ascends the Chrysanthemum Throne in formal ceremony, starting his reign as Japan's emperor. He succeeds soon after death of his father, Emperor Hirohito (1901–89).

1989 Recruit bribery scandal forces resignation of Japanese prime minister Noboru Takeshita (b. 1924). His Liberal Democratic Party successor is soon embroiled in sex scandal and is quickly replaced by yet another Liberal Democrat.

1989 Four additional Sikhs are belatedly charged with involvement in 1984 assassination of Indira Gandhi.

1989 Mass demonstrations are held in Peking by 1 million or more students demanding increased freedoms. Demonstrators, occupying Tienanmen Square, are finally driven off by troops, who shoot at demonstrators and reportedly kill hundreds.

1989 Burma is renamed Myanmar and capital city of Rangoon is renamed Yangon.

Ancient Mesopotamia

B.C.

c. 10,000 Prehistoric peoples in Near East begin to harvest grain.

c. 7000 Village society based on farming is well established in Near East. Villagers practice diversified agriculture.

c. 4000 Sumerians in southern Mesopotamia develop boat for water transportation.

c. 3500 Sumerian cities of Ur, Eridu, Lagash, Erech, Kish, Nippur, and Larsa are built and become thriving commercial centers. These city-states soon begin to expand their control into surrounding countryside.

c. 3500 Early Sumerian sketch shows first known use of wheels, primitive wooden slabs roughly rounded at corners.

c. 3500 Primitive potter's wheel, turned by hand, is in use in Sumeria.

3000 Sumerians have developed advanced agricultural society. They cultivate barley, wheat, flax, grapes, and other crops and raise sheep and cattle.

c. 3000 Sumerians develop written language (cuneiform) by this time.

c. 3000 Babylon is flourishing city in Sumer.

c. 2340 Sargon, ruler of Akkad in northern Mesopotamia, conquers Sumer and other lands, thereby creating first unified kingdom in Mesopotamia.

c. 2180 Kingdom of Akkad is overrun by barbarians from North. Sumerian cities enjoy brief resurgence after fall of Akkad.

c. 2100 Ziggurat of Ur is built by Sumerian king Ur-Nammu. Sumerians call it *temen*, source of later word *temple*.

c. 2100 Mesopotamians construct canals to irrigate arid plains.

▶

2060–1950 City of Ur flourishes. It is finally conquered by Elam.

c. 2000 Simple bellows is developed in Mesopotamia to increase heat for foundries for working metal. True bronze is in use.

c. 2000 Mesopotamian farmers use horses to pull vehicles equipped with spoked wheels.

c. 2000 Metalworkers in Mesopotamia discover glass, probably by accidentally overheating materials. They cut and polish glass cold until c. 1800 B.C.

c. 1792–c. 1750 Hammurabi rules Babylonia. He creates first great Babylonian empire and issues famous law code, Code of Hammurabi.

1530 Babylonia comes under domination of Kassites.

1300–1200 Elam, kingdom to east of Euphrates, reaches height of its power. It conquers Kassites.

c. 1274–c. 1245 Shalmaneser I rules Assyria. He pushes Assyrian borders northward toward Asia Minor.

1247 Assyrians sack Babylon.

c. 1124–c. 1103 Nebuchadnezzar I rules Babylonia. He conquers Elam and creates empire that includes most of Mesopotamia.

c. 1115–c. 1077 Tiglath-pileser I rules Assyria. He establishes short-lived empire that encompasses much of Mideast. Empire collapses after his death.

c. 1050–c. 1032 Ashurnasirpal I rules Assyria. During his reign Assyria suffers widespread famine and constant attacks by desert nomads.

c. 900 Assyrian war chariots are equipped with metal tires (copper or bronze) for greater durability.

c. 883–c. 859 Ashurnasirpal II rules Assyria. His ruthless conquests push empire's borders to Mediterranean.

c. 746–c. 727 Tiglath-pileser III rules Assyria. He conquers Syria and other lands and gains control of Babylon.

c. 726–c. 722 Shalmaneser V rules Assyria. He invades Israel and, following revolt, lays siege to city of Samaria.

c. 722–c. 705 Sargon II rules Assyria. He completes conquest of Israel in 722.

c. 721–c. 710 Marduk-apaliddinia II briefly restores independent Babylon.

c. 704–c. 681 Sennacherib rules Assyria. He captures and destroys city of Babylon, though it is later rebuilt.

680–c. 669 Esarhaddon rules Assyria. He conquers Chaldea and Egypt c. 670.

c. 668–c. 627 Assurbanipal rules Assyria. Assyrian Empire reaches its height during his reign and stretches from Persian Gulf to Egypt. However, Egypt successfully rebels against Assyria during this time, marking start of Assyria's rapid decline.

626–605 Chaldean king Nabopolassar rules Babylonia. He makes Babylon his capital, and his armies capture Ninevah (612), capital of the Assyrian Empire, thus ending that empire.

605–562 Nebuchadnezzar II rules New Babylonian Empire. His reign is marked by period of prosperity. He fights Egyptians at great Battle of Carchemish (605).

586 Nebuchadnezzar II conquers Jerusalem and destroys First Temple. He deports Hebrews en masse to Babylon, where they are forced to remain (Babylonian Captivity) until fall of Babylonia.

c. 575 Hanging Gardens of Babylon, one of Seven Wonders of Ancient World, are constructed by Nebuchadnezzar II.

555–539 Nabonidus rules Babylonia. He is last ruler of independent Babylonia.

539 Persians led by Cyrus the Great (d. 529) conquer Babylonia, thus ending empire.

Ancient Assyrian and Babylonian Rulers

Assyrian Rulers

B.C.	
c. 1813–c. 1781	Shamshi-Adad I
c. 1716–c. 1687	Shamshi-Adad II
c. 1661–c. 1636	Shamshi-Adad III
?–c. 1415	Assurbel-nisesu
c. 1402–c. 1393	Ashur-nadin-ake II
c. 1365–c. 1330	Ashuruballit I
c. 1328–c. 1320	Enlil-Nirari
c. 1319–c. 1308	Arikdenili
c. 1307–c. 1275	Adad-Nirari I
c. 1274–c. 1245	Shalmaneser I
c. 1244–c. 1208	Tukulti-ninurta
c. 1179–c. 1134	Ashurdan I
c. 1115–c. 1077	Tiglath-pileser I
c. 1074–c. 1057	Ashurbel-kala
c. 1050–c. 1032	Ashurnasirpal I
c. 1030–c. 1019	Shalmaneser II
c. 956–c. 934	Tigeth-pileser II
c. 934–c. 912	Assurdan II
c. 911–c. 891	Adad-Nirari II
c. 890–c. 884	Tukulti-Ninurta II
c. 883–c. 859	Ashurnasirpal II
c. 858–c. 824	Shalmaneser III
c. 823–c. 811	Shamshi-Adad V
c. 810–c. 806	Regency of Sammu-ramat
c. 810–c. 783	Adad-Nirari III
c. 783–c. 773	Shalmaneser IV
c. 772–c. 755	Assurdan III
c. 754–c. 746	Ashur-hirari V
c. 746–c. 727	Tiglath-pileser III
c. 726–c. 722	Shalmaneser V
c. 721–c. 705	Sargon II
c. 704–c. 681	Sennacherib
c. 680–c. 669	Esarhaddon
c. 668–c. 627	Assurbanipal

Babylonian Rulers, Old Empire

B.C.	
c. 1792–c. 1750	Hammurabi
c. 1749–c. 1712	Samsuiluna
c. 1646–c. 1626	Ammisaduqa
c. 1600-?	Agum II
c. 1450-?	Ulamburiash
c. 1415–c. 1390	Kurigalzu
c. 1390–c. 1375	Kadashman-Enlil
c. 1375–c. 1347	Burnaburiash II
c. 1345–c. 1324	Kurigalzu II
c. 1242–c. 1235	Kashtiliash IV
c. 1218–c. 1189	Adad-shumnasir
c. 1188–c. 1172	Melishipak
c. 1156–c. 1139	Marduk-kabitakheshu
c. 1124–c. 1103	Nebuchadnezzar I
c. 1098–c. 1081	Marduk-nadinahhe
c. 1080–c. 1068	Marduk-shapikzeri
c. 1024–c. 1007	Simbar-shikhu
c. 930–c. 904	Shamash-mudammiq
c. 904–c. 888	Nabu-shumukin
c. 887–c. 855	Nabu-apaliddin
c. 855–c. 820?	Marduk-zakirshumi
c. 747–c. 734	Nabunasir
c. 734–c. 731	Ukinzer; Babylonia is subsequently under Assyrian rule
c. 721–c. 710	Marduk-apaliddinia II; briefly restores independent Babylon

Babylonian Rulers, New Empire

B.C.	
626–605	Nabopolassar
605–562	Nebuchadnezzar II
561–560	Awil-Marduk (Bibl. Evil Merodach)
559–556	Nergalshar-usur
556	Labashi-marduk (s)
555–539	Nabonidus

Ancient Greece

B.C.

3000 Human settlement on site of present-day Athens begins by this time.

c. 3000–1000 Bronze Age Minoan civilization flourishes on Crete but declines and disappears after destruction of great palace at Knossus c. 1400, perhaps caused by invading Myceneans.

c. 1400–1100 Mycenean civilization flourishes on mainland Greece.

c. 1200 Troy is captured and destroyed. Event is later recounted in *Iliad.*

c. 1100 Dorians invade Greece.

c. 800 Sparta begins to expand its control over surrounding countryside. Rise of other Greek city-states begins.

776 Olympic Games begin.

c. 750 Greek poet Hesiod lives.

c. 735–c. 715 Spartan expansionism results in First Messenian War, between Sparta and Messenia, and ends with conquest of Messenia.

c. 700 Greek phalanx, formation of eight rows of infantrymen, is used throughout Greece by this time.

c. 700–500 Greek city-states actively engage in setting up colonies in Mediterranean, notably those of Syracuse and Tarentum in modern Italy.

c. 650–600 Messenia unsuccessfully rebels against Spartan rule in Second Messenian War.

621 Draco institutes harsh ("Draconian") laws at Athens.

c. 600 Greeks from Ionian city of Phocaea found ancient city of Massilia (modern Marseilles) on distant Atlantic coast of France.

594 Solon (c. 638–c. 559) becomes archon at Athens. His reforms lay basis for Athenian democracy. Serfdom is abolished and property owners govern.

560 Pisistratus (c. 605–527) becomes tyrant of Athens. He promotes building and expansion there.

508 Athenian democracy is established by Cleisthenes (fl. 6th century). Government now is by all Athenian freemen rather than only landed aristocrats.

507 Athenians successfully repel attack by Spartans.

c. 500 Dorian and Ionian Greeks colonize Crimean region of Black Sea, far to east of their homelands.

c. 500 Greeks use sundials. They also discover ore smelting and metal casting.

500s Athens grows rapidly. Athenian Acropolis is rebuilt as religious shrine with many temples.

499–c. 449 Persian Wars. Greeks successfully turn back repeated invasions by Persians, and Athens emerges as major power among Greek city-states. (*See Wars and Military History.*)

488 Athens introduces ostracism as punishment.

480 Athens is sacked during Peloponnesian Wars.

479 Work on rebuilding Athens and its fortifications begins. Walls eventually are extended to enclose city's port, Piraeus.

478 Delian League of Greek city-states is formed to oppose Persians. From 467 Athens is in firm control of league.

464–461 Third Messenian War. Messenia again rebels unsuccessfully against Spartan rule.

460–429 Age of Pericles. During reign of Pericles (c.495–420), Athens becomes political and cultural center of ancient Greece.

447–438 Parthenon is built as part of Pericles's reconstruction of Acropolis.

▶

431–404 Sparta conquers Athens in Peloponnesian War, ending Athens's period of political dominance in Greece. Athens continues to flourish culturally for centuries, however. (*See Wars and Military History.*)

411 Oligarchy is established at Athens. Democracy is reestablished within months, however.

403 Brief Spartan rule in Athens by Thirty Tyrants is overthrown.

399 Philosopher Socrates (470?–399) is forced to poison himself after being accused of corruption and improprieties.

395–87 Corinthian War, between Sparta and Athens and allies (including Persia). Spartan control of Athens is ended.

387 Plato founds Academy at Athens.

371 Thebes defeats Sparta and becomes dominant power in Greece for a time.

355–338 Sacred Wars (third and fourth) occur among various Greek city-states. Two wars (355–346 and 339–338) end in the Macedonians gaining control over all Greece.

c. 335 Lyceum is opened by Aristotle.

334–324 Alexander the Great (356–323), king of Macedonia and Greek peninsula, rapidly conquers vast empire including Persia, Egypt, Media, Scythia, and part of India.

323–281 Wars of the Diadochi, among rival generals for control of Alexander's empire after his death. After Battle of Corupedion (281), empire is permanently divided into Macedonia, Seleucid Asia Minor, and Ptolemaic Egypt.

323–22 Lamian War, between Macedonia and Athens and allies. Athens, after Alexander the Great's death, unsuccessfully rebels against Macedonia's rule.

c. 300 Greek navigator Pytheas explores Atlantic coast of Europe and British Isles. He may have reached Norway or Iceland.

214–148 Macedonian Wars occur between Macedonia and Rome. By these three wars Rome gradually increases its control over Greece until finally Greece is reduced to province of Roman Empire.

Ancient Egypt

B.C.

c. 3400 Early form of hieroglyphic writing is in use in Egypt.

c. 3100 Memphis is founded on Nile River near modern Cairo by Egyptian king Menes, who unites upper and lower Egypt into single kingdom. Memphis becomes his capital.

c. 2700–c. 2200 Old Kingdom period of Egyptian history.

c. 2650 King Zoser builds step pyramid near Memphis to serve as monument. Later pyramids will serve as tombs for rulers.

c. 2600 Egyptians conquer Sinai.

c. 2570 King Khufu, most powerful of early kings, builds largest of three great pyramids at Gizeh. Government administration in Egypt has become highly organized.

c. 2550 King Khafre builds second-largest pyramid at Gizeh. He also orders carving of great statue Sphinx (probably in his own image).

c. 2500 King Menkure builds smallest of pyramids at Gizeh. Power of king, who is worshiped as god and as king, declines in centuries following Menkure's reign.

c. 2500 Egyptians are producing and using papyrus, made from papyrus reeds growing along Nile, for writing paper.

▶

c. 2500 Worship of sun god Ra becomes increasingly important in Egypt. Temple at Heliopolis becomes center of worship.

c. 2500 Pyramid texts appear on inner walls of pyramids dating from this time. They include prayers, spells, and hymns for protection of the dead king during his afterlife.

c. 2294 Pepi II, six years old, begins his long reign as king. Traditional accounts say he reigned for 94 years. However, he was unable to check growing power of provincial rulers, and soon after his death kingdom disintegrated, thus ending Old Kingdom period.

c. 2040 Theban king Mentuhotep II (2010) unites Upper and Lower Egypt under his rule, thus restoring Kingdom of Egypt. He makes Thebes capital, brings about period of prosperity, and annexes northern Nubia during his reign.

c. 2040–1786 Middle Kingdom period. Egyptian arts flourish during this time of commercial prosperity.

c. 1970–1926 Sesostris I reigns as king after his father is assassinated. Sesostris conquers the Libyans and remainder of Nubia, extending his kingdom southward to second Nile cataract. Records show extensive mining in Nubia and Sinai (for copper and gold) during his reign. He also begins work on what becomes great temple complex at Karnak and promotes worship of god Amon.

c. 1890 Egyptians begin to reclaim Fayyum Depression, low area of desert some 50 miles long near modern Cairo. It eventually becomes rich agricultural area.

c. 1800 Egypt has reached height of its prosperity during Middle Kingdom period. Trade with foreign lands is flourishing, and Egyptians have established contacts with peoples in Asia Minor and with developing Greek culture on Crete. Work on irrigation system for Fayyum Depression also is complete.

c. 1700 Nubia gains independence from Egypt for time.

c. 1674 Hyksos, foreign people of uncertain origin, overrun much of Egypt and set themselves up as rulers of land. Horse and chariot are introduced in Egypt by the Hyksos.

c. 1567 Hyksos are driven out of Egypt by indigenous Egyptian military forces from Thebes. Thereafter Thebes is rebuilt and enjoys period of great prosperity as capital of Egypt. New Kingdom period of Egyptian history (1567–c. 1085 B.C.) begins.

c. 1530 Egyptians reconquer Nubia (south to Nile's second cataract). They also reopen copper mines in Sinai.

c. 1526–c. 1512 King Thutmose I rules Egypt. He extends Egypt's border southward to Nile's fourth cataract, thereby gaining territory rich in gold deposits. His armies also drive deep into Mesopotamia, reaching banks of Euphrates. His tomb is first in what becomes Valley of Tombs of Kings.

c. 1504–c. 1450 Thutmose III, greatest of Egyptian kings, reigns and brings Egypt to height of its power. He reconquers territories as far as Euphrates and organizes effective administration of his new empire. He likewise pushes southward in Nubia, well into what is Sudan today.

c. 1373 Religious revolution of King Ikhnaton (reigns c. 1379–62) occurs. He introduces possibly first monotheistic worship (of sun god Aton). He builds new capital, Ikhnaton, which becomes seat of Aton worship.

c. 1361 Boy king Tutankhamen (reigns c. 1361–c. 1352) restores worship of old gods and moves capital back to Thebes.

c. 1355 Tutankhamen's tomb is built. Treasures of King Tut's tomb, including his 250-pound gold coffin, remain intact until modern times.

c. 1304–c. 1237 King Ramses II reigns. He wars against Hittites, fighting inconclusive Battle of Kadesh in Syria. Hebrews are believed to have been enslaved in Egypt during this time.

c. 1232 Egyptians rout an invasion by Sea Peoples and allied Libyans. Records of war make first mention of Israelites, some of whom are among vanquished.

c. 1198–c. 1167 Ramses III, last of the great Egyptian kings, is in power. He successfully defeats three invasions by foreign peoples but is unable

▶

to prevent Sea Peoples from settling in Palestine or to halt growing internal unrest eroding royal authority.

c. 1000 Egypt's Asian empire has disintegrated, and Nubia has gained independence. Royal authority also is greatly diminished, while priests at Thebes are powerful.

c. 945 Egypt comes under rule of succession of foreign powers from this time forward. Libya (945–745 B.C.), Ethiopia (712–663 B.C.), Assyria (663–609 B.C.), and Persia (525–404 B.C.) rule Egypt during time before conquest by Macedonian armies under Alexander the Great.

332 Alexander the Great (356–323 B.C.) conquers Egypt and founds great city of Alexandria; in coming years it becomes leading center of Hellenic Greek culture.

305–285 Ptolemy I rules Egypt. He establishes his rule in Egypt and Libya after Alexander's death. Ptolemy begins long reign of Ptolemaic dynasty in Egypt (305–30 B.C.) and founds great library at Alexandria. Egypt again flourishes under Ptolemies.

51 Ptolemy XIII (63–47 B.C.) and his sister Cleopatra (69–30 B.C.) become rulers of Egypt.

49 Ptolemy XIII deposes Cleopatra to become sole ruler of Egypt.

48 Cleopatra begins her celebrated love affair with Roman leader Julius Caesar (c. 102–44 B.C.), who restores her to power after warring against Ptolemy XIII.

47 Main library at Alexandria burns. Hundreds of thousands of scrolls are lost.

44–30 Cleopatra's son Ptolemy XV (47–30) rules Egypt with his mother, after her younger brother Ptolemy XIV (59–44) is killed.

42 Cleopatra begins her fatal love affair with Roman ruler Marc Antony (c. 83–30). Unpopular in Rome, the affair leads to Antony's defeat at Battle of Actium (31 B.C.) by Octavian (63 B.C.–A.D. 14), who later becomes Augustus.

30 Antony and Cleopatra commit suicide in Alexandria. Octavian later kills Ptolemy XV and reduces Egypt to province of Rome.

Rulers of Ancient Egypt

Old Kingdom

B.C.	
c. 3100?–?	Menes (fl. 3100 B.C.)
c. 2650?–?	Zoser
c. 2613–c. 2589	Snefru (d. c. 2589 B.C.)
c. 2589–c. 2566	Khufu (Cheops)
c. 2555–?	Khafre
c. 2500?–?	Menkure
c. 2345–?	Teti
c. 2343–c. 2294	Pepi I
c. 2294–c. 2200	Pepi II

Middle Kingdom

B.C.	
c. 2040–c. 2010	Mentuhotep II, ruler of reunited Egypt
c. 2000–c. 1970	Amenemhet I (d. c. 1970 B.C.)
c. 1970–c. 1926	Sesostris I (d. c. 1926 B.C.)
c. 1926–c. 1897	Amenemhet II
c. 1897–c. 1878	Sesostris II (d. 1878 B.C.)
c. 1878–c. 1849	Sesostris III (d. 1878 B.C.)
c. 1849–c. 1801	Amenemhet III (d. c. 1801 B.C.)
c. 1801–c. 1792	Amenemhet IV
c. 1789–c. 1786	Sebeknefru

New Kingdom

B.C.	
c. 1570–c. 1546	Amasis I (d. c. 1546 B.C.)
c. 1546–c. 1526	Amenhotep I (d. 1526)
c. 1526–c. 1512	Thutmose I (d. c. 1512 B.C.)
c. 1512–c. 1504	Thutmose II
c. 1504–c. 1450	Thutmose III, coruler with sister Hatshepsut (d. 1481 B.C.), c. 1504–1482 B.C.
c. 1448–c. 1420	Amenhotep II (d. 1425 B.C.)
c. 1420–c. 1411	Thutmose IV (d. c. 1417 B.C.)
c. 1411–c. 1379	Amenhotep III (d. 1379 B.C.)
c. 1379–c. 1362	Ikhnaton (d. c. 1362 B.C.)

Rulers of Ancient Egypt			
c. 1361–c. 1352	Tutankhamen (d. c. 1352 B.C.)	**c. 593–c. 588**	Psamtik II
c. 1348–c. 1320	Horemheb	**c. 588–c. 570**	Apries
c. 1320–c. 1318	Ramses I (d. 1318)	**c. 570–c. 526**	Amasis II
c. 1318–c. 1304	Seti I (d. 1304)	**c. 525**	Psamtik III
c. 1304–c. 1237	Ramses II (d. c. 1237 B.C.)	**305–285**	Ptolemy I (c. 367–c. 283), first of Ptolemies
c. 1236–c. 1223	Merneptah (d. c. 1223 B.C.)	**285–246**	Ptolemy II (c. 308–246 B.C.)
c. 1210–c. 1200	Seti II	**246–222?**	Ptolemy III (d. 222 B.C.)
c. 1200–c. 1198	Setnacht	**221–205**	Ptolemy IV (d. 205 B.C.)
c. 1198–c. 1167	Ramses III (d. 1167 B.C.)	**205–181?**	Ptolemy V (d. 181 B.C.?)
c. 1166–c. 1160	Ramses IV	**180–145**	Ptolemy VI (d. 145 B.C.)
c. 1160–c. 1156	Ramses V	**145**	Ptolemy VII (d. 144 B.C.)
c. 1156–c. 1148	Ramses VI	**145–116**	Ptolemy VIII (d. 116 B.C.)
c. 1148–c. 1147	Ramses VII	**116–107**	Ptolemy IX (d. 81 B.C.)
c. 1147–c. 1140	Ramses VIII	**107–88**	Ptolemy X (d. 88 B.C.)
c. 1140–c. 1121	Ramses IX	**88–81**	Ptolemy IX
c. 1121–c. 1113	Ramses X	**80**	Ptolemy XI (d. 80 B.C.)
c. 1113–c. 1085	Ramses XI	**80–51**	Ptolemy XII (d. 51 B.C.)
c. 945–c. 924	Sheshonk I (d. c. 924 B.C.)	**51–47**	Ptolemy XIII (63–47 B.C.) and Cleopatra (69–30 B.C.)
c. 924–c. 895	Osorkon I	**47–44**	Ptolemy XIV (d. 44 B.C.) and Cleopatra (69–30 B.C.)
c. 874–c. 853	Osorkon II	**44–30**	Cleopatra (69–30 B.C.) and her son Ptolemy XV (47–30 B.C.)
c. 712–c. 700	Shabaka		
c. 689–c. 663	Tirhaka (d. 663 B.C.)		
c. 663–c. 609	Psamtik I (d. 609 B.C.)		
c. 609–c. 593	Necho (d. 593 B.C.)		

Roman Empire

B.C.

753 Traditional date for founding of Rome. At about this time separate villages on the Seven Hills of Rome join.

c. 600 Etruscan (Tarquin) kings gain control of Rome.

509 Romans drive out Tarquin kings and establish republican form of government. Era of Roman Republic begins.

496 Romans defeat neighboring tribe at battle of Lake Regillus. Rome's territorial expansion begins.

450 Roman law code called Twelve Tables is issued.

390 Invading Gauls sack Rome. Romans quickly rebuild city, surrounding it with defensive wall that is not breached for many centuries.

▶

367 Plebeians (ordinary citizens) are granted right to hold rank of consul, right formerly reserved for patricians (aristocrats).

350 Romans develop basic battle formation they use to conquer ancient world. Is based on Roman legion, about 4,500 men arranged in three rows.

343–341 In First Samnite War, Rome begins expanding control into southern Italy.

316–304 Rome gains control of Campania region in southern Italy and puts down revolt by Samnites (298–290).

312 Roman officials begin construction of first stone aqueduct, Aqua Appia (10.3 miles long).

312 Work starts on Appian Way, roadway that eventually extends from Rome to Brundisium.

287 Lex Hortensia reform measure gives plebeians equality with patricians. Senate approval of resolutions voted by plebs is no longer needed.

264–241 First Punic War, between Rome and its rival Carthage. Rome gains Sicily.

218–201 Second Punic War. Carthaginian general Hannibal (247–182?) crosses Alps from Spain to Italy and conquers much of Italy before being halted by lack of supplies. Roman general Scipio Africanus Major (234–183?) in turn invades and defeats Carthage.

214–148 Macedonian Wars, between Macedonia and Rome. By these four wars Rome reduces Greece to province and thereafter completely dominates region.

149–146 Third Punic War. Rome invades Carthage and, after capturing city in bloody fight, razes it and sells survivors into slavery.

101 Consul Gaius Marius (c. 155–86) defeats invading Cimbri in Gaul.

88–82 Period of civil war, unrest, and bloody massacres in Rome follows attempts to impose reforms by force. Lucius Cornelius Sulla (138–78) finally proclaims himself dictator for life (82), kills his opponents, and imposes conservative reforms.

63 Cicero, serving as consul, discovers plot for his overthrow arranged by Catiline (c. 108–62), who escapes Rome. Cicero orders summary execution of other conspirators.

60 Caesar, Marcus Crassus (d. 53), and Pompey (106–48) form First Triumvirate, in which three consuls share power.

58–51 Gallic Wars. Julius Caesar (c. 102–44) leads Roman troops in conquest of Gaul.

53 Consul Crassus is killed in battle. Power struggle between Pompey and Caesar soon begins.

49–46 Roman Civil War. Julius Caesar defies Senate order to disband his armies, makes his famous crossing of Rubicon, and marches into Italy against his rival Pompey. Victorious, Caesar enters Rome and is made consul.

48 Caesar again defeats forces of Pompey, who is later killed in Egypt. In Egypt Caesar defeats King Ptolemy XIII, then installs his lover, Cleopatra (69–30), on Egyptian throne.

46 Caesar is made dictator of Rome.

46 Julian calendar is introduced.

44 Caesar is assassinated (March 15) in Rome's Senate House by Marcus Brutus (85–42), Cassius (d. 42), and others.

43 Octavian (63 B.C.–A.D. 14), Caesar's heir; Marc Antony (c. 83–30); and Marcus Lepidus (d. c. 13) form triumvirate to rule Rome. They defeat forces led by Caesar's assassins.

42 Antony begins his love affair with Egyptian queen Cleopatra. Affair proves unpopular in Rome and leads to Antony's downfall.

36 Lepidus is forced to relinquish power by Octavian.

31 Octavian defeats Antony and Cleopatra at Battle of Actium. He pursues them to Alexandria, where they both commit suicide.

27 Augustus (63 B.C.–A.D. 14) accepts imperial title, thus ending Roman Republic. Rome flourishes during next two centuries of empire.

▶

27 B.C.–A.D. 14 Augustus rules Roman Empire.

c. 8 Jesus of Nazareth (c. 8 B.C.–A.D. 30) is born in Roman Palestine.

A.D.

c. 24–26 Jesus begins ministry. He soon gains devoted following by his teachings and deeds.

c. 30 Jesus is crucified. St. Peter (d. 64?) begins to spread teachings of Christianity.

c. 33 St. Paul (d. 67) begins his missionary work on behalf of Christian Church.

37–41 Caligula (12–41) is in power as Roman emperor. He is assassinated after apparently going mad and instituting notoriously cruel reign.

41–54 Claudius I (10 B.C.–A.D. 54) rules. He adds Great Britain to empire.

c. 47–49 Paul journeys to Asia Minor and Cyprus. He establishes churches and writes earliest epistles.

c. 50 Roman sailors by this time have gained thorough knowledge of eastern coast of Africa south to Zanzibar.

c. 55 Peter travels to Rome. His primacy over church of Rome establishes tradition of papacy.

54–68 Nero (37–68) rules empire. His reign becomes increasingly cruel and ends with revolt and his suicide.

64 Fire destroys Rome. Nero blames Christians for it and begins persecuting them. He rebuilds Rome on grandiose scale, including his lavish palace, Golden House (68).

66–73 Jews mount unsuccessful revolt against Roman rule. Romans destroy Jerusalem.

c. 67 Nero orders executions of Peter and Paul.

68–69 Galba (3 B.C.–A.D. 69) is in power after Nero's overthrow.

80 Colosseum is dedicated by Titus (39–81).

81–96 Domitian (51–96) rules. His cruelty leads to his assassination by his wife.

98–117 Trajan (53–117) reigns. Roman Empire reaches greatest extent during his reign, stretching from Mideast to Spain and from Great Britain in North to Egypt in South.

c. 100 Christian churches are established in Greece, North Africa, Italy, and Asia Minor from this time. Books of New Testament are written during second half of century.

113 Trajan's Column, depicting Trajan's victories over Dacians, is erected in Rome by the Emperor Trajan (53–117).

117–38 Hadrian (76–138) rules. His reign marks beginning of worsening economic problems within empire. He builds 73-mile Hadrian's Wall in Great Britain to secure Roman territories there against invading barbarian tribes.

161–80 Marcus Aurelius (121–80) rules. Rome comes under pressure from invading barbarian tribes, which he successfully repels.

180–92 Commodus (161–92) reigns. He becomes dissolute and ineffective ruler.

c. 200 Romans have constructed over 50,000 miles of roads. Roman vehicles can travel 100 miles in a day.

211–17 Caracalla (188–217) rules. He murders his brother, Geta, who succeeded as coruler with him in 211, and massacres 20,000 of Geta's followers. He grants citizenship to all within borders of empire (212).

249–51 Christians are persecuted throughout Roman Empire under Emperor Decius (201–51).

c. 250 Romans introduce ox-drawn plow and other agricultural methods to farms in Europe.

c. 250 Barbarians, including Franks and Goths, invade Roman frontiers.

270–75 Aurelian (212–75) rules, briefly halting empire's steady decline. He blocks further advance by barbarians at Danube and restores Roman authority in Europe and East.

284–305 Roman emperor Diocletian (245–313) reigns. He divides empire into Western and Eastern empires and institutes reforms but is unable to stop Rome's decline.

▶

307–37 Constantine I the Great (280?–337) rules as emperor. During wars for control of Western Empire he converts to Christianity. By 323 he rules as sole emperor of the entire empire.

313 Constantine issues Edict of Milan, granting legal rights to Christians and restoring their confiscated property.

325 Constantine convenes First Council at Nicaea. Council condemns Arianism and composes Nicene Creed as fundamental statement of Christian doctrine.

330 Constantine rebuilds city of Byzantium as Constantinople and establishes it as new imperial capital of empire. Constantinople develops as "new Rome."

378 Visigoths defeat Roman armies and overrun Europe.

379–94 Theodosius I (346–95) rules as emperor of East.

380 Christianity becomes official religion of Roman Empire.

394–95 Theodosius becomes emperor of East and West, uniting empire for last time. Empire is permanently divided on his death.

407 Romans abandon British Isles.

410 Visigoths sack Rome.

c. 430 Vandals invade and conquer Roman territories in North Africa.

441–53 Attila the Hun invades Roman Empire. Huns advance deep into Italy by 452. Famine and pestilence in Italy at last force Attila to withdraw.

455 Vandals sack Rome.

475–76 Romulus Augustus rules as last emperor of West. He is defeated and deposed by Odoacer (c. 435–493), leader of a Germanic tribe, thus ending Western Roman Empire. Roman rule continues in Eastern Empire. (*See Byzantine Empire*).

Roman Emperors

B.C.	
27 B.C.–A.D. 14	Augustus (63 B.C.–A.D. 14)
A.D.	
14–37	Tiberius (42 B.C.–A.D. 37)
37–41	Caligula (12–41)
41–54	Claudius I (10 B.C.–A.D. 54)
54–68	Nero (37–68)
68–69	Galba (3 B.C.–A.D. 69)
69	Otho (32–62)
69	Vitellius (15–69)
69–79	Vespasian (9–79)
79–81	Titus (39–81)
81–96	Domitian (51–96)
96–98	Nerva (35–98)
98–117	Trajan (53–117)
117–138	Hadrian (76–138)
138–161	Antoninus Pius (86–161)
161–169	Marcus Aurelius and Versus (130–169)
169–180	Marcus Aurelius (121–180)
180–192	Commodus (161–192)
193	Pertinax (126–193)
193	Didius Julianus (133–193)
193–194	Pescennius Niger
193–197	Albinus (d. 197)
193–211	(Septimius) Severus (146–211)
211–212	Caracalla (188–217) and Geta (189–212)
212–217	Caracalla (188–217)
217–218	Macrinus (164–218)
218–222	Elagabalus (Heliogabalus) (204–222)
222–235	Alexander Severus (208–235)
235–238	Maximinus (173–238)
238–244	Gordianus III (224–244)
244–249	Philip the Arabian (d. 249)
249–251	Decius (201–251)
251–253	Gallus (203–253)
253	Aemilianus (206–253)

Roman Emperors

253–260	Valerian (d. 269)
260–268	Gallienus (d. 268)
268–270	Claudius II (214–270)
270	Quintillus (d. 270)
270–275	Aurelian (212–275)
275–276	Tacitus (200–276)
276	Florianus (d. 276)
276–282	Probus (d. 282)
282–283	Carus (223–283)
283–285	Numerianus (d. 284) and Carinus (d. 285); from this time until division of empire, it often has more than one ruler
284–286	Diocletian (245–313)
286–305	Diocletian (245–313) and Maximian (d. 310)
305–306	Constantius (250–306)
306–307	Severus (d. 307)
306–308	Maximian (d. 310)
306–311	Gelerius (d. 311)
306–312	Maxentius (d. 312)
307–313	Maximinus (Maximin) (d. 313)
307–337	Constantine I the Great (280?–337)
308–324	Licinius (270–324)
310–313	Daia
337–340	Constantine II (317?–340)
337–361	Constantius II (317–361)
337–350	Constans I (320–350)
350–353	Magnentius (d. 353)
361–363	Julian the Apostate (331–363)
363–364	Jovian (331–364)
364–367	Valentinian I (321–375), emperor of West
364–378	Valens (328–378), emperor of East
367–375	Valentinian I (321–375) and Gratian (359–383), emperors of West
375–383	Gratian and Valentinian II (372–392), emperors of West
379–394	Theodosius I (346–395), emperor of East
383–388	Maximus (d. 388) and Valentinian II, emperors of West
388–392	Valentinian II (372–392), emperor of West
392–394	Eugenius (d. 394), emperor of West
394–395	Theodosius (346–395), emperor of the East and West; Roman Empire is permanently divided on his death

Roman Emperors of West

395–423	Honorius (384–423)
425–455	Valentinian III (419–455)
455	Petronius Maximus (d. 455)
455–456	Avitus (d. 456)
457–461	Majorian (d. 461)
461–465	Severus (d. 465)
467–472	Anthemius (d. 472)
472	Olybrius
473–474	Glycerius
474–475	Nepos (d. 480)
475–476	Romulus Augustus (c. 461–?)

(For rulers after fall of West, see Byzantine Emperors.)

Byzantine Empire

A.D.

330 Roman emperor Constantine I the Great (280?–337) completes his rebuilding of old city of Byzantium (in modern Turkey). He renames city Constantinople and makes it his capital.

▶

378 Goths, driven into frontiers of eastern half of empire by advancing Huns, revolt and defeat Eastern Roman army. Emperor Theodosius I (346–95) adopts policy of assimilating Goths.

381 City hosts First Council of Constantinople. From this time forward Constantinople is important center of Christianity.

395 Roman Empire is divided between heirs of Emperor Theodosius into eastern and western halves. Constantinople becomes capital of the Eastern Roman Empire.

408–50 Emperor Theodosius II (401–50) reigns. He pays tribute in gold to Huns to prevent their attack in East. When his successor ends payments, Huns under Attila (406?–53) attack Western Roman Empire.

413–47 Great defensive wall, extending 4½ miles to protect landward side of Constantinople from attack, is built. This and seaward wall (built in 439) successfully withstand enemy attacks until 1453.

476 Western Roman Empire falls, at last overwhelmed by various barbarian invasions. Byzantine Empire thus becomes successor state to once mighty Roman Empire. Constantinople becomes leading center of Greek and Roman culture.

527–65 Byzantine emperor Justinian I (483–565) reigns. Empire reaches its height under his rule. He comes to power during period of prosperity and inherits empire that includes Anatolia, Syria, Palestine, Egypt, and Greece.

529–35 Great legal code, *Corpus Juris Civilis*, is compiled during Justinian's reign.

532 Nika riots. Opponents of Justinian unsuccessfully attempt to overthrow him.

537 Great domed church, Hagia Sophia, is completed. It is monumental work of Byzantine architecture.

534 Byzantines, led by Gen. Belisarus (c. 505–65), defeat Vandals in North Africa and thereby gain control of region. They begin reconquest of Italy next year.

554 Byzantine general Narses (c. 478–573) completes conquest of Ostrogoths in Italy. Soon after, however, Lombards invade northern Italy and largely end Byzantine control there.

559 Combined force of Huns and Slavs reaches outskirts of Constantinople before being turned back.

600s Byzantine Empire is at various times threatened or invaded by Persians, Slavs, Avars, Bulgars, and particularly by newly formed Muslim Empire.

636 Muslims gain Syria in first major victory over Byzantines. They later take Jerusalem, Egypt (642), and remainder of North Africa (by 698).

678 Muslim siege of Constantinople is ended by peace settlement.

717 Emperor Leo III the Isaurian (680–741) restores order after period of political unrest lasting over 20 years. He reigns until his death.

817 Bulgars unsuccessfully attack Constantinople. Treaty ends Bulgar threat for many years.

867–86 Basil I the Macedonian (c. 813–86) rules as emperor. He restores Byzantine military strength and begins compilation of new legal code, *Basilica*, completed after his death. His reign marks beginning of Byzantine golden age, lasting until about 1025.

975 Byzantines have retaken Syria and Palestine and have expanded eastward well beyond Euphrates into Mesopotamia.

1018 Bulgarians, who have again been threatening empire (since 889), are subdued.

c. 1050 Seljuk Turks invade Asia Minor, driving Byzantine forces back toward Constantinople. Byzantine Empire enters protracted period of collapse.

▶

1054 Final schism results in permanent division of Eastern and Western Christian churches. Constantinople becomes seat of Eastern Orthodox Church.

1091 Normans complete capture of Byzantine territory in southern Italy and form Kingdom of Two Sicilies.

1096–1291 Era of Crusades. Advancing Seljuks force Byzantine emperors to request aid from Christians in Europe. Popes preach succession of eight Crusades; all but first are militarily unsuccessful.

1185 Bulgars establish independent kingdom after revolting against Byzantine rule.

1202–04 Fourth Crusade. Crusaders begin by intervening in palace coup at Constantinople and end by sacking city and setting up Latin Empire under European rule (1204) in place of Byzantine Empire.

1206 Byzantines set up Empire of Nicaea under Theodore I Lascaris (d. 1222), in opposition to Europeans' Latin Empire.

1261 Michael VIII Palaeologus (1234–82) retakes Constantinople to reestablish Byzantine Empire. He rules it until his death.

1300s Empire comes under attack by Ottoman Turks and by Serbs. It is further weakened by war between rival claimants to throne (1341–47).

1391 Ottoman Turks launch first attacks on Constantinople itself, having reduced empire to this last vestige of Byzantine power.

1453 Turks, using massed artillery, at last breach thick walls defending Constantinople and take city, marking end of Byzantine Empire.

Byzantine Emperors

	Roman Emperors of East
395–408	Arcadius (378?–408)
408–450	Theodosius II (401–450)
450–457	Marcianus (392–457)
457–473	Leo I (400–474)
473–474	Leo I (400–474) and Leo II (d. 474)
474	Leo II (d. 474)
474–491	Zeno (426–491)
491–518	Anastasius I (430–518)
518–527	Justin I (450–527)
527–565	Justinian I (483–565)
565–578	Justin II (d. 578)
578–582	Tiberius (d. 582)
582–602	Maurice (539–602)
602–610	Phocas (d. 610)
610–641	Heraclius (575–641)
641	Constantine III (612–641)
641–668	Constans II (630–668)
668–685	Constantine IV (648–685)
685–695	Justinian II (669–711)
695–698	Leontius (d. 705)
698–705	Tiberius III (d. 705)
705–711	Justinian II (669–711), restored
711–713	Philippicus
713–716	Anastasius II (d. 721)
716–717	Theodosius III (d. 718?)
717–741	Leo III the Isaurian (680–741)
741–775	Constantine V (718–775)
775–780	Leo IV (750–780)
780–797	Constantine VI (c. 770–797)
797–802	Irene (752–803)
802–811	Nicephorus I (d. 811)
811	Stauracius (d. 811)
811–813	Michael I Rhangabe (d. 845)
813–820	Leo V the Armenian (d. 820)
820–829	Michael II Balbus (d. 829)
829–842	Theophilus (d. 842)
842–867	Michael III (d. 867)
867–886	Basil I the Macedonian (c. 813–886)
886–912	Leo VI the Wise (866–912)
913–919	Constantine VII Porphyrogenitus (905–959)
919–944	Romanus I (d. 948)

Byzantine Emperors

944–959	Constantine VII (905–959)
959–963	Romanus II (939–963)
963–969	Nicephorus II Phocus (913–969)
969–976	John I Zimisces (925–976)
976–1025	Basil II (958–1025)
1025–28	Constantine VIII (960–1028)
1028–34	Romanus III (968–1034)
1034–41	Michael IV (d. 1041)
1041–42	Michael V
1042–55	Constantine IX (1000–1055)
1055–56	Theodora (980–1056)
1056–57	Michael VI Stratioticus
1057–59	Isaac I Comnenus (d. 1061)
1059–67	Constantine X Ducas (1007–67)
1067–71	Romanus IV Diogenes (d. 1071)
1071–78	Michael VII Ducas
1078–81	Nicephorus III Botaniates (d. 1081)
1081–1118	Alexius I Comnenus (1048–1118)
1118–43	John II Comnenus (1088–1143)
1143–80	Manuel I Comnenus (1120–80)
1180–83	Alexius II Comnenus (1168–83)
1183–85	Andronicus I Comnenus (1110–85)
1185–95	Isaac II Angelus (d. 1204)
1195–1203	Alexius III Angelus (d. 1210)
1203–4	Alexius IV (d. 1204)
1204	Alexius V (d. 1204)
1204–5	Baldwin I (1171–1205), first of Latin emperors
1205–16	Henry I of Flanders (1174–1216)
1216–17	Peter of Courtenay (d. 1217)
1221–28	Robert of Courtenay (d. 1228)
1228–61	Baldwin II (1217–73)
1206–22	Theodore I Lascaris (d. 1222), first of Nicaean emperors
1222–54	John III Vatatzes (1193–1254)
1254–58	Theodore II Lascaris (d. 1221–58)
1258–61	John IV Lascaris (1250–1300)
1261–82	Michael VIII Palaeologus (1234–82), first emperor of restored Byzantine Empire
1282–1328	Andronicus II Palaeologus (1260–1332)
1328–41	Andronicus III (1296–1341)
1341–47	John V Palaeologus (1332–91)
1347–55	John VI (1292–1383)
1355–76	John V (1332–91), restored

Muslim Empires

See also Middle East.

A.D.

632 Muhammad, Islamic prophet, dies. By this time he has unified much of Arabia, thus forming nucleus for Muslim Empire.

▶

632–34 Abu Bakr reigns as first patriarchal caliph. First period of rapid expansion of Muslim Empire begins.

633 Muslims conquer Syria and Iraq.

634–44 Umar (c.591–644) reigns as second patriarchal caliph.

639–45 Muslims conquer Egypt and Persia.

644–56 Uthman (574–656) reigns as third patriarchal caliph. He launches invasions westward into North Africa.

656–61 Muhammad's son-in-law Ali (d. 661) reigns as fourth patriarchal caliph.

661 Ali's son Hasan becomes caliph upon Ali's assassination by Kharijite zealot.

661–80 Mu'awiya (d. 680), who had been increasing his power in Syria, becomes caliph and founds Umayyad dynasty. He moves capital from Mecca to Damascus. Under his leadership, second wave of Muslim conquests begins.

680–83 Yazid (d. 683), son of Mu'awiya, reigns as caliph. Arabian opposition to Syrian leadership of Umayyad caliphate leads to Yazid's attacks on Mecca and Medina.

683–92 Second civil war. Abd al-Malik (c. 646–705) becomes leader of Muslim world.

696 Arabic is declared official language of Islam, Arabic coinage its official currency.

709 Muslim conquest of Spain begins. Muslims from North Africa push northward up Iberian Peninsula.

724–43 Hisham reigns. Rapid growth of empire ends; empire extends from Afghanistan to east, across Arabian Peninsula and North Africa and up Iberian Peninsula.

732 Muslim advance from Spain into France is halted near Poitiers by Franks.

738 Arab merchant colony is in China.

747–50 Abu Muslim leads revolt that ends in overthrow of Umayyad dynasty.

750 Abbasid dynasty begins when Abbas, a Persian, defeats last Umayyad caliph at Battle of Great Zab. Abbas moves capital from Damascus to Hashimiya in Iraq.

754–775 Abu Jaffar al-Mansur reigns as caliph. He quells uprisings by Shiites disappointed by his treatment of them after they aided his rise to caliphate.

756 Abd al-Rahman, Umayyad refugee, founds Emirate of Cordoba, later an independent caliphate, in Spain.

762 Mansur establishes new capital at Baghdad.

765 School of medicine is founded in Baghdad.

786–809 Harun al-Rashid reigns. Abbasid dynasty reaches its height. Third Muslim civil war breaks out after his death.

788 Morocco gains independence.

799 Tunisia gains independence.

836 Capital is moved to Samarra, on Tigris.

851 First Muslim travelers visit India.

865–925 Razi (Rhazes), Muslim doctor and author of more than 200 books on medicine, alchemy, theology, and astronomy, lives.

869–92 Caliph Mutamid reigns. He makes Baghdad his capital.

910 Shiites conquer North Africa and set up independent caliphate there.

921 First Muslim travelers visit Russia.

929–1030 Independent Umayyad caliphate in Cordoba, Spain flourishes.

945 Buyids, family controlling parts of Persia, occupy Baghdad and control Abbasid caliphate.

969–1171 Fatimids, Shiite dynasty of rulers, conquer Egypt and name Cairo their capital. Ayyubids oust them in 1171.

980–1037 Ibn Sina (Avicenna) lives. Noted Muslim philosopher, he wrote more than 170 books on philosophy, medicine, mathematics, and religion.

998 Ghaznavid Empire emerges in Afghanistan, favoring Sunni theology.

1030 Biruni, scientist and traveler, writes *Description of India,* describing in detail his visits to that country.

▶

1055–92 Seljuk Turks overthrow Buyid rulers in Persia and found Sunni sultanate.

1085 Christians conquer Toledo in Spain and begin regaining country from Umayyad Muslims there.

1095–1291 Christian Crusaders journey to recapture Holy Land from Muslims. After initial successes in First Crusade, they suffer series of defeats and are driven out. (*See Wars and Military History.*)

1154 Idrisi (c. 1099–c. 1155), cartographer in Roger de Hauteville's court in Sicily, writes *The Pleasure of the Ardent Enquirer*, containing circular world map. He holds that earth is round—more than three centuries before Columbus.

1192 Muslims conquer Delhi in India. They found Delhi sultanate.

1219 Mongol armies led by Genghis Khan (1162–1227) invade Islamic territory from central Asia. They reach Persia by 1221.

1260 Mamelukes, Muslims from Egypt, halt Mongol advance in Syria.

1295 Ghazan Khan, Mongol ruler of Persia, is converted to Islam.

1299–1326 Osman, founder of Ottoman Turkish Empire, reigns. He expands his empire by defeating Seljuk dynasty.

1353 Palace of Alhambra is completed in Granada, Spain.

1361 Under reign of Mural I, Ottomans conquer Adrianople, which becomes new Ottoman capital.

1375 Ibn Khaldun, historian and philosopher, begins *Muqaddima*, history of Arabs.

1379 Mongols invaded under Timur. Ottoman Turks are defeated but soon recover their territory after Timur's death.

1453 Ottoman Turks conquer Constantinople, ending Byzantine Empire. They make Constantinople their capital (as Istanbul) and subsequently push westward into Balkans.

1492 Granada, last Muslim stronghold in Spain, is taken by Christians.

1502–1722 Safavid dynasty flourishes in Persia (Iran). Shiism becomes state religion.

1505 Great Mogul Empire in India is founded by Baber (1483–1530).

1517 Ottoman ruler Selim the Grim (1465–1521) proclaims himself successor to caliphs. He takes Cairo from Mamelukes.

1526 Battle of Mohacs in Hungary is won by Ottomans. Ottoman army eventually conquers all of Hungary and lays unsuccessful siege to Vienna.

c. 1530–40 Ottomans conquer Persia.

1551 Ottomans conquer Libya.

1571 Battle of Lepanto. Christian Crusaders defeat Ottoman army in first significant Turkish defeat.

1591–1606 Fifteen Years War takes place between Ottomans and Austrian army.

1605–23 Persians recapture Baghdad in war against Ottomans.

1676–1878 Russo-Turkish Wars occur, long series of conflicts in which Russia greatly expands its borders at expense of weakening Ottoman Empire. Among wars are those in 1676–81, 1695–96, 1710–11, 1736–39, 1768–74, 1787– 92, 1806–12, 1828–29, 1853–56, and 1877–78.

1682–99 Austro-Turkish War. Austrians recapture Hungary and much of Eastern Europe.

1792 Egypt is swept by plague. Some 800,000 perish.

1793 Sultan Selim III (1762–1808) enacts reforms after series of military defeats. He plan reforms and modernizes army, schools, laws, and taxation system.

1798 Napoleon invades Egypt.

1801–5 Tripolitan War. United States launches successful military expedition against Tripoli (in modern Libya) to halt raids by Barbary pirates in Mediterranean.

1805 Egypt successfully rebels against rule by Ottoman Turks, gaining near-complete independence.

▶

1808 Selim III is assassinated by opponents of his reforms.

1812 Russians conquer Bessarabia.

1820–22 Egypt conquers Sudan.

1830 Greece gains independence from Ottomans after gaining support from European powers.

1831–33 Ottomans wage war with Egypt, during which Egypt's Muhammad Ali conquers Syria.

1853–56 Russia sparks Crimean War by invading domains of declining Ottoman Empire. Britain, France, and others aid Turks, launching attack in Crimea at Sevastopol. Russia eventually is driven out.

1857 Afghanistan gains independence.

1859 Construction of Suez Canal begins in Egypt. Canal project and government program to modernize Egypt wreck country's finances.

1869 Suez Canal opens.

1875 Debt-ridden Egypt is forced to sell controlling interest in Suez Canal to British (1875). Soon British and French are in firm control of Egyptian government.

1882 British occupy Egypt after putting down nationalist riots. They reorganize government.

1885 British forces under Sir Charles Gordon (1833–85) are massacred at Khartoum during revolt in Sudan. Sudan is reconquered by British in 1898.

1888 Neutrality of Suez Canal is guaranteed by treaty.

1894–96 Unrest in Armenia leads to widespread massacres of Armenians.

1902 Aswan Dam in Egypt is finished.

1906 Revolt in Persia leads to establishment of constitutional monarchy there.

1908 Bulgaria gains independence.

1911–14 Italians conquer Tripoli (Libya). They develop it as colony.

1912–13 Balkan Wars. Ottoman presence in Eastern Europe is virtually ended.

1914–18 World War I brings on complete collapse of Ottoman Empire; it is dismembered after this war. *(See Middle East for modern history of region.)*

French Colonial Empire

1603 French explorer Samuel de Champlain (1567–1635) establishes colony in Canada called New France.

1608 Champlain establishes settlement at Quebec.

1625 French settlers found first West Indian settlements, at St. Christopher.

1637 Colony of French Guiana is established.

1663 New France, in modern Canada, is established as French colony.

1664 France controls 14 islands in Antilles, including Guadeloupe and Martinique, by this time.

1664 French Company of East Indies is established.

1681 Immigration to New France begins declining. This weakens ties between colonists and mother country.

1682 French explorer Robert Cavelier journeys down Mississippi River to Gulf of Mexico. He claims entire Mississippi River Valley for France.

1707–48 France and England compete for control of India. British dominate following War of Austrian Succession (1740–48).

▶

1718 New Orleans is founded by Louisiana governor Jean-Baptiste Lemoyne (1680–1768).

1754 New France's white population is estimated at 55,000.

1763 Treaty of Paris ends Seven Years War and strips France of many of its colonial possessions. Great Britain takes remaining French colonies in Canada and India, and Spain takes Louisiana Territory.

1800 French emperor Napoleon (1769–1821) conquers Spain and regains Louisiana Territory.

1803 Louisiana Purchase is agreed upon by Napoleon and U.S. President Thomas Jefferson (1743–1926). France cedes to United States more than 800,000 square miles, from Mississippi River to Rocky Mountains, for $27 million.

1830 French invasion of Algeria begins.

1834 Algeria is declared French possession.

1848 Algeria is declared "integral part of France."

1852 Penal colonies are established in French Guiana.

1863 France conquers southern Vietnam.

1867 French colony of Cochin China is established in southern Vietnam.

1875–83 Italian-French explorer Savorgnan de Brazza (1852-1905) explores central Africa and stakes French claims to region.

1883–84 Northern and central Vietnam are made French protectorates.

1885 Madagascar is established as French protectorate.

1891 French Congo is established as colony.

1895 French West Africa is created as federation of territories including Dahomey, French Guinea, French Sudan, Ivory Coast, Mauritania, Niger, Senegal, and Upper Volta.

c. 1900 French Indochina (including Vietnam, Laos, and Cambodia) is consolidated as French possession.

1910 French Equatorial Africa is created as federation of colonies Gabon, Middle Congo, and Ubangi-Shari-Chad. Chad becomes separate colony within federation in 1920.

1911 Morocco is established as French protectorate.

1939 French colonial empire by this time is more than 20 times as large as France itself and one-and-a-half times as populous. It is second in size only to British Empire.

1940 Japan occupies French Indochina during World War II.

1945 France reoccupies Vietnam and Cambodia following defeat of Japan.

1946 French Empire is reorganized as French Union.

1946 Vietnamese leader Ho Chi Minh (1890?–1969) leads nationalistic uprising against French control of Indochina, beginning eight-year Indochinese War.

1954 French defeat at Dien Bien Phu ends Indochinese War. Subsequent Geneva Agreements result in complete French withdrawal from Indochina.

1954 Rebellion breaks out against French control in Algeria. Eight-year Algerian War begins, with France committing 500,000 troops to struggle.

1956 Tunisia and Morocco are given sovereignty.

1958 Riots by French settlers in Algiers lead to election of Charles De Gaulle (1890–1970) as premier and approval of new constitution. Referendum gives French territories right to vote for independence or for self-government within newly formed French Community.

1958 Madagascar votes to become independent republic within French Community.

1958–59 Colonies of French West Africa and French Equatorial Africa are dissolved when their territories elect self-government within French Community.

1962 Algeria is granted independence, ending Algerian War.

▶

1962 French colonial empire is essentially nonexistent by this time.

1970s French Community lapses and becomes gradually defunct as result of growing independence of its members.

British Empire

1609–55 British colonists settle in Caribbean: Bermuda (1609), Barbados (1626), Bahamas (1629), and Jamaica (1655).

1661 St. Helena is granted to British East India Company by Portugal.

1704 Gibraltar is captured from Spanish during War of Spanish Succession.

1713 Newfoundland is ceded to British after Treaty of Utrecht.

1755 Great Britain gains control of Nova Scotia in modern Canada and expels French.

1756–63 Seven Years War. British, in colonial phase of this war, gain control of French colonial territory in India. By this and subsequent conquest of indigenous kingdoms in India, British become masters of all India. Treaty ending war cedes French territories in Canada and east of Mississippi to Great Britain. In Canada, Prince Edward Island, Manitoba, and Ontario thus become British territory.

1770 Eastern part of Australia is claimed by Captain James Cook (1728–79) for Great Britain.

1786 British Honduras is made British superintendency.

1798 Ceylon is made British crown colony.

1800 Malta is captured.

1802 Trinidad is ceded to British by Treaty of Amiens.

1808 Sierra Leone in West Africa is made British colony.

1814 British Guiana in South America and Cape of Good Hope in southern Africa are ceded to Great Britain by Dutch.

1829 Entire continent of Australia is claimed by Great Britain.

1839 Aden in modern Yemen is conquered from Turks by Great Britain.

1840 New Zealand is settled.

1841 Hong Kong is ceded to Great Britain by Chinese.

1843 Natal in southern Africa is made British colony.

1850 Australia is granted limited self-government.

1852 New Zealand is granted self-government.

1858 British government takes direct control of India.

1858 British Columbia is made British crown colony.

1867 In British North America Act, Ontario, Quebec, New Brunswick, and Nova Scotia become provinces of newly created Dominion of Canada.

1867 Singapore, Penang, and Malacca in Southeast Asia are placed under British control.

1870 Manitoba becomes province of Canada.

1871 British Columbia becomes province of Canada.

1873 Prince Edward Island enters Canadian confederation.

1874 Fiji Islands in South Pacific become British colony.

1874 Ghana (Gold Coast) in West Africa becomes British colony after Danish and Dutch settlements are purchased.

1878 Cyprus is placed under British rule.

▶

1881 Sabah (British North Borneo) comes under British rule.

1884 British New Guinea becomes British protectorate.

1897 By new agreement China leases Hong Kong to Great Britain for 99 years.

1900 Transvaal in southern Africa is annexed as British Crown Colony.

1900 Orange Free State in southern Africa is annexed to British dominions.

1905 British New Guinea comes under control of Australia and is renamed Papua.

1910 Union of South Africa is created by uniting former British possessions of Natal, Cape of Good Hope, Transvaal, and Orange Free State.

1931 British Commonwealth of Nations is created, establishing special trade agreements among present and former British dependencies.

1947 India is granted independence.

1949 Newfoundland becomes province of Canada.

1957 Ghana gains independence.

1957 Penang and Malacca become independent states in Federation of Malaya.

1960 Cyprus gains independence.

1961 Sierra Leone gains independence.

1962 Jamaica gains independence.

1964 Malta achieves independence.

1965 Singapore becomes republic.

1966 Barbados gains independence.

1966 British Guiana (now Guyana) gains independence.

1968 Bermuda is granted self-government.

1968 Aden and Socotra become part of independent South Yemen.

1970 Fiji becomes independent.

1973 Bahamas gain independence.

3 Politics and Law

Documents That Shaped Western History

B.C.

c. 1750 Code of Hammurabi is composed by Babylonian emperor Hammurabi (c. 1792–50 B.C.). One of oldest known legal codes, it covers criminal behavior, family life, economics, and ethics and establishes principle of "an eye for an eye."

c. 1000 Oldest books of Old Testament are set into writing.

c. 590 Laws of Solon are established by Athenian statesman Solon (c. 638–c. 559 B.C.). These laws replace severe Draconian code and mark establishment of Athenian democracy.

A.D.

100 New Testament books are composed. They are canonized as sacred writings of Christianity by the fourth century.

313 Edict of Milan is declared by Roman emperors Constantine (c. 280–337) and Licinius (?–325). It grants religious tolerance to Christians throughout Roman Empire.

c. 650 Koran is written as sacred book of teachings of Muhammad.

1215 Magna Carta is signed by English king John (c. 1167–1216). It guarantees such basic liberties as habeas corpus and trial by jury and has major effect on development of English law.

1517 Nailing of 95 theses of Martin Luther (1483–1546) to door of church in Wittenberg, Germany (October 31) begins Protestant Reformation. Theses protest selling of indulgences and other practices of Roman Catholic Church and detail Luther's teachings on faith and works.

1534 Act of Supremacy makes King Henry VIII (1491–1547) head of Church of England. It also establishes church as separate from Roman Catholic Church.

1566 Catechism of the Council of Trent. It details reforms in Roman Catholic Church established by council and begins Counter-Reformation.

1776 U.S. Declaration of Independence is approved by Congress (July 4), declaring 13 American colonies independent of Great Britain.

1787 U.S. Constitution is signed at Constitutional Convention. Constitution goes into effect in 1789 following ratification by states. It establishes effective democratic form of government, providing alternative to then prevailing monarchical system.

1789 Declaration of the Rights of Man and of the Citizen. Approved by French National Assembly (August 26), it is summation of ideals of French Revolution. It declares individual's right to representation and equality and freedom of press, speech, and religion.

1791 U.S. Bill of Rights is ratified as first ten amendments to Constitution. Amendments guarantee freedom of speech, religion, and press and other basic human rights.

1804 Code Napoleon is issued by French emperor Napoleon Bonaparte (1769–1821) as comprehensive compilation of French civil laws. Code forms basis of modern civil law in France and other nations, which voluntarily adopt code after Napoleon's fall.

1848 *Communist Manifesto* is written by Karl Marx (1818–83) and Friedrich Engels (1820–95). It articulates political philosophy of communism and calls for worldwide revolution by workers.

1861 Edict of Emancipation is declared by Russian Czar Alexander II (1818–81), freeing millions of serfs in Russia.

1862 Emancipation Proclamation is made (September 22) by President Abraham Lincoln (1809–65). It declares freedom for slaves in the Confederacy and marks Lincoln's commitment to elimination of slavery.

1918 Fourteen Points are presented by U.S. President Woodrow Wilson (1856–

▶

1924). These proposals amount to plan for world peace following World War I. They lead to founding of League of Nations in 1919.

1925 *Mein Kampf* is written by Adolf Hitler (1889–1945). It details aims of Nazism and calls for German domination of the world. Hitler's later attempted implementation of these plans results in deaths of millions and vast destruction in World War II.

1931 Statute of Westminster grants autonomous government to Great Britain's former colonial possessions, creating British Commonwealth.

1941 Atlantic Charter is adopted (August 14) by U.S. President Franklin D. Roosevelt (1882–1945) and British Prime Minister Winston Churchill (1874–1965). It details policy for world peace and lays groundwork for United Nations Charter.

1945 United Nations Charter is signed (June 26), creating United Nations.

1947 Truman Doctrine is issued by U.S. President Harry S. Truman (1884–1972). It declares U.S. Cold War policy of containing spread of communism.

1948 Universal Declaration of Human Rights is approved by UN General Assembly. It declares essential human rights of all people.

1962–65 Second Vatican Council produces 16 documents detailing wide-ranging reforms and modernization of Roman Catholic practice.

1972 Plaque attached to *Pioneer 10* space probe is attempt to communicate with other possible intelligent life beyond our solar system. It documents significant change in outlook of modern world, brought on by the arrival of space exploration.

Historic Treaties

1814 Treaty of Paris between Allied powers and France ends Napoleonic Wars. France is reduced to roughly its borders of 1792.

1814 Treaty of Ghent between United States and Great Britain ends War of 1812. It provides for return of captured territories and for arbitration of boundary disputes between United States and British Canada but does not address question of impressment, a cause of the war.

1815 Treaty of Paris is effected between Allied Powers and France after final defeat of Napoleon, at Waterloo. France loses additional territory and must pay indemnity.

1817 Rush-Bagot Agreement between United States and British Canada is negotiated by U.S. secretary of state Richard Rush (1780–1843); it limits deployment of military forces on Great Lakes and in other border areas.

1819 In Transcontinential Treaty between United States and Spain, latter cedes Florida to United States, renounces interest in Oregon country, and gains sovereignty over Texas.

1821 Treaty of Cordoba establishes Mexico's independence from Spain.

1824 Treaty of London between British and Dutch defines their respective areas of influence in Malaya (modern Malaysia).

1827 Treaty of London among France, Great Britain, and Russia provides for direct intervention in Greek war for independence from Ottoman Empire.

1829 Treaty of Adrianople ends Russo-Turkish War of 1828–29. Russia gains mouth of Danube and eastern coast of Black Sea. Ottoman Turks recognize autonomy of Greece, Serbia, Moldavia, and Wallachia.

1841 In Straits Convention, agreement sought by European powers, Ottoman Turks agree to close Dardanelles to foreign warships in peacetime.

1842 Webster-Ashburton Treaty between Great Britain and United States is negotiated by Daniel

▶

Webster (1782–1852) and Alexander Baring Lord Ashburton (1774–1848). It grants United States over half disputed Maine–New Brunswick frontier and provides for suppression of African slave trade.

1848 Treaty of Guadalupe Hidalgo between United States and Mexico after Mexican War. United States gains Texas, New Mexico, California, Nevada, Utah, Arizona, and part of Colorado for $15 million.

1850 Clayton-Bulwer Treaty between United States and Great Britain is negotiated by Secretary of State John Clayton (1796–1856) and British minister Henry Bulwer (1801–72). Treaty provides that neither country shall dominate Central America and that any future canal through Isthmus of Panama shall be neutral.

1854 Treaty of Kanagawa between United States and Japan ends Japan's long period of isolation. It opens two ports to U.S. trade and provides for appointment of U.S. consul to Japan.

1856 Treaty of Paris (March 30) ends Crimean War. Russia regains Crimea but gives up mouth of Danube and southern Bessarabia to the Ottoman Empire. Black Sea is made neutral.

1858 Treaty of Aigun between Russia and China settles China's northern boundary. China cedes northern bank of Amur to Russia and agrees to joint possession of other territories. It rejects treaty in 1859 but reaffirms it in 1860 Treaty of Peking.

1866 Peace of Prague between Prussia and Austria after Austro-Prussian War. Austria is excluded from North German Confederation, giving Prussia control of what later becomes modern Germany.

1868 Burlingame Treaty between United States and China is negotiated by Anson Burlingame (1820–70) and Secretary of State William Seward (1801–72). It guarantees Chinese territorial integrity and helps set stage for immigration of Chinese laborers to United States.

1878 Treaty of Berlin between Russia and Ottoman Turkey and European powers ends Russo-Turkish War of 1877–78. Treaty recognizes independence of Montenegro, Serbia, and Rumania and creates autonomous Bulgaria.

1883 Treaty of Ancon between Peru and Chile after War of the Pacific. Chile gains provinces of Tarapaca, Tacna, and Arica.

1895 Treaty of Shimonoseki between China and Japan ends first Sino-Japanese War. Japan forces recognition of Korea's independence, extracts indemnity from China, and gains other concessions.

1898 Treaty of Paris between United States and Spain ends Spanish-American War. Spain gives up control of Cuba and Philippines.

1899 Anglo-Egyptian Condominium Agreement establishes joint British and Egyptian rule in Sudan after unsuccessful Mahdish revolt.

1902 Peace of Vereeniging between Great Britain and defeated Boers ends Boer War and confirms British sovereignty in southern Africa.

1903 Hay-Bunau-Varilla Treaty between United States and Panama is enabling treaty for construction of Panama Canal. It grants United States rights to Canal Zone for $10 million.

1904 Entente Cordiale between Great Britain and France ends tension between them and resolves differences over colonial territories and Africa and Far East.

1905 Treaty of Portsmouth between Russia and Japan ends Russo-Japanese War. U.S. President Theodore Roosevelt negotiates treaty at Portsmouth, New Hampshire.

1915 Treaty of London, secret agreement between the World War I Allies and Italy. Italy agrees to join Allies in return for territorial concessions.

1916 Sykes-Picot Agreement, secret agreement between Great Britain and France (with Russia consenting), provides for division of crumbling Ottoman Empire after World War I.

1919 Treaty of Versailles, between Allies and Germany, ends World War I and establishes League of Nations. Germany protests harsh terms, including responsibility for starting the war, large war-reparations payments, surrender of territory, and demilitarization of Rhineland.

1920 Treaty of Trianon between the Allies and newly established state of Hungary, which cedes much of its territory to surrounding states.

1921 Four-Power Pacific Treaty among United States, France, Great Britain, and Japan recognizes their respective spheres of influence in Pacific.

1922 Five-Power Naval Limitation Treaty among United States, Italy, France, Great Britain, and Japan limits size of their navies.

1925 Locarno Pact among Great Britain, France, Italy, Belgium, and Germany guarantees peace

▶

in Europe and normalizes relations among nations after World War I.

1928 Kellogg-Briand Pact is multilateral treaty renouncing war and agreeing to negotiated settlements of international disputes. Nearly every nation signs treaty.

1929 Lateran Treaty between Italy and Roman Catholic Church recognizes Church's independence and sovereignty over Vatican City. It also recognizes Italy's seizure of Rome and Papal States in 1800s.

1930 London Naval Treaty establishes permanent parity among United States, British, and Japanese navies and continues moratorium on building capital ships to 1936.

1931 Delhi Pact between British authorities and Mohandas Gandhi (1869–1948) provides for end to Gandhi's civil-disobedience campaign and provides for formal talks and release of political prisoners.

1936 Anglo-Egyptian Treaty between Great Britain and Egypt establishes latter as independent state and provides for withdrawal of British military forces except for specified force to protect the Suez Canal Zone.

1938 Munich Pact between Nazi Germany and other European powers provides for gradual German occupation of Czechoslovakia's Sudetenland. Hitler nullifies pact by invading Czechoslovakia in 1939.

1939 Pact of Steel confirms alliance between Nazi Germany and Italy.

1942 Anglo-Soviet Agreement formalizes Soviet Union's switch to Allied side in World War II following Hitler's invasion of USSR.

1945 United Nations Charter, agreement among 50 Allied and associated nations at close of World War II, provides for UN organization.

1948 Organization of American States Charter provides for regional security and economic development.

1949 North Atlantic Treaty between non-Communist European nations and United States provides for regional defense against Soviet aggression.

1951 Japanese Peace Treaty among Japan and 48 other nations formalizes restoration of peaceful relations after World War II.

1954 Southeast Asia Treaty Organization (SEATO) among United States, Great Britain, France, Australia, and other Pacific nations provides for regional defense pact.

1955 Warsaw Treaty between USSR and its East European satellites establishes Warsaw Pact regional defense organization.

1955 Austrian State Treaty is formal peace settlement between Austria and World War II Allies.

1957 Treaty of Rome establishes European Economic Community.

1959 Antarctic Treaty among 12 nations guarantees for 30 years demilitarization of the Antarctic continent and promotion of scientific investigation there.

1963 Nuclear Test-Ban Treaty among United States, USSR, and Great Britain bans atmospheric, underwater, and outer-space testing of nuclear weapons but permits underground tests.

1967 Outer Space Treaty among United States, USSR, and 58 other nations limits military uses of outer space and provides for sharing scientific information.

1968 Treaty on the Nonproliferation of Nuclear Weapons is affected among major powers and many smaller nations to prevent spread of nuclear weapons and capability to nations not already possessing it.

1972 Biological warfare ban is made effective among United States, USSR, and over 100 other nations.

1977 Panama Canal Treaty between the United States and Panama provides for return of Panama Canal to Panama.

1979 Strategic Arms Limitation Treaty (SALT II) between United States and USSR sets a maximum number of long-range missiles and bombers for each nation.

1987 Intermediate Nuclear Forces (INF) Treaty between United States and USSR provides for dismantling of all U.S. and Soviet intermediate-range nuclear weapons.

U.S. Constitution and Amendments

1777 Articles of Confederation are approved by Continental Congress in York, Pennsylvania (November 15). Articles establish government in which individual states rather than central government hold most of power.

1781 Articles of Confederation go into effect (March 1).

1787 Constitutional Congress is held in Philadelphia to draft new constitution to provide for stronger central government. Delegates from every state except Rhode Island attend. Constitution is signed (September 17) and submitted to the original 13 states for ratification (September 28).

1787 Delaware becomes first state to ratify Constitution (December 7).

1787 Pennsylvania and New Jersey become second and third states to ratify Constitution (December 12 and December 18, respectively).

1788 Rhode Island rejects ratification in popular referendum (March 24).

1788 New Hampshire ratifies Constitution (June 21) as last of nine required states.

1789 Constitution goes into effect (March 4) with ratification by 11 of the 13 states.

1789 Bill of Rights is passed by Congress to guarantee certain basic freedoms not explicitly set forth in Constitution (September 25).

1790 Rhode Island ratifies Constitution (May 29) as last of original 13 states to do so.

1791 Bill of Rights is ratified (December 15) as first 10 amendments to Constitution. Amendments provide for (1) freedoms of religion, speech, press, and peaceable assembly, and right to petition; (2) right to bear arms; (3) limits on quartering soldiers; (4) prohibition of unreasonable searches and seizures; (5) right to due process of law; (6) right to speedy trial; (7) right to trial by jury; (8) prohibition of cruel and unusual punishment; (9) powers not in Constitution to be retained by people; and (10) powers not delegated to United States to be retained by states.

1794 Eleventh amendment is proposed by Congress (March 5) to ensure that federal judicial power does not extend to suits between citizens of different states. It is ratified January 8, 1798.

1803 Twelfth amendment is proposed (December 9) to provide for election of President by the Electoral College. It is ratified September 25, 1804.

1865 Thirteenth amendment is ratified (December 2), prohibiting slavery.

1866 Fourteenth amendment is proposed by Congress (June 13) to ensure citizenship rights to former slaves and to prohibit Confederates from holding public office. It is ratified July 28, 1868.

1869 Fifteenth amendment is proposed by Congress (February 26) to ensure that right to vote will not be denied on basis of race, color, or previous condition of servitude. It is ratified March 30, 1870.

1909 Sixteenth amendment is proposed (July 12) to give Congress power to impose income taxes. It is ratified February 25, 1913.

1912 Seventeenth amendment is proposed by Congress (May 16) to institute direct popular election of U.S. senators. It is ratified May 31, 1913.

1917 Eighteenth amendment is proposed by Congress (December 18) to outlaw sale of alcoholic beverages. It is ratified January 29, 1919 and becomes effective in 1920, beginning 13-year. Prohibition.

1919 Nineteenth amendment is proposed by Congress (June) to give women right to vote. It is ratified 1920.

1932 Twentieth amendment is proposed by Congress (March 2) to change terms of office of Congress, President, and Vice President to prevent "lame-duck" sessions of Congress. It is ratified February 6, 1933.

1933 Twenty-first amendment is proposed by Congress to repeal Prohibition (February 20). It is ratified December 5.

1947 Twenty-second amendment is proposed by Congress (March 21) to limit President's service to two terms in office. It is ratified February 26, 1951.

▶

1960 Twenty-third amendment is proposed by Congress to grant rights of voting and representation to District of Columbia. It is ratified in 1961.

1962 Twenty-fourth amendment is proposed by Congress (August 27) to eliminate poll tax. It is ratified January 23, 1964.

1965 Twenty-fifth amendment is proposed by Congress (July 6) to establish system for succession to presidency and for replacement of Vice President. It is ratified February 10, 1967.

1971 Twenty-sixth amendment is approved by Congress (March 23) and ratified (June 30). It lowers legal voting age from 21 to 18.

1972 Equal-rights amendment, specifying equal rights for women, is approved by Congress (March). It is ratified by 35 states but, despite three-year deadline extension in 1979, falls three states short of passage in 1982.

U.S. Presidential Elections

Elections are dated by when the vote was cast. No meaningful popular-vote counts become available until the early 1820s, until which date electors in most states were chosen directly by the state legislatures rather than by popular vote.

1789 George Washington is unanimously elected first President of the United States under newly ratified Constitution. Electoral vote: Washington (unopposed) 69.

1792 George Washington, again unopposed, is reelected to second term with Vice President John Adams. Electoral vote: Washington 132, 3 abstentions.

1796 John Adams (Federalist) is elected President. Electoral vote: Adams 71, Jefferson 68, Thomas Pinckney (Federalist) 59, Aaron Burr (Democratic-Republican) 30, Samuel Adams 15.

1800 Democratic-Republican Thomas Jefferson is elected President after electoral tie forces vote by House of Representatives. Electoral vote: Jefferson 73, Burr (Democratic-Republican) 73, John Adams (Federalist) 65, Charles Pinckney (Federalist) 64, John Jay 1.

1804 Democratic-Republican Thomas Jefferson is reelected to second term as President. Electoral vote: Jefferson 162, Charles Pinckney (Federalist) 14.

1808 Democratic-Republican James Madison is elected President. Electoral vote: Madison 122, Charles Pinckney (Federalist) 47.

1812 Madison is reelected. Electoral vote: Madison 128, De Witt Clinton (Democratic-Republican endorsed by Federalists) 89.

1816 Democratic-Republican James Monroe is elected President in sweeping victory. Electoral vote: Monroe 183, Rufus King (Federalist) 34, 4 abstentions.

1820 James Monroe (unopposed) gains second term. Electoral votes: Monroe 231, 3 abstentions, 1 dissenting vote for John Quincy Adams.

1824 Democratic-Republican John Quincy Adams is elected President. House decides election (February 9, 1825) due to lack of majority winner in electoral votes. Vote: Andrew Jackson 151,271, electoral 99; Adams 113,122, electoral 84; William H. Crawford 40,856, electoral 41; Henry Clay 47,531, electoral 37 (all are Democratic-Republicans).

1828 Democrat Andrew Jackson is elected President. Vote: Jackson 642,553, electoral 178; John Quincy Adams (National Republican) 500,897, electoral 83.

1832 Democrat Andrew Jackson is elected President. Vote: Jackson 701,780, electoral 219; Henry Clay (National Republican) 484,205, electoral 49.

1836 Democrat Martin Van Buren is elected President. Senate decides election of Richard M. Johnson (Democrat) of Kentucky (only Vice President so elected). Vote: Van Buren 764,176, electoral 170; William Henry Harrison (Anti-Masonic) 550,816, electoral 73; Daniel Webster (Whig) 41,201, electoral 14.

1840 Whig William Henry Harrison is elected President. Harrison uses famed campaign slogan "Tippecanoe and Tyler, too." Vote: Harrison

▶

1,275,390, electoral 234; Van Buren (Democrat) 1,128,854, electoral 60.

1844 First "dark-horse" candidate, Democrat James K. Polk, is narrowly elected President. Democrats use slogan "54-40 or Fight." Vote: Polk 1,339,494, electoral 170; Henry Clay (Whig) 1,300,004, electoral 105.

1848 Whig Zachary Taylor is elected President. Vote: Taylor 1,361,393, electoral 163 (for 15 states); Lewis Cass (Democrat) 1,223,460, electoral 127 (for 15 states).

1852 Democrat Franklin Pierce is elected President. Vote: Pierce 1,607,510, electoral 254; Winfield Scott (Whig) 1,386,942, electoral 42.

1856 Democrat James Buchanan is elected President. Vote: Buchanan 1,836,072, electoral 174; John C. Frémont (Republican) 1,342,345, electoral 114; Millard Fillmore (Know-Nothing Party) 873,053, electoral 8.

1860 Republican Abraham Lincoln is elected President. Lincoln's election victory results in start of Civil War. Vote: Lincoln 1,865,908, electoral 180; Stephen A. Douglas (Democrat) 1,380,202, electoral 12; John C. Breckinridge (Southern faction of Democratic Party) 848,019, electoral 72; John Bell (Constitutional Union Party), electoral 3.

1864 Lincoln is reelected. His chances of reelection are in doubt until General Sherman's military victories in Georgia. Vote: Lincoln 2,218,388, electoral 212; George McClellan (Democrat) 1,812,807, electoral 21.

1868 Republican Ulysses S. Grant is elected President. Republicans campaign for radical (harsh) Reconstruction measures in the South; Democrats oppose them. Vote: Grant 3,013,650, electoral 214; Horatio Seymour (Democrat) 2,708,744, electoral 80.

1872 Grant is reelected. This campaign is marked by reaction to radical Reconstruction policies and widespread corruption within Grant's administration. Vote: Grant 3,598,235, electoral 286; Horace Greeley (Democrat/Liberal Republican) 2,834,761. Greeley dies before electoral vote.

1876 Republican Rutherford B. Hayes is eventually declared President by Congress (March 2, 1877) following Election Commission ruling on disputed election results. Vote: Samuel J. Tilden (Democrat) 4,288,546, electoral 184; Hayes 4,034,311, electoral 185. Disputed votes in Florida, Louisiana, South Carolina, and Oregon are all awarded to Hayes.

1880 Republican James A. Garfield is elected President. A dark-horse candidate, he wins popular vote by less than 10,000 ballots. Vote: Garfield 4,449,053, electoral 214; Winfield Scott Hancock (Democrat) 4,442,035, electoral 155.

1884 Democrat Grover Cleveland is elected President. Vote: Cleveland 4,874,621, electoral 219, James G. Blaine (Republican) 4,848,936, electoral 182.

1888 Republican Benjamin Harrison is elected President. Harrison wins by electoral votes, but loses popular vote to Grover Cleveland. Vote: Harrison 5,443,892, electoral 233; Cleveland 5,534,488, electoral 168.

1892 Democrat Grover Cleveland is reelected. Vote: Cleveland 5,551,883, electoral 277; Benjamin Harrison (Republican) 5,179,244, electoral 145.

1896 Republican William McKinley is elected President. William Jennings Bryan delivers his famous "Cross of Gold" speech during this campaign. Vote: McKinley 7,108,480, electoral 271; Bryan (Democrat) 6,511,495, electoral 176.

1900 Republican William McKinley is reelected. Republicans call on government to build Panama Canal during this campaign. Vote: McKinley 7,218,039, electoral 292; William Jennings Bryan 6,358,345, electoral 155.

1904 Republican Theodore Roosevelt is elected to his first full term as President following McKinley's assassination. Vote: Roosevelt 7,626,593, electoral 336; Alton B. Parker (Democrat) 5,082,898, electoral 140.

1908 Republican William H. Taft is elected President. Vote: Taft 7,676,258, electoral 321; William Jennings Bryan (Democrat) 6,406,801, electoral 162.

1912 Democrat Woodrow Wilson is elected President. Theodore Roosevelt's candidacy (opposing incumbent William H. Taft) splits Republican vote. Wilson wins by largest electoral-vote margin to date. Vote: Wilson 6,293,152, electoral 435; Roosevelt ("Bull Moose" or Progressive Party) 4,119,207, electoral 88; Taft 3,486,333, electoral 8; Eugene Debs (Socialist) 900,369.

▶

1916 Wilson is reelected. Vote: Wilson 9,126,300, electoral 277; Charles E. Hughes (Republican) 8,546,789, electoral 254.

1920 Republican Warren G. Harding is elected President. Vote: Harding 16,133,314, electoral 404; James M. Cox (Democrat) 9,140,884, electoral 127; Eugene Debs (Socialist) 913,664.

1924 Republican Calvin Coolidge is elected to his first full term as President. Coolidge wins despite scandal surrounding former President Harding's Republican administration. A woman, Miss Marie C. Brehm, runs as Prohibition Party vice-presidential candidate. Vote: Coolidge 15,717,553, electoral 382; John W. Davis (Democrat) 8,386,169, electoral 136; Robert M. LaFollette (Progressive Party) 4,814,050, electoral 13; Herman P. Faris (Prohibition) 54,833.

1928 Republican Herbert C. Hoover is elected President. Democratic candidate for President, Alfred E. Smith, is first Roman Catholic presidential candidate. Vote: Hoover 21,411,991 electoral 444; Smith 15,000,185, electoral 87.

1932 Democrat Franklin D. Roosevelt is elected President. Campaigning amid worsening Great Depression, Roosevelt pledges to aid "forgotten man" and provide "New Deal" for Americans. Vote: Roosevelt 22,825,016, electoral 472; Hoover 15,758,397, electoral 59; Norman Thomas (Socialist) 883,990.

1936 Roosevelt is reelected by landslide, taking 46 of 48 states. Vote: Roosevelt 27,747,636, electoral 523; Alfred M. Landon (Republican) 16,679,543, electoral 8.

1940 Roosevelt is reelected to third term. Threat of U.S. involvement in World War II overshadows election campaign. Vote: Roosevelt 27,263,448, electoral 449; Wendell L. Willkie (Republican) 22,336,260, electoral 82.

1944 Roosevelt is reelected President for unprecedented fourth term but dies a few months after his election. Vote: Roosevelt 25,611,936, electoral 432; Thomas E. Dewey (Republican) 22,013,372, electoral 99.

1948 Democrat Harry S. Truman is elected to his first full term as President. Southern Democrats opposing strong position on civil rights form Dixiecrats (States' Rights Democrats). Vote: Truman, 24,105,587, electoral 303; Thomas E. Dewey (Republican) 21,970,017, electoral 189; J. Strom Thurmond (Dixiecrats) 1,169,134, electoral 39.

1952 Republican Dwight D. Eisenhower is elected President. Returning hero of war in Europe, General Eisenhower sweeps to easy victory. Vote: Eisenhower 33,936,137, electoral 442; Adlai E. Stevenson (Democrat) 27,314,649, electoral 89.

1956 Eisenhower is reelected President. Vote: Eisenhower 35,585,245, electoral 457; Adlai E. Stevenson (Democrat) 26,030,172, electoral 73.

1960 Democrat John F. Kennedy is elected President by slim margin in popular vote. Television plays important role for first time. Vote: Kennedy 34,221,344, electoral 303; Richard M. Nixon (Republican) 34,106,671, electoral 219; Orval E. Faubus (National States' Rights) 209,314.

1964 Democrat Lyndon B. Johnson is elected to his first full term as President after Kennedy assassination. Johnson campaigns on platform of enlightened social programs and nonaggressive foreign policy. Vote: Johnson 43,126,584, electoral 486; Barry M. Goldwater (Republican) 27,177,838, electoral 52.

1968 Republican Richard M. Nixon is elected President amid turmoil of civil rights unrest, political assassinations, antiwar activism, and protracted, costly war in Vietnam. Vote: Nixon 31,785,148, electoral 301; Hubert H. Humphrey (Democrat) 31,274,503, electoral 191; George C. Wallace (American Independent) 9,901,151, electoral 46.

1972 Nixon is reelected President. His landslide victory is quickly obscured by the unfolding Watergate scandal, which ultimately forces his resignation. Vote: Nixon 47,170,179, electoral 520; George S. McGovern (Democrat) 29,171,791, electoral 17; John G. Schmitz (American) 1,090,673; Benjamin Spock (People's) 78,751.

1976 Democrat Jimmy Carter is elected President. Dark-horse candidate Carter narrowly beats incumbent Gerald R. Ford, then serving out Richard Nixon's term. Vote: Carter 40,830,763, electoral 297; Ford (Republican) 39,147,793, electoral 240; Eugene McCarthy (Independent) 756,691.

1980 Republican Ronald Reagan is elected President in sweeping victory over incumbent Jimmy Carter. Vote: Reagan 43,901,812, electoral 489; Carter (Democrat) 35,483,820, electoral 49; John B. Anderson (Independent) 5,719,722.

▶

1984 Reagan is reelected President, again in landslide victory. Vote: Reagan 54,455,074, electoral 515 (for 49 states); Walter F. Mondale (Democrat) 37,577,137, electoral 13; Lyndon H. LaRouche, Jr. (Independent) 78,807.

1988 Republican George Bush is elected President. Vote: Bush 47,645,225, electoral 425; Michael S. Dukakis (Democrat) 40,797,905, electoral 112.

Composition of U.S. Congresses

The changing fortunes of the two leading U.S. political parties over the years are represented as follows. Total members seated in Senate and House for any given Congress include majority and minority members as well as other seats held by third-party or splinter-group candidates (none before 1831). Vacancies, of course, are not included here.

Senate

1789–91	26 seated: 17 Administration; 9 Opposition
1791–93	29 seated: 16 Federalist; 13 Dem.-Rep.
1793–95	30 seated: 17 Federalist; 13 Dem.-Rep.
1795–97	32 seated: 19 Federalist; 13 Dem.-Rep.
1797–99	32 seated: 20 Federalist; 12 Dem.-Rep.
1799–1801	32 seated: 19 Federalist; 13 Dem.-Rep.
1801–3	31 seated: 18 Dem.-Rep.; 13 Federalist
1803–5	34 seated: 25 Dem.-Rep.; 9 Federalist
1805–7	34 seated: 27 Dem.-Rep.; 7 Federalist
1807–9	34 seated: 28 Dem.-Rep.; 6 Federalist
1809–11	34 seated: 28 Dem.-Rep.; 6 Federalist
1811–13	36 seated: 30 Dem.-Rep.; 6 Federalist
1813–15	36 seated: 27 Dem.-Rep.; 9 Federalist
1815–17	36 seated: 25 Dem.-Rep.; 11 Federalist
1817–19	44 seated: 34 Dem.-Rep.; 10 Federalist
1819–21	42 seated: 35 Dem.-Rep.; 7 Federalist
1821–23	48 seated: 44 Dem.-Rep.; 4 Federalist
1823–25	48 seated: 44 Dem.-Rep.; 4 Federalist
1825–27	46 seated: 26 Administration; 20 Jacksonian
1827–29	48 seated: 28 Jacksonian; 20 Administration
1829–31	48 seated: 26 Democrat; 22 National Republican
1831–33	48 seated: 25 Democrat; 21 National Republican
1833–35	48 seated: 20 Democrat; 20 National Republican
1835–37	52 seated: 27 Democrat; 25 Whig
1837–39	52 seated: 30 Democrat; 18 Whig
1839–41	50 seated: 28 Democrat; 22 Whig
1841–43	52 seated: 28 Whig; 22 Democrat
1843–45	54 seated: 28 Whig; 25 Democrat
1845–47	56 seated: 31 Democrat; 25 Whig
1847–49	58 seated: 36 Democrat; 21 Whig
1849–51	62 seated: 35 Democrat; 25 Whig
1851–53	62 seated: 35 Democrat; 24 Whig
1853–55	62 seated: 38 Democrat; 22 Whig
1855–57	60 seated: 40 Democrat; 15 Republican
1857–59	64 seated: 36 Democrat; 20 Republican
1859–61	66 seated: 36 Democrat; 26 Republican
1861–63	49 seated: 31 Republican; 10 Democrat
1863–65	50 seated: 36 Democrat; 9 Republican
1865–67	52 seated: 42 Unionist; 10 Democrat
1867–69	53 seated: 42 Republican; 11 Democrat
1869–71	67 seated: 56 Republican; 11 Democrat
1871–73	74 seated: 52 Republican; 17 Democrat
1873–75	73 seated: 49 Republican; 19 Democrat
1875–77	76 seated: 45 Republican; 29 Democrat
1877–79	76 seated: 39 Republican; 36 Democrat
1879–81	76 seated: 42 Democrat; 33 Republican
1881–83	75 seated: 37 Republican; 37 Democrat
1883–85	76 seated: 38 Republican; 36 Democrat

▶

1885–87	77 seated: 43 Republican; 34 Democrat
1887–89	76 seated: 39 Republican; 37 Democrat
1889–91	76 seated: 39 Republican; 37 Democrat
1891–93	88 seated: 47 Republican; 39 Democrat
1893–95	85 seated: 44 Democrat; 38 Republican
1895–97	88 seated: 43 Republican; 39 Democrat
1897–99	88 seated: 47 Republican; 34 Democrat
1899–1901	87 seated: 53 Republican; 26 Democrat
1901–3	90 seated: 55 Republican; 31 Democrat
1903–5	90 seated: 57 Republican; 33 Democrat
1905–7	90 seated: 57 Republican; 33 Democrat
1907–9	92 seated: 61 Republican; 31 Democrat
1909–11	93 seated: 61 Republican; 32 Democrat
1911–13	92 seated: 51 Republican; 41 Democrat
1913–15	96 seated: 51 Democrat; 44 Republican
1915–17	96 seated: 56 Democrat; 40 Republican
1917–19	95 seated: 53 Democrat; 42 Republican
1919–21	96 seated: 49 Republican; 47 Democrat
1921–23	96 seated: 59 Republican; 37 Democrat
1923–25	96 seated: 51 Republican; 43 Democrat
1925–27	96 seated: 56 Republican; 39 Democrat
1927–29	96 seated: 49 Republican; 46 Democrat
1929–31	96 seated: 56 Republican; 39 Democrat
1931–33	96 seated: 48 Republican; 47 Democrat
1933–35	96 seated: 60 Democrat; 35 Republican
1935–37	96 seated: 69 Democrat; 25 Republican
1937–39	96 seated: 76 Democrat; 16 Republican
1939–41	96 seated: 69 Democrat; 23 Republican
1941–43	96 seated: 66 Democrat; 28 Republican
1943–45	96 seated: 58 Democrat; 37 Republican
1945–47	95 seated: 56 Democrat; 38 Republican
1947–49	96 seated: 51 Republican; 45 Democrat
1949–51	96 seated: 54 Democrat; 42 Republican
1951–53	96 seated: 49 Democrat; 47 Republican
1953–55	96 seated: 48 Republican; 47 Democrat
1955–57	96 seated: 48 Democrat; 47 Republican
1957–59	96 seated: 49 Democrat; 47 Republican
1959–61	98 seated: 64 Democrat; 34 Republican
1961–63	100 seated: 65 Democrat; 35 Republican
1963–65	100 seated: 67 Democrat; 33 Republican
1965–67	100 seated: 68 Democrat; 32 Republican
1967–69	100 seated: 64 Democrat; 36 Republican
1969–71	100 seated: 57 Democrat; 43 Republican
1971–73	100 seated: 54 Democrat; 44 Republican
1973–75	100 seated: 56 Democrat; 44 Republican
1975–77	99 seated: 60 Democrat; 37 Republican
1977–79	100 seated: 61 Democrat; 38 Republican
1979–81	100 seated: 58 Democrat; 41 Republican
1981–83	100 seated: 53 Republican; 46 Democrat
1983–85	100 seated: 54 Republican; 46 Democrat
1985–87	100 seated: 53 Republican; 47 Democrat
1987–89	100 seated: 54 Democrat; 46 Republican
1989–91	100 seated: 55 Democrat; 45 Republican

House

1789–91	64 seated: 38 Administration; 26 Opposition
1791–93	70 seated: 37 Federalist; 33 Dem.-Rep.
1793–95	105 seated: 57 Dem.-Rep.; 48 Federalist
1795–97	106 seated: 54 Federalist; 52 Dem.-Rep.
1797–99	106 seated: 58 Federalist; 48 Dem.-Rep.
1799–1801	106 seated: 64 Federalist; 42 Dem.-Rep.
1801–3	105 seated: 69 Dem.-Rep.; 36 Federalist
1803–5	141 seated: 102 Dem.-Rep.; 39 Federalist
1805–7	141 seated: 116 Dem.-Rep.; 25 Federalist
1807–9	142 seated: 118 Dem.-Rep.; 24 Federalist
1809–11	142 seated: 94 Dem.-Rep.; 48 Federalist
1811–13	144 seated: 108 Dem.-Rep.; 36 Federalist
1813–15	180 seated: 112 Dem.-Rep.; 68 Federalist
1815–17	182 seated: 117 Dem.-Rep.; 65 Federalist
1817–19	183 seated: 141 Dem.-Rep.; 42 Federalist
1819–21	183 seated: 156 Dem.-Rep.; 27 Federalist
1821–23	183 seated: 158 Dem.-Rep.; 25 Federalist

▶

1823–25 213 seated: 187 Dem.-Rep.; 26 Federalist

1825–27 202 seated: 105 Administration; 97 Jacksonian

1827–29 213 seated: 119 Jacksonian; 94 Administration

1829–31 213 seated: 139 Democrat; 74 National Republican

1831–33 213 seated: 141 Democrat; 58 National Republican

1833–35 260 seated: 147 Democrat; 53 Anti-Masonic

1835–37 243 seated: 145 Democrat; 98 Whig

1837–39 239 seated: 108 Democrat; 107 Whig

1839–41 242 seated: 124 Democrat; 118 Whig

1841–43 241 seated: 133 Whig; 102 Democratic

1843–45 222 seated: 142 Democrat; 79 Whig

1845–47 226 seated: 143 Democrat; 77 Whig

1847–49 227 seated: 115 Whig; 108 Democrat

1849–51 230 seated: 112 Democrat; 109 Whig

1851–53 233 seated: 140 Democrat; 88 Whig

1853–55 234 seated: 159 Democrat; 71 Whig

1855–57 234 seated: 108 Republican; 83 Democrat

1857–59 236 seated: 118 Democrat; 92 Republican

1859–61 237 seated: 114 Republican; 92 Democrat

1861–63 178 seated: 105 Republican; 43 Democrat

1863–65 186 seated: 102 Republican; 75 Democrat

1865–67 191 seated: 149 Unionist; 42 Democrat

1867–69 192 seated: 143 Republican; 49 Democrat

1869–71 212 seated: 149 Republican; 63 Democrat

1871–73 243 seated: 134 Democrat; 104 Republican

1873–75 300 seated: 194 Republican; 92 Democrat

1875–77 292 seated: 169 Democrat; 109 Republican

1877–79 293 seated: 153 Democrat; 140 Republican

1879–81 293 seated: 149 Democrat; 130 Republican

1881–83 293 seated: 147 Republican; 135 Democrat

1883–85 325 seated: 197 Democrat; 118 Republican

1885–87 325 seated: 183 Democrat; 140 Republican

1887–89 325 seated: 169 Democrat; 152 Republican

1889–91 325 seated: 166 Republican; 159 Democrat

1891–93 332 seated: 235 Democrat; 88 Republican

1893–95 356 seated: 218 Democrat; 127 Republican

1895–97 356 seated: 244 Republican; 105 Democrat

1897–99 357 seated: 204 Republican; 113 Democrat

1899–1901 357 seated: 185 Republican; 163 Democrat

1901–3 357 seated: 197 Republican; 151 Democrat

1903–5 386 seated: 208 Republican; 178 Democrat

1905–7 386 seated: 250 Republican; 136 Democrat

1907–9 386 seated: 222 Republican; 164 Democrat

1909–11 391 seated: 219 Republican; 172 Democrat

1911–13 390 seated: 228 Democrat; 161 Republican

1913–15 435 seated: 291 Democrat; 127 Republican

1915–17 435 seated: 230 Democrat; 196 Republican

1917–19 432 seated: 216 Democrat; 210 Republican

1919–21 433 seated: 240 Republican; 190 Democrat

1921–23 433 seated: 301 Republican; 131 Democrat

1923–25 435 seated: 225 Republican; 205 Democrat

▶

1925–27 434 seated: 247 Republican; 183 Democrat

1927–29 435 seated: 237 Republican; 195 Democrat

1929–31 435 seated: 267 Republican; 167 Democrat

1931–33 435 seated: 220 Democrat; 214 Republican

1933–35 432 seated: 310 Democrat; 117 Republican

1935–37 432 seated: 319 Democrat; 103 Republican

1937–39 433 seated: 331 Democrat; 89 Republican

1939–41 429 seated: 261 Democrat; 164 Republican

1941–43 435 seated: 268 Democrat; 162 Republican

1943–45 430 seated: 218 Democrat; 208 Republican

1945–47 434 seated: 242 Democrat; 190 Republican

1947–49 434 seated: 245 Republican; 188 Democrat

1949–51 435 seated: 263 Democrat; 171 Republican

1951–53 434 seated: 234 Democrat; 199 Republican

1953–55 433 seated: 221 Republican; 211 Democrat

1955–57 435 seated: 232 Democrat; 203 Republican

1957–59 433 seated: 233 Democrat; 200 Republican

1959–61 436 seated: 283 Democrat; 153 Republican

1961–63 437 seated: 263 Democrat; 174 Republican

1963–65 435 seated: 258 Democrat; 177 Republican

1965–67 435 seated: 295 Democrat; 140 Republican

1967–69 434 seated: 247 Democrat; 187 Republican

1969–71 435 seated: 243 Democrat; 192 Republican

1971–73 434 seated: 254 Democrat; 180 Republican

1973–75 432 seated: 239 Democrat; 192 Republican

1975–77 435 seated: 291 Democrat; 144 Republican

1977–79 435 seated: 292 Democrat; 143 Republican

1979–81 433 seated: 276 Democrat; 157 Republican

1981–83 435 seated: 243 Democrat; 192 Republican

1983–85 434 seated: 269 Democrat; 165 Republican

1985–87 434 seated: 252 Democrat; 182 Republican

1987–89 438 seated: 259 Democrat; 179 Republican

1989–91 438 seated: 263 Democrat; 175 Republican

The United Nations

1941 U.S. President Franklin D. Roosevelt (1882–1945) coins term *United Nations* to describe nations allied against Axis Powers in World War II.

1942 First official use is made of term when 26 nations sign Declaration of the United Nations, establishing Allied war aims.

1945 UN Charter is drawn up with 51 member nations at San Francisco Conference (June 26). UN organization includes General Assembly, Security Council, Economic and Social Council, Secretariat, International Court of Justice, and Trusteeship Council.

▶

Nobel Peace Prize Winners

1901 Jean H. Dunant (1828–1910, Switzerland) and Frédéric Passy (1822–1912, France)

1902 Élie Ducommun (1833–1906, Switzerland) and Charles A. Gobat (1843–1914, Switzerland)

1903 Sir William R. Cremer (1838–1908, Great Britain)

1904 Institute of International Law

1905 Baroness Bertha von Suttner (1843–1914, Austria)

1906 Theodore Roosevelt (1858–1919, United States)

1907 Ernesto T. Moneta (1833–1918, Italy) and Louis Renault (1843–1918, France)

1908 Klas P. Arnoldson (1844–1916, Sweden) and Fredrik Bajer (1837–1922, Denmark)

1909 Auguste M. F. Beernaert (1829–1912, Belgium) and Baron d'Estournelles de Constant (1852–1924, France)

1910 Permanent International Peace Bureau

1911 Tobias M. C. Asser (1838–1913, Netherlands) and Alfred Fried (1864–1921, Austria)

1912 Elihu Root (1845–1937, United States)

1913 Henri Lafontaine (1854–1943, Belgium)

1914 No prize awarded

1915 No prize awarded

1916 No prize awarded

1917 International Red Cross

1918 No prize awarded

1919 Woodrow Wilson (1856–1924, United States)

1920 Léon V. A. Bourgeois (1851–1925, France)

1921 Karl H. Branting (1860–1925, Sweden) and Christian L. Lange (1869–1938, Norway)

1922 Fridtjof Nansen (1861–1930, Norway)

1923 No prize awarded

1924 No prize awarded

1925 Sir Austen Chamberlain (1863–1937, Great Britain) and Charles G. Dawes (1865–1951, United States)

1926 Aristide Briand (1862–1932, France) and Gustav Streseman (1878–1929, Germany)

1927 Ferdinand E. Buisson (1841–1932, France) and Ludwig Quidde (1858–1941, Germany)

1928 No prize awarded

1929 Frank B. Kellogg (1856–1937, United States)

1930 Nathan Söderblom (1866–1931, Sweden)

1931 Jane Addams (1860–1935, United States) and Nicholas M. Butler (1862–1947, United States)

1932 No prize awarded

1933 Sir Norman Angell (1873–1967, Great Britain)

1934 Arthur Henderson (1863–1935, Great Britain)

1935 Carl von Ossietzky (1888–1938, Germany)

1936 Carlos Saavedra Lamas (1880–1959, Argentina)

1937 Viscount Cecil of Chelwood (1864–1958, Great Britain)

1938 Nansen International Office for Refugees

1939 No prize awarded

1940 No prize awarded

1941 No prize awarded

1942 No prize awarded

1943 No prize awarded

1944 International Red Cross

1945 Cordell Hull (1871–1955, United States)

1946 Emily G. Balch (1867–1961, United States) and John R. Mott (1865–1955, United States)

1947 Friends' Service Council (Great Britain) and American Friends' Service Committee (United States)

1948 No prize awarded

1949 Lord John Boyd Orr (1880–1971, Great Britain)

1950 Ralph J. Bunche (1904–71, United States)

1951 Léon Jouhaux (1879–1954, France)

1952 Albert Schweitzer (1875–1965, France)

1953 George C. Marshall (1880–1959, United States)

1954 Office of UN High Commissioner for Refugees

1955 No prize awarded

1956 No prize awarded

1957 Lester B. Pearson (1897–1972, Canada)

1958 Dominique Georges Pire (1910–69, Belgium)

1959 Philip J. Noel-Baker (1889–1982, Great Britain)

1960 Albert J. Luthuli (1898–1967, South Africa)

1961 Dag Hammarskjöld (1905–61, Sweden)

1962 Linus C. Pauling (b. 1901, United States)

1963 International Red Cross and League of Red Cross Societies

1964 Martin Luther King, Jr. (1929–68, United States)

1965 UN Children's Fund (UNICEF)

1966 No prize awarded

Nobel Peace Prize Winners

1967	No prize awarded
1968	René Cassin (1887–1976, France)
1969	International Labor Organization
1970	Norman E. Borlaug (b. 1914, United States)
1971	Willy Brandt (b. 1913, West Germany)
1972	No prize awarded
1973	Henry Kissinger (b. 1923, United States) and Le Duc Tho (b. 1911, North Vietnam; declined award)
1974	Eisaku Satō (1901–75, Japan) and Sean MacBride (1904–90, Ireland)
1975	Andrei D. Sakharov (1921–90, USSR)
1976	Mairead Corrigan (b. 1944, Northern Ireland) and Betty Williams (b. 1944, Northern Ireland)
1977	Amnesty International
1978	Anwar Sadat (1918–81, Egypt) and Menachem Begin (b. 1913, Israel)
1979	Mother Teresa (b. 1910, India)
1980	Adolfo Pérez Esquivel (b. 1931, Argentina)
1981	Office of UN High Commissioner for Refugees
1982	Alva Myrdal (1902–86, Sweden) and Alfonso Garcia Robles (b. 1911, Mexico)
1983	Lech Walesa (b. 1947, Poland)
1984	Bishop Desmond Tutu (b. 1931, South Africa)
1985	International Physicians for Prevention of Nuclear War (United States)
1986	Elie Wiesel (b. 1928, United States)
1987	Oscar Arias Sánchez (b. 1941, Costa Rica)
1988	UN peace-keeping forces
1989	Tibetan Dalai Lama (14th incarnation 1935–).

1945 UN charter takes effect (October 24). Day is subsequently celebrated as United Nations Day.

1945 World Bank (International Bank for Reconstruction and Development) becomes an agency of United Nations to promote international loans and investment for needy countries.

1946 First official UN act creates Atomic Energy Commission to regulate and control atomic energy.

1946 UNESCO, UN Educational, Scientific, and Cultural Organization, is created to encourage cultural and educational exchanges among nations.

1946–52 Norwegian diplomat Trygve Lie (1896–1968) serves as first secretary-general of United Nations.

1947 Commission on Human Rights, chaired by former first lady Eleanor Roosevelt (1884–1962), drafts Universal Declaration of Human Rights, which is adopted in 1948.

1947 UN flag is adopted.

1948 World Health Organization becomes UN agency to set health standards and promote medical research.

1949 Truce supervision organization from United Nations monitors truce ending first Arab–Israeli War.

1950 India introduces General Assembly resolution to admit China's Communist regime. Resolution is defeated but is reintroduced every year until 1971.

1950–53 Military intervention in Korea (Korean War). UN forces (primarily U.S.) turn back unprovoked North Korean invasion of South Korea.

1951 United Nations moves to permanent headquarters in New York City.

1952 South African policy for apartheid is put on General Assembly agenda for first time. Annual UN resolutions call for change in South Africa's racial policies.

1952 Secretary-General Lie resigns. Stiff Soviet opposition to him, because of UN intervention in Korea, renders his position untenable.

1953–61 Swedish statesman Dag Hammarskjöld (1905–61) serves as second UN secretary-general.

▶

1954 Office of UN High Commissioner for Refugees receives Nobel Peace Prize for refugee work; it also receives prize in 1981.

1956 First UN peace-keeping force is created, to supervise cease-fire and organize troop withdrawal during Suez crisis.

1960 Soviet leader Nikita Khrushchev (1894–1971) loses considerable prestige by his unruly behavior, including banging his shoe on table, at UN General Assembly meeting.

1961 Secretary-General Hammarskjöld dies in suspicious plane crash while on peace-keeping mission related to troubles in war-torn Congo (now Zaire).

1961 Soviet Union introduces unsuccessful resolution to remove Nationalist China from and admit Communist China to United Nations.

1961 UN peacekeeping operation begins in Congo (now Zaire). Troops are sent following Congo's declaration of independence from Belgium.

1962 United States loans $100 million to United Nations to help it resolve financial crisis.

1962–71 Burmese diplomat U Thant (1909–74) serves as UN secretary-general. He is reelected in 1966.

1963 UN Partial Test-Ban Treaty is signed by the Soviet Union, United Kingdom, and United States. It prohibits nuclear testing in atmosphere, in outer space, and under water.

1964 UN peace-keeping troops are sent to Cyprus as conflict between Greek and Turkish Cypriots flares. United Nations fails to prevent partitioning of Cyprus following 1974 Turkish invasion.

1965 United Nations ratifies amendment enlarging Security Council from 11 to 15 members.

1966 Outer Space Treaty is signed by over 60 nations. It controls military use of outer space.

1967 Fund for Population Activities is formed by United Nations to promote family planning.

1971 United Nations admits Communist China as member after many years of debate. Bowing to Communist China's prerequisite for joining, it expels longtime member Nationalist China.

1972 Disaster Relief Office is created following devastating earthquakes and tidal waves of this year.

1972–82 Austrian diplomat Kurt Waldheim (b. 1918) serves as UN secretary-general.

1973 UN Environmental Program (UNEP) is established in Kenya.

1974 UN peace-keeping force returns to Mideast following end of fourth Arab–Israeli War.

1974 Third World nations gain UN support for improvements in trade and industry for underdeveloped countries.

1977 United Nations creates International Fund for Agricultural Development (IFAD).

1979 Special UN committee drafts convention against taking hostages.

1982 UN membership reaches 157 nations.

1982– Peruvian diplomat Javier Pérez de Cuellar (b. 1920) serves as UN secretary-general.

1983 United States withdraws from UNESCO, citing that agency's political bias and mishandling of finances.

1985 Antiterrorism resolution is issued by General Assembly.

1988 PLO leader Yasir Arafat (b. 1929) addresses UN General Assembly in Geneva after United States denies him visa to speak at United Nations in New York City. He proposes peace plan to end conflict between PLO and Israel.

Abolition of Slavery

1671 Quakers in Great Britain and America object to enslavement of blacks.

1700 Massachusetts jurist Samuel Sewall (1652–1730) writes *The Selling of Joseph*, antislavery book.

1726 Virginia colonists attempt to outlaw slave trade, but move is blocked by England.

1775 Society for Relief of Free Negroes Unlawfully Held in Bondage is formed in Philadelphia. Most members are Quaker.

1777 Northern states begin abolishing slavery. By 1804, all states north of Maryland have done so.

c. 1783 English Quakers seek to prohibit importation of African slaves to British colonies and America. They form Abolition Society (1787).

1787 Delegates to U.S. Constitutional Convention discuss but do not include provision against slavery, primarily because of South's influence and its economic dependence on slavery.

1807 Trading of slaves is outlawed by Great Britain and United States.

c. 1815 U.S. abolitionist Benjamin Lundy (1789–1839) starts working for abolition. He succeeds in persuading some slaveholders to free their slaves.

1816 American Colonization Society is founded to return slaves freed in United States to Africa.

1820 First regular journal concerned with abolition is founded in United States.

1820–21 Missouri Compromise is passed by Congress. It provides for admission of Missouri as slave state and Maine as free state to maintain balance between free and slave states.

1823 Antislavery Society forms in England.

1827 About 140 antislavery groups exist in United States by this time.

c. 1829 U.S. abolitionist William Lloyd Garrison (1805–79) founds *Liberator*.

1831 Slave Nat Turner (1800–31) leads only Negro rebellion against slavery in U.S. history. Turner's small band of followers kill more than 50 whites before revolt is quashed.

c. 1831 Term "underground railroad" is in use for network to help escaped slaves get to safety in North.

1833 William Lloyd Garrison founds American Antislavery Society.

1833 British Parliament passes law to free all slaves in British colonies following five- to seven-year apprenticeship to owner.

1836 Gag rule is passed in Congress. It limits consideration of petitions by Congress as means of bypassing petitions by abolitionists.

1837 Abolitionist Rev. Elijah P. Lovejoy (1802–37) is killed when proslavery opponents wreck his printing office in Illinois.

1839 British and Foreign Antislavery Society forms to end slave smuggling.

1844 Underground railroad founder Rev. Charles Torrey (1813–46) is apprehended. He later dies in Maryland prison.

1845 Texas, slave territory, is annexed by United States despite northern fears that move will strengthen South's political stance on slavery.

1848 France abolishes slavery in West Indian colonies.

1848 Quaker Thomas Garrett (1789–1871) is apprehended after assisting some 2,700 slaves escape via underground railroad.

1850 Estimated 50,000 slaves per year are being smuggled from Africa to Cuba, Brazil, and other countries.

▶

1850 Compromise of 1850 is hammered out by U.S. Congress, preventing for time secession of proslavery states. It includes provision for determining free- or slave-state status for new states by test of "popular sovereignty." It also includes passage of Fugitive Slave Act (1850) for return of runaway slaves.

1852 Antislavery novel *Uncle Tom's Cabin* is written by Harriet Beecher Stowe (1811–96).

1857 U.S. Supreme Court rules slavery is legal in U.S. territories by its decision in the Dred Scott case, increasing opposition to slavery in North.

1859 Harper's Ferry Raid (October 16–18). Abolitionist John Brown (1800–59) and band of followers capture U.S. arsenal at Harper's Ferry, West Virginia in unsuccessful attempt to start slave uprising.

1863 During U.S. Civil War, Emancipation Proclamation is issued by President Abraham Lincoln (1809–65). It frees slaves in territory held by Confederacy.

1865 Thirteenth amendment to U.S. Constitution is ratified. It officially abolishes slavery in United States.

1885 International conference in Berlin calls for end of slave trade that supplies African slaves to Middle Eastern countries.

1890 Middle Eastern countries sign General Act of Brussels, which seeks to halt transportation and sale of slaves.

1919 Saint-Germain-en-Laye Convention continues international effort to halt slavery.

1920 League of Nations commission is named to study various forms of slavery and servitude worldwide.

1942 Ethiopia abolishes slavery at behest of League of Nations; Liberia later follows suit.

1955 UN report reveals that slavery persists in some countries in Arabia, Southeast Asia, and South America.

Civil Rights in the United States

1941 President Franklin D. Roosevelt (1882–1945) outlaws discrimination in defense-industry employment by executive order.

1942 Congress of Racial Equality (CORE) forms to work for black equality.

1943 CORE stages sit-in in Chicago to protest segregation.

1943 Race riot in Detroit, Michigan kills nine whites and 25 blacks.

1946 President Harry S. Truman (1884–1972) establishes Committee on Civil Rights. Committee report suggests 27 civil-rights acts to remedy discrimination. None is enacted.

1947 CORE stages "Journey of Reconciliation," first freedom ride to challenge segregation on interstate transit.

1948 Armed forces are desegregated by President Truman's executive order.

1954 *Brown v. Board of Education of Topeka, Kansas.* U.S. Supreme Court holds that long-standing policy of "separate but equal" public-school facilities for blacks is unconstitutional.

1955 Montgomery, Alabama black resident Rosa Parks (b. 1913) is arrested for refusing to move to the black section of a bus in Montgomery, Alabama. Blacks there stage one-day bus boycott to protest her arrest.

1957 Montgomery Improvement Association forms, led by Martin Luther King, Jr. (1929–68). Group organizes bus boycott in Montgomery, Alabama that forces bus company to desegregate its facilities.

▶

1957 Congress passes Civil Rights Act of 1957, creating Commission on Civil Rights to study racial conditions in United States.

1957 Southern Christian Leadership Conference (SCLC) forms. Led by Martin Luther King, Jr., group is dedicated to nonviolent protest to draw attention to discrimination.

1960 Student Nonviolent Coordinating Committee (SNCC) forms during lunch-counter sit-in in Greensboro, North Carolina. It advocates militant direct action instead of nonviolence.

1960 Congress passes Civil Rights Act of 1960, addressing voting-registration practices. As it does not allow for enforcement of act, it is essentially ineffective.

1962 Council of Federated Organizations forms, to register blacks to vote.

1962 Federal troops are sent to University of Mississippi to force school to enroll black James Meredith (b. 1933). Troops remain at university until Meredith is graduated in 1963.

1963 March for racial equality by 200,000 occurs in Detroit.

1963 Protests in Birmingham, Alabama in April and May gain nationwide attention and raise support for civil-rights movement as police resort to dogs and fire hoses to break up demonstrations.

1963 Medgar Evers (1925–63), field secretary for NAACP, is shot and killed in June.

1963 Total of 250,000 participate in March on Washington in August.

1964 Congress passes Civil Rights Act of 1964, banning all discrimination in public facilities, creating the Equal Employment Opportunity Commission and guaranteeing equal voting rights to blacks.

1964 Twenty-fourth amendment to Constitution, outlawing poll taxes, is ratified.

1964 Martin Luther King, Jr. is awarded Nobel Peace Prize.

1964 Civil-rights workers Michael Schwerner, Andrew Goodman, and James Chaney are murdered during voter-registration drive in Mississippi.

1965 March supporting black voting rights is stopped by violence in Selma, Alabama.

1965 Congress passes Voting Rights Act in August, outlawing literacy and other voter-registration tests. By 1968 more than 50 percent of eligible blacks in South are registered voters.

1965 Riots in Watts section of Los Angeles in August cause 34 deaths and more than $35 million in property damage.

1966 James Meredith leads march in Mississippi to support blacks' voting rights. He survives unsuccessful assassination attempt.

1966 Stokely Carmichael (b. 1941) becomes leader of SNCC. He advocates use of violence and promotes Black Power movement.

1966 Chicago race riots take 4,200 National Guardsmen and 533 arrests to quell.

1966 Radical, revolutionary Black Panther Party is organized in Oakland, California by Huey Newton (1942–90) and Bobby Seale (b. 1937).

1966 Floyd B. McKissick (b. 1922) becomes national chairman of CORE. He aligns it with Carmichael's Black Power movement.

1967 H. Rap Brown (b. 1943) succeeds Carmichael at SNCC.

1967 Riots in Newark, Detroit, and other cities cause 100 deaths and more than 2,000 injuries. In first nine months of 1967, total of 164 riots break out.

1967 President Lyndon B. Johnson (1908–73) appoints National Advisory Commission on Civil Disorders, which finds racism is primary cause of recent riots.

1968 Coalition of SNCC, Student Organization for Black Unity (SOBU), and Black Panthers disbands, marking end of organized Black Power movement.

1968 Martin Luther King, Jr. is assassinated in Memphis, Tennessee (April 4). Mass riots breaks out in more than 40 cities. Rev. Ralph D. Abernathy (b. 1926) succeeds King as president of SCLC.

1968 Congress passes Civil Rights Act of 1968, prohibiting housing discrimination and harassment of civil-rights workers.

▶

1968 Congress passes legislation authorizing building or rehabilitation of 1.7 million dwelling units.

1968 July gun battle between black snipers and police in Cleveland, Ohio lasts for five days and does more than $5 million in property damage.

1969 Congressional Black Caucus forms.

1970 Black Panther party leaders Fred Hampton and Mark Clark are killed in Chicago police raid. Panthers by this time have widespread reputation for violent confrontation with police.

1972 Congress passes Equal Employment Opportunity Act, allowing for preferential hiring and promotion of women and minorities.

1979 In *United States v. Weber*, U.S. Supreme Court ruling supports affirmative action.

1986 U.S. Supreme Court upholds affirmative-action hiring quotas, which promote hiring of women and minorities, as remedy for past discrimination.

Women's Rights

1775 American Thomas Paine (1737–1809) writes article for *Pennsylvania Magazine* proposing women's rights.

1777 Abigail Adams (1744–1818), American social reformer and wife of President John Adams (1735–1826), writes that women "will not hold ourselves bound by any laws in which we have no voice."

1792 English feminist Mary Wollstonecraft (1759–97) writes first major feminist tract, *A Vindication of the Rights of Women.*

1833 Oberlin is first U.S. coeducational college.

1840 American women attending World Anti-Slavery Convention are forbidden to participate and must sit in balcony behind curtain.

1841 First U.S. college degree awarded to a woman is presented.

1848 First women's-rights convention is held at Seneca Falls, New York (July 19–20) by American feminists Lucretia Mott (1793–1880) and Elizabeth Cady Stanton (1815–1902).

1848 New York passes legislation to allow married women to own real estate.

1848 Queen's College for Women opens in Great Britain.

1849 Female doctors are allowed to practice in United States.

1850 First national women's-rights convention is held in Worcester, Massachusetts.

1850 Women are gaining acceptance as teachers in United States by this date.

1851 American Amelia Bloomer (1818–94), editor of women's-rights publications, adopts trousers that come to be known as bloomers.

c. 1851 American feminist Susan B. Anthony (1820–1906) becomes active in women's-rights movement. She is later key figure in fight for voting rights for U.S. women.

1867 British economist John Stuart Mill (1806–73) submits first national suffrage bill to Parliament. It does not pass.

1869 National Woman Suffrage Association is founded by Ms. Stanton and Ms. Anthony.

1869 Wyoming Territory becomes first modern political entity to grant vote to women. It becomes first U.S. state to do so when it is admitted to Union (1890).

1869 Female lawyers are licensed in United States.

1870 Married British women who are employed are allowed to spend their own earnings.

1870 British government begins employing women for clerical work.

▶

1875 British Royal College of Surgeons admits women.

1877 Female doctors are allowed to practice in England.

1878 U.S. constitutional amendment to grant full suffrage to women is introduced in Congress. (It is introduced every year until its passage in 1920.)

1879 French Socialist Congress is first political party to demand equal political and economics rights for women. (Rights are not granted.)

1881 French female factory workers are allowed to open bank accounts without consent of their husbands.

1884 Oxford University admits women (but not as full-time students until 1920).

1893 New Zealand is first nation to give vote to women.

1906 Finland is first European country to grant women's suffrage.

1912 National American Women's Suffrage Association organizes congressional committee to campaign for federal amendment to give equal rights to women.

1917 Soviet government of Russia gives vote to women and declares equality between sexes.

1919 Nineteenth Amendment to U.S. Constitution, extending vote to women, passes. (It is ratified in 1920.)

1920 League of Women Voters is founded to educate women in use of their suffrage. Men are admitted after 1974.

1920 All U.S. states allow women to practice law by this date.

1923 American nurse Margaret Sanger (1883–1966) organizes first U.S. birth-control conference; she organizes first international birth-control conference in 1925.

c. 1925 Turkey bans custom requiring women to veil their faces.

1928 Great Britain gives vote to women after 40 years of campaigning.

1944 French women gain right to vote.

1960 Canadian bill of rights prohibits sexual discrimination.

1960s Use of the birth-control pill by women becomes widespread. It contributes to major changes in life-style and outlook of women in 1960s and 1970s.

1963 U.S. author Betty Friedan (b. 1921) publishes *The Feminine Mystique*, seminal work in contemporary women's movement.

1964 U.S. Civil Rights Act of 1964, prohibiting job discrimination on basis of sex, race, religion, or national origin, is passed.

1966 National Organization for Women (NOW) is founded by Friedan. In subsequent years women's liberation movement is dominated by radical feminist agenda, including attempts to eliminate all distinctions between sexes (unisex).

c. 1970 More conservative element of feminist movement begins to develop. It aims at securing legislative and institutional reforms that explicitly benefit women.

1972 Equal Rights Amendment prohibiting sex discrimination passes U.S. Congress. It fails to win ratification by 1982 deadline.

1973 U.S. Supreme Court legalizes abortion, key issue in women's movement.

1975 U.S. Equal Credit Opportunity Act prohibits discrimination against women in granting loans or credit.

1981 Exclusion of women from draft in United States is upheld by U.S. Supreme Court. It is considered setback by women's movement.

1986 U.S. Supreme Court upholds affirmative-action hiring quotas, which promote hiring of women and minorities, as remedy for past discrimination.

1989 Legal status of abortion is challenged by Pro Life groups following reinterpretation of *Roe v. Wade* decision.

Major U.S. Supreme Court Decisions

1793 *Chisholm* v. *Georgia.* Court upholds federal court jurisdiction in cases involving state and citizen of another state, leading to passage of 11th amendment.

1803 *Marbury* v. *Madison.* Court finds act of Congress (Judiciary Act of 1801, later repealed by Congress) unconstitutional in first decision voiding an act of Congress.

1810 *Fletcher* v. *Peck.* Court finds state law unconstitutional for first time.

1816 *Martin* v. *Hunter's Lessee.* Court upholds right of U.S. Supreme Court to hear appeals of state court decisions.

1819 *Dartmouth College* v. *Woodward.* Court limits states' right to abridge property rights.

1819 *McCulloch* v. *Maryland.* Court upholds federal government's right to establish federal bank.

1833 *Barron* v. *Baltimore.* Court decides only federal government, not state governments, must respect protections of Bill of Rights.

1857 *Dred Scott* v. *Sandford.* Court rules against Scott, Negro slave suing for his freedom after being taken to state where slavery is outlawed. Court also strikes down Missouri Compromise (regulating slavery in territories), ruling congressional act unconstitutional.

1866 *Ex Parte Milligan.* Court limits power of military commissions to try civilians to immediate area of theater of war.

1871 *Legal Tender Case* (Second). Reversing earlier ruling, Court decides Legal Tender acts (of 1862 and 1863) are legal exercise of federal government's powers to meet emergencies.

1895 *United States* v. *E. C. Knight Company.* Court sharply limits application of Sherman Antitrust Act (1890), thereby making federal control over monopolies difficult.

1895 *Pollock* v. *Farmers' Loan and Trust Company.* Court voids law establishing federal income tax, resulting in passage of the 16th amendment, authorizing federal income tax.

1896 *Plessy* v. *Ferguson.* Court upholds separate-but-equal facilities for Negroes.

1898 *Holden* v. *Hardy.* Court affirms state's right to regulate labor conditions.

1903 *Champion* v. *Ames.* Court approves federal powers to prohibit as well as regulate commerce, thereby establishing so-called federal police power.

1904 *Northern Securities Company* v. *United States.* Court revives Sherman Antitrust by finding in favor of federal government. In 1905 Court upholds government prosecution of beef trust in *Swift & Co.* v. *United States.*

1908 *Adair* v. *United States.* Court upholds 1898 law forbidding "yellow dog" contracts for railroad workers.

1908 *Danbury Hatters Case (Loewe* v. *Lawler).* Court rules against secondary boycott as restraining trade under Sherman Antitrust Act.

1919 *Schenck* v. *United States.* Court upholds restraint of free speech in wartime by Espionage Act. It applies "clear and present danger" test.

1923 *Massachusetts* v. *Mellon.* Court gives implicit approval to device of federal grants-in-aid to states.

1935 *Sick Chicken Case (Schechter Poultry* v. *United States).* Court voids National Industrial Recovery Act (1933) as unconstitutional. This is one of several rulings in 1930s striking down elements of Roosevelt's New Deal legislation.

1937 *Steward Machine Company* v. *Davis* and *Helvering* v. *Davis.* Court upholds laws establishing Social Security system.

1947 *Friedman* v. *Schwellenback.* Court refuses to review case involving dismissal of federal civil servant on disloyalty charges.

1951 *Dennis et al.* v. *United States.* Court rules in favor of 1946 Smith Act, law against advocating overthrow of government by force.

1954 *Brown* v. *Board of Education of Topeka, Kansas.* Court overturns long-standing practice of

▶

U.S. Supreme Court Justices

1789–95	John Jay (1745–1829), first chief justice
1789–91	John Rutledge (1739–1800)
1789–1810	William Cushing (1732–1810)
1789–98	James Wilson (1742–98)
1789–96	John Blair (1732–1800)
1790–99	James Iredell (1751–99)
1791–93	Thomas Johnson (1732–1819)
1793–1806	William Paterson (1745–1806)
1795	John Rutledge (1739–1800), chief justice
1796–1811	Samuel Chase (1741–1811)
1796–99	Oliver Ellsworth (1745–1807), chief justice
1798–1829	Bushrod Washington (1762–1829)
1799–1804	Alfred Moore (1755–1810)
1801–35	John Marshall (1755–1835), chief justice
1804–34	William Johnson (1771–1834)
1806–23	Henry B. Livingston (1757–1823)
1807–26	Thomas Todd (1765–1826)
1811–45	Joseph Story (1779–1845)
1811–36	Gabriel Duval (1752–1844)
1823–43	Smith Thompson (1768–1843)
1826–28	Robert Trimble (1777–1828)
1829–61	John McLean (1785–1861)
1830–44	Henry Baldwin (1780–1844)
1835–67	James M. Wayne (1790?–1867)
1836–41	Philip P. Barbour (1783–1841)
1836–64	Roger B. Taney (1777–1864), chief justice
1837–52	John McKinley (1780–1852)
1837–65	John Catron (1786?–1865)
1841–60	Peter V. Daniel (1784–1860)
1845–51	Levi Woodbury (1789–1851)
1845–72	Samuel Nelson (1792–1873)
1846–70	Robert C. Grier (1794–1870)
1851–57	Benjamin R. Curtis (1809–74)
1853–61	John A. Campbell (1811–89)
1858–81	Nathan Clifford (1803–81)
1862–77	David Davis (1815–86)
1862–81	Noah H. Swayne (1804–84)
1862–90	Samuel F. Miller (1816–90)
1863–97	Stephen J. Field (1816–99)
1864–73	Salmon P. Chase (1808–73), chief justice
1870–80	William Strong (1808–95)
1870–92	Joseph P. Bradley (1813–92)
1872–82	Ward Hunt (1810–86)
1874–88	Morrison R. Waite (1816–88), chief justice
1877–1911	John M. Harlan (1833–1911)
1880–87	William B. Woods (1824–87)
1881–89	Stanley Matthews (1824–89)
1881–1902	Horace Gray (1828–1902)
1882–93	Samuel Blatchford (1820–93)
1888–93	Lucius Q. C. Lamar (1825–93)
1888–1910	Melville W. Fuller (1833–1910), chief justice
1889–1910	David J. Brewer (1837–1910)
1890–1906	Henry B. Brown (1836–1913)
1892–1903	George Shiras, Jr. (1832–1924)
1893–95	Howell E. Jackson (1832–95)
1894–1910	Edward D. White (1845–1921)
1895–1909	Rufus W. Peckham (1838–1909)
1898–1925	Joseph McKenna (1843–1926)
1902–32	Oliver W. Holmes (1841–1935)
1903–22	William R. Day (1849–1923)
1906–10	William H. Moody (1853–1917)
1910–14	Horace H. Lurton (1844–1914)
1910–16	Charles E. Hughes (1862–1948)
1910–21	Edward D. White (1845–1921), chief justice
1910–37	Willis Van Devanter (1859–1941)
1911–16	Joseph R. Lamar (1857–1916)
1912–22	Mahlon Pitney (1858–1924)
1914–41	James C. McReynolds (1862–1946)
1916–22	John H. Clarke (1857–1945)
1916–39	Louis D. Brandeis (1856–1941)
1921–30	William H. Taft (1857–1930), chief justice
1922–38	George Sutherland (1862–1942)
1922–39	Pierce Butler (1866–1939)
1923–30	Edward T. Sanford (1865–1930)
1925–41	Harlan F. Stone (1872–1946)
1930–41	Charles E. Hughes (1862–1948), chief justice
1930–45	Owen J. Roberts (1875–1955)
1932–38	Benjamin N. Cardozo (1870–1938)
1937–71	Hugo L. Black (1886–1971)
1938–57	Stanley F. Reed (1884–1980)
1939–62	Felix Frankfurter (1882–1965)

U.S. Supreme Court Justices

1939–75	William O. Douglas (1898–1980)
1940–49	Frank Murphy (1890–1949)
1941–42	James F. Byrnes (1879–1972)
1941–46	Harlan F. Stone (1872–1946), chief justice
1941–54	Robert H. Jackson (1892–1954)
1943–49	Wiley B. Rutledge (1894–1949)
1945–58	Harold H. Burton (1888–1964)
1946–53	Frederick M. Vinson (1890–1953), chief justice
1949–56	Sherman Minton (1890–1965)
1949–67	Tom C. Clark (1899–1977)
1953–69	Earl Warren (1891–1974), chief justice
1955–71	John M. Harlan (1899–1971)
1956–90	William J. Brennan Jr. (b. 1906)
1957–62	Charles E. Whittaker (1901–73)
1958–81	Potter Stewart (1915–85)
1962–65	Arthur J. Goldberg (1908-90)
1962–	Byron R. White (b. 1917)
1965–69	Abe Fortas (1910–82)
1967–	Thurgood Marshall (b. 1908), first black named to U.S. Supreme Court
1969–86	Warren E. Burger (b. 1907), chief justice
1970–	Harry A. Blackmun (b. 1908)
1972–87	Lewis F. Powell Jr. (b. 1907)
1972–	William H. Rehnquist (b. 1924); chief justice (1986–)
1975–	John P. Stevens (b. 1920)
1981–	Sandra Day O'Connor (b. 1930), first woman named to U.S. Supreme Court
1986–	Antonin Scalia (b. 1936)
1988–	Anthony M. Kennedy (b. 1936)

separate-but-equal public-school facilities for blacks. Ruling begins U.S. public-school desegregation.

1959 *Abbate* v. *United States* and *Bartkus* v. *Illinois.* Court upholds right of state and federal governments each to prosecute criminal for same crime.

1961 *Mapp* v. *Ohio.* Court applies federal court rule excluding illegally obtained evidence to state courts. This is first of controversial rulings greatly expanding rights of accused criminals.

1963 *Gideon* v. *Wainwright.* Court decides that accused criminals have right to free counsel.

1963 *Abington Township* v. *Schempp.* Court rules public-school prayer unconstitutional.

1964 *Griswold* v. *Connecticut.* Court rules Connecticut law banning contraceptives unconstitutional.

1966 *Sheppard* v. *Maxwell.* Court overturns murder conviction on grounds of prejudicial pretrial publicity.

1966 *Miranda* v. *Arizona.* Court overturns conviction of confessed rapist and establishes requirements for so-called Miranda warnings.

1971 *Phillips* v. *Martin Marietta Corporation.* Court rules in favor of 1963 Equal Pay Act.

1971 *Swann* v. *Charlotte-Mecklenburg Board of Education.* Court upholds school busing to end segregation.

1971 *Griggs* v. *Duke Power Company.* Court orders end of sex discrimination in hiring by removal of all discriminatory barriers to employment not related to job skills.

1972 *Furman* v. *Georgia.* Court rules against current state laws on capital punishment as cruel and unusual punishment.

1973 *Roe* v. *Wade.* Court invalidates state law against abortion for women up to six months pregnant.

1979 *United Steelworkers of America* v. *Weber.* Court upholds affirmative-action programs to end discrimination. It reaffirms decision in 1986.

1980 *Richmond Newspapers, Inc.* v. *Virginia.* Court rules conditions must be extreme before public and press can be barred from criminal trial.

1984 *Nix* v. *Williams.* Court relaxes some restrictions on admissibility of evidence obtained by illegal means.

▶

1984 *Firefighters* v. *Stotts*. Court upholds seniority system in employment, letting stand widespread practice of last hired, first fired.

1987 *United States* v. *Salerno*. Court finds in favor of pretrial preventive detention of dangerous suspects.

1988 *Morrison* v. *Olson*. Court upholds appointment of special prosecutor in cases of suspected misconduct by high government officials.

1989 *Consolidated Rail Corporation* v. *Railway Executives' Association*. Court finds in favor of periodic drug testing of railroad and airline employees, even if no prior negotiations on such testing have taken place with unions.

1989 *U.S.* v. *Peter Monsanto*. Court allows federal government to seize even those assets a criminal defendant could use to pay for his defense.

1989 *Webster* v. *Reproductive Health Services*. Court upholds Missouri law banning abortions by public employees unless life of mother is in danger.

U.S. Criminal Justice System

Prior to 1750, criminal justice involved the regular use of torture, widespread capital punishment, and banishment to foreign lands.

1789 Newly enacted U.S. Constitution includes provision specifying trial by jury for crimes.

1789 Judiciary Act establishes office of U.S. attorney general.

1790 First modern state penitentiary in United States, Philadelphia's Walnut Street Jail, is built. Prisoners are kept in solitary confinement, with Bible their only reading material.

1794 Pennsylvania statute divides murder into degrees, allowing capital punishment for first-degree murder only.

1816 Completion of state prison in Auburn, New York, which becomes model for U.S. prisons. It combines individual sleeping cells with group meals and activities.

1825 Reformatory for juveniles is established in New York. It becomes first institution run under "cottage system."

1841 Boston bootmaker introduces probation. He posts bond for man charged with drunkenness and then helps him get established.

1847 Michigan abolishes death penalty; it is first U.S. state to do so.

c. 1850 Term *penology* is coined by U.S. political philosopher Francis Lieber (1800–1872).

1871 U.S. federal prison system is created, under jurisdiction of newly established Department of Justice.

1876 First U.S. reformatory for young offenders opens, in Elmira, New York.

1890 Electric chair is used for death penalty for first time in United States.

1899 Special courts for juveniles in United States are established.

c. 1900 Writings of Italian scholar Enrico Ferri (1856–1929) provide basis for "social defense" theories of criminal justice: that purpose of criminal law is to protect society from criminal.

1908 Bureau of Investigation, later FBI, is established within U.S. Department of Justice.

1909 Criminal Division of U.S. Department of Justice is established in response to the increase in laws identifying federal criminal offenses.

1925 First U.S. federal reformatory for women opens, in West Virginia.

1930 U.S. Federal Bureau of Prisons is established.

▶

1932 U.S. Supreme Court establishes protection against double jeopardy in cases where accused person has already been acquitted.

1933 Alcatraz, maximum-security federal prison, is opened. Famed for its strict discipline, it remains in operation until 1963.

1945 U.S. Supreme Court adopts federal rules of criminal procedure, defining rules to be followed by courts trying criminal cases.

1961 U.S. Supreme Court rules in *Mapp* v. *Ohio* that federal court rule excluding illegally obtained evidence also applies to state courts.

1963 U.S. Supreme Court rules in *Gideon* v. *Wainwright* that accused criminals have right to free legal counsel.

1964 U.S. Office of Criminal Justice is established, charged with studying and improving criminal-justice process.

1966 U.S. Supreme Court, in *Sheppard* v. *Maxwell*, overturns murder conviction on grounds of prejudicial pretrial publicity.

1966 In *Miranda* v. *Arizona*, U.S. Supreme Court overturns conviction of confessed rapist and establishes requirement for so-called Miranda warnings.

1967 U.S. Supreme Court ruling recognizes juveniles' rights to legal representation.

1968 U.N. General Assembly agrees that there will be no statutes of limitation on war crimes.

c. 1970 Many U.S. states are trying to lower minimum age at which person can be tried as adult, due to extent and seriousness of juvenile crimes by this date.

1971 Riot at Attica, New York prison ends in 43 deaths and marks beginning of increased unrest in U.S. prisons.

1972 U.S. Supreme Court, in *Furman* v. *Georgia*, rules against current state laws on capital punishment as cruel and unusual.

c. 1975 Severe prison riots in United States attest to increasing unrest among inmates.

1980 In *Richmond Newspapers, Inc.* v. *Virginia*, U.S. Supreme Court rules conditions must be extreme before public and press can be barred from criminal trial.

1984 U.S. Supreme Court, in *Nix* v. *Williams*, relaxes some restrictions on admissibility of evidence obtained by illegal means.

1987 U.S. Supreme Court, in *United States* v. *Salerno*, finds in favor of pretrial preventive detention of dangerous suspects.

1988 U.S. Supreme Court, in *Morrison* v. *Olson*, upholds appointment of special prosecutor in cases of suspected misconduct by high government officials.

Political Scandals Since 1900

1906 Dreyfus Affair ends. French appeals court exonerates army officer Alfred Dreyfus (1859–1935) of treason charges, finally ending 12-year political scandal in France. Court overturns Dreyfus's 1894 conviction for offering top-secret military information to Germans, finding it to have been based on forged documents and tainted by anti-Semitism.

1923 Teapot Dome scandal breaks. Earlier, during presidency of Warren Harding (1865–1923), Secretary of the Interior Albert Fall (1864–1944) secretly takes $200,000 bribe and allows Mammoth Oil Co. exclusive leasing rights to Teapot Dome, Wyoming federal oil reserves. Harding's administration ends in disgrace and Fall is convicted (1929) of receiving bribes.

1932 New York mayor Jimmy Walker (1881–1946) resigns following investigation of graft and misuse of funds in city's Tammany Hall political machine.

▶

1948 Hiss case. American editor and confessed Communist spy Whittaker Chambers (1901–61) accuses State Department official Alger Hiss (b. 1904) of spying for Soviet Union. Hiss denies it but is convicted of perjury in 1950 after two dramatic trials, which become focal point for anti-Communist fervor in United States and which provide prominent role for Hiss's leading accusers, Congressman Richard M. Nixon (b. 1913) and Sen. Joseph R. McCarthy (1909–57).

1952 Republican vice-presidential nominee Richard M. Nixon, under attack for alleged misuse of $18,000 campaign fund, successfully defends himself by delivering his famous "Checkers" speech on national television (September 23).

1953 Julius Rosenberg (1918–53) and Ethel Rosenberg (1916–53) are executed for espionage after being convicted for delivering top-secret information on atomic bomb to Soviet Union in 1944–45. Case generates such controversy that many claim Rosenbergs' conviction is result of McCarthy-era anti-Communist hysteria.

1963 Profumo scandal. British war minister John Profumo (b. 1915) resigns amid reports of his affair with 21-year-old Christine Keeler, who also is having affair with Soviet Navy officer.

1963 Robert G. Baker, secretary to U.S. Senate majority, resigns (October 7) in scandal over accepting bribes and other influence-peddling. Baker is convicted in 1967 of tax evasion, theft, and conspiracy to defraud government.

1969 U.S. Senator Edward M. Kennedy (b. 1932) accidentally drives his car off bridge at Chappaquiddick Island, Massachusetts (July 19), drowning his companion, Mary Jo Kopechne (1941–69). Kennedy later admits he waited 10 hours before reporting incident.

1972–74 Watergate scandal. Probably worst political scandal in U.S. history, it starts with break-in and wiretapping of Democratic National Headquarters at Watergate Hotel in Washington (June 17, 1972), which was orchestrated by staffers within Nixon administration. It ends with resignation of President Richard M. Nixon (b. 1913) on August 9, 1974, for attempting cover-up, and with conviction of over 30 Nixon staffers.

1973 U.S. Vice President Spiro Agnew (b. 1918) resigns (October 10) after pleading no contest to charges of failure to report $29,500 in income while serving as governor of Maryland in 1967.

1974 West German chancellor Willy Brandt (b. 1913) resigns (May 6) following revelation that high-ranking member of his staff, Gunter Guillaume, is East German spy.

1980 Abscam ("Arab scam"). This bribery scandal follows two-year undercover operation in which FBI agents posing as Arab businessmen offer bribes in exchange for political favors. Seven U.S. congressmen and various city and state officials are convicted, though operation raises debate over ethics of entrapment.

1983 U.S. representatives Daniel Crane and Gerry Studds are reprimanded by House ethics committee for having sex with teenage congressional pages (July 14).

1986 Former U.N. secretary-general Kurt Waldheim (b. 1918), seeking presidency of his native Austria, is accused of hiding his Nazi past and of involvement in torture of Yugoslavian Jews during World War II. Waldheim denies taking part in cruelties and is elected.

1986 Iran-contra scandal erupts over secret sale of arms by Reagan administration, then seeking release of U.S. hostages. However, attention later focuses on illegal diversion of profits from the arms sale to contra rebels in Nicaragua. Three administration officials are indicated (1987) for conspiracy, including Lt. Col. Oliver North (b. 1943).

1987 Senator Gary Hart (b. 1936) withdraws from U.S. presidential race after newspapers publish reports on his relationship with 27-year-old model Donna Rice. She is seen entering his Washington

▶

townhouse and reportedly traveled with him to Bimini aboard his yacht.

1989 Japanese Prime Minister Noboru Takeshita is forced to resign (April 25) in influence-peddling scandal after admitting he accepted $1 million from Recruit Co. His successor soon is embroiled in scandal involving geisha girl, his alleged mistress.

1989 House ethics scandal. Charges of House rules violations on outside income and other allegations force resignations of Democratic House speaker Jim Wright and (b. 1922) and majority whip Tony Coelho (b. 1942).

Historic Assassinations

B.C.

338 Persian king Artaxerxes III by imperial court eunuch Bagoas.

336 Macedonian king Philip II (382–336 B.C.) by Macedonian youth Pausanias.

330 Persian king Darius III by Persian Satrap Bessus.

44 Julius Caesar (c. 102–44 B.C.) by Marcus Brutus, Cassius, Decimus Brutus, and others (March 15).

A.D.

1170 Thomas à Becket, archbishop of Canterbury, by king's knights (December 29).

1308 German king Albert I (c. 1250–1308) by his nephew.

1437 Scottish king James I (1394–1437) by nobles (February 21).

1483 English king Edward V (1473–83) is deposed and believed killed in Tower of London.

1567 Henry Stewart, Lord Darnley (1545–67) by strangulation at Edinburgh.

1589 French king Henry III (1551–89) by Jacques Clément (August 1).

1610 French king Henry IV (1553–1610) by Ravaillac (May 14).

1634 Austrian general Albrecht Wallenstein (1583–1634) by Irish and Scottish officers.

1792 Swedish king Gustavus III (1746–92) by Ankarström (March 16; dies March 29).

1801 Russian Czar Paul I (1754–1801) by nobles (March 24).

1865 U.S. President Abraham Lincoln (1809–65) is shot by John Wilkes Booth at Ford's Theater, Washington, D.C. (April 14). Lincoln dies April 15; Booth is killed April 26.

1881 Russian czar Alexander II (1818–81) by bomb (St. Petersburg; March 13).

1881 U.S. President James A. Garfield (1831–81) is shot by disgruntled office-seeker Charles J. Guiteau (Washington; July 2). Garfield dies September 19; Guiteau hangs July 30, 1882.

1894 French President Sadi Carnot (1837–94) is stabbed by Italian anarchist Sante Caserio (June 24).

1898 Austrian empress Elizabeth (1837–98) by Italian anarchist Luigi Luccheni.

1900 Italian king Humbert (1844–1900) is shot by anarchist Angelo Bresci (July 30).

1901 U.S. President William McKinley is shot by anarchist Leon Czolgosz (Buffalo, New York; September 6). McKinley dies September 14; Czolgosz is executed October 29.

1908 Portuguese king Carlos I (1863–1908) in Lisbon.

1909 Japanese statesman Prince Hirobumi Ito (1841–1909) by Korean nationalist.

1911 Russian prime minister Peter Stolypin (1863–1911) is shot by socialist lawyer Dmitri Bogroff (September 14).

▶

1913 Greek king George I (1845–1913) by Greek assassin (March 18).

1914 Austrian archduke Francis Ferdinand (1863–1914) is shot by Serbian nationalist Gavrilo Princip (Sarajevo, Bosnia; June 28). Assassination touches off World War I.

1918 Former Russian czar Nicholas II (1868–1918) is executed by Bolsheviks (July 16).

1924 Italian statesman Giacomo Matteotti (1885–1924) by Fascists (Rome; June 10).

1928 Mexican President-elect Alvaro Obregón (1880–1928) is shot by gunman (Mexico City; July 17).

1932 French President Paul Doumer (1857–1932) is shot by Russian émigré Gorguloff (May 7).

1934 Austrian premier Engelbert Dollfuss (1892–1934) is killed by Nazis (Vienna, July 25).

1934 Soviet Communist leader Sergei Kirov (1888–1934) is shot by young party member Leonid Nikolayev (Leningrad; December 1). Assassination becomes pretext for Stalin's bloody purge.

1935 U.S. senator Huey Long (1893–1935) is shot (September 8).

1940 Former Soviet leader Leon Trotsky (1879–1940) by Spaniard (Mexico City; August 20).

1948 Indian nationalist leader Mohandas Gandhi (1869–1948) is shot by Hindu extremist (January 30).

1958 Iraqi king Faisal (1935–58) by rebels (July 14).

1963 South Vietnamese President Ngo Dinh Diem (1901–63) during military coup (November 2).

1963 U.S. President John F. Kennedy (1917–63) is shot by Lee Harvey Oswald (Dallas; November 22). Dallas nightclub owner Jack Ruby shoots and kills Oswald in Dallas jail (November 24).

1968 Civil-rights leader Martin Luther King, Jr. (1929–68) is shot by James Earl Ray (Memphis; April 4).

1968 Senator Robert F. Kennedy (1925–58) is shot by Sirhan Sirhan (Los Angeles; June 5).

1975 Saudi king Faisal (c. 1906–75) is shot by nephew (March 25).

1979 British World War II hero Lord Mountbatten (1900–79) is killed by bomb for which IRA claims responsibility (August 27).

1979 Korean President Park Chung Hee (1917–79) is shot by Korean CIA head Kim Jae Kyu (October 26).

1981 Egyptian President Anwar Sadat (1918–81) is shot by commandos (Cairo; October 6).

1984 Indian prime minister Indira Gandhi (1917–84) is shot by two of her Sikh bodyguards (October 31).

1986 Swedish premier Olaf Palme is shot (Stockholm, February 28).

1987 PLO military head Khalil Wazir is shot by Israeli commandos in Tunisia (April 16).

Uprisings and Revolts

B.C.

90–88 Social War. Other Italian tribes, allies of Rome, rebel successfully to gain rights of Roman citizens.

73–71 Servile War (Third). Unsuccessful revolt by slaves, led by gladiator Spartacus (d. 71). Slaves briefly conquer much of southern Italy.

A.D.

66–70 First Jewish Revolt. Unsuccessful revolt in Judaea against Roman rule results in destruction of the Temple in Jerusalem (70).

132–135 Second Jewish Revolt. Simon Bar Kokhba (d. 135) leads Jews in unsuccessful revolt against Roman rulers.

▶

Jews are defeated and banished from Jerusalem.

184–204 Yellow Turban Rebellion. Taoist faith healer leads this secret society's partly successful revolt against China's Han dynasty rulers.

1358 Jacquerie. French peasants rise against noblemen (May) over increased taxes and other hardships stemming from Hundred Years War.

1378 Revolt of the Ciompi. Short-lived wage-earners' revolt against restrictions on earnings by powerful guilds in Florence, Italy.

1381 Peasants' Revolt. Unpopular poll tax brings on this first major uprising in England. Peasants and workers take London for a time before order is restored.

1440 Praguerie (February). Unsuccessful revolt by French noblemen against King Charles VII to regain lost power in the government.

1524–26 Peasants' War. Inspired by Martin Luther's teachings, peasants revolt and pillage countryside in Holy Roman Empire.

1637–38 Shimabara Rebellion. Roman Catholic peasants from Shimabara area in Japan battle government army of 100,000 before their revolt is crushed.

1649–53 Fronde. Revolt by Parlement and nobles of France against the French monarchy's growing power. Ultimately it fails.

1673–81 Revolt of the Three Feudatories. Briefly successful revolt in southern China against Ch'ing dynasty.

1677–79 Culpeper's Rebellion. John Culpeper leads this uprising of colonists in Carolina colony against British taxation. Rebels govern for two years before being disbanded.

1763–66 Pontiac's Rebellion. Bloody Indian revolt against British control in the Great Lakes region.

1780 Gordon Riots. Anti-Catholic mobs riot for a week in London, burning churches and freeing jailed prisoners. Over 400 die.

1793–96 War of the Vendee. Peasants and royalists in Vendee region of France rebel against radical reforms of the French revolutionary government.

1794 Whiskey Rebellion. Farmers in Pennsylvania rise against liquor tax in first armed revolt against newly constituted U.S. federal government.

1796–1804 White Lotus Rebellion. Uprising in central China against Ch'ing dynasty.

1824 Barrackpore Mutiny (November 2). Native regiment in British India's army refuses to march in dispute over caste taboos. British commanders fire artillery on regiment.

1830 July Revolution. Unpopular ordinances ignite riots (July 27–29) in France, force King Charles X's abdication, and bring "citizen king" Louis-Philippe (1773–1850) to the throne.

1831 Southampton Insurrection (August). Nat Turner (1800–31) leads revolt by some 75 slaves, the only major slave uprising in U.S. history.

1848 Revolutions of 1848. Popular uprisings in France and other European nations against rule by conservative monarchies. In France, a republic replaces the monarchy.

1850–64 Taiping Rebellion. Popular uprising in China against the ruling Ch'ing dynasty. Rebels institute a loosely Christian communal society. Some 20 million die before revolt is quashed.

1857–59 Indian Mutiny. Bloody uprising against British colonial rule in India. It begins when native troops refuse (for religious reasons) to use cartridges greased with tallow and lard.

1863 Draft Riots. Irish and other foreign-born workers in New York City riot for four days over Civil War draft law with paid deferment provision. Over 1,200 die.

▶

1863–64 January Insurrection. Radical elements spark a short-lived general uprising in Poland against their Russian overlords.

1900 Boxer Rebellion. Chinese secret society, called Boxers, rebels against foreign influence in China and gains tacit support of the Empress Dowager. Boxers briefly take control in Peking.

1918 Kiel Mutiny. Rebellion by German sailors at the port of Kiel (October 29–November 3) spreads throughout German navy, bringing about collapse of the German government and World War I armistice.

1921 Kronstadt Rebellion. Soviet navymen at Kronstadt revolt against food shortages and political repression under Bolsheviks. Bolsheviks massacre rebels but institute economic reforms.

1927 Autumn Harvest Uprising. Chinese communists (including Mao Tse-tung; 1893–1976) spark this unsuccessful peasant revolt in China (September).

1944 Warsaw Uprising. At close of World War II, Polish underground unsuccessfully attempts to capture Warsaw from Germans before advancing Soviets can take city (August–October).

1956 Hungarian Revolt. Rebellious students and workers spark an armed revolt against Soviet authority in Hungary. It is put down by Soviet military intervention.

1968 Leftist rebellion in France. Strikes, demonstrations, and riots by leftist students and workers nationwide (May) paralyze France and force change of government.

1978 Iranian Revolution. Muslim fundamentalists rebel against government reforms aimed at further Westernizing and modernizing Iran. After months of unrest, fundamentalist leader Ayatollah Ruhollah Khomeini (1901–89) gains control of national government.

1981 Solidarity. Nationwide strike by workers in Poland briefly brings country to a standstill early in 1981 and leads to growing worker unrest. Imposition of martial law late in the year ends Solidarity for a time.

Socialism and Communism

See also European Nations.

1794 French revolutionary François Babeuf (1760–97), considered first socialist, publishes political journal calling for economic and political equality.

1825 British Utopian socialist Robert Owen (1771–1858), pioneer in cooperative movement, founds self-sufficient cooperative agricultural-industrial community, New Harmony, in Indiana.

1841 Brook Farm, experiment in cooperative living based on ideas of French social philosopher Charles Fourier (1772–1837), is founded in Massachusetts.

1848 German philosopher Karl Marx (1818–83) and businessman Friedrich Engels (1820–95) publish *Communist Manifesto*, predicting violent breakdown of capitalism, revolt of working class, and end to all class struggles.

1863 First workers' party is founded, in Germany, by German socialist Ferdinand Lassalle (1825–64).

c. 1875 Socialist parties exist in most European countries.

1876 Workingmen's Party of America, first U.S. socialist party, is established.

▶

1878 Antisocialist laws are enacted by German chancellor Otto von Bismarck (1815–98); socialists may not hold meetings or distribute literature.

1883 First meeting is held of Fabian Society of Great Britain, advocating patient and peaceful change along with public, cooperative enterprise.

1898 Revisionist School of Socialism emerges in Germany, criticizing some Marxian doctrines and dedicated to peaceful instead of revolutionary change.

1901 Japanese Socialist Party, first socialist party in Asia, is founded.

1903 Modern Communist political movement begins when doctrinal disagreements cause split of Russian Social Democratic Labor Party into revolutionary Bolsheviks, led by Vladimir Lenin (1870–1924), and gradualist Mensheviks.

1906 British Labour Party, federation of trade unions and socialist societies, is formed.

1915 Italian socialist editor Benito Mussolini (1883–1945) breaks with his party and founds Italian Fascist Party (1919). Fascist movements arise in opposition to socialist and communist revolutionary dogmas.

1917 Russian Revolution. Bolsheviks seize power and institute first Communist national government.

1918 U.S. socialist Eugene Debs (1855–1926) is jailed for criticizing government prosecutions under 1917 Espionage Act during World War I.

1918 Russian Bolsheviks form Communist party under Lenin.

1919 Lenin creates Third Communist International (Comintern), formalizing split between Communists and socialists. Lenin uses Comintern to dominate communism worldwide and to spread revolutionary ideology.

1921 Lenin criticizes demands for greater freedom of discussion within party and suppresses opposition political groups.

1922 Most European socialist parties have split by this date; left-wing factions form Communist parties. Right-wing factions adopt term "democratic socialist."

1924 After Lenin's death, Joseph Stalin (1879–1953) becomes Soviet Communist leader. He institutes harsh programs of collectivization and repression and pledges spread of communism worldwide.

1930s German Fascists under Adolf Hitler (1889–1945) suppress socialism and communism. They condemn thousands to concentration camps.

1934 Stalin begins bloody purge of his opponents in party. He establishes "cult of personality" to promote his totalitarian rule over USSR and Communist parties worldwide.

c. 1945 Following World War II, socialist parties in many countries reemerge and are in and out of power throughout postwar years. British Labour Party effects many socialist reforms, including nationalizing banks and power and transportation industries and introducing national health program (1945–51).

1948 Soviet satellite nations are created. Communist takeovers in seven countries by this date form buffer between free Europe and USSR.

1948 Yugoslav Communist leaders, including President Josip Tito (1892–1980), displease Stalin by questioning Soviet authority.

1948 Term "Cold War" is coined by Bernard Baruch (1870–1965).

1949 Chinese Communists are victorious in Chinese Civil War. They establish Communist government in China and aid Communist movements in Southeast Asia.

1956 Destalinization program is begun (February) by Soviet leader Nikita Khrushchev (1894–1971). He calls for moderation of Stalin's repressive rule. Liberalization leads to unrest in Hungary.

1956 Anti-Communist Hungarian revolution is suppressed by invading Soviet troops.

▶

1957 Moscow declaration asserts that socialism worldwide must be based on Marxist-Leninist example and warns against revisionism.

1958 Yugoslav Communists maintain that communism may be achieved through peaceful means.

1959 Revolutionary Fidel Castro (b. 1926) takes power in Cuba, which becomes first Communist state in Latin America.

c. 1960 Split between Chinese and Soviet Communists is widening; Chinese hold to inevitability of revolution and resent Soviet revisionist ideology.

1966–69 Cultural Revolution occurs in China. Mao's Red Guard institutes repressive, hard-line proletarian regime.

1968 In Prague Spring, Czech Communist leader Alexander Dubcek (b. 1921) angers Soviets with liberal socialist theories; Soviets invade Czechoslovakia.

1969 Conference of Communist parties is held in Moscow to discuss Soviet/Chinese differences. China later begins liberalization policy and normalization of relations with West while turning away from USSR.

1975 Khmer Rouge Communists take over in Cambodia, instituting reign of terror. They evacuate cities and force populace to work at hard labor in countryside. A million or more die or are executed.

c. 1975 Communist parties in Italy, France, and Spain choose to reject Soviet dominance and adopt parliamentary means of governing.

1976 Representatives of 29 Communist parties meet in East Berlin; autonomy of individual parties is affirmed.

1977 Leonid Brezhnev (1906–82) becomes President of USSR and takes hard line with satellite countries. He later orders invasion of Afghanistan.

1980 Strikes in Poland reveal growing resentment toward Soviet domination. Independent labor union, Solidarity, demands greater freedoms.

c. 1980 British Labour Party moves sharply to left; moderate and conservative members split to form Social Democratic Party.

1981 Soviet-backed government imposes martial law in Poland. It disbands Solidarity in 1982 but reverses policy and legalizes it in 1989.

1981 French Socialist François Mitterrand (b. 1916) is elected President. He institutes socialization program in France.

1985 Soviet leader Mikhail Gorbachev (b. 1931) is in power. He institutes policy of "openness" (*glasnost*) and "restructuring" (*perestroika*).

1989 Hungary adopts multiparty system, first in any Communist state.

1989 In radical policy change to end chronic food shortages, Soviet government gives farmers control of land and selection of crops they grow. Family farm concept reverses long-standing Communist policy of collectivization of farms.

1990 Soviet Congress enacts sweeping changes in Soviet communist government, including end to one-party government, legalization of private ownership of factories, and creation of a strong presidency. Similar moves toward democracy are taken as well throughout Eastern Bloc nations.

4 U.S. Business, Commerce, and Economics

Origins of Major U.S. Businesses

1802 E.I. Du Pont de Nemours and Co. is founded for manufacture of explosives. It incorporates in 1903 in New Jersey to combine number of businesses and in 1912 is ordered to divest, under Sherman Antitrust Act. Incorporates in Delaware in 1915.

1852 Anheuser-Busch is founded as partnership and incorporates in 1875 as E. Anheuser Co.'s Brewing Assn. Current name is adopted in 1919.

1870 Atlantic Refining Co. incorporates. In 1966 Richfield Oil Co. merges with Atlantic to create Atlantic Richfield Company.

1870 Standard Oil of Ohio, first of several regional Standard Oil corporations, incorporates. John D. Rockefeller (1839–1937) is president.

1875 R.J. Reynolds organizes as tobacco company. It incorporates in 1970 as R.J. Reynolds Industries and adopts name RJR Nabisco in 1986 after merger with Nabisco Brands.

1879 Scott Paper Co. is established. In 1922 it incorporates after merger with Chester Paper Co. (established 1910).

1880 George Eastman establishes his business, which reincorporates in New Jersey in 1901 as Eastman Kodak Co.

1882 Exxon incorporates as Standard Oil Company of New Jersey and in 1892 acquires stocks of other oil companies through Standard Oil Trust. In 1911 U.S. Supreme Court rules that exchange of stocks violates Sherman Antitrust Act, and shares in the other companies are distributed among stockholders. Company adopts name Exxon in 1972.

1882 Mobil Oil Corp. incorporates as Standard Oil Co. of New York. It becomes independent company as result of 1911 antitrust ruling breaking up Standard Oil. In 1976 Mobil Corporation incorporates to act as holding company, and Mobil Oil becomes its subsidiary.

1883 PPG Industries, Inc. forms as Pittsburgh Plate Glass Co. It reincorporates in 1923 after merger with Patton Pitcairn Co. and Columbia Chemical Co. Changes name to PPG Industries in 1968.

1885 American Telephone & Telegraph incorporates. It breaks up in 1982 as result of antitrust suit, forming separate regional telephone companies and a separate AT&T corporation handling long-distance lines and other communications services.

1886 Westinghouse Electric Corporation incorporates as Westinghouse Electric Co. In 1975 Securities & Exchange Commission alleges that Westinghouse made misleading statements to purchase its own stock more profitably.

1887 Johnson & Johnson incorporates in New Jersey.

1887 Clinton Pharmaceutical Company forms; it incorporates as Bristol-Myers in New York in 1900. In 1933 it incorporates in Delaware and acquires capital stock of New Jersey Bristol-Myers Company.

1889 Amoco incorporates in Indiana under name Standard Oil Company as subsidiary of Standard Oil Company of New Jersey. It becomes independent as result of 1911 antitrust action forcing breakup of Standard Oil. Adopts Amoco name in 1985.

1890 Royal Dutch Petroleum Co. incorporates in Netherlands as "Royal Dutch Co. for the Working of Petroleum Wells in the Netherlands Indies."

1890 Procter & Gamble incorporates in New Jersey. In 1905 it joins with a Cincinnati, Ohio partnership (established 1837) to continue the business of producing soap, oil, and glycerine.

1890 Emerson Electric Co. incorporates under name Emerson Electric Manufacturing Co. to produce electrical, and later electronics, systems.

1891 Philips is founded as incandescent lamp factory. It is established under laws of Netherlands in 1920 as holding company and manufacturer of electrical products.

1892 Coca-Cola Co. is established in Georgia, succeeding series of owners who began producing the cola in 1886. Coca-Cola Company incorporates in Delaware in 1919.

1892 General Electric Company incorporates, combining assets of Edison General Electric,

Thomson-Houston Electric, and Thomson-Houston International Electric.

1894 Ralston Purina Co. incorporates as Robinson Danforth Commission Co. It adopts name Ralston Purina in 1902.

1897 Dow Chemical Company forms to build bleach plant to use waste from Midland Chemical Co. In 1900 Dow Chemical acquires Dow Process Co. and Midland Chemical Co., both also formed by Herbert Dow. Dow Chemical reincorporates in Delaware in 1947 and succeeds to all properties and assets.

1898 International Paper Co. incorporates in New York.

1898 Goodyear Tire & Rubber incorporates in Ohio.

1899 NEC Corporation incorporates in Japan as Nippon Electric Company Ltd. It is renamed NEC in 1983.

1899 Borden's Condensed Milk Co. incorporates as successor to New York Condensed Milk Co.

1900 Chas. Pfizer & Co., Inc. incorporates in New Jersey as successor to partnership formed in 1849. In 1970 Pfizer, Inc., research-based corporation, adopts its current name.

1900 Weyerhaeuser Co., originally organized as Weyerhaeuser Timber Co., is founded.

1900 Abbott Alkaloidal Co., which in 1914 is renamed Abbott Laboratories, incorporates, continuing business begun in 1888.

1901 U.S. Steel Co. incorporates. In 1966 it merges with U.S. Steel Corp., and it adopts name USX in 1986. In 1989 USX is sold to subsidiary of Sinochem American Holdings.

1902 Texaco organizes as The Texas Co. It incorporates in 1926. In 1987 it files for bankruptcy as result of $11.1 billion fine in court action by Pennzoil, stemming from Texaco's reported interference in a Pennzoil acquisition. Texaco subsequently reaches settlement with Pennzoil.

1902 Minnesota Mining & Manufacturing Co. incorporates in Delaware.

1903 Ford Motor Company organizes in Michigan with Henry Ford (1863–1947) as principal stockholder. It incorporates in 1919 in Delaware.

1904 Bethlehem Steel incorporates as successor to United States Shipbuilding Co. and acquires several shipbuilders and ironworks.

1904 Spicer Manufacturing Co., which becomes known as Dana Corp. in 1946, incorporates.

1904 American Brands incorporates as The American Tobacco Co. after merger of Consolidated Tobacco Co., American Tobacco Co., and Continental Tobacco Co.

1906 Xerox Corporation is founded as Haloid Co. in New York. It becomes known by its current name in 1961.

1906 Chevron Corporation, known until 1984 as Standard Oil of California, incorporates through merger of Pacific Oil Company and Standard Oil Company (California).

1908 General Motors organizes as auto manufacturer. It incorporates in 1916. Acquires Chevrolet Motor Co. in 1918.

1911 International Business Machines (IBM) incorporates as Computing Tabulating Recording Co. after consolidation of Bundy Manufacturing Co., Tabulating Machine Co., International Time Recording Co. of New York, and Computing Scale Co. of America.

1916 TRW, Inc. is founded as Steel Products Co.; name is changed to TRW in 1965.

1917 Union Carbide Corporation begins business as Union Carbide and Carbon Corporation. In 1989 company reorganizes and becomes holding company known as Union Carbide Corporation, with subsidiary named Union Carbide Chemicals and Plastics Company, Inc.

1919 Philip Morris incorporates in Virginia as tobacco company.

1919 Loft, Inc., which changes its name to PepsiCo in 1965, incorporates. In 1986 PepsiCo, Inc. incorporates in North Carolina to succeed former PepsiCo.

1920 Occidental Petroleum Corporation incorporates in California. In 1961 it begins to acquire a number of companies. Occidental Petroleum Corporation incorporates in Delaware in 1986 to serve as parent company.

1920 Allied Chemical & Dye incorporates. In 1985 it acquires Signal Companies, Inc. and incorporates as Allied-Signal, Inc.

1922 Shell Union Oil Corp., later known as Shell Oil Company, incorporates to merge properties of Royal Dutch-Shell and Union Oil Co.

1923 Eastern Operating Co. incorporates and soon changes its name to Palmolive Co. In 1928 it acquires Colgate and Co. (established in 1806) and changes its name to Colgate-Palmolive-Peet (Peet Brothers is acquired in 1926). In

▶

1953 Colgate-Palmolive is formally adopted as company name.

1923 Archer-Daniels-Midland Co. incorporates as combination of Archer-Daniels Linseed Oil Company, Toledo Seed & Oil Co., and Dellwood Elevator Co.

1925 Caterpillar Tractor Co. incorporates as result of merger between Holt Manufacturing Co. and C.L. Best Tractor Co. It incorporates as Caterpillar, Inc. in 1986.

1925 Chrysler incorporates. In 1980 it receives $1.5 billion loan guarantee from the federal government as bailout for financial crisis; it repays loan by 1983.

1926 American Home Products Corp. incorporates as holding company and acquires Wyeth Chemical Co.

1927 Georgia-Pacific incorporates under name Georgia Hardwood Lumber.

1927 Honeywell, Inc. forms as Minneapolis-Honeywell Regulator Co., merging assets of Minneapolis Heat Regulator Co. and The Honeywell Heating Specialties Co. In 1964 name Honeywell, Inc. is adopted.

1928 BAT Industries registers in England under name Tobacco Securities Trust Co. Ltd. It changes name to BAT Industries Ltd. after merger with British-American Tobacco Co. Ltd. in 1976.

1928 Motorola, Inc. incorporates as Golvin Manufacturing Corp.

1928 General Mills incorporates in Delaware to acquire a number of mills.

1928 Kimberly-Clark Corporation incorporates to absorb Kimberly-Clark Co. (established in 1907) and Kimberly-Clark Corp., Inc., its subsidiary. Kimberly-Clark & Co. was originally established as partnership in 1872 as paper company.

1928 North American Aviation, Inc. begins business. It merges with Rockwell-Standard Corp. in 1967, and in 1973 it adopts name Rockwell International.

1928 Boeing Company is founded for aviation activities. It incorporates in 1934.

1929 SmithKline Beckman, makers of health-care products, incorporates as Smith Kline & French Laboratories. In 1982 it merges with Beckman Instruments (established 1830) and adopts current name.

1931 Baxter International, Inc. incorporates as Baxter Laboratories, Inc. to produce hospital supplies.

1932 Lockheed Aircraft Corp. incorporates in California. In 1977 it changes its name to Lockheed Corp.

1934 United Technologies Corporation incorporates as successor to United Aircraft & Transport Corp. It adopts current name in 1975.

1937 Champion International Corporation incorporates as U.S. Plywood Corp. after consolidation of Aircraft Plywood Co., U.S. Plywood Co. (Delaware), and U.S. Plywood Co. (New York).

1938 The Geophysical Service, Inc., which in 1951 changes its name to Texas Instruments, incorporates.

1939 Northrop Aircraft, Inc. is founded. It reincorporates in 1985 under name Northrop Corp., which it had adopted in 1958.

1939 McDonnell Douglas incorporates as McDonnell Aircraft Corp.

1941 The South Street Co., which in 1985 adopts name Sara Lee, incorporates.

1947 Hewlett-Packard Co. incorporates in California.

1952 General Dynamics incorporates and merges with its parent company, Electric Boat Co., originally formed in 1899.

1953 American Hoechst Cora incorporates in New York. In 1987 it acquires Celanese Corporation and is renamed Hoechst Celanese Corp.

1954 Tenneco incorporates as Cumberland Corp. to produce oil and gas. It adopts Tenneco name in 1960.

1957 Digital Equipment incorporates in Massachusetts.

1961 The Martin Co. (established 1928) and American Marietta Co. (established 1930) join and incorporate as Martin Marietta Corporation, dealing in astronautics, information systems, electronics, and materials.

1967 Textron, Inc. incorporates as American Textron, Inc. In 1968 company merges with company of same name, which was founded in 1928 as Franklin Rayon Corp.

1969 James River Corp. of Virginia is founded.

1969 Hanson Trust Ltd. succeeds Wiles Group Ltd. Renamed Hanson PLC in 1986, company dis-

▶

tributes equipment for construction, agricultural, and industrial uses.

1983 Unocal Corporation incorporates and reorganizes. As part of plan, Union Oil Company of California, established in 1890, becomes subsidiary of Unocal.

1984 Unisys incorporates as successor to Burroughs Corporation. It adopts Unisys name in 1986.

U.S. Business Magnates and Tycoons

1763–1848 John J. Astor lives. He monopolizes fur industry in United States and later invests in real estate. Amasses fortune of over $20 million.

1771–1834 Eleuthère I. Du Pont lives. He makes his fortune manufacturing gunpowder in United States.

1783–1857 William Colgate lives. He founds Colgate & Company (later Colgate-Palmolive Company).

1791–1883 Peter Cooper lives. As head of North American Telegraph Works, he makes fortune manufacturing glue and establishing ironworks.

1794–1887 Cornelius Vanderbilt lives. Steamship and railroad magnate, he amasses estimated $100 million and donates $1 million to what is now Vanderbilt University in Nashville, Tennessee.

1821–1905 Jay Cooke, investment banker and mining magnate, lives.

1822–88 Charles Crocker, railroad magnate, lives. His Crocker & Co. figures in building western half of transcontinental railroad.

1828–1905 Meyer Guggenheim lives. A Swiss immigrant to United States, he makes fortune smelting metallic ores in Colorado and New Mexico.

1832–1901 Philip D. Armour, leader in meat-packing industry, lives. He introduces refrigerated railroad car.

1834–1906 Marshall Field lives. He founds Marshall Field & Co. department stores and leaves estate of over $125 million.

1834–1916 Hetty Green, investor in stocks and real estate, lives. Probably wealthiest woman of her time, she amasses over $100 million.

1835–1919 Andrew Carnegie lives. Steel tycoon and founder of U.S. Steel Company, he donates more than $350 million to philanthropic organizations.

1836–92 Jay Gould lives. Director of Erie Railroad and Union Pacific Railroad, he amasses fortune of over $77 million.

1837–1913 John P. Morgan, financier, lives. He acquires $65 million in gold for U.S. government during Depression of 1895.

1839–1937 John D. Rockefeller lives. Founder of Standard Oil Company (1870) and University of Chicago (1892), he donates more than $500 million to philanthropic organizations.

1841–1900 Marcus Daly lives. An Irishman, he acquires millions mining copper in Montana.

1849–1919 Henry C. Frick, cofounder of U.S. Steel Corporation (1901), lives. He bequeaths $15 million and his Fifth Avenue mansion to New York City to establish Frick Collection of Fine Art.

1852–1919 Frank W. Woolworth, founder of F. W. Woolworth & Co. five-and-dime stores, lives. He builds Woolworth Building in New York City (1913), world's tallest building at time.

1854–1932 George Eastman, founder of Eastman Kodak Company (1884), lives. He donates more than $75 million to MIT and to University of Rochester in New York.

1855–1937 Andrew Mellon, banker, lives. He founds Aluminum Co. of America (Alcoa), Gulf Oil Corporation, and Union Steel Company.

1856–1917 James D. (Diamond Jim) Brady lives. A colorful millionaire, he amasses fortune selling railroad equipment.

▶

1856–1925 James B. Duke, tobacco magnate, lives. Founder of American Tobacco Company (1890), he contributes generously to what becomes Duke University in North Carolina.

1860–1951 Will K. Kellogg lives. Breakfast-cereal tycoon and inventor of the corn flake, he donates more than $47 million to philanthropic organizations.

1861–1947 William C. Durant lives. He founds Buick Motor Car Company (1905) and General Motors (1908).

1863–1914 Richard W. Sears, cofounder and president of Sears, Roebuck & Company department stores, lives.

1863–1947 Henry Ford, businessman who founds Ford Auto Company (1903), lives. Ford's mass-production techniques make him rich and revolutionize manufacturing.

1863–1951 William R. Hearst, newspaper tycoon, lives. He spends more than $30 million building his "castle" in San Simeon, California.

1863–1955 Samuel H. Kress, founder of S. H. Kress & Co. five-and-dime stores (1896), lives. He donates $25 million in art to National Gallery (1939) and $18 million to Bellevue Medical Center at New York University (1949).

1870–1949 Amadeo Giannini, Italian-American businessman, lives. Founder of Bank of America, he makes his fortune providing loans to help rebuild San Francisco after earthquake and fire (1906).

1874–1960 John D. Rockefeller, Jr., oil tycoon, lives. He spends millions to restore Colonial Williamsburg (1926) and donates land worth $9 million for U.N. Secretariat Building in New York City.

1882–1967 Henry J. Kaiser, construction mogul, lives. His companies build Boulder (now Hoover) Dam (1936), San Francisco-Oakland Bay Bridge (1936), and Grand Coulee Dam (1942). He also founds Kaiser Aluminum Corp. (1946).

1885–1956 Charles E. Merrill lives. He cofounds investment firm of Merrill Lynch & Company and amasses fortune of some $25 million.

1888–1979 Conrad N. Hilton, businessman who owns international chain of luxury hotels, lives.

1889–1974 H. L. Hunt lives. Oil tycoon and founder of Hunt Oil Company (1936), he leaves estate of over $2 billion.

1891–1971 Samuel Bronfman, Canadian business executive, lives. In 1920s he takes the helm of Seagrams distillers.

1892–1976 Jean Paul Getty, oil tycoon who amasses $3 billion, lives. He founds J. Paul Getty Museum (1954) with $750 million donation.

1895–1979 Samuel Irving Newhouse, publishing magnate, lives. He begins buying newspapers in 1922, amassing by the 1970s a string of 31 papers and major national magazines, among them *Vogue, Glamour, House and Garden*, and *Mademoiselle*.

1899–1989 August Anheuser Busch, Jr., brewing magnate and owner of St. Louis Cardinals, lives.

1900–1985 John Willard Marriot, hotel and restaurant executive, lives. He establishes Marriot Corp. (formerly Hot Shoppes, Inc.) in 1928.

1901–66 Walt Disney, cartoonist and businessman, lives. He creates entertainment empire out of animated cartoons, including movies, television productions, and multimillion-dollar theme parks.

1901– William S. Paley, multimillionaire broadcasting executive and owner of CBS (1928), lives.

1902–84 Ray A. Kroc lives. He founds McDonald's Restaurant franchise, which eventually produces billions of hamburgers and a fortune besides.

1905–76 Howard Hughes lives. He parlays inherited tool company into $2 billion fortune based on tool manufacture, real estate, airplanes, and other investments.

1908– Walter Hubert Annenberg, billionaire publishing executive, lives. He creates *TV Guide* (1953) and other national publications and is ambassador to the Court of St. James (1969–74).

1909– Harry Brakman Helmsley, real estate billionaire and founder of Helmsley-Spear, Inc., lives.

1909– Ernest Gallo lives. With his brother Julio (b. 1911), he starts billion-dollar wine company with $5,900.

▶

1912– David Packard, billionaire cofounder of Hewlett-Packard Co. (1939), lives. He invents audio oscillator, first used in Walt Disney's *Fantasia*.

1913– William Redington Hewlett, billionaire cofounder (with David Packard) of Hewlett-Packard Co. (1939), lives.

1917– Katharine Graham, multimillionaire inheritor (1963) of Washington Post Company and *Newsweek*, lives.

1919–1990 Malcolm Forbes, flamboyant multibillionaire publisher of *Forbes* magazine, lives.

1920– Sam Moore Walton, multibillionaire discount retail chain-store executive and CEO of Wal-Mart Stores, lives.

1920– Franklin Parsons Perdue, multimillionaire chicken farmer and industrialist, lives. He retires as CEO of Perdue Farms in 1988, with personal net worth over $300 million.

1924– Marvin Harold Davis, billionaire oil, real estate, and entertainment investor, lives. Since 1981 he has been co-owner of 20th Century Fox.

1925– Merv Griffin lives. An entertainer and producer, he owns MGP (Merv Griffin Productions).

1930– H. Ross Perot, billionaire businessman, lives. He founds Electronic Data Systems in 1962 with $1,000 loan from his wife. Sells the business in 1984; founds Perot Sytems in 1988.

1936– Carl Icahn, billionaire arbitrager and options specialist, lives. He is chairman and president of Icahn & Co., chairman and CEO of ACF Industries, and chairman of Trans World Airlines.

1938– Robert Edward (Ted) Turner III, broadcasting and sports billionaire, lives. He founds WTBS-Atlanta.

1939– Ralph Lauren, multimillionaire fashion designer and creator of Polo, Inc., lives.

1940– Francisco A. Lorenzo, airline executive and owner of Eastern Airlines, lives. He is CEO of Texas Air Corp.

1943– Ronald Owen Perelman, billionaire leveraged-buyout specialist, lives.

1946– Donald John Trump, billionaire real-estate magnate and hotel and casino owner, lives.

1956– William Henry Gates III, billionaire computer wizard and cofounder of Microsoft Corporation (1975), lives.

The Developing U.S. Economy

See also chronologies for selected industries.

1723 First commercial corporation is organized in United States. Subsequently insurance companies, banks, and companies for building projects are organized as corporations.

1775–83 American Revolution. Wartime needs produce business boom.

1781 First private commercial bank in United States is chartered.

1784–88 Post-Revolutionary War business recession is exacerbated by British banking crisis.

1789–1807 Business boom, with minor breaks, results from wars in Europe. U.S. Embargo Act and war with Great Britain (War of 1812) ends period of prosperity.

1791 Founding of Bank of United States.

1792 Decimal coinage system is established.

1791 First American securities exchange is established, in Philadelphia; securities exchange is established in New York City in 1792.

1800 Federal government's primary source of income is customs receipts, which account for over 80 percent of revenue.

1811 New York establishes laws governing incorporation, to replace previous practice of charters issued by state legislatures.

▶

System becomes standard practice in other states in coming decades.

1812 First U.S. life-insurance company is founded.

1815–21 Major recession. Collapse of land values in 1819 results in bank failures and country's first banking crisis.

1816 Second chartering of Bank of United States, following fiscal problems brought on by national debt resulting from War of 1812. Its charter also lapses, in 1836.

1820s New York Stock Exchange becomes nation's leading stock exchange.

1819 Savings banks begin system of paying interest on deposits.

1834–36 Brief business boom results from increased railroad building, land speculation, and other factors.

1835 National debt is completely paid off, thanks to budget surpluses in 1820s and early 1830s.

1837 Panic of 1837. Many state banks fail. Business recession lasts until 1843–44.

1846 Independent Treasury is in continuous operation from this time until it is replaced by Federal Reserve system (1913).

1849–56 Business boom is fed by increase in railroad building (made possible by foreign capital influx) and by discovery of gold in California.

1857 Panic of 1857. Speculative boom in real estate and railroads halts briefly with this short-lived financial panic.

1859 Dun & Co. is formed from America's first agency for credit rating, which began operations in 1841.

1861–65 American Civil War. Wartime needs produce business boom. Railroad building after war helps maintain prosperity in North; South suffers period of economic chaos in postwar years.

1862 Federal government issues about $450 million in greenbacks as legal tender during Civil War.

1863 National Banking System is established.

1865 Taxes, enacted to finance Civil War, by this time account for 43 percent of federal government revenue. They include income tax and business taxes. National debt reaches $2.8 billion by this year, due to wartime spending.

1866–94 Federal revenue exceeds public spending each year of this period. Public debt is nearly eliminated again.

1869 Government foils attempt to corner gold, resulting in Black Friday.

1873–78 Major business recession follows failure of Jay Cooke and Co. investment-banking house (this marks end of prolonged speculative boom in railroad securities).

1870s J. P. Morgan emerges as a leading American investment banker.

1880s Investment bankers are active in financing extensive railroad construction.

1890 Sherman Antitrust Act is passed. It becomes tool for breaking up monopolistic combinations during early 1900s.

1893–97 Business recession. Almost 500 banks and 15,000 commercial firms fail.

1897 Period of sustained prosperity begins. Minor setbacks aside, this period lasts through World War I and 1920s.

1900 United States adopts gold standard. $150 million gold reserve is to be maintained to back legal tender.

1903 Federal government's Department of Commerce and Labor is created.

1907 Panic of 1907. Runs on banks force many closings due to insufficient reserves.

1909 Over 2 million Americans now own stocks, as public interest in securities investment grows.

1913 Way is set for establishment of income tax by passage of 16th Amendment to Constitution.

1913 Federal Reserve system is established. It provides commercial banks with emergency source of funds.

1914 Federal Trade Commission is organized.

1914–18 World War I. U.S. business booms as result of orders for war material by Allied nations in Europe and finally by U.S. participation in war.

1916 Public debt is just $62 million prior to U.S. entry into World War I. It skyrockets to $26 billion by 1919.

▶

1919 Business drops off sharply in immediate postwar years, but prosperity returns by 1922. Construction and real estate boom and consumer durable goods industry expands.

1925 Government's main sources of revenue by this time are corporate and individual income taxes, customs receipts, and taxes on tobacco.

1928–29 Securities trading on Wall Street hits all-time highs as broad-based speculative boom in stocks becomes stock-buying frenzy.

1929–33 Great Depression. Crash of stock market in 1929 (begins October 24) touches off worldwide depression and wipes out investments of individual investors and major firms alike. About a third of all U.S. banks suspend operations.

1930 National debt is reduced to $16 billion by this time. It rises sharply in subsequent years owing to public spending programs designed to offset hardships of Great Depression and reaches $37 billion by 1938.

1933 Banking crisis is brought on by wave of failures. Federal government takes emergency action to prop up banking industry. Unites States goes off gold standard.

1935–37 Renewed recession slows recovery from Great Depression.

1938 American life-insurance companies are growing rapidly, with assets up by 500 percent from early 1900s.

1939–45 World War II. Nation enjoys economic boom as industrial production is vastly expanded to meet wartime needs of United States and Allies.

1944–45 American industrial production during war years peaks.

1946 Expenditures of World War II push national debt up to $258 billion.

1953 Federal budget deficit hits high to date for peacetime: $9.4 billion.

1954 U.S. GNP hits high to date of $365 billion.

1955 Sharp drop in stock-market prices leads to biggest one-day losses to date.

1959 Some 13 million Americans own stocks by this time.

1959 Federal debt limit is raised to $295 billion. By 1962, however, public debt is over $300 billion.

1960s Despite sluggish period at beginning of decade, 1960s witness period of sustained prosperity and economic growth, fueled by government spending and wartime business boom created by Vietnam War.

1962 Stock prices drop in sharpest decline since 1929.

1968 Trading volume on New York Stock Exchange tops 16.4 million for first time since 1929.

1970s Recession and inflation, "stagflation," become nation's two leading economic woes. An ongoing problem of inflation, caused by high government spending and Vietnam War, is made much worse by manifold increase in oil prices.

1971 Federal government institutes wage–price freeze to check rising inflation.

1978 United States and Japan set new accord on trade as worsening U.S. trade deficit begins to increase public debt sharply.

1979 Inflation rate is worst in 33 years as inflation becomes nation's leading problem.

1980s Prosperity returns after federal government institutes program of sharply increased interest rates and curtailed federal spending. By mid-1980s economy is booming, though continuing balance of trade deficits and high interest rates create mushrooming public debt.

1980 Banking industry is deregulated.

1981 Public debt hits $1 trillion.

1981–82 Business recession.

1982–87 Sustained bull market sends stock prices soaring. Dow Jones average tops 2,000 for first time in history, in January 1987.

1984 Inflation drops to 1972 levels.

1985 United States becomes net debtor nation for first time since early 1900s, as result of continuing trade deficits.

1987 Crash of 1987. Stock prices plummet 508 points, but repeat of Great Depression following 1929 crash is averted.

►

1989 Federal government successfully pressures banks to new agreement to ease Mexico's massive foreign debt.

1989 Federal government legislates bailout of nation's ailing savings-and-loan banks, to eventually cost $159 billion.

Winners of the Nobel Prize in Economics	
1969	Ragnar Frisch (1895–1973, Norway) and Jan Tinbergen (b. 1903, Netherlands)
1970	Paul A. Samuelson (b. 1915, United States)
1971	Simon Kuznets (1901–85, United States)
1972	Kenneth J. Arrow (b. 1921, United States) and John R. Hicks (b. 1904, Great Britain)
1973	Wassily Leontief (b. 1906, United States)
1974	Gunnar Myrdal (1898–1987, Sweden) and Friedrich A. von Hayek (b. 1899, Great Britain)
1975	Tjalling C. Koppmans (1910–85, Netherlands-United States) and Leonid V. Kantorovich (1912–86, USSR)
1976	Milton Friedman (b. 1912, United States)
1977	Bertil Ohlin (1899–1979, Sweden) and James E. Meade (b. 1907, Great Britain)
1978	Herbert A. Simon (b. 1916, United States)
1979	Theodore W. Schultz (b. 1902, United States) and Sir Arthur Lewis (b. 1915, Great Britain)
1980	Lawrence R. Klein (b. 1920, United States)
1981	James Tobin (b. 1918, United States)
1982	George J. Stigler (b. 1911, United States)
1983	Gerard Debreu (b. 1921, France-United States)
1984	Richard Stone (b. 1913, Great Britain)
1985	Franco Modigliani (b. 1918, Italy-United States)
1986	James M. Buchanan (b. 1919, United States)
1987	Robert M. Solow (b. 1924, United States)
1988	Maurice Allais (b. 1911, France)
1989	Trygve Haavelmo (Norway)

New York Stock Exchange

1792 Twenty-four New York businessmen organize informally to agree on fees and terms of business for exchange. They meet under sycamore tree between 68 and 70 Wall Street.

1795 North Carolina is first state to pass general incorporation laws, allowing business issues to be traded by investors.

1817 New York brokers organize formally indoors. They meet in succession of offices in Wall Street area. Adopt name New York Stock and Exchange Board. Initiation fee is $25.

c. 1820 New York exchange surpasses Philadelphia exchange in volume of trading by successfully floating Erie Canal bond issue. It also has surpassed Philadelphia by this time in field of commercial credit.

1827 NYSE deals in city, country, and state issues, as well as securities of 12 banks, 14 fire-insurance companies, five marine-insurance firms, and several others.

1827 NYSE now pays annual $500 rent for room at Merchants' Exchange, at Wall and Hanover streets.

1830 Unlisted securities are traded for first time.

c. 1830 Industrial stock issues begin to appear.

1835 NYSE moves to temporary quarters on its present site after fire at Merchants' Exchange.

c. 1840 Railroad stocks and bonds (first issued in 1830) make up large portion of investor trading.

▶

c. 1850 Expanded security dealings lead to growth in New York investment houses and brokerage firms. NYSE is firmly established as preeminent in securities trading.

1854 NYSE brokers move into Corn Exchange Building (at William and Beaver streets) but remain only until 1856, due to excessive summer heat.

1863 Official name, "New York Stock Exchange" (NYSE), is adopted.

1867 Stock ticker, device to telegraph and print records of stock transactions, is introduced.

1868 Memberships on NYSE go on sale for first time, changing previous policy of lifetime memberships.

1869 "Black Friday" (September 24). Gold speculation leads to this financial panic. American inventor Thomas A. Edison (1847–1931) later invents improved stock ticker after watching tickers fail to keep up with this day's trading.

1871 NYSE abandons oral "call market" (in which president calls name of each stock and members bid to buy or sell) in favor of continuous auction market.

1873 NYSE closes for 10 day following Panic of 1873, which is caused by failure of several prominent New York banking houses and other firms.

1878 Telephones are introduced at NYSE.

1882 Two Providence businessmen form a company, Dow Jones, to provide hourly financial information.

1886 First day on which volume of over 1 million shares is recorded.

1889 Dow Jones begins publishing daily newpaper, *Wall Street Journal.*

1895 NYSE recommends that member firms issue annual balance sheets to shareholders.

1902 Average closing prices of selected industrial and railroad stocks are published daily by Dow Jones. In 1929, list of public utilities is included.

1910 NYSE no longer trades in unlisted securities.

1914 World War I breaks out. NYSE closes from July 31 to December 11.

1915 Dollar is established as basic currency for quoting and trading in stocks.

1922 Member firms are now required to fill out financial questionnaires on regular basis.

1929 Stock market crashes, with 16.4 million shares traded (October 29). Worldwide financial crisis and Great Depression follow.

1930 Improved stock ticker is introduced. It prints 500 characters per minute.

1933 Regular independent audits are first required for all NYSE member firms.

1933 Securities Exchange Act of 1933 requires full disclosure to investors in attempting to prevent fraud in securities industry.

1934 U.S. Securities and Exchange Commission is established to oversee sales of corporate securities.

1938 First paid president of NYSE is elected.

1964 National Association of Security Dealers (NASD) is established. Modern over-the-counter market takes shape.

1964 Stock ticker able to record 10 million share transactions per day is put into use.

1965 Stock tickers are linked to computers, allowing immediate printing of transaction records.

1967 NYSE has second-largest trading day to date, exceeded only by October 29, 1929.

1974 Price of NYSE seat falls to $70,000, its lowest level since 1958.

c. 1975 NYSE has more than 1,350 members, who trade in listed securities totaling over 1,900 issues of preferred and common stock.

c. 1980s "Merger mania," corporate takeovers financed largely by leveraged buyouts, helps sustain prolonged bull market; $37 billion merger deals are put together in 1985, $39 billion in 1986.

1982 Record 132.7 million shares are traded on August 18.

1982 Earliest computerized trading begins on experimental basis, under auspices of NASD.

▶

1986 SEC announces largest insider-trading scandal. American businessman Ivan Boesky (b. 1946) agrees to pay $100 million in fines. He is barred from securities industry for life and sentenced to three years in jail.

1987 NYSE seat sells for $1.5 million.

1987 Market plunges 508 points (October 19), largest drop to date and at least partly result of computerized trading. Heavy losses lead to collapse of several major brokerage houses. Small investors leave market, despite new limits on computer trading.

1989 "King of junk bonds" Michael Milken is indicted on securities laws violations. He pleads guilty (April 1990) and agrees to pay record $600 million in penalties.

1989 Stock market plunges 190 points (October 17), second-largest drop in one day in history of NYSE. Decline in terms of percentage of market ranks only 12th in history, however.

1990 Major Wall Street investment-banking firm of Drexel, Burnham, Lambert, Inc. closes after suffering heavy losses in market. Firm played key role in developing high-risk junk-bond market as means of financing corporate takeovers during 1980s.

U.S. Labor Movement

1806 First court case brought by employers opposing workers' rights is decided in favor of employers, in Philadelphia. (Decision is reversed in 1842.)

1809 Court pronounces illegal any union activities that injure employers.

1828 Workers establish Workingmen's Party in Philadelphia in reaction to policies of Jeffersonian Republicans. (Eventually, party branches are organized in 33 cities.)

1866 Workers in Baltimore establish National Labor Union (NLU) and work toward eight-hour workday. (NLU disintegrates in 1872.)

1869 Knights of Labor is founded. By 1886 it has 700,000 members and achieves some labor reforms.

1875 Strike by Pennsylvania coal miners, called Molly Maguires, secret and violent organization of Irish-American workers.

1877 Severe depression results in 80-percent unemployment of labor force.

1877 Railroad workers strike and shut down almost all Eastern railroads.

1880 Term *boycott* is coined in Europe after British estate manager Charles Boycott (1832–97) is shunned by local citizens for refusing to lower rents during famine.

1881 Labor leader Samuel Gompers (1850–1924) founds Federation of Organized Trades and Labor Unions.

1886 Haymarket Square Riot (May 4) in Chicago is one of several incidents of violence linked to labor movement in late 1800s. It results in rise of antilabor sentiment.

1886 American Federation of Labor (AFL) is formed in Ohio to succeed Knights of Labor.

1890 Sherman Antitrust Act prohibits trusts that interfere with trade and is used chiefly against organized labor to prohibit strikes.

1892 Homestead strike by Pennsylvania steelworkers erupts into violence when strikers shoot at Pinkerton detectives guarding strikebreakers (July 6).

1894 Treaty with China calls for end to immigration of Chinese laborers to United States.

1894 Coxey's Army, about 400 unemployed demonstrators led by Jacob S. Coxey (1854–1951) and six-piece band, marches into Washington, D.C., demanding federal

▶

help (April 30). The "army" disbands without success.

1894 Pullman strike in Chicago begins (May 11). Labor leader Eugene Debs (1885–1926) organizes nationwide boycott of railroads. Strikers battle with federal troops before strike is finally broken.

1905 U.S. Supreme Court rules that minimum-wage laws are unconstitutional.

1905 Radical union, Industrial Workers of the World (IWW), or "Wobblies," is founded to organize unskilled industrial laborers and to promote overthrow of capitalist system. The union, declining through World War I, is effectively crushed during postwar "red scare."

1916 Congress exempts labor unions from antitrust laws.

1917 Membership in AFL is now over 2 million; significant reforms have been made in salaries and hours.

1919 Steel-industry strike called by AFL leaders is broken by company strikebreakers.

1924 Immigration Act limits numbers of immigrants, reducing competition for jobs.

1932 Injunctions in labor disputes are outlawed by Congress (but later reinstated).

1933 President Franklin Roosevelt (1882–1945) begins New Deal programs to relieve economic hardships of Great Depression.

1933 National Industrial Recovery Act establishes minimum wage for workers and guarantees fair hours, collective bargaining, and right to unionize. (It is found unconstitutional in 1935).

1935 Social Security Act establishes unemployment compensation, retirement benefits, and state welfare programs for workers.

1935 Labor leader John L. Lewis (1880–1969) founds Congress of Industrial Organizations (CIO). (CIO unions are expelled from AFL in 1938.)

1935 Wagner Act requires employers to accept collective bargaining.

1938 Minimum wage for workers is established under Fair Labor Standards Act.

1941 Labor leaders promise to avoid strikes during war years so defense industry remains strong.

1947 Taft-Hartley labor act bans secondary boycotts, closed shops, and jurisdictional strikes.

1948 UAW wins first contract (with General Motors) containing provision for cost-of-living increases.

1949 CIO expels 11 unions thought to be pro-Communist.

1952 Labor leader George Meany (1894–1980) becomes head of AFL. Walter Reuther (1907–70) becomes CIO president.

c. 1953 Increasing automation in factories leads to greater unemployment.

1955 AFL and CIO unify under leadership of George Meany, ending years of rivalry.

1957 Labor leader James "Jimmy" Hoffa (1913–75?) becomes head of Teamsters' Union.

1957 AFL-CIO expels Teamsters' Union following charges of corruption made by Senate committee.

1959 Labor-Management Reporting and Disclosure Act establishes rules to eliminate corruption in use of union funds and procedures for fair union elections.

1962 American labor leader Cesar Chavez (b. 1920) begins organizing California grape pickers. He forms farm workers' association (UFW).

1962 President John F. Kennedy (1917–63) allows federal employees right to organization and collective bargaining but not to calling strikes.

1962 Steelworkers' union develops retraining plan to protect laborers from increased automation.

1963 Federal equal-pay-for-equal-work law is enacted.

1964 Civil Rights Act of 1964 is passed. It creates Equal Opportunity Commission to end discrimination in hiring.

1964 Teamsters' boss Jimmy Hoffa is convicted of charges of corruption. He serves four years in prison.

▶

1970 UFW leader Cesar Chavez is jailed for organizing illegal lettuce boycott.

1972 Equal Employment Opportunity Act is passed, allowing preferential hiring and promotion of women and minorities (affirmative action).

1975 Former Teamster boss Jimmy Hoffa disappears and is presumed murdered.

1981 President Ronald Reagan (b. 1911) orders firing of some 12,000 federal air-traffic controllers after they refuse to end illegal strike.

1982 United Auto Workers, concerned about plant closings and increasing automation, sign contract with Ford that trades off wages and benefits for job security.

1987 Teamsters are readmitted to AFL-CIO, which ousted union in 1957.

U.S. Chemicals Industry

See also Plastics and Synthetics in the Technology section.

1614 First saltworks is established near Jamestown to produce saltpeter, an important ingredient in gunpowder.

1639 Glassworks is established in Plymouth.

1661 English Navigation Acts reserve certain natural resources for England and discourage export by colonies of goods also made in England. Restricted goods include soap, glass, and textiles.

1690 Pennsylvania colony has lime kiln (established in 1681), tannery (1683), and paper mill by this date.

1703 English Parliament lays claim to almost all pitch pines in colonies, citing Royal Navy's need for pitch, tar, and resin.

1716 Manufacture of soap in Boston begins at about this time.

1730 Sugar is refined in New York by this date.

1768 Colonies export 11,000 pounds of soap to England, despite trade restrictions.

c. 1785 Philadelphia becomes center for chemicals manufacturing after Revolution. Factories produce white lead, sulfuric acid, and other substances.

1802 Du Pont Powder Works is established in Delaware. It imports saltpeter for gunpowder until discovery of sufficient supplies in Kentucky and Tennessee.

1816 Baltimore begins first U.S. manufacture of coal gas for street illumination. Eventually almost every American town has a gasworks.

1837 Procter & Gamble is established in Ohio. It makes candles and soap.

1839 U.S. inventor Charles Goodyear (1800–1860) invents vulcanizing, treating rubber with sulfur to prevent it from melting and sticking in warm weather.

c. 1865 Demand for sulfuric acid (for removing impurities from oil) grows with expansion of U.S. petroleum industry. Factories are moved near sites of oil production.

c. 1865 Native sulfur deposits in southern United States are found to be difficult to mine due to quicksand. (Method for pumping liquid sulfur is developed about 1890. United States becomes world's leading sulfur producer by 1910.)

1866 Chemicals concern Duffield, Parke and Company is established. It later is renamed Parke, Davis.

1870 New York inventor John Hyatt (1837–1920) introduces celluloid for manufacture of dentures, billiard balls, and shirt collars.

1871 Carbon black is first produced from natural gas, in Pennsylvania.

1884 Soda-works plant opens in Onondaga salt region of New York. Forerunner of modern

▶

chemicals factory, it includes maze of pipes and tanks with valves for controlling temperatures and pressures.

1885 American businessman George Eastman (1854–1932) manufactures first commercially successful chemically treated photographic film.

1886 American Charles Hall (1863–1914) develops successful process for electrolytic production of aluminum. It begins important industry based on electrochemistry.

1889 Bromine is manufactured on large scale for the first time, by chemist Herbert Dow (1866–1930) in Michigan.

1891 Edward Acheson (1856–1931), assistant to inventor Thomas Alva Edison (1847–1931), discovers carborundum (silicon carbide). It is important for manufacture of abrasive materials.

1897 Construction of huge hydroelectric power sites at Niagara Falls in New York makes enough power available for development of electrochemical industries.

1899 Merger of 12 companies forms General Chemical Company. It becomes Allied Chemical and Dye Company in 1920.

1907 American Cyanamid Corporation is organized in Niagara Falls, Ontario to make calcium cyanamide for fertilizer.

1908 Du Pont Company in Delaware begins production of plastics and paints in addition to gunpowder. It begins to acquire other companies in 1917 and by 1930 has become large, diversified company.

1909 Belgian-born U.S. inventor Leo Baekeland (1863–1944) develops plastic Bakelite. Made from phenol and formaldehyde, it is first synthetic resin.

1914 Corning Glass Works in New York introduces Pyrex, chemically treated glassware for cooking.

1914 At outbreak of World War I, U.S. government encourages chemicals industry to produce goods previously imported from Germany and Great Britain.

1917 Federal government confiscates German dye patents for use by American manufacturers. United States becomes self-sufficient in dyes by 1928.

1917 Union Carbide and Carbon Corporation (later Union Carbide Corporation) is formed by merger of four smaller chemical companies. Company figures in development of what becomes massive U.S. petrochemicals industry.

1931 Successful mining of potash in New Mexico mines leads to U.S. self-sufficiency in potash needed for fertilizer.

1939 Du Pont scientists develop nylon, new plastic fiber for textile materials. It is first genuinely synthetic fiber.

1939 U.S.-manufactured chemical products from about this time are no longer designed to copy natural products. They now are created with desirable features not available naturally.

c. 1945 Manufacture of fluorine is sharply increased for production of uranium hexafluoride needed by Manhattan Project. By-products eventually include fluorocarbons, such as Teflon, which resist corrosion.

c. 1950 Sharp increase in production of agricultural chemicals such as fertilizers and pesticides begins about this time. They are produced for both domestic use and export.

c. 1955 Synthetic diamonds of crystallized carbon are commercially produced for use as drills or cutting tools by this date. New technique developed in 1988 creates synthetic diamonds 85 percent as hard as genuine stones.

1969 DDT, widely used pesticide, is banned in United States because of its adverse environmental effects. This marks beginning of popular concern about environmental effects of many chemical products commonly used in modern world.

1971 American chemists begin research into limiting lifetime of plastics in response to environmental concerns about pollution and litter.

c. 1980 Chemicals industries experiment with batteries and systems for improved conversion of solar energy to electrical energy.

▶

c. 1988 American chemists develop new techniques for chemical-hardening of glass for laser applications.

1989 American chemicals industries continue research into high-temperature ceramic superconductors for use in electronics applications.

U.S. Railroad Industry

1815 First U.S. charter for railroad is granted to American inventor John Stevens (1749–1838), who cannot secure required capital to construct it.

1825 Stevens builds locomotive that operates on short, circular track at his New Jersey home.

1826 Horse-powered 3-mile-long Granite Railway is used in Massachusetts for hauling stone.

1827 Baltimore merchants receive charter for Baltimore & Ohio Railroad. Ground is broken in 1828. First passengers ride in single, horse-drawn cars in 1830.

1830 Experimental locomotive developed by American inventor Peter Cooper (1791–1883) makes its first run on 13 miles of finished track.

1830 South Carolina Canal and Rail Road offers first regular steam railway service. It is first "modern" American railroad to combine tracks, steam-locomotive power, trains of cars, and public passenger service. United States has 23 miles of finished track at this time.

1849 Railroad convention is held at Memphis, Tennessee to promote private investment and public and governmental support for railroads. Many such conventions are held at about this time.

1849–51 Railroad conspiracy against Michigan Central Railroad results from local opposition. Conspirators stone trains, burn stations, and remove rails. Twelve persons eventually are convicted and imprisoned.

1850 Total track mileage exceeds 9,000 miles by this time, despite clergymen's warnings about "the iron horse" and doctors' warnings about railroads' excessive speeds.

1853 New York Central Railroad is formed by consolidation of numerous smaller railroads between Buffalo and Hudson River.

1853 First Texas railroad begins operation.

1855 Railway connections are made from New York to Great Lakes, from Philadelphia to Pittsburgh, and from Baltimore to Ohio River by this date.

1856 Locomotive crosses Mississippi River on first railroad bridge, constructed by Rock Island line (later Chicago, Rock Island, and Pacific).

1856 Railroad building begins in California.

1857 Completion of Illinois Central line, constructed in part with federal government grants. (Government grants railroad land; in return, railroad carries government freight, mail, and troops at low rates.)

1859 First sleeping car is introduced for passenger service.

1860 Some 30,000 miles of railroad track are in use nationwide by this time.

1861–65 Railroads play significant role in warfare for first time, in Civil War. They transport troops and large quantities of food, ammunition, and supplies to battlefront much faster than before. Union states hold over two-thirds of railway mileage during war.

▶

1862 Congress authorizes construction of railroad from Missouri River to California on central route, ending years of debate concerning five possible routes to West. Union Pacific begins building westward; Central Pacific builds eastward from California.

1868 Air brake that can be controlled from locomotive is invented by American George Westinghouse (1846–1914) to replace hand brakes set on each car by brakemen. Westinghouse later invents automatic signal devices.

1869 First U.S. transcontinental railroad is completed. Driving of Golden Spike at Promontory, Utah marks formal completion of construction.

1870 Track mileage nationally exceeds 53,000 miles by this year.

1874–76 Widespread railroad rate war causes rates to fall below amounts required to meet costs, leading to temporary truce among owners and traffic-sharing agreements in 1877.

1880s Railroad building boom peaks. Over 70,000 miles of track are laid in this decade.

1883 Railroads adopt standarized time, with four zones each 1 hour apart. It replaces over 100 different railroad times that previously caused constant confusion.

1886 Railroads move toward standard-gauge track of 4 feet, 8.5 inches.

1887 Interstate Commerce Commission is created by congressional act to establish maximum railroad rates. Public demands federal regulation after numerous rate wars.

1887 Standard design for car couplers is approved following test of many models in Buffalo, New York.

c. 1890 Conversion from iron rail to safer and stronger steel rail is largely complete.

1894 Strike by workers at Pullman Palace Car Company in Chicago over 25-percent pay cut in wake of Panic of 1893 results in nationwide boycott led by Eugene Debs (1855–1926). Strike is broken by federal troops.

1902 Record New York–Chicago run in 20 hours by *20th Century Ltd.*

c. 1905 First electric lamps appear on trains.

1917 President Woodrow Wilson (1856–1924) establishes Railroad Administration to control and coordinate rail services until end of World War I. Government controls end in 1920.

1920 Congressional Transportation Act encourages consolidation of railroads and establishes Railroad Labor Board to determine wages and working conditions.

1922 First nationwide railroad strike is called when shopmen protest Labor Board decision to reduce wages. Strike fails but leads to elimination of Labor Board (1926).

1926 Congress passes Railway Labor Act for mediation of labor disputes.

1929 Railway Labor Executive Association is formed to unite many separate railway unions.

1929 First air-conditioned passenger car. First all-air-conditioned train operates in 1931.

1930 First electric passenger train in United States is tested in New Jersey.

1933 Increasing truck traffic, not federally regulated at this time, becomes new competition for railroads.

1934 First Social Security measure of New Deal is the Railroad Retirement Act. It later is declared unconstitutional. In 1937, acceptable retirement and unemployment insurance system is developed.

1934 Diesel locomotive, developed in 1924, is first used for passenger service. It is introduced for freight transport in 1941.

1934 Streamlining of passenger trains begins by this date.

1946 President Harry S. Truman (1884–1972) seizes the railroads to end nationwide strike by engineers and trainmen. Government seizures occur again in 1948 and 1950.

1946 Congress discontinues grants-in-aid for railroad building.

▶

1950s "Piggyback" system, carrying truck trailers on flatcars, is in use for high-value freight by this time.

1951 Congress ends prohibition against union shops.

1958 Congressional Transportation Act relaxes railroad rate regulations, allowing consideration of competitive cost factors in establishing rates.

1959 "Featherbedding" is issue in railroad labor relations. Management charges that unions require employment of unnecessary workers, such as firemen on diesel locomotives. Firemen on diesel engines are no longer employed after 1972.

1960s Railroad passenger ridership continues sharp decline as result of increasing emphasis on auto and airplane travel. Most rail passenger service becomes uneconomical except in metropolitan areas.

c. 1965 Infrared detection devices placed near tracks are used to detect overheated bearings on freight cars.

1968 Merger of New York Central and Pennsylvania railroads forms 21,000-mile Penn Central system. System files for bankruptcy in 1970.

1969 Four major railroad unions merge, covering railroad conductors, firemen and enginemen, trainmen, and switchmen.

1970 Automatic Car Identification (ACI), nationwide computerized network, provides system for quickly locating rolling stock.

1970 Burlington Northern System of 23,500 miles is formed by merger of Great Northern, Northern Pacific, and Chicago, Burlington and Quincy railroads.

1971 Federally sponsored National Railroad Passenger Corporation, later known as Amtrak, begins. It handles most northeastern passenger service.

1972 Centralized-traffic-control automatic signaling is in use on thousands of miles of track by this time.

1976 Federally funded corporation, Conrail, takes over bankrupt northeastern railroads.

c. 1985 "Doublestack" freight trains carry containers across continent. Cost of crossing is actually cheaper than transport by sea through Panama Canal.

U.S. Steel Industry

1847 American ironmaster William Kelly (1811–88) refines molten pig iron by forcing blasts of air through it to remove impurities.

1856 American patents are granted for Bessemer process (English method developed for refining molten pig iron by blowing air through it), although Kelly's prior claim is recognized.

1864 First U.S. production of steel, using Kelly process, in Michigan.

1865 Bessemer steelworks are constructed at Troy, New York by U.S. businessman A. L. Holley (1832–82). He goes on to alter and improve original Bessemer process.

1866 Bessemer and Kelly steel interests combine, forming Bessemer Steel Association.

1867 U.S. industrialist Abram Hewitt (1822–1903) acquires rights to manufacture steel using Siemens-Martin open-hearth method.

1869 First open-hearth furnace for making steel is built in United States.

▶

1873 Company run by Andrew Carnegie (1835–1919) is producing steel by this date.

1875 Steel manufactured in United States totals 380,000 tons this year. (Huge demand for steel is caused largely by nation's rapidly growing railroad industry.)

1880 Steel production exceeds wrought-iron production in United States for first time. (Steel rails make up 69 percent of total annual production.)

1886 United States becomes world leader in steel production, surpassing Great Britain.

1889 Jones & Laughlin Steel Company produces first U.S.-made Bessemer Steel I-beam. I-beams later are widely used in skyscraper contruction.

1890 Open-hearth process is replacing Bessemer method by this time. It is more precise, and recently invented devices mechanize the job, cutting costs.

1892 Homestead Steel strike against Carnegie Steel Company (June–November). Violence erupts between strikers and Pinkerton guards hired to protect plant. Strikebreakers eventually replace most union men.

1900 High-speed steel, containing tungsten and carbon, is developed. It is exceptionally strong even when heated, making it ideal for manufacturing machine tools.

c. 1900 Enormous ore deposits in Lake Superior area have become primary source for North American steel manufacturing. Ore can be inexpensively transported by water.

1901 Founding of U.S. Steel Company.

1904 Founding of Bethlehem Steel.

1906 Electric-arc furnace is introduced in United States, originally for production of steel for tools and cutlery. It uses electric arc, formed between electrode and metal being heated, to generate heat. Eventually replaces other processes altogether.

1910 Use of alloy steel, carbon steel to which other elements have been added, is increasing rapidly by this time.

1911 U.S. annual production of steel totals 23.6 million tons this year.

1911 Stainless steel, first developed in France in 1904, is recognized for its anticorrosive quality. It is patented in Germany and later is widely used in U.S. industry.

1914 Enormous contracts for war materiel, machine tools, and the like for use in World War I greatly expand U.S. steel industry.

1920 U.S. steel production exceeds 60 million tons annually.

1923 First wide-strip mill opens in United States for producing sheet steel. Sheet stock is needed for manufacture of cars, appliances, and other domestic products. By 1939, U.S. mills produce sheet steel in strips 8 feet wide.

c. 1924 Eight-hour workday is introduced.

1941 U.S. government freezes steel prices as it begins to organize U.S. economy for World War II effort.

1943 Electrodischarge machining, new process for cutting hardened steel using electric impulses, is in use by this date. It becomes widely used in making machine tools and dies.

1950 U.S. manufactures over 100 million tons of steel annually by this date.

c. 1950 Steel industries of Europe and Japan, revived after World War II by U.S. aid, begin to provide stiff competition at about this time.

1952 President Harry S. Truman (1884–1972) orders seizure of U.S. steel mills to prevent strike by Steelworkers of America. U.S. Supreme Court rules seizure unconstitutional, and union strikes. Truman then settles strike by calling all sides to White House conference.

1953 Continuous-casting equipment is introduced to cast rounds of alloy steel, in Pennsylvania.

1959 President Dwight D. Eisenhower (1890–1969) halts steel strike by invoking Taft-Hartley Act (October 9).

▶

1960 President John F. Kennedy (1917–63) forces steel-price rollback after major U.S. companies announce increase to meet increased costs and foreign competition.

c. 1960 Depletion of richer ores leads to use of harder taconite ores found in Lake Superior areas. Technological changes include contruction of new plants to enrich ore and convert it to pellets. Cost increases lead to greater dependence on imported ore and finished steel.

c. 1960 Environmentalists voice concerns about ecological damage caused by strip mining and demand restrictions.

1961 Introduction of steel 10 times stronger than normal, through addition of nickel, cobalt, molybdenum, and titanium to iron. Called maraging steel, it later is used in production of rockets and missiles.

c. 1965 Basic oxygen process of steelmaking developed in Germany and Austria (using pure oxygen instead of air to oxidize impurities) gains ground over open-hearth method.

c. 1965 Electric furnaces are increasingly used. Temperatures can be closely regulated and exact percentages of alloys needed can be precisely determined.

1966 United States has oxygen-process steelmaking equipment sufficient for producing 40 million tons a year.

1968 Spray steelmaking is introduced. Molten iron, sprayed in several directions, is atomized and quickly turns into steel, leading to advances in continuous steelmaking.

1970 Bessemer process for steelmaking has been virtually abandoned by this time.

c. 1970 Foreign competition, high costs, and aging physical plant erode U.S. steel industry's position, though it continues to grow into early 1970s.

c. 1975 To strengthen steel, American manufacturers experiment with blowing gaseous oxygen and powdered lime into hot metal as part of basic oxygen process.

1977 Bethlehem Steel reports losses of $477 million as U.S. steel industry suffers from foreign competition.

1978 Federal government establishes trigger price system to curb dumping of foreign steel in U.S. market at below-minimum prices.

1979 U.S. steelmakers bond glass or metal to steel surfaces to prevent or resist corrosion. They use controlled chemical explosions to weld the materials to steel.

1979 America's largest steel producer, U.S. Steel, lays off 13,000 workers and closes 15 plants as retrenchment in steel industry continues.

1982–84 Federal government restricts imports of foreign steel.

1983 Weirton Steel Works, West Virginia, is bought out by employees who otherwise would have lost their jobs when plant closed down.

1984 Federal government reinstates system of VRA (voluntary restraint agreements) used during 1960s to lessen impact of foreign imported steel on U.S. steel industry.

1985 To decrease concentrations of impurities in finished steel, American manufacturers use "external desulfurization," injecting desulfurizing agents into molten metal in stream of gas.

1986 Increased automation in many U.S. steel plants by or before this date is expected to cut costs by up to 20 percent.

1987 USX reports record high losses.

1989 USX workers end longest steel strike in U.S. history (August 1, 1988–February 2, 1989).

1989 American steel companies report earnings of some $2 billion, marking their strong recovery from 1982–86 lows.

U.S. Oil Industry

1854 Pennsylvania Rock Oil Company forms to recover surface oil from deposits in Cherrytree Township, Pennsylvania, thus becoming the first U.S. oil company.

1859 First commercial oil well is drilled in Titusville, Pennsylvania, marking beginning of modern oil industry.

1861 First cargo ship carrying U.S. oil crosses Atlantic, to London.

1863 First oil pipeline, 5 miles long with 2-inch diameter, is built to deliver Pennsylvania oil to railroad terminal.

1865 Pennsylvania law establishes "rule of capture," allowing well owner in one locale to take all possible oil even if it diminishes flow in neighboring wells.

1870 Standard Oil Company is incorporated in Ohio with U.S. philanthropist John D. Rockefeller (1839–1937) as president. (By 1880s he controls 80 percent of U.S. oil-refining operations and 90 percent of U.S. pipelines.)

1874 U.S. crude-oil output totals 10 million barrels this year.

1899 Rockefeller consolidates his many oil-refining companies into Standard Oil Company of New Jersey.

1900 Domestic oil production this year reaches 60 million barrels.

1901 Discovery of oil fields at Spindletop opens up Texas as greatest U.S. oil-producing region.

1902 Oil-producing firm called The Texas Company incorporates. It later becomes Texaco, Incorporated.

1902 U.S. oil company discovers first oil in Alaska. Gusher rises 200 feet in air.

1907 Gulf Oil incorporates.

1911 U.S. Supreme Court cites antitrust violations and orders breakup of Standard Oil Company. More than 30 smaller companies are formed, including Standard Oil of New Jersey (later Exxon), Standard Oil of New York (later Mobil), and Standard Oil of California (later Chevron).

1911 Gulf Oil drills well in floor of lake in Texas/Louisiana, possibly the earliest offshore drilling.

1913 U.S. allows income-tax depletion allowance for oil companies to offset costs of drilling dry holes. (Tax is terminated in 1975.)

1914 World War I accelerates advances in oil industry as fuel is required for newly developed army transport trucks, airplanes, and submarines.

1914 First commercially "cracked" gasoline is produced by breaking down hydrocarbon molecules under heat and pressure.

1921 U.S. Navy transfers control of strategic oil reserve at Teapot Dome, Wyoming to Department of the Interior. Subsequent bribes connected to leasing of oil reserves result in Teapot Dome scandal.

1923 First antiknock gasoline is sold in United States.

1928 U.S. participation in Middle East oil exploration begins. Five companies form Near East Development Corporation.

c. 1930 Oil industry uses explosion seismology in prospecting for oil. It involves study of seismic waves créated by explosions, to learn about depth and shape of rock strata likely to contain oil.

1932 Federal government gasoline tax (one cent per gallon) begins.

1933 Standard Oil Company of California acquires oil rights in Saudi Arabia. Major discoveries are made in 1935.

1939 World War II brings new demands for oil and oil products, including fuel, explosives, and asphalt. It also results in industry advances, such as new pipeline technol-

▶

ogy and increased refinery production levels.

1943 "Big Inch," longest pipeline to date, opens. It stretches 1,254 miles from Texas to Pennsylvania.

1943 First 50/50 profit-sharing agreement between foreign government and outside oil company is instituted, in Venezuela. It becomes accepted form of operation among oil-producing nations and major companies.

1953 Government establishes U.S. offshore oil deposits as national oil reserve.

1954 Premium-grade gasoline is introduced.

1959 United States establishes oil-import quotas to protect domestic oil for defense purposes. (Quotas are removed in 1973.)

1960 OPEC, Organization of Petroleum Exporting Countries, is established to organize policies of Third World oil-producing nations.

c. 1965 Digital recording of seismological data on magnetic tape vastly improves knowledge gained through explosion seismology. It becomes chief tool in oil exploration.

1967 Major oil deposits are discovered at Prudhoe Bay, Alaska.

1969 Offshore oil-well blowout creates massive oil slick that threatens coastline along Santa Barbara, California. It raises controversy over possible environmental damage connected with offshore drilling.

1970 U.S. oil output this year exceeds 3.5 billion barrels, but American oil consumption is exceeding domestic production. United States is becoming dangerously dependent on foreign oil imports.

1973–74 Arab oil embargo creates shortages in gasoline and other petroleum products and drastically increases prices. Long lines and rationing take place at gas stations in some regions of the country.

1974 Federal Energy Administration issues plan to reduce U.S. dependence on foreign oil. It seeks to cut consumption by imposing gasoline tax, setting fuel-efficiency standards for autos and efficiency standards for electrical appliances, and establishing new standards for building insulation and heating/cooling equipment.

1976 U.S. hydrologists experiment with microbes and algae that can eat spilled oil.

1977 Renewed oil shortages again raise prices and disrupt U.S. economy.

1977 Study of plate tectonics reveals potential locations of petroleum deposits.

1977 Alaska pipeline opens (July 28). It carries oil 799 miles from Prudhoe Bay oil fields to tanker port at Valdez, Alaska.

1978 Satellite data prove useful in oil exploration.

1979 Government orders deregulation of U.S. oil prices.

1980 Crude Oil Windfall Profits Tax Act is passed to tax excess profits made by oil companies as result of increased prices during shortages.

1982 Worldwide glut in oil supply leads to declining gasoline and home-heating-oil prices in United States.

1984 Computer control systems permit instant monitoring and control of oil moving through pipelines.

1985 Environmental Protection Agency begins ban on leaded gasoline.

1986 Price of foreign oil drops below $15 per barrel for first time in years.

1987 Texaco files for bankruptcy after court orders payment of $11.1 billion fine for reportedly interfering in attempted acquisition by Pennzoil. Settlement is subsequently reached.

1989 Exxon tanker *Valdez* runs aground in Prince William Sound, Alaska, spilling 11 million gallons of crude oil. It is worst oil spill in U.S. history.

1989 Introduction of pipeline-corrosion "pig," device that travels through pipeline. Its sensors locate and measure areas of corrosion.

U.S. Automobile Industry

See also Development of Automobiles in the Technology Section.

1895 First U.S. automobile company is founded by Charles Duryea (1862–1938).

1895 American inventor George Selden (1846–1922) receives patent on automobile.

1900 Olds Company in Detroit opens first automobile factory. It produces 400 cars during first year and 4,000 per year by 1903.

1900 One in 9,500 Americans owns a car; 40 percent of cars are steam-powered, 38 percent electric, and 22 percent gasoline-powered.

1900 Fewer than 10,000 cars are produced worldwide during the year.

1903 Henry Ford (1863–1947) forms Ford Motor Co.

1903 Association of Licensed Automobile Manufacturers is formed. Its 10 members agree to pay royalties to George Selden for his patent.

1903 Selden sues Ford for patent infringement after Ford refuses to pay him royalties.

1904 U.S. surpasses France in automobile production to become world's largest car producer.

1904–08 Over 240 auto makers are established in United States.

1908 General Motors Corporation is established by American businessman William C. Durant (1861–1947).

1908 Ford debuts $850 Model T as "a motor car for the great multitude."

1908 Use of interchangeable auto parts is introduced when Cadillac president Henry M. Leland demonstrates that three cars can be taken apart and reassembled using same parts.

1911 Ford triumphs after eight-year legal challenge of Selden's patent. Appeals court restricts the patent to now-defunct engine design. Association of Licensed Automobile Manufacturers is dissolved.

1912 Electric self-starter is marketed by General Motors.

1914 First V-8 engine is produced.

1914 Assembly-line method is introduced by Ford to produce Model T. Method increases Model T production to 500,000 a year while lowering sales price to $440. Ford raises wages for its factory workers to $5 a day, more than twice industry average.

1916 U.S. annual automobile production surpasses 1 million for first time.

1917 There are 4.8 million cars and trucks in United States and 720,000 in rest of world. Average price of new car is $720.

1920 Four-wheel hydraulic brakes are introduced.

1922 Balloon tires are introduced.

1923 Sales of closed cars outpace those of open and touring vehicles for first time in United States.

1925 Chrysler Corporation is established by Walter P. Chrysler (1875–1940).

1927 Model T production ends with replacement by Model A. Some 15 million Model Ts have been sold.

1927 U.S. automakers decrease from 108 in 1923 to 44 this year. Ford, General Motors, and Chrysler establish lasting predominance as "Big Three."

1930s Great Depression causes drastic decline in auto sales.

1931 Fifty million cars have been sold. There is one car for every five Americans.

1934 First diesel-powered car is introduced by Mercedes-Benz.

1935 United Automobile Workers union is formed.

▶

1937 General Motors acknowledges United Auto Workers as bargaining agent.

1939 Automatic transmission is introduced by General Motors.

1939 Big Three automakers account for 90 percent of U.S. car sales.

1941 Ford Motor Co. is forced by National Labor Relations Board to offer unionization to its workers.

1942–45 U.S. auto makers produce 20 percent of American industry's entire contribution to manufacture of war materiel in World War II.

1945 There are 25 million cars in United States The majority are more than 10 years old.

1945 First Volkswagen Beetles are produced. They eventually capture significant share of U.S. import car sales.

1948 Tubeless tire is introduced.

1950 United States produces two-thirds of world's cars and trucks.

1958 Imports account for 8 percent of U.S. auto market. Majority of import sales are of Volkswagen Beetle (produced until 1977).

1965 Vehicle Air Pollution and Control Act regulates automobile emissions. Federal regulations eventually help force auto industry to produce cleaner, more efficient cars.

1970 Clean Air Act is passed, further restricting emission of pollutants.

1973 First Arab oil embargo creates growing market for smaller, more economical cars.

1978 Japanese imports account for 1.5 million sales, half of import market in United States.

1980 Thirty percent of U.S. auto sales are of imports.

1980 Chrysler Corp. narrowly avoids bankruptcy with federal guarantee of $1.5 billion in private loans.

1980 More than 40 million automobiles are produced worldwide this year, one-third of them in United States. One-seventh of U.S. employment is related to automobile production.

1980 Japan surpasses United States for first time to become world's largest auto producer.

1981 Japan agrees to voluntary restriction of auto exports to United States.

1983 Chrysler repays its U.S.-guaranteed loans.

1984 New York State imposes first mandatory seat-belt law.

1985 Chrysler chairman Lee Iacocca (b. 1924) publishes his best-selling autobiography.

1988 United States produces 13 million cars and trucks this year.

U.S. Airline Industry

See also Development of Airplanes in the Technology section.

1918 First U.S. airmail service begins, between New York and Washington, D.C.

1923 Regular night flights begin between Chicago and Cheyenne.

1926 U.S. enacts Air Commerce Act to create the federal airways system and license commercial air operations. It includes government aid for building airports and setting up air navigation aids.

1926 Fledgling U.S. airlines carry just under 6,000 passengers.

1927 First solo nonstop transatlantic flight, New York to Paris, is completed by Charles A.

Lindbergh (1902–74). Lindbergh's feat stirs interest in aviation.

1929 U.S. Army pilot James Doolittle (b. 1896) makes first complete flight by instruments alone, including takeoff and landing. This greatly reduces hazards of flying in bad weather.

1929 U.S. transcontinental air-passenger service begins with three-engine, 12-passenger monoplanes. Trip takes 48 hours, including train trips (over Allegheny Mountains and between Oklahoma and New Mexico).

1929–33 Postmaster General Walter F. Brown (1869–1961) guides development of the U.S. air-carrier network by his awards of airmail contracts (with subsidies) to airline companies.

1930 Founding of TWA.

1930 Start of first transcontinental service completely by airplane. It takes 36 hours.

1931 TWA begins air-freight service.

1934 Scandal over awarding of federally subsidized airmail contracts leads to overhaul of system.

1934 Air travel by credit card is introduced by American Airlines. American also introduces air sleeper service.

1935 TWA introduces first "air hostesses" on their passenger flights.

1935 First scheduled transoceanic service is begun by Pan American Airways, flying from San Francisco to Manila (November 22).

1935 Radar-detection system is developed by British. This and other advances of World War II era are adapted to civilian aviation.

1936 Douglas DC-3 goes into service. Twin-engine, 21-passenger plane becomes workhorse in commercial transportation. It flies almost 160 mph.

1939 Transatlantic air service is begun by Pan American between United States and Great Britain.

1940 Civil Aeronautics Board is created to regulate U.S. commercial air traffic.

1940s Douglas DC-6 and DC-7 passenger airliners are introduced, providing longer range and increased profitability.

1947 Pan American introduces first U.S. round-the-world airline service. First flight of Pan American's world-circling Lockheed Constellation takes off from New York June 17 and returns there on June 30.

1948 Passengers carried by U.S. airlines total almost 13 million this year, over 10 times the number of a decade before.

1952 TWA introduces first tourist-class seating on overseas flights.

1952 British-made de Havilland Comet becomes first turbojet passenger plane to go into regular service. Airlines from various nations, including United States, purchase the British jets.

1955 First executive jet is introduced in United States.

1958 Federal Aviation Agency is created through reorganization of Civil Aeronautics Board. By 1960s, agency provides subsidies to spur growth of local service airlines.

1958 Boeing 707, first American-made jet airliner, begins regular commercial service. It carries 181 economy-class passengers, almost double a prop-driven plane's capacity; maintenance cost is half. Jet capacity also finally makes large-scale air-freight operations practical.

1961 Skyjacking becomes federal crime in United States.

1964 Boeing 727 commercial airliner is introduced. By this time U.S. transcontinental flight takes five hours or less.

1964 Study finds some 64 percent of all domestic passengers flying out of the New York metropolitan area are on business trips.

1970 New supersonic-jet airliners Concorde and Soviet Tu-144 both fly at supersonic speeds for first time during tests.

1970 First "jumbo" jet, Boeing 747, begins passenger service.

▶

1971 U.S. Congress cuts off funding for development of Mach 3 SST jet airliner because of environmental concerns.

1972 TWA and American Airline begin inspecting passengers' baggage to thwart skyjacking attempts.

1976 First supersonic jet airliner, European-made Concorde, goes into service. It cruises at 1,350 mph and carries 100 passengers.

1977 British airline president Freddie Laker inaugurates cut-rate fares between London and New York, providing stiff competition for established American air carriers.

1981 Some 12,000 striking air-traffic controllers are fired by President Ronald Reagan after they refuse to end an illegal strike.

1982 Braniff International is first major U.S. airline to go bankrupt.

1989 Eastern Airlines, once a major U.S. carrier, files for bankruptcy following machinists' strike and walkout by pilots. Soon after, parent company, Texas Air, sells off Eastern's lucrative Northeastern Shuttle service to Donald Trump, who renames it Trump Shuttle.

1989 Anticollision device is introduced for corporate jets. It monitors up to 30 nearby aircraft and features computer voice instructions when collision is imminent.

1989 Major U.S. airlines plan to spend almost $800 million to restore and repair some 1,300 aging Boeing airliners (727, 737, and 747 models).

5 Technology

Basic Technology

B.C.

c. 7000–6000 Sun-dried brick and mortar are used for construction. Pottery and woven baskets are being made. Weaving of cloth appears.

c. 5000 Sailing ships appear in Mesopotamia. Copper ores are being mined and smelted in Egypt.

c. 4000 Kiln-fired bricks appear in Mesopotamia. Gold and silver are smelted in Middle East.

c. 3500 Introduction of bronze. Some early use of iron for making utensils in Egypt.

c. 3500 Early Sumerian sketch shows first known use of wheels, primitive wooden slabs roughly rounded at the corners, on simple funeral vehicle. Assyrians use similar solid wheel by c. 3000 B.C.; people of the Indus Valley use one by c. 2500 B.C.

c. 3500 Primitive potter's wheel, turned by hand, is in use in Sumeria.

c. 2500 Egyptians are producing and using papyrus, made from reedy papyrus plants growing along Nile River.

c. 2500 European craftsmen have developed simple weaving loom.

c. 2000 Mesopotamian craftsmen have developed potter's wheels and kilns.

c. 2000 Simple bellows is developed in Mesopotamia to increase heat for foundries for working metal. Bellows are made from goatskins.

c. 2000 Farmers in Tigris and Euphrates river valleys use horses to pull vehicles equipped with larger, spoked wheels.

c. 2000 Indo-European tribes have developed lighter, more efficient wheels constructed with hub, axle, four to eight wooden spokes, and wooden rim.

c. 2000 Metalworkers in Mesopotamia discover glass, probably by accidentally overheating materials. They cut and polish the glass cold until c. 1800 B.C.

c. 2000 Early Assyrian locks have pin tumblers and are installed in large wooden bolts on palace doors.

c. 2000 Domestic bathroom plumbing appears in Crete.

c. 1500 Egyptian builders now use knotted rope triangles with Pythagorean numbers to construct perfect right angles.

c. 1500 Egyptian farmers irrigate their fields by using leather bucket hanging from horizontal beam. With counterweight fixed to beam's opposite end, farmer can lift water easily.

c. 1500 Egyptians have constructed complex boats that include elaborate rigging, spars, masts, and sails.

c. 1500 Improved ironmaking process is developed by Hittite metalworkers but remains well-guarded secret until c. 1200 B.C.

c. 1450 Seed drill appears in Mesopotamia.

c. 1400 Water clock, "clepsydra," is in use in Egypt. It has bucket-shaped vessel through which water drips. Levels marked on inner surface indicate passage of time.

c. 1050 Primitive compass is developed in China.

c. 1000 Egyptians develop intricate device that measures water flow.

c. 900 Phoenicians are producing purple fabric dyes from certain Mediterranean snails.

c. 900 Jerusalem builders create elaborate water-supply system with reinforced subterranean tunnels.

c. 900 Assyrian war chariots are equipped with metal tires (copper or bronze) for greater durability.

c. 800 Egyptians have developed sundials with six time divisions.

c. 700 Etruscan workers have developed hand cranks.

c. 700 Horseshoes are invented, in Europe.

c. 600 Egyptians begin work on canal connecting Nile River and Red Sea.

c. 600 Soldering of iron has been developed.

c. 600 Assyrians have improved technology for supplying water, by constructing early aqueduct and bucket wells.

▶

566 Invention of iron welding, in Greece.

c. 500 Early form of steel is smelted in India.

c. 500 Babylonians construct tunnel more than half a mile in length. It crosses under Euphrates riverbed.

c. 500 Greek and Chinese peoples use sundials.

c. 500 Greek inventor devises simple lock and key, carpenter's square, and turning lathe.

c. 500 Greek iron smelting uses primitive blast furnace, with charcoal stoked by bellows.

c. 450 Dams are constructed in India.

c. 450 Peoples in Italy and Gaul are practicing viticulture, preserving food, especially grapes, in wooden barrels.

c. 430 Optical telegraph is in use in Greece.

c. 400 Romans develop wagonlike vehicle that can be open for hauling goods or covered with cloth canopy, for commerce and transportation.

c. 300 Cast iron is invented, in China.

c. 300 Simple pump is in use by Greek workers.

c. 257 Greek inventor Archimedes (c. 287–212 B.C.) invents water screw and block and tackle.

c. 200 Concrete is developed by Romans.

c. 200 Ox-driven waterwheel is in use, employing primitive form of gears.

c. 200 Indians of Peru and Ecuador are melting and casting gold, alloying it with silver. They have developed cylindrical clay furnaces, using blowpipes instead of bellows.

c. 200 Simple lathe is in use in many ancient civilizations.

c. 150 First paper is made, by the Chinese. It is manufactured mainly of mulberry bark and hemp fibers, beaten and spread out. Traditionally believed to have been invented in A.D. 105, paper is apparently not used for writing purposes until A.D. c. 110.

c. 110 Horse-collar harness probably is invented, in China; it is unknown in Europe until Middle Ages.

c. 100 Craftsmen of the Fertile Crescent learn to use water power to turn stones in grinding grain. They have developed paddle wheel, shaft, and millstone.

c. 100 Mesopotamians are using early horizonal waterwheel. Romans of about same period have developed vertical waterwheel.

c. 100 Celtic wagonmakers develop swiveling front axles, hardwood rollers and bearings, spokes, and smooth iron tires.

c. 80 Differential gear is invented, by Greeks.

c. 50 Glassblowing technique is introduced in Syria.

c. 20 Chinese apparently invent belt drive. Europeans rediscover it in 15th century.

c. 10 System for drilling deep wells is developed by Chinese.

A.D.

132 Chinese inventor builds first seismograph. Balls, set in raised ivory dragon heads, fall to lower carved frog heads at the slightest tremor.

c. 190 Technique of making porcelain is invented, in China.

271 Mathematicians in China use simple compass.

c. 400 Persian farmers have harnessed wind power, using primitive windmill to generate power.

410 Beginnings of alchemy, as ancients try to find chemical means of converting baser metals into gold.

c. 600 Chinese develop woodblock printing technique.

674 Churches in England are fitted with glass windows for first time.

c. 700 Persian farmers have invented simple horizontal windmill to harness wind power for agricultural purposes.

c. 700 Waterwheels are widely used in Europe to provide power for mills.

700s Vikings develop double-ended warships powered by oarsmen and large, square sail. Hull, 70 to 80 feet long, is made of overlapping oak planks.

c. 725 Earliest known mechanical clock (with escapement mechanism) is built, in China.

c. 740 First printed newspaper, in China.

751 Arabs learn Chinese method for making paper. Paper is introduced in Europe by 12th century.

c. 770 Earliest known surviving printed work, Buddhist prayer for Japanese empress, is produced.

▶

c. 790 Scandinavians use blast furnaces to make cast iron.

c. 800 Byzantines build ships that carry triangular lateen sails.

c. 850 Arabs perfect astrolabe.

c. 850 Chinese probably have developed crude form of gunpowder.

c. 850 Crank for turning grindstone is developed in Netherlands.

984 First workable canal lock for raising and lowering boats is built on China's Grand Canal.

c. 1000 Plow, drawn by horses or oxen, is in use in Europe.

c. 1000 Abacus is used in Europe.

1020 Arab scientist develops parabolic mirror.

c. 1040 Chinese printers use movable type.

1090 Water-driven mechanical clock is invented, in Peking.

1094 Waterwheel clock is built in China.

c. 1100 Multicolor printing to produce paper money is invented, in China.

c. 1100 Horse-collar harness, probably invented c. 110 B.C., in China, is introduced in Europe.

1125 Simple compass is developed for navigational use in China.

c. 1150 European church and monastery workers adapt spring-pole mechanism to their lathes and power them with foot pedals.

1185 English farmers are using vertical windmills for irrigation and other uses.

c. 1200 Bowsprits appear on European sailing ships.

c. 1200 Hinged ship rudders appear.

c. 1200 Chinese develop first arrows propelled by black-powder rockets. Knowledge of rockets quickly spreads to Europe.

c. 1250 Writing is done with quill and ink.

c. 1250 Water-powered machine saw is developed in France.

1288 Small cannon, prototype of gun, is made in China.

1298 German inventor develops earliest spinning wheel, enormously increasing possible output of yarn and thread.

c. 1314 Over 60,000 characters carved in wooden blocks are used for printing by Chinese.

c. 1325 Steel crossbow is introduced.

1335 Earliest recorded reference to European mechanical clock describes one built this year in Milan, Italy chapel tower.

1340 First blast furnace for iron smelting is built in Belgium.

1354 Mechanical clock is built at Strasbourg Cathedral.

1397 Bronze type is used in Korea.

c. 1400 First clocks for use in household appear in Europe.

c. 1400 Dutch use windmill to pump seawater from coastal land being reclaimed.

1423 Earliest dated wood-block print is published in Europe.

1428 Construction of famous Arab observatory at Samarkand (now in south-central USSR) is begun. Finished structure has 180-foot-high quadrant, device for measuring positions of stars.

c. 1438 German printer Johannes Gutenberg (c. 1398–c. 1468) invents printing press (using raised metal type), thus introducing printing in Europe. By 1500, printers have produced more than 8 million books.

1452 Metal plates are used for printing.

1481 First canal lock is installed on European canal.

c. 1500 German locksmith Peter Henlein (1480–1542) invents portable clocks.

1500s Spanish develop galleon ship, narrower than large carracks and with high sterncastle.

1500s First use of jib, a triangular headsail.

1500s *Camera obscura* (dark chamber), believed of Greek origin, is used by artists as drawing tool. It is precursor of modern camera technology.

1502 First watch, "Nuremburg Egg," is built by German clockmaker Peter Henlein (1480–1542).

1510 Italian artist and scientist Leonardo da Vinci (1542–1619) designs horizontal waterwheel. (Other inventions credited to Leonardo include crank-powered armored car, early flying machine, alarm clock, diving suit, and power loom.)

▶

c. 1530 Spinning wheel is in use throughout Europe.

1535 Italians use glass diving bell to explore sunken ships in Lake Nemi.

1551 Theodolite, used in surveying, is invented.

1566 First European seed drill is patented.

1568 Mercator projection is developed for mapmaking.

1569 Wood lathe is in use in France.

c. 1570 *Camera obscura* is studied by Giambattista della Porta (1535–1615).

c. 1570s English shipbuilders develop removable topmast, which can be lowered in storms. It makes taller masts possible.

c. 1589 English curate develops simple knitting machine capable of sewing over 1,000 stitches a minute (as compared to about 100 stitches by hand).

1590 Compound microscope is invented.

1590 Coke is produced from coal for first time, in England.

1592 Italian physicist and astronomer Galileo Galilei (1564–1642) invents primitive thermometer.

1596 Inventor of efficient ribbon loom is strangled in Danzig (now Gdansk) when workers fear his invention will put them out of work.

1606 Italian physicist and astronomer Galileo Galilei (1564–1642) invents proportional compass.

1609 Galileo constructs first of his many telescopes. Telescope has been in secret use for military purposes in Netherlands for 20 years. Galileo, however, turns his telescopes skyward and makes important astronomical discoveries.

1611 First lighthouse with revolving beacon is in operation in France.

1612 Italian physician Sanctorius (Santorio Santorio, 1561–1636) invents first clinical thermometer as part of his work on body metabolism.

c. 1620 Dutch inventor Cornelius van Drebbel (1572–1633) demonstrates first known successful submarine, in Thames River.

1623 Germans build first known mechanical calculator.

1639 Invention of micrometer.

1642 Pascaline, geared adding machine, is built by French mathematician Blaise Pascal (1623–62).

1643 Mercury barometer is invented.

1646 First projection lantern (magic lantern) is constructed by German mathematician Athanasius Kircher (1601–80).

1650 Air pump is invented by German physicist Otto von Guericke (1602–86).

1658 Balance spring is invented for watches by British naturalist Robert Hooke (1635–1703).

1656 Christiaan Huygens (1629–95), Dutch mathematician and physicist, builds first accurate pendulum clock, based on Galileo's theory.

c. 1660 Simple hygrometer is in use.

1661 Huygens develops manometer, for measuring elasticity of gases.

1667 Hooke invents anemometer.

1668 English physicist Isaac Newton (1642–1726) invents reflecting telescope, making possible telescopes of much greater magnification.

1670 Minute hand is developed by British clockmaker William Clement. Before this, clocks had only hour hand.

1670 Jesuit priest living in China develops model cart driven by crude steam turbine.

c. 1670 Stepped reckoner, geared device that adds, subtracts, multiplies, and divides, is invented by Gottfried Wilhelm von Leibniz (1646–1716).

1672 Flexible water hoses are developed for fighting fires by this date.

1672 French scientist N. Cassegrain (1625–?) invents Cassegrain reflecting telescope; design is widely used today.

1675 England's Royal Observatory at Greenwich is established.

1676 Invention of universal joint.

1679 French physicist Denis Papin (1647–c. 1712) invents earliest "pressure cooker," including first known safety valve.

1680 Christiaan Huygens develops an early gunpowder-fueled, internal-combustion engine.

▶

1698 Early steam pump is invented to drain water from mine shafts. It is forerunner of mechanized devices that will herald industrial revolution.

1701 English inventor Jethro Tull (1674–1741) develops mechanical seed drill, considered first modern agricultural machine.

1707 High-pressure boiler is developed by French physicist Denis Papin.

1709 Use of coke in iron smelting is introduced in England.

1714 Mercury thermometer is developed by Gabriel Daniel Fahrenheit (1686–1736).

1716 Invention of double-acting water pump.

1716 Diving bell that uses air replenishment system is built by Englishman Edmund Halley (1656–1742).

1725 Invention of stereotyping, which involves making duplicate plates for printing using hot-metal process.

c. 1730 Sextant, for measuring latitude, is invented.

1733 The "flying shuttle," device for moving weaving shuttle faster than possible by hand, is developed.

1735 First practical marine chronometer enables sailors to calculate longitude accurately.

1738 Caisson is developed to aid construction of bridge piling in Thames River.

1742 Crucible process for making steel is developed by Englishman Benjamin Huntsman (1704–76).

c. 1744 Franklin stove is created by Benjamin Franklin (1706–90).

1745 Leyden jar, device in which electrical charge can be collected and briefly stored, is developed.

1749 Radial ball bearings for use in carriage axles are patented.

c. 1750 Smelting of iron with coke is developed in England.

1758 Machine to knit hose is invented by Jedidiah Strutt (1726–97).

1764 "Spinning jenny" is developed in England.

1769 First water frame, water-powered device for spinning yarn, is produced.

1769 Scottish inventor James Watt (1736–1819) patents his steam engine. Age of steam power begins.

c. 1776 Human-powered, wooden-hulled submarine *Turtle* is built in United States.

1779 English inventor Samuel Crompton (1753–1827) combines water frame and spinning jenny in "spinning mule."

1779 British construct first bridge made of iron.

1780 Invention of fountain pen.

1781 Richard Arkwright's water-powered cotton factory opens, inaugurating factory system of production. Steam-powered plant opens in 1788.

1783 Rolling mill is invented for easier and cheaper production of wrought iron.

1783 First steam-powered vessel, French-made *Pyroscaphe,* is introduced.

1783 Cylindrical printing of cloth is introduced.

1783 Invention of hot-air balloon. Two Frenchmen fly over Paris in earliest recorded hot-air-balloon flight.

1784 Henry Cort patents "puddling furnace," in which molten pig iron is stirred to burn off impurities.

1784 Benjamin Franklin invents bifocal eyeglasses.

1785 Bleaching with chlorine is introduced.

1785 Steam-powered loom is developed.

1792 Coal-gas illumination is invented by British engineer William Murdock (1754–1839). By 1798 it is used to light factory.

1793 U.S. inventor Eli Whitney (1765–1825) invents cotton gin.

1794 Visually based telegraph, the semaphore, is set up in France.

1794 First practical internal-combustion engine is patented.

1796 Lithography, basis for modern offset printing, is invented.

1797 Helmeted diving suit supplied with air pumped from surface is invented. More practical version is fabricated in 1819.

1800 First battery, "voltaic pile," is made of disks of silver and zinc.

c. 1800 Simple machines are developed for cutting chaff and harvesting wheat.

c. 1800 First all-iron printing press is developed in England.

1802 High-pressure steam engine is patented.

▶

1803 World's first steam locomotive, *New Castle,* is introduced.

1804 Sir William Congreve (1772–1828) builds first of his Congreve military rockets.

1804 Method for canning food is developed.

1805 Punched-card method of controlling machinery is invented.

1807 First voyage of *Clermont,* first commercially successful steam-powered boat.

1808 Electric arc light is invented.

1810 Inking roller for printing press is developed. It replaces inking by hand.

1815 Method of road-building employing crushed rock is developed by British engineer John McAdam (1756–1836).

1819 American paddle wheeler *Savannah* crosses Atlantic.

1821 British ship *Vulcan,* first all-iron sailing ship, is launched.

1821 Galvanometers are used for measuring current, its direction, and its strength.

1822 English scientist Michael Faraday (1791–1867) demonstrates principle of electric motor.

1823 First electromagnet is made in England.

1824 Portland cement is patented.

1824–28 Operation of regular steam-powered bus service begins in London.

1825 First regular rail-line service begins in Great Britain.

c. 1825 Combine is invented.

1826 First photographic negative image is produced by using *camera obscura* (pinhole camera).

1827 Polarizing microscope, widely used to view rock and mineral samples in thin sections, is developed.

c. 1827 Microscopes with lens systems free of color distortion (achromatic) are developed.

1829 First reliable locomotive, *Rocket*, capable of traveling 29 mph, is built. It features new-type fire-tube boiler.

1830 Friction matches are invented.

1830 Modern railroad rail (inverted "T") is invented by American Robert L. Stevens (1787–1856).

1831 American inventor Joseph Henry (1797–1878) invents crude electric motor.

1831 Electric telegraph is built by Charles Wheatstone (1802–75) and William Fothergill.

1831 Michael Faraday (1791–1867) builds first electric generators.

1831 U.S. inventor Cyrus McCormick (1809–84) develops reaper.

1835 English scientist William H. F. Talbot (1800–1877) creates first photographic image on sensitized paper.

1835 Telegraph repeater, for sending electric signals over long distances, is invented by Joseph Henry.

c. 1835 Vermont blacksmith Thomas Davenport (1802–51) invents first commercially successful electric motor.

1836 Screw propeller for ships is patented.

1837 Invention of daguerreotypy, process using silver-plated copper with silver iodide emulsion, by French painter Louis Jacques Mandé Daguerre (1789–1851).

1837 Morse patents working model of his telegraph. Telegraph also is patented in England.

1837 John Deere (1804–86) invents steel-bladed plow.

1838 Steam-powered British ship *Sirius* is first to cross Atlantic, without use of sails.

1839 American inventor Charles Goodyear (1800–60) invents process for vulcanizing rubber.

1840 First successful use of chemicals to eliminate insects is recorded in Europe.

1842 Electroplating process is invented by German engineer Werner von Siemens (1816–92).

1842 First photograph is printed in newspaper, in London.

1842 Synthetic-fertilizer industry develops, in England.

1843 Drill powered by compressed air is invented for cutting tunnels through rock.

1844 Samuel Morse introduces first practical telegraph service.

1845 Power loom for carpet manufacture is invented.

▶

1845 *Great Britain* becomes first propeller-driven steamship to cross Atlantic. It is also first to have a hull of wrought iron.

1846 Sewing machine is patented by American Elias Howe (1819–67).

1846 Rotary printing press is invented.

1847 Bridge is constructed with plate girders for first time.

1848–49 Clipper ship *Sea Witch* makes record voyage from Canton, China to New York in 74 days.

1850 First underwater telegraph cable is laid, linking France and England.

1851 Cast-iron frames are used in building construction by American James Bogardus (1800–74).

1852 British Navy ships are issued familiar "admiralty anchor," with curved arms and removable stock (crosspiece).

1852 First flight of dirigible.

1852 Photoengraving is developed.

1852 Development of safety elevator by American inventor Elisha Otis (1811–61).

1853 One of largest wooden sailing ships ever built, 325-foot-long *U.S.S. Great Republic*, is finished.

1856 Bessemer process is developed as first successful method for mass-producing steel.

1859 Drilling of first commercial oil well, in Pennsylvania.

1862 American Richard Gatling (1818–1903) invents first machine gun.

1863 Open-hearth method of producing steel is developed in France.

1863 First rotary printing press fed by continuous roll of paper is invented.

1863 First oil pipeline is built, in Pennsylvania.

1863 Paris Observatory begins publishing first modern weather maps.

1865 Ice-making machine is invented by American Thaddeus Lowe (1832–1913).

1866 Invention of self-propelled torpedo.

1867 Process for making reinforced concrete is patented by Frenchman Joseph Monier (1823–1906).

1867 First practical electric generator to produce AC current is built by Belgian-born French inventor, Zénabe Théophile Gramme (1826–1901).

1867 Dynamite is patented by Swede Alfred B. Nobel (1833–96).

1868 Celluloid is invented.

1869 First bridge made of concrete is built.

1869 Compressed-air brake is patented by American George Westinghouse (1846–1914) for use in railroads.

1871 Dry-plate (with silver halide emulsion) photographic process is invented.

1873 Continuous-ignition combustion engine, basis of turbine engine, is invented.

1874 Ocean liners cross Atlantic in only seven days.

1876 Telephone is patented by Alexander Graham Bell (1847–1922).

1876 First practical refrigerator is built, by German inventor Carl Paul Gottfried von Linde (1842–1934).

1877 American Thomas A. Edison (1847–1931) invents phonograph.

1878 Czech Karl Klietsch (1841–1926) invents rotogravure method for making printing plates to reproduce illustrations.

1879 First successful electric locomotive is demonstrated in Berlin.

1879 Edison demonstrates incandescent light bulb.

1880 First steam-powered plant for generating electricity is built in London.

1881 First electric streetcar system is built, in Berlin.

1881 Improved storage battery is invented.

1883 High-speed internal-combustion engine is invented.

1883 First dirigible that can be steered in flight is developed.

1884 Irishman Charles A. Parsons (1854–1931) develops multistage steam turbine.

1884 French chemist invents first manufactured fiber, viscose rayon.

1884 Electric alternator is invented by American Nikola Tesla (1856–1943).

▶

1884	Early, mechanical scanner type of television is patented in Germany.
1884	Ottmar Mergenthaler (1854–99) invents linotype machine.
1885	Electric transformer is invented.
1885	French inventor Jean J. Étienne Lenoir (1822–1900) develops electrical spark plug.
1886	French and Americans develop cheap method of producing aluminum by electrolysis.
1886	First oil tanker, *Gluckhauf*, is built in Great Britain.
1886	German engineer Karl Benz (1844–1929) patents vehicle powered by gasoline engine.
1887	Monotype typesetting machine is invented.
c. 1887	Celluloid roll film is developed, making motion-picture photography possible.
1888	Adding machine is patented by American William S. Burroughs (1855–98). In 1892 he produces machine that adds, subtracts, and prints.
1888	Eastman Kodak introduces portable $25 camera.
1888	First AC electric motor is patented.
1888	First ball-point pen is patented by American inventor.
1888	Gramophone, which uses flat disks to reproduce sound, is invented.
1888	Radial engine is developed. First in-line cylinder engine also is built this year.
1888	Scot John Dunlop (1840–1921) invents pneumatic tire, thus beginning modern tire industry.
c. 1888	German physicist produces and detects radio waves over short distances.
1891	First AC (three-phase) power transmission system goes into operation, in Germany.
1892	First gasoline-powered tractor is built, in Iowa.
1892	Music is mass-produced on records for first time.
1893	Kinetoscope (peep show) for showing early moving pictures to single viewer is patented.
1895	First commercially successful movie projector is developed in France.
1895	William Roentgen (1845–1923) uses X-rays to photograph bones and internal organs of patients for first time.
1896	Successful model of diesel engine is developed.
1897	First turbine-powered steamship, *Turbinia*, is launched.
1897	First Stanley Steamer automobile is built, in Massachusetts.
1897	Italian scientist Guglielmo Marconi (1874–1937) achieves radio transmission over long distances; his prototype was built in 1894.
1897	World's largest refracting telescope (to date), with 40-inch lens, is in use at Yerkes Observatory.
1898	*Holland VI*, first successful military submarine, is built in United States.
1898	Magnetic recording system is invented.
1900	Photocopying machine is invented in France.
1900	High-speed steel, containing tungsten and carbon, is developed.
1900	Wall-mounted model of telephone with separate earpiece and mouthpiece is introduced.
1900	First vacuum tube is invented.
1900	Electrical ignition system for internal-combustion engines replaces flame ignition system.
1901	Mercury-vapor arc lamp is developed by American engineer Peter C. Hewitt (1861–1921).
1901	Safety razor is patented.
1901	First use of crystal in radio receiver.
1901	Marconi, using improved radio equipment, sends first transatlantic wireless radio message.
1902	Machine to liquefy air is constructed.
1902	Invention of first electrical hearing aid.
1903	Invention of electrocardiograph for diagnosing heart problems.
1903	Wright brothers achieve first successful flight of heavier-than-air craft, at Kitty Hawk, North Carolina.
1904	Flat-disk phonograph is introduced and becomes industry standard.

▶

1904 Invention of silicones.

1904 Invention of diode vacuum tube, rectifier tube that becomes important in electronic equipment.

1904 First practical photoelectric cell is invented.

1904 Development of stainless steel.

c. 1904 American printer accidentally discovers offset printing.

1906 Freeze-drying is developed in France.

1906 Light bulb with tungsten filament is developed.

1908 Invention of steel-toothed drill bit for drilling oil wells.

1908 Gyroscopic compass is introduced.

1908 Electric distributor system for engines is invented.

1908 American Henry Ford (1863–1947) introduces Model T.

1909 First totally synthetic plastic, Bakelite, is invented.

1909 Hydrofoil for boats is invented.

1910 French inventor develops neon light.

1911 British construct first escalator.

1912 Danish ship *Selandia*, first diesel-powered steamship, is launched.

1912 First pressurized, submersible decompression chamber for underwater divers is built.

1912 Electric self-starter for automobiles is introduced.

1913 Germans develop more sensitive photoelectric cell coated with potassium hydride.

1913 Regenerative radio receiver circuit for long-distance reception is patented.

1913 Geiger counter is invented.

1914 Red and green traffic lights are utilized for first time, in Cleveland, Ohio.

c. 1914 First commercial air conditioner is built by American Willis Carrier (1876–1950).

1915 Sonar is invented.

1915 Radiotelephony, long-distance voice communication, is demonstrated, linking Arlington, Virginia to Paris.

1916 Passive sonar system to detect noises coming from subs is introduced.

1916 Turbocharged aircraft engine is built in France.

1917 World's largest reflecting telescope (to date), 100-inch Hooker telescope, is in operation at Mount Wilson.

1917 Freezing of foods is introduced.

1918 Active sonar, using pinging device, is developed.

1919 Introduction of shortwave radio.

1921 Local telephone dialing service is offered by Omaha, Nebraska telephone system.

1923 Iconoscope, electronic television camera tube, is patented.

1923 Experimental movie with sound imprinted directly on film is demonstrated.

1923 First antiknock gasoline is marketed in United States.

1924 Demonstration of wireless transmission of photograph between New York and London.

1924 German Leica camera is introduced.

1924 Self-winding watch is patented.

1926 First successful electric phonograph.

1926 First public demonstration of television.

1926 Liquid-fuel rocket is fired at Auburn, Massachusetts.

1926 First transatlantic conversation via radiotelephone (New York to London).

1926 Talking motion pictures are ushered in with film *The Jazz Singer*.

1927 First iron lung, mechanical breathing apparatus for humans.

1928 American inventors build first quartz clock.

1928 Color motion pictures are first demonstrated.

1928 Color television is demonstrated for first time, by Baird.

1929 First complete flight of airplane by instruments alone, including takeoff and landing.

1929 FM radio transmission begins.

1929 Foam rubber is developed.

1930 Frank Whittle (b. 1907) invents jet engine.

1930 First large-scale analog computer is built.

1930 Newly designed bathysphere successfully reaches record underwater depth of 1,428 feet at end of very long cable.

▶

1931 Freons, later useful for refrigeration and aerosol propellants, are developed.

1932 First successful synthetic rubber becomes available commercially.

1932 Diesel-electric train is introduced.

1934 Electron microscope is developed.

1934 Nylon is invented.

1935 French ocean liner *Normandie* crosses Atlantic in record four days.

1935 Gas turbine engines are separately patented in England and Germany, contributing to development of jet aircraft engine.

1935 Radar is developed by British.

1936 British Broadcasting Corporation begins first public high-definition television broadcasting.

1936 Fluorescent lighting is developed.

1937 First radio telescope (31-foot dish) is put into operation, in Illinois.

1937 First successful, although impractical, helicopter is developed.

1937 Xerography is invented in United States.

1938 Improved ballpoint pen is invented.

1939 FM radio is invented.

1939 DDT is synthesized.

1939 First flight of jet-powered aircraft, Heinkel He-178, in Germany.

1939 Polyethylene and polyvinyl chloride are developed.

1940 RCA demonstrates first electron microscope.

1942 First kidney machine is developed.

1942 First successful test of German V-2 rocket, at Peenemünde.

1942 Polyesters, synthetic textile fibers, become available.

1942 Swiss scientist discovers insecticide properties of DDT.

1942 American scientists achieve first controlled nuclear chain reaction.

1943 First nuclear reactor begins operating.

1943 Continuous casting in steel manufacture is introduced.

1943 Compressed air Aqua-lung is invented for underwater skin diving.

1945 First attempt at manned rocket flight takes place, in Germany. Pilot is killed.

1945 Synchrocyclotron particle accelerator is developed. It is in operation in Berkeley, California in 1946.

1945 First successful test is made of American-made atomic bomb.

1946 Linear particle accelerator is developed.

1946 ENIAC, first all-purpose electronic digital computer, comes on line.

1946 Technique of carbon dating is discovered.

1946 American mathematician John von Neumann (1903–57) works on electronic device to simplify complicated weather computation. By 1950 his computer accurately forecasts weather.

1947 Edwin Land (b. 1909) develops Polaroid camera.

1947 Epoxy glue is developed.

1947 First microwave relay station for long-distance telephone communication begins operation.

1947 Fotosetter, photocomposition machine, is introduced.

1947 The sound barrier is broken for first time in experimental American rocket plane.

1947 First microwave relay station for long-distance telephone communication, between Boston and New York, begins operation. It eliminates need for expensive long-distance trunk lines.

1947 American physicist William Shockley (b. 1910) and others invent transistor, which soon replaces vacuum tube in electronic equipment.

1948 World's largest telescope is now new 200-inch reflecting telescope at Mount Palomar, California.

1948 Atomic clock is developed.

1948 LP (long-playing record) is invented.

1948 American mathematician Norbert Wiener (1894-1964) coins term *cybernetics* to describe science of communication and control in animals and machines.

1949 Vidicon television camera tube, suitable for use in portable television cameras, is introduced.

▶

1949 U.S. scientists build first atomic clock, in which molecular oscillations are used to determine elapsed time.

1950 CATV (Community Antenna Television) system is introduced.

1950 Color television broadcasting begins in United States.

1951 UNIVAC (Universal Automatic Computer) becomes first computer designed for commercial use, ushering in the computer age.

1952 United States explodes first hydrogen bomb.

1952 First turbojet passenger plane goes into regular service.

1952 First implant of artificial heart valve.

1952 Patient's heart action is restarted by electric shock.

1952 First hydrogen bomb is tested by United States.

1952 Cosmotron particle accelerator is built to accelerate protons.

1953 First use of heart-lung machine (developed in 1950) on patient during surgery.

1953 Breeder reactor is developed to produce atomic fuel for reactors.

1953 Redstone missile, American intermediate-range ballistic missile (IRBM), is launched.

1953 Maser, forerunner of laser, is invented.

1954 Photovoltaic cell, which converts sunlight to electricity, is developed at Bell Labs.

1954 United States launches 319-foot *Nautilus*, world's first nuclear submarine.

1954 Deep-diving bathyscaphe *FNRS 3* descends to 13,287-foot depth.

1954 Wankel rotary engine is invented.

1955 RNA is synthesized.

1955 First practical Hovercraft is built.

c. 1955 Synthetic diamonds of crystallized carbon are produced commercially.

c. 1955 Container ships revolutionize cargo-shipping industry.

1955 British scientists construct highly accurate atomic clock using beam of cesium atoms.

1955 First optical fibers are developed.

1955 Field ion microscope is invented. It produces first pictures of single atoms.

1956 DNA is synthesized by American Arthur Kornberg.

1956 Second is redesignated to represent 1/31,556,925,9747 of duration it takes for Earth to orbit sun (solar second).

1956 First nuclear-power generating plant begins operation, in Great Britain.

1956 First transatlantic telephone cable is laid.

1956 Battery-powered wristwatch is introduced.

1957 Stereo recording system is invented.

1957 Technique is developed for welding such metals as titanium and tungsten in vacuum with electron beam.

1957 U.S. Jupiter C, three-stage modified Redstone, is launched. Recoverable nose cone travels 285 miles up into space.

1957 Soviets launch 184-pound *Sputnik I*, first man-made satellite put into space.

1958 First U.S. ICBM missile, Atlas, is fired successfully.

1958 Integrated circuit is invented by American engineer Jack Kilby. Photoengraving process for mass production of integrated circuits is developed in 1959. This further revolutionizes electronics industry and eventually makes possible computer microprocessor chips.

1958 American-built *Savannah*, first nuclear-powered cargo ship, goes into service.

1958 First use is made of ultrasound to examine unborn fetus.

1959 U.S. two-stage ICBM, Titan I, is tested successfully.

1959 System of atomic time is established by U.S. Naval Observatory.

1959 U.S. satellite *Vanguard 2* is first satellite to transmit weather information to Earth.

1959 IBM introduces "second generation" of computers using transistors instead of vacuum tubes. Transistors revolutionize computers, making them smaller, faster, and less expensive.

1960 Felt-tip pen is first marketed.

1960 Artificial kidney is developed.

1960 U.S. scientists build highly accurate atomic clock based on hydrogen maser.

▶

1960 U.S. satellite *Tiros 1*, first meteorological satellite, sends back nearly 23,000 photographs of Earth's cloud cover.

1960 U.S. scientist J. E. Steel coins term *bionics* to describe study of living organisms as models for man-made devices.

1960 First "boiling-water" nuclear reactor is built in United States.

1960 First underwater firing of U.S. IRBM, solid-fuel Polaris.

1960 Laser, high-intensity light beam capable of burning through steel, is discovered. Among uses to which lasers are put is hologram, projection of three-dimensional image.

1960 Pacemaker to control heartbeat is developed.

1961 Soviet Yuri Gagarin (1934–68) is launched into space aboard two-stage *Vostok I* rocket.

1961 Maraging steel, extremely strong steel alloy, is developed.

1962 U.S. satellite *Telstar 1* is in orbit as first commercial communications satellite. In coming years, satellites make possible worldwide communications network, handling telephone, television, and data transmission.

1962 Disk storage system for computer data is introduced by IBM.

1963 Digital Equipment Corporation introduces PDP-8, first successful minicomputer.

1963 First demonstration of home video recorder, in London.

1963 World's largest radio telescope, with fixed spherical dish 1,000 feet across, is completed in Puerto Rico.

1964 Improved rice strain is developed. Higher yields result in "Green Revolution" in underdeveloped nations.

1964 Picturephone system, combining television and telephone, is demonstrated by AT&T at New York World's Fair. Subscriber service begins in 1971.

1964 First public demonstration of satellite television feed via stationary (geosynchronous) satellite. Tokyo Olympic Games are relayed to North America.

1964 First fully automated factory, Sara Lee food-processing plant, opens using computer-operated equipment.

1965 Self-reproducing virus is synthesized.

1965 Digiset, computer-driven, all-electronic photocomposition machine, is introduced.

1965 Soft contact lenses are developed.

1966 United States and Soviet Union both achieve soft landings on moon with unmanned space probes.

1967 Atomic second replaces solar second in time standard.

1967 Dolby device is invented for electronically filtering out background noise in recordings.

1967 Scientists discover metal alloy that acts as superconductor some 20 degrees Kelvin above absolute zero. It stimulates search for superconductors that function at still higher temperatures.

1968 Control Data and NCR introduce the first commercial "third-generation" computers, which use integrated circuits and offer still greater speed and capabilities.

1968 Enzymes that can cut DNA strands at specific point are discovered. They later help make genetic engineering possible.

1969 U.S. *Apollo 11* mission successfully lands first men on moon. The 363-foot-tall, three-stage liquid-fuel rocket develops about 7.7 million pounds of thrust at lift-off. Astronauts reach 24,791 mph during their flight from moon orbit, fastest any humans have ever traveled. First television broadcast from moon reaches 100 million viewers worldwide by satellite feed.

1969 First LASH ship, *Acadia Forest*, is launched.

1969 Experimental TurboTrains, powered by aircraft turbines, are introduced.

1969 Intel 4004, first microprocessor (miniaturized set of integrated circuits on one chip), is invented by American engineer. Microprocessors subsequently make "fourth-generation" computers of 1970s possible.

1969 Scanning electron microscope is built.

1969 Bubble memory system for computer is invented. It retains information even after computer is turned off.

1970 First X-ray satellite observatory, *Uhuru*, is launched.

1970 First complete synthesis of gene occurs.

▶

1970 Floppy disks to store computer data are introduced.

1970 Compact disks (CDs) are developed.

1970 Lasers designed for industrial use are developed.

1970 Supersonic jet airliners French–British Concorde and Soviet Tu-144 both fly at supersonic speeds for first time.

1970s Array technology is developed for electronic equipment, making it possible to put many integrated circuits together on single silicon chip. This leads to further miniaturization of electronic components, notably in computers.

1971 Doppler radar is used by meteorologists to study storm systems.

1972 First of U.S. Landsats (earth resources satellites) is launched. It studies mineral, agricultural, and other Earth resources. (Other Landsats are launched in 1975, 1978, and 1982.)

1972 CAT-scan (computerized axial tomography) imaging system is introduced for medical diagnosis and research work.

1972 Liquid-crystal digital displays for timepieces are introduced.

1973 Continuous-wave, tunable laser is developed in United States.

1973 First genetic engineering is accomplished by American researchers Stanley Cohen (b. 1922) and Herbert Boyer. They successfully implant foreign genetic material into bacterium *Escherichia coli* (*E. coli*).

1973 MRI (magnetic resonance imager) is developed for medical diagnosis.

1973–74 U.S. space probes *Pioneer 11* and *12* transmit first close-up color photos of Jupiter back to Earth.

1974 Pocket calculator is marketed.

1975 First desktop microcomputer become available.

1976 U.S.–West German solar probe *Helios B* sets highest velocity of any space vehicle to date at 149,125 mph.

1976 Scientists completely synthesize functioning gene for first time.

1977 British scientists detail complete genetic structure of organism.

1978 Test-tube fertilization is success with birth of Lesley Brown in England.

1979 Multiple-mirror telescope (MMT), with six 71-inch mirrors operating in unison (equal to 177-inch telescope), is installed at Mount Hopkins in Arizona.

1980 Very Long Array (VLA) radio telescope is completed in New Mexico. It has 27 separate antennae, which operate in unison.

1980 U.S. Magsat satellite completes mapping of Earth's magnetic field.

1980 U.S. spacecraft *Voyager 1* completes first successful flyby of Saturn.

1980 Scanning tunneling microscope is invented. It detects single atoms in surface of scanned material.

1980 Tevatron particle accelerator is built.

1980 Successful transfer of functioning genes is achieved.

1980–81 Two U.S. *GOES* (Geostationary Operational Environmental Satellites) are put in orbit. Each can photograph 7,000 miles of weather patterns in 18 minutes.

1981 First flight of first reusable rocket, U.S. space shuttle *Columbia*.

1981 IBM Personal Computer is marketed. Later IBM-XT is first personal computer with hard-disk storage system (1983).

1981 Solar-powered airplane *Solar Challenger* makes first crossing of English Channel.

1982 Jarvik 7 artificial heart is implanted. Patient, Dr. Barney Clarke, survives 112 days.

1983 More efficient solar cell is invented.

1983 Universe is mapped in infrared wavelength by orbiting infrared telescope of *IRAS* (Infrared Astronomy Satellite). *IRAS* finds first evidence of planetary material around star (Vega) outside our solar system.

1983 First artificially created chromosome.

1984 Laser disk storage system becomes available.

1984 Genetic fingerprinting, based on certain DNA codes unique to each individual, is discovered.

1984 U.S. achieves first midair direct hit of missile by another missile, demonstrating significant advance in antiballistic missile (ABM) technology.

▶

1985 Researchers develop experimental technique for sending 300,000 telephone calls over one optical fiber at same time.

1985 World's largest television set, 80-by-150-foot Sony JumboTron, is first demonstrated in Japan.

1986 U.S. space probe *Voyager 2* flies by Uranus for first time.

1986 First genetically engineered microorganisms are licensed for commercial purposes.

1986 Breakthrough discovery of superconducting oxide that operates at 30 degrees Kelvin. Subsequently, superconductors that work in still warmer temperatures are discovered.

1986 Lightweight airplane *Voyager* completes record round-the-world flight without refueling.

1987 Supercomputer capable of 1,720 billion computations per second goes on line in United States.

1987 Soviet radar satellite weighing 20 tons is launched. It has applications in mapmaking, oceanography, crop predictions, ice monitoring, and prospecting for minerals.

1987 Scientists discover "warm-temperature" superconductor that functions at −321 degrees Fahrenheit, temperature of liquid nitrogen.

1987 Tessa 3 spectrometer is built for more detailed study of nuclear structure.

1987 Electroconductive plastics are developed.

1988 Positron transmission microscope produces its first published images.

1988 Mirror system that multiplies intensity of normal sunlight by 60,000 times is successfully tested.

1988 Parallel processing technique for computers is developed to speed processing of complex problems by 1,000 times.

1988 New, experimental chips called heterojunction devices work hundreds of times faster than conventional computer chips.

1988 Bearing that operates with almost no friction is invented by using warm-temperature superconductor.

1988 Longest suspension bridge to date is constructed in Japan, with span of 6,496 feet.

1988 New process for making synthetic diamonds produces stones with 85 percent of hardness of natural diamonds.

1989 *Voyager 2* space probe makes first flyby of Neptune and discovers planet has third moon.

1990 Powerful Hubble space telescope is put into earth orbit by United States.

Prehistoric Tools and Implements

B.C.

c. 2,400,000 Early stone tools, clubs, and sticks for digging are used by primitive man.

c. 750,000 Primitive man is using fire by this time.

c. 400,000 Axes and spears for hunting are developed.

c. 380,000 Peking man uses wooden spears and primitive pit traps for hunting.

c. 300,000 Primitive humans are developing hammerstone, cleaver, and wooden clubs.

c. 200,000 Bone hand-axes, symmetrically shaped with hammers on both sides, are in use by or before this date.

c. 200,000 Kafuan culture in Africa uses water-worn pebbles for chopping and scraping.

c. 85,000 Neanderthal men are using side scrapers, borers, notched tools, and long blades (possibly earliest knives) by this date.

c. 80,000 Sticks and spears for hunting have fire-hardened points.

c. 80,000 Primitive men make crude stone lamps that burn animal fat.

c. 35,000 Primitive men use tools to make other tools. They use sharp flint blades to

▶

carve many other specific blade implements, such as scrapers and borers.

c. 34,000 Cro-Magnon men develop burin, a pointed, chisel-like tool.

c. 33,000 Cro-Magnon men are carving ivory bracelets, arm rings, necklaces made of teeth, shells, and fish vertebrae, and beads for decorating clothing and headdresses.

c. 33,000 Black manganese and yellow ocher are widely used for coloring faces and bodies.

c. 33,000 Early Cro-Magnon fishhooks are double-pointed, with lines attached in middle.

c. 30,000 First appearance of fired ceramics, in Europe.

c. 28,000 Paleolithic artists carve and paint decorations on tools and weapons made of bone or ivory.

c. 23,000 Bow and arrow and spear-thrower are in use.

c. 20,000 Graver, chisel made by carving narrow cutting edge at right angle to broad blade, is widely used to make tools of ivory, wood, and bone.

c. 20,000 Primitive men modify existing implements for specific uses, developing round scrapers (by rounding a blade's end), end scrapers (by squaring a blade's end), and drills (by working blade to narrow, sharp point).

c. 20,000 Knife with one edge deliberately blunt, presumably to protect fingers, is in use.

c. 20,000 First real spear, with stone point fixed to wooden shaft, is developed.

c. 20,000 Primitives develop flat, symmetrical flint blades that taper to points at each end. They use blades for knives, daggers, and spear points.

c. 20,000 Discovery is made that flint blades can make slots in bone or wood handles and shafts. Blades, points, or knives then transform these handles into new tools.

c. 20,000 Primitives are using bone knives and hammers, chisels, gouges, and smoothing tools.

c. 20,000 Spool-shaped buttons are developed.

c. 20,000 Long, hollow instruments are used to hold dried paint for engraving and painting.

c. 20,000 Small carved statuettes, possible fertility figures, are produced.

c. 20,000 Bone and ivory sewing needles (with eyes for thread) are in use.

c. 18,000 Pressure-flaking, for cutting and trimming blades with greater precision, is developed.

c. 16,000 Barbed harpoons of antler or bone are made with detachable heads attached by strong cords.

c. 13,000 Engravings and carvings of animals on javelin points, harpoons, and other items become common.

c. 12,000 Simple mortar and pestle are used to make pigments for engraving and painting.

c. 10,000 Adze, axelike tool for shaping wood, is developed. It leads to development of carpentry.

c. 10,000 Flint gravers are improved for engraving, as well as stone picks and hammersmiths for carving.

c. 10,000 Rope is developed.

c. 10,000 Prehistoric inhabitants near northern Iraq develop microliths, tiny flint blades often set into wood or bone handles, for precise cutting.

c. 10,000 Near Eastern inhabitants use sickle with flint blade in bone handle for harvesting.

c. 10,000 Clay pottery is hardened in special ovens.

c. 10,000 Prehistoric men make containers of hides and skins for holding liquids.

c. 10,000 American Indians learn precise shaping and flaking of spear and dart points.

c. 10,000 Modeled clay figurines in shapes of people or animals are made.

c. 9000 Sun-dried mud bricks are used to build houses by inhabitants of ancient Palestine.

c. 9000 Barbed fishhooks of bone, shell, or flint are in use.

c. 9000 Nets for fishing are used.

c. 9000 Stone axes for cutting wood now have wooden handles and carved cutting edges.

c. 8000 Inhabitants of northern Iraq create basketry using bone awls and needles.

▶

c. 7500 Prehistoric men in England use paddles with their small boats.

c. 7000 Baskets and woven bags are developed by American Indians in Utah.

c. 7000 Middle Eastern farmers begin to weave mats from reeds.

c. 6500 Inhabitants of northern Iraq spin thread using clay spindles. Earliest known woven cloth is produced c. 6,000 B.C.

c. 6000 Mesolithic inhabitants near Holland develop dugout canoes.

Advances in Agriculture

B.C.

c. 10,000 Prehistoric peoples in Near East begin to harvest grain, using primitive sickles with flint blades and bone handles.

c. 8500 Sheep and goats are domesticated in Iran and Afghanistan.

c. 8500 Potatoes and beans are cultivated in Peru, pumpkins in Middle America, and rice in Indochina.

c. 7000 Village society based on farming is well established in Near East. Villagers practice diversified agriculture.

c. 6750 Pigs are first domesticated, in modern Iraq.

c. 6000 Flax is cultivated in southwestern Asia.

c. 6000 Earliest known domesticated cattle are in Greece.

c. 6000 Chickens are domesticated in southern Asia.

c. 5000 Primitive agricultural villages appear in Egypt. Egyptians have early irrigation systems.

c. 4500 Cotton is cultivated in Mexico.

c. 4000 Domesticated grapes are grown in Turkestan.

c. 4000 Rice is grown as food crop in China.

c. 4000 Horses are domesticated.

c. 3750 Olives are grown on Crete.

c. 3500 Growing of maize becomes widespread in Americas.

c. 3000 Sumerians have developed advanced agricultural society by this time. They cultivate barley, wheat, flax, grapes, and other crops and raise sheep and cattle.

c. 3000 Mules and donkeys are domesticated in Palestine.

c. 2800 Rice paddies are under cultivation in China.

c. 2600 Chinese begin cultivation of silkworms.

c. 2500 Peanuts are cultivated in South America.

c. 1975 Farmers in Palestine use iron plowshares.

c. 1450 Seed drill appears in Mesopotamia.

c. 1450 Soybean becomes farm crop in China.

c. 1000 European farmers grow oats.

c. 600 Olive tree is introduced to Italy by Greeks.

c. 500 Chinese employ advanced agricultural techniques, including row cultivation and applying manure.

c. 400 Chinese develop iron plow.

A.D.

c. 100 Chinese use powdered dried chrysanthemum flowers as first insecticide.

c. 250 Romans introduce to farms in Europe ox-drawn plow and agricultural methods of irrigation, leaving fields fallow; crop rotation; and selective breeding of plants.

304 First recorded use of insects to control other insects that damage crops, in China.

c. 500 Manorial system develops in Europe. Farms that are parts of large estates owned by lords are actually farmed by peasants.

c. 800 Medieval open-field system is adopted. Individual farmers in village have rights to strips of farmland in field, though whole field is worked by local farmers at once.

c. 900 Ditches to drain and reclaim marshlands are used.

c. 1000 Lemon is introduced to Spain and Sicily by Arabs.

▶

c. 1000 Wheeled asymmetrical plow, drawn by horses or oxen, is in use in Europe.

c. 1100 Horse-collar harness, probably invented c. 110 B.C. in China, is introduced in Europe. It replaces old harness, which tends to choke horse and thereby limits work the animal can do. Horse soon replaces ox as work animal on farms.

c. 1220 Wheelbarrow is used in northern Europe.

1520 Corn and turkeys from America and the orange tree from China are brought to Europe.

1565 Potato is introduced into Europe from America.

1645 First text describing crop rotation is published in England.

1701 English inventor Jethro Tull (1674–1741) develops mechanical seed drill, considered first modern agricultural machine. His workers strike in protest.

1724 Possibility of cross-fertilization in corn is discovered.

c. 1730 Crop rotation replaces practice of leaving fields fallow as farmers discover more about soil nutrients from different crops.

1747 Sugar is discovered in beets by German chemist Andreas Marggraf (1709–82).

1775 Farmers in colonial America report serious soil erosion is affecting their farms.

c. 1775 English farms adopt "Norfolk System" of crop rotation, four-field rotation system. Included in system are livestock feed crops, which in turn produce animal manure to be used for fertilizing fields.

c. 1780 Practice of adding ground bones and other waste materials to soil to restore phosphates begins.

1783 First plow-making factory opens, in England.

1785 Cast-iron plow is patented in England. A self-sharpening model is introduced in 1803.

1786 Threshing machine is invented in Scotland.

1793 British Board of Agriculture, to oversee "improvement societies" formed to educate British farmers, is established.

1793 U.S. inventor Eli Whitney (1765–1825) invents cotton gin.

1797 Cast-iron plow is in use in United States by this year.

c. 1800 Simple mechanized machines for cutting chaff and harvesting some food crops are in use on many farms.

1802 Improved strain of sugar beet is used to produce sugar for first time.

1804 Soybean is first cultivated in United States.

1806 First agricultural high school opens, in Germany.

1810 Mowing machine is invented in United States.

1812 Earliest recorded hybridization of corn; U.S. farmer successfully interbreeds two strains.

1820 Guano, a natural fertilizer, is used in Europe.

1820 Cultivator is invented in United States.

c. 1825 Invention of combine.

c. 1830 Early steam-powered farm machinery is in use. It proves expensive, heavy, and difficult to operate.

1831 U.S. inventor Cyrus McCormick (1809–84) develops first harvesting machine, also called reaper.

1834 U.S. inventors John and Hiram Pitts invent efficient thresher. Several early versions of thresher have been in use since c. 1800.

c. 1835 English farm laborers revolt to protest threshing machine, which they say robs them of work.

1836 Grain combine, to cut grain and separate kernels from straw, is developed in United States.

1837 English agriculturalist begins experiments adding manure to soil around plants.

1837 U.S. blacksmith John Deere (1804–86) invents improved plow with steel blade.

1840 First successful insecticide is developed in Europe.

1842 Synthetic-fertilizer industry develops in England.

1845 Potato blight (causing Irish potato famine) and other plant diseases of the decade spur research into effective pesticides.

c. 1850 Austrian botanist Gregor Mendel (1822–84) discovers principles of heredity, thus beginning field of genetics and providing basis for scientific breeding of plants. His

▶

experiments include work with garden peas.

c. 1856 Steel harrow, disk plow, and cultivator are introduced.

1862 U.S. Congress funds federal Department of Agriculture and state colleges for agricultural studies.

c. 1865 Steam-powered farm machinery, such as hay rake, is in use.

1892 First gasoline-powered tractor is built, in Iowa.

1896 U.S. agricultural chemist George Washington Carver (c. 1864–1943) develops methods to make worn-out cropland productive by growing peanuts and sweet potatoes.

c. 1900 U.S. farmers use sprinkler irrigation.

1901 Practical technique for artificial insemination is developed in Russia.

c. 1912 Number of farm horses in use begins to decline due to development of improved gasoline engine. Number of farm mules begins to decline in 1925.

1917 Frozen-food process is developed in United States by Clarence Birdseye (1886–1956).

c. 1918 Earliest crop dusting by airplane is recorded in United States.

1920 Power-operated elevators in barns help relieve farmers of heavy work of loading and moving such things as seed bags and equipment.

1920 Several varieties of genetically bred, high-yielding corn are commercially available by or before this date. By 1960 more than 95 percent of U.S. corn is from hybrid seed.

c. 1920 All-purpose tractors come into use, gradually replacing work animals and steam-powered farm machinery.

c. 1925 Increasing rural electrification provides electrical power to U.S. farms.

1932 Tractors with rubber tires are introduced.

1938 Self-propelled combine is introduced in United States.

1940 Barn fans for ventilation and drying crops come into use.

1942 Swiss scientist Paul H. Müller (1899–1965) discovers insecticide properties of DDT, first synthesized in 1874.

1945 Introduction of herbicide 2,4-D.

1950 First embryos are transplanted in cattle.

1960s Introduction of improved strains of wheat and rice (tailored to regional climates and other special requirements by scientific cross-breeding) result in higher yields. It eases chronic food shortages in some areas, such as India (Green Revolution).

c. 1965 Aerial crop dusting and spraying in United States come under strict federal controls, in response to environmental concerns.

1970 Research in United States into harmful effects of herbicides and pesticides intensifies as result of public concern.

1972 Landsat (earth resources satellite), a monitor of global agricultural production, is launched.

c. 1975 Laser leveling of agricultural fields proves effective in U.S. experiments.

1977 Computer information networks are in use in parts of United States to help farmers with land purchasing, irrigation scheduling, financial budgeting, and other subjects. By 1982 similar systems provide current weather, marketing, and crop information.

1982 New photoelectric device senses optimum ripening of cornfields and applies water and fertilizer automatically. It proves too expensive for many farmers.

1984 U.S. researchers experiment with electrostatic spraying. Droplets are negatively charged and therefore attracted to plant being sprayed, reducing amount of chemicals required.

1987 Test release of genetically altered bacteria (delayed from 1983) for use in protecting crops against frost.

1987 United States announces development of first bioinsecticides, to eliminate insects without harming environment.

1988 Robots are used for fruit-picking in United States experiments.

1988 Computer systems are in use for monitoring and controlling U.S. farm irrigation by or before this date.

Development of Ships

See also Warships in Wars and Military History section.

B.C.

c. 4500 Mesopotamians are building sailing ships by this time.

c. 3300 Reed boats with square sails are used by Egyptians by this time. They are steered with large, rudderlike oars.

c. 2000 Minoans of Crete build wooden ships with log keel, internal ribs, and attached planks to form hull.

c. 1500 Phoenician trading ships ply Mediterranean and beyond. Phoenicians make their wooden ships double-ended with upturned bow and sternposts and single, square sail. Their sailors use rudimentary navigation techniques to sail out of sight of land and to sail at night.

c. 450 Roman merchant ship with two masts (mainmast and shorter sail-bearing mast in front of it) are in use.

A.D.

700s Vikings develop double-ended warships powered by oarsmen and large, square sail. Hull, 70 to 80 feet long, is made of overlapping oak planks and has keel. Rudder and tiller arrangement steers craft.

c. 800 Byzantines build highly maneuverable ships with carvel-planked hulls (planks laid edge to edge and nailed in place) that carry triangular lateen sail.

800s Ungainly-looking but extremely seaworthy Chinese junks are on regular trading voyages to India and back. Junks are built with watertight compartments.

c. 1200 European sailing ships in North are outfitted with bowsprit for first time. Otherwise European sailing ships still are essentially the same as Viking craft.

c. 1200 Hinged rudder appears on European sailing ships. Mediterranean craft continue to use double-oar arrangement.

c. 1300 Italian ships, some over 100 feet long, are built with three decks. They have two masts carrying triangular lateen sails rigged fore and aft.

c. 1450 Three-masted carrack becomes common oceangoing sailing ship. A blend of northern square-sail and Mediterranean lateen-rig designs, it has shorter foremast, a mainmast, and aft mizzenmast. Planks laid edge to edge (carvel construction) form hull, and rudder is mounted on centerline.

1500s Competition among rival European nations leads to building of large carrack-type ships. Spanish develop galleon, narrower than large carracks and with high sterncastle.

1500s First use of jib, triangular headsail.

1513 English King Henry VIII builds large carrack, *Henri Grace à Dieu* (Great Harry), first four-masted ship built in England.

c. 1570s English develop removable topmast, mounted atop mainmast, that can be lowered in storms. It makes taller masts possible.

1581 Englishman William Borough (1536–99) writes treatise on use of magnetic compass.

1659 Early, inaccurate chronometer for sea travel is invented by Dutch mathematician and physicist Christiaan Huygens (1629–95).

c. 1700 East Indiamen, relatively slow but heavily armed cargo vessels, ply trade routes to East India.

1700s U.S. shipbuilders develop the small, fast, two-masted Baltimore clippers that prove worthy in cargo and privateer service.

1735 Englishman John Harrison (1693–1776) builds first of several chronometers accurate enough for use in navigation.

1783 First steam-powered paddleboat *Pyroscaphe* is designed, built, and successfully tested in France by Marquis de Jouffroy (1751–1832).

1787 Steamboat built by American John Fitch (1743–98) is successfully tested on Delaware River.

▶

c. 1800 British perfect copperclad bottom to prevent bottom fouling.

1802 *New American Practical Navigator*, written by American Nathaniel Bowditch (1773–1838), establishes navigation standards for sailing ships.

1802 American John C. Stevens (1749–1839) constructs steamboat that uses propeller.

1805 British experiment with reefing downward to foot, rather than upward to yard. This eventually becomes standard practice in sailing ships.

1807 Fulton launches *Clermont*, first commercially successful steamboat. With 20-horsepower engine, it travels Hudson River from New York to Albany in 32 hours.

1809 American steamboat *Phoenix* becomes first steamboat to make sea passage. It travels Atlantic from Hoboken, New Jersey to Philadelphia.

1817 First Mississippi River steamboat, *Washington*, travels from New Orleans to Louisville, Kentucky in 25 days.

1819 In first transatlantic steamship crossing, *Savannah*, built by American inventor Moses Rogers (1779–1821), crosses from Savannah, Georgia to Liverpool in 29 days. It has 90-horsepower engine.

1821 British ship *Vulcan*, first all-iron sailing ship, is launched.

1832 Term *clipper* is first applied to new type of North Atlantic packet ship, first of which is *Ann McKim*, designed for speed and seaworthiness.

1836 Screw propeller is patented by American John Ericsson (1803–89). Propeller is used in England next year.

1838 British establish regularly scheduled transatlantic steamship operation.

1838 British ship *Sirius* is first to cross Atlantic powered only by steam, without use of sails. It makes crossing in 18 days.

1840 British Transatlantic Cunard Line begins operation with launch of *Britannica*.

1840s American shipyards meet great demand for clipper ships by mass-producing them.

1845 Iron-hulled *Great Britain* becomes first propeller-driven steamship to cross Atlantic. Propellers soon replace paddle wheels.

1845 Clipper ship *Rainbow* is first with what becomes characteristic streamlined clipper-ship hull. She sets record for round trip to Canton, China: six months, 14 days.

1847 Commercially successful Fall River Line begins service. It links New York and Fall River, Massachusetts, using some of the world's largest steamships to date.

1847 Intercoastal service from New York to California begins with passage of paddle wheeler *California*.

1848 Chinese junk makes successful voyage from China to New York and London.

1848-49 Clipper ship *Sea Witch* makes record voyage from Canton, China to New York in 74 days.

1852 British Navy ships are issued familiar admiralty anchor, with curved arms and removable stock (crosspiece). Stockless anchor (developed in 1820s) eventually replaces it.

1853 Clipper ship *Lightning* sails from Melbourne to Liverpool in record 63 days.

1853 One of largest wooden sailing ships ever built, *U.S.S. Great Republic*, is finished. It is 325 feet long.

1853 In first Pacific crossing by steamship, American paddle wheeler *Monumental City* travels from San Francisco to Sydney, Australia.

1854 Clipper ship *Champion of the Seas* makes 462 nautical miles in one day, the world's record for a commercial ship under sail.

1854–57 British iron steamship *Great Eastern* is built. It is world's largest ship for 50 years, with length of 693 feet and room for 4,000 passengers. Designed by I. K. Brunel (1806–59), it weighs 18,000 tons and incorporates numerous technological advances.

1865–66 *Great Eastern* is used in laying first transatlantic cable.

1866 Clipper ships *Serica*, *Taiping*, and *Ariel* in race from China to London complete journey in just 98 days.

1869 *Cutty Sark*, one of last and most famous of all clipper ships, is launched. It remains in service for over 30 years, though by 1870s steamships already are bringing age of sail to close.

▶

1870s Iron-hulled sailing ships are in use, carrying wool from Australia. Use of sailing ships in cargo trade drops off sharply in remaining years of 1800s.

1871 White Star Line's *SS Oceanic*, first modern luxury liner, is launched.

1872 Accurate sounding device is invented.

1874 Ocean liners cross Atlantic in seven days by this time.

1884 Steam turbine is developed by British inventor Charles A. Parsons (1854–1931). It revolutionizes steamships by conserving fuel and increasing passenger space and efficiency.

1886 First oil tanker, *Gluckhauf*, is built in Great Britain.

1889 British steamships *City of Paris* and *City of New York* are built. They have twin-screw propellers. *City of Paris* has more than 14,000 horsepower.

1892 Great Lakes steamship *Christopher Columbus* is launched. It carries 5,600 passengers.

1897 *Turbinia*, first turbine-powered steamship, is launched. It is driven by Parsons' steam turbine engine.

1899 *Oceanic*, at 704 feet, supersedes *Great Eastern* as world's largest ship.

1900 Steamships have almost replaced sailing ships by this year.

1907 Steamship *Lusitania* establishes new transatlantic record of five days, 45 minutes.

1909 Hydrofoil ship is invented.

1911 Gyrocompass is invented by American Elmer Ambrose Sperry (1860–1930).

1912 Danish ship *Selandia*, first diesel-powered steamship, is launched. Its superior fuel conservation is demonstrated when it makes 26,000-mile journey from Bangkok to London without refueling.

1920 Steam turbines begin to be fueled by oil rather than coal.

1935 French ocean liner *Normandie* breaks transatlantic crossing record, making the trip in four days. It weighs nearly 80,000 tons.

1938 *Queen Elizabeth*, at 1,000 feet world's largest ocean liner, is launched. It weighs 83,000 tons.

1952 *United States* is launched. It is world's fastest ocean liner, cruising at 30 knots, and crosses Atlantic in three days, 10 hours.

c. 1955 Container ships revolutionize cargo-shipping industry. Their hulls contain storage compartments to hold truck-size steel boxes, containers filled with merchandise. These ships can be loaded and unloaded much faster than traditional cargo ships.

1958 American-built *Savannah*, first nuclear-powered cargo ship, goes into service. It proves impractical because of its large crew and restricted cargo space.

1968 Introduction of supertankers for transporting oil.

1968 *Ponce de León*, the largest roll-on, roll-off ship, is launched. It holds 200 cargo containers and 30 cars or trucks.

1969 *Acadia Forest*, first LASH ship, is launched. LASH ships carry up to 70 barges, each loaded with 300 tons of cargo.

1969 Transatlantic liner *Queen Elizabeth II* makes its first voyage. It weighs 65,000 tons and measures 963 feet.

1969 U.S. icebreaking tanker *Manhattan* becomes first large ship to travel through Northwest Passage. Journey proves possibility of transporting Alaskan oil to eastern United States.

1973 Supertanker *Brooklyn*, largest U.S. merchant ship, is christened. It weighs 230,000 tons and is 1,094 feet long.

1986 World's largest cargo ship to date, 1,125-foot ore carrier *Berge Stahl*, is launched.

1988 World's largest passenger ship, 73,000-ton Norwegian cruiser *Sovereign of the Seas*, is launched.

Development of Submarines

See also The Underwater Frontier and The Quest for Speed.

c. 1620 Dutch inventor Cornelius van Drebbel (1572–1633) demonstrates first known successful submarine, in Thames River. His 270-foot-long, leather-covered vessel is powered by 12 oarsmen and reportedly submerges to 15 feet below surface.

c. 1776 Human-powered, wooden-hulled submarine *Turtle*, built about this time by Connecticut inventor David Bushnell (1742–1824), unsuccessfully attempts to attach explosive device to hull of British ship.

1801 American inventor Robert Fulton (1765–1815) completes human-powered submarine, *Nautilus*. It has copper hull, precursor of conning tower, and ballast tanks for submerging and uses snorkel for taking in fresh air while running just below surface.

1854 Frenchman designs forerunner of modern periscope. It uses mirrors instead of prisms in its sight tube, however.

1856 Bavarian Wilhelm Bauer (1822–75) builds 52-foot iron-hulled, human-powered submarine, *Sea Devil*.

1863 French submarine *Le Plongeur* becomes first to use compressed air to clear ballast tanks of water for surfacing.

1864 In first sinking of warship in combat by submarine, Confederate iron-hulled sub *David of Hunley*, hand-cranked by crewmen and carrying explosive charge, sinks Union corvette *Housatonic*. Sub sinks with corvette, however.

1866 British inventor Robert Whitehead (1823–1905) invents practical, self-propelled torpedoes.

1870 Novelist Jules Verne (1828–1905) writes his classic tale of submarines of future *Twenty Thousand Leagues Under the Sea*.

1880 British submarine travels several miles underwater powered by steam engine, using reserve steam built up during surface running.

1886 British inventors build submarine using newly invented electric motors. It travels 80 miles before batteries need recharging.

1898 *Holland VI*, first successful military submarine, is built by American inventor John P. Holland (1840–1914). It has steel hull, gasoline engine (for surface running), and battery-powered electric motors (for submerged running), carries crew of seven, and can dive to 75 feet. By early 1900s diesels replace gasoline engines.

1899 French Navy builds 111-foot *Narval*, with steam/electric power. It has double hull with space between holding ballast tanks (present-day system).

c. 1902 Introduction of periscope.

1905 *Unterseeboot No. 1* is built for German Navy. This first U-boat is 139 feet long, is diesel/electric powered, and has single torpedo tube.

1912 German Navy provides submariners with individual rebreather devices for emergency underwater escapes.

1914–18 Germans prove effectiveness of submarines in raiding commerce shipping during World War I. Single German U-boat (*U-35*) sinks all-time record 226 ships.

1914–18 British develop small, fast antisubmarine submarines during this period. They also develop high-speed subs powered by steam turbines for surface running during scouting missions.

1916 Passive sonar system, using hydrophone to detect noises coming from subs, is introduced. Active sonar, using pinging device, is developed in 1918.

1916 First successful use of depth charges in warfare by British, who sink German U-boat.

1918 Deck guns are part of standard submarine weaponry by this time. Role of subs in minelaying operations also has been established.

1928 *Argonaut*, first U.S. Navy long-range sub, is built; 381-foot-long sub has four torpedo tubes.

▶

1933 Dutch Navy officer introduces idea of using snorkel to allow subs to run on diesel power just below surface.

1939–43 Battle of Atlantic. German U-boats operating in wolf packs cripple Allied war effort by sinking supply ships in Atlantic almost at will. Not until 1943 do Allied antisubmarine tactics, including aircraft patrol and radar, end U-boat threat.

1940 Germans build prototype V-80, powered by special turbine developed by German scientist Hellmuth Walter (b. 1900). Turbine runs on oxygen generated from hydrogen peroxide, driving sub at up to 25 knots submerged.

1940s United States uses highly successful Balao and Gato class subs to attack Japanese supply shipping in Pacific Theater. United States equips its subs with radar after 1942.

1940s Japanese build largest-ever conventional-powered subs, 400-foot I-400 subs, which have 102-foot-long aircraft hangar to carry three float planes.

1940s Germans launch first experimental rockets from submarines.

c. 1944 Germans install snorkels on many of their submarines, allowing them to escape radar detection by running just below surface.

1945 Germans complete work on advanced submarine, Type XXI. Postwar submarine designs reflect many Type XXI innovations.

1950s Soviets put short-range ballistic missiles aboard submarines.

1954 United States launches 319-foot *Nautilus*, world's first nuclear submarine. Crusaded for by U.S. Navy admiral Hyman G. Rickover (1900–86), nuclear submarines revolutionize submarine technology.

1958 U.S. Navy launches first of Skipjack class of nuclear submarines. It abandons long, slim U-boat-type hull for teardrop-shaped hull to improve underwater speed (over 30 knots).

1958 *Nautilus* makes first underwater voyage under North Pole icecap.

1958 U.S. Navy successfully launches Polaris missile from submerged sub.

1959 United States launches *George Washington*, first of its fleet of nuclear-powered subs carrying Polaris missiles.

c. 1959 Soviets build their first nuclear sub (November class).

1960 U.S. Navy sub *Triton* makes first round-the-world voyage without resurfacing (36,000 miles, 2½ months).

1961 Steinke Hood, inflatable jacket with attached hood, is successfully tested by U.S. Navy for use in emergency ascents from subs at depths of over 400 feet.

1970 United States begins replacing Polaris missiles with Poseidon missiles on nuclear-powered subs.

Clocks and Other Time-Telling Devices

B.C.

c. 3500 Gnomen, vertical stick or column, is in use for telling time (by length of shadow cast).

c. 1400 Egyptians construct crude water clocks.

c. 800 Egyptians are using sundials, with six time divisions, by this time.

c. 520 Sundial is introduced into Greece by Anaximander (611–547 B.C.).

c. 380 Water clock (clepsydra) complete with alarm is said to have been invented by Greek philosopher Plato (427?–347).

c. 300 Babylonian priest and astronomer devises hemicycle, hemispherical sundial with 12 time divisions. It is used for many centuries.

290 Rome's first sundial, Samnite sundial captured in war, is installed.

▶

c. 270 Greek inventor Ctesibius of Alexandria develops improved water clock using siphon system to replenish reservoir automatically.

c. 100 Greeks build Tower of Winds at Athens. It has eight sundials, each facing different cardinal point.

c. 50 Roman architect Vitruvius describes water clock with a cogwheel arrangement. It has movable pointer to indicate hour.

A.D.

c. 100 Sandglass clocks, which run for an hour or less, are developed. They continue to be used into medieval times.

c. 725 Earliest known mechanical clock (with escapement mechanism) is built in China.

976 Chain drive for mechanical clocks is invented in China.

1094 Waterwheel clock with gong to mark hours is built in China.

c. 1300 Weight-driven mechanical clock is invented in Europe. Called turret clock, it has no hands and is used only to mark proper time for ringing church bells.

1335 First public clock to strike hour is put up in Milan, Italy.

1354 Mechanical clock is constructed at Strasbourg Cathedral.

c. 1400 First small, weight-driven clocks for use in household appear in Europe. Continent soon is taken with novelty of time.

c. 1500 German locksmith Peter Henlein (1480–1542) makes first small, portable clocks. He uses coiled mainspring to drive them.

1574 Clockmaking industry flourishes in Geneva, Switzerland.

c. 1583 Italian scientist Galileo Galilei (1564–1642) discovers that period of pendulum's swing remains almost constant as length of its swing decreases. This later becomes operating principle of pendulum clock.

1656 Christiaan Huygens (1629–95), Dutch mathematician and physicist, builds first accurate pendulum clock, based on Galileo's theory.

1658 English scientist and inventor Robert Hooke (1635–1703) develops balance spring for watches. He also invents improved "recoil" escapement (1660), which is used into modern times.

1670 Minute hand is developed by British clockmaker William Clement. Before this clocks had only hour hand.

c. 1700 Glass cover for clock face is introduced.

1704 Use of jewels in watch movements is introduced.

1715 Improved "deadbeat" escapement for precision timekeeping is invented in Great Britain.

1715 Eight-day clock is invented in England.

1735 British inventor John Harrison (1693–1776) builds first practical marine chronometer. It enables sailors to calculate longitude accurately.

1770 Automatic winding feature for watch is developed.

1780 New York City boasts first town clock in United States. Connecticut-made clock is installed in Old Dutch Church.

1800s Widespread use of pocket watches encourages general concern for punctuality.

1829 Electromagnetically powered clock makes its first appearance.

1859 Big Ben, one of world's most famous clocks, is installed in tower of Westminster in London. Designed by Edmund Beckett (1816–1905), it has 13-ton bell.

c. 1875 Production of watches is estimated at 2.5 million worldwide.

1884 International conference divides world into 24 time zones, changing then-current practice of each locality keeping its own time.

1892 Electric timers are used to clock sports events in Japan.

1904 Radio stations begin beaming time signal to ships at sea as aid to navigation.

1906 First self-contained, battery-operated electric clock is introduced.

1918 First clock driven by electric motor appears.

1924 Mechanism for self-winding watch is patented.

1928 American inventors build first quartz clock. Vibrations of quartz crystal regulate synchronous motor to produce precision timekeeping device.

1948 U.S. scientists build first atomic clock, in which molecular oscillations are used to determine elapsed time.

▶

1949 Timex introduces cheap, durable watch. It later offers shockproof, waterproof, self-winding watches.

1955 British scientists construct highly accurate atomic clock using beam of cesium atoms.

1956 Battery-operated watch is introduced in France.

1956 Second is redesignated to represent 1/31,556,925.9747 of duration it takes for earth to orbit sun (solar second).

1959 System of atomic time is established by U.S. Naval Observatory.

1960 U.S. scientists build highly accurate atomic clock based on hydrogen maser.

1962 Telstar becomes first satellite used to synchronize time internationally (between United States and Great Britain, to 1 microsecond).

1967 Atomic second replaces solar second as time standard.

1968 Quartz watches are marketed. Earliest models cost about $1,000.

1969 Seiko of Japan introduces its first electronic wristwatch. American-made Pulsar soon follows.

1972 Texas Instruments and Eubauches S.A. together make liquid-crystal digital-display timepieces.

1981 Talking wristwatch is introduced by Sharp.

1989 One-of-a-kind Swiss watch, called world's most complicated, is put up for auction. At about 2½ pounds and 3½ inches in diameter, watch has 26 separate hands and performs 31 more functions than standard watch.

Printing and Copying

c. 150 Chinese begin carving religious texts on stone tablets, inking surfaces to make impressions.

c. 400 Chinese make special ink for printing from carved stones.

600 Chinese develop woodblock printing method.

751 Chinese prisoner taken in battle near Samarkand reveals secret of making paper to Arabs. Paper is not finally introduced in Europe until 12th century.

c. 770 First known printed work, Buddhist prayer for Japanese empress, is produced.

868 Chinese print oldest known printed book in existence, *The Diamond Sutra.* It contains only six pages and one woodcut illustration.

c. 1040 Chinese printers use movable type made from earthenware.

1107 Multicolor printing is used by Chinese in making paper money.

c. 1314 Over 60,000 characters carved in wooden blocks are used for printing by Chinese.

1397 Printer in Korea creates text using bronze type.

1423 Earliest dated woodblock print is published in Europe.

c. 1438 German printer Johannes Gutenberg (c. 1398–c. 1468) invents printing press (using raised metal type), thus introducing printing in Europe.

c. 1440 Inks created for printing consist of linseed-oil varnish and carbon black.

1454 A 42-line Bible is printed with movable type by Gutenberg at Mainz.

c. 1457 First two-color printing system is developed.

1475 First book in English is printed.

1487 Printing presses are in use in all European countries by or before this year.

1492 Book publishing emerges as profession.

1500 Italic type is used, in Italy.

1530 French type designer sells italic, Roman, and Greek type styles.

1539 First North American print shop opens, in Mexico City.

1638 First printing press is set up in American colonies, in Cambridge, Massachusetts.

▶

1704 Regularly published newspaper is printed in American colonies.

1719 Painter in England takes out patent for color printing process using primary colors (red, yellow, and blue) plus black. Printing is done with four separate, engraved printing plates, each inked separately.

1725 Stereotyping is invented; it involves making duplicate plates for printing using hot-metal process.

1737 Point system of type measurement is developed by Frenchman Pierre Simon Fournier (1712–68).

1757 First use of woven paper for printing, in England.

1772 Metal parts are replacing wood in presses, increasing power and speed of presses.

1790 Cylinder press is invented, in England.

1796 Lithography is invented in Munich by Aloys Senefelder (1771–1834). It becomes basis for modern offset printing.

c. 1800 First all-iron printing press is developed, in England.

1806 Machine to print serial numbers on bank notes is invented by English engineer Joseph Bramah (1748–1814).

c. 1810 Steam-powered flatbed cylinder printing press is invented.

1810 Inking roller is developed to replace inking by hand.

1822 Casting machine is patented. It produces 3,000 pieces of type per hour.

1842 First photograph is printed in a newspaper, in London.

1844 British inventor W. H. Fox Talbot (1800–77) experiments with gravure printing of photographs; in 1852 he invents photoengraving.

1846 Rotary press, using revolving cylinders to hold type, distribute ink, and move paper, is invented. Machine is forerunner of modern newspaper presses.

1863 American printer William Bullock (1813–67) invents first rotary press fed by continuous roll of paper.

1873 Remington Company begins manufacturing modern typewriter.

1878 Czech Karl Klietsch (1841–1926) invents rotogravure method for making printing plates to reproduce illustrations.

1884 Linotype machine is invented by German-born American Ottmar Mergenthaler (1854–99). Device creates entire line of type in single casting.

1887 Monotype typesetting machine is patented.

1900 Photocopying machine is invented in France. It provides first facsimile-copying capability.

1904 American printer discovers offset printing by accident when he inadvertently transfers inked images to cylinder instead of to paper.

1906 Photostats, photographs of documents, are developed.

1907 English printer introduces use of silk frames in printing, forerunner of modern screen-process printing.

1920 Dry offset printing is developed, in United States.

c. 1929 Teletypesetting system is in use in United States. Typesetters produce punched tapes, which are then fed into automated typesetting machine. Machine produces finished type at high rate of speed.

1937 Xerography, electrostatic copying process, is invented in United States. It does not become commercially available until 1950.

1947 Fotosetter, photocomposition machine, is introduced.

1959 Xerox introduces its first commercial copier.

c. 1960 Three-dimensional printing is developed (printing two views of text, superimposed).

1965 Digiset, computer-driven, all-electronic photocomposition machine, is introduced. It theoretically can set up to 10 million characters per hour.

1967 *London Daily Express* is transmitted electronically, via telephone lines and communications satellite, for printing in Puerto Rico. (Transmission takes 15 minutes.)

1968 First novel to be typeset entirely by computers is published, in United States.

c. 1968 Over 500,000 copy machines are in use.

c. 1975 Xerographic machines are able to duplicate colored materials.

1976 Ink-jet printer is introduced by IBM.

1988 Some 1 million fax machines are bought during year as popularity of electronic transmission and reproduction of documents surges.

Age of Steam

See also Railroad Development.

c. 100 Aeolipile, primitive steam turbine, is invented by Hero of Alexandria (fl. c. 100).

1690 French physicist Denis Papin (1647–1712) develops concept of using steam to move piston.

1698 English engineer Thomas Savery (c. 1650–1715) patents his steam pump, cumbersome precursor of steam engines designed to pump water out of mines.

1712 English inventor Thomas Newcomen (1663–1729) builds practical steam pump that is more efficient than horse-powered water pumps.

1769 Scottish inventor James Watt (1736–1819) patents his vastly improved steam engine. Watt's engine includes such basic advances as separate steam condenser and steam jacket. Watt builds and installs his first successful engines in 1776.

1770 First steam-powered wagon is built by French artillery officer Nicholas-Joseph Cugnot (1725–1804). Vehicle creeps along at 2 miles per hour.

1781 Watt develops means to convert steam piston's back-and-forth (reciprocal) motion to rotary motion. New engine's ability to turn shaft creates many new applications for steam power.

1782 Watt patents his double-acting piston engine (steam power on both upstroke and downstroke).

1783 French-made *Pyroscaphe* becomes first steam-powered vessel. It completes 15-minute voyage on Saône River near Lyons.

1784 William Murdock (1754–1839), engineer for British firm Boulton & Watt, builds steam carriage.

1786 Steamboat built by American inventor John Fitch (1743–98) goes into regular operation on Delaware River, but boat is not commercially successful.

1789 James Watt invents governor to control speed of steam engine.

1789 First U.S. patent for land vehicle driven by steam is granted to Oliver Evans (1755–1819).

1789 First steam-powered cotton factory opens, in Manchester, England.

1801 Successful test of steam carriage built by British inventor Richard Trevithick (1771–1833) is made.

1802 High-pressure steam engine built in 1800 is patented by Trevithick. (Watt builds only low-pressure engines, which are ineffective for such self-propelled steam vehicles as locomotives and steamboats.)

1803 Trevithick builds world's first steam railway locomotive.

1807 First voyage of *Clermont*, first commercially successful steamboat. Built by American inventor Robert Fulton (1765–1815), *Clermont* travels up Hudson River at 5 miles per hour.

1814 British inventor George Stephenson (1781–1848) builds new locomotive that can pull loads faster than horse team.

1814 *Fulton*, first steam-powered warship, is designed by Robert Fulton.

1814 Cylinder printing press powered by steam is used to publish *The Times* of London.

1819 First Atlantic crossing by steamship, American paddle wheeler *Savannah*.

1824–28 Regular steam-powered bus service operates in London.

1825 First regular rail-line service begins in Great Britain.

1829 *Rocket* locomotive, capable of traveling 29 miles per hour, is built by George Stephenson. It is first reliable steam locomotive and heralds opening of railway era.

1849 Efficient, four-valve steam engine is patented by American George Henry Corliss (1817–88).

1870s Manufacture of early four-stroke gasoline engines begins. Development of powerful, lightweight internal-combustion power

▶

plants (gasoline and diesel) marks the beginning of end of age of steam.

1880 Steam-powered plant for generating electricity is built in London (New York City plant starts operating later in year). Steam-generating plants soon become major suppliers of electricity for lighting and manufacturing.

1884 Irishman Charles A. Parsons (1854–1931) develops multistage steam turbine. His turbines eventually are used to drive electrical generators and propeller shafts on large ships, applications that continue to present day.

1896 Unmanned steam-powered flying machine designed by American airplane pioneer Samuel Pierpont Langley (1834–1906) flies 0.75 miles before crashing.

1897 *Turbinia*, first turbine-powered steamship, is launched. It uses Parsons's turbine power plant.

1897 First Stanley Steamer is built, in Massachusetts. By 1906 a Stanley Steamer, probably best-remembered of all steam cars, reaches top speed of 127 miles per hour. Steam cannot compete with quick-starting gasoline engines in burgeoning automobile market, however.

Railroad Development

c. 1550 Forerunner of railroad, wagons run on wooden tracks in mines are used in Holy Roman Empire.

1801 Surrey Iron Railway starts service between Wandsworth and Croydon. First commercial freight line, it uses horse-drawn wagons.

1803 World's first steam locomotive, *New Castle*, is built by British inventor Richard Trevithick (1771–1833). It runs but is too heavy for rails.

1812 First practical locomotive, built by British inventor John Blenkinshop (1783–1831), runs on tooth-rack railway.

1813 *Puffing Billy*, successful traction locomotive, hauls mining wagons without tooth-rack system.

1814 George Stephenson (1781–1848), British railroad inventor, builds his first locomotive, *Blücher*, which hauls loads faster than horse team can.

1829 Stephenson's locomotive *Rocket*, capable of traveling at 29 mph, wins contest in Great Britain for best design. *Rocket*, whose main advantage is new-type fire-tube boiler, becomes first reliable locomotive.

1830 Service begins on the Baltimore & Ohio Railroad, America's first steam-powered railroad. Locomotive is *Tom Thumb*, built by American manufacturer Peter Cooper (1791–1883).

1830 American inventor Robert L. Stevens invents modern railroad rail (inverted "T"). He also introduces railroad spike and roadbed of wooden ties set in gravel.

1832 France's first steam-powered railway service begins, on St. Étienne–Andrüezieux line.

1835 First German steam-powered rail service begins on Nürnberg–Fürth line.

1836 Canada's first railway service begins operations, in Quebec, between Laprairie and St. John.

1837 Early version of sleeping car is in service in United States.

1840 Railroad building boom begins in United States. Nation has 2,800 miles of track now; by 1860 it has 30,000 miles.

1841 Semaphore signal—for decades universal railroad signaling device—is in use for first time.

1844–46 Height of Great Britain's railroad-building boom. Parliament approves construction of over 400 new rail lines in this period.

1857 First steel railroad rails are in use in Great Britain.

1859 Pullman sleeper cars, invention of American George M. Pullman (1831–97), are in service for first time.

▶

1869 U.S. transcontinental railroad is completed when Union Pacific tracks being laid from east and west link up at Promontory, Utah.

1879 Germans demonstrate first successful electric locomotive, in Berlin.

1880s United States reaches peak of railroad building in this decade, with over 70,000 miles of track laid in 10 years.

1882 Knuckle-type coupler for railroad cars becomes available.

1885 First Canadian transcontinental railway is complete.

1892 Diesel engine is patented; soon after, inventors begin trying to adapt diesel to rail uses.

1893 Compressed-air brake, patented by American inventor George Westinghouse (1846–1914) in 1869, becomes mandatory on all U.S. rolling stock.

1895 Two British trains average more than 60 mph on competitive run between London and Scotland.

1902 Italian railroad completes electrification of main line. In following years many railroads in cities and suburbs are electrified.

1913 First use of diesel-electric railway engines, in Sweden.

1916 Completion of 5,787-mile Trans-Siberian Railroad linking Moscow and Vladivostok, Russia.

1925 First successful diesel switching engine is put into service.

1932 Germans introduce two-car, diesel-electric train on Berlin–Hamburg line; train averages 77 mph.

1933 U.S. railroad tries welding segments of rail together for first time. Years later, 440-yard lengths of welded rail become standard.

1939 First diesel-electric freight locomotive is built, by General Motors.

1941 World's largest steam locomotive, Union Pacific's *Big Boy*, is built. Designed for mountain service, it develops 6,000 horsepower.

1955 Two French-made electric locomotives set world record by reaching 205 mph.

1964 Japan's high-speed commuter line, New Tokaido Line, begins service. Its electric passenger trains hit top speeds of 132 mph on 320-mile line connecting Osaka and Tokyo.

1969 Experimental TurboTrains, powered by aircraft turbines, are put in service between New York and Boston and between Mon treal and Toronto.

c. 1970 By this time diesel-electric locomotives have replaced steam locomotives in most parts of world.

1970s Trains using principle of magnetic levitation are being developed. Japanese test model achieves speed record of 321.2 mph in 1979.

Key Developments in the Industrial Revolution

See also The Age of Steam.

1698 Earliest steam pump is invented in England. It pumps water from mine shafts and is called "miner's friend."

1712 English ironmonger Thomas Newcomen (1663–1729) invents early steam engine. Unfortunately, it burns enormous quantities of coal.

1733 English machinist John Kay (1704–c. 64) develops "flying shuttle," device for moving weaving shuttle faster than possible by hand. However, machine often goes out of control.

1738 English inventor develops roller spinning machine, to replace spinning wheels.

▶

c. 1750 Coke smelting is developed in England, providing more economical and efficient means of producing iron.

1764 English carpenter and engineer James Hargreaves (c. 1720–78) develops spinning jenny, mechanical device to do work of hand spinner.

1769 English inventor Richard Arkwright (1732–92) produces first water frame, water-powered device for spinning yarn.

1769 Scottish inventor James Watt (1736–1819) patents more efficient steam engine. With later improvements, Watt's steam engines become efficient power sources for wide variety of machines and play important role in bringing about Industrial Revolution.

1775 Boring mill is developed to bore holes more precisely, thus eliminating steam leakage from steam engines.

1779 English inventor Samuel Crompton (1753–1827) combines water frame and spinning jenny in "spinning mule," which greatly increases output of spun thread and effectively ends home spinning industry.

1783 English inventor Henry Cort (1740–1800) develops rolling mill for easier and cheaper production of wrought iron. In 1784 he patents "puddling furnace," in which molten pig iron is stirred to burn off impurities.

1785 Steam-powered loom is developed by English inventor Edmund Cartwright (1743–1823).

1786 U.S. inventor develops early steam-powered vessel. Machinery required to power it fills entire vessel, however.

1789 Leblanc process for producing soda cheaply helps revolutionize chemicals industry.

1793 U.S. inventor Eli Whitney (1765–1825) develops cotton gin to mechanically remove seeds from picked cotton, tripling amount of cotton a worker can process.

1795 Hydraulic press is invented in England.

1797 Sliderest is developed to simplify precise cutting of metals. It leads to screwmaking machine.

1800 Eli Whitney's use of interchangeable parts in gun manufacture later provides basis of modern mass production.

1800 Engine lathe is developed, in England.

1802 High-pressure steam engine is built in England by Richard Trevithick (1771–1833), five years after one is patented in United States.

1803 Metal steam-powered loom is developed.

1804 French invent mechanical loom for weaving designs. Workers riot to protest ensuing unemployment.

1806 Gas lighting for interior illumination is introduced.

1807 Wood-mortising machine is developed in France.

1807 U.S. inventor Robert Fulton (1765–1815) develops the first commercially successful steamboat. It simplifies transportation of goods and natural resources.

1822 English scientist Michael Faraday (1791–1867) invents electric motor. In 1831 he invents dynamo, forerunner of modern dynamos and generators.

1824 Portland cement is patented by English bricklayer and stonemason Joseph Aspdin (1799–1855).

1825 Steam locomotion proves successful in England for overland transportation of coal from mines to waterways.

1827 Water-powered turbine is developed.

1829 British inventor George Stephenson (1781–1848) builds first reliable locomotive, *Rocket*, making large-scale commercial rail service practical.

c. 1835 First commercially successful electric motor is invented by Vermont blacksmith John Davenport (1802–51).

1839 Steam hammer is invented in Scotland.

1846 Sewing machine is invented by Elias Howe (1819–67).

1855 Turret lathe, American invention, further revolutionizes machine tool production.

1856 Synthetic-dye industry begins with discovery of aniline dye mauve, coal-tar derivative.

1856 Bessemer process is developed, allowing mass producion of steel.

1860 Internal-combustion engine is invented in France.

1861 Process for mass-producing ammonia is patented by Belgian industrial chemist Ernest Solvay (1838–1922).

▶

1873 Electricity is first used to drive industrial machinery, in Austria; it is quickly adopted elsewhere.

1876 First practical four-stroke engine appears.

1876 Belgian industrialist develops steam-powered machine to generate electricity.

1876 Centennial Exposition in Philadelphia exhibits over 8,000 machines.

1884 Steam turbine is patented.

1898 Diesel engine, perfected by German inventor Rudolf Diesel (1858–1913) in 1896, becomes commercial success. Diesel engines later find wide array of applications in transportation, power generation, and other industries.

Plastics and Synthetics

1828 French scientists discover vinyl and polystyrene plastics, but development does not progress.

1862 English chemist develops early plastic substance, Parkesine.

1868 New York merchant firm offers $10,000 prize for material to replace costly ivory in billiard balls. U.S. printer and inventor John Hyatt (1837–1920) mixes nitrated cellulose with camphor, inventing celluloid. (Celluloid proves ideal for making false teeth, piano keys, and other products as well.)

1868 Eyeglass frames are being made from cellulose nitrate.

1878 Patent is granted in England for making rayon, which also is developed in France in 1879.

1882 First motion-picture film is made, of celluloid.

1884 Eastman Kodak Company begins making photographic film from celluloid.

1884 W. H. Walker invents roll film.

1892 Englishman Charles Frederick Cross (1855–1935) invents viscose rayon, superior form of this synthetic.

1897 Discovery of plastic casein.

1903 Method of producing viscose "silk" is developed by W. H. Stearn and F. Topham.

1904 Frederick Kipping discovers silicone.

1908 Swiss chemist discovers cellophane.

1909 First totally synthetic plastic is invented by Belgian-American chemist Leo H. Baekeland (1863–1944). Called Bakelite, it is used to make electrical equipment, coatings, and handles.

1909 Cold-molded plastics are developed (little heat is required to shape them).

1919 Researchers develop casein, cold-molded plastic made from skimmed milk and formaldehyde. It later is used for knitting needles, buttons, and buckles.

c. 1926 German scientist Hermann Staudinger (1881–1965) accurately describes molecular makeup of polymers (chemical compounds made up of chainlike molecules), leading to greater understanding and development of plastics. (He wins 1953 Nobel Prize in Chemistry.)

1927 Cellulose acetate is developed. It is far less flammable material than celluloid.

1927 Chemists discover polyvinyl chloride. Flexible and elastic, its many uses include raincoats and phonograph records.

1928 Diene synthesis (Diels-Alder reaction) is developed in Germany, vital process for producing synthetics.

1929 Urea formaldehyde is introduced for use in United States. Products eventually include lighting fixtures.

1930 Polystyrene production is developed in Germany by I. G. Farberindustrie.

1932 First successful synthetic rubber becomes available commercially, marketed by Du Pont in United States.

1935 Nylon is discovered by U.S. chemist Wallace Carothers (1896–1937). It is first completely synthetic fiber and finds many uses in manufacture of fabrics.

▶

c. 1935 Chemical firms produce acrylics such as Lucite and Plexiglas, now standard for airplane windows.

1936 Acrylic resin, transparent coating, is developed.

1936 Researchers discover polyvinyl acetate, adhesive in some paints.

1937 Polyurethane is patented, in Germany. This synthetic is produced in many forms, and products made of it range from stretch clothing to furniture.

1937 Polystyrene becomes commercially available. Kitchen utensils, wall tile, and toys are made from it.

1938 Perlon, a synthetic polyamide fiber, is developed.

1938 Nylon goes on sale.

1938 Polyvinyl acetal, a water-insoluble resin in adhesives, is developed.

1938 Scientists at Du Pont discover Teflon (polytetrafluoroethylene). Used for electrical insulation and for coatings to produce nonstick surfaces, it becomes commercially available in 1948.

1939 Polyethylene, invented in Great Britain, becomes commercially available. It is at first primarily used in radar equipment.

1939 Polyvinyl chloride, discovered in 1927, is developed. It finds many uses, including manufacture of automobile seat covers, brush bristles, and window screens.

1940 Stockings made of nylon become commercially available in United States.

c. 1940 U.S. scientist manages to convert soybeans to plastic parts, for use in automotive industry.

1941 First all-plastic car body is manufactured in United States, by Ford. It proves costly, however.

1942 Polyesters, synthetic textile fibers, become available. Strong and wrinkle-resistant, they are widely used for manufacturing clothing. (Dacron is patented by U.S. Du Pont Company in 1946; Orlon, in 1948.)

1943 Chemists develop silicones, which were discovered in 1904. Chemically very stable, silicones make good electrical insulation and waterproof coatings for fabric and other materials and are component of hydraulic fluid.

1945 U.S. chemist introduces first commercial product made of polyethylene, a plastic water glass. He soon begins large-scale marketing of plastic products called Tupperware.

1947 Epoxy glue is developed as especially strong adhesive.

1948 Researchers develop acrylonitrile-butadiene-styrene (ABS). Products using it include luggage, football helmets, and skates.

c. 1952 German scientist Karl Ziegler (b. 1898) finds process (using catalyst) to create harder and more rigid polyethylene. (New polyethylene is first used to make Hula Hoops.)

1953 Automotive industry experiments with plastic strengthened by glass fibers (forerunner of fiberglass-reinforced plastics).

1953 Development of isostatic polymers, plastics containing metal ions.

c. 1955 Italian scientist Giulio Natta (b. 1903) discovers that Ziegler's earlier use of catalysts to produce stronger polyethylene works with many other plastic materials as well. (Ziegler and Natta share 1963 Nobel Prize in Chemistry for work on polymers.)

1956 Acetal resins, stable formaldehyde polymers, are in commercial use for automotive parts. (Previously, formaldehyde polymers depolymerized quickly, before reaching temperatures required for processing.)

1956 Polycarbonates are developed. Products made of it later include film and unbreakable windows.

1957 Polypropylene, lightweight plastic now used for luggage, baby bottles, and water-resistant coatings, is developed.

1965 Polysulfones are developed. Uses today range from smoke alarms to shower heads.

c. 1968 Anionic polymerization is being used in plastics manufacture. It allows precise control of molecular size when synthesizing plastic materials.

1970s Widespread use of plastics becomes target of environmental movement because of difficulty in disposing of the material.

1980s High insulating property of certain plastics leads to increased use in building insulation and even (experimentally) as structural building material.

1985 Lanxides, composite materials of ceramics and metal, are declassified by U.S. military.

1987 Electroconductive plastic is developed by Herbert Naarmann and N. Theophilou.

Photography

See also Motion Pictures in the Media and Entertainment section for information on the development of motion-picture photography.

1500s *Camera obscura* (dark chamber), believed of Greek origin, is used by artists as drawing tool. Pinhole admits light inside enclosed box, throwing inverted image on inside of box.

1777 Swedish chemist Karl Wilhelm Scheele (1742–86) finds that when silver nitrate is exposed to light, blackening effect results.

1802 Englishmen Thomas Wedgewood (1771–1831) and Humphry Davy (1778–1829) coat paper with silver salts to reproduce silhouette images temporarily.

1810 German physicist Thomas Johann Seebeck (1770–1831) experiments with wet silver chloride and temporarily reproduces color spectrum. His work pioneers later color film development.

1816 Joseph Nicéphore Niepce (1765–1833), French chemist, experiments with silver nitrate, producing grayish images on paper from negative. He is unable, however, to remove unexposed silver salts and obtain permanent image.

1826 Niepce produces negative image by using *camera obscura* and light-sensitive emulsion. "Heliograph" image is on pewter plate and can be printed on ordinary printing press.

1837 French painter Louis Jacques Mandé Daguerre (1789–1851) develops daguerreotypy, process using silver-plated copper with silver iodide emulsion.

1838 William H. F. Talbot (1800–1877), English scientist, creates first image on sensitized paper. He later is first to use hyposulfate of soda to "fix" images permanently.

1839 Daguerre announces his image-making process to public. Soon after, portrait salons in Europe and United States open.

1839 France's *Le Lithographe* magazine publishes daguerreotype for first time. In late 1880s halftone images are used for facsimile reproduction.

1840 Earliest extant photograph of person is taken.

1840 First picture of moon is taken.

1851 Collodion process for making "wet-plate" glass negatives is introduced. Many good paper prints are to be made with this negative.

1851 First photograph is taken using high-speed flash to stop action.

1853 Photographic Society is founded in London. In 1854 Société Française de Photographie is founded in Paris. These organizations help establish photography as profession.

1855 Roger Fenton (1819–69), one of pioneer news photographers, leaves London to photograph Crimean War.

1858 French balloonist takes first aerial photograph.

1861 Mathew Brady (1823–96) organizes first photographic coverage of actual combat in American Civil War. He and others hired by him compile photographic record of this war.

1869 French chemist Louis D. du Hauron (1837–1920) builds camera that shoots three negatives per exposure, to be developed using different pigments. One crude color image combines pigments.

1871 Dry-plate (with silver halide emulsion) process is invented. Coated plates are marketed by 1876.

1873 German chemist Hermann Wilhelm Vogel (1834–89), experimenting with film emulsion, discovers that some dyes increase film sensitivity. His work pioneers development of high-speed, light-sensitive film.

1873 Photographic print paper (silver bromide gelatin-coated) is introduced.

1878 First moving-picture technology emerges. Zoopraxiscope, step beyond

▶

earlier Zoetrope, casts series of photographs nonstop at viewer's eye.

1880 Anthony Company (now GAF) begins selling Eastman dry-plate negatives for photo images.

c. 1880 Flash powder (magnesium and potassium chlorate) is used to create extra light for photographs.

1881 First color photograph is produced, by Frederic E. Ives (1856–1937).

1885 High-speed photograph of bullet in flight is taken, using sparks to provide flash.

1887 Celluloid roll film is developed by Hannibal W. Goodwin (1822–1900).

c. 1887 First leaf-type shutter is invented.

1888 Eastman Kodak introduces portable $25 camera loaded at factory with 100-exposure roll of celluloid film. Factory later must unload camera and develop exposed roll of film.

1890 American Alfred Stieglitz (1864–1946) publishes *Camera Notes*, quarterly journal of Camera Club of New York. Stieglitz helps to establish aesthetic standards for photography.

1891 Photographic paper for contact printing is marketed.

1891–1903 Kodak improves celluloid roll film, introducing paperbacked, daylight-loading roll. It later makes gelatin-coated film that allows easier handling and printing of negatives.

1900 Kodak introduces its first Brownie camera for children.

1907 Color photography is developed by French chemist Louis Lumière (1864–1948).

1918 First "schadograph" is created by Dadaist artists in Zurich. Light-sensitive paper is exposed with objects placed on it.

1920 Color camera with sensitized plates that reproduce color images is marketed.

1924 German Leica camera is introduced, first commercially successful camera using 35mm film and prototype for many similar cameras.

1924 Demonstration is made of wireless transmission of photograph between New York and London.

1925 Prototype of flash bulb is introduced.

1927 World's first underwater color photographs are published, by *National Geographic*.

1928 Introduction of the Rolleiflex, double-lens reflex cameras.

1930s Frenchman Henri Cartier-Bresson (b. 1908) begins his career in photography; he sets aesthetic standards for photojournalism.

1930s First appearance of single-lens reflex cameras.

1935 American Ansel Adams (1902–84) publishes *Making a Photograph*. Many of Adams's famous photographs are high-contrast photos of American West.

1935 Kodachrome, first commercial color film, is introduced by Kodak. Film has three layers to produce color image.

1935 Farm Security Administration authorizes historical documentation of Great Depression.

1939 Transfer-diffusion process for photographic prints is developed. It becomes basis for Polaroid process.

1940 Electronic flash is first marketed.

1947 American Edwin Herbert Land (b. 1909) develops first black-and-white Polaroid camera with one-step development of photographs.

1955 "The Family of Man," noted photography exhibit, is first shown at Museum of Modern Art in New York City.

1963 Polaroid introduces color version of popular Polaroid camera.

1966 First pictures of moon are sent back. Photographs became important part of space exploration.

1982 Kodak introduces 15-picture film disk.

Electricity

1600 English physicist William Gilbert (1540–1603) describes substances that can produce static electricity, in his treatise, *De magnete*.

1630 Attraction and repulsion of electrically charged bodies is observed.

c. 1720 British scientists experiment with conducting and insulating materials. They conduct electricity from rubbed glass along 886-foot string.

1733 French scientist discovers that materials charged with similar kinds of electricity repel one another and that those charged with opposite kinds of charge attract each other.

1740 Machines for producing electrical charges and electroscopes for measuring charges are in existence by this date.

1746 Leyden jar, device in which electrical charge can be collected and briefly stored, is developed.

1749 American inventor Benjamin Franklin (1706–90) installs lightning rod on his home after discovering that pointed conductor can take charge from charged body.

1752 Franklin experiments with kite in thunderstorm to prove that lightning is electricity. He gives terms *positive* and *negative* to different electrical charges.

1760 Swiss professor Daniel Bernoulli's (1700–82) experiments reveal that electricity is governed by an inverse-square law.

1775 Italian physicist Alessandro Volta (1745–1827) describes device to produce and store static electrical charges.

1785 French physicist Charles Coulomb (1736–1806) describes mathematically how positive and negative electrical charges attract and repel one another. His theory is known as Coulomb's law.

1786 Italian physician Luigi Galvani (1737–98), observing leg contractions in frog he is dissecting, mistakenly theorizes that animal tissues produce electricity. His research leads to study of electric currents, and many electrical devices (such as galvanometer) incorporate his name.

1800 Volta disproves Galvani's animal-electricity theory, showing that chemical action of moisture and two different metals can generate flow of electricity. He makes world's first battery, a voltaic pile, using disks of silver and zinc. Electric volt is named for him.

1807 English chemist Humphry Davy (1778–1829) makes important strides in new field of electrochemistry, following Volta's discovery that electric current can decompose water. He identifies elements of sodium and potassium. In 1808 he isolates barium, boron, calcium, and magnesium and also produces first electric (carbon arc) lamp.

1820 Danish physicist Hans Christian Oersted (1777–1851) demonstrates important relationship between electricity and magnetism (that compass needle is deflected when placed near electric current), beginning study of electromagnetism.

c. 1820 French physicist André-Marie Ampère (1775–1836) continues Oersted's work, demonstrating that two parallel wires carrying electrical current in same direction will attract each other and if currents travel in opposite directions, wires repel one another. Electric current unit ampere is named after him.

1821 Ampère and others develop galvanometers for measuring current, its direction, and its strength.

1821 English scientist Michael Faraday (1791–1867) invents instruments to demonstrate effects of magnetic "field of force" around current and magnet; this becomes basic principle of electric motor.

1823 British invent first electromagnet, iron bar with a current-carrying wire wrapped around it.

1826 German physicist Georg Simon Ohm (1787–1854) describes electrical conduction in solids, from which he formulates

▶

Ohm's law (for measuring electrical current). Ohm, unit of electrical resistance, is named for him.

1831 American inventor Joseph Henry (1797–1878) invents first electric motor, consisting of electromagnet and rocker arm that moves up and down.

1831 Faraday discovers electromagnetic induction (also discovered independently by Henry), phenomenon by which moving magnet can generate electric current. He builds first crude electric generators.

1833 Faraday formulates laws of electrolysis.

1835 American inventor Samuel F. B. Morse (1791–1872) builds working model of his electrical telegraph. He patents it in 1837 in both United States and England.

1835 Joseph Henry invents electric relay, facilitating long-distance transmission of current.

c. 1835 Vermont blacksmith Thomas Davenport (1802–51) invents first commercially successful electric motor.

1839 Electric fuel cell is invented; it is based on combination of hydrogen and oxygen.

1843 English physicist Charles Wheatstone (1802–75) invents devices to measure large electrical currents and electrical resistance accurately.

1844 Morse introduces first practical telegraph service, beginning operation between Washington, D.C. and Baltimore. It marks beginning of use of electricity for communications.

1858 First lighthouse with electric arc lamp goes into operation.

1859 Rechargeable storage battery is invented.

1864 Scottish physicist James Maxwell (1831–79) unifies known facts about electricity into concise, four-equation mathematical theory, providing framework for understanding all electric and magnetic phenomena in nature. He publishes *Electricity and Magnetism* in 1873.

1867 First practical AC generator is built by Belgian-born French inventor Zénobe Théophile Gramme (1826–1901). Two years later he builds practical DC generator.

1876 American inventor Alexander Graham Bell (1847–1922) patents telephone, second great communications device based on electrical current. Telephone soon surpasses telegraph as primary communications system.

1879 U.S. inventor Thomas Alva Edison (1847–1931) demonstrates incandescent light bulb, constructed with carbonized cotton filament, in New Jersey. (Bulb introduced by British inventor in 1878 was unsuccessful.)

1880 First electric generating station, built by Edison in London, England to provide power for streetlights, goes into operation.

1881 Invention of improved storage battery.

1881 System for distributing AC current, allowing longer-distance power transmission, is patented.

1882 Modern three-wire system of electrical transmission is patented by Edison.

1882 First electric generating plant in United States opens in New York City, providing power for nation's first system of electric lighting.

1884 Electric alternator is invented by Croatian-born American Nikola Tesla (1856–1943).

1884 First turbogenerator for producing hydroelectric power is built by Irishman Charles A. Parsons (1854–1931).

1885 Electric transformer is invented.

1886 American George Westinghouse (1846–1914) demonstrates AC-current power system, revealing advantages over DC systems then in use.

1888 First AC motor is patented. Over next decades, it becomes efficient and widely used means for converting electricity to mechanical power for both industrial and home applications.

1890 First electric-chair execution takes place, in New York.

1891 Tesla invents coil bearing his name; it produces high voltage at high frequency.

1891 First AC (three-phase) power transmission system goes into operation in Germany. First U.S. AC system begins service in 1896.

1891 Irish physicist theorizes that electric current is composed of small moving particles, which he calls electrons.

▶

1897 Italian scientist Guglielmo Marconi (1874–1937) achieves long-distance radio transmission (short-distance transmission is demonstrated in 1894), thus pioneering new "wireless" form of electronic communications based on electromagnetic radiation. He achieves transatlantic transmission in 1901.

1897 English physicist Joseph Thomas (1856–1940) confirms electron theory and discovers that all atoms contain electrons.

c. 1900 First vacuum tube is invented. Discovery that electrons can be moved at great speeds through vacuum tubes begins electronics age.

1900s Electric home appliances become available and gain increasing popularity during this century. Electric ovens are marketed by c. 1910, electric refrigerators by c. 1913.

1904 First practical photoelectric cell is invented by Johann P. L. Elster.

1904 British scientist John Fleming (1849–1945) invents diode vacuum tube, rectifier tube that becomes important in radios and other electronic equipment.

1906 Tungsten-filament light bulb is introduced.

1906 American inventor Lee De Forest (1873–1961) patents Audion vacuum tube, which amplifies electronic signal. It becomes key part of electronic equipment for radio, television, telephone, radar, and other communications systems.

1907 Suspension-type insulators for high-voltage lines are invented, allowing transmission of much higher voltages (up to 150,000 in next few years) over longer distances.

1910 Introduction of neon light.

1910 First underground hydroelectric power plant is constructed, in Sweden.

1921 Invention of magnetron, vacuum tube that produces microwaves.

1922 World's largest hydroelectric power plant to date is built in Ontario, Canada using Niagara Falls water to drive its turbines.

1926 First public demonstration is made of television.

1929 FM radio is introduced.

1935 Development of radar in England.

1936 Fluorescent lighting is introduced.

1947 American physicist William Shockley (b. 1910) and others invent transistor, which soon replaces vacuum tube in electronic equipment. It marks beginning of miniaturization and increasing complexity of electronic devices.

1951 First commercial electronic computers (UNIVAC) are introduced by Remington Rand.

1954 Development of photovoltaic cell at Bell Labs.

1954 Sweden builds first EHV (extra-high-voltage) power transmission lines, which carry 380,000 volts. By 1965, lines carrying 750,000 volts are in operation in Canada.

1956 First nuclear-power generating plant begins operation in Great Britain. Nuclear-power generation is expected to meet rapidly expanding need for electrical power in industrialized world. High plant construction costs, potential safety hazards, and nuclear-waste disposal are not widely recognized as major problems until later, however.

1958 Integrated circuit is invented. Methods for mass producing integrated circuits further revolutionize electronics industry and eventually make possible computer microprocessor chips.

1960 Increased efficiency of steam-turbine electric generating plants, lower initial cost, and low cost of fossil fuel lead to shift from hydroelectric to thermal generation in United States. Largest turbines produce 450 megawatts in 1960, 1,000 megawatts by 1965, and almost 1,450 megawatts by 1980s.

1960s–80s Many high-tech electronic consumer devices become popular, including calculators, audio and video equipment, computers, and timepieces.

1967 Construction of largest hydroelectric power plant to date is completed, on Yenisei River in Siberia. It generates three times power produced by Grand Coulee Dam in United States.

1970s Array technology is developed, making it possible to put many integrated circuits together on single silicon chip. This leads

▶

to further miniaturization of electronic components, notably in computers.

1980s Scientists develop array of new superconducting materials, which offer virtually no resistance to the flow of electricity.

1981 Completion of world's largest solar electrical generating plant, Solar One, with capacity of 10 megawatts.

1984 Paraná River power plant on border of Brazil and Paraguay begins producing electricity. Total output when all 16 turbines are operating will reach 12.6 million kilowatts.

Telegraph and Telephone

B.C.

c. 430 Primitive optical telegraph using torches on hilltops is used in Greece.

A.D.

1790s French coin word *télégraphe*, meaning "to write far," to refer to system for relaying messages by semaphore along system of signal towers (in operation in 1794).

1831 Crude electric telegraph is built by Charles Wheatstone (1802–75) and William Fothergill (1771–1867). Magnetized needle is made to point at specific letter by passing current through nearby wire.

1831 American inventor Joseph Henry (1797–1878) uses recently invented electromagnet (1825) to ring bell at end of 1-mile-long wire.

1832 American painter and inventor Samuel F. B. Morse (1791–1872) begins work on his telegraph apparatus.

1833 Electromagnetic telegraph that functions at 1.2-mile distance is developed by German mathematician Carl Friedrich Gauss (1777–1855) and German physicist Wilhelm Eduard Weber (1804–91).

1835 Repeater, for sending electric signals over long distances, is invented by American Joseph Henry (1797–1878).

1837 Charles Wheatstone (1802–75) and William Cooke (1806–79) invent and patent telegraph independently of Morse.

1837 Morse patents working model of his telegraph.

1844 First practical telegraph service begins on Morse's 37-mile line connecting Washington, D.C. and Baltimore. First message, sent in Morse's special code of dots and dashes, is "What hath God wrought?"

1846 Telegraph transmitter using prepunched paper tape is invented. It greatly increases speed and efficiency of message transmission.

1850 First underwater telegraph cable is laid, linking France and England.

1850–70 Many new telegraph systems are built, in United States and abroad, to link cities and to provide rapid communications within large cities.

1851 Western Union becomes leading telegraph company in United States.

1852 Word *telegram* is coined.

1855 Printing telegraph is invented by David E. Hughes (1831–1900).

1866 Beginning of regular transatlantic cable service, connecting United States and Europe, after failure of first working transatlantic cable (1858).

1874 Frenchman Émile Baudot (1845–1903) invents five-unit code later used in teletypewriter system. He develops prototypes of teletypewriter in 1870s.

1876 American inventor Alexander Graham Bell (1847–1922) patents his telephone, using one-piece transmitter-receiver that induces current when spoken into.

1878 First telephone switchboard is in operation, in New Haven, Connecticut.

▶

1878 Carbon-type microphone telephone transmitter developed by Edison is in commercial use, making long-distance telephone possible.

1879 Great Britain's first telephone exchange is built, in London.

1887 U.S. telephone companies by this time serve total of 150,000 customers. From this time forward, telephone begins to replace telegraph.

1892 Automatic telephone switchboard is first used.

1893 Boston and Chicago are linked by long-distance telephone lines for first time.

1894 Guglielmo Marconi (1874–1937) develops his prototype wireless telegraph.

1899 M. I. Pupin facilitates long-distance telephone transmission through use of loading coils (inductors).

1900 Wall-mounted model of telephone with separate earpiece and mouthpiece is introduced.

1901 First transatlantic radio telegraphic transmission is achieved by Marconi.

1910 Nearly 125,000 phones are in use in Great Britain.

1915 Researchers discover that vacuum tube successfully amplifies telephone voice signal, making cross-country and transatlantic telephone calls possible.

c. 1915 Printing telegraphy using teletypewriters is replacing manual telegraphy in telegraph-company operations.

1921 First completely automatic local dialing service is offered by Omaha, Nebraska telephone system. Dial telephone is first developed in 1905.

1926 First transatlantic conversation is made via radiotelephone (New York to London).

1930s Telex system of privately rented teletypewriters begins.

1930s Dry chemical copying process is developed, making possible desk-fax system.

1947 First microwave relay station for long-distance telephone communication, between Boston and New York, begins operation. It eliminates need for expensive long-distance trunk lines.

1950 Traditional black, rotary-dial desk-model telephone is introduced.

1951 Direct long-distance dialing service begins in United States.

1956 First transatlantic telephone cable is laid, made possible by newly designed amplifiers with 20-year life.

1956 "Visual telephone" is under development in United States.

1962 World's first telecommunications satellite, *Telstar*, is in orbit. In coming years, satellites make possible worldwide communications network, handling telephone, television, and data transmission.

1963 First push-button telephones are introduced.

1966 First successful transatlantic direct-dial phone calls are made. Regular service from New York to Paris and London begins in 1967.

1970 First commercial Picturephone service begins, in Pittsburgh.

1977 Testing of new fiber-optic system begins in Chicago.

1982 Court order breaks up AT&T, U.S. telephone monopoly, into AT&T Long Lines and regional telephone companies.

1985 Single optical fiber carries equivalent of 300,000 simultaneous phone calls in Bell Labs test.

c. 1987 AT&T completes "digitalization" of all its long-distance facilities. Voice transmissions are converted to computer codes for more efficient transmission through lines, then reassembled at receiving end.

Radio and Television

See also Radio and Television in Media and Entertainment section.

1864 Scottish physicist James Maxwell (1831–79) develops mathematical formulations on electromagnetic radiation, providing theoretical basis for what becomes basic principle of radio-wave transmission.

1866 Radio waves are successfully used to transmit telegraphic messages in West Virginia by Mahlon Loomis (1826–86).

1873 Selenium is found to have property that later is important to development of television. Amount of light striking selenium varies its ability to conduct electricity.

1878 English scientist Sir William Crookes (1832–1919) develops Crookes tube, precursor of cathode-ray tube, vital component of television.

c. 1880 Basic principle of television, line-by-line scanning of image, is discovered in United States by W. E. Sawyer and in France by Maurice Leblanc (1857–1923).

1884 German inventor Paul Nipkow (1860–1940) patents an early type of television, using mechanical scanning system (rotating disk).

c. 1888 German physicist Heinrich Hertz (1857–94) produces and detects radio waves (predicted by Maxwell) over short distances.

1897 Italian scientist Guglielmo Marconi (1874–1937) achieves radio transmission over long distance (12 miles), with equipment of his design (including ground and overhead wire antennae).

1900 Human speech is transmitted via radio waves by American R. A. Fessenden.

1901 First use of crystal to detect radio waves by German physicist Ferdinand Braun (1850–1918).

1901 Marconi, using improved equipment, sends first transatlantic wireless radio message, from Cornwall, England to St. John's, Newfoundland. He sends Morse code letter "s."

1902 Discovery of a radio-wave reflection layer in upper atmosphere.

1904 Diode vacuum tube is invented by British scientist John Fleming (1849–1945).

1905 Marconi develops directional radio antenna.

1906 AM radio is invented by Canadian-American Reginald A. Fessenden (1866–1932).

1906 American inventor Lee De Forest (1873–1961) patents Audion vacuum tube, which amplifies electronic signal. It is key element in electronic equipment for radio, television, telephone, radar, and other communications systems.

1906 Russian scientist makes crude television system using newly developed fluorescent-screen cathode-ray tube as receiver.

1908 Basic scheme for what becomes modern television system is outlined by Scottish engineer. His ideas are too advanced for current technology, however.

1910 First radio communication from airplane to ground station is achieved.

1911 Locating radio source by direction-finding is developed.

1913 Heterodyne and cascade-tuning radio receivers are invented.

1913 Germans develop photoelectric cell coated with potassium hydride. It is many times more sensitive to changes in light intensity than one made with selenium.

1913 Regenerative receiver circuit is patented by American Edwin H. Armstrong (1890–1954), making long-distance reception practical.

1915 First major demonstration of long-distance voice communication, radiotelephony, from Arlington, Virginia to Paris.

1916 David Sarnoff (1891–1971) outlines the basic scheme of the radio broadcasting industry. He recommends building stations to broadcast voice and music and specifies manufacture of a "radio music box" to be bought by potential listeners.

▶

1918 Development of superheterodyne radio receiver.

1919 Shortwave radio is developed.

1921–25 Experimentation reveals shortwave radio transmission is best for long-distance voice communication.

1923 Iconoscope electronic television camera tube is patented by Russian-born American electronics engineer Vladimir K. Zworykin (1889–1982). With improved receiver patented in 1924, Zworykin makes possible modern, electronic scanning television system.

1925 Zworykin develops color television system (patented 1928).

1926 First public demonstration of television, by British inventor J. L. Baird (1888–1946). System uses mechanical scanner.

1928 Color television is demonstrated for first time, by Baird. He again relies on mechanical scanner.

1929 FM radio is introduced.

1929 Low-definition color images are transmitted between New York and Washington.

1932 RCA demonstrates all-electronic television system, operating on 120-line scan, built under Zworykin's direction.

1936 British Broadcasting Corporation begins its first public television broadcasting service (405 scanning lines).

1938 Sideband filter used in television broadcasting is developed, doubling resolution of pictures.

1939 Researchers begin to develop orthicon camera tube. It ultimately produces better pictures than earlier iconoscope tube.

1940s Development of printed circuit, which greatly simplifies manufacture of radio and television components.

1941 Start of regular television broadcasting in United States.

1948 Invention of transistor, longer-lasting and much less bulky substitute for vacuum tube. It helps make miniaturization of radio and television components possible.

1949 Vidicon camera tube is introduced. Compact size eventually makes it suitable for use in portable television cameras.

1950 CATV (Community Antenna Television) system is introduced. It initially provides cable-television feed for locales with poor reception, but by late 1970s, rapidly expanding cable-television systems compete with regular broadcast networks.

1951 Transcontinential television broadcasts are inaugurated in United States.

1952 Sony introduces pocket-size radio using transistors.

1953 "Shadow-mask" color television tube is marketed. It is standard design for color receiver tubes for next decades.

1953 Technical standards for "compatible" television broadcast signal are set in United States by National Television Systems Committee (NTSC), allowing black-and-white and color receivers to operate from the same broadcast signal.

1956 Videotape recording system is demonstrated.

1958 First monolithic integrated circuit is demonstrated. Integrated circuits revolutionize manufacture of electronic components in radio and television.

1960s Special color camera tube, plumbicon, is developed.

1960s–70s Instant-replay feature, playing back magnetic videotape recorded simultaneously with live broadcast, is introduced in sports broadcasts. Split-screen images are also introduced.

1962 First experimental transoceanic television broadcast by *Telstar* satellite, in color.

1963 First demonstration of home video recorder, in London.

1964 Color television sets become popular in United States. Sales eventually outstrip those of black-and-white receivers.

1964 Picturephone system, combining television and telephone, is demonstrated by AT&T at New York World's Fair. Subscriber service begins in 1971.

1964 First public demonstration of satellite television feed via "stationary" (geosynchronous) satellite. Tokyo Olympic Games are relayed to North America.

1969 First live television broadcast from moon, by U.S. *Apollo 11* moon-landing mission, is transmitted to some 100 million viewers worldwide by satellite feed.

▶

1970s High-definition television (HDTV) is under development in Japan. By 1981, working system is demonstrated.

1970s Sony introduces new Chromatron color picture tube. It has improved "metal grille" system for controlling illumination of color dots on screen, making brighter color picture possible than with standard shadow-mask design.

1982 Japanese introduce wristwatch-size television with 1.2-inch screen.

1985 World's largest television set, 80-by-150-foot Sony JumboTron, is first demonstrated, in Japan.

Development of Automobiles

B.C.

c. 800 Chinese develop steam-propelled cart.

A.D.

1670 A cart driven by crude steam turbine is developed by Jesuit priest living in China.

1769 French Army officer Nicholas-Joseph Cugnot (1725–1804) invents steam-powered gun tractor, which travels 2 mph but is unmaneuverable. He suffers what may be first automobile accident.

1789 First U.S. patent for land vehicle driven by steam is granted to Oliver Evans (1755–1819).

1801 British inventor Richard Trevithick (1771–1833) creates the first passenger-carrying vehicle powered by steam.

c. 1840 Regularly scheduled "road locomotives," steam-powered vehicles carrying 12 to 16 passengers, exist in England. However, they prove slower and less efficient than water or rail transportation.

1860 French engineer Jean J. Étienne Lenoir (1822–1900) develops two-stroke internal-combustion engine powered by illuminating gas.

1864 Lenoir sells one of his vehicles to Russian czar Alexander II (1818–81), in what is possibly first automobile export.

1865 German engineer Karl Benz (1844–1929) designs first automobile built as motor vehicle (rather than being converted from carriage), a three-wheeled, gas-driven model.

1865 Pressure from rail and horse-drawn carriage industries leads to British law requiring flagman to walk ahead of all steam-powered carriages. Law is not revoked until after invention of gasoline-powered automobile.

1876 First successful, lightweight, four-stroke engine is invented by German engineer Nikolaus Otto (1832–91). He originally intends it as stationary power source. Among German manufacturers who recognize engine's potential are Karl Benz and Gottlieb Daimler (1834–1900).

1879 First U.S. automobile patent is granted, to George B. Selden.

1888 Scot John Dunlop (1840–1921) invents pneumatic tire, thus beginning modern tire industry.

1891 French inventor develops the first four-cylinder gas engine with mechanical valve operation, for use in boat. (He also builds first six-cylinder engine.)

1893 First practical American automobile is built by businessmen Charles (1861–1938) and J. Frank Duryea (1869–1967). They begin manufacturing in 1895. Their second car wins first U.S. automobile race, between Chicago and Evanston, Illinois.

1896 American businessman Henry Ford (1863–1947) produces his first car, two-cylinder "quadricycle."

1896 U.S. manufacturer Ransom Olds (1864–1950) builds his first car. He begins production of first Oldsmobiles in 1899.

▶

1896 French car is manufactured with pneumatic tires as standard feature. It is called *voiturette* (little car).

c. 1898 Several U.S. companies begin large-scale manufacture of freight-carrying trucks.

1900 Electric cars, despite their limited power, make up 38 percent of American market.

1901 Fire at factory of Ransom Olds leads to his purchasing car components from outside suppliers, marking first step toward future mass production of automobiles.

1902 Frenchman Louis Renault (1877–1944) invents first drum brake, which is soon in use on most vehicles.

1905 Society of Automobile Engineers is established in United States, encouraging standardization of auto parts.

1906 Stanley steam car travels at 127 mph in Florida race.

1908 Henry Ford introduces inexpensive, well-built Model T. It becomes enormously popular worldwide.

1909 V-8 engine is standard in French production-line automobile manufactured this year.

1911 British Cadillac Company introduces model with electric self-starter, which eliminates difficult hand-cranking. First electric lights also are featured on Cadillacs.

1914 Ford begins assembly-line production of Model T, further lowering its price.

1915 U.S. trucks total 158,000 this year (up from 700 in 1905).

1915 Ford's Michigan assembly-line factory produces over 500,000 Model Ts this year. Each sells for $440. Ford offers first car rebates on his Model T, $50.

1920 Four-wheel hydraulic brakes are introduced by U.S. Deusenberg company.

1922 Electric windshield wipers are introduced.

1923 Closed sedans become more popular than open models for first time.

1924 Benz introduces first diesel truck.

1925 Headlights with two beam levels, operated by foot switch, are introduced.

1926 First power steering is installed on Pierce-Arrow Runabout. It is not available on passenger cars and trucks until 1950.

1929 Electric gauges are developed to determine amount of fuel left in tank.

1930 First windshield washer, manual pump, is invented.

c. 1930 Hydraulic brakes, balloon tires, and self-starting devices are becoming common on most American-made cars and trucks.

c. 1930 Trucking gains advantage over railroad transportation due to small size of trucking companies and to deficiencies in railroad industry.

1933 Auto manufacturers such as Pierce-Arrow and Chrysler begin redesigning their cars for increased aerodynamics and smoothness of line; 1934 Chrysler Airflow also features first overdrive transmission.

1934 Rigid, four-axle trucks in use in England carry loads of up to 15 tons.

1934 First diesel-powered passenger car is introduced, by Mercedes-Benz.

c. 1934 Adolf Hitler (1889–1945) challenges German automakers to develop car for the people. He suggests name Volkswagen.

c. 1935 Front-wheel-drive automobiles are introduced by manufacturers such as Audi, Citroën, and Cord.

1937 Oldsmobile introduces first automatic transmission.

1938 Prototype of Volkswagen beetle is introduced in Germany by Ferdinand Porsche.

1939 Earliest car air conditioning is offered by U.S. manufacturers (by fitting small refrigerator in ventilating system).

1939 Outbreak of World War II brings technological advances in automotive and trucking industries but causes scarcity of gasoline and materials.

c. 1940 Power brakes are now available for trucks.

1947 Goodyear introduces tubeless tire.

1948 First four-wheel-drive Land-Rovers are sold.

1950 First gas turbine engine is installed in truck in United States.

c. 1950 Mercedes-Benz introduces fuel injection in its racing models. It becomes popular for passenger vehicles.

1953 Radial ply tires are introduced in Germany and Italy.

1957 British Lotus Elite sports car with fiberglass body is introduced.

►

1958 Aluminum car engine is developed in United States. It is more resistant to wear and 30 percent lighter than previous, cast-iron engines.

1960 Individual front suspension is introduced for trucks.

1960s American auto makers design compact cars in response to consumer interest in small, imported models.

1966 First electronic fuel-injection system is developed, in England. Small computer checks engine operations, then injects appropriate fuel into each cylinder.

1968 Antipollution devices are required on U.S. cars to control hydrocarbon emissions.

c. 1970 Disk brakes are standard on front wheels of cars.

1979 U.S. scientists develop DISC (Direct-Injected Stratified Charge) engine to improve pollution control and fuel economy.

c. 1980 Electronic antilock braking systems are introduced to prevent skidding.

1981 Computer chips installed in automobiles can control air conditioning and monitor cruising speed, radio, and engine performance.

c. 1985 Automobile and truck manufacturers are developing microchip-powered devices to display maps and efficient travel routes on tiny computer screens in cars.

Internal Combustion Engine

c. 1680 Dutchman Christiaan Huygens (1629–95) develops one of earliest internal-combustion engines. Fueled by gunpowder, it is difficult to control.

1791 English inventor designs gas turbine engine, but it is never built.

1794 First practical internal-combustion engine is patented by British inventor Robert Street.

1823 Water-cooled automotive engine is invented in Great Britain.

1824 French physicist Sadi Carnot (1796–1832) publishes book *Reflections on the Motive Power of Heat*, which outlines theory behind internal-combustion engine.

1838 Englishman William Barnett develops continous-flame ignition system.

1841 French physicists Antoine Masson (1806–60) and Louis Breguel (1804–84) introduce induction coil, which enables development of electrical ignition system.

1859 French engineer Jean J. Étienne Lenoir (1822–1900) designs two-stroke internal-combustion engine modeled after steam engine. It uses illuminating gas as fuel and includes sliding valves and pathway for exhaust.

1862 Alphonse Beau de Rochas (1815–93) develops principle of four-stroke engine. His engine (never built by him) compresses fuel/air mixture before ignition.

1873 American engineer George Brayton invents continuous-ignition combustion engine that later becomes basis of turbine engine.

1875 Air-cooling system for engine is developed in France.

1876 German engineer Nikolaus Otto (1832–91) uses de Rochas's principle to develop first successful four-stroke engine. Reliable and efficient four-stroke engines soon provide first practical alternative to steam power.

1883 German inventor Gottlieb Daimler (1834–1900) invents first high-speed internal-combustion engine.

1885 Daimler develops carburetor that makes it possible to use gasoline instead of illuminating gas as fuel.

1885 Étienne Lenoir (1822–1900) develops electrical spark plug.

1886 German engineer Karl Benz (1844–1929) patents vehicle driven by gasoline engine.

1888 Radial engine is developed. First in-line cylinder engine also is built this year.

1889 First V-type engine is constructed by Daimler and German engineer Wilhelm Maybach (1846–1929).

1893 Maybach invents improved carburetor that is used widely.

1893 Benz improves carburetor by adding throttle valve.

1896 German inventor Rudolf Diesel (1858–1913) uses de Rochas's principle to develop successful model of his compression-ignition diesel engine, which he described in 1893.

1900 Electrical ignition system replaces flame ignition by this time.

Internal Combustion Engine	
1908	American engineers Charles Kettering (1876–1958) and Edward A. Deeds invent electric distributor system for engines.
1912	Electric self-starter, invented by Kettering, is introduced.
1914	V-8 engine is introduced; 12-cylinder engine is unveiled in 1915.
1916	Turbocharged aircraft engine is built by engineer Auguste Rateau (1863–1930) in France.
1924	High-compression engines are introduced.
1935	Englishman Frank Whittle (b. 1907) and German Hans von Ohain (b. 1911) separately patent gas turbine engines.
1938	Chrysler installs gas turbine engine in passenger car.
1940s	World War II promotes development of first jet aircraft, powered by improved gas turbine engines.
c. 1950	Mercedes-Benz introduces fuel-injection system for auto engines used in racing.
1953	Volvo manufactures first turbocharged diesel trucks.
c. 1954	German Felix Wankel (b. 1902) successfully develops Wankel rotary engine, patented by him in 1929. Rotary power rather than reciprocating piston action turns crankshaft.
1965	First legislation limiting exhaust emissions is passed in California. Subsequent air-pollution legislation and fuel shortages force major changes in engine design during next decades.
1966	British develop first electronic fuel-injection system for auto engines.
1967	Japenese car manufacturer Mazda builds first cars using rotary engine.
1980s	Computer control and monitoring systems are introduced on auto engines in United States.
1984	In West Germany, Bosch develops platinum spark plug.

Development of Airplanes

See also Warplanes in Wars and Military History section.

1492 Leonardo da Vinci (1452–1519) sketches a proposed flying machine. In 1500 he designs prototype helicopter.

1875 British engineer demonstrates steam-powered monoplane. It rises 6 inches into air.

1896 German aeronautical pioneer Otto Lilienthal (1848–96) is killed while flying one of his experimental gliders, after making hundreds of successful flights. His pioneering work heavily influences the Wright brothers' airplane design.

1896 First successful powered flight of unmanned, heavier-than-air plane built by American scientist Samuel P. Langley (1834–1906). Craft, weighing 26 pounds, is powered by small steam engine.

1902 Prototype radial engine for airplanes is invented by American engineer Charles M. Manly (1876–1927).

1902 Wilbur (1867–1912) and Orville (1871–1948) Wright successfully test at Kitty Hawk, North Carolina new glider design based on aeronautical data they have compiled experimentally.

1903 Wright Brothers achieve first successful powered flight of heavier-than-air craft at Kitty Hawk. Flight lasts just 12 seconds, covering 120 feet, but marks start of age of aviation.

1904 Wright Brothers design new model of aircraft, capable of sustained flight and doing turning and banking maneuvers.

▶

1906 Brazilian Alberto Santos-Dumont completes first successful flight of heavier-than-air craft in Europe.

1909 First successful flight across English Channel, by Frenchman Louis Blériot (1872–1936).

1911 First U.S. transcontinental flight, by American flier Calbraith Rogers. Rogers crashes 15 times during 49-day flight.

1911 First successful takeoffs and landings by airplane on U.S. Navy ship, marking beginning of development of aircraft-carrier operations.

1914–18 Airplane design advances rapidly during World War I. By 1918 planes make top speed of 145 mph, reach altitude of 30,000 feet, and can carry payloads of over 3,000 pounds.

1919 First transatlantic flight is completed by U.S. Navy seaplane. It makes two stops (Newfoundland and Azores) between New York and Plymouth, England in 25 days.

1920 Altitude record of 33,113 feet is set by U.S. Army pilot.

1923 First nonstop transcontinental flight, from New York to San Francisco, is made in just under 27 hours (May 2–3).

1923 First midair refueling of a plane is accomplished by U.S. Army pilots.

1924 U.S. Army pilots complete first round-the-world flight, in 175 days, aboard two Douglas World Cruisers.

1926 First polar flight is made by Americans Richard Byrd (1888–1957) and Floyd Bennett (1890–1928).

1927 New altitude record of 37,995 feet is set in United States.

1927 American pilot Charles A. Lindbergh (1902–74) completes first nonstop solo transatlantic flight (May 20–21) aboard his Ryan monoplane *Spirit of St. Louis* in just under 33½ hours. "Lucky" Lindbergh's flight creates sensation that makes him national hero and provides new momentum for development of aviation.

1929 Endurance record of 150 hours, 40 minutes is set by Air Corps pilots in Fokker. Flight is aided by midair refueling.

1929 First rocket-powered flight is made, in Germany.

1929 U.S. Army pilot James Doolittle (b. 1896) makes first complete flight by instruments alone, including takeoff and landing. It greatly reduces hazards of flying in bad weather.

1930 British inventor Frank Whittle (b. 1907) patents the first jet engine.

1930 U.S. transcontinental air passenger service begins with three-engine, 12-passenger monoplanes.

c. 1930 By this time aluminum is replacing fabric on wood as material for constructing airplanes, and new planes are being given streamlined, aerodynamic designs.

1933 Solo round-the-world flight is completed by American Wiley Post (1900–35) in 7¾ days.

1935 American pilot Howard Hughes (1905–76) sets new world speed record of 352 mph while flying plane he had built.

1935 Radar detection system is developed by British and proves valuable in tracking aircraft. Advances in radar and other navigational equipment during World War II are later adapted to air-traffic control in civilian aviation.

1936 Douglas DC-3 goes into service. It becomes workhorse in commercial and military transport service.

1937 German Focke-Achegelis Company develops first successful helicopter.

1937 Howard Hughes sets new U.S. transcontinental flight record of 7 hours, 28 minutes.

1939 American Igor Sikorsky (1889–1972) builds first practical helicopter.

1939 First flight of jet-powered aircraft, the Heinkel He-178, in Germany.

1939 B-29 bomber prototype becomes world's largest plane aloft to date during successful test flights.

▶

1939 German Messerschmitt BF-109 sets new world speed record of 469 mph.

1941 First flight of British-made jet, Gloster E.28/39, using Whittle jet engine.

1947 U.S. test pilot Charles E. Yeager (b. 1923) becomes first man to break sound barrier, in Bell X-1 experimental rocket plane.

1947 First (and only) flight of Howard Hughes's *Spruce Goose*, eight-engine, propeller-driven flying boat with world's largest wing span (just under 320 feet). Hughes flies $40 million prototype about 1,000 yards and then mothballs it.

1950 Jet makes its first transatlantic flight.

1952 First transatlantic helicopter flight is made.

1952 British-made de Havilland Comet becomes first turbojet passenger plane to go into regular service.

1953 Yeager becomes first to reach Mach 2.5, in Bell X-1A rocket plane.

1955 Practical hovercraft is invented.

1955 First executive jet is introduced in United States.

1956 Soviet jet airliner Tupolev TU-104 becomes second jet passenger craft to go into regular service.

1957 Three Boeing B-52 bombers complete first nonstop round-the-world flight (Jan. 18).

1958 Boeing 707, first American jet airliner, begins regular commercial service.

1959 First flight of X-15, high-flying research plane capable of making Mach 6. In test flights it eventually reaches speeds of over 4,500 mph (1967) and flies above 350,000 feet (1963).

1964 Boeing 727 commercial airliner is introduced.

1968 First supersonic jet airliner, Soviet Tupolev TU-144, makes its debut; it breaks sound barrier in 1969.

1969 Largest military transport plane, Lockheed C-5 Galaxy, goes into service.

1970 Supersonic French–British jet airliner Concorde flies at supersonic speeds for the first time during tests.

1970 First "jumbo" jet, Boeing 747, begins passenger service. (It was first flown in 1969.)

1971 U.S. Congress cuts off funding for development of Mach 3 SST jet airliner because of environmental concerns.

1976 Supersonic jet airliner, Concorde, goes into service. It cruises at 1,350 mph and carries 100 passengers.

1977 California engineer Paul MacCready achieves first successful man-powered flight in *Gossamer Condor* (August 23).

1979 First man-powered aircraft crossing of the English Channel is made in *Gossamer Albatross*.

1981 First solar-powered airplane, *Solar Challenger*, makes its first crossing of English Channel. Lightweight plane uses 16,000 solar cells to power its electric motor.

1986 Specially designed lightweight airplane *Voyager* completes record round-the-world flight without refueling in just over 9 days.

1989 Anticollision device is introduced for corporate jets. It monitors up to 30 nearby aircraft and features computer voice instructions when collision is imminent.

1990 New radar system to detect wind shear is installed for test purposes at Denver's Stapleton International Airport.

Rockets

B.C.

c. 360 Greek philosopher-scientist Archytas of Tarentum (400-365 b.c.) builds first known device to operate on reaction principle. Thrusts from jets of steam move pigeon-shaped device.

A.D.

c. 1200 Chinese develop the first rocket-propelled weapons—arrows propelled by black-powder rockets. Knowledge of rockets quickly spreads to Europe.

1379 Word *rocket* is first used (in Italian, *rocketta*).

1500 Chinese mandarin named Wan-Hoo supposedly attaches seat and 47 rockets to bamboo stakes to make rocket-powered vehicle. According to account, probably legend, Wan-Hoo is killed when rockets are ignited.

1650 Writer Cyrano de Bergerac (1619–55) suggests rockets as means of traveling from Earth to moon.

1687 Isaac Newton (1642–1726) publishes his laws of motion. His third law—for every action there is an equal and opposite reaction—explains the basic motive force of rocket engines.

1804 William Congreve (1772–1828) builds first of his Congreve military rockets, which later are used to bombard (among other places) Fort McHenry in Baltimore (1814) during War of 1812 (thus "the rockets' red glare" in "The Star Spangled Banner").

1883 Pioneering work in application of rocketry to space travel is begun by Russian schoolteacher-scientist Konstantin Tsiolkovsky (1857–1935). He lays theoretical groundwork for space flight (such as calculating escape velocity and use of liquid fuel rocket).

1909 American scientist Robert H. Goddard (1882–1945), father of modern rocketry, begins his work on liquid-propellant rockets. By building and testing his rockets, Goddard makes many basic advances in rocketry.

1919 Goddard gains notoriety after publishing technical report on his rocketry work that includes speculation on sending rocket to moon.

1923 German scientist Hermann Oberth (b. 1894) publishes his first book on rocket and space travel. Theoretician Oberth subsequently does much to promote development of rocketry.

1926 First launch of rocket carrying instruments (camera, barometer, and thermometer), by Robert Goddard.

1926 First successful firing (March 16) of liquid-fuel rocket, at Auburn, Massachusetts. Built by Goddard, rocket travels just 184 feet.

1929 German automaker achieves 75-second, 2-mile flight in rocket-powered craft.

1930–41 Goddard conducts extensive, privately funded liquid-propellant rocket testing program at New Mexico ranch.

1931 Official study groups are established at Moscow and Leningrad for development of Soviet rockets and rocketry.

1933 Soviet liquid-propellant sounding rocket reaches 3-mile altitude.

1937 Germans establish Peenemünde as center for their rocket research. Wernher von Braun (1912–77) becomes its technical director.

1942 First successful test of German V-2 rocket at Peenemünde. Powered by liquid oxygen and alcohol, it travels 120 miles (reaching over 50-mile altitude) and lands 2½ miles from target.

1945 United States establishes rocket testing grounds at White Sands, New Mexico.

▶

1945 Wernher von Braun and over 100 other German rocket scientists surrender to Allied armies as World War II ends. Soviets eventually make use of some 4,000 German rocket specialists.

1949 Cape Canaveral, Florida becomes rocket proving ground.

1953 Redstone, American intermediate-range ballistic missile (IRBM), is launched. Developed by von Braun and other former Nazi rocket scientists, it is first large missile developed in United States after World War II.

1956 Soviets reportedly launch their first IRBM. Early Soviet one-stage IRBMs eventually include the Shyster, Sandal, and Skean. Soviets deployed the Sandal in Cuba in 1962.

1957 First Soviet intercontinental ballistic missile (ICBM) is tested, year ahead of first U.S. ICBM tests. Modified Soviet ICBMs are later used for space launches.

1957 U.S. Jupiter C, three-stage modified Redstone, is launched. Recoverable nose cone travels 285 miles up into space.

1957 Soviets launch 184-pound *Sputnik I*, first man-made satellite put into space.

1958 United States launches (January 31) its first artificial satellite, *Explorer 1*, into space, aboard modified Jupiter C rocket (fourth stage added).

1958 National Aeronautics and Space Administration (NASA) is created to develop U.S. space program.

1958 First U.S. ICBM missile, Atlas, is fired successfully. Atlas ICBM has maximum range of up to 9,000 miles.

1959 U.S. two-stage ICBM, Titan 1, is tested successfully. It has maximum range of 6,000 miles.

1960 First underwater firing of U.S. IRBM, solid-fuel Polaris (1,400-mile range).

1961 Soviets launch first person into space, Yuri Gagarin (1934–68), aboard two-stage *Vostok I* rocket, probably modified SS-6 ICBM.

1966 Titan 3C, designed for military space satellite launches, is successfully tested.

1966 First flight of Saturn 5 rocket for Apollo moon mission.

1969 Saturn 5 rocket launches 96,000-pound *Apollo 11* command module and lunar lander on successful manned voyage to moon. The 363-foot-tall, three-stage liquid-fuel rocket develops about 7.7 million pounds of lift-off thrust from five first-stage engines. Five smaller second-stage engines produce about 1 million pounds of thrust, while single third-stage engine provides enough thrust to achieve Earth orbit and later to break away for moon.

1981 First flight of first reusable rocket, U.S. space shuttle *Columbia* (April 12–14). Shuttle is powered by three variable-thrust liquid fuel engines (burning liquid oxygen and liquid hydrogen), which are designed to be reused for up to 7½ hours of firing. Shuttle's two solid-fuel rocket boosters (for lift-off) develop 2.65 million pounds of thrust.

1984 United States achieves first midair direct hit of missile by another missile, demonstrating significant advance in antiballistic missile (ABM) technology.

1986 Tragic explosion of space shuttle *Challenger* during lift-off (January 28) apparently is caused by faulty O-ring seal on one of solid-fuel rocket boosters. United States begins redesign of equipment and overhaul of space-shuttle program to increase safety factors.

1988 Space shuttle *Discovery*, with redesigned boosters, is launched successfully.

1988 Soviets successfully launch their first reusable space vehicle.

Nuclear Technology

See also Physics chronology.

1895 Cloud chamber for tracking charged particles is invented.

1913 Radiation detector is invented by German Hans Geiger (1882–1945).

1925 First cloud-chamber photographs of nuclear reactions are made in England by Patrick M. S. Blackett.

1930 First crude particle accelerator is built by British scientists John Crockroft (1897–1967) and E. Walton (b. 1903).

1930 Princeton University researcher Robert J. van de Graaff (1901–66) develops another early particle accelerator, Van de Graaff generator.

1930 American physicist Ernest Lawrence (1901–58) invents cyclotron to accelerate atomic particles. His device uses magnetic field to speed particles along spiraling path.

1932 Atom is split by scientists for first time with particle accelerator.

1933 University of California at Berkeley scientists are constructing cyclotron with a 27-inch magnet.

1934 Italian-born American physicist Enrico Fermi (1901–54) proposes use of neutrons to split atoms.

1938 First fission of uranium is achieved by German scientist Otto Hahn (1879–1978).

1939 Fermi begins creating first nuclear pile. He uses thin layers of uranium oxide and graphite to create environment where chain reaction will occur.

1940 Betatron is built. It accelerates electrons on circular path.

1942 Fermi and colleagues achieve first controlled, self-sustaining nuclear reaction, at University of Chicago. Their work leads to building of first atomic bomb and in subsequent years to nuclear-power-generating industry.

1944 Electromagnetic method of "enrichment," separation of uranium isotopes U-235 and U-238, is achieved in Tennessee with calutron. Enrichment is necessary to produce fissionable material for atomic bomb, then being developed.

1944 Gas-diffusion enrichment also is developed. It soon replaces calutron.

1945 First atomic bomb is successfully exploded by United States, in secret test (July 16, 1945).

1945 Synchrotron method of controlling particle orbits by using varying magnetic fields or radio frequencies is developed.

1946 Linear proton accelerators are developed in United States and Great Britain.

1951 First nuclear reactor is built by U.S. Atomic Energy Commission.

1952 United States explodes first H-bomb in successful test (November 6).

1952 Most powerful particle accelerator to date, cosmotron, is in operation at Brookhaven, New York.

1953 First breeder reactor is operated by the United States. In reaction, U-235 is split to produce energy, while U-238 is changed into plutonium, which can be used to fuel nuclear reactors.

1954 First submarine powered by a nuclear reactor, *Nautilus*, is launched.

1955 Nuclear-generated electricity is used for first time in United States at Schenectady, New York.

1957 French develop new electron-beam method for welding metals in vacuum. It makes fuel fabrication of nuclear fuel rods safer, since welding of titanium and tungsten is involved.

1960 World's first "boiling-water" nuclear reactor is built in United States. Nuclear core heats water under pressure, producing steam to drive generator.

1962 First nuclear-powered surface ship, *Savannah*, puts to sea.

▶

1962 First advanced gas-cooled reactor (AGR) is built, in England. It is fueled by enriched uranium.

1963 First privately owned commercial nuclear reactor for generating electricity is built, in New Jersey.

1963 Canada's CANDU nuclear reactors begin operating. They use natural uranium fuel in pressure tubes surrounded by heavy water.

1968 Fermi National Accelerator Laboratory (Fermilab) is founded in Illinois to study elementary particles. It has synchrotron particle accelerator capable of speeding protons to energies of 500 billion electron volts.

1971 Nuclear power station in Canada uses ordinary water for cooling instead of more costly heavy water.

1972 World's largest atom-smashing device to date is built in United States, at U.S. National Accelerator Laboratory. Its rapid-cycling synchrotron accelerator moves protons through circular track 4 miles in circumference and achieves energies of 200 billion electron volts.

1975 Technique for laser separation of uranium isotopes is developed. It may greatly reduce cost of nuclear energy.

1980 Lasers are used to create device for study of spin phenomena in nuclear physics.

1982 Spin spectrometer is developed by American scientists to study collisions between nucleus-heavy ions.

1984 New Fermilab tevatron accelerator accelerates protons to energies of 0.8 trillion electron volts.

1985 Scientists develop lead-iron phosphate glass, new material that will be used to create more durable containment medium for storing nuclear wastes.

1987 TESSA 3 spectrometer is built in Great Britain. It allows more detailed study of atomic structure.

Computers

1623 German professor Wilhelm Schickard builds first known mechanical calculator. His calculating clock remains unknown until 1935, however.

1642 Blaise Pascal (1623–62) creates Pascaline, geared adding machine.

c. 1670 Stepped reckoner, geared device that adds, subtracts, multiplies, and divides, is invented by Gottfried Wilhelm von Leibniz (1646–1716).

1820 Arithmometer becomes first calculator marketed commercially.

1834 English mathematician Charles Babbage (1792–1871), father of modern computers, devises analytical engine. Never built, engine has in mechanical form most of basic components found in electronic computers.

c. 1850 English mathematician George Boole (1815–64) develops Boolean algebra, which later provides basis for computer-logic circuitry.

1890 U.S. Census Bureau uses electromechanical tabulating machine for tallying 1890 census. Machine automatically reads, counts, and sorts information on punched cards.

1892 Adding–subtracting machine with its own printer is produced by American William S. Burroughs (1857–98).

1896 American Herman Hollerith (1860–1929), who worked on the 1890 census tabulation, founds Tabulating Machine Company, forerunner of IBM.

1924 International Business Machines (IBM) incorporates.

1930 First large-scale analog computer is built at MIT by American Vannevar Bush (1890–1974).

▶

1943 Colossus, Alan Turing's (1912–54) all-electronic calculator, helps crack German codes.

1943 Prototype digital computer, Harvard Mark I, is built by Harvard mathematician Howard Aiken (1900–73). Computer, 50 feet long, uses electromechanical switches.

1946 ENIAC (Electronic Numerical Integrator and Calculator), first all-purpose electronic digital computer, comes on line. Built by John Mauchly (1907–80) and J. Presper Eckert (b. 1919), 80-foot-long computer has 18,000 vacuum tubes.

1947 Transistor is invented by American physicist William Shockley (b. 1910) and others at Bell Labs.

1948 Concept of cybernetics is introduced by American mathematician Norbert Wiener (1894–1964) as analysis of feedback and automated processes.

1948 Stored-program electronic computer, Mark I, begins operation in England. First in United States, BINAC, is in service following year.

1951 Mauchly and Eckert's UNIVAC (Universal Automatic Computer) becomes first computer designed for commercial use. Computer, 15 feet long, has 1.5K of memory. Magnetic tape is used for data storage.

1953 IBM introduces 701, its first computer.

1955 IBM's 705 business-model computer becomes first commercially available computer to use magnetic-core memory, innovation of MIT engineer Jay W. Forrester (b. 1918).

1956 Development of FORTRAN, the first programming language.

1958 Integrated circuit is invented by American engineer Jack Kilby. Photoengraving process for mass production of integrated circuits is developed in 1959.

1959 Business-use computer language COBOL is developed.

1959 IBM introduces "second generation" of computers, using transistors instead of vacuum tubes. Transistors revolutionize computers, making them smaller, faster, and less expensive.

1962 Disk-storage system for computer data is introduced by IBM.

1962 First computerized airline-reservation system is introduced by American Airlines, as on-line computer services are put to new uses.

1963 Downsizing of computers begins with Digital Equipment Corporation's PDP-8, first successful minicomputer.

1964 Number of computers in United States tops 18,000, up from just over 2,500 in 1958, thanks largely to transistorization.

1964 First fully automated factory using computer-operated equipment, Sara Lee food-processing plant, opens in Deerfield, Illinois.

1964 IBM introduces its highly successful System/360 family of compatible computers.

1965 BASIC, first all-purpose computer-programming language, is demonstrated by inventors John Kemeny and Thomas Kurtz.

1968 Control Data and NCR introduce first commercial "third-generation" computers, which use integrated circuits and offer greater speed and capabilities.

1969 Intel 4004, first microprocessor (miniaturized set of integrated circuits on one chip), is invented by American engineer Marcian E. Hoff. Microprocessors subsequently make "fourth-generation" computers of 1970s possible.

1969 "Bubble memory" is invented for computer data storage.

1970 Floppy-disk system for data storage is introduced.

1971 Texas Instruments produces first pocket calculator.

1971 Pascal computer language is introduced.

1975 First desktop microcomputer becomes available. MIT's Altair computer is instant success. By 1977 TRS-80, Apple II, and PET microcomputers are being marketed.

1979 VisiCalc, first spreadsheet software for microcomputers, goes on the market. It helps

▶

redefine general perception of microcomputer as hobby machine to that of office machine.

1981 IBM Personal Computer is marketed. IBM-XT becomes first personal computer with hard-disk storage system in 1983.

1981 Commodore VIC-20 microcomputer is introduced. It will become first microcomputer to sell over 1 million units.

1983 Computer "mouse" is introduced by Apple.

1984 Laser disk-storage system is marketed.

1986 Compaq markets computers based on Intel 80386 32-bit chip.

1987 Advanced supercomputer, which can perform 1.72 billion computations per second, goes on line.

1988 Parallel processing, technique first proposed in 1960s, is being used to speed processing of complex problems by 1,000 times.

1988 New, experimental chips called heterojunction devices work hundreds of times faster than conventional computer chips.

The Quest for Speed

1935 French ship *Le Terrible* sets destroyer speed record of 45.25 knots (51.84 mph).

1938 Highest confirmed speed for steam locomotive is set near Peterborough, England. Locomotive *Mallard*, pulling seven coaches, reaches 125 mph. (Engine is slightly damaged.)

1941 Italian Fiat CR42B biplane sets speed record for biplanes, 323 mph.

1955–59 U.S. Navy four-engine jet Xp6m-1 Seamaster becomes world's fastest flying boat, reaching speeds of up to 646 mph.

1964 British-made *Bluebird* becomes fastest wheel-driven automobile to date. Powered by turbine engine, it reaches 429.311 mph.

1966 U.S. pilot in Hawker Sea Fury achieves fastest measured speed on record in piston-engine plane, 520 mph. Official, confirmed record is 517.055 mph, flown in North American P-51D Mustang in 1983.

1967 USAF Major William Knight (b. 1930) sets speed record for fixed-wing aircraft, flying X-15A-2 rocket at up to 4,520 mph.

1969 Fastest speed ever traveled by humans is set when U.S. astronauts aboard *Apollo 10* reach 24,791 mph, maximum speed during their space voyage.

1971 American driver sets speed record for compression-ignition-engined cars, 190.344 mph.

1973 Containership *Sea-Land Commerce* sets speed record for fastest Pacific Ocean crossing, averaging 33.27 knots during voyage lasting 6 days, 1 hour, 27 minutes from Yokohama, Japan to Long Beach, California.

1975 Fastest rotary speed is achieved in ultracentrifuge at Birmingham, England, 4,500 mph.

1976 USAF Lockheed SR-71 reconnaissance jet sets official speed record for jets when it reaches 2,193.167 mph.

1976 Supersonic airliner used for passenger service, French–British *Concorde*, cruises at speeds of over 1,200 mph.

1976 Highest velocity of any space vehicle to date is set at 149,125 mph by U.S.–West German solar probe *Helios B*.

1976 American driver sets world record for highest land speed achieved by woman, 525.016 mph.

1977 Australian Kenneth Warby (b. 1930) reaches highest speed achieved on water,

▶

345 mph, in his hydroplane *Spirit of Australia.* In 1978 he sets official world water speed record of 319.627 mph.

1978 Diesel-engine car speed record is set in Italy, 203.3 mph.

1978 Californian Donald Vesco (b. 1939) sets speed record for motorcycling when he travels at average speed of 318.598 mph on *Lightning Bolt*, streamlined motorcycle powered by two Kawasaki engines.

1979 Japanese Maglev (magnetic levitation) test train sets record for rail speed with passengers, 321.2 mph.

1979 Fastest land speed ever claimed (but not confirmed), 739.666 mph, is set by U.S. *Budweiser Rocket*, rocket-engined wheeled car, during test at Edwards Air Force Base, California. It becomes only land vehicle to break sound barrier to date.

1980 Fastest warship on record is U.S. Navy test vehicle SES-100B, which travels at 103.9 mph during tests in Maryland.

c. 1980 Fastest submarines are believed to be Soviet Alfa-class vessels, which reportedly reach maximum speed of 42 knots (48 mph).

1981 Highest speed on national railway is 236 mph, set by French SNCF high-speed train.

1982 U.S. driver sets world record for snowmobile speed, 148.6 mph, in Budweiser-Polaris snowmobile.

1983 World speed record for propeller-driven water craft is set by an American supercharged hydroplane traveling 215.33 mph.

1983 World's official 1-mile land speed record, 633.468 mph, is set by American driver Richard Noble (b. 1946), in jet-powered *Thrust 2*.

1984 Highest tested speed for factory-stock road car is 179.9 mph, achieved by Lamborghini Countach LP500S.

1985 American bicyclist John Howard (b. 1950) reaches speed of 152.284 mph riding directly behind car whose tail section is specially equipped to decrease air resistance.

1986 British flier in Somerset, England averages record-making speed of 249.1 mph in his helicopter.

1986 65-foot motorboat, *Virgin Atlantic Challenger II*, makes fastest west-to-east Atlantic crossing by any boat or ship on record, completing voyage in three days, eight hours, 31 minutes.

1986 American sailor circumnavigates globe in 60-foot sloop, setting speed record of 150 days, 1 hour, 6 minutes. He averages 171.1 miles per day.

1986 U.S. cyclist sets unofficial speed record when he travels 200 meters (656 feet) in 6.832 seconds. (His speed is 65.484 mph.)

1987 Ferrari claims record for fastest road car, 201 mph, achieved by $250,000 F40.

1987 World's fastest tank, British Scorpion AFV, can reach speed of 50 mph.

1988 Fastest transatlantic supersonic passenger flight is set by French–British *Concorde*, which averages 1,215 mph.

6 The Arts

Novelists, Poets, and Essayists

Noted writers with selected major works.

B.C.

c. 800 Greek poet Homer is said to have written ancient Greek epics *Iliad* and *Odyssey* at this time. The works are probably based on much older stories, passed on orally, concerning war with Troy (1100s B.C.).

c. 600 Aesop, Greek fabulist and probably legendary figure, lives. Earliest known collection of fables credited to Aesop appears in fourth century B.C. and contains materials that date back to 14th-century B.C. Egypt.

522–439 Pindar, greatest of Greek lyric poets, lives. Works include 44 *Epinicia* (victory odes), *Hymns, Choral Dithyrambs to Dionysus, Encomia* (laudatory odes), *Scolia* (festive songs), *Dirges* (for flute and choral dance), and *Songs.*

c. 484–c. 420 Greek historian Herodotus lives. Called father of history, his major work is *History of the Persian Wars to 479 B.C.*, in nine books.

470–404 Thucydides, Greek historian, lives. Works include *History of the Peloponnesian War* in eight books, ending in 411 B.C.—21st year of that war.

443–359 Xenophon lives. Greek soldier, historian, and essayist, his works include *Anabasis*, history of a military expedition; *Cyropaedia*; and *Memorabilia.*

382–322 Demosthenes, greatest of Greek orators, lives. Among his works are the orations *Philippics Concerning the Crown* (351 B.C.) against Philip II of Macedon.

310–250 Theocritus, Greek idyllic poet, creates pastoral poetry. His *Idylls* (32 survive) include *Adoniazusae* and *Dirge on Daphnis.*

c. 295–230 Greek poet Apollonius of Rhodes lives. He writes four-volume epic *Argonautica.*

239–169 Roman epic poet Quintus Ennius, father of Latin literature, lives. Works include *Annales* (epic on Roman history in 18 books) and *Saturae.*

234–149 Cato the Elder, Roman statesman and prose writer, lives. Extant works include *De Agri Cultura, Origines* (history), and *Praecepta.*

200–118 Polybius, Greek historian, lives. He writes 40-volume history of Punic Wars (five survive intact).

106–43 Roman orator and philosopher Marcus Tullius Cicero lives. He is best known for his orations, including *Orations Against Catiline, First and Second Philippics, On Behalf of Archias*, and *Against Verres.*

100–44 Julius Caesar, Roman emperor, writes military commentaries, notably *Commentarii de Bello Gallico (Commentaries on the Gallic War).*

c. 96–55 Lucretius, Roman poet and philosopher, lives. Works include poem *De Rerum Natura* (On the Nature of Things), first Latin treatment of Greek philosophy.

c. 87–54 Catullus (Caius Valerius), Roman poet, lives. His best-known works are 116 love poems written to Lesbia (real name Clodia), his love and sister of Cicero's enemy, Clodins.

70–19 Virgil, great Roman poet, lives. Works include epic *Aeneid*, which tells of wanderings of Aeneus and the founding of Rome, and shorter collections, *Eclogues* and *Georgics.*

65–8 Horace, great Roman lyric poet, lives. Works include *Satires* (35 B.C.), *Epodes* (30 B.C.), *Odes, Epistles,* and *Ars Poetica.*

64 B.C.–A.D. 23 Greek geographer Strabo lives. His great work *Geographica*, in 17 books, is an important source of information on ancient times.

59 B.C.–A.D. 17 Livy, Roman historian and most important prose writer of Augustan age, lives. Works include comprehensive *History of Rome.*

43 B.C.–A.D. 17 Ovid, Roman poet, lives. Works include *Amores, Letters from Heroines, Art of Love* (in three books), *Metamorphosis*

▶

Pulitzer Prize for Poetry

1922 *Collected Poems* by Edwin Arlington Robinson

1923 *The Ballad of the Harp-Weaver, A Few Figs from Thistles and Other Poems* by Edna St. Vincent Millay

1924 *New Hampshire: A Poem with Notes and Grace Notes* by Robert Frost

1925 *The Man Who Died Twice* by Edwin Arlington Robinson

1926 *What's O'Clock* by Amy Lowell

1927 *Fiddler's Farewell* by Leonora Speyer

1928 *Tristram* by Edwin Arlington Robinson

1929 *John Brown's Body* by Stephen Vincent Benét

1930 *Selected Poems* by Conrad Aiken

1931 *Collected Poems* by Robert Frost

1932 *The Flowering Stone* by George Dillon

1933 *Conquistador* by Archibald MacLeish

1934 *Collected Verse* by Robert Hillyer

1935 *Bright Ambush* by Audrey Wurdemann

1936 *Strange Holiness* by Robert P. Tristram Coffin

1937 *A Further Range* by Robert Frost

1938 *Cold Morning Sky* by Marya Zaturenska

1939 *Selected Poems* by John Gould Fletcher

1940 *Collected Poems* by Mark Van Doren

1941 *Sunderland Capture* by Leonard Bacon

1942 *The Dust Which Is God* by William Rose Benét

1943 *A Witness Tree* by Robert Frost

1944 *Western Star* by Stephen Vincent Benét

1945 *V-Letter and Other Poems* by Karl Shapiro

1946 No prize awarded

1947 *Lord Weary's Castle* by Robert Lowell

1948 *The Age of Anxiety* by W. H. Auden

1949 *Terror and Decorum* by Peter Viereck

1950 *Annie Allen* by Gwendolyn Brooks

1951 *Complete Poems* by Carl Sandburg

1952 *Collected Poems* by Marianne Moore

1953 *Collected Poems 1917–52* by Archibald MacLeish

1954 *The Waking* by Theodore Roethke

1955 *Collected Poems* by Wallace Stevens

1956 *Poems, North and South* by Elizabeth Bishop

1957 *Things of This World* by Richard Wilbur

1958 *Promises: Poems 1954–56* by Robert Penn Warren

1959 *Selected Poems 1928–58* by Stanley Kunitz

1960 *Heart's Needle* by W. D. Snodgrass

1961 *Times Three: Selected Verse from Three Decades* by Phillis McGinley

1962 *Poems* by Alan Dugan

1963 *Pictures from Brueghel* by William Carlos Williams

1964 *At the End of the Open Road* by Louis Simpson

1965 *77 Dream Songs* by John Berryman

1966 *Selected Poems* by Richard Eberhart

1967 *Live or Die* by Anne Sexton

1968 *The Hard Hours* by Anthony Hecht

1969 *Of Being Numerous* by George Oppen

1970 *Untitled Subjects* by Richard Howard

1971 *The Carrier of Ladders* by William S. Merwin

1972 *Collected Poems* by James Wright

1973 *Up Country* by Maxine Winokur Kumin

1974 *The Dolphin* by Robert Lowell

1975 *Turtle Island* by Gary Snyder

1976 *Self-Portrait in a Convex Mirror* by John Ashbery

1977 *Divine Comedies: Poems* by James Merrill

1978 *The Collected Poems* by Howard Nemerov

1979 *Now and Then: Poems 1976–78* by Robert Penn Warren

1980 *Selected Poems* by Donald R. Justice

1981 *The Morning of the Poem* by James Schuyler

1982 *The Collected Poems* by Sylvia Plath (posthumous)

1983 *Selected Poems* by Galway Kinnell

1984 *American Primitive* by Mary Oliver

1985 *Yin* by Carolyn Kizer

1986 *The Flying Change* by Henry Taylor

1987 *Thomas and Beulah* by Rita Dove

1988 *Partial Accounts: New and Selected Poems* by William Meredith

1989 *New and Collected Poems* by Richard Wilbur

(collection of Greek, Latin, and Eastern myths), and *Letters from the Black Sea.*

A.D.

23–79 Pliny the Elder, Roman naturalist, lives. He writes complete history of German wars, *Bellorum Germanial Libri* (in 20 books—no longer extant), as well as elaborate *Historia Naturalis* (Natural History), encyclopedia in 37 books.

39–65 Roman poet Lucan lives. Works include *Pharsalia*, epic poem on warfare between Caesar and Pompey.

c. 46–c. 120 Plutarch, Greek biographer and essayist, lives. Works include essays *On Virtue and Vice, On Superstition, On Rivers*, and celebrated *46 Parallel Lives.*

c. 55–c. 117 Roman historian Tacitus lives. Works include *Germania* (ethnology of Germans); *Historiae*, events in reigns of emperors Galba to Domitian; and *Annales*, history of Julian dynasty.

c. 60–c. 140 Juvenal, Roman satiric poet, lives. Works include 16 well-known satires written to expose flagrant vices of Rome in his day.

c. 69–c. 140 Suetonius, Roman biographer, lives. He writes *Lives of the Caesars*, anecdotal tales of first 12.

120–180 Lucian, Greek satirist and rhetorician, lives. Works include the mocking *Dialogues of the Gods* and *Dialogues of the Dead* and a travelogue, *True History.*

121–80 Roman emperor Marcus Aurelius lives. He writes noted *Meditations* on Stoic philosophy.

315–430 St. Augustine of Hippo, noted theologian and bishop of Roman Catholic Church, lives. Works include *Confessions* and *City of God.*

c. 370–c. 404 Roman poet Claudian, last major classical poet, lives. Works include *Claudianus Major* and *Claudianus Minor.*

c. 475–525 Roman philosopher and statesman Boethius lives. Works include *The Consolation of Philosophy.*

c. 673–735 Venerable Bede, English historian and scholar, lives. Works include *Ecclesiastical History of the English People* (731).

700s Unknown Anglo-Saxon author composes *Beowulf.*

1050–1123 Omar Khayyám, Persian astronomer and poet, lives. Works include *Rubáiyát of Omar Khayyám.*

1100–55 Geoffrey of Monmouth, English historian, lives. Major works include *History of the Kings of Britain* (1137).

fl. 1165–90 Chrétien de Troyes, French poet, lives. Works include *Erect et Enide, Launcelot, or the Knight of the Cart* and *Perceval, or the Story of the Grail.*

c. 1170–c. 1220 Wolfram von Eschenbach, German lyric poet, lives. Works include *Parzival, Willehalm*, and *Titurel.*

1265–1321 Dante Alighieri, Italian poet, lives. Works include *La Vita Nuova* (1209), *The Banquet* (1304–7), *De velgari eloquentia* (On the Vernacular Tongue) (1304–6), *De monarchia* (On Monarchy) (c. 1319), and *The Divine Comedy* (1321).

1304–74 Petrarch, Italian poet and scholar, lives. Works include *De viris illustribus* (On Illustrious Men), *Africa, Secretum, De vita solitaria* (On the Solitary Life), *De otio religioso* (On the Virtue of the Religious Life), and *De remediis utriusque* (Remedies Against One and the Other Fortune).

1313–75 Giovanni Boccaccio, Italian prose writer and poet, lives. Works include *Caccia di Diana* (Diana's Hunt) (1334–36), *Filocolo, Filostrato, Teseida, Fiammetta*, and *Decameron* (1351–53).

c. 1330–1408 John Gower, English poet, lives. Works include *Cinkanti Balades* (c. 1374), *Vox Clamantis* (c. 1379–81), and *Confessio Amantis* (c. 1393).

▶

Pulitzer Prize for Fiction

1917	No prize awarded
1918	*His Family* by Ernest Poole
1919	*The Magnificent Ambersons* by Booth Tarkington
1920	No prize awarded
1921	*The Age of Innocence* by Edith Wharton
1922	*Alice Adams* by Booth Tarkington
1923	*One of Ours* by Willa Cather
1924	*The Able McLaughlins* by Margaret Wilson
1925	*So Big* by Edna Ferber
1926	*Arrowsmith* by Sinclair Lewis (prize declined)
1927	*Early Autumn* by Louis Bromfield
1928	*The Bridge of San Luis Rey* by Thornton Wilder
1929	*Scarlet Sister Mary* by Julia Peterkin
1930	*Laughing Boy* by Oliver LaFarge
1931	*Years of Grace* by Margaret Ayer Barnes
1932	*The Good Earth* by Pearl S. Buck
1933	*The Store* by T. S. Stribling
1934	*Lamb in His Bosom* by Caroline Miller
1935	*Now in November* by Josephine W. Johnson
1936	*Honey in the Horn* by Harold L. Davis
1937	*Gone with the Wind* by Margaret Mitchell
1938	*The Late George Apley* by John P. Marquand
1939	*The Yearling* by Marjorie Kinnan Rawlings
1940	*The Grapes of Wrath* by John Steinbeck
1941	No prize awarded
1942	*In This Our Life* by Ellen Glasgow
1943	*Dragon's Teeth* by Upton Sinclair
1944	*Journey in the Dark* by Martin Flavin
1945	*A Bell for Adano* by John Hersey
1946	No prize awarded
1947	*All the King's Men* by Robert Penn Warren
1948	*Tales of the South Pacific* by James A. Michener
1949	*Guard of Honor* by James Gould Cozzens
1950	*The Way West* by A. B. Guthrie, Jr.
1951	*The Town* by Conrad Richter
1952	*The Caine Mutiny* by Herman Wouk
1953	*The Old Man and the Sea* by Ernest Hemingway
1954	No prize awarded
1955	*A Fable* by William Faulkner
1956	*Andersonville* by MacKinlay Kantor
1957	No prize awarded
1958	*A Death in the Family* by James Agee
1959	*The Travels of Jaimie McPheeters* by Robert Lewis Taylor
1960	*Advise and Consent* by Allen Drury
1961	*To Kill a Mockingbird* by Harper Lee
1962	*The Edge of Sadness* by Edwin O'Connor
1963	*The Reivers* by William Faulkner
1964	No prize awarded
1965	*The Keepers of the House* by Shirley Ann Grau
1966	*Collected Stories of Katherine Anne Porter* by Katherine Anne Porter
1967	*The Fixer* by Bernard Malamud
1968	*Confessions of Nat Turner* by William Styron
1969	*House Made of Dawn* by M. Scott Momaday
1970	*Collected Stories* by Jean Stafford
1971	No prize awarded
1972	*Angle of Repose* by Wallace Stegner
1973	*The Optimist's Daughter* by Eudora Welty
1974	No prize awarded
1975	*The Killer Angels* by Michael Shaara
1976	*Humboldt's Gift* by Saul Bellow
1977	No prize awarded
1978	*Elbow Room* by James Alan McPherson
1979	*The Stories of John Cheever* by John Cheever
1980	*The Executioner's Song* by Norman Mailer
1981	*A Confederacy of Dunces* by John Kennedy Toole
1982	*Rabbit Is Rich* by John Updike
1983	*The Color Purple* by Alice Walker
1984	*Ironweed* by William Kennedy
1985	*Foreign Affairs* by Alison Lurie
1986	*Lonesome Dove* by Larry McMurtry
1987	*A Summons to Memphis* by Peter Taylor
1988	*Beloved* by Toni Morrison
1989	*Breathing Lessons* by Anne Tyler

c. 1343–1400 Geoffrey Chaucer, English poet, lives. Works include *Book of the Duchess* (1369), *The House of Fame* (c. 1379), *Parliament of Fowls* (c. 1382), *Troilus and Criseyde* (c. 1385), *Legend of Good Women* (c. 1386), and *Canterbury Tales* (c. 1387–1400).

c. 1360 Anonymous poet writes Middle English romance *Gawain and the Green Knight.*

1391–1465 Charles duc d'Orleans, French prince and poet, lives. Works include "My Ghostly Father," "The Smiling North," and "Oft in My Thought."

c. 1408–71 Sir Thomas Malory, English writer, lives. Works include *Le Morte D'Arthur* (c. 1469).

c. 1460–c. 1525 William Dunbar, Scottish poet, lives. Works (c. 1503–8) include "The Thrissil and the Rose," "The Golden Targe," and "The Lament for the Makaris" (poems).

1460–1529 John Skelton, English poet and humanist, lives. Works include "The Bowge of Court" (1499), "The Tanning of Elynor Rummyng" (c. 1517), "Why Come Ye Not to Court" (c. 1522), and "The Garland of Laurel" (1523).

c. 1460–1553 François Rabelais, French physician and author, lives. Works include *Gargantua and Pantagruel* (1532–64).

1466–1536 Desiderius Erasmus, Dutch scholar and philosopher, lives. Works include *The Praise of Folly* (1509).

1469–1527 Niccolò Machiavelli, Italian statesman and political philosopher, lives. Works include *Discorsi (Discourses)* (1513–17), *The Prince* (1517), and *La mandragola* (The Mandrake) (c. 1518).

1478–1529 Baldassare Castiglione, Italian diplomat and writer, lives. Major works include *The Book of the Courtier* (1518).

1478–1535 Sir Thomas More, English author and statesman, lives. Works include *The History of Richard III* (c. 1513), *Utopia* (1516), and *A Dialogue Concerning Heresies* (1518).

1533–92 Michel E. de Montaigne, French moralist and creator of personal essays, lives. Works include *Essay* (1580).

1544–95 Torquato Tasso, Italian poet, lives. Works include *Rinaldo, Gerusalemme Liberata* (c. 1559–75), *Torrismondo* (1586), and *Mondo creato* (1607).

1547–1616 Miguel de Saavedra Cervantes, novelist and dramatist, lives. Spain's greatest literary figure, his works include novels *La Galatea* (1585), *Don Quixote* (Part 1, 1605; Part 2, 1615), and *Los trabajos de Persiles y Sigismunda* (1617) and plays *Ocho comedias, y ocho entremeses nuevos* (1615) and *La Numancia* (discovered in 1784).

1552?–99 Edmund Spenser, English poet, lives. Works include *The Shepheardes Calendar* (1579), *The Faerie Queene* (1590–96), *Astrophel* (1586), *Colin Clout's Come Home Again* (1595), *Amoretti, Epithalamion* (1595), and *Prothalamion* (1596).

1554–86 Sir Philip Sidney, English poet and scholar, lives. Major works include *Arcadia* (c. 1580), *Astrophel and Stella* (1591), and *The Defence of Poesie* (1595).

1561–1626 Sir Francis Bacon, English philosopher and essayist, lives. Works include *Essays or Counsels* (1597–1625), *The Advancement of Learning* (1605), *Novum Organum* (1620), and *The New Atlantis* (1627).

1563–1631 Michael Drayton, English poet, lives. Major works include a series of sonnets (1593–1619), *England's Heroical Epistles* (1597–99), *Poly-Olbion* (1612–22), *Nymphidia* (1627), and *Muses' Elysium* (1630).

1567–1601 Thomas Nash, English satirist, lives. Major works include *Pierce Penilesse, his Supplication to the Divell* (1592), *The Unfortunate Traveller or the Life of Jack Wilton*, (1594), and *The Isle of Dogs* (1597).

▶

1572?–1631 John Donne, first English metaphysical poet, lives. Works include *Holy Sonnets, Biathanatos*, and *Essays in Divinity*. Donne's works were published, for the most part, posthumously, beginning in 1633 with a collection of poems.

1591–1674 Robert Herrick, English poet, lives. Works include *To the Virgins, to Make Much of Time* (1648), *Delight in Disorder* (1648), *Hesperides* (1648), and *Corinna's Going A-Maying* (1648).

1593–1633 George Herbert, English poet, lives. Major works include *The Collar* (1633), *The Sacrifice* (1633), "Easter Wings" (1633), and *The Temple* (1633).

1606–87 Edmund Waller, English poet, lives. Works include "On a Girdle" (1645), "Go, Lovely Rose" (1645), and *Poems* (1645).

1608–74 John Milton, English poet, lives. Works include *L'Allegro* (1632), *Il Penseroso* (1632), *Lycidas* (1637), *Areopagitica* (1644), *Eikonoklastes* (1649), *Paradise Lost* (1667), *Paradise Regained* (1671), and *Samson Agonistes* (1671).

1609–42 Sir John Suckling, English poet, lives. Major works include "Why So Pale and Wan, Fond Lover?" (1637), *Session of the Poets* (1637), and *Brennoralt, or The Discontented Colonel* (1639).

c. 1612–72 Anne (Dudley) Bradstreet, early American poet, lives. Works include anthologies *The Tenth Muse Lately Sprung Up in America* (1650) and *Several Poems* (1678), containing "Contemplations."

1612–80 Samuel Butler, English poet, lives. Works include *Hudibras* (1663–78) and *The Rehearsal* (1671).

1618–57 Richard Lovelace, English poet, lives. Works include *To Althea from Prison, To Lucasta, Going Beyond the Seas* (1649), and "To Lucasta, Going to the Wars" (1649).

1621–78 Andrew Marvell, English poet, lives. Works include "To His Coy Mistress" (1650), *The Definition of Love* (1681), *The Garden* (1681), and "A Dialogue Between the Soul and Body."

1622–95 Henry Vaughan, English poet, lives. Major works include *The Retreat* (1650–55), *The World* (1650–55), and *They Are All Gone into the World of Light* (1650–55).

1628–88 John Bunyan, English writer, lives. Works include *Grace Abounding to the Chief of Sinners* (1666) and *Pilgrim's Progress* (1678).

1631–1700 John Dryden, English poet, dramatist, and critic, lives. Major works include *All for Love* (1677), *Achitophel* (1681), *Religio Laici* (1682), *Absalom and Macflecknoe* (1682), *Essay of Dramatic Poesy*, "A Song for St. Cecilia's Day" (1687), *The Hind and the Panther* (1687), and "Alexander's Feast" (1697).

1633–1703 Samuel Pepys, English diarist, lives. His major work is a diary (kept 1660–69).

1637–74 Thomas Traherne, English metaphysical poet and prose writer, lives. Works include *Roman Forgeries* (1673) and *Christian Ethicks* (1675).

1660–1731 Daniel Defoe, English novelist, lives. Works include *The Shortest Way with the Dissenters* (1702), *Robinson Crusoe* (1719–20), *Moll Flanders* (1722), and *Journal of the Plague Year* (1722).

1667–1745 Jonathan Swift, English satirist, poet, and political writer, lives. Works include *The Tale of a Tub* (1704), *Gulliver's Travels* (1726), *A Modest Proposal* (1729), and *Verses on the Death of Dr. Swift* (1739).

1672–1719 Joseph Addison, English poet and essayist, lives. Works include *The Campaign* (1704), *The Tatler* (1709–11), *The Spectator* (1711–12), "The Spacious Firmament on High" (1712), and *Cato* (1713).

1672–1729 Sir Richard Steele, Irish-born English playwright and essayist, lives. Works include *The Christian Hero* (1701), *The Tatler* (1709–11), and *The Spectator* (1711–12).

1685–1732 John Gay, English poet and playwright, lives. Works include *The Shepherd's Week* (1714) and *The Beggar's Opera* (1728).

▶

Pulitzer Prize for History

1917 *With Americans of Past and Present Days* by J. J. Jusserand

1918 *History of the Civil War* by James F. Rhodes

1919 No prize awarded

1920 *The War with Mexico* by Justin H. Smith

1921 *The Victory at Sea* by William S. Sims

1922 *The Founding of New England* by James T. Adams

1923 *The Supreme Court in United States History* by Charles Warren

1924 *The American Revolution: A Constitutional Interpretation* by Charles H. McIlwain

1925 *A History of the American Frontier* by Frederick L. Paxton

1926 *History of the United States* by Edward Channing

1927 *Pinckney's Treaty* by Samuel F. Bemis

1928 *Main Currents in American Thought* by Vernon L. Parrington

1929 *The Organization and Administration of the Union Army, 1861–65* by Fred A. Shannon

1930 *The War of Independence* by Claude H. Van Tyne

1931 *The Coming of the War, 1914* by Bernadotte E. Schmitt

1932 *My Experiences in the World War* by Gen. John J. Pershing

1933 *The Significance of Sections in America History* by Frederick J. Turner

1934 *The People's Choice* by Herbert Agar

1935 *The Colonial Period of American History* by Charles M. Andrews

1936 *A Constitutional Period of American History* by Andrew C. McLaughlin

1937 *The Flowering of New England* by Van Wyck Brooks

1938 *The Road to Reunion, 1865–1900* by Paul H. Buck

1939 *A History of American Magazines* by Frank L. Mott

1940 *Abraham Lincoln: The War Years* by Carl Sandburg

1941 *The Atlantic Migration, 1607–1860* by Marcus L. Hansen

1942 *Reveille in Washington* by Margaret Leech

1943 *Paul Revere and the World He Lived In* by Esther Forbes

1944 *The Growth of American Thought* by Merle Curti

1945 *Unfinished Business* by Stephen Bonsal

1946 *The Age of Jackson* by Arthur M. Schlesinger, Jr.

1947 *Scientists Against Time* by James P. Baxter III

1948 *Across the Wide Missouri* by Bernard De Voto

1949 *The Disruption of American Democracy* by Roy F. Nicholas

1950 *Art and Life in America* by O. W. Larkin

1951 *The Old Northwest: Pioneer Period 1815–1840* by R. Carlyle

1952 *The Uprooted* by Oscar Handlin

1953 *The Era of Good Feelings* by George Dangerfield

1954 *A Stillness at Appomattox* by Bruce Catton

1955 *Great River: The Rio Grande in North American History* by Paul Horgan

1956 *The Age of Reform* by Richard Hofstadter

1957 *Russia Leaves the War* by George F. Kennan

1958 *Banks and Politics in America—From the Revolution to the Civil War* by Bray Hammond

1959 *The Republican Era: 1869–1901* by Leonard D. White and Jean Schneider

1960 *In the Days of McKinley* by Margaret Leech

1961 *Between War and Peace: The Potsdam Conference* by Herbert Feis

1962 *The Triumphant Empire: Thunderclouds Gather in the West* by Lawrence H. Gibson

1963 *Washington, Village and Capital, 1800–1878* by Constance Green

1964 *Puritan Village: The Formation of a New England Town* by Sumner C. Powell

1965 *The Greenback Era* by Irwin Unger

1966 *Life of the Mind in America* by Perry Miller

1967 *Exploration and Empire: The Explorer and Scientist in the Winning of the American West* by William H. Goetzmann

1968 *The Ideological Origins of the American Revolution* by Bernard Bailyn

1969 *Origins of the Fifth Amendment* by Leonard W. Levy

1970 *Present at the Creation: My Years in the State Department* by Dean Acheson

1971 *Roosevelt: The Soldier of Freedom* by James M. Burns

1972 *Neither Black Nor White* by Carl N. Degler

1973 *People of Paradox: An Inquiry Concerning the Origin of American Civilization* by Michael Kammen

Pulitzer Prize for History

1974 *The Americans: The Democratic Experience* by Daniel J. Boorstin

1975 *Jefferson and His Time* by Dumas Malone

1976 *Lamy of Santa Fe* by Paul Horgan

1977 *The Impending Crisis: 1841–61* by David M. Potter

1978 *The Visible Hand: The Managerial Revolution in American Business* by Alfred D. Chandler, Jr.

1979 *The Dred Scott Case: Its Significance in American Law and Politics* by Don E. Fehrenbacher

1980 *Been in the Storm So Long* by Leon F. Litwack

1981 *American Education: The National Experience, 1783–1876* by Lawrence A. Cremin

1982 *Mary Chestnut's Civil War* by C. Vann Woodward

1983 *The Transformation of Virginia, 1740–1790* by Rhys L. Issac

1984 No prize awarded

1985 *The Prophets of Regulation* by Thomas K. McCraw

1986 *The Heavens and the Earth* by Walter A. McDougall

1987 *Voyagers to the West* by Bernard Bailyn

1988 *The Launching of Modern American Science, 1846–1876* by Robert V. Bruce

1989 *Parting the Waters: America in the King Years, 1954–63* by Taylor Branch; *Battle Cry of Freedom: The Civil War Era* by James M. McPherson

1688–1744 Alexander Pope, English poet and satirist, lives. Works include *Pastorals* (1709), *An Essay on Criticism* (1711), *The Rape of the Lock* (1714), "Epistle from Eloisa to Abelard" (1717), *The Dunciad* (1728, 1742), *An Essay on Man* (1733–34), and *Epistle to Dr. Arbuthnot* (1735).

1694–1778 Voltaire, French satirist, philosopher, historian, dramatist, and poet, lives. Works include *Lettres philosophiques* (1734), *Zadig* (1748), *Candide* (1759), and *Dictionnaire philosophique* (1764).

1703–58 Jonathan Edwards, American theologian and essayist, lives. Major work is sermon *Sinners in the Hands of an Angry God* (1741).

1706–90 Benjamin Franklin, American statesman and author, lives. Works include *Poor Richard's Almanack* (1732–47) and *Autobiography* (1771–88).

1707–54 Henry Fielding, English novelist and playwright, lives. Works include *Apology for the Life of Mrs. Shamela Andrews* (c. 1742), *Joseph Andrews* (1742), *Jonathan Wilde* (1743), *Tom Jones* (1749), and *Amelia* (1751).

1709–84 Samuel Johnson, English lexicographer, essayist, and poet, lives. Works include *London* (1738), *The Vanity of Human Wishes* (1749), *The Rambler* (1750–52), *Dictionary of the English Language* (1755), *The Idler* (1758–60), *Rasselas* (1759), and *Lives of the Poets* (1779–81).

1709–89 John Cleland, English novelist and journalist, lives. His major work is *Fanny Hill: Memoirs of a Woman of Pleasure* (1748–49).

1712–78 Jean-Jacques Rousseau, Swiss-born French philosopher and author, lives. Works include *The Social Contract* (1762), *Émile* (1762), *Confessions* (1788), and *Julie ou la Nouvelle Héloïse* (1791).

1713–68 Laurence Sterne, English novelist, lives. Works include *Tristram Shandy* (1759–67) and *A Sentimental Journey* (1768).

1713–84 Denis Diderot, French encyclopedist and author, lives. Works include *Pensées philosophiques* (1746), *Encyclopédie* (1751–72), *Le Neveu de Rameau* (1762), and *Le Rêve de d'Alembert* (1769).

1717–97 Horace Walpole, English author and historian, lives. Works include *Castle of Otranto* (1764) and *Historic Doubts on Richard III* (1768).

▶

1729–97 Edmund Burke, Irish-born statesman and author, lives. Works include *A Philosophical Inquiry into the Origin of Our Ideas of the Sublime and Beautiful* (1756), *On American Taxation* (1774), *On Conciliation with the Colonies* (1775), and *Reflections on the French Revolution* (1790).

1730–74 Oliver Goldsmith, Irish-born English poet, playwright, and novelist, lives. Works include *The Citizen of the World* (1762), *The Vicar of Wakefield* (1766), *The Deserted Village* (1770), and *She Stoops to Conquer* (1773).

1731–1800 William Cowper, English pre-Romantic poet, lives. Works include *The Diverting History of John Gilpin* (1784) and *The Task* (1785).

1737–1809 Thomas Paine, American pamphleteer and political radical, lives. Works include *Common Sense* (1776), *The American Crisis* (1776–83), *The Rights of Man* (1791–92), and *The Age of Reason* (1794–96).

1740–1814 Marquis (Donatien Alphonse François) de Sade, French author, lives. Works include *Justine ou les Malheurs de la vertu* (1791) and *Juliette* (1797).

1740–95 James Boswell, English biographer and diarist, lives. He is best known for *Life of Samuel Jonson* (1791).

1745–1833 Hannah More, English writer, lives. Works include *Percy* (1777), *The Fatal Falsehood* (1779), and *Coelebs in Search of a Wife* (1809).

1749–1832 Johann Wolfgang von Goethe, German poet, playwright, and novelist, lives. Works include *The Sorrows of Young Werther* (1774), *Iphigenia in Tauris* (1779), *Egmont* (1788), and *Die Wahlver-Wandtschaften* (1809).

1752–70 Thomas Chatterton, English poet, lives. Works include *The Revenge* (1770), *Bristowe Tragedie* (1772), and *An Excelente Balade of Charitie* (1777).

1752–1840 Fanny Burney (Frances, Madame d'Arblay), English novelist, lives. Works include *Evelina* (1778), *Cecilia* (1782), and *Camilla* (1796).

1754–1832 George Crabbe, English poet, lives. Works include *The Village* (1783), *The Newspaper* (1785), *The Parish Register* (1807), and *The Borough* (1810).

1757–1827 William Blake, English poet and artist, lives. Works include *Songs of Innocence* (1789), *The Book of Thel* (1789), *The Marriage of Heaven and Hell* (1790), *Songs of Experience* (1794), *Jerusalem* (1804), and *Milton* (1804).

1759–96 Robert Burns, Scottish poet, lives. Works include *The Holy Fair* (1786), *The Cotter's Saturday Night* (1786), *Tam O'Shanter* (1791), *A Red, Red Rose* (1794), *Scots* (1794), and *Auld Lang Syne* (1796).

1766–1817 Baronne Anne Louise de Staël-Holstein, French writer of belles-lettres, lives. Works include *Delphine* (1802), *Corinne* (1807), and *De l'Allemagne* (1813).

1767–1849 Maria Edgeworth, Irish novelist, lives. Works include *Castle Rackrent* (1800), *Belinda* (1801), and *Ormond* (1817).

1768–1848 François René de Chateaubriand, French writer, lives. Works include *Atala* (1801), *Le Genie du Christianisme* (1802), and *Memoires d'Outre-Tombe* (1848–50).

1770–1850 William Wordsworth, English poet, lives. Works include *Lyrical Ballads* (1798), "The Solitary Reaper" (1798), "The World Is Too Much with Us" (1798), "Lines Composed a Few Miles Above Tintern Abbey" (1798), *Ode: Intimations of Immortality* (1798), "The Daffodils" (1807), *The Excursion* (1814), and *The Prelude* (1850).

1771–1810 Charles Brockden Brown, American novelist and editor, lives. Works include *Wieland* (1798), *Ormond* (1799), *Clara Howard* (1801), and *Jane Talbot* (1801).

1771–1832 Sir Walter Scott, Scottish novelist and poet, lives. Works include poems *The Lay of the Last Minstrel* (1805), *Marmion* (1808), and *The Lady of the Lake* (1810) and novels *Waverley* (1814),

▶

Guy Mannering (1815), *The Antiquary* (1816), *Rob Roy* (1817), *Ivanhoe* (1819), *Peveril of the Peak* (1822) and *The Talisman* (1825).

1772–1834 Samuel Taylor Coleridge, English poet, lives. Works include *Kubla Khan* (1797), *The Rime of the Ancient Mariner* (1798), "Christabel," and "Dejection: An Ode" (1802).

1773–1853 Ludwig Tieck, German author, lives. Works include *Puss in Boots* (1797), *Franz Sternbald's Wanderings* (1798), and *Life's Overflow* (1837).

1774–1843 Robert Southey, English romantic poet, lives. Works include "The Battle of Blenheim" (1798) and biographies *Nelson* (1813), *Wesley* (1830), and *Cowper* (1833).

1775–1817 Jane Austen, English novelist, lives. Works include *Sense and Sensibility* (1811), *Pride and Prejudice* (1813), *Emma* (1816), *Northanger Abbey* (1818), and *Persuasion* (1818).

1775–1834 Charles Lamb, English essayist and critic, lives. Major works include *Tales from Shakespeare* (with Mary Lamb, 1704) and *Essays of Elia* (1823).

1776–1822 E.T.A. Hoffman, German author, lives. Works include *Fantasy-pieces* (1814–15), *The Devil's Elixirs* (1816), and *Night-pieces* (1817).

1778–1830 William Hazlitt, English essayist and literary critic, lives. Works include *The Characters of Shakespeare's Plays* (1817), *Lectures on the English Poets* (1818), and *The Spirit of the Age* (1825).

1782–1846 Esaias Tegner, Swedish poet, lives. Works include *Hjalten* (1813), *Nattuardsbarnen* (1820), and *Frithiofs Saga* (1825).

1783–1842 Stendhal (Marie Henri Beyle), French novelist, lives. Works include *The Red and the Black* (1830) and *The Charterhouse of Parma* (1839).

1783–1859 Washington Irving, American writer and historian, lives. Works include *Salmagundi* papers (1807–8), *A History of New York from the Beginning of the World to the End of the Dutch Dynasty* (1809), *The Sketch Book of Geoffrey Crayon* (including "Legend of Sleepy Hollow" and "Rip Van Winkle," 1820), *Bracebridge Hall* (1822), and *The Legend of the Alhambra* (1832).

1785–1873 Count Alessandro Manzoni, Italian novelist and dramatist, lives. Works include *Inni sacri* (Sacred Hymns) (1822) and *I Promessi Sposi* (1827).

1787–1862 Ludwig Uhland, German romantic poet, lives. Works include *Duke Ernst of Swabia* (1818), *Ludwig the Bavarian* (1819), and *Old South and North German Folk-Songs* (1844–45).

1788–1824 George Gordon, Lord Byron, English poet, lives. Works include *English Bards and Scotch Reviewers* (1809), *The Giaour* (1813), *The Bride of Abydos* (1813), *Corsair* (1814), *Lara* (1814), *The Prisoner of Chillon* (1816), *Childe Harold's Pilgrimage* (1816–18), *Manfred* (1817), *Beppo* (1818), and *Don Juan* (1819–24).

1788–1866 Friedrich Ruckert, German poet, lives. Works include *Harnessed Sonnets* (1814), *The Wisdom of the Brahman* (1836), and *Songs on Children's Deaths* (1872).

1789–1851 James Fenimore Cooper, American novelist, lives. Works include *The Pioneers* (1823), *The Last of the Mohicans* (1826), *The Prairie* (1827), *The Pathfinder* (1840), and *The Deerslayer* (1841).

1789–1862 Bernard Severin Ingemann, Danish poet and novelist, lives. Works include *Waldemar, Surnamed Seir, or the Victorious* (1826), *The Childhood of King Eric Menved* (1828), *King Eric and the Outlaws* (1833), and *Morgen og Aften sange* (1839).

1790–1869 Alphonse Marie Louis de Lamartine, French poet and writer, lives. Works include *Les Premières Méditations* (1820), *Les Nouvelles Méditations*

▶

(1823), *Les Harmonies poétiques et religieuses* (1830), *Jocelyn* (1836), and *La Chute d'un ange* (1838).

1792–1822 Percy Bysshe Shelley, English poet, lives. Works include *Ozymandias* (1818), *The Cenci* (1819), *The Cloud* (1819), "To a Skylark" (1819), *Ode to the West Wind* (1819), *The Sensitive Plant* (1819), *Prometheus Unbound* (1820), *Epipsychidion* (1821), *Adonais* (1821), and *Defence of Poetry* (1821).

1793–1864 John Clare, English poet, lives. Major works are included in volumes *Descriptive of Rural Life and Scenery* (1820), *The Village Minstrel* (1821), and *Rural Muse* (1835).

1794–1878 William Cullen Bryant, American poet and critic, lives. Works include *To a Waterfowl* (1815), *Thanatopsis* (1817), and *Inscription for the Entrance to a Wood* (1817).

1795–1821 John Keats, English poet, lives. Works include *Endymion* (1818), *Hyperion* (1818), "To Autumn" (1818), *The Eve of St. Agnes* (1819), "La Belle Dame sans Merci" (1819), *Ode on a Grecian Urn* (1819), and *Ode to a Nightingale* (1819).

1795–1881 Thomas Carlyle, English essayist and historian, lives. Works include *Sartor Resartus* (1833–34), *The French Revolution* (1837), and *On Heroes, Hero-worship, and the Heroic in History* (1841).

1797–1851 Mary Wollstonecraft Shelley, English novelist, lives. Works include *Frankenstein, or the Modern Prometheus* (1818).

1797–1856 Heinrich Heine, German poet and writer, lives. Works include *Die Harzreise* (1826), *Buch der Lieder* (Book of Songs) (1827), *Die romantische Schule* (1833), *Atta Troll* (1843), *Deutschland, ein Wintermarchen* (Germany, a Winter Tale) (1844), and *Romanzero* (1851).

1797–1863 Comte Alfred Victor de Vigny, French poet, writer, and playwright, lives. Works include novel *Cinq-Mars* (1826), play *Chatterton* (1835), and poems *Des Destinées* (1864).

1798–1837 Count Giacomo Leopardi, Italian poet, lives. Works include *Canzoni* (1824), *Versi* (1826), *Canti* (1836), and *Pensieri* (1837).

1798–1855 Adam Mickiewicz, Polish poet, lives. Works include *Ballady i romanse* (1822), *Dziady* (1823–32), and *Pan Tadeusz* (1834).

1799–1837 Aleksandr Sergeyevich Pushkin, Russian poet, author, and playwright, lives. Works include poems *Kislan and Ludmilla* (1820) and *Bronze Horeseman* (1833), play *Boris Godunov* (1825), and novel *Eugene Onegin* (1831).

1799–1850 Honoré de Balzac, French novelist, lives. Works include *Les Chouans* (1829), *Eugénie Grandet* (1833), and *Le Père Goriot* (1834).

1800–59 Thomas Babington Macaulay, English statesman and writer, lives. Works include poems *Lays of Ancient Rome* (1842) and *History of England from the Accession of James the Second* (1848–61).

1802–70 Alexandre Dumas *père*, French novelist and dramatist, lives. Works include *The Count of Monte Cristo* (1844), *The Three Musketeers* (1844), *Twenty Years After* (1845), *The Viscount of Bragelonne* (1848–50), and *The Black Tulip* (1895).

1802–76 Harriet Martineau, English author, lives. Works include *Illustrations of Political Economy* (1832–34) and *Poor Laws and Paupers Illustrated* (1833).

1803–39 José Maria de Heredia, Cuban poet, lives. Works include "En el teocalli de Cholula" (1820) and "Niagara" (1824).

1803–70 Prosper Mérimée, French novelist, lives. Works include *La Chronique du temps de Charles IX* (1829), *Colomba* (1841), and *Carmen* (1852).

▶

Pulitzer Prize for Biography or Autobiography

1917 *Julia Ward Howe* by Laura E. Richards and Maude Howe Elliott

1918 *Benjamin Franklin, Self-Revealed* by William C. Bruce

1919 *The Education of Henry Adams* by Henry Adams

1920 *The Life of John Marshall* by Albert J. Beveridge

1921 *The Americanization of Edward Bok* by Edward Bok

1922 *A Daughter of the Middle Border* by Hamlin Garland

1923 *The Life and Letters of Walter H. Page* by Burton J. Hendrick

1924 *From Immigrant to Inventor* by Michael Pupin

1925 *Barrett Wendell and His Letters* by DeWolfe Howe

1926 *Life of Sir William Osler* by Harvey Cushing

1927 *Whitman: An Interpretation in Narrative* by Emory Holloway

1928 *The American Orchestra and Theodore Thomas* by Charles E. Russell

1929 *The Training of an American: The Earlier Life and Letters of Walter H. Page* by Burton J. Hendrick

1930 *The Raven (Sam Houston)* by Marquis James

1931 *Charles W. Eliot* by Henry James

1932 *Theodore Roosevelt* by Henry F. Pringle

1933 *Grover Cleveland* by Allan Nevins

1934 *John Hay* by Tyler Dennett

1935 *R. E. Lee* by Douglas S. Freeman

1936 *The Thought and Character of William James* by Ralph B. Perry

1937 *Hamilton Fish: The Inner History of the Grant Administration* by Allan Nevins

1938 *Pedlar's Progress* by Odell Shepard; *Andrew Jackson* by Marquis James

1939 *Benjamin Franklin* by Carl Van Doren

1940 *Woodrow Wilson, Life and Letters* by Ray S. Baker

1941 *Jonathan Edwards* by Ola E. Winslow

1942 *Crusader in Crinoline* by Forrest Wilson

1943 *Admiral of the Ocean Sea* by Samuel E. Morison

1944 *The American Leonardo: The Life of Samuel F. B. Morse* by Charleton Mabee

1945 *George Bancroft: Brahmin Rebel* by Russell B. Nye

1946 *Son of the Wilderness* by Linnie M. Wolfe

1947 *The Autobiography of William Allen White* by William A. White

1948 *Forgotten First Citizen: John Bigelow* by Margaret Clapp

1949 *Roosevelt and Hopkins* by Robert E. Sherwood

1950 *John Quincy Adams and the Foundations of American Foreign Policy* by Samuel F. Bemis

1951 *John C. Calhoun: American Portrait* by Margaret L. Coit

1952 *Charles Evans Hughes* by Merlo J. Pusey

1953 *Edmund Pendleton, 1721–1803* by David J. Mays

1954 *The Spirit of St. Louis* by Charles A. Lindbergh

1955 *The Taft Story* by William S. White

1956 *Benjamin Henry Latrobe* by Talbot F. Hamlin

1957 *Profiles in Courage* by John F. Kennedy

1958 *George Washington*, vols. I–VI by Douglas S. Freeman and *George Washington*, vol. VII by John A. Carroll and Mary W. Ashworth

1959 *Woodrow Wilson: American Prophet* by Arthur Walworth

1960 *John Paul Jones* by Samuel E. Morison

1961 *Charles Sumner and The Coming of the Civil War* by David Donald

1962 No prize awarded

1963 *Henry James*: Vol. II: *The Conquest of London*, 1870–1881; Vol. III: *The Middle Years*, 1881–1895 by Leon Edel

1964 *John Keats* by Walter J. Bate

1965 *Henry Adams* by Ernest Samuels

1966 *A Thousand Days* by Arthur M. Schlesinger, Jr.

1967 *Mr. Clemens and Mark Twain* by Justin Kaplan

1968 *Memoirs (1925–1950)* by George F. Kennan

1969 *The Man from New York* by B. L. Reid

1970 *Huey Long* by T. Harry Williams

1971 *Robert Frost: The Years of Triumph, 1915–1938* by Lawrence Thompson

1972 *Eleanor and Franklin* by Joseph P. Lash

1973 *Luce and His Empire* by W. A. Swanbert

1974 *O'Neill, Son and Artist* by Louis Sheaffer

1975 *The Power Broker: Robert Moses and the Fall of New York* by Robert A. Caro

1976 *Edith Wharton: A Biography* by R.W.B. Lewis

Pulitzer Prize for Biography or Autobiography	
1977	*A Prince of Our Disorder, The Life of T. E. Lawrence* by John E. Mack
1978	*Samuel Johnson* by Walter J. Bate
1979	*Days of Sorrow and Pain: Leo Baeck and the Berlin Jews* by Leonard Baker
1980	*The Rise of Theodore Roosevelt* by Edmund Morris
1981	*Peter the Great: His Life and World* by Robert K. Massie
1982	*Grant: A Biography* by William S. McFeely
1983	*Growing Up* by Russell Baker
1984	*Booker T. Washington* by Louis R. Harlan
1985	*The Life and Times of Cotton Mather* by Kenneth Silverman
1986	*Louise Bogan: A Portrait* by Elizabeth Frank
1987	*Bearing the Cross: Martin Luther King, Jr., and the Southern Christian Leadership Conference* by David J. Garrow
1988	*Look Homeward: A Life of Thomas Wolfe* by David Herbert Donald
1989	*Oscar Wilde* by Richard Ellmann

1803–73 Edward George Earle Bulwer-Lytton, English novelist, lives. Works include *The Last Days of Pompeii* (1834), *Rienzi* (1835), *The Last of the Barons* (1843), and *King Arthur* (1849).

1803–82 Ralph Waldo Emerson, American poet, essayist, and philosopher, lives. Works include *Essays, First and Second Series* (1841, 1844), *Representative Men* (1850), *The Conduct of Life* (1860), and *English Traits* (1865).

1804–64 Nathaniel Hawthorne, American novelist and short-story writer, lives. Works include *Twice-Told Tales* (1837), *The Scarlet Letter* (1850), *The House of the Seven Gables* (1851), *The Blithedale Romance* (1852), and *The Marble Faun* (1860).

1804–76 George Sand (Baroness Dudevant), French writer, lives. Works include *Indiana* (1832), *Consuelo* (1842), and *La Mare au diable* (1846).

1804–77 Johan Ludvig Runeberg, Finnish poet, lives. Works include *The Elk Hunters* (1832), *King Fjalar* (1844), *Tales of Ensign Stal* (1848, 1860), and "Vart Land" (Our Land).

1804–81 Benjamin Disraeli, English statesman and novelist, lives. Works include *Vivian Grey* (1826), *Coningsby, or the New Generation* (1844), *Sybil* (1845), and *Tancred, or The New Crusade* (1847).

1805–75 Hans Christian Andersen, Danish writer of fairy tales, lives. Works include "The Red Shoes," "The Ugly Duckling," "The Emperor's Clothes," "The Tinder Box," and "The Fir Tree."

1805–85 Victor Hugo, French poet, novelist, and dramatist, lives. Works include poems *Les Feuilles d'automne* (1831) and *Contemplations* (1856) and novels *The Hunchback of Notre Dame* (1831) and *Les Misérables* (1862).

1806–61 Elizabeth Barrett Browning, English poet, lives. Works include *Cry of the Children* (1843), *Poems* (1844), *Sonnets from the Portugese* (1850), and *Aurora Leigh* (1856).

1807–82 Henry Wadsworth Longfellow, American poet, lives. Works include *The Village Blacksmith* (1839), *Wreck of the Hesperus* (1841), *Evangeline* (1847), *Hiawatha* (1855), *The Courtship of Miles Standish* (1858), *The Children's Hour* (1860), and *Tales of a Wayside Inn* (1863).

1807–92 John Greenleaf Whittier, American poet, lives. Works include *Maud Muller* (1854), *The Barefoot Boy* (1855), *Telling the Bees* (1858), *Barbara Frietchie* (1863), and *Snow-Bound* (1866).

1809–49 Edgar Allan Poe, American poet and short-story writer, lives. Works include *The Narrative of A. Gordon Pym* (1837), poems *The Raven* (1845), *Eureka* (1848), and *Annabel Lee* (1849), and stories "The Fall of the House of

▶

Usher" (1840), "Murders in the Rue Morgue" (1841), "The Pit and the Pendulum" (1842), "The Tell-Tale Heart" (1843), and "The Purloined Letter" (1845).

1809–52 Nikolai Vasilyevich Gogol, Russian author and playwright, lives. Works include *Arabesques* (1835), *Taras Bulba* (1835), "The Nose" (1836), *Dead Souls* (1842), and *The Overcoat* (1842).

1809–92 Alfred Lord Tennyson, English poet, lives. Works include *Locksley Hall* (1842), *In Memoriam* (1850), *Charge of the Light Brigade* (1854), *Idylls of the King* (1859–85), *Enoch Arden* (1864), and *Crossing the Bar* (1889).

1809–94 Oliver Wendell Holmes, American poet, lives. Works include *Old Ironsides* (1830), *The Autocrat of the Breakfast-Table* (1858), *The Chambered Nautilus* (1858), *The Deacon's Masterpiece* (1858), and *Elsie Venner* (1861).

1810–65 Elizabeth Cleghorn Gaskell, English novelist, lives. Works include *Mary Barton* (1848), *Cranford* (1853), *North and South* (1855), and *Wives and Daughters* (1864–66).

1811–63 William Makepeace Thackeray, English novelist, lives. Works include *The Memoirs of Barry Lyndon* (1844), *The Rose and the Ring* (1845), *Vanity Fair* (1848), *Henry Esmond* (1852), and *The Virginians* (1857).

1811–72 Théophile Gautier, French poet and novelist, lives. Works include *Poesies* (1830), *Albertus* (1832), *La Comédie de la mort* (1833), *Émaux et camées* (1852), *La fettatura* (1856), and *Le Captaine Fracass* (1861–63).

1811–96 Harriet Beecher Stowe, American novelist, lives. Works include *Uncle Tom's Cabin* (1852), *The Pearl of Orr's Island* (1862), *Old-Town Folks* (1869), and *Poganuc People* (1878).

1812–70 Charles Dickens, English novelist, lives. Works include *Pickwick Papers* (1836–37), *Oliver Twist* (1837–39), *Nicholas Nickleby* (1838–39), *The Old Curiosity Shop* (1840), *Barnaby Rudge* (1841), *American Notes* (1842), *A Christmas Carol* (1843), *Martin Chuzzlewit* (1844), *Dombey and Son* (1848), *David Copperfield* (1849–50), *Bleak House* (1852), *Little Dorrit* (1855–57), *A Tale of Two Cities* (1859), *Great Expectations* (1860–61), *Our Mutual Friend* (1864), and *Edwin Drood* (1870).

1812–88 Edward Lear, English painter and writer, lives. Works include *A Book of Nonsense* (1846), "The Owl and the Pussycat" (1871), and *Laughable Lyrics* (1877).

1812–89 Robert Browning, English poet, lives. Works include *Sordello* (1840), *Pippa Passes* (1841), *My Last Duchess* (1842), *Fra Lippo Lippi* (1855), *Rabbi Ben Ezra* (1855), *Prospice* (1864), and *The Ring and the Book* (1868–69).

1814–41 Mikhail Yuryevich Lermontov, Russian poet and novelist, lives. Works include poems *Son*, *Angel*, *Demon*, *The Novice*, and *The Testament* and novel *A Hero of Our Time* (1840).

1815–82 Anthony Trollope, English novelist, lives. Works include *Barchester Towers* (1857), *The Claverings* (1867), and *The Eustace Diamonds* (1873).

1816–55 Charlotte Brontë, English novelist, lives. Works include *Jane Eyre* (1847), *Shirley* (1849), and *Villette* (1853).

1817–62 Henry David Thoreau, American essayist and poet, lives. Works include *Civil Disobedience* (1849), *Walden* (1854), and *The Maine Woods* (1864).

1818–48 Emily J. Brontë, English novelist, lives. Works include *Wuthering Heights* (1847).

1818–83 Ivan Sergeyevich Turgenev, Russian novelist, lives. Works include *Sportsman's Sketches* (1852), *A Nest of Gentlefolk* (1859), *On the Eve* (1860), *Fathers and Sons* (1862), and *Smoke* (1867).

▶

1819–75 Charles Kingsley, English novelist and reformer, lives. Works include *Westward Ho!* (1855), *The Water Babies* (1863), and *Hereward the Wake* (1865).

1819–80 George Eliot (Mary Ann Evans), English novelist, lives. Works include *Adam Bede* (1859), *The Mill on the Floss* (1860), *Silas Marner* (1861), *Middlemarch* (1871–72), and *Daniel Deronda* (1874–76).

1819–91 James Russell Lowell, American poet and writer, lives. Works include *The Bigelow Papers* (1848), *The Vision of Sir Launfal* (1848), and *A Fable for Critics* (1848).

1819–91 Herman Melville, American writer and poet, lives. Works include *Typee: A Peep at Polynesian Life* (1846), *Omoo* (1847), *Moby-Dick, or the Whale* (1851), *Bartleby the Scrivener* (1853), *The Confidence-Man, His Masquerade* (1857), and the posthumously published *Billy Budd, Foretopman* (1924).

1819–92 Walt Whitman, American poet, lives. Works include *Leaves of Grass* (1855), *Song of Myself* (1855), *Crossing Brooklyn Ferry* (1856), *Out of the Cradle Endlessly Rocking* (1859), *Children of Adam* (1860), *Calamus* (1860), *O Captain! My Captain!* (1865), and *When Lilacs Last in the Dooryard Bloom'd* (1867).

1819–1910 Julia Ward Howe, American poet and reformer, lives. Works include *The Battle Hymn of the Republic* (1862).

1820–49 Anne Brontë, English novelist, lives. Works include *Agnes Grey* (1847) and *The Tenant of Wildfell Hall* (1848).

1821–67 Charles Pierre Baudelaire, French poet, lives. Works include *Les Fleurs du mal* (1857).

1821–81 Fyodor Dostoyevsky, Russian novelist, lives. Works include *The House of the Dead* (1861–62), *Notes from the Underground* (1864), *Crime and Punishment* (1866), *The Idiot* (1868), *The Possessed* (1871–72), and *The Brothers Karamazov* (1879–80).

1821–88 Gustave Flaubert, French novelist, lives. Works include *Madame Bovary* (1856), *Salammbô* (1862), *L'Education sentimentale* (1869), *The Temptation of St. Anthony* (1874), and *Bouvard et Pecuchet* (1881).

1822–88 Matthew Arnold, English critic and poet, lives. Works include *The Strayed Reveler and Other Poems* (1849), *Empedocles on Etna and Other Poems* (1852), *The Scholar-Gipsy* (1853), "Stanzas from the Grande Chartreuse" (1855), "Thyrsis" (1866), *Dover Beach* (1867), and *Culture and Anarchy* (1869).

1824–89 (William) Wilkie Collins, English novelist, lives. Works include *The Woman in White* (1860) and *The Moonstone* (1868).

1825–1900 Richard Doddridge Blackmore, English novelist, lives. Works include *Clara Vaughan* (1864), *Craddock Nowell* (1866), *Lorna Doone* (1869), and *Christowell* (1882).

1828–82 Dante Gabriel Rossetti, English poet, lives. Works include *The House of Life* (1848–81) and *The Blessed Damozel* (1850).

1828–1905 Jules Verne, French writer, lives. Works include *A Voyage to the Center of the Earth* (1864), *The Mysterious Island* (1870), *Twenty Thousand Leagues Under the Sea* (1870), and *Around the World in Eighty Days* (1873).

1828–1909 George Meredith, English writer and critic, lives. Works include *The Ordeal of Richard Feverel* (1859), *Modern Love* (1862), and *The Egoist* (1879).

1828–1910 Count Leo Tolstoy, Russian novelist, lives. Works include *The Cossacks* (1863), *War and Peace* (1864–69), *Anna Karenina* (1873–76), and *The Death of Ivan Ilyich* (1886).

1830–86 Emily Dickinson, American poet, lives. Collections of her works are first published posthumously (1890–94), beginning with *Poems by Emily Dickinson.*

▶

1830–94 Christina Georgina Rossetti, English poet, lives. Works include *Goblin Market and Other Poems* (1862), *The Prince's Progress* (1866), "A Birthday," and "When I Am Dead."

1830–1914 Paul J. Heyse, German novelist, lives. Works include *L'Arrabbiata* (1855) and *Kinder de Welt* (1873).

1830–1914 Frédéric Mistral, French poet, lives. Works include *Mirèio* (1859), *Lis Isclo d'or* (1875), and *Lou Poèmō Jóu Rose* (1897).

1832–88 Louisa May Alcott, American novelist, lives. Works include *Hospital Sketches* (1863), *Little Women* (1868), and *Little Men* (1871).

1832–98 Lewis Carroll (Charles Dodgson), English author, lives. Works include *Alice's Adventures in Wonderland* (1865), *Through the Looking Glass* (1872), and *The Hunting of the Snark, an Agony in Eight Fits* (1876).

1832–99 Horatio Alger, American writer, lives. Works include *Ragged Dick Series* (1867), *Luck and Pluck Series* (1869), and *Tattered Tom Series* (1871).

1832–1910 Björnstjerne Björnson, Norwegian novelist and poet, lives. Works include *Trust and Trial* (1881), *In God's Way* (1890), *Flags Are Flying in Town and Port* (1892), and *A Happy Boy* (1896).

1833–91 Pedro A. de Alarcon, Spanish novelist, lives. Works include *The Three-Cornered Hat* (1874), *El escandalo* (1875), *El niño de la bola* (1880), and *Captain Poison* (1881).

1834–91 Jan Neruda, Czech poet and writer, lives. Works include *Cemetery Flowers* (1858), *Books of Verses* (1867), *Tales of the Little Quarter* (1878), *Cosmic Songs* (1878), *Simple Motifs* (1883), *Ballads and Romances* (1883), and *Friday Songs* (1896).

1834–96 William Morris, English artist and poet, lives. Works include *The Defense of Guinevere and Other Poems* (1858), *The Life and Death of Jason* (1867), *The Earthly Paradise* (1868–71), and *Sigurd the Volsung* (1876).

1835–1902 Samuel Butler, English satirist and novelist, lives. Works include *Erewhon* (1872), *Erewhon Revisited* (1901), and *The Way of All Flesh* (1903).

1835–1910 Mark Twain (Samuel L. Clemens), American writer, lives. Works include *The Celebrated Jumping Frog of Calaveras County* (1865), *The Innocents Abroad* (1869), *Tom Sawyer* (1876), *The Prince and the Pauper* (1881), *Life on the Mississippi* (1883), *Huckleberry Finn* (1884), *A Connecticut Yankee in King Arthur's Court* (1889), *Tragedy of Pudd'nhead Wilson* (1894), *The Mysterious Stranger* (1898), and *The Man That Corrupted Hadleyburg* (1900).

1836–1902 Bret Harte, American writer, lives. Works include *The Luck of Roaring Camp* (1868), *The Outcasts of Poker Flat* (1869), and *Plain Language from Truthful James* (1870).

1836–1938 Gabriele D'Annunzio, Italian poet, novelist, and dramatist, lives. Works include novels *The Intruder* (1892) and *The Flame of Life* (1900) and play *La Gioconda* (1898).

1837–1909 Algernon Charles Swinburne, English poet, lives. Works include *Atalanta in Calydon* (1865), *A Song of Italy* (1867), *Songs Before Sunrise* (1871), *Tristram of Lyonesse* (1882), *Marino Faliero* (1885), *Astrophel* (1894), *A Tale of Balen* (1896), and *A Channel Passage* (1904).

1837–1920 William Dean Howells, American novelist, lives. Works include *A Modern Instance* (1881) and *The Rise of Silas Lapham* (1885).

1838–1918 Henry (Brooks) Adams, American historian and man of letters, lives. Works include *Mont-Saint-Michel and Chartres* (1904) and *The Education of Henry Adams* (1907).

1840–97 Alphonse Daudet, French novelist, lives. Works include *Lettres de Mon Moulin* (1866), *Le Nabab* (1877), and *Sapho* (1884).

▶

1840–1902 Émile Zola, French writer, lives. Works include *La Confession de Claude* (1865), *La Fortune des Rougons* (1871), *Les Rougon-Macquart* (1871–93), *Les Trois Villes* (1894–98), and *Quatre Evangiles* (1899–1903).

1840–1928 Thomas Hardy, English novelist and poet, lives. Works include *Far from the Madding Crowd* (1874), *The Return of the Native* (1878), *The Mayor of Casterbridge* (1886), *Tess of the D'Urbervilles* (1891), and *Jude the Obscure* (1895).

1842–98 Stephane Mallarmé, French poet, lives. Works include *The Afternoon of a Faun* (1876) and *Le Cygne* (1885).

1842–c. 1914 Ambrose G. Bierce, American journalist, short-story writer, and poet, lives. Works include *Cobwebs from an Empty Skull* (1874), *In the Midst of Life* (1892), *Can Such Things Be?* (1893), and *The Devil's Dictionary* (1911).

1843–1916 Henry James, American novelist, lives. Works include *Daisy Miller* (1878), *Washington Square* (1881), *The Portrait of a Lady* (1881), *The Bostonians* (1886), *The Aspern Papers* (1888), *What Maisie Knew* (1897), *The Turn of the Screw* (1898), *The Ambassadors* (1903), *The Golden Bowl* (1904), and *The Art of the Novel* (1934).

1844–89 Gerard Manley Hopkins, English poet, lives. Works include *The Wreck of the Deutschland* (1875).

1844–96 Paul Verlaine, French poet, lives. Works include *Poèmes saturniens* (1866), *Fêtes galantes* (1869), *Romances sans paroles* (1874), and *Sagesse* (1881).

1844–1924 Anatole France, French novelist and poet, lives. Works include *The Crime of Sylvester Bonnard* (1881), *Thais* (1890), *The Red Lily* (1894), and *Penguin Island* (1908).

1846–1916 Henryk Sienkiewicz, Polish novelist, lives. Works include *With Fire and Sword* (1884), *Krzyzacy* (1900), and *Quo Vadis?* (1896).

1847–1912 Bram Stoker, English writer, lives. Works include *Dracula* (1897).

1848–1908 Joel Chandler Harris, American journalist and author, lives. Works include *Uncle Remus, His Songs, and His Sayings* (1880) and *Nights with Uncle Remus* (1883).

1849–87 Emma Lazarus, American poet, lives. Works include *Songs of a Semite* (1882), "The New Colossus" (1883), and *By the Waters of Babylon* (1887).

1849–1916 James Whitcomb Riley, American poet, lives. Works include "When the Frost Is on the Punkin" (1883), *Little Orphan Annie* (1885), and *The Raggedy Man* (1890).

1849–1924 Frances Hodgson Burnett, English-born American author, lives. Works include *Little Lord Fauntleroy* (1886), *Sara Crewe* (1888), and *The Secret Garden* (1911).

1850–93 Guy de Maupassant, French short-story writer and novelist, lives. Works include "En famille," "Le Rendezvous," "The Necklace," "The Umbrella," *Une Vie* (1883), *Bel-Ami* (1885), and *Pierre et Jean* (1888).

1850–94 Robert Louis Stevenson, Scottish novelist and poet, lives. Works include *Treasure Island* (1883), *A Child's Garden of Verses* (1885), *The Strange Case of Dr. Jekyll and Mr. Hyde* (1886), *Kidnapped* (1886), *The Master of Ballantrae* (1888), and *Weir of Hermiston* (1894).

1850–1923 Pierre Loti (Louis-Marie-Julien Viaud), French novelist, lives. Works include *Pecheur d'islande* (1866), *Mon Frère Yves* (1883), *Madame Chrysanthème* (1888), and *Matelot* (1893).

1851–1904 Kate Chopin (O'Flaherty), American novelist, lives. She is best known for *The Awakening* (1899).

1852–1933 George Moore, Irish novelist, playwright, and poet, lives. Works include novels *Confessions of a Young Man* (1888), *Héloise and Abélard* (1921), and *Aphrodite in Aulis* (1931) and poem *Flowers of Passion* (1878).

▶

1853–95 José Martí, Cuban poet, lives. Works include *Ismaelillo* (1882), *Versos libres* (1882), and *Versos sencillos* (1891).

1854–91 Arthur Rimbaud, French poet, lives. Works include *The Drunken Boat* (1871), *Une Saison Enfer* (1873), and *Les Illuminations* (1886).

1854–1900 Oscar Wilde, Irish dramatist, poet, novelist, and essayist, lives. Works include novel *The Picture of Dorian Gray* (1891), play *The Importance of Being Earnest* (1895), poem *Ballad of Reading Gaol* (1898), and essay *The Critic as Artist* (1891).

1855–1916 Émile Verhaeren, Belgian poet, lives. Works include *Les Flamandes* (1883), *Les Villes tentaculaires* (1895), *Les Heures claires* (1896), *Les Heures d'après-midi* (1905), and *Les Heures du soir* (1911).

1856–1919 L. Frank Baum, American author, lives. His best-known work is *The Wonderful Wizard of Oz* (1900).

1856–1927 Matilde Serao, Italian novelist, lives. Works include *Il paeso di Cuccagna* (1891).

1857–1919 Karl A. Gjellerup, Danish writer, lives. Works include *En Idealist* (1878), *Germanernes Laerling* (1882), and *Der Pilger Kamanoto* (1906).

1857–1924 Joseph Conrad, Polish-born English novelist, lives. Works include *The Nigger of the Narcissus* (1897), *Lord Jim* (1900), *Heart of Darkness* (1902), and *The Secret Sharer* (1912).

1858–1940 Selma Lagerlof, Swedish novelist and short-story writer, lives. Works include *Gosta Berlings Saga* (1891) and *Jerusalem* (1901–2).

1859–1930 Sir Arthur Conan Doyle, English novelist, lives. Works include *A Study in Scarlet* (1887), *The Adventures of Sherlock Holmes* (1892), and *The Hound of the Baskervilles* (1902).

1859–1932 Kenneth Grahame, English author, lives. Works include *The Wind in the Willows* (1908).

1859–1936 A. E. Housman, English poet, lives. Works include *A Shropshire Lad* (1896).

1859–1940 Verner von Heidenstam, Swedish poet and novelist, lives. Works include *Vallfart och vandringsar* (1888), *Dikter* (1895), and *Nya dikter* (1915).

1859–1952 Knut Hamsun, Norwegian novelist, poet, and playwright, lives. Works include *Mysterier* (1892) and *Markens grode* (1917).

1860–1940 Hamlin Garland, American short-story writer and novelist, lives. Works include *Main-Travelled Roads* (1891), *A Son of the Middle Border* (1917), and *A Daughter of the Middle Border* (1921).

1862–1910 O. Henry (William Sydney Porter), American short-story writer, lives. Works include collections *The Four Million* (1906), *The Voice of the City* (1908), and *Options* (1909) and stories "The Gift of the Magi" (1906) and "The Last Leaf" (1907).

1862–1923 Maurice Barres, French novelist, lives. Works include *Le Culte du moi* (1888–91) and *Colette Baudoche: The Story of a Young Girl of Metz* (1918).

1862–1937 Edith Wharton, American novelist and short-story writer, lives. Works include *The House of Mirth* (1905), *Ethan Frome* (1911), *The Custom of the Country* (1913), and *The Age of Innocence* (1920).

1863–1924 Gene Stratton Porter, American novelist, lives. Works include *Freckles* (1904) and *A Girl of the Limberlost* (1909).

1863–1929 Arno Holz, German poet, lives. Works include *Phantasus* (1898) and *Revolution of Lyric Poetry* (1899).

1863–1952 George Santayana, Spanish-born American poet and novelist, lives. Works include *The Last Puritan* (1945).

1864–1931 Erik A. Karlfeldt, Swedish poet, lives. Works include *Fridolins visor* (1898), *Fridolins lustgard* (1901), and *Hosthorn* (1927).

▶

1864–1936 Miguel de Unamuno y Jugo, Spanish poet and novelist, lives. Works include novels *Tres novelas ejemplares* (1920) and *San Manuel Bueno, martir* (1931) and poems *El Cristo de Velazquez* (1920).

1865–1936 Rudyard Kipling, English short-story writer, poet, and novelist, lives. Works include story "The Man Who Would Be King" (1889), *Barrack Room Ballads* (1892, including *Gunga Din*), *The Jungle Book* (1894), novels *Captains Courageous* (1897) and *Kim* (1901), and poem "The White Man's Burden" (1899).

1865–1939 William Butler Yeats, Irish poet and dramatist, lives. Works include collected poems *The Green Helmet* (1910), *The Wild Swans at Coole* (1917), *The Tower* (1928), and *The Winding Stair* (1929).

1866–1946 H. G. Wells, English novelist, lives. Works include *The Time Machine* (1895), *The Invisible Man* (1897), *The War of the Worlds* (1898), and *Tono-Bun-gay* (1909).

1867–1925 Wladyslaw Reymont, Polish novelist, lives. Works include *Ziemia obiecana* (1899), *Chlopi* (1904–9), and *Rok 1794* (1913–18).

1867–1928 Vicente Blasco Ibáñez, Spanish novelist, lives. Works include *La barraca* (The Cabin) (1898), *Canas y barro* (Reeds and Mud) (1902), and *The Four Horsemen of the Apocalypse* (1916).

1867–1933 John Galsworthy, English novelist, lives. His best-known work is *The Forsyte Saga* (1906–21).

1868–1936 Maxim Gorky, Russian writer, lives. Works include *Twenty-six Men and a Girl* (1899), *Mother* (1907) and the trilogy *Childhood* (1913–14), *In the World* (1915–16), and *My Universities* (1923).

1869–1935 Edwin Arlington Robinson, American poet, lives. Works include *The Children of the Night* (1897), *The Man Against the Sky* (1916), *The Man Who Died Twice* (1924), and *Tristram* (1927).

1869–1946 Booth Tarkington, American novelist and playwright, lives. Works include *Penrod* (1914), *Seventeen* (1916), *The Magnificent Ambersons* (1918), and *Alice Adams* (1921).

1869–1950 Edgar Lee Masters, American poet and novelist, lives. Works include *Spoon River Anthology* (1915) and *Lincoln, the Man* (1931).

1869–1951 André Gide, French writer, lives. Works include *The Immoralist* (1902), *Strait Is the Gate* (1909), *The Pastoral Symphony* (1919), and *The Counterfeiters* (1926).

1870–1902 Frank Norris, American novelist, lives. Works include *McTeague* (1899), *The Octopus* (1901), and *The Pit* (1903).

1870–1953 Hilaire Belloc, French-born English writer, lives. Works include *The Bad Child's Book of Beasts* (1896) and *Cautionary Tales* (1907).

1871–1900 Stephen Crane, American writer, lives. Works include *The Red Badge of Courage* (1895), *The Open Boat* (1898), *The Bride Comes to Yellow Sky* (1898), *The Blue Hotel* (1899), and *Whilomville Stories* (1900).

1871–1922 Marcel Proust, French novelist, lives. Works include *Remembrance of Things Past* (1913–27), *Jean Santeuil* (1952), and *Contre Sainte-Beuve* (1954).

1871–1945 Paul Valéry, French poet, lives. Works include *La Jeune Parque* (1917) and *Le Cimitière Marin* (1932).

1871–1945 Theodore Dreiser, American novelist, lives. Works include *Sister Carrie* (1900), *Jennie Gerhardt* (1911), *The Financier* (1912), and *An American Tragedy* (1925).

1872–1906 Paul Laurence Dunbar, American poet, lives. Works include *Lyrics of Lowly Life* (1896) and *The Uncalled* (1898).

1872–1963 John Cowper Powys, English novelist, essayist, and critic, lives. Works include

▶

A Glastonbury Romance (1932), *Weymouth Sands* (1934), *The Art of Happiness* (1935), and *The Pleasure of Literature* (1938).

1872–1970 Bertrand Russell, English philosopher, lives. Works include *Principia Mathematica* (1910–13), *An Inquiry into Meaning and Truth* (1940), and *A History of Western Philosophy* (1945).

1873–1939 Ford Madox Ford, English novelist, lives. Works include *The Good Soldier* (1915) and *Parade's End* (1924–28).

1873–1954 Colette (Sidonie-Gabrielle Colette), French novelist, lives. Works include *La Maison de Claudine* (1922), *Sido* (1929), and *Gigi* (1945).

1873–1956 Walter De la Mare, English poet and novelist, lives. Works include *The Listeners* (1912) and *Early One Morning* (1935).

1874–1925 Amy Lowell, American poet, lives. Works include "Patterns" (1916), *Sword Blades and Poppy Seeds* (1914), "Lilacs" (1925), and *What's o'Clock* (1925).

1874–1942 L. M. Montgomery, Canadian writer, lives. His best-known work is *Anne of Green Gables* (1908).

1874–1946 Gertrude Stein, American poet, lives. Works include *Three Lives* (1909), *The Making of Americans* (1925), and *The Autobiography of Alice B. Toklas* (1933).

1874–1963 Robert L. Frost, American poet, lives. Works include poems *Mending Wall* (1914), *The Death of the Hired Man* (1914), *The Road Not Taken* (1916), *Birches* (1916), and *Stopping by Woods on a Snowy Evening* (1923).

1874–1965 Sir Winston Churchill, English statesman and author, lives. Major work is *A History of the English-Speaking Peoples* (1956–58).

1874–1965 W. Somerset Maugham, English novelist and short-story writer, lives. Works include *Of Human Bondage* (1915), *The Moon and Sixpence* (1919), *Cakes and Ale* (1930), and *The Razor's Edge* (1944).

1875–1926 Rainer Maria Rilke, German poet, lives. Works include *Das Stundenbuch* (1905), *The Duino Elegies* (1923), and *The Sonnets to Orpheus* (1923).

1875–1939 Zane Grey, American writer, lives. Works include *The Last of the Plainsmen* (1908) and *Riders of the Purple Sage* (1912).

1875–1955 Thomas Mann, German novelist, lives. Works include *Buddenbrooks* (1901), *Death in Venice* (1912), *The Magic Mountain* (1924), and *Doktor Faustus* (1947).

1876–1916 Jack London, American writer, lives. Works include *The Call of the Wild* (1903), *The Sea Wolf* (1904), *White Fang* (1905), and *Martin Eden* (1909).

1876–1931 Ole E. Rolvaag, Norwegian-born American novelist, lives. Works include *Giants in the Earth* (1924), *Peder Victorious* (1929), and *Their Father's God* (1931).

1876–1941 Sherwood Anderson, American short-story writer and novelist, lives. Works include novel *Winesburg, Ohio* (1919), collected poems *The Triumph of the Egg* (1921), and novel *Dark Laughter* (1925).

1876–1947 Willa S. Cather, American novelist, lives. Works include *O Pioneers!* (1913), *My Antonia* (1918), and *Death Comes for the Archbishop* (1927).

1877–1962 Hermann Hesse, German novelist and poet, lives. Works include *Demian* (1919), *Siddhartha* (1922), *Steppenwolf* (1929), and *Magister Ludi* (1943).

1878–1967 John Masefield, English poet, dramatist, and novelist, lives. Works include *Salt-Water Ballads* (1902) and *The Everlasting Mercy* (1911).

1878–1967 Carl Sandburg, American poet, lives. Works include *Smoke and Steel* (1920), *Rootabaga Stories* (1922),

▶

Abraham Lincoln (1926, 1939), and *The American Songbag* (1927).

1878–1968 Upton Sinclair, American novelist, lives. Works include *The Jungle* (1906), *Oil!* (1927), and *Boston* (1928).

1879–1931 Vachel Lindsay, American poet, lives. Works include *General William Booth Enters into Heaven* (1913), *The Congo* (1914), and *The Santa Fe Trail* (1914).

1879–1935 Will Rogers, American humorist, lives. Works include *The Cowboy Philosopher on Prohibition* (1919) and *The Illiterate Digest* (1924).

1879–1955 Wallace Stevens, American poet, lives. Works include *Notes Toward a Supreme Fiction* (1942) and *Collected Poems* (1954).

1879–1970 E. M. Forster, English novelist, lives. Works include *Where Angels Fear to Tread* (1905), *The Longest Journey* (1907), *A Room with a View* (1908), *Howard's End* (1910), *A Passage to India* (1924), and the posthumous *Maurice* (1971).

1880–1956 H. L. Mencken, American essayist and short-story writer, lives. Works include *The American Language* (1919) and *Prejudices* (1919–27).

1880–1957 Sholem Asch, Polish-born American novelist, lives. Works include *The Nazarene* (1939), *The Apostle* (1943), and *Mary* (1949).

1881–1975 P. G. Wodehouse, English writer and humorist, lives. Works include *Leave It to Psmith* (1923), *Jeeves* (1925), and *The Code of the Woosters* (1938).

1882–1941 James Joyce, Irish novelist, short-story writer, and poet, lives. Works include *Dubliners* (1914), *Ulysses* (1922), *A Portrait of the Artist as a Young Man* (1916), and *Finnegans Wake* (1939).

1882–1941 Virginia Woolf, English novelist and essayist, lives. Works include *Mrs. Dalloway* (1925), *To the Lighthouse* (1927), *Orlando* (1928), *A Room of One's Own* (1929), and *The Waves* (1931).

1882–1949 Sigrid Undset, Norwegian novelist, lives. Works include *Kristin Lavransdatter* (1920–22) and *The Snake Pit* (1929).

1882–1956 A. A. Milne, English novelist, lives. Works include *Winnie-the-Pooh* (1926) and *The House at Pooh Corner* (1928).

1883–1924 Franz Kafka, Austrian novelist, lives. Works include *The Metamorphosis* (1915), *The Castle* (1922), *The Trial* (1925), and *Amerika* (1927).

1883–1931 Kahlil Gibran, Syrian-born American writer, lives. His best-known work is *The Prophet* (1923).

1883–1957 Nikos Kazantzakis, Greek novelist, lives. Works include *The Odyssey: A Modern Sequel* (1938), *Zorba the Greek* (1946), *The Greek Passion* (1951), and *The Last Temptation of Christ* (1951).

1883–1963 William Carlos Williams, American poet, lives. Works include *In the American Grain* (1925), *Pictures from Brueghel and Other Poems* (1963), and *Paterson* (1946–58).

1885–1930 D. H. Lawrence, English novelist, short-story writer, and poet, lives. Works include *Sons and Lovers* (1913), *The Rainbow* (1915), *Women in Love* (1920), and *Lady Chatterley's Lover* (1928).

1885–1933 Ring Lardner, American humorist and short-story writer, lives. Works include *You Know Me, Al* (1916) and *How to Write Short Stories (with Samples)* (1924).

1885–1951 Sinclair Lewis, American novelist, lives. Works include *Main Street* (1920), *Babbitt* (1922), *Arrowsmith* (1925), and *Elmer Gantry* (1927).

1885–1962 Isak Dinesen, Danish short-story writer, lives. Works include *Seven Gothic Tales* (1934), *Out of Africa* (1937), and *Winter's Tales* (1942).

1885–1972 Ezra L. Pound, American poet, lives. Works include *Personae* (1909), *Lustra* (1916), "Homage to Sextus Propertius" (1917), and *Cantos* (1925–68).

▶

Nobel Prize for Literature

1901 René F. A. Sully Prudhomme (1839–1907, France)

1902 Theodor Mommsen (1817–1903, Germany)

1903 Björnsterne Björnson (1832–1910, Norway)

1904 Frédéric Mistral (1830–1914, France) and José Echegaray y Eizaguirre (1832–1916, Spain)

1905 Henryk Sienkiewicz (1846–1916, Poland)

1906 Giosuè Carducci (1835–1907, Italy)

1907 Rudyard Kipling (1865–1936, Great Britain)

1908 Rudolf C. Eucken (1846–1926, Germany)

1909 Selma Lagerlöf (1858–1940, Sweden)

1910 Paul J. von Heyse (1830–1914, Germany)

1911 Maurice Maeterlinck (1862–1949, Belgium)

1912 Gerhart Hauptmann (1862–1946, Germany)

1913 Rabindranath Tagore (1861–1941, India)

1914 No prize awarded

1915 Romain Rolland (1866–1944, France)

1916 Verner von Heidenstam (1859–1940, Sweden)

1917 Karl A. Gjellerup (1857–1919, Denmark) and Henrik Pontoppidan (1857–1943, Denmark)

1918 No prize awarded

1919 Carl F. G. Spitteler (1845–1924, Switzerland)

1920 Knut Hamsun (1859–1952, Norway)

1921 Anatole France (1844–1924, France)

1922 Jacinto Benavente y Martinez (1866–1954, Spain)

1923 William Butler Yeats (1865–1939, Ireland)

1924 Wladyslaw S. Reymont (1868–1925, Poland)

1925 George Bernard Shaw (1856–1950, Ireland)

1926 Grazia Deledda (1875–1936, Italy)

1927 Henri Bergson (1859–1941, France)

1928 Sigrid Undset (1882–1949, Norway)

1929 Thomas Mann (1875–1955, Germany)

1930 Sinclair Lewis (1885–1951, United States)

1931 Erik A. Karlfeldt (1864–1931, Sweden)

1932 John Galsworthy (1867–1933, Great Britain)

1933 Ivan A. Bunin (1870–1953, France)

1934 Luigi Pirandello (1867–1936, Italy)

1935 No prize awarded

1936 Eugene O'Neill (1888–1953, United States)

1937 Roger Martin du Gard (1881–1958, France)

1938 Pearl S. Buck (1892–1973, United States)

1939 Frans E. Sillanpää (1888–1964, Finland)

1940 No prize awarded

1941 No prize awarded

1942 No prize awarded

1943 No prize awarded

1944 Johannes V. Jensen (1873–1950, Denmark)

1945 Gabriela Mistral (1889–1957, Chile)

1946 Hermann Hesse (1877–1962, Switzerland)

1947 André Gide (1869–1951, France)

1948 T. S. Eliot (1888–1965, Great Britain)

1949 William Faulkner (1897–1962, United States)

1950 Bertrand Russell (1872–1970, Great Britain)

1951 Pär F. Lagerkvist (1891–1974, Sweden)

1952 François Mauriac (1885–1970, France)

1953 Sir Winston Churchill (1874–1965, Great Britain)

1954 Ernest Hemingway (1899–1961, United States)

1955 Halldór K. Laxness (b. 1902, Iceland)

1956 Juan Ramón Jiménez (1881–1958, Spain)

1957 Albert Camus (1913–60, France)

1958 Boris L. Pasternak (1890–1960, USSR; declined award)

1959 Salvatore Quasimodo (1901–68, Italy)

1960 Saint-John Perse (1887–1971, France)

1961 Ivo Andríc (1892–1975, Yugoslavia)

1962 John Steinbeck (1902–68, United States)

1963 George Seferis (1900–1971, Greece)

1964 Jean-Paul Sartre (1905–80, France; declined award)

1965 Mikhail Sholokhov (1909–84, USSR)

1966 Shmuel Yosef Agnon (1888–1970, Israel) and Nelly Sachs (1891–1970, Sweden)

1967 Miguel Angel Asturias (1899–1974, Guatemala)

1968 Yasunari Kawabata (1899–1974, Japan)

1969 Samuel Beckett (1906–89, Ireland)

1970 Aleksandr I. Solzhenitsyn (b. 1918, USSR)

1971 Pablo Neruda (1904–73, Chile)

1972 Heinrich Böll (1917–85, West Germany)

1973 Patrick White (b. 1912, Australia)

1974 Eyvind Johnson (1900–1976, Sweden) and Harry E. Martinson (1904–78, Sweden)

Nobel Prize for Literature

Year	Laureate
1975	Eugenio Montale (1896–1981, Italy)
1976	Saul Bellow (b. 1915, United States)
1977	Vicente Aleixandre (1898–1984, Spain)
1978	Isaac Bashevis Singer (b. 1904, United States)
1979	Odysseus Elytis (b. 1911, Greece)
1980	Czeslaw Milosz (b. 1911, United States/Poland)
1981	Elias Canetti (b. 1905, Great Britain)
1982	Gabriel Garcia Márquez (b. 1928, Colombia)
1983	William Golding (b. 1911, Great Britain)
1984	Jaroslav Seifert (1902–86, Czechoslovakia)
1985	Claude Simon (b. 1913, France)
1986	Wole Soyinka (b. 1934, Nigeria)
1987	Joseph Brodsky (b. 1940, United States)
1988	Naguib Mafouz (b. 1911, Egypt)
1989	Camilo José Cola (b. 1916, Spain)

1886–1962 H. D. (Hilda Doolittle), American poet, lives. Works include anthologies *Sea Garden* (1916), *Red Shoes for Bronze* (1931), *The Walls Do Not Fall* (1944), and *Bid Me to Live* (1960).

1887–1959 Edwin Muir, English author and poet, lives. Works include anthologies *Chorus of the Newly Dead* (1926), *The Labyrinth* (1949), and *Collected Poems* (1952).

1887–1962 Robinson Jeffers, American poet and dramatist, lives. Works include anthologies *Tamar and Other Poems* (1924), *Dear Indas* (1929), *Such Counsels You Gave to Me* (1937), and *Hungerfield and Other Poems* (1954).

1887–1964 Dame Edith Sitwell, English poet, lives. Works include collected poems *Clown's House* (1918).

1887–1968 Edna Ferber, American novelist and short-story writer, lives. Works include *The Girls* (1921), *So Big* (1924), *Show Boat* (1926), and *Giant* (1952).

1887–1972 Marianne C. Moore, American poet, lives. Works include *Observations* (1924), *Pangolin, and Other Verse* (1936), and *Nevertheless* (1944).

1888–1923 Katherine Mansfield, New Zealand-born British short-story writer, lives. Works include *Bliss* (1920), *A Dill Pickle* (1920), and *The Garden Party* (1922).

1888–1965 T. S. Eliot, American-born English poet, critic, and dramatist, lives. Works include *The Love Song of J. Alfred Prufrock* (1915), *The Wasteland* (1922), *The Hollow Men* (1925), *Ash Wednesday* (1930), *Old Possum's Book of Practical Cats* (1939), and *Four Quartets* (1943).

1888–1974 John Crowe Ransom, American poet and critic, lives. Works include anthologies *Poems About God* (1919), *Chills and Fever* (1924), *Gentlemen in Bonds* (1926), *The World's Body* (1938), *Selected Poems* (1969), and *Beating the Bushes: Selected Essays* (1941–70).

1889–1945 Robert Charles Benchley, American humorist, critic, and actor, lives. Collected works from his *New Yorker* columns include *Of All Things* (1921), *Inside Benchley* (1942), and *Benchley Beside Himself* (1943).

1889–1957 Gabriela Mistral, Chilean poet, lives. Works include *Sonetos de la muerte* (1914), *Ternura* (1924), and *Tala* (1938).

1889–1963 Jean Cocteau, French poet and novelist, lives. Works include *Les Enfants terribles* (1929), *La Voix humaine* (1930), and *The Infernal Machine* (1934).

1889–1973 Conrad P. Aiken, American writer, lives. Works include *The Jig of Forslin* (1916), *The Charnel Rose* (1918), and *Brownstone Eclogues* (1942).

1890–1945 Franz Werfel, Austrian novelist, poet, and playwright, lives. Works include *Die Weltfreund* (1911) and *The Song of Bernadette* (1941).

▶

1890–1948 Claude McKay, American poet and novelist, lives. Works include anthologies *Spring in New Hampshire* (1920), *Harlem Shadows* (1922), and novels *Home to Harlem* (1927), *Banjo* (1929), and *Banana Bottom* (1933).

1890–1960 Boris L. Pasternak, Russian lyric poet and novelist, lives. His best-known work is *Doctor Zhivago* (1957).

1890–1980 Katherine Anne Porter, American short-story writer and novelist, lives. Works include *Pale Horse, Pale Rider* (1939), *Noon Wine* (1939), *Old Mortality* (1939), and *Ship of Fools* (1962).

1891–1970 Nelly Sachs, German poet and playwright, lives. Works include *In the Dwellings of Death* (1947), *Flight and Metamorphosis* (1959), and *Journey into a Dustless Room* (1961).

1891–1980 Henry Miller, American autobiographer and poet, lives. Works include *Tropic of Cancer* (1934) and *Tropic of Capricorn* (1939).

1892–1950 Edna St. Vincent Millay, American poet, lives. Works include *A Few Figs from Thistles* (1920) and *The Harp Weaver and Other Poems* (1923).

1892–1973 J. R. R. Tolkien, English novelist, lives. Works include *The Hobbit* (1937), *The Fellowship of the Ring* (1954), *The Two Towers* (1955), and *The Return of the King* (1956).

1892–1973 Pearl S. Buck, American novelist, lives. Works include *The Good Earth* (1931) and *The Spirit and the Flesh* (1944).

1892–1982 Archibald MacLeish, American poet, lives. Works include *Songs for a Summer Day* (1915), *Nobodaddy* (1926), *Conquistador* (1932), and *J.B.* (1958).

1892–1983 Dame Rebecca West, English novelist, lives. Works include *The Judge* (1922), *The Thinking Reed* (1936), *The Fountain Overflows* (1956), and *The Birds Fall Down* (1966).

1892– Hugh MacDiarmuid (Christopher Murray Grieve), Scottish poet and critic, lives. Works include essays *At the Sign of the Thistle* (1934), poem *A Drunk Man Looks at the Thistle* (1962), *Collected Poems* (1962), and *More Collected Poems* (1971).

1893–1957 Dorothy L. Sayers, English detective-story writer, lives. Works include *Strong Poison* (1930), *The Nine Tailors* (1934), and *Busman's Honeymoon* (1937).

1893–1967 Dorothy R. Parker, American short-story and verse writer, lives. Works include *Enough Rope* (1926), *After Such Pleasurers* (1932), and *Here Lies* (1939).

1894–1961 Dashiell Hammett, American detective-story writer, lives. Works include *The Maltese Falcon* (1930), *The Glass Key* (1931), and *The Thin Man* (1932).

1894–1961 James Thurber, American humorist and short-story writer, lives. Works include *The Seal in the Bedroom and Other Predicaments* (1932), *The Middle-Aged Man on the Flying Trapeze* (1935), and *The Thurber Carnival* (1945).

1894–1962 e e cummings, American poet, lives. Works include *The Enormous Room* (1922), *Tulips and Chimneys* (1923), *1 × 1* (1944), and *Ninety-five Poems* (1958).

1894–1963 Aldous L. Huxley, English novelist, essayist, and satirist, lives. Works include *Crome Yellow* (1921), *Antic Hay* (1923), *Point Counter Point* (1928), and *Brave New World* (1932).

1894–1984 J. B. Priestley, English novelist and playwright, lives. Works include *The Good Companions* (1929), *Angel Pavement* (1930), and "An Inspector Calls" (1946).

1895–1985 Robert Graves, English poet and novelist, lives. Works include *Fairies and Fusiliers* (1917), *Goodbye to All That* (1929), *I, Claudius* (1934), *Claudius the God* (1934), and *The White Goddess* (1948).

1896–1940 F. Scott Fitzgerald, American novelist and short-story writer, lives. Works include *This Side of Paradise* (1920), *The*

▶

Great Gatsby (1925), *Tender Is the Night* (1934), and *The Last Tycoon* (1941).

1896–1970 John R. Dos Passos, American writer, lives. Works include *Three Soldiers* (1921), *Manhattan Transfer* (1925), and the *USA* trilogy *The 42nd Parallel* (1930), *1919* (1932), and *The Big Money* (1936).

1896–1984 Liam O'Flaherty, Irish novelist and short-story writer, lives. Works include *The Black Soul* (1924), *The Informer* (1925), and *The Assassin* (1928).

1897–1962 William Faulkner, American novelist and short-story writer, lives. Works include *The Sound and the Fury* (1929), *As I Lay Dying* (1930), *Sanctuary* (1931), *Light in August* (1932), *Absalom, Absalom!* (1936), *The Unvanquished* (1938), *The Hamlet* (1940), *A Fable* (1954), and *The Reivers* (1962).

1897–1975 Thornton N. Wilder, American novelist and playwright, lives. Novels include *The Bridge of San Luis Rey* (1927) and *The Eighth Day* (1967).

1898–1936 Federico García Lorca, Spanish poet and playwright, lives. Works include *Romancero Gitano* (1928), *Poeta en Nueva York* (1929, published in 1940), and *Bodas de sangre* (1933).

1898–1943 Stephen Vincent Benét, American poet, lives. Works include *King David* (1923) and *John Brown's Body* (1928).

1898–1963 C. S. Lewis, English novelist, lives. Works include *The Screwtape Letters* (1942) and *Perelandra* (1943).

1898–1970 Erich Maria Remarque, German novelist, lives. Works include *All Quiet on the Western Front* (1929) and *A Time to Love and a Time to Die* (1954).

1899–1932 Hart Crane, American poet, lives. Works include *White Buildings* (1926), *The Bridge* (1930), and *Collected Verse* (1933).

1899–1961 Ernest M. Hemingway, American novelist and short-story writer, lives. Works include *The Sun Also Rises* (1926), *A Farewell to Arms* (1929), *To Have and Have Not* (1937), *The Snows of Kilimanjaro* (1938), *For Whom the Bell Tolls* (1940), and *The Old Man and the Sea* (1952).

1899–1973 Elizabeth D. C. Bowen, Anglo-Irish novelist and short-story writer, lives. Works include *The Hotel* (1927), *The House in Paris* (1935), *The Death of the Heart* (1938), and *The Heat of the Day* (1949).

1899–1977 Vladimir Nabokov, Russian-born American novelist, lives. Works include *The Gift* (1937), *Lolita* (1955), *Pale Fire* (1962), and *Ada, or Ardor: A Family Chronicle* (1969).

1899–1985 E. B. White, American humorist and essayist, lives. Works include *Stuart Little* (1945), *Charlotte's Web* (1952), and *The Trumpet of the Swan* (1970).

1899–1986 Jorge Luis Borges, Argentinian short-story writer, essayist, poet, and man of letters, lives. Works include *Ficciones* (1944), *El Aleph* (1949), and *Sietas noches* (1980).

1900–1938 Thomas Wolfe, American novelist, lives. Works include *Look Homeward, Angel* (1929), *Of Time and the River* (1935), *The Web and the Rock* (1939), and *You Can't Go Home Again* (1940).

1900–1944 Antoine de Saint Exupéry, French novelist, lives. Works include *Night Flight* (1931), *Wind, Sand, and Stars* (1939), and *The Little Prince* (1943).

1900–1949 Margaret Mitchell, American novelist, lives. Her single novel is *Gone with the Wind* (1936).

1900– Sean O'Faolain, Irish novelist and short-story writer, lives. Works include *A Nest of Simple Folk* (1933) and *King of the Beggars* (1938).

1901–68 Salvatore Quasimodo, Italian poet, lives. Works include *Oboe sommerso* (1932), *Poesie nuove* (1942), and *La terra impareggiabile* (1958).

1901–76 André Malraux, French novelist and man of letters, lives. Works include *Les Conquérants* (1928), *Man's Fate*

▶

(1933), *La voie royale* (1930), and *Le Temps du mépris* (1935).

1902–67 Langston Hughes, American poet, lives. Works include *The Weary Blues* (1926), *The Ways of White Folks* (1934), *Shakespeare in Harlem* (1942), and *Something in Common* (1963).

1902–68 John Steinbeck, American novelist and short-story writer, lives. Works include *Tortilla Flat* (1935), *Of Mice and Men* (1937), *The Grapes of Wrath* (1939), *Cannery Row* (1945), *The Pearl* (1947), and *East of Eden* (1952).

1902–71 Ogden Nash, American poet and humorist, lives. Works include collections *Free Wheeling* (1931), *Everyone but Thee and Me* (1962), and *Bed Riddance* (1970).

1903–40 Nathanael West, American novelist, lives. Works include *A Cool Million* (1934) and *The Day of the Locust* (1939).

1903–46 Countee Cullen, American poet, lives. Works include *Color* (1926), *Copper Sun* (1927), and *The Medea and Some Poems* (1929).

1903–50 George Orwell (Eric Arthur Blair), English novelist, lives. Works include *The Road to Wigan Pier* (1937), *Animal Farm* (1944), and *1984* (1949).

1903–66 Evelyn Waugh, English novelist, lives. Works include *Decline and Fall* (1928), *Black Mischief* (1932), *A Handful of Dust* (1937), *The Loved One* (1948), and *Brideshead Revisited* (1945).

1903–77 Anaïs Nin, French-born American diarist and essayist, lives. She is best known for her 10-volume *Diary* (1966–83).

1903– Alan Paton, South American novelist, lives. Works include *Cry, the Beloved Country* (1948) and *Tales from a Troubled Land* (1961).

1903–87 Erskine Caldwell, American novelist and short-story writer, lives. Works include *Tobacco Road* (1932) and *God's Little Acre* (1933).

1904–73 Pablo Neruda, Chilean poet, lives. Works include *Veinte poemas de amor y una cancion desesparada* (1924), *Residencia en la tierra* (1932–35), and *Estravagario* (1959).

1904–79 James T. Farrell, American novelist, lives. His best-known work is *Studs Lonigan* (1935).

1904–79 S. J. Perelman, American journalist and humorist, lives. Works include *Westward Ha! or Around the World in Eighty Clichés* (1948) and *The Most of S. J. Perelman.*

1904–86 Christopher Isherwood, English novelist and dramatist, lives. Works include *All the Conspirators* (1928) and *Goodbye to Berlin* (1939).

1904– (Henry) Graham Greene, English novelist, short-story writer, and playwright, lives. Works include *The Power and the Glory* (1940), *The Heart of the Matter* (1948), *The Third Man* (1950), *Our Man in Havana* (1959), and *The Human Factor* (1978).

1904– Isaac Bashevis Singer, Polish-born American novelist and short-story writer, lives. Works include *Gimpel the Fool* (1957), *Shosa* (1978), and *The Manor* (1989).

1905–70 John O'Hara, American novelist, short-story writer, and screenwriter, lives. Works include *Appointment in Samarra* (1934), *Butterfield 8* (1935), *A Rage to Live* (1949), *Elizabeth Appleton* (1963), and *The Ewings* (1972).

1905–75 Lionel Trilling, American critic and essayist, lives. Works include *Matthew Arnold* (1939), *E. M. Forster* (1943), *A Gathering of Fugitives* (1956), and *Sincerity and Authenticity* (1972).

1905–80 Jean-Paul Sartre, French dramatist, novelist, and philosopher, lives. Major works include novel *Nausea* (1938), treatise *Being and Nothingness* (1943), and play *No Exit* (1944).

1905–80 C. P. Snow, English novelist, lives. Works include *Strangers and Brothers*

▶

(1940–70), a series of novels, and *Corridors of Power* (1964).

1905–83 Arthur Koestler, Hungarian-born English novelist and essayist, lives. Major works include *Spanish Testament* (1937), *Scum of the Earth* (1941), *Darkness at Noon* (1941), and *The Ghost in the Machine* (1967).

1905–84 Mikhail A. Sholokhov, Russian novelist, lives. Works include *The Quiet Don* (1928–40).

1905– Robert Penn Warren, American poet and novelist, lives. Works include novels *All the King's Men* (1946), *World Enough and Time* (1950), *Wilderness* (1961), and *A Place to Come To* (1977) and collected poems *Promises* (1957) and *Now and Then: Poems* (1978).

1905– Elias Canetti, Bulgarian-born American novelist, lives. Works include *Auto da Fé* (1984), *The Human Province* (1985), and *The Wedding* (1986).

1906–64 T. H. White, English writer, lives. Works include *The Sword in the Stone* (1939) and *The Once and Future King* (1958).

1907–63 Louis MacNeice, Irish poet and critic, lives. Works include *The Poetry of W. B. Yeats* (1941) and *Collected Poems* (published posthumously in 1979).

1907–73 W. H. Auden, English poet and dramatist, lives. Works include *Letters from Iceland* (1937), *The Quest* (1941), and *The Age of Anxiety* (1947).

1907–89 Daphne Du Maurier, English novelist and short-story writer, lives. Works include *Rebecca* (1938) and *The Scapegoat* (1957).

1907– James A. Michener, American novelist, lives. Works include *Tales of the South Pacific* (1947), *Hawaii* (1959), *Centennial* (1974), *Chesapeake* (1978), *Space* (1982), and *Texas* (1985).

1908–60 Richard Wright, American novelist, lives. Works include *Native Son* (1940) and *Eight Men* (1961).

1908–63 Theodore Roethke, American poet, lives. Works include *Collected Poems* (1954), *Words for the Wind* (1958), and *The Far Field* (1964).

1908–81 William Saroyan, American short-story writer, novelist, and playwright, lives. Works include collections *The Daring Young Man on the Flying Trapeze* (1934) and *My Name is Aram* (1940), play *The Time of Your Life* (1939), and novel *The Human Comedy* (1943).

1908–86 Simone de Beauvoir, French writer, lives. Works include novel *Les Mandarins* (1954) and treatises *The Second Sex* (1949–50) and *The Coming of Age* (1970).

1909–55 James Agee, American writer, lives. Works include novels *The Morning Watch* (1951) and posthumously published *A Death in the Family* (1957) and film commentary *Agee on Film*.

1909–57 Malcolm Lowry, English novelist, lives. His best-known work is *Under the Volcano* (1947).

1909– Eudora Welty, American novelist and short-story writer, lives. Works include *The Robber Bridegroom* (1942), *The Ponder Heart* (1954), and *The Collected Stories of Eudora Welty* (1980).

1911–79 Elizabeth Bishop, American poet, lives. Works include *Selected Poems* (1967) and *Poems* (1979).

1911– William Golding, English novelist, lives. Works include *The Lord of the Flies* (1954), *The Inheritors* (1955), and *Rites of Passage* (1980).

1912–82 John Cheever, American novelist and short-story writer, lives. Works include *The Wapshot Chronicle* (1937) and *Falconer* (1977).

1912– Mary McCarthy, American critic, novelist, and short-story writer, lives. Works include memoir *Memories of a Catholic Girlhood* (1957) and novel *The Group* (1963).

1912– Patrick White, Australian novelist, dramatist, and poet, lives. Works include *The Happy Valley* (1939), *The Tree of Man* (1955), and *Voss* (1957).

▶

1913–60 Albert Camus, Algerian-born French novelist and dramatist, lives. Works include *The Stranger* (1942), *The Myth of Sisyphus* (1942), *Caligula* (1944), and *The Plague* (1947).

1914–53 Dylan M. Thomas, Welsh poet and prose writer, lives. Works include collected stories *Portrait of the Artist as a Young Dog* (1940) and poem "Do not go gentle into that good night" (1952).

1914–72 John Berryman, American poet, lives. Works include *The Dream Songs* (1964), *Delusions* (1972), and *The Freedom of the Poet* (1976).

1914–86 Bernard Malamud, American novelist and short-story writer, lives. Works include *The Natural* (1952) and *The Fixer* (1966).

1914– William S. Burroughs, American novelist, lives. Works include *The Naked Lunch* (1959) and *The Ticket That Exploded* (1962).

1914– Ralph Ellison, American essayist, short-story writer, and novelist, lives. His best-known work is novel *The Invisible Man.*

1914– John Hersey, American novelist, lives. Works include *A Bell for Adano* (1944), *Hiroshima* (1946), and *The Wall (1950).*

1915– Saul Bellow, Canadian-born American novelist, lives. Works include *The Adventures of Augie March* (1953), *Herzog* (1964), and *Humboldt's Gift* (1975).

1916–90 Walker Percy, American novelist, lives. Works include *The Moviegoer* (1961), *The Last Gentleman* (1966), *Love in the Ruins* (1971), and *Second Coming* (1986).

1917–67 Carson S. McCullers, American novelist and short-story writer, lives. Works include *The Heart Is a Lonely Hunter* (1940) and *The Member of the Wedding* (1946).

1917–77 Robert Lowell, American poet, lives. Works include *Lord Weary's Castle* (1946), *Life Studies* (1959), *The Dolphin* (1973), and *Day by Day* (1977).

1917– Anthony Burgess, English novelist and critic, lives. Works include *A Clockwork Orange* (1962), *Earthly Powers* (1980), and *The Piano Players* (1986).

1918– Aleksandr Solzhenitsyn, Russian novelist, lives. Works include *One Day in the Life of Ivan Denisovich* (1962), *The Gulag Archipelago* (1973–75), and *The Oak and the Calf* (1980).

1918– Muriel Spark, English novelist and poet, lives. Works include *Memento Mori* (1959), *The Prime of Miss Jean Brodie* (1961), and *A Far Cry from Kensington* (1988).

1919–65 Shirley Jackson, American novelist and short-story writer, lives. Her best-known work is "The Lottery."

1919– Doris Lessing, English novelist and short-story writer, lives. Works include *The Grass is Singing* (1950), *The Golden Notebook* (1962), *African Stories* (1964), *The Sentimental Agents* (1983), and *The Fifth Child* (1988).

1919– Iris Murdoch, English novelist and philosopher, lives. Works include *Under the Net* (1954), *The Sandcastle* (1957), *A Severed Head* (1961), and *The Book and the Brotherhood* (1987).

1919– J. D. Salinger, American novelist and short-story writer, lives. Works include *The Catcher in the Rye* (1951) and *Franny and Zooey* (1961).

1920– Howard Nemerov, American poet and essayist, lives. Works include *The Image and the Law* (1947), *Guide to the Ruins* (1950), and *The Salt Garden* (1955).

1922–69 Jack Kerouac, American novelist and poet, lives. Works include *On the Road* (1957), *Big Sur* (1962), and *Desolation Angels* (1965).

1922– Kingsley Amis, English novelist and poet, lives. Works include *That Uncertain Feeling* (1955), *The Egyptologists* (1965), *The Alteration* (1976), and *The Old Devils* (1986).

▶

1922– Kurt Vonnegut, American novelist, lives. Works include *Cat's Cradle* (1963), *Slaughterhouse Five* (1969), and *Deadeye Dick* (1982).

1923– James Dickey, American poet, critic, and novelist, lives. Works include novel *Deliverance* (1970) and poetry volumes *The Zodiac* (1979) and *Puella* (1982).

1923– Norman Mailer, American writer, lives. Works include *The Naked and the Dead* (1948), *The Armies of the Night* (1968), and *The Executioner's Song* (1979).

1923– Nadine Gordimer, South African novelist, lives. Works include *A Guest of Honor* (1970), *Burger's Daughter* (1979), and *A Sport of Nature* (1987).

1923– Joseph Heller, American novelist, lives. Works include *Catch-22* (1961) and *Good as Gold* (1979).

1924–84 Truman Capote, American writer, lives. Works include *Other Voices, Other Rooms* (1948), *Breakfast at Tiffany's* (1958), *In Cold Blood* (1966), and *Music for Chameleons* (1983).

1924–87 James A. Baldwin, American novelist, lives. Works include *Go Tell It on the Mountain* (1953), *Notes of a Native Son* (1955), and *Evidence of Things Not Seen* (1986).

1925–64 Flannery O'Connor, American novelist and short-story writer, lives. Works include *Wise Blood* (1952) and *A Good Man Is Hard to Find* (1955).

1925–70 Yukio Mishima, Japanese novelist, lives. Works include *Confessions of a Mask* (1949) and *The Sailor Who Fell from Grace with the Sea* (1963).

1925– William Styron, American novelist, lives. Works include *Lie Down in Darkness* (1951), *Set This House on Fire* (1960), *The Confessions of Nat Turner* (1967), and *Sophie's Choice* (1979).

1925– Gore Vidal, American novelist and essayist, lives. Works include *Myra Breckenridge* (1968), *Burr* (1973), and *Lincoln* (1984).

1926– John Fowles, English novelist, lives. Works include *The Magus* (1965), *The French Lieutenant's Woman* (1969), and *A Maggot* (1985).

1926– J. P. Donleavy, American novelist, lives. Works include *The Ginger Man* (1955), *The Beastly Beatitudes of Balthazar B.* (1968), and *The Destinies of Darcy Dancer, Gentleman* (1977).

1926– Allen Ginsberg, American poet, lives. Works include *Kaddish and Other Poems* (1961).

1927– John Ashbery, American poet and art critic, lives. He is best known for his poetry collection *Self-Portrait in a Convex Mirror* (1975).

1927– Günter Grass, German novelist, lives. Works include *The Tin Drum* (1963) and *The Flounder* (1978).

1928–74 Anne Sexton, American poet, lives. Works include anthologies *To Bedlam and Part Way Back* (1960) and *The Awful Rowing Towards God* (1975).

1928– Maya Angelou, American novelist, poet, playwright, and short-story writer, lives. Her best-known work is novel *I Know Why the Caged Bird Sings* (1970).

1928– Gabriel Garcia Márquez, Colombian novelist, lives. Works include *One Hundred Years of Solitude* (1967) and *Love in the Time of Cholera* (1988).

1929– Adrienne Rich, American poet and writer, lives. Works include *Of Woman Born* (1976) and *On Lies, Secrets and Silence* (1976).

1929– Thom Gunn, English poet, lives. Works include *The Sense of Movement (1957)*, *Moly* (1971), and *The Passages of Joy* (1982).

1930– John Barth, American novelist and short-story writer, lives. Works include *The Floating Opera* (1956), *Giles Goat-Boy* (1966), and *Lost in the Funhouse* (1968).

1930– Chinua Achebe, Nigerian novelist and poet, lives. Works include *Things Fall Apart* (1958), *Arrow of God* (1964),

▶

and *Christmas in Biafra and Other Poems* (1973).

1930– Ted Hughes, English poet, lives. Works include *Meet My Folks* (1961) and *Crow* (1970).

1931–89 Donald Barthelme, American short-story writer, lives. Works include *Come Back, Dr. Caligari* (1964), *Unspeakable Practices, Unnatural Acts* (1968), and *Amateurs* (1976).

1931– E. L. Doctorow, American novelist, lives. Works include *Ragtime* (1975), *Loon Lake* (1980), and *World's Fair* (1985).

1931– Toni Morrison, American novelist, lives. Works include *Song of Solomon* (1977), *Tar Baby* (1981), and *Beloved* (1988).

1931– Fay Weldon, English novelist, lives. Works include *The Fat Woman's Joke* (1967) and *The Life and Loves of a She-Devil* (1983).

1932–63 Sylvia Plath, American poet and novelist, lives. Works include poetry collections *The Colossus* (1960) and *Ariel* (1965) and novel *The Bell Jar* (1963).

1932– V. S. Naipaul, Trinidadian novelist of Indian descent, lives. Works include *A House for Mr. Biswas* (1961) and *A Bend in the River* (1979).

1932– John Updike, American novelist, short-story writer, and poet, lives. Works include trilogy *Rabbit, Run* (1960), *Rabbit Redux*, and *Rabbit Is Rich* (1981) and *The Witches of Eastwick* (1985).

1933– Philip Roth, American novelist and short-story writer, lives. Works include *Goodbye, Columbus* (1959), *Portnoy's Complaint* (1969), *The Ghost Writer* (1979), and the *Anatomy Lesson* (1983).

1933– Susan Sontag, American critic and novelist, lives. Works include *The Benefactor* (1963), *Trip to Hanoi* (1969), and *Under the Sign of Saturn* (1980).

1934– Joan Didion, American novelist and essayist, lives. Works include *Slouching Toward Bethlehem* (1968) and *The White Album* (1979).

1935–89 Richard Brautigan, American poet and novelist, lives. Works include *A Confederate General from Big Sur* (1964) and *Trout Fishing in America* (1967).

1935– Ken Kesey, American novelist, lives. His best-known work is *One Flew Over the Cuckoo's Nest* (1962).

1937– Thomas Pynchon, American novelist, lives. Works include *V* (1963), *The Crying of Lot 49* (1966), *Gravity's Rainbow* (1973), and *Vineland* (1989).

1938– Joyce Carol Oates, American novelist, poet, short-story writer, and critic, lives. Works include novels *A Garden of Earthly Delights* (1967), *them* (1969), *Bellefleur* (1980), and *You Must Remember This* (1989) and short-story collection *The Wheel of Love* (1970).

1939– Margaret Atwood, Canadian novelist, poet, and short-story writer, lives. Works include *The Edible Woman* (1969), *The Handmaid's Tale* (1986), and *Cat's Eye* (1989).

1939– Margaret Drabble, English novelist and short-story writer, lives. Works include *A Summer Bird-Cage* (1962), *The Needle's Eye* (1972), and *The Ice Age* (1977).

1939– Seamus Justin Heaney, Irish poet, lives. Works include *Death of a Naturalist* (1966), *The Field Work* (1979), and *The Haw Lantern* (1987).

1941– Anne Tyler, American novelist, lives. Works include *Dinner at the Homesick Restaurant* (1982), *The Accidental Tourist* (1985), and *Breathing Lessons* (1988).

1944– Alice Walker, American poet, novelist, and short-story writer, lives. Works include poetry collection *Once: Poems* (1968) and novel *The Color Purple* (1982).

1947– Salman Rushdie, novelist, lives. Works include *Midnight's Children* (1981) and *The Satanic Verses* (1989).

Painters and Sculptors

Noted artists with selected major works.

c. 1220–c. 1280 Nicola Pisano, Italian Gothic sculptor, lives. Works include pulpit scenes for the Baptistery in Pisa and reliefs for the fountain at Perugia.

c. 1240–c. 1302 Cimabue (Cenni de Peppi), Gothic Florentine artist known as Father of Italian Painting, lives. Work includes famed altar panel *Madonna Enthroned* (c. 1280–90).

c. 1255–c. 1319 Duccio di Buoninsegna, Gothic Sienese painter, lives. Works include several panels from the Maestà Altar (1308–11), among them *Madonna Enthroned, Christ Entering Jerusalem*, and *Annuciation of the Death of the Virgin.*

1266–1337 Giotto, Florentine Gothic painter, lives. Works include *Death of St. Francis, Betrayal of Christ* (1306), *Meeting of Joachim and Anna* (c. 1306), and *The Madonna in Glory* (1310).

1285–1344 Simone Martini, Italian Gothic painter, lives. Works include *Maestà* (1315), *St. Martin Abandoning His Arms* (c. 1326), and *Annunciation* (1333).

c. 1370–c. 1440 Jan van Eyck, Late Gothic Flemish painter, lives. Works include *Annunciation* (1434), *Madonna of Chancellor Rolin* (1436), *Giovanni Arnolfini and His Bride* (1434), *A Man in a Red Turban* (c. 1433), *Cardinal Albergati* (1435), *The Lucca Madonna* (c. 1436), *St. Barbara* (1437), and *Madonna at the Fountain* (1439).

c. 1370–c. 1427 Gentile da Fabriano, Italian Late Gothic painter, lives. Works include *The Adoration of the Magi* (1423) and *The Nativity* (1423).

1386–1466 Donatello (Donato di Niccolo di Betto Bardi), Florentine Renaissance sculptor, lives. Works include *David* (1430–32), *General Gattamelata* (1443), and *Mary Magdalen* (1445).

1397–1475 Paolo Uccello, Florentine Renaissance painter, lives. Works include *The Creation of Adam, The Flood, The Recession of the Flood, The Fall* (1430s), *Battle of Sant' Egidio* (1432), *The Battle of San Romano* (1457), *St. George and the Dragon* (c. 1460), and *A Hunt in a Forest* (c. 1460).

1400–55 Fra Angelico, Florentine Renaissance painter, lives. Works include *The Martyrdom of St. Peter* (1429), *Madonna and Child* (1430), *Madonna of the Star* (1430), *The Last Judgment* (c. 1433), *Death and Assumption of the Virgin* (1434), *Coronation of the Virgin* (1435), *Lamentation* (1436), *Annunciation* (c. 1438), and *Madonna and Child Enthroned* (c. 1438).

c. 1400–64 Rogier van der Weyden, early Flemish painter, lives. Works include *Deposition* (c. 1435), *Annunciation* (c. 1435), *Madonna and Child in a Niche* (c. 1435), *Descent from the Cross* (1440), *Altar of the Crucifixion* (1450), and *Madonna and Child with Four Saints* (c.1450).

1401–28 Masaccio, Florentine Renaissance painter, lives. Works include *Virgin and Child* (1426), *Expulsion from Paradise* (c. 1426), *Adoration of the Magi* (1426), *The Trinity* (1427), *SS. John the Baptist and Jerome* (c. 1428), and *The Tribute Money* (c. 1428).

c. 1406–69 Fra Filippo Lippi, Florentine Renaissance painter, lives. Works include *Madonna and Child* (1437), *Madonna and Child Enthroned* (c. 1437), *The Annunciation* (1443), *Adoration of the Magi* (c. 1457), and *Seven Saints* (1460).

▶

Art of Ancient Greece and Rome

B.C.

c. 1100–700 Geometric style, beginning with simple linear patterns, develops. First distinctly Greek artistic scheme, it appears on pottery and small sculptures. Most sophisticated examples appear in ninth and eighth centuries.

c. 700 Athenians adopt black-figure style of vase painting, originated by Corinthian artisans.

c. 700–600 Oriental influence makes Greek Geometric style more fluid. Straight lines give way to curvilinear forms, and human and animal representations appear.

c. 600 Earliest statues of Kouros (standing youth) are sculpted, marking beginning of emerging naturalistic style of sculpture in Archaic period.

c. 550 Famed Greek vase painter Exekias produces some of the finest surviving examples of black-figure pottery, including kylix *Dionysus in a Boat.*

c. 525 Red-figure style of vase painting emerges, showing three-dimensional space and naturalism.

c. 510 Greek painter and potter Euphronius produces his painted vase of Hercules wrestling Antaeus.

c. 510 Euthymides, noted Athenian red-vase painter, produces his amphora decorated with "Revelers."

c. 480–450 Early Classical period: Humanistic elements take precedence over formalistic design.

c. 470 *Charioteer*, rare life-size bronze in early Classical style, is crafted.

c. 456 Doric Temple of Zeus is built and adorned with some of best examples of early Classical sculpture.

450 Bronze *Poseidon* is completed. Lost in shipwreck off Cape Artemision, it is rediscovered in modern times.

c. 450 Bronze statue *Discus Thrower* is completed by Greek sculptor Myron (fl. c. 450–440 B.C.), noted for his studies of athletes in action.

c. 450–400 Golden age in Greek arts.

437–c. 432 Parthenon is built in Athens. Phidias (c. 500–432 B.C.), great Classical sculptor, supervises decorative sculptures. His frieze and one of several caryatids are among "Elgin Marbles," removed in modern times by a British collector and now on view at British Museum, London.

c. 359–351 Scopas creates frieze *Battle of the Greeks and Amazons* for Mausoleum at Helicarnassus. Scopas is noted for introducing powerful emotion into Greek sculpture.

c. 330 Praxiteles (c. 370–330 B.C.) sculpts *Aphrodite of Cnidus.* He blends realism with graceful elegance in his figures, and his style is imitated in later centuries.

c. 310 Lysippus produces his renowned bronze of athlete (known from Roman copy *Apoxyomenos*). He is said to have produced over 1,500 bronzes.

c. 230–220 *Dying Gaul* is sculpted by unknown artist. It shows concern for realism that marks the Hellenistic period, the last great phase of ancient Greek culture.

c. 200–190 Production of Hellenistic masterpiece *Victory of Samothrace.*

c. 200–190 *Venus de Milo*, most famous female figure in Greek sculpture, is produced; it probably is modeled on Praxiteles's *Aphrodite of Cnidus.*

c. 200–100 Incrustation style of wall painting prevails in Roman houses. Plaster is painted to resemble marble.

c. 100 Exceptional floor mosaic *The Defeated Persians Under Darius* is installed in house in Pompeii.

c. 80 Portrait of a Roman is sculpted. Roman sculptors perfect portrait heads with truly Roman character, revealing generalized national character in guise of individual portrait.

c. 38 Noted Hellenistic Greek statue *Laocoön*, celebrating a legend, is produced by Greek sculptors Agesander, Athenodorus, and Polydorus.

c. 20 Roman sculptor invests portrait of his emperor, *Augustus of Prinicipata*, with appearance of divinity.

c. 25 Series of eight wall paintings, *Odyssey Landscapes*, illustrates Roman architectural style of painting, in which scenes are framed, stagelike, by columns or other architectural detail.

c. 10 Ornate style of painting replaces architectural style in Roman wall paintings. Architectural

Art of Ancient Greece and Rome

	borders are now secondary to painting within them.
A.D.	
c. 50	Intricate style dominates wall painting at Pompeii, with architectural motifs fading and brilliantly colored scenes dominating.
81	*Triumphal Arch of Titus* is erected in Rome, commemorating foreign victories of Emperor Titus. Its relief sculpture marks apex of illusionism in this art form.
c. 90	Completion of marble *Portrait of a Lady*, perhaps most finely executed female character study in Roman art.
c. 100	Life-size marble head *Trajan* is sculpted. It is another Roman masterpiece of character study.
106–133	*Trajan's Column* is erected in Rome to celebrate emperor's victory against Dacians. It is famous for great height (125 feet) and for its continuous spiraling relief sculpture depicting events of campaign.
161–180	Bronze *Equestrian Statue of Marcus Aurelius* is produced. It reflects Rome's renewed interest in Greek ideals.
312–315	Completion of relief sculpture *The Arch of Constantine* marks end of purely Roman, non-Oriental, monumental sculpture. Sculptural style here points toward developing Christian artworks.

1416–92 Piero della Francesca, Italian Renaissance painter, lives. Works include *The Baptism of Christ* (c. 1445), *Flagellation of Christ* (1450s), *Resurrection* (c. 1463), *Battle of Constantine* (1466), *The Legend of the True Cross* (1466), and *The Nativity* (c. 1480).

1426?–1516 Giovanni Bellini, Venetian Renaissance painter, lives. Works include *Agony in the Garden* (1460s), *Dead Christ with Angels* (c. 1468), *Madonna and Child* (1470s), *Madonna of the Lake* (c. 1488), *Madonna and Child with St. Paul and St. George* (1490), *Portrait of a Young Man* (1490s), *Miracle of the True Cross* (1500), *Doge Leonard Loredano* (c. 1503), and *Madonna* (1510).

1430?–94 Hans Memlinc, Flemish painter, lives. Works include *The Last Judgment* (c. 1473), *The Deposition* (c. 1475), *Adoration of the Magi* (1479), *The Descent from the Cross* (1480), *Bathsheba* (c. 1485), *Madonna and Child* (1487), and *Resurrection* (c. 1490).

1430–95 Carlo Crivelli, Italian painter, lives. Works include *Virgin and Child* (c. 1470) and *Madonna and Child with Angels* (1489).

1431–1506 Andrea Mantegna, Italian Renaissance painter, lives. Works include *Agony in the Garden* (c. 1450), *Virgin with Sleeping Child* (c. 1450), *St. James Led to His Execution* (c. 1452), *The Crucifixion* (1459), *Portrait of a Man* (c. 1460), *Dead Christ* (1466), *St. George* (c. 1462), *Saint Sebastian* (c. 1465), *The Madonna of the Caves* (c. 1484), *Parnassus* (1497), and *Wisdom Overcoming the Vices* (1502).

1435–88 Andrea del Verrocchio, Florentine Renaissance sculptor and painter, lives. Works include *Equestrian Monument of Colleoni* (c. 1483–88).

1444–1510 Sandro Botticelli, Florentine Renaissance painter, lives. Works include *Fortitude* (1470), *Primavera* (c. 1478), *Pallas and the Centaur* (c. 1485), *Mars and Venus* (c. 1485), *Madonna of the Pomegranate* (1487), *Birth of Venus* (1487), *The Coronation of the Virgin* (1490), *St. Augustine in His Cell* (c. 1495), *The Nativity* (c. 1500), and *Three Miracles of St. Zenobius* (1505).

c. 1450–1516 Hieronymous Bosch, Dutch late Gothic painter, lives. Works include *Ship of Fools* (c. 1500), *Adoration of the Magi, Christ Bearing the Cross, Garden of Delights, Temptation of St. Anthony,* and *The Haywain.*

▶

1452–1519 Leonardo da Vinci, Italian Renaissance painter, lives. Works include *Ginevra de'Benci* (1474), *The Benois Madonna* (1480), *Adoration of Kings* (1482), *The Last Supper* (1497), *Adoration of the Magi* (c. 1481), *Portrait of Cecilia Gallerani* (c. 1483), *St. Jerome* (c. 1483), *Virgin of the Rocks* (c. 1490), *Virgin and Child with St. Anne* (1499), *Virgin with Her Son and St. Anne* (c. 1503), *Mona Lisa* (1503), *Virgin of the Rocks* (1508), *Holy Family* (1506), *Virgin of the Balances* (c. 1506), *Madonna of the Rocks* (c. 1508), and *St. John the Baptist* (c. 1512).

1460–1523 Gerard David, Flemish painter, lives. Works include *Judgment of Cambyses* (1498), *Virgin and Child with Donors and Patron Saints* (c. 1505), *Baptism of Christ* (1507), *Madonna with Angels and Saints* (1509), and *The Crucifixion* (1515).

1460–1526 Vittore Carpaccio, Venetian painter, lives. Works include *Blood of the Redeemer* (1496), *Dream of St. Ursula* (c. 1495), *Life of St. Jerome* (c. 1502), *Presentation in the Temple* (1510), and *Two Courtesans on a Balcony* (c. 1510).

1460–1524 Hans Holbein the Elder, German painter, lives. His masterpiece is the altar of St. Sebastian in Munich (1515).

c. 1460–1528 Matthias Grünewald, German painter, lives. Works include *Mocking of Christ* (1503), *The Crucifixion* (1511), *Miracle of the Snow* (1519), and *SS. Erasmus and Mauritius* (1523).

1471–1528 Albrecht Dürer, German painter and engraver, lives. Works include *Adoration of the Magi* (1504), *Salvator Mundi* (1502), *Adam and Eve* (1507), *Adoration of the Trinity* (1511), and *Virgin and Child* (1516).

1473–1553 Lucas Cranach the Elder, German painter, lives. Works include *The Crucifixion of Christ* (1502), *Christ on the Cross* (1503), *John, Duke of Saxony* (1518), *Adam and Eve* (1526), *Judgment of Paris* (1530), *Paradise* (1530), *Venus* (1532), and *Nymph of the Spring* (c. 1537).

1475–1564 Michelangelo Buonarroti, Italian Renaissance painter and sculptor, lives. Works include *Holy Family* (1505), *Delphic Sibyl* (1509), *Jeremiah* (1512), *Last Judgment* (1541), *Conversion of St. Paul* (1545), and *Crucifixion of St. Peter* (1550).

1477–1511 Giorgio Giorgione, Venetian Renaissance painter, lives. Works include *Judgment of Solomon* (1500), *The Tempest* (1505), *Fugger Youth* (c. 1505), *Laura* (1506), *Epiphany* (1508), *The Three Philosophers* (c. 1510), *Sleeping Venus* (c. 1510), and *Pastoral Symphony* (c. 1510).

c. 1487–1576 Titian, Venetian Renaissance painter, lives. Works include *The Gypsy Madonna* (1510), *Sacred and Profane Love* (1515), *Assumption of the Virgin* (1518), *The Worship of Venus* (1518), *Man with the Glove* (1520), *Bacchus and Ariadne* (1523), *Madonna with the Rabbit* (1530), *Presentation of the Virgin* (1538), *The Pardo Venus* (1540), *Christ Crowned with Thorns* (1550), *Venus and Cupid* (1550), *Self-Portrait* (1550), *Venus and Adonis* (1554), *Venus with the Mirror* (1555), *Perseus and Andromeda* (1556), *Martyrdom of St. Lawrence* (1557), *Diana and Actaeon* (1559), *Diana and Callistro* (1559), *Venus and the Lute Player* (1560), *The Rape of Europa* (1562), *The Last Supper* (1564), *Annunciation* (1565), *Portrait of Jacopo Strada* (1568), and *Pietà* (1576).

1494–1534 Antonio da Correggio, Italian Renaissance painter, lives. Works include *Madonna of Saint Frances* (1514), *Four Saints: Peter, Martha, Mary Magdalen,* and *Leonard* (1517), *Jupiter and Antiope* (1522), *Madonna of St. Jerome* (c. 1523), *Jupiter and Io* (c. 1530), and *School of Love* (1532).

▶

1497–1543 Hans Holbein the Younger, German Renaissance painter, lives. Works include *The Last Supper* (1520), *Christ in the Tomb* (1521), *Erasmus of Rotterdam* (1523), *Sir Thomas More* (1526), *Sir Henry Guilford* (1527), *Ambassadors* (1533), *Derick Born* (1533), *Henry VIII* (1536), *Jane Seymour* (1537), *King Henry VIII* (1539), and *Edward VI as Prince of Wales* (c. 1543).

1500–71 Benvenutto Cellini, Italian Mannerist sculptor, lives. Works include bust of Cosimo I and *Perseus with the Head of Medusa* (1546–54).

c. 1515–69 Pieter Bruegel the Elder, Flemish painter, lives. Works include *Battle of Carnival and Lent* (1559), *Tower of Babel* (1563), *Adoration of the Kings* (1564), *Wedding Dance* (c. 1566), *Conversion of St. Paul* (1567), *Birds Nesting* (1568), *The Cripples* (1568), *Children's Games* (1569), and *Massacre of the Innocents* (1569).

1518–94 Tintoretto, Venetian Mannerist painter, lives. Works include *Apollo and Marsyas* (1545), *Self-Portrait* (c. 1546), *The Last Supper* (1547), *Christ Washing the Feet of the Apostles* (1547), *Miracle of St. Mark* (1548), *Susanna Bathing* (c. 1550), *Knight of Malta* (c. 1551), *The Finding of Moses* (1555), *Portrait of Gentlemen with a Gold Chain* (c. 1560), *The Last Judgment* (c. 1562), *The Crucifixion* (1565), *Bacchus and Ariadne* (1578), and *Paradise* (1590).

1528–88 Paolo Veronese, Venetian Realist painter, lives. Works include *Temptation of St. Anthony* (1552), *Transfiguration* (1556), *Portrait of a Lady and Her Daughter* (c. 1556), *Baptism of Christ* (1561), *Marriage at Cana* (1563), *Feast in the House of Levi* (1573), *Adoration of the Kings* (1573), and *The Magdalen* (1583).

1541–1614 El Greco (Domenico Theotocopoulos), Spanish Mannerist painter, lives. Works include *Christ Healing the Blind* (1570), *Annunciation* (1576), *Trinity* (1579), *Knight Taking an Oath* (1580), *St. Martin and the Beggar* (1590), *Christ Carrying the Cross* (1590), *View of the City of Toledo* (1595), *The Burial of Count de Orgaz* (1588), *Purification of the Temple* (c. 1610), and *Adoration of the Shepherds* (1612).

1564–1638 Pieter Bruegel the Younger, Flemish painter, lives. Works include *The Whitsun Bride, Village Fair,* and *Crucifixion.*

1568–1625 Jan Bruegel, Flemish painter, lives. Works include *Adam and Eve in Paradise* and *Windmill.*

1573–1609 Michelangelo Caravaggio (Merisida), Italian Baroque painter, lives. Works include *Rest on the Flight to Egypt* (1590), *Boy with Fruit* (1593), *The Fortune-Teller* (1594), *The Lute-Player* (1596), *Sacrifice of Isaac* (1596), *The Calling of St. Matthew* (1597), *St. Matthew and the Angel* (1600), *The Conversion of St. Paul* (1601), *The Crucifixion of St. Peter* (1601), *The Deposition* (1602), *Madonna De Loretto* (1606), *St. Jerome* (1604), *Death of the Virgin* (1606), *Madonna Dei Palafrenieri* (1606), *Salome with the Head of St. John* (1607), *Cupid Asleep* (1608), *Beheading of St. John the Baptist* (1608), and *Raising of Lazarus* (1609).

1577–1640 Peter Paul Rubens, Flemish Baroque painter, lives. Works include *Elevation of the Cross* (1611), *The Descent from the Cross* (1614), *The Holy Family* (c. 1615), *Arrival of Marie de Medici at Marseilles* (1625), *Marie de Medici* (c. 1625), *Philip II of Spain* (c. 1629), *The Ildefonso Altarpiece* (1632), *The Garden of Love* (1634), *The Rape of the Sabines* (1635), *Venus and Adonis* (c. 1635), *Sunset* (c. 1638), *The Judgment of Paris* (1639), *The Massacre of the Innocents* (c. 1639), and *Portrait of Himself* (1640).

c. 1580–1666 Franz Hals, Dutch portrait painter, lives. Works include *Banquet of the Officers of the Civil Guard of St. George* (1616), *Yonker Ramp and His Sweetheart*

▶

(1623), *The Laughing Cavalier* (1624), *The Merry Toper* (1630), *Gypsy Girl* (1630), *A Burgomaster* (c. 1633), *Man with a Beer Keg* (1635), *Portrait of a Young Man* (c. 1645), *Women Regents of the Old Men's Home* (1664), and *Man in a Slouch Hat* (c. 1666).

1594–1665 Nicholas Poussin, French classical painter, lives. Works include *Orpheus and Eurydice* (c. 1626), *Triumph of David* (c. 1626), *Triumph of Flora* (1635), *Kingdom of Flora* (1635), *Triumph of Neptune* (1635), *Bacchanalian Revel Before a Term of Pan* (1638), *Seven Sacraments* (1642), *The Funeral of Phocion* (1648), *Self-Portrait* (1660), and *Four Seasons* (1664).

1598–1680 Giovanni Lorenzo Bernini, Italian Baroque sculptor and architect, lives. Works include *David* (1623) and *The Ecstasy of St. Theresa* (1645–52).

1599–1641 Anthony van Dyck, Flemish portrait and religious painter, lives. Works include *Carrying of the Cross* (c. 1617), *Wife of Peter Paul Rubens* (1621), *The Taking of Christ* (c. 1621), *Self-Portrait* (c. 1621), *Portrait of François Dequesnoy* (1623), *Madonna of the Rosary* (1628), *Rinaldo and Armida* (1629), *Christ on the Cross* (c. 1630), *Samson and Delilah* (1630), *Prince Ruprecht of the Palatinate* (c. 1631), *Lamentation for Christ* (1634), *Portrait of Thomas Killigrew and Thomas Carew* (1638), and *King Charles on Horseback* (1638).

1599–1660 Diego Rodriguez de Silva y Velázquez, Spanish Baroque painter, lives. Works include *Christ in the House of Martha and Mary* (1619), *Triumph of Bacchus* (1629), *The Forge of Vulcan* (1630), *Prince Baltasar Carlos on His Pony* (1634), *Philip IV of Spain* (1635), *The Surrender of Breda* (1635), *Lady with a Fan* (1640), *The Coronation of the Virgin* (1641), *Mars* (1648), *The Toilet of Venus* (1651), *Infant Margarita in a Pink Dress* (1654), *The Maids of Honor* (1656), *The Fable of Arachne* (1655), *Mercury and Argos* (1659), and *Prince Felipe Prospero* (1659).

1600–82 Claude Lorrain, French landscape painter, lives. Works include *Harbor Scene* (1634), *Landscape: The Marriage of Isaac and Rebekah* (1648), *Pastoral Landscape* (1651), and *Coast Scene with Perseus* (1674).

1607–69 Paul Rembrandt van Rijn, Dutch Baroque painter, lives. Works include *The Money-Changer* (1627), *Christ at Emmaus* (1629), *The Anatomy Lesson of Dr. Nicolaes Tulp* (1632), *The Blinding of Samson* (1636), *Angel Leaving Tobias and His Family* (1637), *The Stone Bridge* (c. 1638), *Maria Trip* (1638), *The Painter's Mother* (1639), *Holy Family* (1640), *A Girl at the Window* (1645), *The Man with the Golden Helmet* (c. 1650), *Large Self-Portrait* (1652), *Aristotle Contemplating the Bust of Homer* (1653), *Woman Bathing* (1654), *Bathsheba* (1654), *Polish Rider* (1655), *The Flayed Ox* (1655), *Portrait of an Old Man* (1657), *Portrait of a Man Holding a Manuscript* (1658), *The Small Self-Portrait* (1658), *Self-Portrait* (1659), *St. Peter's Denial* (1660), *Board of the Clothmakers' Guild* (1662), *Family Portrait* (1669), and *The Return of the Prodigal Son* (c. 1669).

1617–82 Bartolomé Estéban Murillo, Spanish religious and portrait painter, lives. Works include *Virgin of the Rosary* (1642), *Immaculate Conception* (1652), *St. Leandro* (1655), *St. Isidoro* (1655), *Vision of St. Anthony* (1656), *The Pie-Eater* (1675), *Soult Immaculate Conception* (1678), and *Two Trinities* (1681).

1626–79 Jan Steen, Dutch genre painter, lives. Works include *The Lovesick Maiden* (1655), *Skittle Players Outside an Inn* (1660), *Morning Toilet* (1663), *Merry Company* (1663), *Dissolute Life* (1663), *The World Upside Down* (1663), *Feast of St. Nicholas Day* (1667), and *Serenade* (c. 1675).

1632–75 Jan Vermeer, Dutch genre and landscape painter, lives. Works include *Officer and Laughing Girl* (1657), *View of Delft* (1660), *Young Woman with a*

▶

Water Jug (1660), *Lady Weighing Gold* (1663), *Girl with a Flute* (1665), *Girl with a Red Hat* (1665), *Head of a Girl* (1665), *The Lace Maker* (1665), *Lady Writing a Letter with her Maid* (1667), and *Painter in His Studio* (1670).

1638–1709 Meyndert Hobbema, Dutch landscape painter, lives. Works include *The Watermill* (1665), *Entrance to a Village* (c. 1665), *Water Mill with the Great Red Roof* (c. 1670), and *Ruins of Brederode Castle* (1671).

1669–1779 Jean-Baptiste-Siméon Chardin, French painter, lives. Works include *Le Buffet* (1728), *Scouring Maid* (1738), *Jar of Olives* (1760), *Dessert* (1763), *Wild Duck* (1764), *Attributes of the Ants* (1765), *Madame Chardin* (1775), and *Portrait of the Painter* (1775).

1684–1721 Jean-Antoine Watteau, French Baroque painter, lives. Works include *La Conversation* (1715), *Jupiter et Antiope* (1716), *The French Comedy* (1716), *Les Deux Cousines* (1717), *Embarkation for the Isle of Cythera* (1717), *Champs-Élysées* (1719), *Italian Comedians* (1720), and *La Toilette* (1720).

1696–1770 Giovanni Tiepolo, Venetian Baroque painter, lives. Works include *Repudiation of Hagar* (1719), *The Force of Eloquence* (c. 1725), *The Adoration of the Christ Child* (1732), *Time Revealing Truth* (1745), *Story of Anthony and Cleopatra* (1750s), *Olympus* (1753), and *Apotheosis of the Pisani Family* (1762).

1697–1764 William Hogarth, English painter, satirist, and engraver, lives. Works include *Beggar's Opera* (1728), *Fishing Party* (c. 1730), *Good Samaritan* (1735), *The Rake's Progress* (1735), *Captain Thomas Coran* (1740), *The Graham Children* (1742), *Self-Portrait with His Dog Trump* (1745), *Ascension* (1756), *Hogarth's Servants* (c. 1758), and *The Shrimp Girl* (1759).

1703–70 François Boucher, French Rococo painter, lives. Works include *The Interrupted Sleep* (1750), *Toilet of Venus* (1751), *Vulcan Presenting to Venus the Arms of Aeneas* (1757), and *Rustic Landscape with Figures by a Stream* (1768).

1712–93 Francesco Guardi, Venetian landscape and architectural painter, lives. Works include *Piassetta, Venice*, *Arsenal at Venice* (1780s), *Piazza St. Marco in Venice* (1780s), and *Santa Maria Della Salute* (1793).

1723–92 Joshua Reynolds, English portrait painter, lives. Works include *Lady Chambers* (1752), *Nelly O'Brien* (1762), *Dr. Johnson* (1772), *William Robertson* (1772), *Self-Portrait* (1773), *St. John in the Wilderness* (1776), *John Musters* (1780), and *Heads of Angels* (1787).

1727–88 Thomas Gainsborough, English portrait painter, lives. Works include *The Charterhouse* (1748), *Cornard Wood* (1748), *Mr. and Mrs. Andrews* (1750), *Peasants Returning from Market* (1767), *The Blue Boy* (1770), *The Harvest Wagon* (1770), *The Honorable Mrs. Graham* (1777), *The Cottage Door* (1780), *Mrs. Robinson* (1781), *Queen Charlotte* (1781), *Duchess of Devonshire* (1783), *Mrs. Siddons* (1784), *Mrs. Richard Brinsley Sheridan* (c. 1785), *The Morning Walk* (1785), *The Market Cart* (1786), and *Cottage Children* (1787).

1732–1806 Jean Fragonard, French Rococo painter, lives. Works include *The See-Saw* (c. 1752), *Blindman's Buff* (1752), *The Washerwoman* (1765), *The Storm* (1765), *Women Bathing* (1765), *Inspiration* (1766), *The Music Lesson* (c. 1770), *The Swing* (c. 1766), *The Cascade* (1773), *Storming the Citadel* (1772), *The Schoolmistress* (1778), and *The Stolen Kiss* (1788).

1737–1815 John Singleton Copley, American portrait painter, lives. Works include *John Hancock* (1765), *Boy with a Squirrel* (1765), *Nathaniel Hurd* (1765), *Mrs. Thomas Boylston* (1766), *Colonel William Montresor* (1772), *Watson and the Shark* (1778), and *The Copley Family* (1785).

▶

1738–1820 Benjamin West, American painter, lives. Works include *Penn's Treaty with the Indians* (1771), *Saul and the Witch of Endor* (1777), *Franklin Drawing Electricity from the Sky* (1805), *Robert Fulton* (1806), and *Death on the Pale Horse* (1817).

1741–1827 Charles William Peale, American portrait painter, lives. Works include *Peale Family Group* (1773), *The Staircase Group* (1795), *Exhuming of the Mastodon* (1806), and *James Peale* (1822).

1746–1828 Francisco de Goya y Lucientes, Spanish Neo-Baroque painter, lives. Works include *The Nude Maja* (c. 1797), *The Family of Charles IV* (1800), and *The Milkmaid of Bordeaux* (1827).

1748–1825 Jacques Louis David, French classical painter, lives. Works include *The Death of Socrates* (1787), *The Oath of the Hora III, Death of Marat* (1793), and *Charlotte De Val D'Ognes* (c. 1795).

1755–1828 Gilbert Stuart, American portrait painter, lives. Works include *George Washington* (1793), *Mrs. Richard Yates* (1793), *Thomas Jefferson* (1799), *Washington at Dorchester Heights* (1806), and *Admiral Sir Isaac Coffin* (1810).

1756–1823 Sir Henry Raeburn, Scottish portrait painter, lives. Works include *The Archers* (c. 1788), *Sir John and Lady Clerk* (1790), *Sir John Sinclair* (1795), *Lord Newton* (1811), and *The McNab* (1813).

1757–1827 William Blake, English Romantic painter and poet, lives. He is best known for his illustrations of Milton and the Book of Job (c. 1790–1805).

1775–1851 Joseph Mallord William Turner, English Romantic painter, lives. Works include *Calais Pier* (1803), *Frosty Morning* (1813), *Regatta at Cowes* (1828), *Evening Star* (c. 1830), *Grand Canal, Venice* (c. 1835), *The Slave Ship* (1839), *The Fighting Temeraire* (1839), and *Rain, Steam, and Speed* (1844).

1776–1837 John Constable, English landscape painter, lives. Works include *Wivenhoe Park, Essex* (1816), *The White Horse* (1819), *The Hay Wain* (1821), *Salisbury Cathedral from the Bishop's Garden* (1823), *The Leaping Horse* (1825), *View at Stoke-by-Nayland* (c. 1830), and *Salisbury Cathedral* (1831).

1779–1843 Washington Allston, American painter, lives. Works include *Bellshazzar's Feast* (1804), *Rising of a Thunderstorm at Sea* (1804), *Classical Landscape* (1808), *Coast Scene on the Mediterranean* (1811), *Dead Man Revised* (1813), *Elijah Fed by the Ravens* (1818), *Flight of Florimell* (1819), and *Moonlight Landscape* (1819).

1780–1867 Jean-Auguste Dominique Ingres, French painter, lives. Works include *Louis Bertin* (1832), *La Source* (1856), *Madame Moitessier, Seated* (1856), *Odalisque* (1858), and *Turkish Women at the Bath* (1862).

1790–1824 Jean Louis André Théodore Géricault, French painter, lives. Works include *Mounted Officer of the Imperial Guard* (1812) and *The Raft of the "Medusa"* (c. 1817).

1796–1875 Jean-Baptiste Camille Corot, French landscape painter, lives. Works include *The Inn* (1831), *Hagar in the Wilderness* (1835), *Bacchante by the Sea* (1865), *The Sleep of Diana* (1868), and *The Gypsies* (1872).

1799–1863 Eugène Delacroix, French Romantic painter, lives. Works include *Dante and Virgil in Hell* (1821), *Massacre at Scio* (1824), *Death of Sardanapalus* (1827), *Liberty Leading the People* (1830), *Oriental Lion Hunt* (1834), *George Sand's Garden at Nohant* (1841), *Abduction of Rebecca* (1846), and *Christ on the Sea of Galilee* (1854).

1801–48 Thomas Cole, English-born American landscape painter, lives. Works include *Voyage of Life: Manhood, Landscape with Tree Trunks* (c. 1827), *Expulsion from the Garden of Eden* (1828), *The Architest's Dream* (1840), and *Roman Campagna* (1843).

▶

1808–79 Honoré Daumier, French painter, caricaturist, and engraver, lives. Works include *The Drinkers* (1856), *Third-Class Carriage* (c. 1862), *Washwoman* (c. 1863), and *Don Quixote* (c. 1868).

1811–77 George Caleb Bingham, American painter, lives. Works include *Raftsmen Playing Cards* (1850), *Shooting for the Beef* (1850), *The Trappers' Return* (1851), and *Verdict of the People* (1855).

1814–75 Jean Françoise Millet, French Realist painter, lives. Works include *The Gleaners* (1850), *The Sower* (1850), *Angelus* (1859), and *Man with the Hoe* (1863).

1819–77 Gustave Courbet, French Realist painter, lives. Works include *After the Hunt* (1846), *Man with a Pipe* (1846), *The Stone Breakers* (1850), *Les Demoiselles de Village* (1852), *The Winnowers* (1854), *The Artist's Studio* (1855), *Young Women on the Banks of the Seine* (1856), *The Bride at Her Toilet* (1859), *Sleeping Woman* (1866), *Woman with a Parrot* (1866), *The Source* (1868), *Calm Sea* (1869), and *Marine—the Waterspout* (1876).

1822–99 Rosa Bonheur, French painter, lives. Works include *The Horse Fair* (1855).

1824–98 Eugène Boudin, French landscape painter, lives. Works include *On the Beach at Trouville* (1863), *Canal in Brussels* (1871), *Beaulieu Baie de Fourmis* (1892), and *Port of Dieppe* (1896).

1828–82 Dante Gabriel Rossetti, English Pre-Raphaelite painter, lives. Works include *Found* (1854), *Beata Beatrix* (1863), *Lily Lileth* (1863), and *Proserpine* (1874).

1829–96 Sir John Everett Millais, English Pre-Raphaelite painter, lives. Works include *Christ in the Carpenter's Shop* (1850).

1831–1903 Camille Pissarro, French Impressionist painter, lives. Works include *Still Life* (1867), *The Road* (1870), *The Little Bridge* (1875), *Towpath at Pontoise* (1882), *Street in Rouen* (1898), and *Bridge at Bruges* (1903).

1832–83 Édouard Manet, French pre-Impressionist painter, lives. Works include *The Old Musician* (1862), *Luncheon on the Grass* (1863), *Olympia* (1863), *The Fifer* (1866), *Blond Half-Nude* (1878), *Bar at the Folies-Bergère* (1881), and *Springtime* (1881).

1834–1903 James Abbott McNeill Whistler, American painter, lives. Works include *At the Piano* (1854), *Last of Old Westminster* (1862), *The Little White Girl* (1862), *Battersea Reach* (1865), *Valparaiso* (1866), *The Ocean* (1867), *Arrangement in Gray and Black* (1871), *Thomas Carlyle* (1872), *Nocturne in Black and Gold* (c. 1874), *The Lagoon, Venice* (1880), *The Lifeboat* (c. 1884), and *The Angry Sea* (1884).

1834–1917 Edgar Degas, French Impressionist painter, lives. Works include *Woman with Chrysanthemums* (1865), *Woman Drying Her Foot* (1866), *Portrait of a Young Woman* (1867), *James Tissot* (1868), *Le Foyer* (1872), *The Dancing Class* (1872), *Pouting* (1875), *The Rehearsal* (1877), *Dancer Practicing at the Bar* (1877), *Blue Dancer* (1883), *Two Laundresses* (1884), *Woman Bathing* (1885), *The Dancers* (1899), and *Ballet Scene* (1907).

1836–1910 Winslow Homer, American landscape painter, lives. Works include *Prisoners from the Front* (1866), *The Croquet Match* (1866), *Long Branch, New Jersey* (1869), *Country School* (1871), *Breezing Up* (1876), *Rowing Home* (1890), *Huntsman and Dogs* (1891), *All's Well* (1896), *Gulf Stream* (1899), and *Early Morning After a Storm at Sea* (1902).

1839–99 Alfred Sisley, French Impressionist landscape painter, lives. Works include *Banks of the Seine* (1863), *Land Near a Small Town* (1863), *Street at Marly* (1863), *Banks of the Oise* (1863), *Island of La Grande-Jatte* (1873), and *The Watering Place* (1874).

▶

1839–1906 Paul Cézanne, French painter, lives. Works include *Portrait of Boyer* (1887), *The Blue Vase* (1887), *The Bather* (1887), *The Card Players* (1892), *Group of Bathers* (1894), *Still Life with Basket of Apples* (1894), *The Clock Maker* (1896), *Still Life* (1900), and *Woman Bathers* (1906).

1840–1917 Auguste Rodin, French Impressionist sculptor, lives. Works include *The Thinker* (1879–89), *The Kiss* (1886–98), and *Balzac* (1892–97).

1840–1926 Claude Monet, French Impressionist painter, lives. Works include *Still Life* (c. 1859), *Terrace at the Seaside Near Le Havre* (1865), *Women in the Garden* (1867), *The River* (1868), *Basin at Argenteuil* (1872), *Impression: Fog* (1872), *Wild Poppies* (1873), *Gare St. Lazare* (1877), *Haystacks at Giverny* (1884), *The Haystacks* (1891), and *Water-Lily Pond* (1906).

1841–1919 Pierre-Auguste Renoir, French Impressionist painter, lives. Works include *Lise* (1867), *Odalisque* (1870), *Dancer* (1874), *Le Moulin de la Galette* (1876), *Oarsmen at Chatou* (1879), *Little Blue Nude* (1880), *The Loge* (1883), *Dance at Bougival* (1883), *Bather* (1892), *Sleeping Bather* (1897), and *Seated Bather* (1913).

1844–1910 Henri Rousseau, French primitive painter, lives. Works include *Myself, Portrait-Landscape, Prague* (1890), *Storm in the Jungle* (1891), *The Sleeping Gypsy* (1897), and *The Dream* (1910).

1845–1926 Mary Cassatt, American Impressionist painter, lives. Works include *Young Mother Sewing* (1880), *Woman Reading in a Garden* (1880), *La Toilette* (1894), *The Family* (c. 1887), and *Woman with a Dog* (c. 1889).

1848–1903 Paul Gauguin, French Symbolist painter, lives. Works include *Belle Angele* (1889), *The Yellow Christ* (1889), *Ia Orana Maria: Hail Mary* (1891), *Revière* (1891), *Hina Te Fatou: The Moon and the Earth* (1893), *Poèmes Barbares* (1896), *Te Arii Vahine: The Woman with Mangoes* (1896), *Whence Do We Come? What Are We? Where Are We Going?* (1898), *The White Horse* (1898), *Two Tahitian Women* (1899), *Golden Bodies* (1901), *L'Appel* (1902), and *Apparition* (1902).

1853–90 Vincent van Gogh, Dutch Expressionist painter, lives. Works include *The Loom* (1884), *Potato-Eaters* (1885), *Self-Portrait* (1886), *Skull with Cigarette* (1886), *Bread* (1888), *La Mousme* (1888), *The Night Cafe* (1888), *Flowers in a Blue Vase* (1888), *Self-Portrait with Straw Hat* (1888), *Sunflowers* (1888), *The Chair of Gauguin* (1888), *Vase with Flowers* (1888), *View of Arles* (1888), *Postman Roulin* (1888), *Self-Portrait with Pipe* (1889), *Green Corn* (1889), *Bedroom at Arles* (1889), *The Starry Night* (1889), *Church at Auvers* (1890), and *Old Man in Sorrow* (1890).

1856–1925 John Singer Sargent, American painter, lives. Works include *Madame X* (1884), *Robert Louis Stevenson* (1885), *Mountain Fire* (1895), *Lord Ribblesdale* (1902), and *Gassed* (1918).

1859–91 Georges Seurat, French Pointillist painter, lives. Works include *Sunday Afternoon on the Island of La Grande Jatte* (1886), *The Models* (1888), *Side Show* (1889), and *The Circus* (1890).

1860–1949 James Baron Ensor, Belgian painter, lives. Works include *Music in the Rue de Flandre* (1891) and *Old Woman with Masks* (1899).

1860–1961 Anna M. R. (Grandma) Moses, American primitive painter, lives. Works include *Catching the Thanksgiving Turkey, July 4th* (1941), *Old Oaken Bucket* (1941), *The McDonel Farm* (1943), *Hoosick Falls in Winter* (1943), *Look, It's a New Little Colt* (1945), *Out for Christmas Trees* (1946), and *The Mailman Has Gone* (1949).

1861–1944 Aristide Joseph Maillol, French painter and sculptor, lives. Works include *Statue at the Jardin des Tuileries* (1906).

▶

1862–1918 Gustav Klimt, Austrian Symbolist painter, lives. Works include *Judith and Holofernes* (1903), *Rosebushes Under Trees* (1905), and *The Kiss* (1907).

1863–1944 Edvard Munch, Norwegian Symbolist painter, lives. Works include *Portrait of a Lady in Black* (1892), *The Scream* (1893), *The Kiss* (1895), *Red Vine* (1900), *The Bridge* (1905), *The Farm* (1905), and *Model by the Armchair* (1929).

1864–1901 Henri de Toulouse-Lautrec, French painter and lithographer, lives. Works include *Portrait of the Artist's Mother Reading* (1887), *The Ringmaster* (1888), *À La Mie* (1891), *At the Moulin de la Galette* (1891), *Quadrille at the Moulin Rouge* (1892), *At the Moulin Rouge* (1892), *Salon in the Rue des Moulins* (1894), *La Visite: Rue des Moulins* (1894), *Women in a Brothel* (1895), *Chilperic* (1896), *The Toilette* (1896), and *The Grand Loge* (1897).

1866–1944 Wassily Kandinsky, Russian abstract painter, lives. Works include *The Blue Rider* (1903), *Blue Mountain* (1908), *Picture with Three Spots* (1914), *Improvisation No. 30* (1913), *Rain* (1913), *Improvisation* (1915), *White Line* (1920), *Yellow-Red-Blue* (1925), *Affirmed Pink* (1932), *Little Circles, No. 555* (1935), *Dominant Curve* (1936), *Moderation* (1940), and *Tempered Élan* (1944).

1867–1947 Pierre Bonnard, French Nabi painter, lives. Works include *Interior* (1898), *Girl in a Straw Hat* (1903), *Regatta* (1912), *Woman and Dog* (1923), *Still Life* (1924), *Interior with a Boy* (1925), *Standing Nude* (1927), *The Breakfast Room* (1931), *Bowl of Fruit* (1933), and *Flowering Almond Tree* (1946).

1868–1940 Édouard Vuillard, French Nabi painter, lives. Works include *Seated Woman* (c. 1898), *Portrait of Lugne Poe* (1891), and *Portrait of Theodore Duret* (1912).

1869–1954 Henri Matisse, French Fauve painter, lives. Works include *Woman with the Hat* (1905), *Joy of Life* (1906), *The Dessert* (1908), *The Red Studio* (1911), *The Blue Window* (c. 1912), *Acrobats in the Circus* (1918), *White Plumes* (1919), *Flowers* (1923), *Odalisque with Magnolias* (1924), *The Pink Nude* (1935), and *Sorrows of the King* (1952).

1871–1951 John Sloan, American painter, lives. Works include *Hairdresser's Window* (1907), *The Haymarket* (1907), *The Wake of the Ferry* (1907), *Madame Berthe Bady* (1912), and *Sixth Avenue and Third Street, New York* (1928).

1871–1956 Lyonel Feininger, American painter, lives. Works include *Glorious Victory of the Sloop "Maria"* (1926) and *The Steamer Odin II* (1927).

1871–1958 Georges Rouault, French Fauve painter, lives. Works include *Ordeal of Samson* (1893), *Circus* (1906), *Circus Woman* (1906), *Parade* (1907), *Pierrot* (1911), *Portrait of a Man* (1911), *Three Clowns* (1917), *The Crucifixion* (1918), *Christ Mocked by Soldiers* (1932), *The Old King* (1936), and *Flight into Egypt* (1948).

1872–1944 Piet Mondrian, Dutch Stijl painter, lives. Works include *The Red Tree* (1908), *Composition* (1915), *Composition Checkerboard* (1919), *Fox Trot A* (1927), *Vertical Composition with Blue and White* (1936), and *Victory Boogie Woogie* (1944).

1876–1957 Constantin Brancusi, Romanian abstract sculptor, lives. Works include *The Kiss* (1908), *Bird in Space* (1919), *Adam and Eve* (1921), *The Seal* (1924–36), and *Endless Column* (1937).

1876–1958 Maurice de Vlaminck, French Fauve painter, lives. Works include *The River* (c. 1908) and *Still Life* (1928).

1877–1953 Raoul Dufy, French Fauve painter, lives. Works include *View of Marseilles* (1908), *Port of Marseilles* (1925), *Tightrope* (1925), *Casino at Nice* (1929), *The Blue Train* (1935), *Harlequin* (1939), *Racetrack* (1940), *Blue Mozart*

▶

(1951), and *Moulin de la Galette* (1951).

1878–1935 Kazimir Malevich, Russian painter, lives. Works include *The Return of the Harvests* (1912) and *Black Square on White* (1918).

1879–1940 Paul Klee, Swiss abstract painter and lithographer, lives. Works include *Demon Above the Ships* (1916), *Arctic Thaw* (1920), *Dance You Monster to My Sweet Song* (1922), *Twittering Machine* (1922), *Around the Fish* (1926), *Pastoral Landscape* (1927), *The Jester* (1928), *The Mocker Mocked* (1930), *Mixed Weather* (1930), *Mask of Fear* (1932), *Child Consecrated to Suffering* (1935), *Demonry* (1939), and *They Persuaded the Child* (1940).

1879–1953 Francis Picabia, French painter, lives. Works include *Udnie* (1913) and *Parade amoureuse* (1917).

1880–1916 Franz Marc, German Blaue Reiter painter, lives. Works include *The Tiger*, *Blue Horses* (c. 1911), *Red Horses* (c. 1911), *The Blue Horse* (c. 1911), *The Gazelle* (1912), and *Deer in the Forest* (1913).

1880–1938 Ernst Ludwig Kirchner, German Expressionist painter, lives. Works include *Self-Portrait with Model* (1907) and *Bare-Breasted Woman with Hat* (1911).

1880–1954 André Derain, French Fauve painter, lives. Works include *Blackfriars Bridge, London* (1906), *Window on the Park* (1912), *The Old Bridge* (c. 1910), *Woodland Scene* (c. 1922), *Portrait of Madame Hessling* (1925), *Southern France* (1927), *Back of Woman* (1928), *Guitar Player* (1928), *Manot, the Dancer* (1928), and *Monastery* (1928).

1881–1955 Fernand Léger, French Cubist painter, lives. Works include *Nudes in the Forest* (1910), *Three Figures* (1911), *City* (1919), *The Mechanic* (1920), *Three Women* (1921), *The Vase* (1927), *Divers on Yellow Background* (1941), *The Great Julie* (1945), *The Builders* (1950), *The Card Players* (1950), and *The Great Parade* (1954).

1881–1961 Max Weber, American painter, lives. Works include *The Geranium* (1911), *The Deer* (1913), *Chinese Restaurant* (1915), *The Visit* (1919), *Gesture* (1921), *Whither Now?* (1939), and *Three Literary Gentlemen* (1945).

1881–1973 Pablo Picasso, Spanish painter and sculptor, lives. Works include *Le Moulin de la Galette* (1900), *The Mourners* (1901), *La Vie* (1903), *The Old Guitarist* (1903), *Family of Saltimbanques* (1905), *Woman with a Fan* (1905), *Gertrude Stein* (1906), *La Toilette* (1906), *Les Demoiselles d'Avignon* (1907), *Girl with a Mandolin* (1910), *Accordionist* (1911), *Harlequin* (1915), *Guitar* (1919), *Two Seated Women* (1920), *Three Musicians* (1921), *Pipes of Pan* (1923), *Woman in an Armchair* (1927), *Seated Women* (1930), *Seated Bather* (1930), *Acrobat* (1930), *Girl Writing* (1934), *Guernica* (1937), *Girl with Dark Hair* (1939), *Night Fishing at Antibes* (1939), *Still Life with a Guitar* (1942), *The Charnel House* (1945), and *Las Meninas* (1957).

1882–1963 Georges Braque, French Cubist painter, lives. Works include *Man with a Guitar* (1911), *Soda* (1911), *Musical Forms* (1913), *Boat on the Beach* (1928), *Still Life* (1928), *The Table* (1928), *The Mantle* (1929), *The Bathers* (1931), *Le Duo* (1937), *Le Billiard* (1945), and *The Blackbirds* (1957).

1882–1967 Edward Hopper, American painter, lives. Works include *House by the Railroad* (1925), *Dawn in Pennsylvania* (1942), *Nighthawks* (1942), and *Second-Story Sunlight* (1960).

1883–1955 Maurice Utrillo, French painter, lives. Works include *Abbey of St. Denis* (1910), *Bistros in a Suburb* (1910), *Suburban Landscape* (1910), *White Château* (1911), *Chartres Cathedral* (1913), *Notre-Dame* (1913), *Place du Tertre, the Fourteenth of July* (1914),

▶

Street in Paris (1914), *Compiègne Barracks* (1915), *Factory Street* (1917), *Renoir's Garden* (1919), *Country Church* (1920), *Montmartre* (1938), *Flower Still Life* (1946), and *Windmills of Montmartre* (1949).

1884–1920 Amedeo Modigliani, Italian painter, lives. Works include *Gypsy Woman and Baby* (1916), *Boy in Red Sweater* (1917), *Portrait of a Girl* (1917), *Woman with a Necklace* (1917), *Gypsy Woman with a Baby* (1919), *Self-Portrait* (1919), *Yellow Sweater* (1919), and *Reclining Nude* (1920).

1884–1950 Max Beckmann, German Expressionist painter, lives. Works include *The Shore* (1936) and *The Rainbow* (1942).

1885–1918 Egon Schiele, Austrian Expressionist painter, lives. Works include *Dual Self-Portrait* (1910) and *Cover* (1913).

1886–1957 Diego Rivera, Mexican painter, lives. Works include *Man and Machinery* (c. 1933), *Flower Vendor* (1935), *Mother Mexico* (1935), and *Market Place* (1947).

1886–1962 Antoine Pevsner, Russian Constructivist painter and sculptor, lives. Works include *Portrait de Marcel Duchamp* (1926), *Surface developpable* (1938), and *Emool l'oiseau* (1955).

1886–1980 Oskar Kokoschka, Austrian Expressionist painter, lives. Works include *The Betrothed of the Wind* (1914) and *The Charles Bridge in Prague* (1934).

1886–1966 Jean Arp, French abstract painter and sculptor, lives. Works include *Head of Tzara* (1916) and *Ptolemy I* (1953).

1887–1948 Kurt Schwitters, German abstract artist, lives. Works include *The First Merzbau* (1933) and *Chocolate* (1947).

1887–1964 Alexander Archipenko, Russian abstract sculptor, lives. Works include *Torso* (1909) and *Medrano* series.

1887–1968 Marcel Duchamp, French Dadaist artist, lives. Works include *The Bride* (1912), *Nude Descending a Staircase* (c. 1912), *Bride Stripped Bare by Her Bachelors* (1923), *The Green Box* (1934), *Pocket Chess Set* (1943), *Female Fig Leaf* (1950), *Self-Portrait in Profile* (1958), and *Torture-morte* (1959).

1887–1985 Marc Chagall, Russian painter, lives. Works include *I and My Village* (1911), *Half-past Three* (1911), *The Musician* (1913), *The Praying Jew* (1914), *Birthday* (1923), *Circus Riders* (1931), *The Cellist* (1939), *The Juggler* (1943), *The Cow with a Parasol* (1946), *King David* (1951), *The Roofs* (1953), and *The Circus* (1956).

1887–1987 Georgia O'Keefe, American painter, lives. Works include *American Radiator Building* (1924), *Dark Corn* (1924), *White Barn, Canada* (1932), and *Sky Above Clouds IV* (1965).

1888–1978 Giorgio de Chirico, Italian Surrealist painter, lives. Works include *Nostalgia of the Infinite* (1911), *Delights of the Poet* (c. 1913), *Evangelical Still Life* (1916), *The Regret* (1916), *Horses* (1927), *Lion and Gladiators* (1927), *Gladiators* (c. 1928), and *Antique Era* (1948).

1890–1976 Man Ray, American Surrealist artist and photographer, lives. Works include *Admiration of the Orchestrelle for the Cinematograph* (1919), *Clock Wheels* (1925), and *Observatory Time—the Lovers* (1934).

1890–1977 Naum Gabo, Russian Constructivist sculptor, lives. Works include *Kinetic Sculpture* (1920) and *Tower* (1923).

1891–1973 Jacques Lipchitz, Lithuanian-born French abstract sculptor, lives. Works include *Marin à la guitare* (1914), *Arlequin et guitare* (1926), and *Prometheus* (1936).

1891–1976 Max Ernst, German Surrealist painter, lives. Works include *Two Children Are Menaced by a Nightingale* (1924), *Night and Day* (1942), *Echo-Plasm* (1944), and *Mundus Est Fabula* (1959).

1892–1942 Grant Wood, American painter, lives. Works include *Woman with Plants* (1929), *American Gothic* (1930), and *Midnight Ride of Paul Revere* (1931).

▶

1892–1966 Jean Lurçat, French painter and decorator, lives. Works include *La chant du monde* (1956).

1893–1943 Chaim Soutine, Lithuanian-born French Expressionist painter, lives. Works include *Auto-portrait* (1918) and *Le boeuf écorché* (1925).

1893–1959 George Grosz, German Expressionist painter, lives. Works include *Dr. Max Hermann-Neisse* (c. 1927), *Man of Opinion* (1928), *No Letup* (1940), and *The Pit* (1946).

1893–1983 Joan Miro, Spanish Surrealist painter, lives. Works include *The Goldsmith* (1918), *The Farm* (1922), *The Harlequin's Carnival* (1925), *Dog Barking at the Moon* (1926), *Dutch Interior* (1928), *The Farmer's Dinner* (1935), *The Circus* (1937), *Figures Under the Moon* (1938), *Woman and Stars* (1944), *The Harbor* (1945), *Red Disk in Pursuit of the Lark* (1953), *Blue II* (1961), *Ski Lesson* (1966), and *Red Accent in the Quiet* (1968).

1894–1964 Stuart Davis, American painter, lives. Works include *Summer Landscape* (1930), *Red Cart* (1932), *New York Under Gaslight* (1941), and *Ursine Park* (1942).

1895–1946 László Moholy-Nagy, Hungarian abstract artist, lives. Works include *Space Modulator in Plexiglas Ch 4* (1941).

1898–1967 René Magritte, Belgian Surrealist painter, lives. Works include *Rape* (1934), *Les marches de l'été* (1938), *The Call of the Summit* (1942), and *The Fire Ladder* (1949).

1898–1969 Benjamin Shahn, American painter, lives. Works include *Passion of Sacco and Vanzetti* (1932), *Handball* (1939), *Fourth of July Orator* (1943), *Father and Child* (1946), and *Miners' Wives* (1948).

1898–1976 Alexander Calder, American abstract sculptor, lives. He is best known for his mobiles and constructions in wood, metal, and wire, such as *The Circus*.

1898–1986 Henry Moore, English abstract sculptor, lives. Works include *Composition* (1933), *Reclining Figure* (1938), and *Falling Warrior* (1956).

1900–88 Louise Nevelson, Russian-born American abstract sculptor, lives. Works include *Dawn's Wedding Feast* (1959); *Wedding Chapel IV* (1960), and *Dawn Shadows* (1982).

1901–66 Alberto Giacometti, Swiss sculptor and painter, lives. Works include *Head of the Mother of the Artist* (1947), *The Street* (1953), *Walking Man* (1960), and *Caroline* (1961).

1901–85 Jean Dubuffet, French painter, lives. Works include *Sals et terrains* (1952), *Fleur de barbe* (1960), and *J'accours* (1964).

1903–70 Mark Rothko, American abstract painter, lives. Works include *Omen of the Eagle; No. 2* (1948), *Light, Earth, and Blue* (1954), and *Black on Gray* (1970).

1904–48 Arshile Gorky, Armenian-born American abstract painter, lives. Works include *Betrothal II* (1947) and *The Leaf of the Artichoke Is an Owl* (1944).

1904–89 Salvador Dali, Spanish Surrealist painter, lives. Works include *Illumined Pleasures* (1929), *Persistence of Memory* (1931), *Burning Giraffe* (1936), *Spanish Dancer* (1938), *Dali Atomicus* (1948), *Last Supper* (1955), and *Tuna Fishing* (1966).

1904– Willem de Kooning, Dutch-born American Expressionist painter, lives. Works include *Man* (1939) *Standing Man* (1942), *Queen of Hearts* (1946), *Excavation* (1950), *Woman and Bicycle* (1953), *Door to the River* (1960), and *The Visit* (1967).

1906–65 David Smith, American abstract sculptor, lives. Works include *Australia* (1951) and the Agricola series.

1908– Balthus (Count Bathazar Klossowski de Rola), French painter, lives. Works include *The Street* (1933), *The Room* (1952), and *The Good Days* (1944–46).

▶

1910– Francis Bacon, Irish painter, lives. Works include *Study After a Portrait of Pope Innocent X By Valasquez* (1953), *Reclining Person with Hypodermic Syringe* (1963), and *Two Persons Lying on a Bed with Witnesses* (1968).

1912–56 Jackson Pollock, American Abstract Expressionist painter, lives. Works include *The Flame* (1937), *Male and Female* (1942), *She-Wolf* (1943), *The Blue Unconscious, Number Ten* (1949), *Portrait and a Dream* (1953), and *Sleeping Effort* (1953).

1913–67 Ad Reinhardt, American abstract painter, lives. Works include *Abstract Painting, Red* (1952) and *Abstract Painting* (1956).

1914–55 Nicolas de Staël, Russian-born French abstract painter, lives. Works include *La vie deire* (1946) and *Composition* (1950).

1914–73 Asger Jorn (Asger Olaf Jorgensen), Danish abstract painter, lives. Works include *Disfigurations* (1959 and 1962) and *Quiet Look* (1971).

1915– Robert Motherwell, American Abstract Expressionist painter, lives. Works include *The Spanish Prison* (1944), *Five in the Afternoon* (1950), *Africa* (1965), and *Open* (1969).

1917– Andrew Wyeth, American painter, lives. Works include *The Hunter* (1943), *Christina's World* (1948), *River Cove* (1958), *Tenant Farmer* (1961), *The Patriot* (1964), and *Grapevine* (1966).

1921–46 Karel Appel, Dutch painter, lives. Works include *The Cry of Freedom* (1948) and *Two Heads in a Landscape* (1968).

1921– Georges Mathieu, French painter, lives. Works include *Batailles* (1954) and *Hommage aux poètes du Monde extier* (1956).

1923– Roy Lichtenstein, American Pop painter, lives. Works include *Look Mickey* (1961), *Whaam!* (1963), and *Temple of Apollo* (1964).

1923– Antoni Tàpies, Spanish painter, lives. Works include *Desconsuelo lunar* (1949), *Negro y ocre* (1955), and *Grand triangle marron* (1963).

1924– George Segal, American sculptor, lives. Works include *Cinema* (1963) and *Times Square* (1970).

1925– Robert Rauschenberg, American painter and sculptor, lives. Works include *Charlene* (1954), *Rebus* (1955), *Revolvers* (1967), and *Venetian Series* (1973).

1925– Jean Tinguely, Swiss sculptor, lives. Works include *Metamechanicals* (1953) and *Homage to New York* (1959).

1928–62 Yves Klein, French abstract painter, lives. Works include *Monochromes* (1946) and *Portrait relief d'Arman* (1962).

1928– Bernard Buffet, French painter, lives. Works include *Horreur de la guerre* (1955).

1928– Friedrich Hundertwasser (Fritz Stowasser), Austrian painter, lives. Works include *The Tower of Babel Punctures the Sun* (1959) and *Irinaland over the Balkans* (1971).

1929– Claes Oldenburg, Swedish-born American Pop sculptor, lives. Works include *The Street, The Store, The Home*, and *Lipstick on Caterpillar Tracks* (1969).

1930–87 Andy Warhol, American Pop artist, lives. Works include *The Twenty Marilyns* (1962) and *Self-Portrait* (1966).

1930– Jasper Johns, American painter, lives. Works include *Flag* (1954), *Painted Bronze* (1960), and *Double White Map* (1965).

1936– Frank Stella, American abstract painter and sculptor, lives. Works include *Newstead Abbey* (1960) and *Aybotana II* (1968).

1937– David Hockney, English Pop artist, lives. Works include *Painting of Tea in Illusionist Style* (1961) and *Portrait of an Artist (Pool with Two People)* (1971).

Playwrights

B.C.

c. 251–c. 184 Plautus, Roman comic dramatist, lives. Twenty-one of his plays survive, including *The Comedy of Asses* and *The Pot of Gold.*

c. 185–c. 159 Terence, Roman comic dramatist, lives. Works include *Andria* (166), *Hecyra* (165), *Heauton Timoroumenos* (163), *Eunuchus* (161), *Phormio* (161), and *Adelphoe* (160).

3 B.C.–A.D. 65 Seneca, Roman philosopher and dramatist, lives. Works include *Phaedra, Agamemnon,* and *Medea.*

A.D.

1469–1527 Niccolò Machiavelli, Florentine statesman and political philosopher, lives. Plays include *La mandragola* (c. 1518).

1505–56 Nicholas Udall, English dramatist, lives. Plays include *Ralph Roister Doister* (1566).

1538–1612 Battista Guarini, Ferrarese poet, lives. Plays include *Il pastor fido* (1590).

1544–95 Torquato Tasso, Italian poet, lives. Plays include *Aminta* (1573) and *Torrismondo* (1586).

1554–1606 John Lyly, English prose writer, poet, and playwright, lives. Plays include *Alexander and Campaspe* (1584), *Endimion* (1591), and *Midas* (1592).

1557?–94 Thomas Kyd, English dramatist, lives. Plays include *The Spanish Tragedy* (1584–89).

1558?–97 George Peele, English playwright, lives. Plays include *The Arraignment of Paris* (1584), *The Love of King David and Fair Bethsabe* (c. 1588), and *The Old Wives' Tale* (1595).

1562–1635 Lope de Vega, Spanish dramatist, lives. Plays include *Peribáñez* (1614), *El caballero de Olmendo, Fuenteovejuna* (1619), and *El mejor alcalde, el rey* (1623).

1564–93 Christopher Marlowe, English dramatist and poet, lives. Plays include *Tamburlaine the Great* (1587), *The Jew of Malta* (1589), *Edward II* (1590), *Dido, Queen of Carthage* (c. 1593) with Thomas Nash, *The Massacre at Paris* (1593), and *The Tragical History of Dr. Faustus* (1601).

1564–1616 William Shakespeare, English dramatist, lives. *See Shakespeare's Plays.*

1567–1601 Thomas Nash, English satirist and dramatist, lives. Plays include *Summer's Last Will and Testament* (1593).

c. 1570–1641 Thomas Heywood, English dramatist, lives. Plays include *A Woman Killed with Kindness* (1603), *The Fair Maid of the West* Parts I and II, *The Wise Woman of Hogsden* (c. 1604), *The Golden Age* (c. 1610), *The Captives* (1624), *A Maidenhead Well Lost* (c. 1633), and *Love's Mistress* (1634).

1571–1648 Tirso de Molina (Gabriel Telley), Spanish playwright, lives. Works include some 300 to 400 plays, among them *The Love Rogue*, perhaps the original version of the Don Juan legend.

1572?–1632 Thomas Dekker, English dramatist, lives. Plays include *Old Fortunatus* (1599), *The Shoemaker's Holiday* (1600), *The Honest Whore* Parts I and II (1604, 1605), *The Roaring Girl* (1610) with Thomas Middleton, *The Virgin Martyr* (c. 1620) with Philip Massinger, *The Witch of Edmonton* (1621) with John Ford and William Rowley, and *The Sun's Darling* (1624) with John Ford.

1573–1637 Ben Jonson, English dramatist and poet, lives. Plays include *Every Man in His Humour* (1598), *Every Man Out of His Humour* (1599), *Cynthia's Revels*

▶

The Greek and Roman Dramatists

B.C.

525–456 Aeschylus, Greek tragedian, lives. Major works include *The Persians* (472); *Seven Against Thebes* (467); *Suppliants* (463); *Oresteia* (458); trilogy *Agamemnon, The Libation-Bearers,* and *The Furies*; and *Prometheus Bound.*

c. 496–406 Sophocles, Greek tragedian, lives. Works include *Ajax* (c. 441), *Antigone* (c. 441), *Oedipus Rex* (c. 430), *Trachiniai* (after 430), *Electra* (after 418), *Philoctetes* (408), and *Oedipus at Colonus* (c. 406).

c. 480–406 Euripides, Greek, tragedian, lives. Works include *Alcestis* (438), *Medea* (431), *Hippolytus* (428), *Heraclidae* (c. 428), *Andromache* (426?), *Hecuba* (423?), *Suppliants* (c. 420), *Trojan Women* (415), *Electra* (413), *Helen* (412), *Heracles* (408), *Orestes* (408), *Ion* (408), and *Iphigeneia in Tauris* (408).

c. 448–c. 388 Aristophanes, Greek comedian, lives. Works include *The Knights* (425), *Hippias* (424), *The Clouds* (423), *The Wasps* (422), *The Peace* (421), *The Birds* (414), *Lysistrata* (411), *Ladies' Day* (411), and *The Frogs* (405).

342–292 Menander, Greek comedian, lives. Works include more than 100 pieces, among them *Anger* (321), *Dyscolus* (c. 316), *Perikeiromene*, and *Second Adelphoe.*

(1600), *The Poetaster* (1601), *Richard Crookbacke* (1603), *Sejanus* (1603), *Eastward Ho* (1605) with John Marston and George Chapman, *Volpone, or the Fox* (1606), *Epicene, or the Silent Woman* (1609), *The Alchemist* (1610), *Bartholomew Fair* (1614), *The Devil Is an Ass* (1616), *The Staple of News* (1625), *The Magnetic Lady* (1632), and *Tale of a Tub* (1633).

c. 1575–1625 Cyril Tourneur, English dramatist, lives. Plays include *Revenger's Tragedy* (1607) and *Atheist's Tragedy* (1611).

1579–1625 John Fletcher, English dramatist, lives. Plays include *The Faithful Shepherdess* (1609), *The Wild Goose Chase* (1621), *The Scornful Lady* (c. 1614), *Without Money* (1614), and *Rule a Wife and Have a Wife* (1624).

1580–1627 Thomas Middleton, English dramatist, lives. Plays include *A Trick to Catch the Old One* (1606), *The Roaring Girl* (1610) with Thomas Dekker, and *The Changeling* (1623) with William Rowley.

c. 1581–1639 Juan Ruiz de Alarcón y Mendoza, Spanish dramatist, lives. Plays include *Las paredes oyen, Mudarse por mejorarse, La prueba de las promesas,* and *La verdad sospechosa* (1619).

1583–1640 Philip Massinger, English dramatist, lives. Plays include *A New Way to Pay Old Debts* (1625), *The Duke of Milan* (1620), *The Maid of Honor* (c. 1621), *The Virgin Martyr* (1622), *The Great Duke of Comedy* (1627), and *The City Madam* (1632).

1584–1616 Francis Beaumont, English dramatist, lives. Plays include *The Woman-Hater* (c. 1606) and *The Knight of the Burning Pestle* (1607). Major plays written with John Fletcher (1579–1625) include *The Nice Valour* (1606?), *Philaster* (1609), *A King and No King* (1611), and *The Maid's Tragedy* (1611).

1585–c. 1642 William Rowley, English dramatist, lives. Plays include *All's Lost by Lust* (1622), *A Match at Midnight* (1633), and *The Changeling* (1623) with Thomas Middleton.

1586–c. 1640 John Ford, English dramatist, lives. Plays include *'Tis a Pity She's a Whore* (1627), *The Lovers' Melancholy* (1628), and *The Broken Heart* (1629).

1596–1666 James Shirley, English dramatist, lives. Plays include *The Witty Fair One* (1628), *The Traitor* (1631), *Hyde Park* (1632), *The Gamester* (1633),*The Triumph of Peace* (1634), *The Lady of*

▶

Pleasure (1635), and *The Cardinal* (1641).

?–1652 Richard Brome, English dramatist, lives. Plays include *The City Wit, or The Woman Wears the Breeches* (c. 1628), *The Weeding of Covent Garden* (1632), *The Sparagus Garden* (1635), and *A Jovial Crew* (1641).

1600–1681 Pedro Calderón de la Barca, Spanish dramatist, lives. Plays include *El alcalde de Zalamea* (c. 1640), *La devoción de la cruz, El cisma de Inglaterra, El médico de su honra* (c. 1649), *La vida es sueño* (c. 1630), and *El mágico prodigioso* (1637).

1606–84 Pierre Corneille, French dramatist, lives. Plays include *Mélite* (1629), *Clitandre* (1632), *Médée* (1635), *Le Cid* (1637), *Horace* (1640), *Polyeucte* (1641), *Le Menteur* (1643), *La Toison d'or* (1660), and *Pulchérie* (1672).

1616–64 Andreas Gryphius, German baroque poet and dramatist, lives. Plays include *Absurda Comica oder Herr Peter Squentz* (1633), *Katharina von Georgien* (1657), *Carolus Stuardus* (1657), and *Cardenio und Celinde* (1657).

1622–73 Jean-Baptiste Poquelin Molière, French comic dramatist, lives. Plays include *Le Dépit amoureux* (1659), *Les Précieuses Ridicules* (1659), *Sganarelle* (1660), *L'École des femmes* (1662), *Le Mariage forcé* (1664), *La Princesse d'Élide* (1664), *Le Tartuffe* (1664), *Dom Juan ou le festin de pierre* (1665), *L'Étourdi* (1665), *Le Misanthrope* (1666), *Les Femmes savantes* (1672), and *Le Malade imaginaire* (1673).

1625–1709 Thomas Corneille, French playwright, lives. Plays include *Timocrate* (1656) and *Ariane* (1672).

1631–1700 John Dryden, English poet, dramatist, and critic, lives. Plays include *The Indian Emperor* (1665), *Almanzor and Almahide, or The Conquest of Granada* (1670), *Aurengzebe* (1675), *All for Love* (1678), and *Don Sebastian* (1690).

1639–1699 Jean B. Racine, French playwright, lives. Plays include *La Thébaïde* (1664), *Alexandre le Grand* (1665), *Andromaque* (1667), *Les Plaideurs* (1668), *Bajazet* (1672), *Mithridate* (1673), *Phèdre* (1677), *Iphigénie en Aulide* (1679), *Esther* (1689), and *Athalie* (1691).

1640–1689 Aphra J. Behn, English novelist and dramatist, lives. Plays include *The Rover, or The Banished Cavaliers* Parts I and II (1677, 1680), *Sir Patient Fancy* (1678), *Oroonoko, or The History of the Royal Slave* (c. 1678).

1640?–1716 William Wycherley, English comic dramatist, lives. Plays include *Love in a Wood, or St. James' Park* (1671), *The Gentleman Dancing Master* (1672?), *The Country Wife* (1675), and *The Plain Dealer* (1676).

1655–1709 Jean F. Régnard, French comic dramatist, lives. Plays include *Le Joueur* (1696), *Les Folies amoureuses* (1704), *Les Ménechmes* (1705), and *Le Légataire universel* (1708).

1670–1729 William Congreve, English dramatist, lives. Plays include *The Old Bachelor* (1693), *The Double Dealer* (1693), *Love for Love* (1695), *The Mourning Bride* (1697), and *The Way of the World* (1700).

1674–1718 Nicholas Rowe, English poet and dramatist, lives. Plays include *The Fair Penitent* (1703), *Tamerlane* (1701), *Jane Shore* (1714), and *Lady Jane Grey* (1715).

1685–1732 John Gay, English poet and playwright, lives. Plays include *The Beggar's Opera* (1728).

c. 1700–1797 Charles Macklin, English dramatist, lives. Plays include *Love à la Mode* (1759) and *The Man of the World* (1781).

1707–93 Carlo Goldoni, Italian dramatist, lives. Plays include *La bottega del caffè, La locandiera, La vedova scaltra, La famiglia dell'antiquario, I rusteghi*, and *Le barufe chiozote* (1748–60).

▶

Development of the Theater from Elizabethan Times

See also Playwrights.

c. 1580–90 English dramatist and poet Christopher Marlowe introduces blank verse to playwrighting.

1564–1616 English dramatist and poet William Shakespeare, considered greatest of all playwrights, lives. (*See Shakespeare's Plays.*)

c. 1575 Commedia dell'arte flourishes in Italy.

1576 First British public theater, The Theatre, is built near London.

1599 Globe Theatre in London, where Shakespeare's plays are performed, opens. It burns in 1613 and is rebuilt by 1614.

1606–84 French dramatist Pierre Corneille establishes neoclassicism in France with such plays as *El Cid* (1637) and *Horace* (1640).

c. 1610 Tragicomedy becomes popular in Great Britain for first time.

c. 1660–80 French playwright and actor Jean Molière creates high French comedy.

c. 1600–80 French playwright Jean Racine's plays exemplify French Classicism.

1642 British Puritan government closes all theaters and bans theatrical performances. Globe Theatre is torn down in 1644.

c. 1660 New theaters include galleries and private boxes; previously spectators often sat on stage.

1660–80 Heroic plays, written in rhymed couplets and dealing with love, flourish in England.

c. 1661 Women's roles, previously played by men and boys, are now played by women.

1690–1700 English dramatist William Congreve's plays are considered height of Restoration comedy.

1680 Comédie-Française is established in Paris; it is oldest state theater still in existence.

1752 First professional acting troupe performs in American Colonies, in Virginia.

1775 Continental Congress bans all theatrical performances in Colonies.

c. 1820 Introduction of gas lighting replaces candlelight in theaters. (In c. 1850 limelight is developed, and in 1880 electricity is introduced in theaters.)

1828–1906 Norwegian playwright Henrik Ibsen, often called founder of modern drama, lives.

c. 1835 Showboats are in use for transporting theatrical sets and equipment in major river towns in U.S. Midwest.

1843 First matinee performance in a U.S. theater is offered in New York City.

1850–70 U.S. minstrel troupes, in which white actors perform in blackface, are popular.

c. 1850 Burlesques, musical parodies of plays and personalities, are popular in United States.

c. 1855 U.S. touring companies, called "tommers," perform dramatic versions of *Uncle Tom's Cabin* in small towns and rural areas.

1863–1938 Russian theatrical director and teacher Constanin Stanislavsky lives. His innovations in acting, known as Stanislavsky method, revolutionize modern theater.

1865–1939 Irish poet and playwright William Butler Yeats, a leader of Irish Literacy Renaissance, lives. Under patronage of Annie Horniman, he and Lady Gregory found the Abbey Theatre in Dublin (1903).

c. 1875 U.S. theaters build elaborate machinery to handle complex stagecraft and sets.

c. 1875 Directors begin to appear, taking the places of actors who staged plays while also acting.

c. 1885 Symbolist drama, in which appearance is considered only small part of reality, develops in France.

c. 1900 Community theaters begin to appear in United States.

1905 First U.S. motion-picture theater opens, in Pennsylvania. (Motion pictures eventually draw audiences away from established theater.)

1910 Expressionist drama, in which inner vision is more important that outward appearance, begins in Germany.

c. 1925 Epic theater, revolutionary political dramatic style, appears in the work of such playwrights as German Bertolt Brecht (1898–1956).

1938 Playwrights' Company is founded in United States. It attempts to stage dramatists' productions despite economic hardships of Great Depression.

1941 Experimental theater is founded in New York City.

Development of the Theater from Elizabethan Times

1946	Soviet government withdraws its artistic subsidies and imposes strict theatrical censorship. (After Stalin's death, censorship is relaxed and dramatic production increases.)
c. 1950	Theater of the absurd, in which illogical and chaotic aspects of life are emphasized, appears in France.
c. 1950	Development of off-Broadway movement in New York City, low-budget response to the financial problems of Broadway theater. (In 1965, off-off-Broadway theaters stage experimental dramatic productions.)
1962	Lincoln Center for the Performing Arts is built in New York City.
1964	Shakespeare's 400th anniversary is celebrated.
1965	U.S. Congress establishes National Endowment for the Arts to make grants to theatrical groups and projects.
1973	Broadway theater revenues hit 10-year low. Although revenues bounce back by the 1974–75 season, financial worries of New York City's theater district are far from over.
1976	Australia's Perth Entertainment Centre is built with seating capacity of 8,003, making it world's largest theater in regular operation.
1983	*A Chorus Line* becomes Broadway's longest-running show on completing its 3,389th performance. Still running in 1988, it tops 5,300 performances by midyear.
1984	American Academy of Dramatic Arts marks its 100th anniversary.
1985	Sagging ticket sales and high cost of production in New York City's theater district leads to cutbacks in number of shows staged. Year 1985 has lowest number of shows staged since 1900.
1988	Longest-running play to date is Agatha Christie's *The Mousetrap*. First performed in London in 1952, it tops 14,500 performances in 1988.
1990	*A Chorus Line* closes to a capacity crowd of 1,500 at New York's Shubert Theatre.

1713–84 Denis Diderot, French writer and dramatist, lives. Plays include *Le Fils naturel* (1757) and *Le Père de famille* (1758).

1731–74 Oliver Goldsmith, Irish-born English playwright and novelist, lives. Plays include *The Good Natur'd Man* (1768) and *She Stoops to Conquer* (1773).

1732–99 Pierre-Augustin Carn de Beaumarchais, French dramatist, lives. Plays include *Eugénie* (1767), *The Barber of Seville* (1775), and *The Marriage of Figaro* (1784).

1749–1803 Vittorio Alfieri, Italian dramatist and poet, lives. Plays include *Cleopatra, Antigone, and Agamemnone* (1787).

1749–1832 Johann von Goethe, German poet, playwright, and novelist, lives. Plays include *Gotz von Berlichingen* (1773), *Clavigo* (1774), *Stella* (1774), *Torquato Tasso* (1790), and *Faust* (1808).

1751–1816 Richard Brinsley Sheridan, Irish-born English dramatist, lives. Plays include *The Rivals* (1775), *St. Patrick's Day* (1775), *The Duenna* (1775), *The School for Scandal* (1777), *The Critic* (1779), and *Pizarro* (1799).

1759–1805 Friedrich (J. C.) von Schiller, German dramatist and poet, lives. Works include *The Robbers* (1782), *Kabale und Liebe* (1784), *Don Carlos* (1787), *Wallenstein* (1798–99), *The Maid of Orleans* (1801), and *Wilhelm Tell* (1804).

1768–1823 Zacharias Werner, German dramatist, lives. Plays include *Martin Luther; oder, Die Weihe der Kraft* (1807) and *Der 24. Februar* (1810).

1776–1839 William Dunlap, early American playwright, lives. Works include *The Father, or American Shandyism* (1789) and *André* (1798).

1777–1811 Heinrich von Kleist, German dramatist and novella writer, lives. Plays include *Robert Guiskard* (1801–2), *Das Käthchen von Heilbronn* (1810), *Penthesilea* (1808), *Der zerbrochene Krug*

▶

(1811), and *Prinz Friedrich von Homburg* (1821).

1791–1861 Augustin E. Scribe, French dramatist, lives. Plays include *Le Mariage de raison* (1826), *Le Mariage d'argent* (1828), *Le Camaraderie ou la courte échelle* (1837), and *Une Chaîne* (1841).

1799–1837 Aleksandr Pushkin, Russian writer, poet, and playwright, lives. Plays include *Boris Godunov* (1825) and *Little Tragedies* (1830).

1802–70 Alexandre Dumas père, French novelist and dramatist, lives. Plays include *Henri III et sa cour* (1829), *Napoléon Bonaparte* (1831), *Antony* (1831), and *La Tour de Nesle* (1832).

1802–85 Victor Hugo, French writer and dramatist, lives. Plays include *Cromwell* (1827), *Hernani* (1830), *Le Roi s'amuse* (1832), and *Angelo* (1835).

1810–57 Alfred de Musset, French writer and dramatist, lives. Plays include *On ne badine pas avec l'amour* (1834), *Barberine* (1835), *Un Caprice* (1837), and *Il faut qu'une porte soit ouverte ou fermée* (1845).

1813–37 Georg Büchner, German dramatist, lives. Plays include *Dantons Tod* (1835), *Leonace und Lena* (1836), and *Woyzeck* (1879).

1813–63 Friedrich Hebbel, German dramatist, lives. Plays include *Judith* (1841), *Maria Magdalena* (1844), *Herodes and Marianne* (1850), *Agnes Bernaner* (1855), *Gyges und sein Ring* (1856), and *Die Nibelungen* (1862).

1817–93 Don José Zorrilla y Moral, Spanish poet and dramatist, lives. Plays include *El zapatero y el rey* (1840–41), *El puñal del Godo* (1842), *Don Juan Tenorio* (1844), and *Traidor, inconfeso y martir* (1849).

1824–95 Alexandre Dumas fils, French dramatist, lives. Plays include *La Dame aux camélias* (1852), *Le Demi-Monde* (1855), *L'Ami des femmes* (1864), and *Les Idées de Mme. Aubray* (1867).

1828–1906 Henrik Ibsen, Norwegian poet and playwright, lives. Plays include *Lady Inger of Östrat* (1855), *The Vikings of Helgoland* (1858), *Love's Comedy* (1862), *The Pretenders* (1864), *Brand* (1866), *Peer Gynt* (1867), *A Doll's House* (1879), *Ghosts* (1881), *An Enemy of the People* (1882), *The Wild Duck* (1884), *The Lady from the Sea* (1888), *Hedda Gabler* (1890), and *John Gabriel Borkman* (1896).

1849–1912 (Johan) August Strindberg, Swedish playwright, novelist, and poet, lives. Plays include *Master Olof* (1872), *Lucky Per's Travels* (1880), *Sir Bengt's Wife* (1882), *The Father* (1887), *Comrades* (1888), *Miss Julie* (1888), *The Creditor* (1890), and *The Ghost Sonata* (1907).

1852–1932 Lady Augusta P. Gregory, Irish dramatist, lives. Plays include *Spreading the News* (1904) and *The Rising of the Moon* (1907).

1854–1900 Oscar Wilde, Irish, dramatist, and novelist, lives. Plays include *Lady Windermere's Fan* (1892), *A Woman of No Importance* (1893), *Salomé* (1893), and *The Importance of Being Earnest* (1895).

1855–1934 Sir Arthur W. Pinero, English dramatist, lives. Plays include *The Second Mrs. Tanqueray* (1893), *The Notorious Mrs. Ebbsmith* (1896), *Trelawney of the Wells* (1898), *The Gay Lord Quex* (1899), *Iris* (1901), and *Mid-Channel* (1909).

1856–1950 George Bernard Shaw, Irish dramatist, lives. Plays include *Mrs. Warren's Profession* (1893), *Arms and the Man* (1894), *The Devil's Disciple* (1897), *Candida* (1898), *The Philanderer* (1898), *Caesar and Cleopatra* (1899), *Captain Brassbound's Conversion* (1900), *Man and Superman* (1905), *John Bull's Other Island* (1904), *Major Barbara* (1905), *Fanny's First Play* (1912), *Androcles and the Lion* (1912), *Pygmalion* (1913), *Heartbreak House* (1913), *Back to Methuselah* (1921), and *St. Joan* (1923).

▶

Shakespeare's Plays

Plays are arranged below by the date they were written. Exact dates for the writing and first performance of Shakespeare's plays are not known, however, and dates given here are based on available historical evidence.

1590 *Henry VI*, Part I (history). Performed 1589–92.

1590 *Henry VI*, Part II (history). Performed 1589–92.

1590 *Henry VI*, Part III (history). Performed 1589–92.

1592 *Richard III* (history). Performed 1592–93.

1592 *The Comedy of Errors* (comedy). Performed 1592–93.

1593 *The Taming of the Shrew* (comedy). Performed 1593–94.

1593 *Titus Andronicus* (tragedy). Performed 1593–94.

1594 *Love's Labor Lost* (comedy). Performed 1594–95.

1594 *Romeo and Juliet* (tragedy). Performed 1594–95.

1594 *The Two Gentlemen of Verona* (comedy). Performed 1594–95.

1595 *A Midsummer Night's Dream* (comedy). Performed 1595–96.

1595 *Richard II* (history). Performed 1595–96.

1596 *King John* (history). Performed 1596–97.

1596 *The Merchant of Venice* (comedy). Performed 1596–97.

1597 *Henry IV*, Part I (history). Performed 1597–98.

1597 *Henry IV*, Part II (history). Performed 1597–98.

1598 *Henry V* (history). Performed 1598–99.

1598 *Much Ado About Nothing* (comedy). Performed 1598–99.

1599 *As You Like It* (comedy). Performed 1599–1600.

1599 *Julius Caesar* (tragedy). Performed 1599–1600.

1599 *Twelfth Night* (comedy). Performed 1601–2.

1600 *Hamlet* (tragedy). Performed 1600–1.

1600 *The Merry Wives of Windsor* (comedy). Performed 1600–1601.

1601 *Troilus and Cressida* (comedy). Performed 1600–1602.

1602 *All's Well That Ends Well* (comedy). Performed 1602–3.

1602 *Measure for Measure* (comedy). Performed 1604–5.

1604 *Othello* (tragedy). Performed 1604–5.

1605 *King Lear* (tragedy). Performed 1605–6.

1605 *Macbeth* (tragedy). Performed 1605–6.

1606 *Antony and Cleopatra* (tragedy). Performed 1606–7.

1607 *Coriolanus* (tragedy). Performed 1607–8.

1607 *Timon of Athens* (tragedy). Performed 1607–8.

1608 *Pericles* (comedy). Performed 1608–9.

1609 *Cymbeline* (comedy). Performed 1609–10.

1610 *The Winter's Tale* (comedy). Performed 1610–11.

1611 *The Tempest* (comedy). Performed 1611–12.

1612 *Henry VIII* (history; possibly a collaboration). Performed 1612–13.

1612 *Two Noble Kinsmen* (comedy; possibly a collaboration). Performed 1612–13.

1859–1927 Jerome K. Jerome, English novelist and playwright, lives. Plays include *The Passing of the Third Floor Back* (1908).

1860–1904 Anton Chekhov, Russian playwright and short-story writer, lives. Plays include *Na bolsoi doroge* (1884), *Ivanov* (1887), *Lebedinaya pesnya* (1887), *Medved* (1888), *Predlozheniye* (1888), *Tragik ponevole* (1889), *Sevadba* (1889), *Yubilei* (1891), *The Seagull* (1896), *Uncle Vanya* (1899), *The Cherry Orchard* (1904), and *The Three Sisters* (1901).

1862–1921 Georges Feydeau, French dramatist, lives. Plays include *L'Hôtel du Libre Exchange* (1894), *Un Fil à la patte* (1899), and *Occupe-toi d'Amélie* (1911).

1862–1946 Gerhart Hauptmann, German dramatist, lives. Plays include *Vor Sonnenaufgang* (1889).

1862–1949 Maurice Maeterlinck, Belgian poet, dramatist, and essayist, lives. Plays include *La Princesse Maleine* (1889), *L'Intruse, les Aveugles* (1890), *Pelléas et Mélisande* (1892), and *L'Intérieur* (1894).

▶

Pulitzer Prize for Drama

1917 No prize awarded

1918 *Why Marry?* by Jesse Lynch Williams

1919 No prize awarded

1920 *Beyond the Horizon* by Eugene O'Neill

1921 *Miss Lulu Bett* by Zona Gale

1922 *Anna Christie* by Eugene O'Neill

1923 *Icebound* by Owen Davis

1924 *Hell-Bent for Heaven* by Hatcher Hughes

1925 *They Knew What They Wanted* by Sidney Howard

1926 *Craig's Wife* by George Kelly

1927 *In Abraham's Bosom* by Paul Green

1928 *Strange Interlude* by Eugene O'Neill

1929 *Street Scene* by Elmer L. Rice

1930 *The Green Pastures* by Marc Connelly

1931 *Alison's House* by Susan Glaspell

1932 *Of Thee I Sing* by George S. Kaufman, Morrie Ryskind, and Ira Gershwin

1933 *Both Your Houses* by Maxwell Anderson

1934 *Men in White* by Sidney Kingsley

1935 *The Old Maid* by Zoe Akins

1936 *Idiot's Delight* by Robert E. Sherwood

1937 *You Can't Take It with You* by Moss Hart and George S. Kaufman

1938 *Our Town* by Thornton Wilder

1939 *Abe Lincoln in Illinois* by Robert E. Sherwood

1940 *The Time of Your Life* by William Saroyan (declined)

1941 *There Shall Be No Night* by Robert E. Sherwood

1942 No prize awarded

1943 *The Skin of Our Teeth* by Thornton Wilder

1944 No prize awarded

1945 *Harvey* by Mary Chase

1946 *State of the Union* by Russel Crouse and Howard Lindsay

1947 No prize awarded

1948 *A Streetcar Named Desire* by Tennessee Williams

1949 *Death of a Salesman* by Arthur Miller

1950 *South Pacific* by Richard Rodgers, Oscar Hammerstein II, and Joshua Logan

1951 No prize awarded

1952 *The Shrike* by Joseph Kramm

1953 *Picnic* by William Inge

1954 *The Teahouse of the August Moon* by John Patrick

1955 *Cat on a Hot Tin Roof* by Tennessee Williams

1956 *The Diary of Anne Frank* by Albert Hackett and Frances Goodrich

1957 *Long Day's Journey into Night* by Eugene O'Neill

1958 *Look Homeward, Angel* by Ketti Frings

1959 *J.B.* by Archibald MacLeish

1960 *Fiorello!* book by Jerome Weidman and George Abbott, music by Jerry Bock, and lyrics by Sheldon Harnick

1961 *All the Way Home* by Tad Mosel

1962 *How to Succeed in Business Without Really Trying* by Frank Loesser and Abe Burrows

1963 No prize awarded

1964 No prize awarded

1965 *The Subject Was Roses* by Frank D. Gilroy

1966 No prize awarded

1967 *A Delicate Balance* by Edward Albee

1968 No prize awarded

1969 *The Great White Hope* by Howard Sackler

1970 *No Place to Be Somebody* by Charles Gordone

1971 *The Effect of Gamma Rays on Man-in-the-Moon Marigolds* by Paul Zindel

1972 No prize awarded

1973 *That Championship Season* by Jason Miller

1974 No prize awarded

1975 *Seascape* by Edward Albee

1976 *A Chorus Line* by James Kirkwood and Nicholas Dante

1977 *The Shadow Box* by Michael Cristofer

1978 *The Gin Game* by Donald L. Coburn

1979 *Buried Child* by Sam Shepard

1980 *Talley's Folly* by Lanford Wilson

1981 *Crimes of the Heart* by Beth Henley

1982 *A Soldier's Play* by Charles Fuller

1983 *'Night, Mother* by Marsha Norman

1984 *Glengarry Glen Ross* by David Mamet

1985 *Sunday in the Park with George* by Stephen Sondheim and James Lapine

1986 No prize awarded

1987 *Fences* by August Wilson

1988 *Driving Miss Daisy* by Alfred Uhry

1989 *The Heidi Chronicles* by Wendy Wasserstein

1864–1918 Frank Wedekind, German playwright, lives. Plays include *Frühlings Erwachen* (1891), *Der Erdgeist* (1895), and *Die Büchse der Pandora* (1903).

1865–1939 William Butler Yeats, Irish poet and dramatist, lives. Plays include *The Countess Cathleen* (1891), *The Land of Heart's Desire* (1894), *Cathleen ni Houlihan* (1902), *Deirdre* (1907), *The Herne's Egg* (1938), and *Purgatory* (1939).

1866–1954 Jacinto Benavente, Spanish playwright, lives. Plays include *Los intereses creados* (1907), *Señora ama* (1908), and *La malquerida* (1913).

1867–1933 John Galsworthy, English novelist and playwright, lives. Plays include *The Silver Box* (1906).

1867–1936 Luigi Pirandello, Italian dramatist, lives. Plays include *Six Characters in Search of an Author* (1921), *Enrico IV* (1923), and *La nuova colonia* (1928).

1868–1918 Edmond Rostand, French playwright, lives. Plays include *La Princesse lointaine* (1895), *Cyrano de Bergerac* (1897), *La Samaritaine* (1897) and *L'Aiglon* (1900).

1868–1936 Maxim Gorky, Russian novelist and playwright, lives. His best-known work is *The Lower Depths* (1902).

1868–1955 Paul Claudel, French dramatist and poet, lives. Plays include *Tête d'or* (1889), *La Cité* (1889), *L'Échange* (1893), *La Jeune Fille Violaine* (1901), *The Tidings Brought to Mary* (1912), and *The Satin Slipper* (1928–29).

1871–1909 John M. Synge, Irish dramatist, lives. Plays include *In the Shadow of the Glen* (1903), *Riders to the Sea* (1904), *Playboy of the Western World* (1907), *The Tinker's Wedding* (1907), and *Deirdre of the Sorrow* (1910).

1874–1965 W. Somerset Maugham, English novelist, short-story writer, and playwright, lives. Plays include *The Circle* (1921), *Our Betters* (1923), and *The Constant Wife* (1926).

1878–1952 Ferenc Molnár, Hungarian playwright, novelist, and short-story writer, lives. Plays include *Az odog* (1907) and *Liliom* (1909).

1878–1967 John Masefield, English poet and dramatist, lives. Plays include *The Tragedy of Nan* (1908).

1882–1944 Jean Giraudoux, French dramatist, novelist, and essayist, lives. Plays include *Amphitryon 38* (1929), *Judith* (1931), *Tiger at the Gates* (1935), *Electre* (1937), *Ondine* (1939), *Sodome et Gomorre* (1943), and *The Madwoman of Chaillot* (1945).

1882–1951 Henri-René Lenormand, French dramatist, lives. Plays include *Le Temps est un songe* (1919), *Les Ratés* (1920), *Le Mangeur des rêves* (1922), and *L'Homme et ses fantômes* (1924).

1884–1965 Sean O'Casey, Irish dramatist, lives. Plays include *Juno and the Paycock* (1924), *The Plough and the Stars* (1926), *The Silver Tassel* (1928), and *Within the Gates* (1933).

1888–1953 Eugene O'Neill, American playwright, lives. Plays include *Beyond the Horizon* (1920), *The Emperor Jones* (1920), *Anna Christie* (1922), *The Hairy Ape* (1922), *All God's Chillun Got Wings* (1924), *Desire Under the Elms* (1926), *The Great God Brown* (1926), *Lazarus Laughed* (1927), *Marco Millions* (1928), *Mourning Becomes Electra* (1931), *Ah, Wilderness!* (1933), *Days Without End* (1934), *The Iceman Cometh* (1939), *A Moon for the Misbegotten* (1943), and *Long Day's Journey into Night* (1956).

1888–1959 Maxwell Anderson, American playwright, lives. Works include *What Price Glory?* (1924) and *Valley Forge* (1934).

1888–1965 T. S. Eliot, American-born English poet, critic, and dramatist, lives. Plays include *Murder in the Cathedral* (1935), *The Family Reunion* (1939), *The Cocktail Party* (1949), *The Confidential Clerk* (1953), and *The Elder Statesman* (1958).

▶

1889–1961 George S. Kaufman, American playwright, lives. Plays include *Dulcy* (1921) and *Beggar on Horseback* (1924) with Marc Connelly, *Dinner at Eight* (1932) and *Stage Door* (1936) with Edna Ferber, *June Moon* (1929) with Ring Lardner, and *You Can't Take It with You* (1930) and *The Man Who Came to Dinner* (1939) with Moss Hart.

1889–1963 Jean Cocteau, French poet, novelist, and dramatist, lives. Plays include *Antigone* (1922), *Orphée* (1926), *Oedipe-Roi* (1928), *La Machine infernale* (1934), *Les Chevaliers de la Table Ronde* (1937), *La Voix humaine* (1930), and *Les Parents terribles* (1938).

1890–1938 Karel Capek, Czech writer and dramatist, lives. Plays include *R.U.R.* (1920).

1890–1945 Franz Werfel, Austrian novelist, poet, and playwright, lives. Plays include *Goat Song* (1921), *Juarez und Maximilian* (1924), *Der Weg der Verheisslung* (1936), and *Jacobowsky und der Oberst* (1936).

1892–1967 Elmer Rice, American playwright, lives. Plays include *The Adding Machine* (1923), *Street Scene* (1929), *The Left Bank* (1931), *We, the People* (1933), and *Dream Girls* (1945).

1893–1939 Ernst Toller, German dramatist, lives. Plays include *Die Wandlung* (1919), *Masse Mensch* (1920), *Die Maschienenstürmer* (1922), *Hinkemann* (1924), and *Hoppla, wir leben!* (1927).

1896–1955 Robert E. Sherwood, American playwright, lives. Plays include *The Road to Rome* (1927), *Reunion in Vienna* (1931), *The Petrified Forest* (1935), *Idiot's Delight* (1936), *Abe Lincoln in Illinois* (1938), and *There Shall Be No Night* (1940).

1897–1975 Thorton Wilder, American novelist and playwright, lives. Plays include *The Angel That Troubled the Waters* (1928), *Our Town* (1938), *The Skin of Our Teeth* (1942), *The Matchmaker* (1954), and *Plays for Bleecker Street* (1962).

1898–1936 Federico García Lorca, Spanish poet and playwright, lives. Plays include *La zapatera prodigiosa* (1930), *Yerma* (1934), and *La casa de Bernarda Alba* (1936).

1898–1956 Bertolt Brecht, German poet and dramatist, lives. Plays include *Trommeln in der Nacht* (1922), *Im Dickicht der Städte* (1924), *Mann ist Mann* (1926), *The Three-Penny Opera* (1928) and *Rise and Fall of the City of Mahagonny* with Kurt Weill (1929), *The Private Life of the Master Race* (1938), *The Life of Galileo* (1938–39), *Mother Courage and Her Children* (1939), *The Good Woman of Setzuan* (1939–41), and *The Caucasian Chalk Circle* (1944–45).

1899–1973 Sir Noel Coward, English playwright and actor, lives. Plays include *The Vortex* (1924), *Hay Fever* (1925), *Bitter Sweet* (1929), *Private Lives* (1930), and *Blithe Spirit* (1941).

1904–61 Moss Hart, American playwright, lives. Plays include *You Can't Take It with You* (1936), *The American Way* (1939), *The Man Who Came to Dinner* (1939), and *George Washington Slept Here* (1940), all with George S. Kaufman, and *Winged Victory* (1943), *Light Up the Sky* (1948), and *The Climate of Eden* (1952).

1905–80 Jean-Paul Sartre, French philosopher and dramatist, lives. Plays include *The Flies* (1943), *No Exit* (1944), *The Respectful Prostitute* (1946), *Dirty Hands* (1948), *The Devil and the Good Lord* (1951), and *The Condemned of Altona* (1959).

1905–84 Lillian Hellman, American playwright and memoirist, lives. Plays include *The Children's Hour* (1939), *The Little Foxes* (1939), *Watch on the Rhine* (1941), *The Searching Wind* (1944), *Another Part of the Forest* (1946), and *Toys in the Attic* (1960).

1906–63 Clifford Odets, American playwright, lives. Plays include *Waiting for Lefty* (1935), *Awake and Sing!* (1935),

▶

Golden Boy (1937), and *The Country Girl* (1950).

1906–89 Samuel Beckett, Irish-born French novelist and dramatist, lives. Plays include *Waiting for Godot* (1952), *End-game* (1957), *Krapp's Last Tape* (1958), *Happy Days* (1961), *Breath* (1971), and *Company* (1980).

1907–73 W. H. Auden, English poet and dramatist, lives. Plays include *The Dance of Death* (1933), *The Dog Beneath the Skin* (1935) with Christopher Isherwood, *The Ascent of F6* (1936), and *On the Frontier* (1938).

1907– Christopher Fry, English dramatist, lives. Plays include *A Phoenix Too Frequent* (1946), *The Lady's Not for Burning* (1948), *Venus Observed* (1950), and *The Dark Is Light Enough* (1954).

1908–81 William Saroyan, American novelist, short-story writer, and playwright, lives. Plays include *My Heart's in the Highlands* (1939) and *The Time of Your Life* (1939).

1910–86 Jean Genet, French dramatist and novelist, lives. Plays include *The Maids* (1948), *Haute Surveillance* (1949), *The Balcony* (1956), *Les Nègres* (1959), and *Les Paravents* (1961).

1910–87 Jean M. L. P. Anouilh, French dramatist and screenwriter, lives. Plays include *Antigone* (1942), *The Lark* (1953), *La Culotte* (1978), and *Le Nombril* (1981).

1911–83 Tennessee Williams, American playwright, lives. Plays include *The Glass Menagerie* (1944), *A Streetcar Named Desire* (1947), *Summer and Smoke* (1947), *The Rose Tattoo* (1951), *Camino Real* (1953), *Cat on a Hot Tin Roof* (1955), *Suddenly Last Summer* (1959), *Sweet Bird of Youth* (1959), and *The Night of the Iguana* (1961).

1911– Max Frisch, Swiss playwright, novelist, and dramatist, lives. Plays include *Now They're Singing Again* (1944), *The Great Wall of China* (1969), *When the War Was Over* (1967), and *Triptychon* (1978).

1912– Eugène Ionesco, Rumanian-born French dramatist, lives. Plays include *The Bald Soprano* (1950), *The Lesson* (1950), *Amédée* (1954), *Rhinoceros* (1959), and *Exit the King* (1962).

1913–73 William M. Inge, American playwright, lives. Plays include *Come Back, Little Sheba* (1949), *Picnic* (1953), *Bus Stop* (1955), and *The Dark at the Top of the Stairs* (1957).

1915– Arthur Miller, American playwright, short-story writer, and novelist, lives. Plays include *All My Sons* (1947), *Death of a Salesman* (1949), *The Crucible* (1953), *A View from the Bridge* (1955), and *After the Fall* (1964).

1916–82 Peter Weiss, German-born Swedish dramatist, novelist, and film director, lives. Plays include *The Tower* (1948), *The Insurance* (1952), *Marat/Sade* (1964), *Discourse on Vietnam* (1967), and *Trotsky in Exile* (1972).

1921– Friedrich Dürrenmatt, Swiss playwright, lives. Plays include *The Marriage of Mr. Mississippi* (1952) and *The Visit* (1956).

1923–64 Brendan Behan, Irish playwright, lives. Plays include *The Quare Fellow* (1954) and *The Hostage* (1958).

1926– Peter Shaffer, English playwright, lives. Plays include *Equus* (1973), *Amadeus* (1974), and *Lettice and Lovage* (1987).

1927– Ann Jellicoe, English playwright and director, lives. Plays include *The Knack* (1961) and *The Giveaway* (1969).

1927– Neil Simon, American playwright, lives. Plays include *Barefoot in the Park* (1963), *The Odd Couple* (1965), *Chapter Two* (1978), *I Ought to Be in Pictures* (1980), *Brighton Beach Memoirs* (1984), *Biloxi Blues* (1985), and *Broadway Bound* (1986).

1928– Edward Albee, American playwright, producer, and director, lives. Plays include *The Zoo Story* (1959), *Who's Afraid of Virginia Woolf?* (1962), and *Tiny Alice* (1964).

▶

1929– John Osborne, English playwright, lives. Plays include *Look Back in Anger* (1957), *The Entertainer* (1957), *Luther* (1961), and *Inadmissible Evidence* (1964).

1930– Harold Pinter, English dramatist, lives. Plays include *The Room* (1957), *The Birthday Party* (1958), *The Dumb Waiter* (1960), *The Caretaker* (1960), *The Collection* (1962), *The Homecoming* (1965), and *Betrayal* (1980).

1934– Imamu Amiri Baraka (formerly LeRoi Jones), American poet and playwright, lives. Plays include *Dutchman* (1964) and *Slave Ship* (1967).

1937– Tom Stoppard, Czech-born English playwright, lives. Plays include *Rosencranz and Guildenstern Are Dead* (1967) and *Travesties* (1971).

1937– Lanford Wilson, American playwright, lives. Plays include *The Hot L Baltimore* (1973), *The Fifth of July* (1978), and *Talley's Folly* (1979).

1939– Alan Ayckbourne, English playwright, lives. Plays include *The Norman Conquests* (1973), *A Chorus of Disapproval* (1984), and *Henceforward* (1988).

1940– David Rabe, American playwright, lives. Plays include *The Basic Training of Pavlo Hummel* (1971) and *Streamers* (1978).

1943– Sam Shepard, American playwright, lives. Works include *The Curse of The Starving Class* (1976) and *Fool for Love* (1984).

1945– August Wilson, American playwright and poet, lives. Plays include *Fences* (1986) and *The Piano Lesson* (1988).

1947– David Mamet, American playwright, lives. Works include *American Buffalo* (1975), *Glengarry Glen Ross* (1984), and *Speed the Plough* (1987).

1947– David Hare, American playwright, lives. Works include *Plenty* (1978).

1950– Wendy Wasserstein, American playwright, lives. Works include *Uncommon Women and Others* (1978) and *The Heidi Chronicles* (1988).

1952– Beth Henley, American playwright, lives. Works include *Crimes of the Heart* (1981).

Composers

An Overview

c. 1300–77 Guillaume de Machaut, French poet and composer, lives. He is considered greatest French musician of the ars nova.

1325–97 Francesco Landini, Italian composer and instrumentalist, lives. He is considered a master of Florentine ars nova.

c. 1440–1521 Josquin des Prés, Flemish composer, lives. Best known for his chansons and motets, he was considered greatest composer of his age.

c. 1505–85 Thomas Tallis, English organist and composer, lives. Works include choral and instrumental church and secular music.

c. 1525–94 Giovanni Pierluigi da Palestrina, Italian composer, organist, and choirmaster, lives. Works include sacred and secular madrigals, masses, motets, and lamentations.

1532–94 Orlando di Lasso, Belgian composer of liturgical music, lives. Works include masses, motets, and magnificats.

▶

Development of Western Music

B.C.

c. 800 Mesopotamians use string and wind instruments and develop system of musical notation.

500s Greek philosopher and mathematician Pythagoras (c. 582–507 B.C.) studies ratio of vibrations. It provides basis of Western system of musical notation.

c. 400 Music is used extensively in Greek tragedies. Lyre called *kithara* and reed-blown instrument known as *aulos* are in use.

A.D.

300s Roman bishop Ambrose (c. 339–97) catalogs growing body of church chants, which come to be known as Ambrosian chants.

c. 600 Pope Gregory I (c. 540–604) collects church chants known as plainsong or Gregorian chants, monophonic melodies without instrumental accompaniment.

800s Troping develops, addition of embellishing melody above existing melody.

c. 900 *Musica enchiriadis* is written as handbook for singers. It describes practice of polyphony, or use of two or more melodies over one another.

1100s Troubador tradition of wandering poet-singers develops in southern France. Secular songs are written in vernacular.

c. 1200 Motet is developed, adding text to the voice parts above tenor.

c. 1300 *Ars Nova* (New Art), treatise by Philippe de Vitry, is written. It catalogs advances in harmony and metrics and marks turning point in developing musical sophistication.

1364 French composer Guillaume de Machaut (c. 1300–77) writes earliest known Mass to be composed by one person.

1426 French composer Guillaume LeGrant composes first Mass to combine choral and solo sections.

c. 1450 Distinction of vocal parts for tenor, soprano, bass, and contralto is completed.

c. 1450 Beginning of Renaissance in music.

1475 First dictionary of music terms, compiled by Johannes Tinctoris (1436–1511), is published.

1500s Major and Minor scales are developed.

1500s Emergence of instrumental music separate from vocal. Lute, organ, harpsichord, and clavicord are commonly used.

1501 First printed musical composition, by Josquin des Prés (c. 1440–1521), is published, in Venice.

c. 1550 Emergence of final form of violin, probably by Andrea Amati. His grandson Nicolò is teacher of Antonio Stradivari, most famous of Cremonese instrument-makers.

1597 First opera, *Dafne*, by Jacopo Corsi (1561–1622) and Jacopo Peri (1561–1633), is performed, in Florence.

c. 1600 Baroque period begins with development of monodic style, which combines sung melody with underlying bass played by two instruments.

1607–8 Italian composer Claudio Monteverdi (1567–1643) sophisticates opera form in *Orfeo* and *L'Arianna* by introducing overture and recitative.

1678–1741 Italian composer Antonio Vivaldi lives. He develops three-movement concerto form much imitated by his contemporaries, including J. S. Bach.

1685–1750 German composer Johann Sebastian Bach lives. Considered greatest Baroque composer, he writes fugues, concertos, organ works, harpsichord compositions, and choral cantatas, including *Brandenburg Concertos* (1721), *The Well-Tempered Clavier* (1722), *St. Matthew's Passion* (1729), and *Mass in B Minor* (1738).

1685–1759 German composer Georg Friedrich Handel lives. He develops oratorio. Handel's 20 oratories include his masterpiece *Messiah* (1742). He also composes operas, concertos, and orchestral works.

c. 1700 First pianoforte is constructed by Bartolomeo Cristofori, in Florence.

c. 1730 Orchestral symphony begins to develop into its eventual classical four-movement form.

1732–1809 Austrian composer Franz Joseph Haydn, a pivotal figure in development of symphony, lives. He composes more than 100 symphonies, 80 string quartets, and 60 piano sonatas as well as concertos, operas, oratorios, and Masses.

Development of Western Music

c. 1750	Classical period, characterized by form, order, balance, and restraint in composition, begins. Instrumental music eclipses vocal in importance for first time.
1770–90	Austrian composer Wolfgang Amadeus Mozart, considered one of greatest of all composers, contributes to all musical forms, writing about 650 compositions. He writes over 50 symphonies, over 40 concertos, 20 operas, and hundreds of other works, both sacred and secular.
c. 1770	Development of chamber music form, played by small group (trio, quartet, etc.) of musicians, as opposed to full orchestra.
1800–27	German composer Ludwig van Beethoven's compositions mark culmination of Classical period and beginning of Romantic period. His masterpieces include nine symphonies, 32 piano sonatas, 16 string quartets, 10 violin sonatas, and two Masses.
1813–83	German composer Richard Wagner lives. His highly influential operas mark height of German Romanticism.
1850–80	Italian composer Giuseppe Verdi, often considered greatest composer of operas, composes his major works.
c. 1825	Romantic period, characterized by emotional exuberance, begins.
1910–60	Russian-born American composer Igor Stravinsky (1882–1971) composes highly innovative work: ballet music characterized by dissonance, abrupt juxtapositions, and imbalance that leads to development of atonal music.
1924	Austrian composer Arnold Schoenberg (1874–1951) composes his revolutionary *Quintet for Winds*, giving equal validity to all 12 tones of chromatic scale and rejecting traditional harmony.
c. 1950	Development of electronic music, using synthesizers and digital computers, begins.

1543–1623 William Byrd, English composer and organist, lives. He is best known for sacred polyphonic compositions.

1557–1602 Thomas Morley, English composer, theorist, and organist, lives. He is best known for his comprehensive treatise on musical composition (1597).

c. 1563–1626 John Donland, English lutenist and composer, lives. Works include *Lachrimae* (1604) and collection of songs for lute.

1567–1643 Claudio Monteverdi, Italian composer of liturgical and secular music, lives. Works include *Vespers* (1639), 12 operas, madrigals, and motets.

1583–1625 Orlando Gibbons, English organist and composer, lives. Works include madrigals, chamber music, and keyboard pieces.

1632–87 Jean-Baptiste Lully, French composer of comedy ballets, opera, and church music, lives.

1653–1713 Arcangelo Corelli, Italian violinist and composer, lives. Works include solo sonata and trio sonatas. Corelli establishes the concerti grossi style for the violin.

c. 1659–1695 Henry Purcell, English organist and composer, lives. He is known for compositions for plays, including *The Fairy Queen* (1692) and *The Tempest* (1695), as well as songs, church music, and chamber music.

1660–1725 Alessandro Scarlatti, Italian conductor and composer, lives. Works include 115 operas, 20 oratorios, 600 cantatas, 10 masses, motets, and chamber music.

1668–1733 Françoise Couperin, French organist and composer, lives. Works include organ and harpsichord pieces and secular and church music.

1678–1741 Antonio Vivaldi, Italian violinist and composer, lives. Master of Italian Baroque, his works include church music, oratorios, and the concertos known as *The Four Seasons*.

▶

Development of Western Musical Instruments

B.C.	
c. 800	String and wind instruments are used by Mesopotamians, who develop system of musical notation.
200s	First mechanically blown instrument, ancient Greek *hydraulis* (water organ), is invented by Ctesibios of Alexandria, Egypt and is used into Middle Ages.
A.D.	
1000	Rebabs (simple bowed string instruments) from North Africa enter Europe by way of wandering troubadors.
1000	Harp is developed.
1000s	Early slide organs develop into lever-operated organs.
1200s	Three types of drums are in use in medieval Europe by this time: paired kettle drums (*nakers*); tabor; and tambourine, ancestral to modern timpani.
1300s	Metal horns, in imitation of animal horns, begin to be made.
1385	*Eschiquier*, type of keyboard instrument, is in use in Europe by this time.
c. 1400	Development of one-piece, wooden recorder.
1400s	Keyed organs are built, using broad keys played by fists, not fingers; mechanism of stops is included.
1400s	*Scheitholt*, primitive zither, is developed in Switzerland and southern Germany.
1400s	Keyboard instruments clavichord and harpsichord are developed. Clavichord's strings are struck by brass blade, while those of harpsichord are plucked.
1400s	Curved, looped trumpets begin to replace straight trumpets.
1400s	Early attempt to alter natural trumpets. Loose mouthpiece permits player to alter pitch by sliding it in and out.
1400s	Sackbut, forerunner of trombone, first appears. By the end of century it evolves into its modern form.
1470s	*Vihuela*, lutelike guitar, is developed in Spain.
1470s	Introduction of Italian viola, then known as *lira da braccio*.
1500	Harpsichords are being built with two keyboards instead of one.
c. 1500	Precursor of the modern violin first appears, in Italy.
1500s	Swiss infantry fife is introduced into France and England.
1500s	Finger-keyed organs are built.
c. 1530	*Curtal*, one-piece bassoon, is developed in Italy.
c. 1550	Final form of violin emerges from workshop of Cremonese master Andrea Amati.
c. 1550	Precursor of cello appears along with first double bass, which developed from violone, member of viol family.
1644–1737	Antonio Stradivari, apprentice to Amati family workshop, makes over 1,100 violins, violas, cellos, and guitars in his lifetime. He is considered maker of the world's finest violins.
c. 1660	Oboe, developed from shawm, an early Middle Eastern woodwind, is introduced in France.
c. 1660	Four-piece bassoon develops out of *curtal*.
1670s	Banjo first appears, in Jamaica.
1670s	Bagpipes, ancient instruments used all over Near East and continental Europe, first appear in Scotland, where they become national instruments.
c. 1675	Early, one-keyed transverse flute is developed, in Paris.
1700s	Development of Neapolitan mandolin from earlier lute form *mandola*.
c. 1700	First clarinet is developed in Nuremberg.
c. 1700	Slide trumpet is invented in England. Two-legged slide alters pitch and permits playing in minor keys.
c. 1700	Florentine harpsichord-maker Bartolomeo Cristofori (1655–1731) builds first piano. Its four-octave range will eventually be expanded into modern seven and a half.
c. 1718	Crooks are introduced into design of French horn, allowing instrument to play different pitches.
1720	Pedal harp is invented.
1739	First contrabassoon (double bassoon) is built, in London.
1754	French horn with padded lever keys similar to woodwind keys is built in St. Petersburg.
1760s	Organ hurdy-gurdy is developed in France.

Development of Western Musical Instruments

1761	First electric instrument, electric harpsichord, is invented in Paris.
c. 1770	Violin bow currently in use is designed by François Tourte; it is characterized by slight inward curve and allows for virtuosic playing.
c. 1770	Grand piano is developed in London.
1793	Development of bass clarinet.
1796	Lever-keyed trumpet debuts at performance of Haydn's *Concerto in E flat* in Vienna.
c. 1800	Violin is modernized: Sound post and base bar are enlarged to produce louder tones for performances in large concert halls; bridge, neck, and fingerboard are altered to allow artist to play more virtuosic compositions.
1810	Keyed bugle is patented in London.
1820	Development of alto clarinet.
1820	Double-pedal harp is invented by Sebastian Erard.
c. 1820	Violin chinrest is invented by violinist and composer Louis Spohr to help keep instrument steady for virtuosic playing.
1820s	First cylindrical piston valves for brass instruments, Stolzen valves, are introduced in Paris.
1821	Harmonica is developed in Germany.
1822	Keyed accordion is invented in Germany.
1825	First one-piece cast-iron frame for piano is introduced to replace wooden frame. It allows use of heavier strings and greatly increases piano's loudness.
1828	Valved cornet appears in Paris.
1830	Valved trombone is invented in Vienna.
1830s	Intricate system of keys is developed for flute by Theobald Böhm (1794–1881).
1835	Bass tuba is patented in Berlin.
1839	Modern piston valves, Perinet valves, are introduced in Paris.
1840	Saxophone is invented by Adolphe Sax (1814–94) and patented in 1846.
1847	Introduction of first cylindrical-bored silver flute. Previous flutes had had conical bores.
1849	Circular helicon tuba, designed to rest partially on shoulder, is perfected in Vienna.
1880s	Frets and metal strings are first used on banjos, in United States.
1899	Sousaphone, helicon tuba with large removable bell, is designed by John Philip Sousa (1854–1932).
1906	Telharmonium, built by Thaddius Cahill (1867–1934) in Massachusetts, is first instrument to actually generate sound electronically, rather than simply reproducing it.
1920s	Early music synthesizers are produced: ondes Martenot, by Maurice Martenot, and theremin, invented by Lev Theremin.
1935	Hammond organ, first commercially successful electric instrument, is invented.
1935	Electric acoustic guitar is invented by Gibson Guitar Company.
1946	Solid-body electric guitar is invented by Fender Guitar Company.
1955	First electronic music synthesizer is built, in Princeton, New Jersey.
1964	Moog synthesizer, a keyboard instrument built by Robert Moog, is introduced; first with chord capabilities is introduced by Moog in 1976.

1683–1764 Jean-Philippe Rameau, French music theorist and composer, lives. His works include operas, opera ballets, and treatises outlining modern harmony.

1685–1757 Domenico Scarlatti, Italian composer and son of Alessandro Scarlatti, lives. Works include 550 sonatas for harpsichord, as well as concertos, operas, cantatas, and sacred music.

1685–1759 George Friedrich Handel, German-born English composer, lives. Works include *The Water Music* (1717), *Messiah* (1742), and *Music for the Royal Fireworks* (1749).

1695–1750 Johann Sebastian Bach, German organist and composer, lives. Works include two- and three-part inventions, fugues, suites, and instrumental and religious works.

▶

1710–36 Giovanni Battista Pergolesi, Italian composer of operas and comic intermezzos, lives. Works include the renowned *Stabat Mater*, as well as other church music and sonatas.

1714–87 Christoph Willibald von Gluck, German composer, lives. His best-known works are operas, including *Orfeo ed Euridice* (1762) and *Alceste* (1767).

1732–1809 Franz Joseph Haydn, Austrian composer, lives. Well-known works include over 100 symphonies; he is known as the father of the symphony. Mature works include *Surprise Symphony* (no. 94 in G Major, 1791) and *London Symphony* (no. 104 in D Major, 1795). After 1800 Haydn wrote oratorio *The Seasons* (1801), Masses, and string quartets.

1743–1805 Luigi Boccherini, Italian composer, lives. Well-known works include *Cello Concerto* (in B Flat, c. 1780s), in addition to 20 symphonies, much chamber music, *Stabat Mater* (1781) for three voices and strings, and *Christmas Cantata* (1802).

1752–1832 Muzio Clementi, Italian-born British composer, lives. Works include *Gradus ad Parnassum* (1817) and much early piano music (galants, sonatas, and sonatinas), chamber music, and symphonies.

1756–91 Wolfgang Amadeus Mozart, Austrian composer, lives. Works include 41 symphonies, among them the *Jupiter*, No. 41 (1788); 20 operas, including *The Marriage of Figaro* (1786), *Don Giovanni* (1787), and *The Magic Flute* (1791); 21 concertos, among them *Elvira Madigan*, No. 23; church music; fantasias; piano sonatas; and chamber music.

1760–1842 Luigi Cherubini, Italian composer, lives. Works include *Mass in F Major* (1809) and *Requiem in D Minor* (1836), in addition to operas, orchestral music, and chamber music.

1770–1827 Ludwig van Beethoven, German composer, lives. Well-known works include *Symphony Number 1* (C Major, opus 21, 1800), *Moonlight Sonata* (Sonata in C Sharp Minor, opus 27, no. 2 for piano, 1801), *Symphony No. 2* (D Major, opus 36, 1802), *Kreutzer Sonata* (in A Major for piano and violin, 1803), *Eroica (Symphony No. 3* in E Flat Major, opus 55, 1804), *Symphony No. 4* (B Flat Major, opus 60, 1806), *Violin Concerto in D Major* (opus 61, 1806), *Symphony No. 5* (C Minor, opus 67, 1808), *Pastoral Symphony* (No. 6, in F Major, opus 68, 1808), *Symphony No. 7* (A Major, opus 92, 1812), *Symphony No. 8* (F Major, opus 93, 1812), *Wellington's Victory* (opus 91, 1813), and nine Classical-Romantic symphonies (1800–1824).

1782–1840 Niccolò Paganini, Italian composer, lives. Well-known works include violin caprices (1801–7); concertos for violin and guitar; and six *Violin Concertos*, most virtuosic of his violin music.

1786–1826 Carl Maria von Weber, German composer, conductor, and pianist, lives. Works include operas *Der Freischutz* (1821) and *Oberon* (1826) and other choral and orchestral pieces.

1791–1857 Karl Czerny, Austrian composer, lives. He is best known for technical studies for piano.

1791–1864 Giacomo Meyerbeer, German composer, lives. He is best known for operas, including *Robert le Diable* (1831) and *Les Huguenots* (1836).

1792–1868 Gioacchino Rossini, Italian composer, lives. Works include *William Tell* and *The Barber of Seville* as well as cantatas, songs, and chamber music.

1797–1828 Franz Schubert, Austrian composer, lives. Works include *Tragic Symphony* (no. 4 in C Minor, 1816), *Symphony No. 5* (in B-flat Major, 1816), *Trout Quintet* (1819), *Unfinished Symphony* (in B Minor, 1822), *String Quartet* (in D Minor, 1824), and *The Great Symphony* (in C Major, 1828), as well as song cycles *Müllerin* (1823), *Die*

▶

Winterreise, (1827) and *Schwanengesang* (1828).

1797–1848 Gaetano Donizetti, Italian composer, lives. His best-known works are operas, including *Lucrezia Borgia* (1833) and *Don Pasquale* (1843).

1801–35 Vincenzo Bellini, Italian composer of operas, lives. Works include *La Straniera* (1829) and *Norma* (1831).

1803–1869 Hector Berlioz, French composer, lives. Works include *Symphonie Fantastique* (1831), overture to *Le Corsaire* (1831–52), symphonies *Harold en Italie* (1834) and *Romeo et Juliette* (1839), and oratorio *L'Enfance du Christ* (1854).

1804–49 Johann Strauss (the Elder), Austrian composer, lives. His best-known work is *Radetzky March*. His son Johann Strauss (the Younger; 1825–99) wrote operettas and waltzes, including *The Blue Danube* and *Tales from the Vienna Woods*.

1804–57 Mikhail Glinka, Russian composer, lives. Works include *Capriccio Brillante* (1845), *Summer Night in Madrid* (1848), *Kamarinskaya* (1849), and *Festival Polonaise* (1855).

1809–47 Felix Mendelssohn, German composer, lives. Works include *Symphony No. 1* (in C Minor, opus 1, 1824), *A Midsummer Night's Dream* (1826), *Reformation Symphony* (no. 5 in D Major, opus 107, 1830), *The Hebrides Overture* (1832), *Scottish Symphony* (no. 3 in A Minor-Major, opus 56, 1842), *Italian Symphony* (no. 4 in A Major-Minor, opus 90, 1833), and *Violin Concerto in E Minor* (opus 46, 1844).

1810–49 Frédéric Chopin, Polish composer, lives. His works, written primarily for piano, include *Piano Concerto No. 2 in F Minor* (opus 21, 1829), *Twelve Grandes Études* (opus 10, 1832), *Piano Concerto No. 1* (in E Minor, opus 11, 1830), *Twelve Études* (opus 25, 1836), *Twenty-four Préludes* (opus 28, 1839), *Sonata in B-flat Minor* (opus 35, 1839), *Sonata in B Minor* (opus 58, 1844), *Polonaise-Fantasie* (opus 61, 1845–46), and *Minute Waltz* (1847).

1810–56 Robert Schumann, German composer, lives. Works include *Carnaval* (opus 9, 1834–35) and *Kreisleriana* (1838), both written for piano, and four symphonies: no. 1, *Spring Symphony* (1841); no. 2 (1846); no. 3, *Rhenish* (1850); and No. 4, *Manfred Overture* (1849, rewritten 1851).

1811–86 Franz Liszt, Hungarian composer, lives. Works include piano *Concerto No. 1* (in A Major, 1849); *Grandes Études de Paganini* (1838) and *Fantasia on Hungarian Rhapsodies* (1846–85), both for piano; 13 tone poems, including *Les Préludes* (1854) and *A Faust Symphony* (1854–57); and the oratorio *Christus* (1862–67).

1813–1901 Giuseppe Verdi, Italian composer, lives. In addition to his 27 operas, including *Rigoletto* (1851), *La Traviata* (1853), and *Aïda* (1871), his works include *Messa da Requiem* (1874), *Ave Maria* (1889), *Te Deum* (1896), and *Stabat Mater* (1897).

1813–83 Richard Wagner, German composer, lives. He is best known for his operas, including four-part *Ring of the Nibelungen* (1853–74), *Tristan and Isolde* (1859), and *Parsifal* (1882).

1818–93 Charles Gounod, French composer of operas, church music, and symphonies, lives.

1819–80 Jacques Offenbach, French composer, lives. Works include 90 operettas and opera *The Tales of Hoffmann* (1881).

1822–90 César Franck, Belgian-born French composer, lives. Works include tone poem *Les Djinns* (1884), *Sonata in A Major for Violin and Piano* (1886), *Symphony in D Minor* (1888), *String Quartet in D Major* (1889), and *Trois Chorals* (1890) for organ.

1824–84 Bedrich Smetana, Bohemian composer, lives. In addition to his operas, including *The Bartered Bride* (1866), his

▶

symphonic poems *Má Vlast* (My Country) (1874–79, including *The Moldau*) and string quartet *From My Life* (1876).

1824–96 Anton Bruckner, Austrian composer, lives. Well-known works include nine symphonies (1886–96), *Mass No. 1* (in D Minor, 1864), *Mass No. 2* (in E Minor, 1866), *Grosse Messe* (no. 3 in F Minor, 1867–68), *Te Deum* (1881–84), chamber music, and organ works.

1825–99 Johann Strauss the Younger, Austrian composer, lives. Works include over 400 waltzes, including "The Blue Danube" (1867) and "Tales from the Vienna Woods" (1868).

1833–87 Aleksandr Borodin, Russian composer, lives. Works include *Symphony No. 1 in E-flat Major* (1862–67), popular "Polovtsian Dances" from his opera *Prince Igor* (1869), and tone poem *In the Steppes of Central Asia* (1880).

1833–97 Johannes Brahms, German composer, lives. In addition to 214 solo songs with piano and chamber music, works include *Ein deutsches Requiem* (opus 45, 1868), *Hungarian Dances* (1869), four symphonies (1876–85), *Violin Concerto in D Major* (opus 77, 1878), *Academic Festival Overture* (1880), and *Piano Concerto No. 2 in B-flat Major* (opus 83, 1881).

1835–1921 Camille Saint-Saëns, French composer, lives. Works include popular *Danse Macabre* (1874), *Carnival des Animaux* (1886), and two cello concertos (1873 and 1902).

1838–75 Georges Bizet, French composer, lives. In addition to operas, well-known works include *Symphony in C Major* (1855), orchestral suite *Roma* (1860–68), piano suite *Jeux d'enfants* (1871), and *L'Arlesienne Suite* (1872).

1838–1920 Max Bruch, German composer, lives. Works include three violin concertos (in G Minor, 1868; in D Minor, 1878; and in D Minor, 1891), the *Scottish Fantasy* (1880) for violin and orchestra, and *Kol Nidre* (1881) for cello and orchestra.

1839–81 Modest Mussorgsky, Russian composer, lives. Works include orchestral work *Night on Bald Mountain* (1860–66), opera *Boris Godunov* (1868), and piano piece *Pictures at an Exhibition* (1874).

1840–93 Peter Ilyich Tchaikovsky, Russian composer, lives. Works include overture *Romeo and Juliet* (1869), *Little Russian Symphony* (no. 2 in C Minor, opus 17, 1872), *Piano Concerto No. 1* (in B-flat Minor, opus 23, 1875), *Symphony No. 3* (in D Major, opus 29, 1875), ballet *Swan Lake* (1877), *Violin Concerto in D Major* (opus 35, 1878), *Manfred Symphony* (1885), ballets *The Sleeping Beauty* (1890) and *The Nutcracker* (1892), and *Symphonies Pathétique* (no. 6 in B Minor, opus 74, 1893).

1841–1904 Antonin Dvorák, Bohemian composer, lives. Works include *Slavonic Dances* (1878–86); *Slavonic Rhapsodies* (1878); *Violin Concerto in A Minor* (opus 53, 1880); nine symphonies (1874–93), including *From the New World, Ninth Symphony* (1893); and much chamber music, including *Dumky Trio* (1891).

1843–1907 Edvard Grieg, Norwegian composer, lives. Works include *Lyric Pieces* (1867–1901), *Piano Concerto in A Minor* (1868), *Slatter, Norwegian Peasant Dances* (opus 72), and incidental music to *Peer Gynt* (1874–75), and the *Holberg Suite* (1884).

1844–1908 Nikolai Rimsky-Korsakov, Russian composer, lives. In addition to operas, works include three symphonies (1865–74), *Capriccio Espagnol* (1887), symphonic suite *Sheherazade* (1888), and *Russian Easter Festival Overture* (1888).

1845–1924 Gabriel Fauré, French organist and composer, lives. Works include operas, orchestral music, chamber music, piano pieces, and church music, including *Requiem* (1888).

▶

History of European and American Opera

1597 First opera, *Dafne*, by Jacopo Corsi (1561–1602) and Jacopo Peri (1561–1633), libretto by Ottavio Rinuccini (1562–1621), is performed, in Florence. No copy of musical score survives.

1600 *Euridice*, also by Peri, is earliest opera whose libretto and musical score survive. It is first performed at Pitti Palace in Florence.

1607 *La favola d'Orfeo* by Claudio Monteverdi (1567–1643) is presented during Carnival in Mantua, Italy. Monteverdi improves upon Peri's operatic form by increasing number of instruments in orchestra, introducing operatic overture, and expressively using recitative.

1627 First German opera, *Dafne*, by Heinrich Schütz (1585–1672), is performed at Torgau.

1637 First public opera house, San Cassiano, opens in Venice.

1639 First completely comic opera, *Chi soffre speri*, by Virgilio Mazzocchi (1597–1646) and Marco Mazzoli (1619–62), debuts in Rome. Its libretto is by Giulio Cardinal Rospigliosi (1600–1669), later Pope Clement IX.

1671 First French opera, *Pomone*, by Robert Cambert (1628–77), inaugurates Académie Royal de Musique, now Paris Opera.

1678 *Adam und Eva*, by Johann Theile (1646–1724), is first *Singspiel*, comic opera with spoken dialogue.

1689 First English opera, *Dido and Aeneas*, by Henry Purcell (1659–95), is performed by students at girls' school in London.

1711 *Rinaldo*, first opera of Georg Friedrich Handel (1685–1759), is performed in London. He becomes England's leading opera composer. Among his operas are *Julius Caesar* (1724), *Tamerlano* (1724), *Rosalinda* (1725), *Sosarme* (1732), *Orlando* (1733), *Ariodante* (1735), *Alcina* (1735), *Berenice* (1737), and *Xerxes* (1738).

1720 Italian composer Benedetto Marcello (1686–1739) publishes satire on current state of opera. It prompts reform movement among composers to improve dramatic structure and give greater importance to orchestra and chorus.

1728 *The Beggar's Opera*, with text by Englishman John Gay (1685–1732) and music by Pepusch, becomes overnight sensation in London.

1762 *Orfeo ed Euridice*, by Christoph Willibald Gluck (1714–87), debuts in Venice. Gluck's works, influenced by reform movement, stress importance of music and story equally. His later works include *Alceste* (1767), *Paris and Helen* (1770), *Iphigenia in Aulis* (1774), and *Iphigenia in Tauris* (1779).

1767 Wolfgang Amadeus Mozart (1756–91) writes his first opera, *Apolla and Hyacinth*. Among his later works are *Bastien and Batienne* (1768); *Idomeneo* (1781), his first serious (as opposed to comic) opera; *The Abduction from the Seraglio* (1782); *The Marriage of Figaro* (1786); *Don Giovanni* (1787); *Cosi fan tutti* (1790); *The Magic Flute* (1791); and *La Clemenza di Tito* (1791).

1805 Ludwig van Beethoven (1770–1827) writes his first and only complete opera, *Fidelio*.

1816 *The Barber of Seville*, most famous opera of Gioacchino Rossini (1792–1868), is first performed. Among Rossini's other operas are *The Italian Girl in Algiers* (1813), *La Cenerentola* (1817), *Semiramide* (1823), and *William Tell* (1829).

1821 *Der Freischütz*, by Carl Maria von Weber (1786–1826), introduces gloomily heroic German Romanticism to opera.

1830 *Anna Bolena*, first opera by Italian Gaetano Donizetti (1797–1848), is produced in Milan. His later works include *L'elisir d'amore* (1832), *Lucia di Lammermoor* (1835), and *The Daughter of the Regiment* (1840).

1831 Italian Vincenzo Bellini (1801–35), known for his impressive and difficult arias in *bel canto* style, produces his two greatest works: *Norma* and *La Sonnambula*.

1831 *Robert le Diable*, by German composer Giacomo Meyerbeer (1791–1864), is performed in Paris. Its large orchestra, huge stage, and spectacular pageantry qualify it as the first "grand opera."

1853 Two of greatest operas of Italian composer Giuseppe Verdi (1813–1901), *La Traviata* and *Il Trovatore*, premiere. Among Verdi's other masterpieces are *Rigoletto* (1851), *La Forza del destino* (1862), *Aïda* (1871), *Otello* (1887), and *Falstaff* (1893).

1853 German composer Richard Wagner (1813–83) begins 21 years of work on his operatic tetralogy *Der Ring des Nibelungen*, consisting of *Das Rheingold, Die Walküre, Siegfried*, and *Götterdammerüng*. Among Wagner's other operas are *The Flying Dutchman*, first performed in 1843; *Tannhäuser*, which premiered in 1845; *Lohengrin* (1850); *Tristan und Isolde* (1865); and *Parsifal* (1882).

1858 French composer Jacques Offenbach's (1819–80) *Orpheus in the Underworld* debuts in Paris. His later works include *La Vie Parisienne* (1866) and *The Tales of Hoffman*, staged posthumously in 1881.

History of European and American Opera

1859 Charles Gounod's (1818–93) *Faust* debuts in Paris.

1866 Bohemian Bedrich Smetana (1824–84) composes his most famous opera, *The Bartered Bride.*

1871 Verdi composes *Aïda* for opening of new Cairo Opera House.

1874 Viennese composer Johann Strauss the Younger (1825–99) produces his most famous comic operetta, *Die Fledermaus.*

1874 Russian Modest Mussorgsky (1839–81) composes his most famous opera, *Boris Godunov.*

1875 *Carmen*, last and most famous opera of Georges Bizet (1838–75), debuts in Paris. Among Bizet's other works are *The Pearl Fishers* (1863) and *The Fair Maid of Perth* (1867).

1879 Debut of *Eugene Onegin,* first opera by Russian Peter Ilyich Tchaikovsky (1840–93). He composes 10 operas, including *The Queen of Spades* (1890) and *Iolanta* (1891).

1884 Jules Massenet (1842–1912) becomes most popular opera composer in France with *Manon.* Among his later works are *Werther* (1892) and *Thäis* (1894).

1888 Italian Pietro Mascagni's (1863–1945) *Cavalleria rusticana* introduces Italian *verismo*, or realism of human passions, to opera.

1892 *I Pagliacci*, most famous opera of Ruggiero Leoncavallo (1858–1919), is produced in Milan.

1893 Debut of *Hänsel und Gretel*, first and most popular of the operas of German Engelbert Humperdinck (1854–1921).

1893 Italian Giacomo Puccini (1858–1924) composes his first opera, *Manon Lescaut.* His enormously popular later works include *La Bohème* (1896), *Tosca* (1900), and *Madama Butterfly* (1904). His last opera, *Turandot*, is produced posthumously in 1926.

1902 French composer Claude Debussy (1862–1918) produces his only completed opera, *Pélleas et Mélisande.*

1905 German Richard Strauss (1864–1949) composes his first opera, *Salome.* His later works include *Elektra* (1909), *Der Rosenkavalier* (1911), *Ariadne auf Naxos* (1912), *Die Frau ohne Schatten* (1919), and *Capriccio* (1942).

1907 Debut of *The Legend of the Invisible City of Kitezh*, considered to be finest work of Russian Nikolai Rimsky-Korsakov (1844–1908). Other works include *The Snow Maiden* (1881), *Sadko* (1898), and *The Golden Cockerel* (1907).

1914 Debut of Russian-born Igor Stravinsky's (1882–1971) first opera, *The Nightingale.* He later composes *The Rake's Progress* (1951), with English libretto by poets W. H. Auden (1907–73) and Chester Kallman.

1919 Soviet Sergei Prokofiev (1891–1953) composes his most popular opera, *The Love for Three Oranges.* Among his later works are *War and Peace* (1943) and *Angel of Fire* (1954).

1934 American Virgil Thomson (1896–1989) composes *Four Saints in Three Acts* to text by Gertrude Stein (1874–1946).

1935 American George Gershwin (1898–1937) composes American folk opera *Porgy and Bess.*

1937 Austrian Alban Berg (1885–1935) composes *Lulu*, atonal opera including film clips and spoken dialogue. His only other opera, also atonal, is *Wozzeck*, dating from 1925.

1937 German Carl Orff (1895–1982) incorporates medieval songs into *Carmina Burana*, "scenic oratorio."

1945 British composer Benjamin Britten (1913–76) writes his most popular opera, *Peter Grimes.* His later works include *Billy Budd* (1951) and *The Turn of the Screw* (1954).

1951 Italian Gian Carlo Menotti (b. 1911) composes his popular Christmas opera, *Amahl and the Night Visitors*, for television performance.

1956 American Douglas Moore (1893–1969) composes *Ballad of Baby Doe.*

1957 Debut of *Dialogues of the Carmelites*, by French composer Francis Poulenc (1899–1963).

1958 American Samuel Barber's (1910–81) *Vanessa* is awarded Pulitzer Prize.

1970 Debut of British composer Sir Michael Tippett's (b. 1905) *The Knot Garden.* Among his other operas are *The Midsummer Marriage* (1955) and *King Priam* (1962).

1980s American opera houses begin placing screens above stage to display English "supertitles" to operas sung in foreign languages.

1854–1921 Engelbert Humperdinck, German composer, lives. Works include six operas, including *Hansel and Gretel* (1893), and other vocal works.

1854–1928 Leoš Janáček, Czech composer, lives. Works include 10 operas, among them *Jenúsa* (1904); orchestral, choral, and piano works, including *Glagolithic Mass* (1926) and *Sin Sonietta* (1926); chamber music; and songs.

1857–1934 Sir Edward Elgar, British composer, lives. Works include *Enigma Variations* (1899) for orchestra, oratorio *The Dream of Gerontius* (1900), and his five *Pomp and Circumstance* marches (1901–7, 1930).

1858–1924 Giacomo Puccini, Italian composer of operas, lives. Works include *La Bohême* (1896), *Tosca* (1900), and *Madame Butterfly* (1904).

1860–1909 Isaac Albéniz, Spanish composer, lives. Works include *Suite Española*, *Cantos de Espana*, and *Iberia* (1906–9).

1860–1911 Gustav Mahler, Austrian composer, lives. Works include *Resurrection Symphony* (no. 2, 1894), song cycle *Kindertotenlieder* (1901–4), *Giant Symphony* (no. 5, 1902), *Symphony of a Thousand* (no. 8 in E-flat Major, 1907), *Das Lied von der Erde* (1908), and *Symphony No. 9* (1910).

1860–1941 Ignace Paderewski, Polish composer and statesman, lives. Works include many piano pieces, opera *Manru* (1901), and *Symphony in B Minor* (1909).

1861–1908 Edward A. MacDowell, American composer, lives. Works include *Indian Suite* (1892), *Woodland Sketches* (1896), and *Sea Pieces* (1898), all for piano, and tone poem *Hamlet and Ophelia* (1885).

1862–1918 Claude Debussy, French composer, lives. Works include *Suite Bergamasque* (1890–1905) and *Estampes* (1903) for piano; *Prélude à l'Après-midi d'un Faune* (1894) and *Nocturnes* (1899) for orchestra; and orchestral suite *La Mer* (1905).

1862–1934 Frederick Delius, British composer, lives. Works include *Brigg Fair* (1907) and *On Hearing the First Cuckoo in Spring* (1912).

1864–1949 Richard Strauss, German composer, lives. Works include symphonic poems *Don Juan* (opus 20, 1889); *Till Eulenspiegel* (opus 28, 1894–95); *Also sprach Zarathustra* (opus 30, 1896); and his final works, *Metamorphoses* (1945) and *Four Last Songs* (1948).

1865–1935 Paul Dukas, French composer, lives. Works include *The Sorcerer's Apprentice* (1897), opera *Ariane et Barbebleue* (1907), and ballet *La Peri* (1912).

1865–1936 Alexander Glazunov, Russian composer, lives. Works include ballets *Raymonda* and *Les Saisons* (1890s) and *Violin Concerto in A Minor* (1904).

1865–1957 Jean Sibelius, Finnish composer, lives. Works include symphonic tone poem *The Swan of Tuonela* (1893–95), *Finlandia* (1899), *Violin Concerto in D Minor* (1903), and *Valse Triste* (1903).

1866–1925 Erik Satie, French composer, lives. Works include *Trois Gymnopédies* (1888) for piano; ballet *Parade* (1917); and symphonic drama *Socrate* (1918), for voices and orchestra.

1872–1915 Alexandr Scriabin, Russian composer, lives. Works include Third Symphony, *Le Divin Poème* (1905), *Poème de L'extase* (1908), and *Prometheus* (1911).

1872–1958 Ralph Vaughan Williams, British composer, lives. Works include *Sea Symphony* (1903–9), *A London Symphony* (1914), *A Pastoral Symphony* (1921), and *Symphonia Antarctica* (1952).

1873–1916 Max Reger, German composer, lives. Works include *Bocklin Suite* and *Phantasie und Fuge über Bach*.

1873–1943 Sergei Rachmaninoff, Russian composer, lives. Works include *Piano Concerto in C Minor* (1901); tone poem *The Isle of the Dead* (1909); choral work *The Bells* (1910); and *Rhapsody on a Theme of Paganini* (1934), for piano and orchestra.

▶

1874–1934 Gustav Holst, British composer, lives. Works include orchestral suite *The Planets* (1915), *Choral Symphony* (1925), and *Egdon Heath* (1928).

1874–1951 Arnold Schoenberg, Austrian composer, lives. Works include *Verklärte Nacht* (1899), *Pelléas and Mélisande* (1903), *First Chamber Symphony* (1906), and *Pierrot Lunaire* (1912).

1874–1954 Charles Ives, American composer, lives. Works include four symphonies, chamber and choral music, songs, and piano pieces.

1875–1937 Maurice Ravel, French composer, lives. Works include *Habanera* (1896); *Sheherazade* (1899); *Rhapsodie espagnole* (1907); ballet *Daphnis et Chloé* (1912); *Le Tombeau de Couperin* (1917), for piano; and ballets *La Valse* (1920) and *Bolero* (1928).

1876–1946 Manuel de Falla, Spanish composer and pianist, lives. Works include operas; piano pieces; and ballets, including *The Three-Cornered Hat* (1919).

1879–1936 Ottorino Respighi, Italian composer, lives. Works include *Fountains of Rome* (1916); *Pines of Rome* (1924); and *Antique Dances and Arias* (1917–31), based on music of Italian Renaissance.

1881–1945 Béla Bartók, Hungarian composer, lives. Works include *Mikrokosmos* (1926–39), for piano; *Music for Strings, Percussion, and Celesta* (1936); and *Concerto for Orchestra* (1943).

1882–1961 Percy Aldridge Grainger, Australian pianist and composer, lives. His best-known works are orchestral and band arrangements of traditional folk melodies.

1882–1967 Zoltán Kodály, Hungarian composer, lives. Works include *Psalmus Hungaricus* (1923), *Háry János* (1926), *Dances of Galanta* (1934), and *Symphony in C* (1961).

1882–1971 Igor Stravinsky, Russian composer, lives. Works include ballets *The Firebird* (1910), *Petrouchka* (1911), and *The Rite of Spring* (1913); stage work *The Soldier's Tale* (1918); *Violin Concerto in D Major* (1931); *Dunbarton Oaks Concerto* (1938), for 16 wind instruments; *Symphony in C Major* (1940); and *Threni* (1958).

1883–1945 Anton von Webern, Austrian avant-garde composer, editor, and conductor, lives. Works include a symphony; cantatas; string quartets, including *Five Pieces* (1909); and songs, including *Lieder*, Opus 3 and 4 (1908–9).

1885–1935 Alban Berg, Austrian composer, lives. Works include *Piano Sonata* (opus 1, 1908), *String Quartet* (opus 3, 1910), opera *Wozzeck* (1921), *Chamber Concerto* (opus 8, 1925), *The Lyric Suite* (1926), and 12-tone *Violin Concerto* (1935).

1887–1959 Heitor Villa-Lobos, Brazilian composer, lives. Works include ballet *Amazonas* (1917), *Bachianas Brasilieras* (1930–44), and *Genesis* (1954).

1891–1953 Sergei Prokofiev, Russian composer, lives. Works include *Classical Symphony* (1917) and six other symphonies (1924–52), *Lieutenant Kije Suite* (opus 60, 1934), symphonic tale *Peter and the Wolf* (opus 67, 1936), and choral work *Alexander Nevsky* (1938).

1892–1974 Darius Milhaud, French composer, lives. Well-known works include ballets *L'Homme et son désir* (1918), *Saudades do Brasil* (1921–22), and *La Création du monde* (1923).

1894–1976 Walter Piston, American composer, lives. Works include ballet *The Incredible Flutist* (1938), Pulitzer Prize-winning *Third Symphony* (1948) and *Seventh Symphony* (1961), and orchestral suite *Three New England Sketches* (1959).

1895–1963 Paul Hindemith, German-born American composer, lives. Works include *Das Marienleben* (1923; 1948), radio cantata *Lindberghflug* (1929), and ballet *The Four Temperaments* (1940).

1895–1982 Carl Orff, German composer of stage works, lives. His best-known work is *Carmina Burana* (1936).

1896–1985 Roger Sessions, American composer, lives. Works include eight symphonies,

▶

a violin concerto, piano and organ works, and songs.

1896–1989 Virgil Thomson, American composer, lives. Well-known works include collaboration with Gertrude Stein on *Four Saints in Three Acts* (1928), film score *The Plow that Broke the Plains* (1936), and ballet *The Filling Station* (1937).

1898–1937 George Gershwin, American composer, lives. Classical compositions include *Rhapsody in Blue* (1924), *An American in Paris* (1928), *Second Rhapsody* (1931), and opera *Porgy and Bess* (1935).

1899–1963 Francis Poulenc, French composer, lives. Works include *Trois Mouvements Perpétuels* (1918), for piano; ballet *Les Biches* (1924); *Organ Concerto* (1941); and choral work *Gloria* (1960).

1899–1978 Carlos Chávez, Mexican composer, lives. Works include *Sinfonía de Antígona* (1933) and ballet *Xochipilli Macuilxochitl* (1940).

1900– Aaron Copland, American composer, lives. Well-known works include *El Salòn Mexico* (1936), for orchestra; ballets *Billy the Kid* (1938) and *Rodeo* (1942); *Lincoln Portrait* (1942), for narrator and orchestra; ballet *Appalachian Spring* (1944); and three symphonies (1928, 1933, and 1946).

1903–78 Aram Khachaturian, American composer, lives. Works include operas; orchestral works; and ballets, including *Gayani* (1942) and its popular "The Saber Dance."

1904–75 Luigi Dallapiccola, Italian composer of atonal music, lives. Works include opera *The Prisoner* (1944) and *Christmas Concerto* (1956).

1906–1975 Dmitri Shostakovich, Soviet composer, lives. Works include *Piano Concerto No. 1* (1933), *Piano Quintet* (1940), *Leningrad Symphony No. 7* (1941), and 14 other symphonies (1924–71), as well as chamber works and scores for film, theater, and ballet.

1908– Oliver Eugène Massiaen, French organist and composer, lives. Works include symphonic poems and piano, organ, and vocal pieces, among them a series of church works (1932–51).

1910–1981 Samuel Barber, American composer, lives. Works include *Violin Sonata* (1928), overture to *The School for Scandal* (1933), *Adagio for Strings* (1938), *Violin Concerto* (1939), and *Knoxville: Summer of 1915* (1948).

1910– William H. Schuman, American composer, lives. Works include *A Free Song* (Secular Cantata no. 2, 1942), for which he won first Pulitzer Prize in music (1943); *New England Triptych* (1956); and *Concerto on Old English Rounds* (1976).

1911– Gian Carlo Menotti, Italian composer, lives. Works include ballets; a piano concerto; and several operas, among them *Amahl and the Night Visitors* (1951) and *The Saint of Bleecker Street* (1954).

1912– John Cage, American composer, lives. Works include *Imaginary Landscape No. 4* (1951), scored for 12 performers and 24 randomly tuned radios; *4′ 33″* (1952), also known as *Silence*, for any instrument or instruments; and *Apartment Building 1776* (1976).

1913–76 Benjamin Britten, English composer, lives. Works include *A Ceremony of Carols* (1942), *Peter Grimes* (1945), and *A Young Person's Guide to the Orchestra* (1945).

1913– Norman Dello Joio, American composer, pianist, and organist, lives. Works include piano sonatas, chamber music, ballets, and orchestral and choral pieces.

1918– Leonard Bernstein, American composer and conductor, lives. Works include symphony *Age of Anxiety* (1949), musical scores for *Candide* (1956) and *West Side Story* (1957), and *Mass* (1971).

7 Religion, Philosophy, and Education

The Bible

B.C.

c. 1225	Exodus from Egypt of Hebrews, led by Moses.
c. 1000	Earliest Old Testament writings are composed from highly developed oral tradition.
c. 1000–950	Biography of David is composed. It eventually forms narrative of David's life that is included in parts of I and II Samuel and I Kings.
c. 850	Composition of the Yahwistic (J) document, one of two documents believed to have been edited together to form basis of Books of Moses (Pentateuch)—Genesis, Exodus, Leviticus, Numbers, and Deuteronomy.
c. 750	Elohistic (E) document of Pentateuch is composed.
c. 740–700	Composition of prophetic books of Isaiah.
c. 600s	First Proverbs are written.
c. 650	J and E documents are believed to have been edited together to form core of Books of Moses.
c. 626–585	Composition of Jeremiah.
621	Law of Moses (Deuteronomy 5:26 and 28) is discovered in Temple at Jerusalem and adopted as law.
c. 600	I and II Kings are composed.
c. 600–400	Book of Job is composed.
c. 592–570	Prophetic Book of Ezekiel is written.
586	Destruction of Jerusalem by Babylonian king Nebuchadnezzar (c. 630–532) marks end of golden age of Old Testament literature.
c. 550	Books of Genesis through Kings are edited to form history of Hebrews from creation through c. 560 B.C. Collection spans books of Genesis to Kings.
c. 538	Composition of Second Isaiah.
c. 400	Torah, or Law, is established as canon. It includes books of Genesis, Exodus, Leviticus, Numbers, and Deuteronomy.
c. 300s	Book of Jonah is composed.
c. 200s	Composition of I and II Chronicles, Song of Songs, Ezra, and many psalms and proverbs.
c. 200s	Greek translation of Hebrew Old Testament writings known as the Septuagint is begun.
c. 200	Books known as "The Prophets" are established as canon. These include Joshua, Judges, Samuel, Kings, Isaiah, Jeremiah, Ezekiel, and 12 minor prophets.
c. 165	Book of Daniel is composed.
c. 150	Ecclesiastes is written. By this time, the Old Testament is complete.

A.D.

c. 30	Jesus of Nazareth is crucified in Jerusalem.
c. 49	Paul composes his epistle to Thessalonians, earliest New Testament writing.
c. 51	Paul writes epistle to Galatians.
c. 55	Paul writes epistles to Corinthians.
c. 60	Paul's epistle to Romans is composed.
c. 68	Mark, earliest of four gospels, is written.
c. 81–96	Revelation is written.
c. 85–95	Gospel of Luke and Book of Acts are written.
c. 90	Old Testament books known as The Writings are established as Christian canon. These include Psalms, Proverbs, Job, Song of Songs, Ruth, Lamentations, Ecclesiastes, Esther, Daniel, Ezra, and Chronicles.
c. 95–105	Pastoral Epistles, falsely attributed to Paul, are written. These include Hebrews, I and II Timothy, Titus, and I Peter.
c. 100	Gospel of Matthew is composed.
c. 100–125	Gospel according to John is written.
c. 150	The four canonical gospels are collected.

▶

200 New Testament canon is largely fixed in its present form.

367 Festal Epistle of St. Athanasius (c. 293–373) gives earliest listing of New Testament canon in its present form.

c. 405 St. Jerome (c. 347–419) completes Vulgate, Latin translation of Old and New Testaments that remains Latin Bible of Roman Catholic Church into modern times.

c. 1380 English priest John Wycliffe (c. 1320–84) starts first English translation of Bible.

1582 Douay Version (English translation) of New Testament is completed. With Old Testament translation (completed c. 1610), it becomes authorized English translation of Bible for Roman Catholics.

1604 English king James I (1566–1625) commissions King James translation of Bible.

Roman Catholicism

Roman Catholicism is practiced by 880 million people, representing 18 percent of the world's population.

B.C.

c. 8 Jesus of Nazareth (c. 8 B.C.–A.D. 30) is born in Roman Palestine.

A.D.

c. 24–26 Jesus begins ministry. He soon gains devoted following by his teachings and deeds.

c. 30 Jesus is crucified. His followers declare that he has been resurrected from dead and appeared among them. St. Peter (d. c. 64) begins to spread teachings of Christianity.

c. 33 Saul of Tarsus (d. 67), rabbi and bitter opponent of Christianity, is converted to Christianity. Thereafter known as Paul, he becomes one of early church's most important missionaries.

c. 47–49 Paul journeys to Asia Minor and Cyprus. He establishes churches and writes earliest epistles that eventually become part of New Testament.

c. 49 Early church council at Jerusalem agrees to expand missionary work to Gentiles. Christianized Jews are to be excused from full obedience to Jewish law.

c. 53–57 Paul writes epistles to Romans and Corinthians.

c. 55 Peter travels to Rome. His primacy over church of Rome establishes tradition of papacy. He is regarded as first bishop of Rome (pope).

64 Roman emperor Nero (A.D. 37–68) charges Christians with starting fire that destroyed much of Rome. He initiates persecution of Christians.

c. 67 Nero orders executions of Peter and Paul.

c. 95 Clement of Rome (c. 30–100), among earliest of popes, writes letter arguing that church leaders have divine authority inherited from Christ and apostles.

c. 100 Christian churches are established in Greece, North Africa, Italy, and Asia Minor. Books of New Testament are written during second half of century.

100–165 St. Justin Martyr lives. He is one of first Christian apologists who write to defend and explain Christianity.

c. 140 *Shepherd of Hermas* is written. It reveals highly developed ecclesiastical system of bishops, deacons, and priests.

c. 144 Marcion founds influential sect arguing for existence of two gods—one good and one evil—and rejecting Old Testament.

c. 150 School of Alexandria is founded in Egypt. It becomes center for early Christian theology

▶

and Greek philosophy. Its prominent teachers include theologians Clement (d. c. 215) and Origen (c. 185–254).

c. 180 Catholic theologian Irenaeus (125–c. 202) writes *Against Heresies*. It argues against spreading Gnostic belief and claims that "every church must agree" with church of Rome because of its apostolic authority.

249–51 Christians are persecuted throughout Roman Empire under Emperor Decius (201–251).

284–305 Roman emperor Diocletian reigns and orders further persecution of Christians.

312 Roman emperor Constantine the Great (c. 280–337) converts to Christianity following vision before Battle of Milvian Bridge. He triumphs over his rival for throne and ascribes his victory to the conversion.

313 Constantine issues Edict of Milan, granting legal rights to Christians and restoring their confiscated property.

325 Constantine convenes first ecumenical council, at Nicaea, to settle dispute over teachings of Arius (d. 336) that Christ was not equal in divinity to God. Council condemns Arianism and composes Nicene Creed as fundamental statement of Christian doctrine.

330 Constantine rebuilds city of Byzantium as Constantinople and establishes it as new imperial capital of empire. Constantinople develops as "new Rome."

367 First evidence that New Testament canon is established in its lasting form.

387 Augustine of Hippo (354–430) converts to Christianity. He becomes a leading theologian in church's history through such influential works as *The Confessions, City of God*, and his polemics against numerous heresies.

380 Christianity becomes official religion of Roman Empire under reign of Theodosius I (c. 346–95).

395 Division of Roman Empire between East and West sets stage for growing split between churches of Constantinople and Rome.

418 British monk Pelagius (c. 354–420) is excommunicated as heretic. He denies original sin and need for baptism, teaching that man can achieve righteousness through good works. Augustine condemns Pelagianism in series of writings.

c. 430 St. Patrick (c. 385–461) introduces Christianity in Ireland.

431 Council of Ephesus (ecumenical) denounces teaching of Nestorius (d. 451) that Christ had separate human and divine natures. It declares that Mary is mother of God as well as of Christ.

451 Council of Chalcedon (ecumenical) declares Christ to be simultaneously "truly man and truly God."

476 Fall of Western Roman Empire leaves church as primary authority in West.

c. 529 Italian monk Benedict establishes monastery at Monte Cassino, Italy. His Benedictine order initiates Western monasticism and spreads rapidly across Europe.

590–604 Pope Gregory the Great (c. 540–604) initiates reforms in liturgy and church administration. He enhances power and prestige of papacy.

596 Pope Gregory sends Augustine of Canterbury (d. c. 605) on missionary journey to spread Christianity to England.

756 Donation of Pepin. Papal domains in Italy (reduced to Vatican in modern times) are established as territory under sovereignty of popes.

800 Charlemagne (c. 742–814) is crowned emperor by Pope Leo III (750–816) at St. Peter's in Rome. Coronation marks new relationship between church and state, with emperor's authority dependent on blessing of pope.

c. 950 Catholicism is prevalent throughout Europe.

1054 Schism of 1054 between Roman Catholic and Eastern Orthodox churches finalizes long-standing split between the two centers of Christiandom. Pope Leo IX (1002–54) and Patriarch Michael Cerularius exchange anathemas of excommunication.

1059 College of Cardinals is established as elector of popes.

1073 Pope Gregory VII (c. 1020–85) initiates widespread reform in church practice, liturgy, and administration.

▶

1075 Investiture controversy breaks out between Pope Gregory VII and Holy Roman emperor Henry IV (1050–1106). Gregory denies emperor's right to appoint bishops. He excommunicates and deposes Henry (1077) but finally is driven into exile following Henry's invasion of Italy (1081–83).

1095 First of eight crusades against Moslems in Holy Land is launched by Pope Urban II (c. 1035–99). Crusades continue for almost two centuries.

1122 Diet of Worms ends investiture controversy with compromise that maintains authority of church over the empire.

1198–1216 Power of papacy is at its height during reign of Innocent III (1161–1216). He excommunicates Holy Roman emperor Otto IV (1182–1218) in 1210 and English king John (c. 1167–1216) in 1209.

1208 Albingensian Crusade is launched by Pope Innocent III against Cathar and Waldensian sects in southern France. Total of 20,000 people are massacred by crusaders at Béziers in 1209.

1209 Italian monk Francis of Assisi (1182–1226) founds Franciscan order of mendicant monks.

1216 Spanish theologian Dominic (1170–1221) establishes Dominican order, dedicated to preaching, scholarship, and teaching.

1225–74 Italian theologian Thomas Aquinas (1225–74) lives. He codifies Catholic theology in such monumental works as *Summa Theologica*. His work marks high point of Scholastic movement in theology.

1233 Inquisition is created by Pope Gregory IX (c. 1155–1241) to abolish heresy.

1302 Pope Boniface VIII (c. 1235–1303) issues papal bull *Unum Sanctum*, declaring supreme authority of pope in civic and spiritual matters.

1303 Boniface VIII is kidnapped by agents of French king Philip IV (1268–1314) after threatening to depose Philip. Boniface dies within a month.

1305 Seventy-year "Babylonian Captivity" of papacy begins when Pope Clement V (1260–1314) moves papacy to Avignon in France.

1377 Pope Gregory XI (1329–78) returns papal seat to Rome.

1378 Great Schism begins when election of Italian pope Urban VI (c. 1318–89) is challenged by French cardinals, who elect Clement VII (d. 1394) as antipope in Avignon. Rome and Avignon have rival popes for following 40 years.

1409 Council in Pisa elects Alexander V pope to end the schism. Popes in Rome and Avignon refuse to step down, leaving three rival popes.

1417 Council of Constance (ecumenical) ends schism with election of Pope Martin V (1368–1431) and deposing of all rival popes.

c. 1450 First papal indulgences for dead are granted. Remission of punishment for dead in purgatory is promised in exchange for good works.

1478 Spanish Inquisition is established by Spanish monarchs. Using Dominican inquisitors, it persecutes thousands of Muslims and Jews over following three centuries.

1506 Construction begins on St. Peter's Basilica in Rome.

1517 German theologian Martin Luther (1483–1546) begins Protestant Reformation with protest against sale of indulgences and other corrupt church practices.

1520 Luther is excommunicated as heretic.

1521 Diet of Worms condemns Luther after he refuses to recant his teachings against supremacy of pope, sale of indulgences, priesthood, five of the seven Catholic sacraments, and doctrine of justification by works.

1534 Church of England separates from Roman Catholic Church following dispute between King Henry VIII (1491–1547) and pope, who refused to annul Henry's marriage.

1534 Society of Jesus (Jesuits) is established by Spanish churchman St. Ignatius of Loyola (1491–1556).

1545–63 Council of Trent (ecumenical) initiates Catholic Reformation. Council reacts to Protestant Reformation with reforms in

▶

dogma, indulgences, sacraments, and practices of clergy.

1555 Peace of Augsburg gives princes right to determine whether their state should be Catholic or Protestant.

1562–98 Wars of religion are fought between French Catholics and Protestants.

1600 Lutheranism, Calvinism, and other Protestant movements dominate much of northern Europe by about this time.

c. 1622 Congregation for the Propagation of the Faith is founded as Catholic missionary organization.

1773 Society of Jesus is abolished by Pope Clement XIV (1705–74).

1781 Last executions are carried out by Spanish Inquisition.

1798 French general Napoleon (1769–1821) imprisons Pope Pius VI (1717–99) after conquering papal territories in Italy.

1809 Papal domains become part of French Empire. Napoleon holds Pope Pius VII (1742–1823) captive for five years.

1814 Society of Jesus is refounded.

1815 Congress of Vienna restores papal domains to church.

1854 Doctrine of immaculate conception of Mary is declared by Pope Pius IX (1792–1878).

1859 Most of Papal States are conquered during drive to unify Italian peninsula into Kingdom of Italy. Papal control over Vatican City is not officially recognized until 1929.

1864 *Syllabus of Errors* is published by Pope Pius IX. It places church in opposition to socialism, pantheism, secular education, civil marriage, and other "errors" of 19th century.

1869–70 First Vatican Council (ecumenical) is called by Pope Pius IX. It defines dogma of papal infallibility. Italian troops conquer Rome during council.

1879 Thomism, based on theology of Thomas Aquinas, becomes official philosophy of church.

1900 U.S. Catholic Church has 12 million members following rapid growth among 19th-century immigrants.

1907 Pope Pius X (1835–1914) declares Modernism "the résumé of all the heresies."

1917 Code of Canon Law is established.

1929 Lateran Treaty between church and Italy cedes Vatican City to church as independent republic and makes Catholicism state religion of Italy.

1937 Encyclicals condemning Nazism and communism are issued by Pope Pius XI (1857–1939).

1950 Mary's bodily assumption into heaven is declared church dogma by Pope Pius XII (1876–1958).

1961 Pope John XXIII (1881–1963) issues his famous encyclicals *Mater et Magistra* and *Pacem in Terris* on world peace and social policy of church.

1962–65 Second Vatican Council (ecumenical) is convened by Pope John XXIII. It institutes vast range of reforms in church practice and encourages ecumenical advances. It also denounces antisemitism, introduces use of vernacular in Mass, and increases role of laity in church.

1964 Pope Paul VI (1897–1978) and Eastern Orthodox patriarch Athenagoras meet in Jerusalem.

1965 Anathemas of excommunication between Roman Catholic and Eastern Orthodox churches are officially removed.

1968 Pope Paul VI issues encyclical *Humanae Vitae,* condemning use of artificial birth control.

1978 Polish-born pope John Paul II (b. 1920) is first non-Italian elected to papacy in 455 years.

1981 Pope John Paul II survives attempted assassination by Turkish terrorist (May 13).

1982 World population of Roman Catholics is estimated at 580 million.

1987 Vatican takes stand against such current medical practices relating to conception as *in vitro* fertilization, artificial insemination, and surrogate motherhood.

Popes			
c. 41?–c. 67?	Saint Peter (d. c. 67)	**402–17**	Saint Innocent I (d. 417)
67–c. 79	Saint Linus (d. c. 79)	**417–18**	Saint Zosimus (d. 418)
79–c. 91	Saint Anacletus (d. c. 91)	**418–22**	Saint Boniface I (d. 422)
91–100	Saint Clement (c. 30–c. 100)	**422–32**	Saint Celestine I (d. 432)
100–c. 107	Saint Evaristus (d. 107?)	**432–40**	Saint Sixtus III (d. 440)
107–c. 16	Saint Alexander I (d. c. 116)	**440–61**	Saint Leo I (The Great) (c. 390–461)
116–c. 25	Saint Sixtus I (d. c. 125)	**461–68**	Saint Hilary (d. 468)
125–c. 36	Saint Telesphorus (d. c. 136)	**468–83**	Saint Simplicius (d. 483)
136–c. 40	Saint Hyginus (d. c. 140)	**483–92**	Saint Felix II (or III) (d. 492)
140–c. 54	Saint Pius I (c. 80–c. 157)	**492–96**	Saint Gelasius I (d. 496)
154–c. 65	Saint Anicetus (d. 165?)	**496–98**	Saint Anastasius II (d. 498)
165–74	Saint Soter (d. 174)	**498–514**	Saint Symmachus (d. 514)
174–89	Saint Eleutherius (d. 189)	**514–23**	Saint Hormisdas (d. 523)
189–98	Saint Victor I (d. 198)	**523–26**	Saint John I (c. 470–526)
198–217	Saint Zephyrinus (d. 217)	**526–30**	Saint Felix III (d. 530)
217–22	Saint Calixtus I (d. 222)	**530–32**	Boniface II (d. 532)
222–30	Saint Urban (d. 230)	**533–35**	John II (d. 535)
230–35	Saint Pontian (d. 235)	**535–36**	Saint Agapetus I (d. 536)
235–36	Saint Anterus (d. 236)	**536–c. 37**	Saint Silverius (d. 537?)
236–50	Saint Fabian (d. 250)	**c. 537–55**	Vigilius (d. 555)
251–53	Saint Cornelius (d. 253)	**556–61**	Pelagius I (d. 561)
253–54	Saint Lucius I (d. 254)	**561–74**	John III (d. 574)
254–57	Saint Stephen I (d. 457)	**575–79**	Benedict I (d. 579)
257–58	Saint Sixtus II (d. 258)	**579–90**	Pelagius II (d. 590)
259–68	Saint Dionysius (d. 268)	**590–604**	Saint Gregory I the Great (c. 540–604)
269–74	Saint Felix I (d. 274)	**604–606**	Sabinian (d. 606)
275–83	Saint Eutychianus (d. 283)	**607**	Boniface III (d. 607)
283–96	Saint Caius (d. 296)	**608–15**	Saint Boniface IV (d. 615)
296–304	Saint Marcellinus (d. 304)	**615–18**	Saint Deusdedit (d. 618)
308–9	Saint Marcellus I (d. 309)	**619–25**	Boniface V (d. 625)
309	Saint Eusebius (d. 309)	**625–38**	Honorius I (d. 638)
c. 310–14	Saint Miltiades (d. c. 310)	**640**	Severinus (d. 640)
314–35	Saint Sylvester (d. 335)	**640–42**	John IV (d. 642)
336	Saint Marcus (d. 336)	**642–49**	Theodore I (d. 649)
337–52	Saint Julius I (d. 352)	**649–55**	Saint Martin I (d. 655)
352–66	Saint Liberius (d. 366)	**654–57**	Saint Eugene I (d. 657)
366–84	Saint Damasus I (c. 304–84)	**657–72**	Saint Vitalian (d. 672)
384–99	Saint Siricius (d. 399)	**672–76**	Adeodatus II (d. 676)
399–401	Saint Anastasius I (d. 401)	**676–78**	Donus (d. 678)

Popes

678–81	Saint Agatho (d. 681)	**903**	Leo V (d. 903)
682–83	Saint Leo II (d. 683)	**904–11**	Sergius III (d. 911)
684–85	Saint Benedict II (d. 685)	**911–13**	Anastasius III (d. 913)
685–86	John V (d. 686)	**913–14**	Lando (d. 914)
686–87	Conon (d. 687)	**914–28**	John X (c. 860–928)
687–701	Saint Sergius I (d. 701)	**928**	Leo VI (d. 928)
701–5	John VI (d. 705)	**928–31**	Stephen VII (d. 931)
705–7	John VII (d. 707)	**931–35**	John I (906–35)
708	Sisinnius (d. 708)	**936–39**	Leo VII (d. 939)
708–15	Constantine (d. 715)	**939–42**	Stephen VIII (d. 942)
715–31	Saint Gregory II (d. 731)	**942–46**	Marinus II (d. 946)
731–41	Saint Gregory III (d. 741)	**946–55**	Agapetus II (d. 955)
741–52	Saint Zacharias (d. 752)	**955–64**	John XII (938–64)
752–57	Stephen II (d. 757)	**964–65**	Leo VIII (d. 965)
757–67	Saint Paul I (d. 767)	**965–72**	John XIII (d. 972)
768–72	Stephen III (d. 772)	**973–74**	Benedict VI (d. 974)
772–95	Saint Adrian I (d. 795)	**974–983**	Benedict VII (d. 983)
795–816	Saint Leo III (750–816)	**983–84**	John XIV (d. 984)
816–17	Stephen IV (V) (d. 817)	**985–96**	John XV (d. 996)
817–24	Saint Paschal I (d. 824)	**996–99**	Gregory V (972–99)
824–27	Eugene II (d. 827)	**999–1003**	Sylvester II (c. 945–1003)
827	Valentine (d. 827)	**1003**	John XVII (d. 1003)
827–44	Gregory IV (d. 844)	**1004–9**	John VXIII (d. 1009)
844–47	Sergius II (d. 847)	**1009–12**	Sergius IV (d. 1012)
847–55	Saint Leo IV (c. 800–855)	**1012–24**	Benedict VIII (d. 1024)
855–58	Benedict III (d. 858)	**1024–32**	John XIX (d. 1032)
858–67	Saint Nicholas I (c. 800–867)	**1032–45**	Benedict IX (1012–56)
867–72	Adrian II (d. 872)	**1045**	Sylvester III
872–82	John VIII (c. 820–82)	**1045**	Benedict IX
882–84	Marinus I (d. 884)	**1045–46**	Gregory VI (d. 1047)
884–85	Saint Adrian III (d. 885)	**1046–47**	Clement II (d. 1047)
885–91	Stephen V (d. 891)	**1047–48**	Benedict IX
891–96	Formosus (c. 816–96)	**1048**	Damasus II (d. 1048)
896	Boniface VI (d. 896)	**1049–54**	Saint Leo IX (1002–54)
896–97	Stephen VI (d. 897)	**1055–57**	Victor II (1018–57)
897	Romanus (d. 897)	**1057–58**	Stephen IX (d. 1058)
897	Theodore II (d. 897)	**1059–61**	Nicholas II (c. 1010–61)
897–900	John IX (d. 900)	**1061–73**	Alexander II (d. 1073)
900	Benedict IV (d. 903)	**1073–85**	Saint Gregory VII (c. 1020–85)

Popes			
1086–87	Victor III (1027–87)	**1334–42**	Benedict XII (c. 1285–1342)
1088–99	Urban II (c. 1035–99)	**1342–52**	Clement VI (1291–1352)
1099–1118	Paschal II (c. 1050–1118)	**1352–62**	Innocent VI (1282–1362)
1118–19	Gelasius II (d. 1119)	**1362–70**	Urban V (1310–70)
1119–24	Calixtus II (1050–1124)	**1370–78**	Gregory XI (1329–78)
1124–30	Honorius II (d. 1130)	**1378–89**	Urban VI (1318–89)
1130–43	Innocent II (d. 1143)	**1389–1404**	Boniface IX (c. 1350–1404)
1143–44	Celestine II (d. 1144)	**1404–6**	Innocent VII (c. 1336–1406)
1144–45	Lucius II (d. 1145)	**1406–15**	Gregory XII (c. 1325–1417)
1145–53	Eugene III (d. 1153)	**1417–31**	Martin V (1368–1431)
1153–54	Anastasius IV (d. 1154)	**1431–47**	Eugene IV (1383–1447)
1154–59	Adrian IV (c. 1100–1159)	**1447–55**	Nicholas V (1397–1455)
1159–81	Alexander III (d. 1181)	**1455–58**	Calixtus III (1378–1458)
1181–85	Lucius III (c. 1110–85)	**1458–64**	Pius II (1405–64)
1185–87	Urban III (d. 1187)	**1464–71**	Paul II (1417–71)
1187	Gregory VIII (c. 1110–87)	**1471–84**	Sixtus IV (1414–84)
1187–91	Clement III (d. 1191)	**1484–92**	Innocent VIII (1432–92)
1191–98	Celestine III (c. 1105–98)	**1492–1503**	Alexander VI (1431–1503)
1198–1216	Innocent III (1161–1216)	**1503**	Pius III (1439–1503)
1216–27	Honorius III (d. 1227)	**1503–13**	Julius II (1453–1513)
1227–41	Gregory IX (c. 1155–1241)	**1513–21**	Leo X (1475–1521)
1241	Celestine IV (d. 1241)	**1522–23**	Adrian VI (1459–1523)
1243–54	Innocent IV (d. 1254)	**1523–34**	Clement VII (1478–1534)
1254–61	Alexander IV (d. 1261)	**1534–49**	Paul III (1468–1549)
1261–64	Urban IV (c. 1200–64)	**1550–55**	Julius III (1487–1555)
1265–68	Clement IV (c. 1195–1268)	**1555**	Marcellus II (1501–55)
1271–76	Gregory X (1210–76)	**1555–59**	Paul IV (1476–1559)
1276	Innocent V (1224–76)	**1559–65**	Pius IV (1499–1565)
1276	Adrian V (1205–76)	**1566–72**	Saint Pius V (1504–72)
1276–77	John XXI (d. 1277)	**1572–85**	Gregory XIII (1502–85)
1277–80	Nicholas III (c. 1216–80)	**1585–90**	Sixtus V (1520–90)
1281–85	Martin IV (c. 1210–85)	**1590**	Urban VII (1521–90)
1285–87	Honorius IV (1210–87)	**1590–91**	Gregory XIV (1535–91)
1288–92	Nicholas IV (1227–92)	**1591**	Innocent IX (1519–91)
1294	Saint Celestine V (c. 1210–96)	**1592–1605**	Clement VIII (1536–1605)
1294–1303	Boniface VIII (c. 1235–1303)	**1605**	Leo XI (1535–1605)
1303–04	Benedict XI (1240–1304)	**1605–21**	Paul V (1552–1621)
1305–14	Clement V (1260–1314)	**1621–23**	Gregory XV (1554–1623)
1316–34	John XXII (1244–1334)	**1623–44**	Urban VIII (1568–1644)

Popes			
1644–55	Innocent X (1574–1655)	**1800–23**	Pius VII (1742–1823)
1655–67	Alexander VII (1599–1667)	**1823–29**	Leo XII (1760–1829)
1667–69	Clement IX (1600–69)	**1829–30**	Pius VIII (1761–1830)
1670–76	Clement X (1590–1676)	**1831–46**	Gregory XVI (1765–1846)
1676–89	Innocent XI (1611–89)	**1846–78**	Pius IX (1792–1878)
1689–91	Alexander VIII (1610–91)	**1878–1903**	Leo XIII (1810–1903)
1691–1700	Innocent XII (1615–1700)	**1903–14**	Saint Pius X (1835–1914)
1700–21	Clement XI (1649–1721)	**1914–22**	Benedict XV (1854–1922)
1721–24	Innocent XIII (1655–1724)	**1922–39**	Pius XI (1857–1939)
1724–30	Benedict XIII (1649–1730)	**1939–58**	Pius XII (1876–1958)
1730–40	Clement XII (1652–1740)	**1958–63**	John XXIII (1881–1963)
1740–58	Benedict XIV (1675–1758)	**1963–78**	Paul VI (1897–1978)
1758–69	Clement XIII (1693–1769)	**1978**	John Paul I (1912–78)
1769–74	Clement XIV (1705–74)	**1978–**	John Paul II (b. 1920)
1775–99	Pius VI (1717–99)		

Eastern Orthodoxy

Eastern Orthodoxy is practiced by 130.8 million people, representing 3 percent of the world's population.

330 Roman emperor Constantine I (c. 280–337) builds city of Constantinople at site of Byzantium and makes it capital of Roman Empire.

395 Roman Empire is split between Eastern and Western empires. Latin Christianity, under leadership of pope, is based in Rome, while Eastern Orthodoxy develops in Constantinople under leadership of patriarch. Cultural friction and doctrinal disputes develop between the two Christian churches.

476 Western Roman Empire collapses. Byzantine Empire, based in Constantinople, prevails for next 977 years.

532 Church of Hagia Sofia (Divine Wisdom), largest church in Christian world, is built by Emperor Justinian (483–565) in Constantinople as focal point of Byzantine Christianity.

726 Emperor Leo III (c. 680–741) begins Iconoclastic Controversy by banning worship of religious images (icons).

787 Second Nicaean Council is last of seven church councils commonly accepted as authoritative by both Roman Catholic and Eastern Orthodox churches. Council allows veneration, but not worship, of icons.

843 Worship of icons is restored in the East by Emperor Michael III (839–867), thus ending Iconoclastic Controversy.

▶

Ecumenical Councils

The Roman Catholic Church recognizes the following 21 councils as ecumenical; the Orthodox Eastern Church recognizes only the first seven. *See dates in the Roman Catholicism chronology for more on the major ecumenical councils.*

325	First Council of Nicaea (1st ecumenical)
381	First Council of Constantinople (2nd)
431	Council of Ephesus (3rd)
451	Council of Chalcedon (4th)
553	Second Council of Constantinople (5th)
680–81	Third Council of Constantinople (6th)
787	Second Council of Nicaea (7th)
869–70	Fourth Council of Constantinople (8th)
1123	First Lateran Council (9th)
1139	Second Lateran Council (10th)
1179	Third Lateran Council (11th)
1215	Fourth Lateran Council (12th)
1245	First Council of Lyons (13th)
1274	Second Council of Lyons (14th)
1311–12	Council of Vienne (15th)
1414–18	Council of Constance (16th)
1431–45	Council of Basel and Ferrara-Florence (17th)
1512–17	Fifth Lateran Council (18th)
1545–63	Council of Trent (19th)
1869–70	First Vatican Council (20th)
1962–65	Second Vatican Council (21st)

867 Photian Schism breaks out between Eastern and Western churches during first patriarchy of Photius (c. 820–91). Photius excommunicates Pope Nicholas I (c. 800–67) during dispute over various issues, including use in Western church of *filioque* clause in Nicene Creed (phrase "and from the Son" added to the statement that the holy Spirit proceeds from the Father). Schism leads to Photius's deposition, although he later effects reconciliation with Rome.

988 Vladimir I (c. 956–1015), grand duke of Kiev, declares Eastern Orthodox Christianity state religion of Russia.

1054 Permanent schism between Latin and Byzantine churches begins. Disagreements finally result in mutual excommunications by Pope Leo IX (1002–54) and Patriarch Michael Cerularius. Eastern church rejects *filioque* clause and primacy of pope.

1071 Turkish victory over Byzantine Empire at Battle of Manzikert leads to sharp decline in Byzantine power.

1096–99 First Crusade is organized by Christians in Europe to aid Byzantines against Muslim invaders. Crusaders conquer Antioch and Jerusalem and put Latin prelates in power.

1204 Sack of Constantinople by Fourth Crusade leads to establishment of Latin Empire and Latin domination of Eastern church. Thomas Morosini of Venice is made patriarch of Constantinople, intensifying rivalry between Eastern and Western churches.

1261 Michael Palaeologus (1224–82) drives Latin rulers out of Constantinople, reestablishing Eastern Orthodox rule.

1439 Decree of Union between East and West results from Council of Basel and Ferrara-Florence (1431–45). Patriarch Joseph accedes to Roman positions on *filioque* clause, primacy of pope, and other issues in return for Western military aid against invading Muslim Turks.

1453 Ottoman Turks finally conquer Constantinople itself and end Byzantine Empire. They establish Hagia Sophia as Moslem mosque. Union of Florence becomes obsolete.

1551 Russian emperor Ivan IV (the Terrible) leads council of 100 bishops in Moscow to canonize saints and settle issues of liturgy. Moscow becomes known as a "third Rome," after Rome and Constantinople.

1589 Russian church becomes independent patriarchate.

▶

1666 Russian patriarch Nikon (1605–81) is deposed for his declaration that the patriarch is greater than the czar and for his attempts to reform church liturgy.

1721 Russian czar Peter I (1672–1725) eliminates office of patriarch in Russia and establishes church administration as division of state. He issues Spiritual Regulation, governing all church activities in Russia.

1755 Synod of Constantinople denies validity of all sacraments of Western church.

1794 Eastern Orthodox Church is first established in North America by Russian monks on Kodiak Island, Alaska.

1850 Orthodox Church of Greece becomes independent and self-governing.

1900 Diocese of Aleutian Islands and North America is established in San Francisco.

1914 Church of Russia has 50,000 priests, 21,000 monks, and 73,000 nuns.

1918 Revolutionary Bolshevik government in Russia denies Orthodox Church all rights.

1922 Bolshevik government seizes all church treasures.

1939 Fierce persecution of Russian church under Joseph Stalin (1879–1953) leaves only 100 churches in operation.

1943 Stalin halts persecution of church, and 25,000 churches open during next 15 years.

1959 Soviet prime minister Nikita Khrushchev (1884–1971) begins new persecution of church. Fewer than 10,000 churches survive.

1964 Pope Paul VI (1897–1978) and Constantinople Patriarch Athenagoras (1880–1972) formally abolish anathemas of 1054 between Roman Catholic and Eastern Orthodox churches. Both pledge to work toward reestablishment of communion between the churches.

1985 Membership of Eastern Orthodox churches reaches 85 million worldwide. There are 15 autocephalous (self-governing) churches worldwide.

Protestantism

Protestantism is practiced by 455.5 million people, representing 9.5 percent of the world's population.

1514 Albrecht becomes archbishop of Mainz and offers indulgences in return for contributions to building of St. Peter's Basilica in Rome.

1517 German theologian Martin Luther (1483–1546) nails his 95 theses, protesting sale of indulgences and other long-standing corrupt church practices, to a church door in Wittenberg (October 31). This marks beginning of Protestant Reformation.

1518 Swiss Reformation begins independently under leadership of Huldreich Zwingli (1484–1531).

1519 Luther debates German Catholic theologian Johann Eck (1486–1543) in Leipzig. Luther argues against any divine establishment of papacy and rejects authority of popes and church councils.

1520 Luther is excommunicated as heretic. He burns both papal bull ordering him to submit to church and copy of church Canon Law. Luther argues against papal infallibility and articulates essential Protestant doctrine that Scriptures are the only true bases of authority.

1520 Luther writes *The Babylonian Captivity*, *Address to the Christian Nobility of the German Nation*, and *Liberty of a Christian Nation*. Works condemn church abuses and announce his positions on justification by faith rather than works, priesthood of all believers, and authority of Scriptures.

▶

c. 1520 Anabaptist movement develops in Switzerland and Germany. Anabaptists deny efficacy of infant baptism and renounce violence. They are widely persecuted by both Catholics and Lutherans.

1521 Luther refuses to recant his teachings at Diet of Worms. Edict of Worms condemns him as heretic.

1521 Luther translates New Testament into German while in hiding under protection of Frederick III, elector of Saxony (1463–1525).

1524–25 Peasants' Revolt breaks out in Austria and southern Germany, led by radical Protestant Thomas Münzer (c. 1490–1525). Luther condemns uprising. It is suppressed and Münzer is executed.

1529 Term *Protestant* originates at Diet of Speyer when supporters of Luther formally protest imperial attempt to limit spread of Lutheranism.

1529 Lutheranism is declared sole religion of Denmark by royal decree.

1530 Augsburg Confession, major statement of Lutheran beliefs, is written at Diet of Augsburg. Luther and Swiss Protestant leader Zwingli agree on 14 of 15 articles of faith but disagree on significance of Lord's Supper.

1531 Schmalkaldic League is formed as alliance of Lutheran states against Holy Roman Empire.

1534 English Reformation begins when King Henry VIII (1491–1547) becomes supreme head of Church of England and is excommunicated by Rome. Schism follows dispute with church on annulment of Henry's marriage to Katharine of Aragon.

1536 French theologian John Calvin (1509–64) publishes first edition of his *Institutes of the Christian Religion*, one of most influential works of Protestantism. It details his theology of man's depravity, necessity of grace, and predestination.

1541–64 Calvin leads theocratic Protestant government in Geneva.

1544 Catholic worship is forbidden in Sweden as Lutheranism becomes official state religion.

1544–47 Schmalkaldic War results in defeat of Schmalkaldic League by Holy Roman emperor Charles V (1500–58).

1546 Death of Luther. His more than 400 published works exert enormous influence throughout Europe.

1549 *Book of Common Prayer* is adopted in England. It establishes liturgy and practice of Church of England.

1550 Calvinism spreads rapidly in France, Switzerland, and Netherlands.

1553–58 Catholic queen Mary I (1516–58) reigns in England and persecutes Protestants.

1555 Peace of Augsburg reduces religious conflict in Holy Roman Empire by allowing princes to choose Catholicism or Lutheranism for their subjects.

1558 Elizabeth I (1533–1603) succeeds Mary in England and restores Protestantism. Puritan movement develops among those dissatisfied with Elizabeth's reform of church.

1560 Scottish churchman John Knox (1514–72), disciple of Calvin, leads Scottish reformation and plays pivotal role in founding Presbyterian Church of Scotland.

1562 Wars of Religion begin in France between Catholics and French Calvinists (known as Huguenots).

1563 Thirty-nine Articles are adopted as orthodox beliefs of Anglican Church.

1572 Massacre of St. Bartholomew's Day in France results in death of thousands of French Calvinists.

1577 *Book of Concord* is accepted by Lutheran churches. It establishes orthodox creeds for Lutheranism.

1598 Edict of Nantes by French king Henry IV (1553–1610) marks end of bloody Wars of Religion in France and grants religious freedom to Protestants there.

1608 Baptist Church is founded by John Smyth (c. 1570–1612).

1618 Thirty Years' War begins with Protestant uprising in Bohemia.

▶

1620 English Puritans known as Pilgrims establish American colony at Plymouth Rock in pursuit of religious freedom.

1642–49 English Civil War is fought between Puritans and Presbyterians on one side and Anglicans and Catholics on the other.

c. 1647 Society of Friends (Quakers) is founded in England by George Fox (1624–91). Quakers repudiate violence and oath-swearing and worship without ministers or liturgy. They teach an "inner light" of divine revelation.

1648 Peace of Westphalia ends Thirty Years' War. Germany is divided between Catholic and Protestant states.

1670 German Pietism movement begins. Led by German pastor Philipp Jacob Spener (1635–1705), movement emphasizes pious living and personal experience of faith.

1685 Edict of Nantes is renounced by French king Louis XIV (1638–1714), denying rights to French Protestants. Mass exodus of Huguenots from France follows.

1689 Act of Toleration in Great Britain grants rights to dissenters from Church of England.

1706 Presbyterian Church is established in America by Irish churchman Francis Makemie (c. 1658–1708).

1720–60 Great Awakening revival in American colonies. Widespread conversions are stimulated by charismatic Protestant evangelists such as Jonathan Edwards (1703–58).

1729 Methodist movement is founded by English evangelist John Wesley (1703–91).

1789 Protestant Episcopal Church is founded as independent American branch of Anglicanism.

1799 German theologian Friedrich Schleiermacher (1768–1834) publishes *Religion: Speeches to Its Cultured Despisers*. His influential theology initiates Romantic movement, emphasizing religious emotion over 18th-century rationalism.

1896 American Presbyterian evangelist Billy Sunday (1863–1935) begins preaching. He attracts huge crowds to his revival meetings.

1900 Protestantism is now established on every continent.

1900 U.S. Protestants include 6 million Methodists, 5 million Baptists, 1.5 million Lutherans, and 1.5 million Presbyterians.

1948 World Council of Churches is formed to encourage cooperation among denominations.

c. 1965 There are 70 million Lutherans worldwide, including 35 million in Germany, 18 million in Scandinavia, and 7 million in United States.

1986 Worldwide Protestant population is estimated at 332 million, or 6.6 percent of world population. U.S. Protestants number 53 million in more than 23,000 churches.

Anglicanism/Episcopalianism

300s Christianity takes root in England after being introduced by Romans.

400s Invasions by Anglo-Saxon pagans all but obliterate Christianity in England.

597 St. Augustine of Canterbury (d. c. 605) begins his mission to reintroduce Christianity to England and once again bring country under auspices of Roman Catholic Church.

601 First cathedral is built in England.

1170 Thomas à Becket (1118–70) is martyred by King Henry II (1133–89) after his long

▶

struggle to maintain church against expansion of royal authority.

1533 Pope Clement VII (1478–1534) refuses to allow King Henry VIII (1491–1547) to divorce Katharine of Aragon (1485–1536) and marry Anne Boleyn (1505–36).

1534 Henry VIII issues Act of Supremacy in response to pope's refusal. Repudiates papal authority and makes England's king head of Church of England. No changes are made in Roman Catholic Church doctrine, and Henry goes ahead with his divorce.

1539 Six Articles, intended to suppress Lutheranism, are enacted as Henry VIII also seizes monasteries in Great Britain.

1547 Edward VI (1537–53) repeals Six Articles and introduces Protestant doctrine into church service.

1549 First *Book of Common Prayer* is introduced by Archbishop Thomas Cranmer (1489–1556).

1552 Archbishop Cranmer publishes Forty-two Articles of Anglican faith and doctrine.

1553 Mary I (1516–58) ascends throne and restores Catholicism. During her six-year reign over 300 Protestant men and women are burned at stake as heretics.

1559 Upon ascension of Protestant Elizabeth I (1533–1603), Parliament officially reestablishes national Church of England.

1563 Thirty-nine Articles of Religion replace Forty-two Articles, incorporating primarily Protestant doctrine with Catholic liturgy.

1570 Pope Pius V (1504–72) excommunicates Elizabeth I.

1604 Hampton Court conference, organized by James I (1566–1625), discusses but does not resolve differences between Protestant and Catholic doctrines.

1611 First edition of King James Bible is completed.

1646 Long Parliament establishes Presbyterianism as official national religion during English Civil War and subsequent interregnum.

1660 Episcopacy is restored as Stuart dynasty regains throne.

1661 Modern version of *Book of Common Prayer* is adopted.

1661–65 Harsh Clarendon Codes are enacted, suppressing Catholicism and Presbyterianism in effort to promote uniformity of worship.

1662 Act of Uniformity declares *Book of Common Prayer* only legal form of worship.

1689 Act of Toleration gives Englishmen right to worship in other churches.

1689 English Bill of Rights excludes Catholics from throne.

1699 Society for Promotion of Christian Knowledge is founded by Anglican Church to promote education and missionary work.

1701 Act of Settlement is passed, stipulating that all British monarchs must be members of Church of England.

1701 Society for the Propagation of the Gospel in Foreign Parts is founded.

1729 Evangelical Methodist movement begins to coalesce within Church of England. Its leader, John Wesley (1703–91), is Anglican priest.

1789 Episcopal Church is officially founded as successor to Anglican Church in America after United States gains independence from Great Britain. New church revises Anglican constitution and *Book of Common Prayer.*

1799 Anglican Church Missionary Society is formed.

1829 Catholic Emancipation Act repeals repressive regulations against Catholics in Great Britain.

c. 1833 Oxford Movement. Anglicans at Oxford University unsuccessfully attempt to return to doctrines and worship of early Christian times.

1867 Lambeth Conference, first Pan-Anglican conference, is held (and has been held nearly once every 10 years since). This and subsequent conferences strengthen ties with Anglican churches in other parts of British Empire and establish basis for worldwide Anglican Communion.

▶

1874 Public Worship Regulation Act permits ecclesiastical courts to punish Anglican clergymen who deviate too far from *Book of Common Prayer* in their services.

1893 Church of England in Canada establishes national organization separate from Church of England.

1896 Pope Leo XIII (1810–1903) declares clerical orders within Church of England invalid.

1920 Act of Disestablishment. Churches in Wales and Monmouthshire (in western England) become separate archbishopric.

1959 Church of England in Canada changes its name to Anglican Church of Canada.

1966 Archbishop of Canterbury, Arthur Ramsey (b. 1904), meets with Pope Paul VI (1897–1978) in Rome. He begins process of establishing closer relations between the two churches.

1970 Confirmed membership of Church of England is about 9.9 million.

1982 Pope John Paul II (b. 1920) visits England. His meeting with Anglican church leaders marks significant step toward reconciliation between the two churches.

Archbishops of Canterbury

601–4	Saint Augustine of Canterbury (d. 605)
604–19	Saint Laurentius
619–24	Mellitus
624–27	Justus
627–53	Honorius
655–64	Deusdedit
668–90	Theodore of Tarsus (602–90)
692–731	Beorhtweald
731–34	Tatwin
735–39	Nothelm
740–60	Cuthbeorht
761–64	Breguwine
765–92	Jaenbeorht
793–805	Ethelhard
805–32	Wulfred
832	Feologild
833–70	Ceolnoth
870–89	Ethelred
890–914	Plegmund
914–23	Aethelhelm
923–42	Wulfhelm
942–59	Oda
959	Aelfsige
959	Beorhthelm
959–88	Saint Dunstan (c. 910–88)
988–90	Ethelgar
990–94	Serio
995–1005	Aelfric
1005–12	Saint Aelfheah (954–1012)
1013–20	Living
1020–38	Ethelnoth
1038–50	Eadsige
1051–52	Robert of Jumieges (fl. 1037–52)
1052–70	Stigand (d. 1072)
1070–89	Lanfranc (1005–89)
1093–1109	Saint Anselm (1033–1109)
1114–22	Ralph de Turbine
1123–36	William de Corbeil
1138–61	Theobald (d. 1161)
1162–70	Thomas à Becket (1118–70)
1174–84	Richard
1185–90	Baldwin
1191	Reginald Fitzjoceline
1193–1205	Hubert Walter (d. 1205)
1207–28	Stephen Langton (d. 1228)
1229–31	Richard le Grant
1234–40	Saint Edmund (Edmund Rich) (1175–1240)
1245–70	Boniface of Savoy (d. 1270)
1272–78	Robert Kilwardby (d. 1279)
1279–92	John Pecham (d. 1292)
1293–1313	Robert Winchelsea (d. 1313)
1313–27	Walter Reynolds (d. 1327)
1328–33	Simon Mepeham
1333–48	John de Stratford (d. 1348)

Archbishops of Canterbury			
1349	Thomas Bradwardine (1290–1349)	**1663–77**	Gilbert Sheldon (1598–1677)
1349–66	Simon Islip (d. 1366)	**1678–90**	William Sancroft (1617–93)
1366–68	Simon Langham (1310–76)	**1691–94**	John Tillotson (1630–94)
1368–74	William Whittlesey (d. 1374)	**1695–1715**	Thomas Tenison (1636–1715)
1375–81	Simon of Sudbury (d. 1381)	**1716–37**	William Wake (1657–1737)
1381–96	William Courtenay (1342–96)	**1737–47**	John Potter
1396–97	Thomas Arundel (1353–1414)	**1747–57**	Thomas Herring
1398	Roger Walden	**1757–58**	Matthew Hutton
1399–1414	Thomas Arundel (1353–1414), restored	**1758–68**	Thomas Secker
1414–43	Henry Chichele (1364–1443)	**1768–83**	Frederick Cornwallis
1443–52	John Stafford	**1783–1805**	John Moore
1452–54	John Kemp (1380–1454)	**1805–28**	Charles Manners-Sutton (1755–1828)
1454–86	Thomas Bourchier (1404–86)	**1828–48**	William Howley
1486–1500	John Morton (1420–1500)	**1848–62**	John Bird Sumner
1501–3	Henry Dean	**1862–68**	Charles Thomas Longley
1503–32	William Warham (1450–1532)	**1868–83**	Archibald Campbell Tait (1811–82)
1533–56	Thomas Cranmer (1489–1556)	**1883–96**	Edward White Benson (1829–96)
1556–58	Reginald Pole (1500–58)	**1896–1902**	Frederick Temple (1821–1902)
1559–75	Matthew Parker (1504–75)	**1903–28**	Randall Thomas Davidson (1848–1930)
1576–83	Edmund Grindal (1519–83)	**1928–42**	Cosmo Gordon Land (1864–1945)
1583–1604	John Whitgift (1530–1604)	**1942–44**	William Temple (1881–1944)
1604–10	Richard Bancroft (1544–1610)	**1945–61**	Geoffrey Francis Fisher (1945–61)
1611–33	George Abbot (1562–1633)	**1961–74**	Arthur Michael Ramsey (b. 1904)
1633–45	William Laud (1573–1645)	**1974–80**	Frederick Donald Coggan (b. 1909)
1660–63	William Juxon (1582–1663)	**1980–**	Robert Alexander Kennedy Runcie (b. 1921)

Presbyterian Church

1536 French theologian John Calvin (1509–64) publishes first edition of his *Institutes of the Christian Religion*. Earliest statement of Reformed Christianity, it promotes idea that every true believer is equal before God and that only Scriptures are authority for belief.

1536 Calvin establishes theocracy at Geneva based on his doctrine. Here he sets basic organization of church: Single congregation is governed by elected officeholders, or elders (presbyters). Councils at various levels, culminating in synods, govern congregations. Calvin and associates later establish congregations at Zurich, Strasbourg, and elsewhere.

▶

1555 First Reformed congregation is organized in France.

1559 Calvin's *Gallican Confession* is adopted by French Protestants, who come to be known as Huguenots. Despite persecution in France, their numbers grow.

1559 John Knox (1515–72) returns to Scotland from exile in Switzerland to preach Reformed Christianity.

1560 Scottish Parliament abolishes authority of pope in Church of Scotland, which adopts Presbyterian organization.

1598 Henry IV (1553–1610) issues Edict of Nantes, granting equal rights and privileges to Huguenots.

1611 First congregation in American colonies is organized at Jamestown, Virginia.

1643–49 Presbyterians dominate Westminster Assembly, appointed by Parliament to reform English Church during English Civil War. *Directory for the Worship of God* (1645), *Westminster Confession* (1646), and *Larger Catechism* and *Shorter Catechism* (1647) are written by Assembly. They eventually become standards of doctrine and liturgy for all Presbyterians.

1646 English Long Parliament establishes Presbyterianism as official national religion during English Civil War and subsequent interregnum.

1648 Peace of Westphalia in Germany ends Thirty Years' War and grants members of Reformed churches same rights that Lutherans were granted in 1555.

1685 Edict of Nantes is revoked in France. Thousands of Huguenots flee France for England, Germany, Netherlands, Switzerland, and North America.

1689 Toleration Act in England grants Presbyterians religious freedom.

1690 Act of Settlement establishes Presbyterian Church in Scotland. Church's position is strengthened by unification of kingdoms of Scotland and England (1707).

1733 First Secession Church breaks off from Church of Scotland over question of separation of church and state. Other seceding churches are Relief Church (1761) and Free Church (1843).

1789 First General Assembly of Presbyterian denominations in colonial America is held.

1811 Welsh Calvinistic Methodist Church (Calvinist-Presbyterian) is formed.

1840 Two Irish synods unite to form Presbyterian Church in Ireland.

1847 Secessionist churches in Scotland unite to form United Church of Scotland. It later (1900) joins Free Church of Scotland to form United Free Church of Scotland.

1858 United Presbyterian Church of North America is formed, following conservative Scottish Presbyterian traditions.

c. 1861 Presbyterian Church in United States, also known as Southern Presbyterian Church, is formed as result of split with northern branch over slavery.

1875 Four Canadian Presbyterian churches unite to form Presbyterian Church in Canada.

1876 Various branches of English Presbyterianism join to form Presbyterian Church of England.

1900 Church membership in United States is 1.5 million.

1925 In Canada, Presbyterian, Methodist, and Congregational churches merge to form United Church of Canada.

1929 United Free Church of Scotland rejoins Church of Scotland.

1955 General Assembly in United States approves ordination of women.

c. 1965 Church membership in United States reaches high of 4 million.

1967 Confession of 1967 is adopted by Presbyterian Church in United States, marking first adoption of major Presbyterian confession since 1647.

1972 Presbyterian Church of England merges with Congregational Church in England and Wales. United Reformed Church is formed.

▶

1983 United Presbyterian Church of North America and Presbyterian Church in United States merge to form Presbyterian Church (U.S.A.).

1989 Church membership has fallen to 2.9 million. There are 11,573 churches.

Baptist Church

c. 1608 British clergyman John Smyth (c. 1570–1612) leads group of separatists from Church of England to Amsterdam to escape religious persecution. Smyth teaches that baptism should be administered to believers only, not to infants.

1609 Smyth publishes *The Character of the Beast*, detailing his view on baptism for believers only. Baptizing 37 people (including himself), he forms a Baptist church.

c. 1611 Smyth advocates union of his church with Mennonite Church in Amsterdam, having come to believe that latter constitutes a legitimate faith.

1611 Thomas Helwys (c. 1550–c. 1616) objects to Mennonite union and leads group from Smyth's church back to London. The church he founds gives rise to General Baptist Movement, based on the "general atonement" doctrine of Dutch theologian Jacobus Arminias (1560–1609). It holds that Christ died to atone for the sins of all, not just for those predestined for salvation.

1638 Particular Baptist movement is established by Henry Jacobs outside London. This group follows doctrine of French theologian John Calvin (1509–64), called "particular atonement." Members hold Christ's death is atonement only for those predestined for salvation, a doctrine that later becomes dominant among Baptists.

1639 Roger Williams (c. 1603–83) establishes Baptist Church in America. He founds church in Providence, Rhode Island after being expelled from Massachusetts Bay Colony in 1635 for expressing "newe [sic] and dangerous opinions." Williams soon concludes that all churches lack apostolic legitimacy, however, and abandons his new church.

1640–58 Particular Baptists win many converts in England during English Civil War, notably among followers of Puritan leader Oliver Cromwell (1599–1658). General Baptists lose many of their members to Quakers.

1660 Restoration of Stuart monarchy in England drives both General and Particular Baptists into hiding for fear of persecution.

1666 English author John Bunyan (1628–88), Baptist minister imprisoned for his beliefs, composes *Grace Abounding to the Chief of Sinners* while in jail.

1689 Act of Toleration is passed in England, granting greater religious freedom to Baptists and other dissenters from Church of England.

1707 Philadelphia Baptist Association is established in Philadelphia, Pennsylvania, bringing together five churches in Pennsylvania, Delaware, and New Jersey. The association later founds many new Baptist churches in the American colonies.

1755 Shubael Stearns (1706–71) founds Baptist church in Sandy Creek, North Carolina during American religious revival known as Great Awakening. Stearns's church helps spread Baptist revival movement throughout South, laying groundwork for lasting Baptist domination of southern Protestantism.

1764 Brown University is founded by Baptists as Rhode Island College in Warren, Rhode Island. College is moved to Providence in 1770.

▶

1792 William Carey (1761–1834) organizes English Baptist Missionary Society and serves as its first missionary to India.

1800 Some 48 different associations of Baptist churches are extant in United States.

1814 Baptist General Convention is founded as first U.S. national association of Baptist churches. Convention coordinates growing foreign missionary activity of churches.

1845 Southern Baptist Convention splits from General Convention. Founded in Augusta, Georgia, new group results from friction between northern and southern churches over slavery and doctrinal and procedural disputes.

1863 First Baptist church in Canada is organized in Nova Scotia.

1880 National Baptist Convention of America is established to organize the growing number of black Baptist churches formed by freed slaves after Civil War.

1884 Russian Baptist Union is organized.

1891 University of Chicago is established as national Baptist university.

1891 Baptist Union of Great Britain is founded.

1900 Membership of Baptist Church in United States reaches 5 million.

1905 Baptist World Alliance is founded in London as multinational cooperative association for all Baptists. It results from growth of Baptist churches throughout Asia, Africa, Europe, and Latin America.

1907 Northern Baptist Convention is established, completing split between regional factions of U.S. Baptist churches.

1908 Union of Evangelical Christians is established in Russia.

1927 Baptist church membership in Soviet Union reaches high of 4.5 million. Membership decreases sharply thereafter, although Baptists remain largest Protestant denomination in Soviet Union.

1950 In United States, the Northern Baptist Convention is renamed American Baptist Convention.

c. 1970 Worldwide membership of Baptist churches reaches about 28 million. U.S. Baptist churches alone have membership of 24.5 million.

1979 Fundamentalists gain control of Southern Baptist Convention, previously dominated by more moderate factions.

1984 Membership in U.S. Southern Baptist Convention totals over 14.3 million.

1988 Baptist preacher Reverend Jerry Falwell resigns from Moral Majority, a group that promotes fundamentalist Christian values through political action, to rededicate himself to his church in Lynchburg, Virginia. He notes adverse reaction among his followers resulting from his attempt to rescue PTL Club during Jim and Tammy Bakker scandal.

1988 Fundamentalists again win presidency of Southern Baptist Convention, continuing their decade-long control of the organization.

Methodist Church

1729 John Wesley (1703–91), attending Oxford with his brother Charles (1707–88), leads religious study and prayer group known as The Holy Club. They are later called Methodists because of their strict daily routine.

1736 John and Charles Wesley arrive in Georgia as missionaries to colonists and local Indians. Mission fails, however, and both return to England.

▶

1738 John Wesley and George Whitefield (1714–70) meet with unexpected success while preaching in rural areas of England, due to Anglican Church's neglect of poor. John begins his lifelong devotion to circuit preaching.

1738 Wesley organizes first of Methodist societies for Bible study, prayer groups, and charitable deeds. These societies remain within Church of England.

1739 Bishop of London denounces George Whitefield for preaching to congregations of 20,000 or more. Whitefield continues preaching.

1740 Whitefield denounces Wesley's doctrine of "free grace." He breaks with Wesley to preach a Calvinist Methodism, thereby splitting Methodist doctrine.

1742 Welsh Calvinist Methodists are formed by Howel Harris (1714–73).

1744 Wesley calls first annual conference in England. He meets with lay preachers he has recruited to help him.

1766 Philip Embury (1729–75), one of first unofficial Methodist lay preachers in United States, begins preaching in New York.

1768 First Methodist church in United States is dedicated in New York City (October 30). Wesley Chapel on John Street is used by Philip Embury's congregation and is rebuilt in 1817 and 1840.

1769 Methodist Society in Philadelphia purchases St. George's Church, now oldest Methodist church in continuous service.

1771 Francis Asbury (1745–1816), a blacksmith, arrives in United States. He becomes only lay preacher to remain through American Revolution.

1773 First annual conference of American Methodists is held at St. George's Church in Philadelphia (July 14).

1784 Wesley ordains Richard Boardman (1738–82), Joseph Pilmoor (1738–1825), and Thomas Coke (1747–1814) as Methodist preachers and sends them to United States after Church of England refuses to do so. Coke is designated superintendent of American Methodist Society.

1784 Wesley's *Deed of Declaration*, charter of Wesleyan Methodism, appoints 100 men to govern Methodist societies after his death.

1784 Christmas conference in Baltimore. American Methodist Society breaks from Church of England, taking name Methodist Episcopal Church and establishing Methodist Church in America (December 24). Francis Asbury becomes its bishop.

1787 Founding of Cokesbury College, first Methodist college in United States, in Abingdon, Maryland.

1788 Charles Wesley dies. During his lifetime he composed more than 6,000 hymns for Methodist societies, including "Hark, the Herald Angels Sing," "Christ the Lord Is Ris'n Today," and "Love Divine, All Loves Excelling."

1789 Methodist Book Concern is organized in New York City to propagate religious works.

1791 John Wesley dies at 88, having traveled 250,000 miles and preached 40,000 sermons.

1794 First independent Methodist church for blacks is established in Philadelphia.

1795 British Methodist Society breaks from Church of England to establish Methodist Church in Great Britain.

1796 African Methodist Episcopal Zion Church is formed in New York.

1800 First recorded Methodist camp meeting is held in Logan County, Kentucky.

1816 African Methodist Episcopal Church becomes independent church in Philadelphia.

1816 Methodists in United States now total 214,235, outnumbering those in England.

1830 Methodist Protestant Church is established by reformers in United States.

1843 Wesleyan Methodist Church is formed by abolitionists in church.

1844 Methodist Episcopal Church splits over issue of slavery. Proslavery faction creates Methodist Episcopal Church, South.

1844 American Methodists now total more than 1,170,000.

▶

1860 Free Methodist Church is established by members seeking return to more conservative theology.

1870 Christian (formerly Colored) Methodist Episcopal Church is established (December 16) by blacks in Methodist Episcopal Church, South.

1900 American Methodist population now numbers 6 million.

1924 Ban on dancing and theatergoing is lifted by Methodist Episcopal General Conference.

1939 Methodist Episcopal Church; Methodist Episcopal Church, South; and Methodist Protestant Church unite to form Methodist Church.

1956 General Conference of Methodist Church bans racial segregation.

1956 Women become eligible to be ministers.

1968 United Methodist Church is created by merger of Methodist Church and Evangelical United Brethren Church.

1968 Methodists in United States number over 13 million.

1984 General Conference of United Methodist Church prohibits ordination of noncelibate homosexuals.

1986 Opposition to use of nuclear weapons is voted unanimously by Council of Bishops of United Methodist Church.

1987 Methodists in over 100 countries number 50 million.

Judaism

Judaism is practiced by 17.8 million people, representing 0.4 percent of the world's population.

B.C.

c. 2000 Abraham, first Hebrew patriarch, recognizes existence of one supreme God. He leaves family to spread his teachings. Dies in Canaan.

c. 2000–1600 In Nomadic Period, primitive form of Judaism develops. Prevailing concerns are with spirits, unclean and clean animals, and rites of dead. Period marks beginnings of ethical concepts, later developed by Moses.

c. 1300 In Exodus, Moses leads Hebrew people out of slavery in Egypt to edge of Canaan. He declares Yahweh (God) to be liberator of his people. A body of law (Ten Commandments, criminal code, and liturgical code) to govern people is shaped while Hebrews are wandering in desert. Graven images—any representations of the divine—are strictly forbidden.

c. 1000 Torah, or five books of Moses, containing basic laws of Judaism, is first set into writing. (Other two parts of Jewish Bible, Prophets and Writings, are compiled between 1000 and 200 B.C.) Holy days (Sabbath, Passover, Shavuot, and Sukkoth), religious rites, and laws relating to conduct, diet, and social justice are established.

c. 940 First Temple is constructed in Jerusalem during reign of Solomon (973–933). Religious ceremonies are conducted by priests, who also execute laws.

c. 750 Amos, Hebrew prophet, lives. He accurately predicts destruction of Israel's northern kingdom and preaches absolute sovereignty of the divinity, who demands social justice for both rich and poor.

742 Isaiah prophesies invasion of Israel by Assyria as sign of divine disapproval of Israel's godlessness. He preaches that God is more concerned with people than with performance of ritual.

c. 650–570 Prophet and reformer Jeremiah lives. He writes book of Bible bearing his

▶

	name, condemns social injustice, and prophesies destruction of Temple.
c. 621	Deuteronomic reform under Josiah (c. 640–609), king of Judah. He abolishes Jewish practice of pagan rites and demands absolute monotheism.
c. 592	Ezekiel (fl. early sixth century B.C.) receives call to prophesy; he prophesies destruction of Israel. After Babylonian exile, he preaches hope for restoration of Israel and Temple.
c. 586	Kingdom of Judah and Temple are destroyed by Babylonian invaders. Babylonians forcibly resettle prominent Jews in Babylonia.
c. 586–538	Babylonian Captivity, period of forced resettlement in Babylon. Religious rites not tied to Temple are emphasized, such as Sabbath observance and ritual circumcision. Jewish practice becomes more democratic. Persian conquest of Babylonia ends captivity (538).
c. 538	Jews begin to return to Israel. Temple is rebuilt. Jews divorce foreign wives to eliminate pagan influence and preserve Jewish community.
c. 500–200	Ezra and scribes who succeed Nehemiah complete three-part canon of Bible.
c. 165	Judas Maccabeus defeats Syrians sent to destroy him, occupies Jerusalem, and rededicates Temple. Hanukkah festival celebrates this victory.
c. 135	Pharisees advocate strict adherence to both written and oral law. They are opposed by priestly aristocracy known as Sadducees, who accept only written law as valid. Strife between the two groups leads to military intervention by Romans.
c. 100 B.C.–A.D. 200	Tannaim, or period in which Hebraism is transformed into modern Judaism.

A.D.	
70	Second Temple is destroyed by Romans during First Jewish Revolt (66–73) against Roman rule.
132–35	Bar Kokba leads Second (and final) Jewish Revolt against Roman rule and is defeated. Jerusalem is destroyed and Jews are no longer allowed to live there.
c. 135–500	Jews disperse from land of Israel. Synagogue replaces Temple as center of worship, and rabbis replace priests as spiritual leaders. Beginning of Rabbinic Judaism.
c. 200	Mishna, systematization of traditions, ideologies, and law, is compiled.
c. 500	Talmud, compilation of Jewish oral law, is put in its definitive form in Babylonia. It consists of text of oral law (Mishna) in Hebrew and commentaries on Mishna (Gemara) in Aramaic.
600–1050	Gaonate Period. Heads of Jewish academies in Sura and Plumbeditha in Babylonia explicate Jewish law and expand influence of Talmud.
c. 800	Anan Ben David and Karaite sect challenge authority of Talmud, calling for strict adherence to law of Bible. Talmudists increase scholarly efforts, and Karaite influence is halted.
c. 800–1200	Jews experience "golden age" in Spain under Moorish rule. Great writers and thinkers flourish.
1040–1105	Rashi (Rabbi Solomon ben Isaac of Troyes) lives. He writes famed commentaries on Talmud and Torah.
1085–1142	Judah Halevi, Spanish-Jewish philosopher and poet, lives. He opposes Karaite sect in favor of rabbinic Judaism.
1135–1204	Moses Ben-Maimon, or Maimonides, lives. Most influential Jewish philosopher of his time, he produces famous compilation of all laws of Judaism, *Mishne Torah*, and famous philo-

▶

sophical work *A Guide for the Perplexed.* An Aristotelian, Maimonides uses reason to fight growing Jewish interest in mysticism.

c. 1200 Cabala, Jewish mystic movement, arises in France and spreads to Spain. Cabalists believe that every word, letter, and number in Bible can reveal mysteries through cabalistic interpretation.

1250–1305 Spanish Jew, Moses de Leon, lives. He compiles *Zohar*, which becomes most sacred cabalistic work.

c. 1290–1500 European persecutions of Jews. Dating from Crusades, Jews are denied land, forced into city ghettos, and restricted to certain occupations such as moneylending. In 1290, Jews are expelled from England; in 1392, from France; in 1492, from Spain; in 1497, from Portugal. Those who survive find asylum in Netherlands and in Turkish possessions. German Jews escape to Poland.

1626–76 Sabbatai Zevi, Jewish mystic and cabalist, lives. He proclaims himself messiah in 1648 and gains thousands of followers. He is captured (1666) in Constantinople and converts to Islam to escape death.

1700–60 Baal Shem Tov lives. Founder of modern Hasidism, he teaches that a pure heart is more valued by God than a scholarly mind and that God should be worshiped joyfully.

c. 1700–1900 European Enlightenment frees Jews from ghettos, prescribed occupations, and exclusive study of Talmud.

1720–97 Elijah ben Solomon, influential scholar and commentator on Bible and Talmud, lives. He actively opposes spread of Hasidism in Lithuania and Poland, fearing weakening of Jewish community.

1729–86 Moses Mendelssohn lives. He translates Torah into German, opens first Jewish parochial school in Berlin, and advocates a Judaism more assimilated into general culture.

1800–1900 Reform movement arises in Germany. Prompted by already-developing assimilationism, movement is led mainly by laymen, with religious grounding provided by rabbis such as Abraham Geiger (1810–74) and Samuel Holdheim (1806–60). Stressing reason and scientific approach to Jewish law, it deemphasizes idea of Jewish "chosenness."

1801–75 Zacharias Frankel lives. He founds Conservative Judaism. Seen as compromise between extremes of Orthodox and Reform Judaism, it emphasizes historical approach to Judaism and preservation of Jewish ritual.

1801–88 Samson Raphael Hirsch lives. In response to Jewish assimilationism, he founds new orthodoxy in Germany. Hirsch proposes that Judaism and humanism are compatible but that strict observance of ritual is central to Jewish moral mission.

c. 1880–1917 Pogroms, violent attacks against Jews, take place in Russia. They lead to mass emigrations, especially to United States and Palestine.

1881–1983 Mordechai Kaplan lives. He founds Reconstructionist Judaism. Tied to Conservative Judaism, it holds that Judaism is an evolving religious culture and that its practices are determined by its people, not by historical documents.

1886 Term *Zionism* is first used in reference to Jewish political movement that seeks to reestablish Jewish state in Palestine.

1897 Theodore Herzl (1860–1904) convenes first World Zionist Congress in Basel, Switzerland. Rise of European antisemitism gives impetus to movement for Jewish homeland.

1917 British government issues Balfour Declaration, supporting creation of Jewish state in Palestine.

▶

1933–45 Holocaust occurs. Persecution and mass extermination of estimated 6 million Jews begins with rise of Adolf Hitler to power and ends with defeat of Germany by Allies in World War II.

1948 State of Israel is established.

1967 Old City of Jerusalem, annexed by Jordan during 1948–49 Arab–Israeli War, is captured by Israel during Six-Day War. Jews gain access to remains (Eastern Wall) of Temple, which becomes place for Jewish pilgrimage.

1969 Oldest depiction of seven-branched candelabrum, the menorah, is discovered near site of Temple. Carving in rock wall is some 2,200 years old.

1973 Conservatives approve counting of women to establish quorum of 10 adults necessary for worship.

1975 Union of American Hebrew Congregations publishes first volume of *The Torah—A Modern Commentary.*

1977 Translation of "Temple Scroll," one of the Dead Sea Scrolls, is completed. Scroll largely treats reconstruction of Temple.

1979 National Havurah Committee is formed as a network of religious fellowship groups.

1985 Ordination of women rabbis is approved by conservative movement after long debate.

Buddhism

Buddhism is practiced by 295 million people, representing 6.2 percent of the world's population.

B.C.

521 Siddhartha Gautama (c. 563–c. 483), Buddha, begins wanderings in search of spiritual truth.

c. 528 Siddhartha, meditating at Buddha Gaya in northwestern India, discovers enlightenment.

c. 528 Four Noble Truths are revealed by Siddhartha Buddha during his first sermon after his enlightenment. The Truths become basic Buddhist teaching (existence is suffering; the cause of suffering is within the self; an end to suffering is possible; and following the Eightfold Path ends suffering and leads to nirvana).

c. 483 Buddha dies, having dedicated his later life to spreading his teachings. His followers continue to spread Buddhism orally, in opposition to then-dominant Hinduism.

c. 483 First Buddhist Council is held, at Rajagrha, India. Buddha's teachings *(Sutta)* and a text on monastic discipline *(Vinaya)* are written.

c. 400 Buddhism spreads to Nepal.

383 Second Buddhist Council is held, at Vesali.

c. 340 Buddhist movement splits into rival factions. Eventually two major sects form, the Mahayana (Greater Vehicle) and the more orthodox Theravadin (Lesser Vehicle).

259–232 King Asoka (273–232) reigns. A Buddhist convert, he sends out Buddhist missionaries and promotes Buddhism in his kingdom. "Rock Edicts," stones and pillars engraved with Buddhist principles, are created during his reign.

c. 251 Mahinda (d. c. 204), Asoka's son, introduces Buddhism to Ceylon.

250 Third Buddhist Council, at Patna, is held under King Asoka in northeastern India.

▶

	At the council, Buddhist canon *(Tipitaka)* is completed.
c. 200	Buddhism spreads to central Asia.
c. 24	Two important monasteries are founded in Ceylon: the Theravadin monastery, called Mahaviranhara, and the Mahayana monastery, Abhayagiri. Long-lasting controversy develops between them.

A.D.

c. 61	Mahayana Buddhism spreads to China.
c. 100	Fourth Buddhist Council, at Jalandhar, is held in northern India.
c. 150	Nagarjuna founds "Middle Way" (Madhyamika), a major school of Mahayana Buddhism. It holds that salvation is reached by shedding all knowledge until only a void remains.
c. 300–400	Brothers Vasubandhu and Asanga found second major Mahayana school, Yoga (Yogocara) school. It holds that absolute reality can be described as mind or consciousness and that thought creates its objects out of itself.
334–413	Kumarajiva lives. He translates more than 100 Buddhist texts from Sanskrit to Chinese.
372	Buddhism spreads to Korea.
c. 400	Buddhaghosa writes *Visuddhimagga*, a major treatise on Theravadin philosophy.
515–97	Chih-i, founder of T'ien-t'ai sect in China, lives. He holds that apparent contradictions in Buddha's teachings are actually different levels of one truth.
520	Indian scholar Bodhidharma goes to China. Zen school of Buddhism eventually evolves from his teachings.
538 or 552	First written account of Buddhism in Japan.
549–623	Chi-tsang lives. He founds Madhyamika school in China.
596–664	Hsuan Tsang lives. He founds Yoga school (Fa-hsiang) in China.
600–800	Mantrayana Buddhism develops in India. It uses sacred chants (mantras) to reach enlightenment.
613–81	Shan-tao lives. He founds Pure Land sect (Ching-t'u) in China and postulates Pure Land, or paradise, to which Buddha will lead faithful followers.
625	Middle Way school forms in Japan (Sanron).
628–700	Dosho lives. He founds Yogacara school (Hosso) in Japan.
632	Tantric Buddhism becomes state religion of Tibet.
638–713	Hui-neng lives. He founds Ch'an sect in China. Sect is reaction to intense scholasticism of traditional Buddhism.
668–749	Gyogi, Korean priest, lives. He unites Buddhism and Shintoism.
700	Buddhist influence in India declines.
787	First Tibetan monastery is built.
803	Saicho (767–822) founds T'ien-t'ai sect (Tendai) in Japan. It grows into three important Japanese schools: Pure Land, Zen, and Nichiren.
817–836	Buddhism reaches high point in Tibet during reign of King Ralpa-can. His successor, however, begins persecuting Buddhists.
958–1055	Buddhism in Tibet revives. Scholar Rinchen bzang-po translates many Indian Buddhist texts.
972	Printing of Buddhist canon *Tipitaka* in China.
1030	Atisa (982–1054), Tantric master, arrives in Tibet. He begins tradition of Tibetan Lamaism headed by Dalai Lama; it becomes dominant form of Buddhism in Tibet.
1140–1390	Koryo Dynasty reigns in Korea. Buddhism reaches its height there.
1160	Council of Anuradhapura in Ceylon ends long-standing dissension between Mahavirahara and Abhayagiri monasteries.

▶

c. 1200 True Sect of Pure Land is founded in Japan by Shinran Shonin (1173–1262). Believers rely on Buddha's grace for salvation rather than on individual effort.

c. 1200 Muslim conquests in India effectively eliminate Buddhism there, the nation of its origin.

1244 Dogen (1200–53) founds Soto Zen sect in Japan. It emphasizes gradual enlightenment, rejecting sudden enlightenment.

1253 Nichiren (1222–82) founds sect named after him in Japan. Patriotic sect holds that all other schools are leading Japan to disaster.

1260 Lamaism is declared national religion of Mongols.

c. 1500 Buddhism declines in Japan.

1731 *Kanjur*, first part of Tibetan Buddhism's canonical sutras, is printed in sNarthang in Tibet.

c. 1800 Buddhist revival occurs in Ceylon, where Buddhist Theosophical Society forms with help from American Theosophical Society.

1890 Buddhist revival in Japan.

1891 D. H. Hewavitarne (1865–1933) founds Maha Bodhi Society in Ceylon to spread Buddhism to other English-speaking lands.

1929 T'ai-hsu (1889–1947) organizes Chinese Buddhist Society, which by 1947 has 4.5 million members.

1950 World Fellowship of Buddhists forms in Ceylon.

1954 Buddhist Council is held in Rangoon to mark 2,500th anniversary of Buddha's death according to Theravadin reckoning.

1960 Buddhist Congress is convoked in Cambodia to combat undermining of Buddhism there by Communists.

1964 Buddhaghosa's *Visuddhimagga* is translated into English.

Hinduism

Hinduism is practiced by 650 million people, representing 13.5 percent of the world's population.

B.C.

c. 1700–c. 500 The six basic texts of Hindu religious philosophy and mysticism are written: *Rig Veda, Sama Veda, Yajor Veda, Atharva Veda, Aranyakas,* and *Upanishads.*

c. 550 Ascetic sects break away in opposition to authority of the priests (Brahmins). Notable dissidents are Siddhartha Gautama (c. 563–c. 483 B.C.), founder of Buddhism, and Nataputta Vardhamana (c. 599–527 B.C.), founder of Jainism.

A.D.

c. 200–400 Vaisnava cult, devoted to Hindu god Vishnu, develops. Important texts for Vaisnavas: *Mahabharata*, Hindu epic encouraging Vishnu cult, and *Ramayana*, story of Rama, ideal king and follower of Vishnu.

c. 300–800 Development of *Puranas*, sacred texts devoted to Hindu gods of Brahma, Vishnu, and Shiva.

320–540 Gupta Dynasty flourishes. Temple architecture develops during this period. Saivite cult, devoted to destructive Hindu god Shiva, develops.

▶

423 Temple to Mothers (wives of Hindu pantheon) is built at Gangdhar, Rajasthan.

c. 500 Mother Goddess cult (Saktism) develops. Followers are devoted to female creative aspect of Brahma.

c. 500–c. 28 Reign of Mihirakula, Saivite king who persecutes Buddhists.

c. 600–c. 35 King Sasanka (d. c. 635) reigns. He is a Saivite and persecutes Buddhists.

c. 700 Hindu caste system becomes increasingly complex from this time forward. It eventually consists of some 3,000 hereditary castes and subcastes.

788–820 Shankara, Saivite philosopher, lives. He addresses issues of monasticism and asceticism and opposes Buddhism and other popular cults.

c. 1000 Gorakhnath founds Natha sect of Yogis and spreads practice of Yoga in India.

1192 Muslims conquer Delhi, establish kingdom there, and later expand control over much of India. Hindus subsequently endure periods of persecution by Muslim rulers.

1400–70 Ramananda lives. He founds sect devoted to Hindu god Rama. Rejects caste system and image worship and is first to write in popular Hindi, not scholarly Sanskrit.

1449–1569 Sankardeb lives. He spreads Vaisnava cult.

1485–1533 Caitanya lives. He founds Bengali Vaisnava sect. Establishes popular celebrations to Vishnu, including hymn singing, dancing, and parades.

1556–1605 Akbar (1542–1605), king of Muslim India, reigns. Hindu and Muslim cultural influences blend during his reign.

1623 Tulsidas, Hindu poet, dies. He is author of important text written in Hindi, *The Lake of the Deeds of Rama.*

1647–80 Hindu rebel Sivaji (1627–80), opposing Muslim rule of India, establishes and expands by conquest Hindu Maratha Kingdom.

1659–1707 Aurangzeb (1618–1707), king of Muslim India, reigns. He persecutes Hindus.

1784 First English translation of Hindu *Bhagavadgita.*

1828 Rammohan Ray (1772–1833), Hindu reformer, founds Society of God ashram. It opposes image worship and holds that Hinduism is root of universal religion.

1836–86 Hindu saint Ramakrishna lives. He spreads Shankara's philosophy that all religions are equally valid.

1856–1920 Bal Gangadhar Tilak lives. A key figure early in India's nationalist movement for independence, he writes commentary on Hindu *Bhagavadgita.*

1869–1948 Indian independence leader Mohandas K. Gandhi lives. He promotes among Hindus concepts of nonviolence, passive resistance *(satyagraha)*, manual labor, end to caste system, equality of women, vegetarianism, and celibacy. His campaign of nonviolent civil disobedience against British rule helps win India's independence.

1872–1950 Sri Aurobindo lives. He forms ashram encouraging translation of religious sentiments into political action for Indian independence.

1879–1950 Ramana Maharishi lives. Nearly mute ashram leader has tremendous influence on disciples. He teaches asceticism as best route to truth.

1897 Vivekanada (1862–1902) founds Ramakrishna mission. He eventually develops offshoots in Europe and North America known as Vedanta societies. Based on Ramakrishna's teachings, mission opposes caste system and encourages social service.

1949 Indian law abolishes "untouchable" class, lowest of old Hindu hereditary castes.

1951 Vinoba Bhave (b. 1895), follower of Gandhi, introduces idea of donating

▶

	fields and money to India's poor as devotional act and means to salvation.
1955–56	Indian Congress approves legislation tending to undermine traditional Hinduism. It gives women equal rights to inheritances, enforces monogamy, and facilitates divorce.
1965	Swami Pradhupada founds in Los Angeles, California the International Society of Krishna Consciousness, Vaisnava devotional movement.

Islam

Islam is practiced by 817 million people, representing 17.1 percent of the world's population.

c. 610	Muhammad (c. 570–632), merchant from Mecca, has vision on Mount Hira telling him he is God's prophet, marking beginning of his rise as religious leader.
c. 613	Muhammad begins preaching publicly. He teaches that devotees should submit to will of God.
c. 616	Abu Jahl, leader of mounting opposition to Muslims in Mecca, organizes unsuccessful boycott of merchants in Muhammad's clan, Hashim.
622	Two Pledges of al-Aqaba. Some 75 converts from Medina take oath agreeing to profess Islam and to protect Muhammad from opposition.
622	Hegira occurs. Increasing opposition to Islamic teachings in Mecca forces Muhammad and some 70 followers to immigrate to Medina (September 22). Hegira marks traditional start of Islamic calendar.
624	Muhammad breaks with Jews. He emphasizes Arabness of Islam and changes prayer ritual so that devotees face Mecca, not Jerusalem, when praying.
624	Battle of Badr (March 15). Muslims decisively defeat Meccans. Muhammad's chief rival in Mecca, Abu Jahl, is killed, and the victory strengthens Muhammad's position in Medina.
627	Battle of the Trench (April). Meccan Abu Sufyan (c. 567–c. 655) lays siege to Muhammad's forces in Medina for 15 days with 10,000 men but ultimately is unsuccessful.
c. 627	Constitution of Medina is created, forming confederation between Muhammad's followers from Mecca and the eight Arab clans in Medina.
628	Muhammad leads some 1,600 men on pilgrimage to Mecca. Meccans block their passage but then negotiate with Muhammad. They agree to Pact of Hudaibiya, ending hostilities and providing for Muslim pilgrimage to Mecca in 629.
629	Muhammad dissolves Pact of Hudaibiya after attack on group of Muslims. He prepares attack on Mecca.
630	Muhammad leads army of 10,000 against Mecca. Mecca surrenders with little resistance, whereupon Muhammad takes control of the city, further strengthening his position. Mecca thereafter is spiritual center of Islam.
632	Muhammad dies near Medina after final pilgrimage to Mecca. A close follower, Abu Bakr (573–634), is appointed caliph (successor), thus beginning line of successors to Muham-

▶

mad. His selection later becomes a matter of contention between orthodox Sunni Muslims and Shiites (Legitimists).

650 Official version of Koran is established.

657 Kharijite (Seceder) sect forms. Followers hold that sinners who do not sincerely repent are no longer Muslims and that militant rebellion is a fundamental principle of Islam.

670 Great Mosque of Kairouan in Tunisia is built.

680 Tragedy of Karbala (in modern Iraq). Troops of Umayyad caliph Yazid I (645–83) murder Husayn (c. 626–80), who, according to Shiites, was legitimate successor to caliphate. This begins Shiites' open opposition to Sunni Umayyad caliphs. Anniversary of Husayn's death becomes Islamic day of mourning.

691 Oldest surviving Islamic mosque, Dome of the Rock, is built in Jerusalem. Mosque is built on rock where Muhammad is said to have ascended to heaven.

705 Great Mosque in Damascus, considered one of seven wonders of Muslim world, is built.

c. 722 Development of Hadith, elaborate method of validating religious theories and Koranic commentaries by linking them to Muhammad and his companions.

785 Great Mosque is built in Cordoba.

c. 800–900 Sufism, an ascetic, mystical sect of Islam, grows as result of opposition to tyrannical caliphate and trend toward rationalistic philosophy.

810–70 Muhammad ibn Isma'il al-Bukhari lives. He writes Sunni canonical compilation of Koranic commentary, *Book of the Authentic Collection* (in 97 volumes).

827 Rationalistic philosophy of Mu'tazilite school, based on earlier Greek philosophy, becomes state doctrine. Sunni opposition forces reversal of the policy in 849, however.

858–922 Sufi mystic Hallaj lives. He is executed for heresy for his religious riddles and claim that he is "the truth."

c. 873–c. 936 Islamic theologian Ash'ari lives. He successfully melds Greek rationalism and Muslim thought. He holds that revelation is superior to reason but uses logic to support faith in revelation.

876 Great Mosque of Cairo is built.

1067 Founding of Mizamiyya mosque-university in Baghdad. Ash'arism becomes orthodox theology.

1096 Islamic theologian Abu Hamid Mohammed al-Ghazali (1058–1111) begins his book *The Revival of the Religious Sciences*, which helps make Sufi mysticism part of Islamic orthodoxy.

1126–98 Spanish-born Islamic philospher Averroës lives.

1149–1290 Fakhr al-Din al-Razi lives. His works discuss important philosophical themes in orthodox Sunni theology.

c. 1250 Nasir al-Din Tusi flourishes. His teachings inject philosophical elements into Shiite theology. Holds man is "creator of his own actions."

c. 1300 Fundamentalist leader Ahmad ibn Taymiyah flourishes. He teaches "science of the oneness of God," among other fundamentalist doctrines, and greatly influences later Wahabi movement.

1465 Famous book of prayers, *The Signs of the Blessings* is written by Muhammad ibn Sulayman al-Jazuli.

1703–92 Muhammad ibn Abd al-Wahhab lives. He preaches against popular religious practices in Arabia and in 1744 wins support of Arabian potentate. Thus begins Wahabi movement, which becomes part of the movement to unify Saudi Arabia.

▶

c. 1838–97 Reformer Jamal ad-Din al-Afghani lives. He holds that Islam must discover its own values and unite against foreign influences.

1899 Muhammad Abduh (1849–1905) becomes mufti of Egypt (its most powerful Islamic religious leader). A follower of the reformer al-Afghani, he advocates educational, religious, and linguistic reforms designed to maintain strength of Islam in face of growing Western influence.

1899 Qasim Amin, disciple of Muhammad Abduh, publishes book advocating emancipation of women.

1926 First international Muslim organization, Al-Mutamar al-Alam al-Islami, is formed.

1976 World of Islam festival takes place in London.

1978 Islamic fundamentalist revolution takes place in Iran, fueled by opposition to shah's program of modernization and Westernization.

Confucianism

B.C.

551–479 Confucius (551–479 B.C.) lives. He develops teachings on virtue, perfectability of humanity, obedience to will of heaven, and role of the sage ruler in bringing reform. Attracts many disciples. Five classics of Confucianism (*Wu Ching*) are traditionally attributed to Confucius.

550–250 Classical Age, or "age of a hundred philosophers," in China. This period sees growth of Confucianism, Taoism, and other philosophies.

479 Death of Confucius. *Analects,* or *Lun Yu,* a record of his life and teachings, is compiled by his disciples and becomes the basic text of Confucianism.

371–289 Mencius (Meng-tzu) lives. Known as the "second sage," after Confucius, he teaches that human nature is inherently good.

c. 289–230 Hsun-tzu lives. He argues that human nature is evil but that goodness can be acquired. Believes heaven is a natural force that does not concern itself with human affairs.

221 Ch'in Dynasty, a repressive regime, unifies China but ends period of philosophical growth of Confucianism.

213 Burning of Confucian classics is ordered by Ch'in Emperor Shi Huang Ti (255–210 B.C.).

206 B.C.–A.D. 25 Informal canons of Confucian classics gradually develop.

202 B.C.–A.D. 220 Han dynasty rules. Confucian thought generally stagnates during this period.

179–104 Tung Chung-shu lives. He categorizes the five cardinal virtues of love, righteousness, decorum, wisdom, and trustworthiness. Plays important role in winning imperial acceptance of Confucianism.

141–87 Emperor Wu Ti (156–87 B.C.) reigns. He makes Confucianism the basis of imperial government political ideology and creates the state cult of Confucianism (136 B.C.). Confucius is subsequently elevated to patron saint of all scholars, and teaching of Confucianism is introduced in public schools.

c. 124 Chinese Examination System, requiring detailed knowledge of Confucian classics, is instituted by Wu Ti. It serves as required test for all Chinese civil servants until its abolition in 1905.

▶

124 First imperial university is founded as a center for study of the Five Confucian Classics. Its enrollment reaches 30,000 by A.D. 220.

A.D.

c. 8 Old Text school forms by this time in reaction to New Text school, which supports infusing Confucianism with such new elements as cosmological speculations (Yin and Yang) and numerology.

c. 100 Confucian temples are built in each of China's 2,000 counties.

220 Han dynasty falls. Former unity of China, maintained by Ch'in and Han dynasties, gives way to division. Taoism grows in influence as Confucianism declines. Buddhism spreads rapidly in China.

581–618 Sui dynasty reigns and restores Chinese unity.

618–907 T'ang dynasty is in power. Civil-service examinations, based on Confucian classics, are expanded and a new edition of the Confucian classics, *Wu Ching*, is completed.

768–824 Han Yu lives. His polemical attacks on Buddhism are highly influential.

960–1279 Neo-Confucian revival takes place during reign of Sung dynasty.

1130–1200 Chu Hsi lives. Widely regarded as the preeminent figure of the neo-Confucian revival, he codifies Confucianism into comprehensive system of thought. His commentaries on the Confucian classics become official interpretation for use in civil-service examinations.

1527–1602 Li Chih lives. Famous Confucian thinker and heretic is part of an ongoing reaction to rigid interpretation of Confucianism that has developed during Ming dynasty.

c. 1850 Weakening central government and growing influence of Western powers in China lead to decline of traditional Chinese society based on Confucianism. By end of 1800s, formal efforts are made to westernize Chinese society.

1905 Examination system is abolished. Confucian classics are gradually eliminated from school curricula.

1911 Chinese Revolution ends imperial rule.

1928 Sacrifices at Confucian temples and widespread study of Confucian classics are no longer practiced.

1949 Communist rule is established in China. Confucianism declines greatly in influence as Marxism replaces it as state ideology.

1974 Chinese government officially condemns Confucius and his teachings as reactionary.

1988 Followers of Confucianism worldwide now number about 5.9 million.

Philosophy

B.C.

c. 640–546 Thales of Miletus lives. Often called the first philosopher, he inquires into underlying constituents of all things, proposing that all is water.

c. 610–547 Anaximander of Miletus lives. He attempts to explain all things as arising from *apeiron*, or indeterminacy.

c. 550 Anaximenes, Greek pre-Socratic, lives. Like Thales, he believes that one sub-

▶

stance underlies all things, though names air as the one universal substance.

c. 580–c. 500 Pythagoras of Samos lives. He inquires into mathematical harmonies upon which world is ordered. Founds religious-philosophical school in Crotona, Italy, preaching purification and transmigration of souls.

c. 540–c. 475 Heraclitus of Epheseus, called "dark philosopher," argues that all is in state of flux and that because everything is constantly changing, it is impossible "to step into the same river twice."

c. 515–450 Parmenides of Elea lives. He inquires into nature of "Being," arguing in his poem *On Nature* that it is changeless and not perceptible by senses. Venerated by Plato, whom he greatly influenced, Parmenides is portrayed in Plato's dialogue *Parmenides.*

c. 500–428 Anaxagoras, Greek pre-Socratic thinker, lives. He rejects Empedocles's theory of four elements and proposes that infinite numbers of unique particles compose all objects in nature.

c. 490–c. 430 Empedocles, Greek pre-Socratic philosopher, lives. He believes that universe consists of four elements: earth, air, fire, and water. According to legend, he ultimately throws himself into the volcano Mount Etna to prove to his followers he is a god.

c. 490–410 Protagoras of Abdera lives. Most important of Sophists, he preaches extreme relativism in his works *On Truth* and *On the Gods.* Argues that "Man is the measure of all things."

c. 470–399 Socrates lives. Exclusively an oral teacher, he spends his life in Athens inquiring into nature of virtue. Teaching his Socratic method of inquiry, he attracts such brilliant students as Plato and Xenophon (431–350). Socrates is represented in many of Plato's dialogues, including *Apology, Crito,* and *Phaedo.* Eventually given a death sentence on charges of impiety and corrupting the young, Socrates takes poison.

c. 460–c. 370 Demoritus lives. He proposes a mechanistic rather than a supernatural explanation of the world and holds that all things are made up of atoms.

428–348 Plato lives. Great systematic philosopher and student of Socrates, he writes many famous dialogues and founds a philosophical school, the Academy, which lasts until A.D. sixth century. His teaching centers on theory of forms or ideas, presented in *Republic.* This work also reveals his political theory by describing an ideal city ruled by philosopher-kings.

c. 400–c. 325 Diogenes lives. He is the founder and leading figure of the Cynics, a Greek philosophical school that defies social conventions. He lives the life of a beggar, promoting self-sufficiency and eccentric protests against society.

384–322 Aristotle lives. Greek philosopher from Macedonia and one of the most influential thinkers of all time, he spends 20 years at Plato's Academy, eventually founding his own school in Athens, the Lyceum. He makes major contributions to wide range of academic fields in such works as *Metaphysics, Politics, Poetics,* and *Nicomachean Ethics.*

341–270 Epicurus from Samos lives. A follower of Democritus, he founds Epicurean school and atomism. He preaches materialism and a moderate hedonism in *On Nature* and other works.

335–264 Zeno of Citium lives. He founds Stoic school, which emphasizes living according to nature and natural law.

c. 99–c. 55 Lucretius, Roman Epicurean philosopher and poet, lives. He holds that the soul is comprised of atoms and ceases to exist when the body dies, expanding this thesis in *De rerum nahira.*

A.D.

121–180 Marcus Aureluis, Roman emperor, lives. He writes classic work on stoic philosophy, *Meditations.*

▶

205–70 Plotinus, founder of Neoplatonism and Egyptian by birth, lives. He founds his school in Rome. His main teachings are collected by his pupil Porphyry in *Enneads.*

354–430 St. Augustine lives. Residing mainly in North Africa, he becomes bishop of Hippo. Synthesizes elements of Greek philosophy and Christian theology in such works as *The City of God* (413–26) and *Confessions* (c. 400).

c. 475–525 Boethius lives. A Roman translator of Aristotle, he writes *The Consolation of Philosophy*, a much-read treatise of the Middle Ages.

980–1037 Avicenna lives. A leading Islamic philosopher who interprets Aristotle, his work eventually influences such medieval philosophers as St. Thomas Aquinas.

1033–1109 St. Anselm lives. Italian by birth, he eventually becomes archbishop of Canterbury in England (1093–1109). He is best known for his proof of God's existence, the Ontological Argument, contained in *Proslogion.*

1079–1142 Peter Abelard, French theologian, logician, and teacher, lives. He is condemned by the church for heresy because of his nominalist views but is best remembered for tragic events that befell himself and his wife, Héloïse.

1225–74 St. Thomas Aquinas lives. He synthesizes Christian thought with Aristotelian teachings, only recently rediscovered in West. Italian by birth, he serves in Dominican order while writing *Summa Theologica* (c. 1265–74), *Summa Contra Gentiles* (c. 1258–64), and other works.

1126–98 Averroës, Spanish-born Arabian philosopher, lives. His commentaries on Aristotle's works are widely read. He holds philosophical knowledge is derived from reason, not faith.

c. 1266–1308 Duns Scotus lives. He pursues questions of logic, metaphysics, and epistemology in scholastic tradition in England and France.

c. 1285–c. 1349 William of Occam, English scholastic philosopher, lives. He devises "Occam's razor."

1561–1626 Sir Francis Bacon, English statesman and philosopher, lives. He advances his theory of the scientific method of inquiry in his best-known work, *Novum Organum* (1620).

1588–1679 Thomas Hobbes lives. He develops materialist account of nature and politics. In *Leviathan* (1651) he argues for absolute government as remedy to chaos wrought by Civil War in England.

1596–1650 René Descartes lives. A French philosopher in rationalist tradition, he postulates dualism of mind and body and devises famous philosophical proposition "I think, therefore I am" (*Cogito, ergo sum*). Descartes constructs materialist account of world, an essential philosophical concomitant to progress in natural sciences. His "method of doubt" is presented in his *Discourse on Method* (1637) and *Meditations* (1641).

1632–77 Baruch (Benedict) Spinoza lives. Dutch by birth and of Jewish descent, he works out implications of Cartesian metaphysics in his *Ethics* (1677) and other works.

1632–1704 John Locke, English philosopher, lives. His great contribution to English empiricism is *Essay Concerning Human Understanding* (1690). He presents classical statement of liberal political theory and social contract of government in *Two Treatises on Government* (1690).

1646–1716 Gottfried Wilhelm Leibniz, German philosopher, lives. He adopts rationalist approach to metaphysical questions in *The Monadology* (1714) and other works.

1685–1753 George Berkeley, idealist critic of English empiricism, lives. He writes *Three Dialogues Between Hylas and Philonous* (1713) and other works. Devises philosophical proposition *esse est*

▶

percipi—material world exists only in being perceived by the mind.

1711–76 Scottish empiricist David Hume lives. He draws skeptical implications of British empiricism in *The Treatise of Human Nature* (1737–40). Criticizes view that certain knowledge can be based on sense perception. His posthumously published *Dialogues Concerning Natural Religion* (1779) vigorously questions extent to which religious beliefs can be founded on reason.

1712–78 Jean-Jacques Rousseau lives. French philosopher and political theorist, he writes *The Social Contract* (1762). Rousseau is a sometimes erratic figure in romantic reaction to current rationalist view of man.

1724–1804 Immanuel Kant, great German philosopher, lives. He spends his life attempting grand synthesis of empiricist and rationalist traditions, to respond to Hume's skepticism. Kant publishes his monumental *Critique of Pure Reason* (1781), *Critique of Practical Reason* (1781), and *Critique of Judgment* (1790) late in life. *Groundwork Concerning the Metaphysics of Morals* (1785) presents his influential moral philosophy, centering on categorical imperative: "Act as if the maxim of thy act were to become by thy will a universal law of nature."

1748–1832 Jeremy Bentham, English philosopher and lawyer, lives. One of the founders of utilitarianism, he won worldwide recognition with his *Introduction to the Principles of Morals and Legislation* (1789).

1770–1831 German philosopher G.W.F. Hegel lives. He creates his idealist philosophical system based on dialectic. His teachings, presented in such works as *Encyclopedia of Philosophy* (1817), *Phenomenology* (1807), and *Philosophy of Right* (1821), exert enormous influence on subsequent thinkers, including Karl Marx (1818–83) and Søren Kierkegaard (1813–55).

1788–1860 Arthur Schopenhauer, German pessimistic and aesthetic philosopher, lives. He develops his teaching under the influence of Kant and Plato; it is presented most fully in *The World as Will and Idea* (1819).

1798–1857 Auguste Comte, founder of the 19th-century philosophical system called positivism and of sociology, lives.

1806–73 John Stuart Mill lives. He works out implications of English utilitarian and empiricist traditions in logical, moral, and political philosophy and political economy. His *Utilitarianism* (1863) defends utilitarian moral philosophy. *On Liberty* (1859) probably is best-known defense of individual liberty in Western philosophy.

1813–55 Danish religious philosopher Søren Kierkegaard, founder of existentialist philosophy, lives. His works include *Either/Or* (1843) and *Fear and Trembling* (1843).

1818–83 German philosopher, revolutionist, and economist Karl Marx lives. He theorizes that his "dialectical materialism" explains economic basis of historical development. Argues for inevitable collapse of capitalism and rise of communism in *Das Kapital* (1867–94) and other works.

1839–1914 American philosopher Charles S. Peirce lives. He founds American school of pragmatism, writing papers in epistemology and metaphysics. He coins term *pragmatism* (1878) to describe his theory of meaning, which emphasizes practical consequences of conceptions.

1842–1910 William James, American philosopher and psychologist, lives. He is a leader of Pragmatism movement. Among his philosophical works are *The Varieties of Religious Experience* (1902) and *Essays in Radical Empiricism* (1912).

1844–1900 Friedrich Nietzsche, German philosopher, lives. He argues for death of God and will to power in *Thus Spake Zarusthustra* (1883–92), *The Will to Power* (1901), and other works. Originally trained as classicist, Nietzsche completes *The Birth of Tragedy* (1872),

▶

noted in this field, before undertaking his studies in philosophy.

1848–1925 Gottlob Frege, German philosopher, lives. He makes pioneering contributions to mathematical logic that will greatly influence Bertrand Russell and other thinkers.

1859–1938 Edmund Husserl, German philosopher, lives. He founds phenomenology and writes such influential works as *Logical Investigations* (1900–01) and *Ideas: General Introduction to Pure Phenomenology* (1913).

1859–1952 John Dewey, American pragmatist, philosopher, psychologist, and educational theorist, lives. He explores questions in science, education, and political philosophy in *Democracy and Education* (1916), *Experience and Nature* (1925), *The Quest for Certainty* (1929), and other works. He maintains that knowledge is not passive, that it must be pursued.

1863–1952 George Santayana, American philosopher and poet, lives. He asserts that "animal faith" (impulse) is the basis of reason and belief. His works include *The Sense of Beauty* (1896) and *Scepticism and Animal Faith* (1923).

1872–1970 Bertrand Russell, English philosopher, lives. He does influential work in mathematical logic and other areas of philosophy. Writes *Principia Mathematica* (1910–13, with A. N. Whitehead) and other works. Receives Nobel Prize for Literature (1952).

1884–1976 Martin Heidegger, German philosopher, lives. He studies under Husserl and then becomes a central figure in existentialism. Publishes *Being and Time* (1927) and numerous other works.

1889–1951 Ludwig Wittgenstein, Austrian philosopher, lives. He influences development of logical positivism with his *Tractatus Logico-Philosophicus* (1922). He retires from philosophy to become elementary schoolteacher and architect. He later analyzes use of ordinary language, however, in *Philosophical Investigations* (published posthumously in 1953).

1902– Karl Popper, British philosopher, lives. He argues for scientific explanation based on "falsification" in *The Logic of Scientific Discovery* (1935). Criticizes totalitarian political philosophy in *The Open Society and Its Enemies* (1945).

1905–80 French philosopher Jean-Paul Sartre lives. Influenced by Husserl and Heidegger, he becomes the major French existentialist with *Being and Nothingness* (1943) before embracing Marxian tradition in his *Critique of Dialectical Reason* (1960). Declines Nobel Prize for Literature (1964).

1908– American philosopher Willard Quine lives. He is noted for his work in mathematical logic, especially in applying it to language. His works include *Word and Object* (1938) and *Philosophy of Logic* (1969).

1921– John Rawls, American philosopher, lives. In *A Theory of Justice* he criticizes utilitarianism and reformulates device of the social contract to justify deontological moral principles.

Education

B.C.

c. 3000 Sumerians and Egyptians establish schools to teach reading and writing to boys.

c. 500 Emergence of Greek Sophists, itinerant teachers who receive fees for their teaching.

▶

c. 400 Athenian gymnasiums are centers of advanced learning.

387 Plato founds the Academy, the most influential school of ancient world.

c. 100 Romans develop extensive educational system, for girls as well as boys.

A.D.

840 Under rule of Charlemagne (c. 742–814), monastic schools become widespread.

1160 University of Paris, generally considered to be finest university of Middle Ages, is founded.

1376 Nonmonastic teaching order, dedicated to democratic teaching principles, develops in Holland.

c. 1438 Invention of printing press by German metalworker Johannes Gutenberg (c. 1400–c. 68) leads to publication of inexpensive books and pamphlets. It stimulates interest in education.

c. 1500 Almost 80 universities have been established in Europe.

c. 1500 Primary schools begin to appear in Europe.

1534 Spanish nobleman Ignatius of Loyola (1491–1556) organizes Society of Jesus, known as Jesuits.

1628 Czech educator John Comenius (1592–1670) publishes *Didactica Magna*, in which he expounds teaching in vernacular (rather than in Latin) and universal education for men and women.

1636 Harvard, first college in American colonies, is founded.

1647 Massachusetts law requires that every town of 50 families or more must hire teacher to teach reading and writing.

1663 First Canadian university is founded.

1689 Founding of the first U.S. secondary school, in Pennsylvania.

1690 *The New England Primer*, widely used colonial textbook, is published.

1723 Pauper schools are established in Maryland.

1745 Pennsylvania Quakers establish first elementary schools for black children.

1770 Publication of the first book on teaching, by German educator Christoph Dock.

1779 Thomas Jefferson (1743–1826) proposes first modern state school system: free tuition to all free children, attendance voluntary.

1783 Noah Webster's *American Spelling Book* is published.

1785 U.S. Land Ordinance reserves land in every western territorial township for establishment of public schools.

1802 France establishes national system of secondary schools and universities.

1805 Swiss educational reformer Johann Pestalozzi (1746–1827) founds experimental school in which he opposes memory-learning techniques and strict discipline of students.

1816 First infant school is established in Boston.

1818 Publication of first teachers' journal, *The Academician*.

1821 First U.S. high school is established in Boston.

1837 First permanent women's college in United States, Mount Holyoke, is founded.

1837 American educator Horace Mann (1796–1859) establishes major school reforms in Massachusetts.

c. 1840 Blackboards are introduced, prompting educators to predict revolution in education.

1852 Massachusetts is first U.S. state to mandate compulsory school attendance.

1861 First Ph.D. degree in United States is awarded, at Yale University.

1862 Federal land grants are provided to aid agricultural colleges in United States.

1867 First U.S. commissioner of education, Henry Barnard (1811–1900), begins term.

1873 Beginning of Chautauqua Movement in U.S. adult education.

1873 First public-school kindergarten is established, in Missouri.

1874 Michigan Supreme Court upholds taxes for public high schools.

▶

1890 Educational testing is begun in some U.S. schools.

1896 U.S. Supreme Court decision approves "separate but equal" schooling, legal precedent for segregated schools.

1896 American philosopher and educator John Dewey (1859–1952) founds elementary school in Chicago to experiment with progressive education ideas.

1897 Founding of National Congress of Mothers, precursor of PTA.

1907 Establishment of first day-care center, in Rome, by Italian educator and physician Maria Montessori (1870–1952), whose teaching approach becomes known as Montessori Method.

c. 1920 Establishment of first U.S. nursery school.

1944 G.I. Bill provides payments for tuition, books, and living expenses to World War II veterans. (Similar programs follow Korean and Vietnam wars.)

1946 Fulbright scholarships are established for U.S. teachers, researchers, and students to encourage exchange programs with other nations.

1954 U.S. Supreme Court decision requires desegregation of public schools. School desegregation becomes highly controversial issue in subsequent years.

1955 By this date, many schools have adopted "core curricula."

1958 U.S. National Defense Education Act promotes teaching of sciences, foreign languages, and mathematics.

c. 1960 Introduction of "new math" in U.S. schools.

1961 France moves to liberalize education.

1962 U.S. Supreme Court upholds ban on public-school prayer.

1964 Head Start, U.S. educational program for disadvantaged preschool children, is established.

1968 Open-classroom system of informal learning areas and activities is becoming popular in United States.

1970 Many U.S. schools have adopted team-teaching methods of instruction.

1971 U.S. Supreme Court approves controversial school-busing plans as means of achieving desegregation.

1971 Punishing students by caning is banned in London schools.

1975 Many U.S. schools are moving toward reviving traditional approaches to teaching.

1980 Microcomputers begin to appear in U.S. classrooms.

1983 Government report on education terms United States "a nation at risk" educationally in comparison to other industrialized nations.

1984 Average SAT scores of U.S. high-school students improve by four points, ending steady decline in the 1970s and early 1980s.

1987 Elementary- and secondary-school teachers in United States earn average salary of $26,700.

1989 New studies comparing math abilities of students in six countries show South Koreans first and U.S. students last.

U.S. and Canadian Colleges and Universities (founding dates)

1636 Harvard University, Cambridge, Massachusetts

1693 College of William and Mary, Williamsburg, Virginia

1701 Yale University, New Haven, Connecticut

1740 University of Pennsylvania, Philadelphia

1746 Princeton University, Princeton, New Jersey

1749 Washington and Lee University, Lexington, Virginia

1754 Columbia University, New York, New York

1764 Brown University, Providence, Rhode Island

1766 State University of Rutgers at New Brunswick, New Jersey

1769 Dartmouth College, Hanover, New Hampshire

1773 Dickinson College, Carlisle, Pennsylvania

1784 St. John's College, Annapolis, Maryland

U.S. and Canadian Colleges and Universities (founding dates)

1785	University of Georgia, Athens
1785	University of New Brunswick, Fredericton, Canada
1787	Franklin and Marshall College, Lancaster, Pennsylvania
1787	University of Pittsburgh, Pennsylvania
1789	Georgetown University, Washington, D.C.
1789	University of North Carolina, Chapel Hill
1791	University of Vermont, Burlington
1793	Williams College, Williamstown, Massachusetts
1794	Bowdoin College, Brunswick, Maine
1794	University of Tennessee, Knoxville
1795	Union College, Schenectady, New York
1798	University of Louisville, Louisville, Kentucky
1800	Middlebury College, Middlebury, Vermont
1801	University of South Carolina, Columbia
1802	St. Mary's University, Halifax, Nova Scotia, Canada
1802	U.S. Military Academy, West Point, New York
1804	Ohio University, Athens
1809	Miami University, Oxford, Ohio
1812	Hamilton College, Clinton, New York
1812	Lycoming College, Williamsport, Pennsylvania
1813	Colby College, Waterville, Maine
1817	University of Michigan, Ann Arbor
1818	Dalhousie University, Halifax, Nova Scotia, Canada
1818	St. Louis University, Missouri
1819	Colgate University, Hamilton, New York
1819	University of Cincinnati, Ohio
1819	University of Virginia, Charlottesville
1820	Indiana University, Bloomington
1821	Amherst College, Amherst, Massachusetts
1821	George Washington University, Washington, D.C.
1821	McGill University, Montreal, Quebec, Canada
1823	Trinity College, Hartford, Connecticut
1824	Rensselaer Polytechnic Institute, Troy, New York
1826	Case Western Reserve University, Cleveland, Ohio
1826	Furman University, Greenville, South Carolina
1826	Lafayette College, Easton, Pennsylvania
1827	University of Toronto, Ontario, Canada
1830	University of Richmond, Virginia
1831	La Grange College, Georgia
1831	New York University, New York, New York
1831	University of Alabama, Tuscaloosa
1831	Wesleyan University, Middletown, Connecticut
1833	Haverford College, Pennsylvania
1833	Oberlin College, Ohio
1833	University of Delaware, Newark
1834	Tulane University of Louisiana, New Orleans
1834	University of the Ozarks, Clarksville, Arkansas
1834	Wake Forest University, Winston-Salem, North Carolina
1834	Wheaton College, Norton, Massachusetts
1836	Emory University, Atlanta, Georgia
1837	Mount Holyoke College, South Hadley, Massachusetts
1838	Acadia University, Wolfville, Nova Scotia, Canada
1838	Duke University, Durham, North Carolina
1839	Boston University, Massachusetts
1839	University of Missouri at Columbia
1841	Fordham University, New York , New York
1841	Queen's University at Kingston, Ontario, Canada
1842	University of Notre Dame, Notre Dame, Indiana
1842	Villanova University, Villanova, Pennsylvania
1843	College of the Holy Cross, Worcester, Massachusetts
1844	State University of New York (first of 64 SUNY colleges and centers)
1844	University of Mississippi, University
1845	U.S. Naval Academy, Annapolis, Maryland
1846	Bucknell University, Lewisburg, Pennsylvania
1847	University of Iowa, Iowa City
1848	Rhodes College, Memphis, Tennessee
1848	University of Ottawa, Ontario, Canada
1848	University of Wisconsin at Madison
1850	University of Rochester, New York
1850	University of Utah, Salt Lake City
1850	Oregon State University, Corvallis
1851	Northwestern University, Evanston, Illinois
1851	University of Minnesota, Minneapolis-St. Paul
1852	Antioch University, Yellow Springs, Ohio

U.S. and Canadian Colleges and Universities (founding dates)

1852	Laval University, Quebec City, Canada
1852	Tufts University, Medford, Massachusetts
1853	University of Florida, Gainesville
1853	Washington University, St. Louis, Missouri
1855	Michigan State University, East Lansing
1855	Pennsylvania State University, University Park
1855	Trenton State College, New Jersey
1856	Auburn University, Alabama
1856	University of Maryland, College Park
1857	Illinois State University, Normal
1857	University of the South, Swanee, Tennessee
1858	Iowa State University, Ames
1860	Louisiana State University and Agricultural and Mechanical College, Baton Rouge
1861	Massachusetts Institute of Technology, Cambridge
1861	University of Washington, Seattle
1861	Vassar College, Poughkeepsie, New York
1862	University of South Dakota, Vermillion
1863	Boston College, Chestnut Hill, Massachusetts
1864	Marquette University, Milwaukee, Wisconsin
1864	Swarthmore College, Pennsylvania
1864	University of Denver, Colorado
1865	Cornell University, Ithaca, New York
1865	Lehigh University, Bethlehem, Pennsylvania
1865	University of Maine, Orano
1865	University of Kentucky, Lexington
1866	Drew University, Madison, New Jersey
1866	Carleton College, Northfield, Minnesota
1866	University of Kansas, Lawrence
1866	University of New Hampshire, Durham
1867	Howard University, Washington, D.C.
1867	University of llinois at Urbana-Champaign
1867	West Virginia University, Morgantown
1868	University of California, Berkeley
1868	Wayne State University, Detroit, Michigan
1869	Purdue University, West Lafayette, Indiana
1869	Southern Illinois University at Carbondale
1869	University of Nebraska, Lincoln
1870	Colorado State University, Fort Collins
1870	Ohio State University, Columbus
1870	St. John's University, New York, New York
1870	Syracuse University, New York
1870	Wellesley College, Massachusetts
1871	Smith College, Northampton, Massachusetts
1871	University of Arkansas, Fayetteville
1873	Vanderbilt University, Nashville, Tennessee
1874	Macalester College, St. Paul, Minnesota
1875	Brigham Young University, Provo, Utah
1876	Johns Hopkins University, Baltimore, Maryland
1876	Texas A&M University, College Station
1876	U.S. Coast Guard Academy, New London, Connecticut
1876	University of Colorado at Boulder
1876	University of Montreal, Quebec, Canada
1876	University of Oregon, Eugene
1877	University of Manitoba, Winnipeg, Canada
1878	University of Western Ontario, London, Canada
1880	University of Southern California, Los Angeles
1881	Tuskegee Institute, Alabama
1881	University of Connecticut, Storrs
1883	Jacksonville State University, Alabama
1883	University of Texas, Austin
1884	University of North Dakota, Grand Forks
1884	Temple University, Philadelphia, Pennsylvania
1885	Arizona State University, Tempe
1885	Bryn Mawr College, Pennsylvania
1885	Georgia Institute of Technology, Atlanta
1885	Stanford University, Stanford, California
1885	University of Arizona, Tucson
1886	University of Wyoming, Laramie
1887	Clark University, Worcester, Massachusetts
1887	Pomona College, Claremont, California
1887	Troy State University, Alabama
1888	University of Rhode Island, Kingston
1889	Barnard College, New York, New York
1889	University of Idaho, Moscow
1889	University of New Mexico, Albuquerque
1890	Oklahoma State University, Stillwater

U.S. and Canadian Colleges and Universities (founding dates)	
1890	University of Chicago, Illinois
1890	University of Oklahoma, Norman
1890	Washington State University, Pullman
1891	California Institute of Technology, Pasadena
1891	Rice University, Houston, Texas
1893	American University, Washington, D.C.
1897	San Diego State University, California
1899	San Francisco State University, California
1906	University of Alberta, Edmonton, Canada
1907	University of Hawaii, Honolulu
1909	Reed College, Portland, Oregon
1911	Connecticut College, New London
1911	Southern Methodist University, Dallas, Texas
1915	University of British Columbia, Vancouver, Canada
1919	New School for Social Research, New York, New York
1919	University of California at Los Angeles
1923	Texas Tech University, Lubbock
1925	University of Miami, Coral Gables, Florida
1942	Carleton University, Ottawa, Ontario, Canada
1942	Fairfield University, Fairfield, Connecticut
1945	University of Calgary, Alberta, Canada
1946	Claremont-McKenna College, Claremont, California
1948	Brandeis University, Waltham, Massachusetts
1954	U.S. Air Force Academy, Colorado Springs, Colorado
1954	University of Alaska, Anchorage
1957	University of Waterloo, Ontario, Canada
1959	York University, North York, Ontario, Canada

8 Architecture and Engineering Feats

Architecture

An Overview

B.C.

6500	Ancient builders at Jarmo, Mesopotamia, are already using precursor of brick, called pisé block. Some houses here also have stone foundations.
4000	Communal houses 160 feet long are being built in Europe.
4000	First use of mortar to bind bricks, in Mesopotamia. Bitumen slime is used to bind sun-dried bricks. Mesopotamians also later discover kiln-drying as means of making long-lasting brick.
c. 2664–2155	Temple of Sphinx at Giza employs clerestory roof.
2600–2500	Egyptian pyramids are constructed at Giza. These massive structures are created as burial places for the Egyptian pharaohs. Largest is more than 480 feet high.
2400	Mesopotamian builders use arch and vault construction.
c. 1800–1600	Great palace at Knossos in Crete is built. One of most monumental palaces in ancient world, it is built around massive central court. It features well-planned suites and stairways with light. Well system of baths and water pipes demonstrates sophisticated knowledge of sanitation.
1570–1070	Egyptian temples are constructed during Empire period, using entrance pylons, peristyle with colonnades, large hall, and sanctuary.
c. 1400	Lion's Gate at Mycenae, Greece, is built. Two enormous doorjambs are supported by lintel in early substitute for arch.
c. 700	Palace of Sargon II is constructed of brick in Khorsabad, Mesopotamia. It is 60 feet tall and covers more than 100,000 square feet.
600	Doric order is developed, possibly in Corinth. It features columns without bases, curved capitals, and sharp friezes over column.
c. 600	Ionic order first appears in Smyrna. Its characteristics include thin stone columns with curved capitals carved in floral pattern.
c. 600	Temple of Bel and Hanging Gardens are built in Babylon by King Nebuchadnezzar (c. 630–562 B.C.). Temple is a multicolored brick ziggurat, or pyramid. The terraced gardens are considered among the seven wonders of the ancient world.
c. 500	Persian palace at Persepolis is built. It features colonnades, gates, vast staircases, and decorative friezes.
477–432	Parthenon is constructed in Athens. One of greatest works of Greek architecture, it employs Doric capitals, monumental steps, and hanging cornice.
c. 460	Temple of Zeus at Olympia, Greece is built as one of first classically Doric temples.
435–408	Erechtheum is built on Acropolis in Athens. It is considered supreme example of Ionic order.
c. 430	Temple of Apollo Epicurius at Bassae is built. It incorporates perhaps first use of Corinthian order.
214–204	Great Wall of China is constructed. Designed for military defense, it eventually spans 1,500 miles.
c. 100	Roman architect Vitruvius writes *De Architectura*, comprehensive handbook of architectural styles.

A.D.

70–82	Roman Colosseum is built. One of greatest amphitheaters ever constructed, it occupies six acres and seats 50,000.
c. 120–24	Roman Pantheon is built. It incorporates a massive concrete dome (world's largest until modern times) and vault supported by walls 20 feet thick. The vault allows for a vast interior unbroken by columns.
211–17	Construction of baths of Caracalla in Rome. It includes hot, cold, and

▶

Mounds and Megaliths

B.C.

c. 7500 Ancient city of Jericho is enclosed by 14-foot wall 10 feet thick and some 800 yards in circumference. Cylindrical tower of 25 feet protects city.

c. 3500 Grand Menhir of Locmariaquer, 380-ton megalith, is set up in complex of other megalithic structures near Gulf of Morbihan in Brittany. It probably is used for astronomical observations.

c. 3250 Tomb made of giant rock slabs and covered with earthen mound is constructed at New Grange, Ireland. It contains beehivelike passages and central chamber of elaborately carved stones. Arranged according to astronomical and mathematical principles, tomb is made up of 200,000 tons of stone.

c. 3000 Seven-story brick ziggurat of Etemenanki, probably biblical Tower of Babel, is constructed along Euphrates in ancient Babylon.

c. 3000 Taxien, temple among Europe's largest Stone Age monuments, is constructed at Malta. Megalithic complex of 20 acres contains maze of chambers and corridors, giant statues, and sacrificial altars.

c. 3000 Stonehenge is built in southern England. Circle of giant megalithic stones surround altar stone. Stone structure probably is used for religious rituals and may have astronomical significance as well.

c. 2670 Maes Howe, passage grave in the Orkney Islands, is constructed between two groups of megalithic stones: Circle of Stenness and Ring of Brodgar (possibly lunar observatory). Astronomical alignment of tomb's placement in relation to sun and megaliths makes it extraordinary feat.

c. 2500 Temples of Mnajdra at Malta are constructed in huge, fan-shaped arrangement overlooking Mediterranean. Giant stone blocks make up elaborate structures.

c. 2500 Temple of Hagar Qim in Malta, one of most complex of ancient megaliths, is built of giant stone blocks. It contains sanctuary, statuettes, and crude altars.

c. 2500 Inhabitants of Gozo, island near Malta, construct Ggantija ("Work of the Giants"), imposing 26-foot temple of giant limestone slabs.

c. 2200 Avebury, one of biggest ceremonial megalithic structures in Europe, is constructed in British Isles. Great circle of standing stones, some weighing over 60 tons, encloses entire village. It probably is work of many generations of people.

c. 2100 Largest artificial hill in Europe, Silbury Hill, is constructed in the United Kingdom. Hill is 130 feet high and 550 feet in diameter and is built of clay, gravel, and earth covered with chalk. It possibly is ancient burial place.

c. 2100 Ziggurat of Ur is built by Sumerian king Ur-Nammu. Temple, of 30,000 square feet, is constructed of fired bricks set in black tar. Sumerians call it "temen," source of later word *temple*.

c. 2000 Megalithic structure of Hal Saflieni at Malta is built of enormous flat stone slabs. It contains huge underground labyrinth where some 7,000 people are interred.

c. 2000 Megalithic monument of Palamos in southern Spain is constructed. It consists of menhirs and dolmen (large flat blocks supported on raised stones).

c. 1500 Soul Statues in Corsica are erected, possibly as war trophies or memorials. Single standing menhirs, they are roughly shaped to represent dead souls.

c. 1500 Some 2,934 giant stones have been set upright in parallel lines at Carnac, France. Legend says they represent pursuing warriors turned to stone.

c. 1500 Shepherds of Palaggui in southern Corsica erect enormous line of giant menhirs to mark ceremonial location.

---

lukewarm baths. System of crossed axes provides logical architectural plan for sprawling complex.

c. 400 Earliest medieval architects borrow basilica church design from Roman public meeting halls. Christian basilicas are made of brick or stone. Features include atrium (open courtyard); narthex (vestibule); nave (central aisle of church's interior); transept (section perpendicular to and at top of nave); and apse (semicircular altar area often covered by half-dome). Entryway to basilica, called galilee, is framed by columns.

►

Temples, Palaces, and Monuments of the Ancient World

B.C.

c. 2750 Step Pyramid of King Zoser at Suqqara is completed. Unlike later pyramids, it is completely solid and serves as landmark, not tomb.

c. 2650 Great Pyramid of Khufu at Gizeh, one of seven wonders of ancient world and largest pyramid ever built, is erected. Constructed of limestone, it covers 13 acres and stands near pyramids of Menkure (c. 2575 B.C.) and Khafre (c. 2600 B.C.).

c. 2600 Great Sphinx at Gizeh, one of seven wonders of ancient world, is carved out of rock. It represents pharaoh as sun god Ra.

c. 1475 Cleopatra's Needles, two red granite obelisks covered with hieroglyphics, are erected in Heliopolis, Egypt. They are transported to Alexandria c. 14 B.C. In 1878 Egypt gives one to England; other is given to United States in 1880 and now stands in New York's Central Park.

1411–1371 Famous Temple of Luxor is built at Thebes during reign of Amenhotep III. It is 623 feet long and is altered by successive pharaohs, especially Ramses II.

c. 1350 Tomb of Tutankhamen is erected. Only undisturbed pharaonic tomb unearthed in modern times, it contains great riches, including the king's 250-pound gold coffin.

957 First Temple at Jerusalem is completed. It consists of building with five altars, one containing ark, and large courtyard. Nebuchadnezzar (c. 605–562 B.C.) destroys it in 586 B.C.; it is rebuilt in 515 B.C.

c. 575 Hanging Gardens of Babylon are constructed by Nebuchadnezzar. One of seven wonders of ancient world, they are connected series of roof gardens irrigated by water pumped from Euphrates. They do not survive to modern times.

c. 575 Ishtar Gate from Nebuchadnezzar's sacred precinct in Babylon is built. It is reconstructed in modern times using thousands of glazed bricks that once covered it.

c. 550 Temple of Artemis at Ephesus, often counted among seven wonders of the ancient world, is erected. Goths destroy it in A.D. 262.

c. 448–432 Parthenon is built under Pericles (c. 495–429 B.C.), according to designs by Ictinus and Callicrates. Masterpiece of Greek architecture, its western portion remains intact.

c. 435 Statue of Zeus is sculpted by Phidias (c. 500–432 B.C.) for temple of Olympia. Temple is carved out of ivory and gold and counts among seven wonders of ancient world because of its enormous size and elaborate ornamentation. It does not survive.

c. 432 Propylaea, monumental gateway to the Acropolis, is erected. Designed by Mnesicles, it has five doors.

c. 425 Temple of Athena Nike is built on Acropolis.

c. 421–405 Erectheum is built on Acropolis. Possibly designed by Mnesicles, it is finest surviving example of Ionic order.

c. 352 Tomb of Mausolus, one of seven wonders of ancient world, is built at Halicarnassus. It is huge white marble structure topped by stepped pyramid, which in turn is topped by four-horse chariot. Building is later destroyed, but some sculpture from it survives.

c. 350 Greek Theater of Epidaurus is built. It is noted surviving example of Greek open-air theater with concentric seating.

c. 292–280 Colossus of Rhodes, one of seven wonders of ancient world, is erected in Rhodes Harbor. Probably over 100 feet tall, statue is made of bronze and is toppled by earthquake in 224 B.C.

c. 280 Pharos lighthouse at Alexandria, Egypt stands 200 to 600 feet tall. Usually numbered among seven wonders of ancient world, it is destroyed by earthquake in A.D. 1375.

27 Pantheon is built in Rome by Agrippa (63–12 B.C.). Later destroyed, it is rebuilt by Hadrian (A.D. 76–138) between A.D. 118 and 125. Built of brick, it is best preserved of all surviving Roman edifices.

20 B.C.–A.D. 26 Second Temple is built at Jerusalem. Focal point of Israelite life, it houses Holy Scriptures and court of Sanhedrin, high court of Jewish law. Romans destroy it in A.D. 70. All that remains is Western Wall.

A.D.

65–68 Golden House palace is constructed by Nero (A.D. 37–68). It consists of elaborate gardens, pavilions, and baths. Unpopular even in its own day, little of it survives today.

Temples, Palaces, and Monuments of the Ancient World

80	Colosseum in Rome is dedicated by Titus (A.D. 39–81) and completed in 82 by Domitian (A.D. 51–96), who adds top story. It is freestanding structure (not built into surrounding hill) and seats 50,000.
106–113	Trajan's Column is erected in Rome by Emperor Trajan (A.D. 53–117). This monument stands 125 feet high and is covered with carving in low relief, depicting Trajan's victories over Dacians.

532 Hagia Sophia, domed cathedral in Constantinople, is under construction. Fine example of medieval Byzantine architecture, its central dome with square base becomes common in architecture of that time. New dome supports, called pendentives, support higher, larger dome than previously possible.

c. 550 Church of San Vitale, built in Ravenna, Italy, uses another popular medieval church design, many-sided centralized building plan.

c. 600 Byzantine architecture of time shows attention to dangers of earthquakes. Lead sheets and metal bands are used to strengthen columns and shafts. Walls are bonded with timber. Iron ties protect vast domes.

c. 700 Stone church buildings with glass windows replace wooden churches in England.

c. 800 Carolingian church architects add high towers, elaborate chapels, and tombs to early Christian designs. New styles for entrances, called westworks, include porches, intricate chapels, and small towers, or turrets.

c. 800 Carolingian monks create new designs for monasteries with covered walkways to the church and other main facilities. Monasteries of period are like small cities of unconnected buildings built around main church. Each has its own cupola and bell tower.

c. 800 Islamic mosques are influenced by Persian and Byzantine styles. Columns and cupolas become essential parts of buildings' designs. Minarets, lofty towers sometimes surrounded with balconies, take place of bell towers. Fountains are constructed near mosque for bathing before prayer.

c. 850 Secular architecture of time is influenced by need for strong defense. Castles are built with stone and earthworks and are surrounded by moats.

c. 1000 Roman bells are introduced in medieval churches. Housed in small tower on galilee or in separate building, they are rung to call faithful to services. Architects also reserve and enclose section of courtyard for burial of church members.

c. 1000 Romanesque churches now under construction in many European countries combine Roman with Byzantine influences. They also introduce thick walls with columns closely positioned and tower rising from roof. Nave is separated from side aisles by arcade on which is built gallery called triforium. Clerestory, row of windows within arches, is constructed over triforium.

c. 1000 Romanesque churches feature doorways that widen outward in several steps, each covered by separate arch. Tympanum, space between arch and doortop, is decorated with biblical reliefs and carvings.

c. 1000 Indian Hindu architecture of period is characterized by elaborate sculptural ornamentation. Many temples are constructed with massive and graceful towers.

c. 1050 Romanesque churches undergo continuous development from this period, as populations increase and additional space is required. Chapels and new apses are added. Columns become increasingly solid to support taller buildings and to allow window openings.

▶

Castles and Cathedrals

c. 500 Chinon, among oldest castles in France, is constructed on rocky cliff above Vienne River. It is composed of three fortresses within enclosure complete with moats and ramparts. Here Joan of Arc (c. 1412–31) later reveals her plan to rally France to victory in Hundred Years War for uncrowned King Charles VII (1403–61).

532 Hagia Sophia is under construction in Istanbul, Turkey. Largest domed structure built outside Rome, it is magnificent engineering feat.

617 Castle-fort, earliest of buildings that constitute majestic Edinburgh Castle, is erected. Castle is eventually surrounded by Scotland's capital city, Edinburgh.

708 Benedictine abbey at Mont-Saint-Michel is founded on island in northwestern France. Rising 164 feet above shoreline on Normandy coast, it is notable example of Gothic architecture.

914 Warwick Castle, medieval fortress, is built by Ethelfleda, daughter of Alfred the Great (849–99), in England. It is enlarged by William the Conqueror (c. 1027–87) in 1068.

c. 1000 La Brede Château is built near Bordeaux, France. Gothic in form, castle is constructed in middle of lake. Notable for its drawbridges, towers, turrets, and buttresses, La Brede becomes home of Charles Louis, baron de Montesquieu (1689–1755), French political philosopher.

c. 1000 Foundations are laid for construction of Vitre, massive feudal fortress on border between Brittany and France. Built on promontory above town, castle's fortified walls and high towers form triangular enclosure.

c. 1000 Corinthian Monastery of Daphni is constructed in Greece. Containing finest mosaics in Greece, monastery is outstanding example of Byzantine architecture.

1008 Château de Josselin is built on cliff above France's Oust River. Thick outer stone walls and massive towers contrast with beautiful architecture within. It is rebuilt in 1173 after being damaged in war with Great Britain.

1024 Original construction is complete on French castle of Fougéres in northern Brittany. Red masonry structure is composed of turrets, gables, highly fortified towers, and strong walls. It later inspires French literary figures Victor Hugo (1802–85) and Honoré de Balzac (1799–1850).

c. 1050 Arundel Castle is constructed in Sussex, England. It is repaired and enlarged successively after being damaged in sieges by Henry I (c. 1008–60) and others.

c. 1075 Construction is begun, under William the Conqueror, on Windsor Castle, chief royal residence in England. Castle contains beautiful Gothic church, massive Round Tower, and numerous chapels and apartments. It is largest inhabited castle in Europe.

c. 1078 Work begins on historic fortress Tower of London, along Thames River. Central tower of white limestone is built at this time; several concentric outer walls are added later. Fortress serves various purposes but becomes notorious as royal prison where such noted persons as Anne Boleyn (1507–36) and Sir Walter Raleigh (1552?–1618) are jailed or executed.

1140 Cathedral of San Ruffino is built in Italy. It is site where Holy Roman emperor Frederick II (1194–1250) is baptized.

1140 Abbey of St. Denis, near Paris, is constructed in Gothic style. Beautiful buttresses, rose window, high nave, and pointed arches make it remarkable for its time.

1142 Krak des Chevaliers, most massive and impressive of "Crusader castles," is built in Syria by Knights Hospitalers.

1145 Gothic cathedral of Chartres is erected in northwestern France. Magnificent spires and stained-glass windows are renowned.

1160 Scotland's largest cathedral, St. Andrew's Cathedral, is founded. It is consecrated in 1318 in presence of King Robert I (1274–1329).

1163 Cornerstone is laid for renowned Gothic cathedral Notre-Dame de Paris, famous for its flying buttresses, rose window, and sculptures.

1180 Modeled after Syrian Crusader castles, Castle of Counts of Flanders is erected in Ghent, Belgium. It has statuesque towers and parapets.

c. 1180 Construction begins on remains of Roman fortress of Dover Castle, overlooking English Channel at Dover, England. Massive stone buildings include castle, ancient Roman lighthouse, and Anglo-Saxon church. Fortress is referred to as "key of England" because of its strategic location.

1196 Château Gaillard, along France's Seine River, is constructed by Richard I (1157–99) of England to prevent French king Philip Augustus (1165–

Castles and Cathedrals

	1223) from passing through valley to Rouen. Fortress is considered finest example of early Crusader castle design.
c. 1200	Huge island-fortress, Kalmar Castle, is constructed in Sweden. It withstands more than 20 sieges (14th–15th centuries) and later serves as distillery, then prison.
c. 1200	Château of Cherveux, constructed in Brittany, France, is known for exquisite stonework and sculpture and frequently comes under attack because of strategic location.
1222	Castle Tornese is erected in the Peloponnesus to defend Frankish port of Clarenza from Byzantines. Hexagonal in shape, castle is exceptional example of French medieval architecture.
1245	British king Henry III (1207–72) begins construction of Westminster Abbey, originally abbey church of monastery. Magnificent English national shrine where monarchs are crowned and noted subjects are buried, its nave is highest in England and its architecture is among most spectacular.
1263	Cathedral at Pisa, Italy is constructed to commemorate naval victory over Saracens at Palermo. It serves as model for future Tuscany cathedrals.
1284	Conway Castle is erected in Wales. Most picturesque of six castles built by Edward I (1239–1307) to subjugate Welsh, it has 15-foot-thick walls and eight cylindrical towers.
1285	Caernarvon Castle is constructed in Wales. Designed as fortress and royal residence, it has octagonal towers.
1296	Florentine clothmakers' guild commissions construction of Cathedral Santa Maria del Fiore over remains of Florence's earliest cathedral. Multicolored marble structure is topped by great 350-foot dome.
1352	Spanish king builds Alcazar, Romanesque-style fortress and castle, at Segovia, Spain. It serves as royal residence and stronghold of city.

c. 1050 Romanesque influence on secular construction leads to elimination of moats and enlargement and elaboration of castle towers.

1066 Norman Conquest leads to increased construction of churches, cathedrals, and castles (for defense and political reasons) in northern France, England, and Italy. Normans finance extensive rebuilding and encourage return to classical styles.

1077 Church of St. Étienne is built in Romanesque Norman style. It has three levels of arches in interior; lowest separates nave from aisles.

1080 Construction begins on Church of St. Sernin in Toulouse, France. "Pilgrimage church," it houses bones or remains of saints and is built large enough to accommodate many visitors. Church has two aisles alongside nave, with intricate chapels adjacent to semicircular aisle around apse, enabling pilgrims to move around without disrupting main church services.

c. 1100 Gothic style is introduced in Europe. Architects use new construction techniques to build thinner walls with lighter supports. Piers extending into roof areas of churches are designed to curve out, like separate ribs. Arched ceiling is called ribbed vault.

c. 1100 Conversion of Russia to Orthodox Christianity leads to development of elaborate religious architecture there. Stone becomes important building material.

1145 Cathedral at Chartres, France is built. Gothic architecture is further advanced by substituting windows for nonbearing walls to increase interior light, adopting pointed arch in place of rounded arch, and perfecting flying buttresses to provide support for vaults.

c. 1145 Abbey of St. Denis, whose construction begins near Paris, contains rounded and pointed arches. It reveals gradual evolution of pointed arch.

▶

Earthworks, Dams, and Canals

B.C.

c. 2700 Construction of the earliest known dam, Sadd el-Kafara in Egypt, rock-fill dam on Nile River.

c. 2100 Early Mesopotamian rulers supervise construction of canals to irrigate arid plains. Earliest known land disputes occur over water rights.

c. 1000 Largest North American village of time is constructed in Louisiana. Huge concentric semicircles of piled dirt (15–25 feet high and 50–150 feet wide) surround town.

c. 750 Rulers of Saba (now Yemen) complete Marib Dam, one of great ancient dams, to provide mountain water for irrigation. It is made of boulders without mortar.

c. 600 Babylonian ruler Nebuchadnezzar oversees construction and restoration of irrigation canals in Tigris–Euphrates Valley, which makes agrarian economy possible.

c. 600 Canal from Nile River in Egypt to Red Sea is begun by Egyptian ruler Necho (d. 593 B.C.). Ancient forerunner of Suez Canal, it later is completed by Persian king Darius (c. 549–486 B.C.).

530 Water-supply tunnel built on Greek island of Samos is dug through 3,300 feet of solid stone. It contains clay pipe that carries fresh water to city.

518 Construction of *qanaats* (long tunnels, gently sloped, for transporting water) begins in Egypt, where more than 9 miles of tunnels are completed this year. Rediscovered in 1900, they still carry water.

c. 500 Work begins on Grand Canal to link China's major river systems. World's oldest and longest artificial waterway, it eventually extends 1,000 miles, from Tianjin in northern China to Hangzhou and Yangtze River Valley in south.

312 Roman officials begin construction of first stone aqueduct, Aqua Appia (10.3 miles long), which includes channels, piping, and supporting structures. It leads to development of support arches.

c. 250 Maiden Castle is built in southern England by ancient Britons. Designed for defense, massive earthworks consist of triple ditches, steep scarps, and slopes.

A.D.

50 Romans finish tunnel to drain Lake Fucino after 11 years of work by some 30,000 workers. The tunnel, dug from 40 separate shafts, is 3.5 miles long.

54 Portus, new harbor in Ostia, is constructed by Roman emperor Claudius (10 B.C.–A.D. 54). It includes new harbor basin and 247-acre water area protected by breakwater.

115 Roman emperor Hadrian (76–138) builds 15-mile aqueduct in Athens. It still supplies part of city.

c. 600 Valleys in northern Peru have extensive irrigation canals.

c. 1300 Remarkable irrigation canals are constructed in Peruvian city of Chan Chan. One section has 40-foot, 1-mile-long aqueduct. Canals link city's arid farmlands with interior mountain areas 40 miles away.

c. 1450 Aztec capital of Tenochtitlán, built on man-made islands in lake, is supplied with water by masonry aqueducts on causeways. Stone-slab dike 9 miles long provides flood protection. Spanish destroy all this in conquest of 1521.

c. 1450 Land-reclamation projects begin in northern Holland. Eventually many large lakes are drained. Ditches, banks, and drainage channels are built.

1482 Italian engineer and artist Leonardo da Vinci (1452–1519) designs and directs construction of canals in Milan following appointment as that city's engineer.

1692 Languedoc Canal in France, one of greatest engineering accomplishments of that time, is completed. It connects Bay of Biscay with Mediterranean. Project includes aqueducts and tunnels. Tunneling involves first recorded use of gunpowder for blasting.

1718 Elaborate system of earthen levees are built along Mississippi River at New Orleans, Louisiana to control floodwaters.

1825 United States opens New York Erie Canal, linking Great Lakes with New York City by way of Hudson River. It leads to increased development of western New York State.

1832 Swedish engineers construct Gota Canal, which crosses Sweden, by connecting lakes and rivers.

1843 Construction is completed on 41-mile-long Old Croton Aqueduct in New York, built to supply water for New York City. Another aqueduct is added in 1893.

Earthworks, Dams, and Canals

1843	Tunnel under Thames River in Great Britain finally opens after failed attempts and many construction problems. It is converted to railroad use in 1865.
1866	Construction is completed on French Furens Dam, 170 feet high. It is first dam built on modern dam-engineering principles that deal with internal stresses.
1869	Suez Canal, 100 miles long, is complete, built by French engineer Ferdinand de Lesseps (1805–94) to connect Mediterranean and Red seas. It is enlarged in 1980 to enable passage of supertankers.
1870	Tower subway tunnel in London is complete, marking first use of cast-iron rings for lining tunnel.
1895	Kiel Canal in Germany, originally Kaiser Wilhelm Canal, connects North and Baltic seas.
1902	Aswan Dam is built on Nile River in Egypt. Considered one of finest dams of all time, it has record-setting length of 6,400 feet.
1904–14	Panama Canal across Isthmus of Panama connects Atlantic and Pacific oceans. It is built by U.S. military engineers on land leased from Republic of Panama. Canal Zone is returned to Panama in 1978.
1905	Twin rail tunnels below Hudson River in New York are complete after 25 years of unsuccessful attempts. They are among earliest tunnels excavated using compressed air to prevent influx of water.
1906	Simplon Tunnel, longest railroad tunnel of its time (12.3 miles), is complete. It links France and Italy.
1917	Catskill Aqueduct, 93 miles long, is constructed to provide water to New York City. In many places it is cut through solid rock.
1928	Rove Tunnel is built in southern France. World's largest canal tunnel, it connects port of Marseilles with Rhône River. Tunnel is 4.5 miles long and 72 feet wide.
1934	Tunnel under Mersey River in England opens. It requires excavation of over 1 million tons of sandstone.
1936	Hoover Dam is constructed on Colorado River between Nevada and Arizona. Arched dam is 726 feet high and 1,244 feet long.
1940	Tennessee Valley Authority, created in 1933 to develop Tennessee River basin, constructs system of dams to produce hydroelectric power.
1941	Grand Coulee Dam, built for electric generation and irrigation, is completed on Columbia River in Washington State. At 550 feet high and 4,173 feet long, dam is world's largest concrete structure.
1944	World's longest tunnel, Delaware Aqueduct, is complete. It is 105 miles long and supplies water to New York City.
1954	Gage Dam in France is complete. It is modern "thin-arch" dam, made possible by advances in arch design.
1959	United States and Canada complete construction of St. Lawrence Seaway. It provides access to Lake Ontario for oceangoing traffic by way of St. Lawrence River.
1970	Aswan High Dam, on Nile River in Egypt, is complete. Dam is 364 feet high and 12,565 feet long.
1985	Construction on world's longest railroad tunnel is complete in Japan. Almost 33.5 miles long, tunnel connects islands of Hokkaido and Honshu.

c. 1200 Gothic architects practice "stereotomy," art of constructing windows so any one of 90 or more pieces of stained glass can be removed for replacement without damaging whole. (Horizontal and vertical iron bars reinforce structure.)

c. 1200 Castle architecture in Great Britain is changing as keeps (innermost central towers) are largely eliminated. Many existing castles add stronger fortifications, such as moats or surrounding walls.

c. 1200 Russian churches, strongly influenced by Byzantine styles, begin to develop specific architectural characteristics, such as onion-shaped domes.

c. 1200 In Southeast Asia, great Cambodian temple cities of Angkor Wat and Angkor

▶

Thom are constructed. Vast temple "mountains," they represent impressive architectural accomplishments.

c. 1200 British architects devise hammer-beam truss to support and strengthen timber roof without blocking view through east and west windows.

c. 1200 Many French Gothic cathedrals provide water drainage from roofs through elaborate stone gargoyles, thus eliminating need for unsightly downpipes.

c. 1250 German Gothic "hall" churches have nave and aisles equal in height, making columned interiors look like halls.

c. 1250 Elaborate rayonnant (from radiant rose windows that characterize this style) architecture of High Gothic period influences church construction in Western Europe.

c. 1350 "Perpendicular" style appears briefly in Great Britain. It is characterized by timber-roof construction, vertical tracery, and intricate stonework.

c. 1815 Classical and Renaissance architectural styles (notably Greek and Gothic) are revived. Revival partly results from increased interest in archaeological discoveries.

1819 First Greek Revival structure in United States, Philadelphia's Second Bank of the United States, is designed. It resembles Greek Doric temple.

1823 London's British Museum, resembling huge Ionic temple, is constructed.

1840 London's Houses of Parliament are designed in Gothic Revival style (popular from the 1700s to c. 1850).

1845 Work begins on library of Ste. Geneviève in Paris, combining traditional and modern architectural styles. Ste. Geneviève is first public building in which iron is part of architectural style, with uncovered iron vaults and columns. Interiors combine utility with aesthetically pleasing architecture.

1850 Crystal Palace, first prefabricated structure, is designed by British architect Sir Joseph Paxton (1801–65). Modern timber, iron, and plate-glass building covers over 18 acres and influences 20th-century skyscraper design.

c. 1850 Arts and Crafts Movement in England attracts artists and architects who oppose poor-quality machine-made products of industrial revolution. Members create high-quality furniture, glass, and textiles.

c. 1860 Keble College buildings at England's Oxford University are designed in Gothic style. They feature impressive façades, ornate towers, and arches built with stone and colored brick.

1861 Paris Opera House is designed. Effective modern version of classical styles, with grand foyers, ornate staircases, and boxes, it typifies movement toward combining several architectural styles.

1871 Great Chicago Fire destroys much of city. Rebuilt city becomes center for modern American architecture.

1884 First modern metal-frame skyscraper, Chicago's 10-story Home Insurance Building, is designed by U.S. architect William Jenney (1832–1907). It features metal skeleton of cast-iron columns and nonsupporting curtain walls, which become characteristic of modern design.

1885 American architect Henry Richardson (1838–86), among earliest of U.S. architects to experiment with modern designs, constructs several buildings in Chicago. He successfully blends medieval Romanesque architectural elements in modern geometric style.

1889 "Machine Hall" is designed in Paris. Arches are constructed of steel girders, which, with glass panels, allow enormous interior space.

c. 1890 Architects in Netherlands, Austria, and Germany favor simplicity and lack of decoration, in reaction to elaborate ornamentation of earlier Revival styles. Amsterdam Stock Exchange, built in 1896, typifies this trend.

▶

Soaring Spans: Bridges Since 1800

1801 First U.S. suspension bridge is built at Uniontown, Pennsylvania.

1812 Colossus, longest wooden bridge, is constructed in Philadelphia, Pennsylvania.

1826 Menai Strait Bridge, greatest suspension bridge of its time, is constructed over Menai Strait in Wales. It is rebuilt in 1940.

c. 1830 London Bridge in England, first built of stone in 1176, is replaced. In 1968, work begins on another replacement and parts of old bridge are shipped to Arizona.

1840 First U.S. iron-truss bridge is built at Frankford, New York.

1845 First iron railway bridge is built in United States at Manayunk, Pennsylvania.

1846 U.S. Army adopts inflatable India-rubber pontoon for military use. Between world wars I and II, aluminum replaces rubber and wooden pontoons.

1847 First plate-girder bridge is constructed.

1849 American civil engineer Charles Ellet (1810–62), pioneer in long-span suspension bridges, builds 1,010-foot span bridge at Wheeling (now West Virginia). Bridge is rebuilt in 1854 after storm destroys it.

1850 British engineer Robert Stephenson (1803–59) builds Britannia Railway Bridge across Menai Strait in Wales, longest railway bridge of its time.

1850 Worst recorded bridge accident occurs in France when suspension bridge falls, killing 200 soldiers.

1855 John Roebling builds Niagara Bridge, first railway suspension bridge. It is replaced in 1896, by which time trains have become heavier and faster.

1869 First bridge using concrete is built.

c. 1870 Introduction of modern cantilever bridge.

1874 U.S. engineer James Eads (1820–87) uses steel for first time in U.S. bridge. He constructs two-level (railway and highway) tubular steel arch bridge across Mississippi River.

1877 Firth of Tay Bridge is completed. Considered one of seven wonders of modern world, windstorm destroys it in 1879.

1883 Brooklyn Bridge over East River, New York City, is completed. It breaks previous record for span length (1,595 feet) and minimizes bridge vibrations by use of "stay cables." Construction, begun by John Roebling (1806–69), is finished by his son, Washington Roebling (1837–1926).

1890 Firth of Forth Bridge is built in Scotland. A cantilever railway bridge, it is 8,248 feet long and strong enough to support fast-moving trains safely.

1894 Tower Bridge, drawbridge with two Victorian masonry towers, is constructed in London.

1894 First notable vertical lift bridge is completed, in Chicago. It spans 130 feet and lifts to height of 140 feet.

1898 Eads Bridge, constructed in 1874, is the first bridge to appear on a commemorative stamp.

1900 Construction begins on longest steel cantilever bridge (1,100 feet), Quebec Railway Bridge over St. Lawrence River in Canada. During construction in 1907, bridge collapses, killing 75 workers; a 5,000-pound span falls in 1916, killing over 10 more men. Bridge finally opens in 1917.

1927 First bridge specially constructed to withstand earthquakes, Carquinez Strait Bridge at Vallejo, California, is finished.

1929 Grand 'Mère Bridge is constructed in Quebec, Canada. It is the first to use prestressed twisted-strand cables, introduced by U.S. civil engineer David Steinman (1886–1960).

1931 George Washington Bridge across Hudson River, connecting New York City and New Jersey, is finished. A suspension bridge, its main span is 3,500 feet, double any previously built.

1931 Bayonne Bridge is constructed over Kill van Kull between New York and New Jersey. It has longest span of any arch bridge (1,652 feet).

1933 Swiss engineer Robert Maillart (1872–1940) completes Schwandbach Bridge in Switzerland. A reinforced-concrete arch bridge, its roadway is curved, which is unusual for this time.

1933 Aluminum alloys are used in bridge construction for first time. They significantly reduce bridge weight but prove costly to use.

1936 Prestressed-concrete bridge completed at Aue, Gemany is one of the first of its kind.

1937 Golden Gate Bridge is completed in San Francisco, California. It is 700 feet longer than George Washington Bridge and holds record for 28 years. Its 746-foot steel towers are tallest in world.

Soaring Spans: Bridges Since 1800

1937 La Roche-Guyon Bridge is built in France. Its span of 528 feet makes it longest overhead-arch bridge of its time.

1940 Tacoma Narrows Bridge in Washington opens to traffic but vibrates up and down, earning nickname of "Galloping Gertie." Bridge eventually rips from its supports and falls, killing only a luckless dog; it is rebuilt in 1951.

c. 1940 Floating Bailey Bridge, prefabricated for easy assembly and removal, is in military use. It is named after its designer, British engineer Sir Donald Baily (1901–85).

1955 Garibaldi Bridge is constructed in Rome. First arched-shell bridge made of reinforced concrete, its roadway curves in two directions.

1955 Stayed girder bridge is introduced. Cables are used to support girders.

1957 MacKinac Bridge in Michigan, designed by David Steinman, is constructed. It has longest total suspended span of any bridge (7,400 feet).

1958 Commodore John Perry Bridge opens, linking Pennsylvania and New Jersey. It has longest total length of any cantilever bridge (13,915 feet).

1964 Verrazano-Narrows Bridge, possibly most expensive structure ever built ($325 million), is constructed in New York City. Its main span is 4,260 feet.

1964 Construction is completed on the 17.6-mile-long Chesapeake Bay Bridge-Tunnel. Roadway dips down into tunnels in two places to allow open waterway for ship passage.

1964 Largest concrete arch span (1,000 feet) is constructed for Gladseville Bridge, crossing Paramatta River in Sydney, Australia. Some of the precast arch segments weigh 50 tons or more.

1965 Suez Canal Railway bridge in Egypt is built. It is swing bridge with 548-foot span.

1972 Europe's longest bridge, Oland Island Bridge (19,882 feet) is constructed in Sweden.

1973 Bosporus Bridge opens in Istanbul. Fifth-longest suspension bridge, it links Asia and Europe.

c. 1980 Humber Bridge in England, with total length of 7,283 feet, is completed. Its towers are reinforced concrete instead of steel, and it has world's longest main span to date (4,626 feet).

1988 Construction begins on suspension bridge Akaishi-Kaikyo between Honshu and Shikoku, Japan, world's longest span to date (6,496 feet).

1890s Chicago School of architects becomes famous for construction of modern steel-frame buildings and stores.

c. 1895 Austrian teacher and architect Otto Wagner (1841–1918) designs buildings to emphasize horizontal aspect. His buildings have flat, straight roofs projecting beyond wall and little or no ornamentation.

1898 Secession Building is designed in Vienna for group of architects and artists known as Vienna Secession, which opposes Revival styles. Building has simple geometric form and stark, flat walls characteristic of much modern architecture.

1900 U.S. architect Frank Lloyd Wright (1869–1959) becomes famous for designing houses in "prairie style," characterized by low, horizontal lines and use of natural earth colors. Wright believes buildings should complement settings.

1905 Palais Stoclet is designed in Brussels, Belgium. Considered among most architecturally advanced buildings in Europe, it features cubelike shapes and unadorned white walls.

1909 German architect Peter Behrens (1868–1940) designs AEG Turbine factory in Berlin. With glass walls and steel and concrete construction, it is one of first factories built according to modern architectural designs. Behrens's students include Ludwig Mies van der Rohe (1886–1969), Walter Gropius (1882–1969), and Charles Jeanneret-Gris (known as le Corbusier, 1887–1965).

▶

Taller than Ever Before—Developing the Skyscrapers

1885 Home Insurance Building, Chicago (130 feet tall). It is world's first skyscraper and has 10 stories.

1889 Tacoma Building, Chicago (165 feet). One of first buildings to use steel construction with concrete curtain-wall façade, it is demolished in 1929.

1889 Eiffel Tower, Paris (984 feet). Built for French Revolution centennial celebration, it remains world's tallest structure for more than 40 years.

1889 Auditorium Building, Chicago (270 feet). It is one of first buildings to have forced-air ventilation system.

1890 Rand McNally Building, Chicago (125 feet). First tall building constructed with all-steel load-bearing frame and terra-cotta walls, it is demolished in 1911.

1895 Reliance Building, Chicago (200 feet). It is first building to make use of glazed terra cotta over its steel frame.

1903 Ingals Building, Cincinnati (210 feet). It is first skyscraper built with reinforced concrete.

1903 Flatiron Building, New York (286 feet). This wedge-shaped skyscraper is among first constructed on a triangular foundation.

1913 Woolworth Building, New York (792 feet). Although almost 200 feet shorter than Eiffel Tower, it remains world's tallest office building for more than 25 years. It is first skyscraper with a high-speed elevator.

1930 Chrysler Building, New York (1,046 feet). First building taller than Eiffel Tower, it briefly holds title of world's tallest building.

1931 Empire State Building, New York (1,250 feet). Most famous American skyscraper, it is world's tallest building for more than 40 years.

1933 RCA Building, New York (850 feet). It is first skyscraper to consist of a complex of towers.

1958 Seagram Building, New York (525 feet). It is first highrise to make use of glass and metal flat-topped design.

1962 Marina Towers, Chicago (588 feet). It is tallest reinforced concrete building in the United States to date.

1965 Brunswick Building, Chicago (475 feet). Its innovative design leaves interior space free of columnar supports by placing all weight on load-bearing concrete outer structure.

1968 John Hancock Center, Chicago (1,127 feet). It is tallest multi-use building in world.

1972 Transamerica Building, San Francisco (853 feet). Its distinctive pyramid shape quickly becomes hallmark of San Francisco skyline.

1972 World Trade Center, New York (two towers, 1,368 feet and 1,362 feet). Both towers surpass Empire State Building and become world's two tallest buildings.

1974 Sears Tower, Chicago (1,454 feet). It becomes world's tallest building and remains so to date.

1977 Citicorp Center, New York (919 feet). Distinctive slanted roof is designed for solar panels.

1919 Gropius founds Bauhaus, German school of design, to combine arts and architecture with modern industrial technology. Bauhaus styles are notable for geometric lines and use of steel, glass, and concrete.

1928 Noted American architect (Richard) Buckminster Fuller (1895–1983) designs self-contained "4-D" technological house. Fuller becomes known for his "Dymaxion" principle of trying to get most from least amount of material and energy.

c. 1930 Phrase "International Style" is used to describe changes in American architecture since 1920s. Style features geometric design, white walls, use of reinforced concrete, large windows, and lack of decoration. It is popular until c. 1950.

1931 New York City's Rockefeller Center is designed. Utilitarian pedestrian space is provided within its simple, towering structure. It exemplifies American Modernism in urban architecture.

1937 Gropius moves to United States. He greatly influences American architecture.

1937 Mies van der Rohe emigrates to United States and becomes leader in glass-and-

▶

steel architecture. He pioneers rectangular lines in design, including cubelike brick structures, uncovered steel columns, and large areas of tinted glass.

1946 Mies van der Rohe designs Seagram Building in New York. This office skyscraper is known for its bronze walls and bronze-tinted glass.

1947 Frenchman Le Corbusier, famed International Style architect, designs Unité d'Habitation apartment complex in Marseilles, France. Buildings are notable for white reinforced concrete construction, raised on pillars called "pilotis."

1951 Finnish-born American architect Eero Saarinen (1910–61) becomes known for innovative designs for various buildings in United States. His sweeping style features soaring roof lines, extensive use of glass, and curved lines.

1955 Saarinen designs MIT's Kresge Auditorium as spherical triangle. It is one of many buildings of this time designed in monumental, single forms for cities, campuses, and corporations.

1970 New York's World Trade Center opens. Its tubular design eliminates steel skeleton of previous skyscrapers. Exterior tubular walls, achieved by exterior columns built close together and/or diagonal bracing, are strong enough to withstand wind and other pressures.

1979 New York's Citicorp Building is completed. It contains open plaza and sunlit atrium and has direct access to subways.

c. 1985 Architectural Postmodernism develops. Movement tends away from glass-and-metal structures as architects restore and redesign existing buildings in harmony with natural environment.

9 Science

Medicine

B.C.

c. 3000	Unknown Sumerian physician has seal produced. Sumerians practice mixture of primitive medicine and magic.
c. 3000	First physician is mentioned by name in ancient records. Sekhet'enanach, an Egyptian, apparently cures malady of "king's nostrils" and by that achieves lasting place in history.
c. 2980–2950	Imhotep, famed Egyptian physician, lives. He is worshiped as medical demigod in ancient Egypt.
2500	Physicians perform primitive surgical procedures in Egypt by this time or earlier, including splinting and possible surgery of limbs and neck.
c. 2500	Chinese probably develop practice of acupuncture by this time.
c. 1700s	Hammurabi promulgates his famous code in Babylonia. It includes regulations on medical fees and penalties for malpractice.
c. 1600	Unknown Egyptian physicians compile oldest known medical book. *Smith Papyrus*, as it is called today, contains 48 clinical descriptions of surgical cases, including injuries to head, spine, and chest.
c. 1550	Egyptian physicians write *Ebers Papyrus*, encyclopedic work that lists remedies for many diseases, including deformative arthritis, tumors, and conjunctivitis.
c. 1550	Egyptians claim perhaps earliest known "cures" for baldness. One is pomade containing equal parts of fat from lion, hippopotamus, crocodile, goose, serpent, and ibex; other is mix of writing ink and cerebrospinal fluid.
c. 650	Assyrians compile Mesopotamian medical knowledge in collection of clay tablets at Royal Library in Nineveh. Over 300 medications and such diseases as paralytic stroke and rheumatism are described.
c. 500	Greek physiologist Alcmaeon (fl. 500 B.C.) performs first recorded dissection of human body for research purposes. He identifies brain as center of intelligence.
c. 500	First cataract operations are performed, in India.
c. 460–c. 370	Hippocrates, father of medicine, lives. His writings later provide scientific and ethical basis for modern Western medicine. Works describe many common diseases, recommend dry wound dressings, and propose incorrect humoral theory of disease—that diseases result from disorders in bodily fluids.
437	Hospital, possibly earliest, is built in Sri Lanka (Ceylon).
c. 300	Herophilus, father of anatomy, and Erasistratus, first experimental physiologist, reside and work at Alexandria. Herophilus distinguishes cerebrum from cerebellum and describes retina, liver, spleen, and other body parts. Erasistratus theorizes that heart is pump and studies respiration, muscle action, secretion, and nutrition.
c. 300	*Nei Ching*, classic Chinese medical book, is written in its present version. It advances idea of balancing *yin* and *yang*, upon which all subsequent Chinese medicine is based.
c. 255	*Mo Ching*, so-called pulse classic, is written in China. (Patient's pulse is used to diagnose imbalances in forces of *yin* and *yang*.)
c. 250	Mauryan emperor Asoka constructs hospitals in India.
c. 100	Chinese physicians form accurate theory of circulation of blood.

A.D.

c. 178	Roman Aulus Cornelius Celsus (fl. A.D. 178) writes his classic encyclopedic work on medicine. It contains first description of sewing an artery and of heart disease and also describes hernia operation.

▶

The Fight Against Disease

Following is a listing of well-known diseases. The first date shows when the disease was first identified (or described with accuracy); the second, when its cause was discovered; and the third, when a cure or effective treatment was found.

	Disease	Cause and Treatment
B.C.		
c. 650	tuberculosis	1882, bacillus
		1943, antibiotic streptomycin
c. 650	gonorrhea	1879, gonococcus
		1937, sulfanilamide and later penicillin
c. 650	leprosy	1873, bacillus
		1943, DDS (a sulfonamide shown to be partly successful)
c. 400	malaria	1880, plasmodium parasite
		by early 1900s, mosquito-control programs sharply reduced incidence
c. 400	gout	1848, excess uric acid
		1948, caranamide (penicillin by-product)
A.D.		
c. 100	pneumonia	1881, pneumococcus
		c. 1935, sulfa drugs, followed by penicillin in 1940s
c. 100	diabetes	1889, pancreatic dysfunction
		1922, insulin
c. 100	tetanus	1889, bacillus toxin
		1890, antitoxin serum
c. 100	diphtheria	1883, bacillus
		1892, serum antitoxin widely available
c. 900	smallpox	virus
		1796, effective vaccination by English physician Edward Jenner (1749–1823)
c. 900	measles	1914, virus
		1963, effective vaccine
1496	syphilis	1905, spirochete
		1907, arsphenamine by German bacteriologist Paul Ehrlich (1854–1915)
c. 1500	cholera	1883, bacillus
		late 1800s, isolation of infection and improved sanitation
1643	typhoid fever	1884, bacillus
		1888, vaccine
1675	scarlet fever	1924, bacillus
		1940s, penicillin
1741	African sleeping sickness	1895, protozoa
		early 1900s, arsenic-based chemotherapy
1840	polio	1947, virus
		1953, vaccine

c. 100 Writings of Greek Soranus of Ephesus, founder of obstetrics and gynecology, influence medical practice in these fields until 15th century.

c. 130–c. 199 Galen, noted Greek physician and physiologist, lives. Though flawed and based on animal dissections, his voluminous writings become definitive medical work for 12 centuries. Galen correctly explains respiration and discovers sympathetic nervous system.

369 Hospital of St. Basil is built at Caesarea. It heralds start of hospital-building movement by Christians.

375 Charity hospital with 300 beds for plague victims is set up at Edessa in Eastern Roman Empire.

590 Epidemic of St. Anthony's Fire (ergotism) in France.

651 Hôtel-Dieu is founded in Paris to care for sick and poor.

794 St. Albans Hospital is founded in England.

c. 800 Text reveals recipe for soporific sponge, a medieval anesthetic that contains opium, hemlock, mandrake, and other ingredients.

c. 848 Medical School of Salerno gains notice. First important medieval medical school, it helps stimulate medical advances between ninth and eleventh centuries.

▶

Medications

1775 Digitalis is first used in cases of heart disease by English physician William Withering (1741–99).

1805 Morphine is first identified and named by German researcher Friedrich Sertüner (1783–1841).

1898 Ben-Gay ointment for sore muscles, discovered by French pharmacist Jules Bengué, is marketed.

1899 Aspirin is introduced in powdered form in Germany (tablets are introduced in 1915).

1905 Vick's VapoRub is invented and marketed by North Carolina druggist Lunsford Richardson.

1907 Salvarsan (arsphenamine), cure for syphilis, is discovered by German bacteriologist Paul Ehrlich (1854–1915).

1912 Phenobarbital, sedative and anticonvulsant, is introduced.

1928 Penicillin, the first antibiotic, is discovered by Scottish bacteriologist Alexander Fleming (1881–1955).

1931 Alka-Seltzer is marketed as remedy for headache and upset stomach.

1937 Pyrilamine, first antihistamine, is discovered by Swiss pharmacologist Daniel Bovet.

1941 Sulfadiazine is discovered in United States. It becomes most widely used sulfa drug.

1943 Streptomycin, antibiotic, is discovered by U.S. biochemist Selman Waksman (1888–1973).

1943 Benadril, antihistamine, is discovered by George Rieveschl, Jr.

1944 Quinine, remedy for malaria, is completely synthesized.

1946 Cortisone is synthesized.

1948 Dramamine is found to prevent seasickness.

1948 Aureomycin, antibiotic, is discovered by U.S. botanist Benjamin M. Duggar (1872–1956).

1949 Neomycin, antibiotic, is discovered by Waksman.

1950 Miltown, widely used tranquilizer, is introduced.

1952 Reserpine, tranquilizer for mental patients, is isolated from root long used in India to calm them.

1952 Isoniazid is introduced for tuberculosis therapy.

1952 Polio vaccine is discovered by American virologist Jonas Salk (b. 1914).

1955 Oral polio vaccine is developed by American virologist Albert Sabin (b. 1906).

1957 Interferon, substance produced by body to fight viruses, is discovered. It is reportedly produced artificially by bacteria in 1982.

1963 Valium, tranquilizer, is developed. It becomes world's most widely used tranquilizer.

1968 Propanolol, for treatment of angina and hypertension, is introduced.

1982 Human insulin produced by genetically engineered bacteria gains FDA approval.

1983 Cyclosporine, immunosuppressant to facilitate organ transplants, is approved by FDA.

1986 Vaccine for hepatitis-B is approved by FDA.

1988 Minoxidil, successful remedy for baldness, is approved by FDA.

1200s Height of leprosy epidemic occurs in Europe. Spread by returning Crusaders, leprosy is finally controlled by isolating lepers from community in "lazar" houses.

1240 Imperial decree in Holy Roman Empire permits human dissection, ending longstanding ban of it early in Christian era and opening way for advances in human anatomy.

c. 1280 Alessandro di Spina of Florence introduces eyeglasses in Europe. They appear in China at about same time.

1305 City Hospital of Siena is founded, marking beginning of trend toward building nonsectarian hospitals.

1319 First body snatcher is prosecuted as criminal.

1348–50 Bubonic plague, which has already ravaged Asia and Africa, sweeps over Europe. This is but one of eight major epidemic diseases afflicting medieval Europe, including leprosy, anthrax, scabies, and tuberculosis. Influenza also appears in Middle Ages.

1361–91 Bubonic plague epidemic occurs in Great Britain.

▶

Modern Dentistry

Dentistry as a practice distinct from surgery emerged in Europe in the 1500s. Before that, barber surgeons routinely performed extractions.

1530	Publication of first textbook on dentistry in Leipzig.
1570	England's Queen Elizabeth I receives gift of six gold toothpicks. Toothpicks are preferred for cleaning teeth at this time, toothbrushes not becoming common until much later. The Chinese invent modern toothbrush (bristles at right angle to handle) earlier, in c. 1490.
c. 1575	French surgeon Ambroise Paré (1510–90) publishes writings on dentistry. He introduces reimplantation of teeth, devises workable bridges attached by wires, and treats cavities by cauterizing with acid (but not filling them).
1678–1761	French dentist Pierre Fauchard lives. He introduces mounting of crowns by silver post in root canal, full dentures with teeth of carved ivory, and filling of drilled-out cavities with lead or tin.
c. 1700	Dutch scientist Anton van Leeuwenhoek (1682–1723) investigates "toothworm," popularly believed the cause of cavities. Using his microscope, he finds that worms lodged in cavities come from eating worm-infested overripe cheese.
1771	Noted English surgeon John Hunter (1728–93) publishes his important work on dental anatomy and disease. He coins the terms *cuspid*, *bicuspid*, and *incisor*.
1788	French dentist introduces artificial teeth made from porcelain.
1790	Construction of the first dental chair, by American dentist Josiah Flagg (c. 1763–1816).
1832	First reclining dental chair is constructed. Specially manufactured dental chairs become available after mid-1800s.
1833	Introduction of new amalgam of silver shavings and mercury for filling cavities, improvement over gold foil then in use.
1840	World's first dental college opens, in Baltimore.
1844	First use of nitrous oxide as anesthetic during tooth extraction. Connecticut dentist Horace Wells (1815–48) has one of his own teeth pulled after breathing this gas.
1855	Vulcanite is developed. It proves effective in denture bases, replacing gold.
1858	First practical powered dental drill, driven by foot treadle, is introduced.
1872	First electric-powered dental drill is introduced.
1874	Fluoride's decay-preventative property is discovered.
1879	Forerunner of modern dental cements, containing zinc oxyphosphate, becomes available.
1880	Norman W. Kingsley (1829–1913), founder of orthodontics, publishes outline of procedures for treating irregularly positioned teeth.
1880–84	Porcelain crowns with metal mounting pins are developed.
1882	Modern saliva ejector to drain saliva from patient's mouth is introduced.
1890	American dentist Willoughby D. Miller (1853–1907) accurately describes biological process of tooth decay.
1895	Modern filling material containing silver, copper, tin, and zinc becomes available.
1896	American dentist takes first dental X-rays.
1904	German chemist develops Novocain. It soon replaces cocaine as dentists' preferred anesthetic.
1918	Tooth implants, driven directly into patient's jaw, are introduced.
1940s	Large-scale testing is done of water fluoridation in selected U.S. cities. Subsequently many cities fluoridate public drinking water to help prevent cavities.
1957	First clinically successful high-speed (300,000 rpm) dental drill is introduced. The drill reduces patient discomfort.
1967	Composite resins are introduced, making possible modern bonding procedures for filling and restoring teeth.
1974	Dentists are warned about exposing patients to excessive radiation through too-frequent use of dental X-rays.
1983	Greater use of plastic sealants on children's teeth is urged to help prevent cavities.
1986	FDA approves Caridex system for removing tooth decay. Most tooth decay is cleaned out with mild acid instead of by drilling.
1987	Transparent braces are introduced.
1987	New plaque-removing mouth rinse is patented.

Psychology

1600s Work on mind/body problem by noted philosophers in this century (René Descartes, Baruch Spinoza, and Gottfried Leibniz) later provides theoretical basis for emerging field of modern psychology.

1749 British philosopher David Hartley (1705–57) publishes *Observations on Man*, which attempts to explain thought processes by associationism.

1770 First American public mental hospital is opened, in Williamsburg, Virginia.

1776–1841 German philosopher Johann Friedrich Herbart develops theories concerning the psychology of learning.

1778 German physician Friedrich Anton Mesmer (1734–1815) is forced to leave Austria for practicing magic after he develops therapy called "mesmerism"—forerunner of hypnosis.

1793 French physician Philippe Pinel (1745–1826), a founder of psychiatry, pioneers humane treatment of mental patients by removing chains from patients and setting up program of counseling at Bicetre asylum in Paris. He debunks popular belief that mental illness derives from demonic possession and describes various mental disorders.

1829 British philosopher James Mill (1773–1836) publishes *Analysis of the Phenomenon of the Human Mind*, which deals with association of ideas.

1841 British physician James Braid (1795–1860) begins his pioneering work in scientific study of hypnotism. He coins term *hypnosis*.

1869 British scientist Sir Francis Galton (1822–1911), founder of eugenics, publishes *Hereditary Genius*, concerning hereditary aspects of talent.

1882 French clinician Jean-Martin Charcot (1825–93) opens what will become Europe's foremost neurological clinic of that time. Charcot's work inspires his student Sigmund Freud.

1873 British philosopher Alexander Bain (1818–1903) publishes *Mind and Body: The Theory of Their Relation*. With this and other works on relationship of physiology to psychology, he lays foundation for physiological psychology.

1878 German physiologist and psychologist William Max Wundt (1832–1920) establishes first experimental psychology laboratory, in Leipzig.

1881 First journal of experimental psychology is founded.

1885 Austrian psychologist Sigmund Freud (1856–1939) learns techniques of hypnotism, which later will help him formulate his ideas on psychoanalysis, form of therapy based largely on exploration of unconscious.

1890 American philosopher and psychologist William James (1842–1910) publishes *Principles of Psychology*, pioneering work in physiological psychology. His studies also include work on hypnosis, religion, and psychic phenomena.

1893 Freud and Josef Breuer (1842–1925) publish landmark paper on cause of hysteria. This and later work by Freud provides foundation for modern psychoanalysis.

1896 First psychology laboratory in Great Britain is opened by James Sully (1842–1923), who writes *Outlines of Psychology and the Human Mind* and introduces child guidance clinic.

1898 Russian physiologist Ivan Pavlov (1849–1936) begins his famous research on animals that results in concept of conditioned reflex.

1900 Freud publishes *The Interpretation of Dreams*, generally considered his major work. His *The Psychopathology of Everyday Life* (1901) advances, among other things, now famous idea of Freudian "slip."

c. 1902–06 Freud, whose emerging theories of psychoanalysis are widely opposed, is joined by Austrian psychologist Alfred Adler (1870–1937), Swiss psychologist Carl Jung (1875–1961), and others. Collaboration eventually leads to international recognition of system of analysis based on Freud's theories.

1904–40 American educator and psychologist Edward L. Thorndike (1874–1949) teaches at Columbia University Teachers College. By his research and teaching, he greatly influences development of educational psychology.

1905 First use is made of organized group therapy. This method of treatment does not come into wide use until World War II, when greatly increased number of patients requires group sessions.

Psychology

1905–11 Alfred Binet (1857–1911), French psychologist, devises first intelligence tests.

1908–20 American psychologist John B. Watson (1878–1958) presides over laboratory for psychological studies at Johns Hopkins University. He founds U.S. school of behaviorism (behavior as response to stimuli from environment).

1910 Freud and collaborators form International Psychoanalytic Association.

1912 Carl Jung publishes *Psychology of the Unconscious*, marking formal break with Freud. He opposes Freud's emphasis on infantile sexuality in his psychoanalytic framework, discusses theory of collective unconscious, and establishes that archetypes represent a way of symbolizing universal human experience.

1912 Publication of paper by psychologist Max Wertheimer (1880–1943) marks founding of *Gestalt* psychology. Wertheimer is joined by German-born American psychologist Kurt Koffka (1886–1941) and others.

1913 Alfred Adler (1870–1937) also breaks with Freud over question of emphasis. Adler bases his personality theories on human desire for power.

1914 Influential American psychiatrist Adolf Meyer (1866–1950) becomes director of Johns Hopkins Henry Phipps Psychiatric Clinic. He introduces concepts of psychobiology and mental hygiene.

1921 Swiss psychiatrist Hermann Rorschach (1884–1922) introduces his famous Rorschach inkblot test for diagnosing mental disorders.

1923 Freud publishes *The Ego and the Id*, his final major work directly concerning psychoanalytic theory.

1933 Insulin shock therapy is developed for treating psychotic patients. Electric shock therapy is developed later in Italy for treating severe depressive disorders (1938). (Both are later replaced by treatment with tranquilizers.)

1935 First projective test is published.

1943 Mind-altering properties of LSD are accidentally discovered in Switzerland.

1948 Controversial, quasiscientific Kinsey Report on male sexual behavior is published. Kinsey Report on female sexual behavior is published in 1953.

1949 Lithium is used to treat mental patients in Australia.

1950 *Love Is Not Enough* by Bruno Bettelheim (b. 1903) details his work with severely disturbed children; his methods later are adopted for study and education of normal children.

1954 Reserpine is used as antipsychotic agent. From this time forward, large-scale use of tranquilizing agents for mental patients begins, as part of more general psychopharmacological revolution.

1955 Thorazine is used in treating mental disorders.

1958 Eugene Aserinsky and Nathaniel Kleitman of Chicago report Rapid Eye Movement (REM) sleep, making possible advances in sleep research.

1964 Noted Swiss child psychologist Jean Piaget (1896–1980) publishes *The Early Growth of Logic in the Child*, one of his many works on cognitive development.

1965 Psychological experiments with monkeys show that young monkeys deprived of social contact become emotionally impaired.

1974 *About Behaviorism* is published by Burrhus (B. F.) Skinner (b. 1904), advancing his theories on behaviorism.

1452–1519 Leonardo da Vinci lives. His lifetime work includes notable anatomical drawings.

1457 Europe's first medical publication, Gutenberg *Purgation Calendar*, is printed.

c. 1493–1541 Paracelsus, medieval physician, alchemist, and astrologer, lives. Despite his occult interests, he pioneers chemical therapeutics in treating disease and

▶

adds sulfur, mercury, iron, copper sulfate, and other elements to physicians' pharmacopoeia.

1496–1500 Syphilis epidemic starts spreading throughout Europe. Disease continues to be major public health problem for centuries.

1500 First living patient undergoes Caesarean section.

1510–90 French physician Ambroise Paré lives. Father of modern surgery, he develops (1545) new treatment for gunshot wounds (replacing boiling oil method), invents hernia truss, and introduces artificial limbs.

1514 First medical description of gunshot wounds.

c. 1520–74 Bartolommeo Eustachio lives. His discoveries include Eustachian tube and nerve and blood vessels inside teeth.

1543 Andreas Vesalius (1514–64) publishes his landmark anatomy text *De humani corpis fabrica* and thereby founds modern science of anatomy.

1557–58 Influenza epidemic in Europe.

1561 Gabrielle Fallopius (1523–62) publishes discovery of Fallopian tubes.

1579 Manufacture of the first glass eyes.

1612 Italian physician Sanctorius (Santorio Santorio, 1561–1636) invents first clinical thermometer, as part of his work on body metabolism.

1624–89 Thomas Sydenham, greatest 17th-century clinician, lives. "English Hippocrates," he revives emphasis on direct observation of patient and on experience.

1628 William Harvey (1578–1657) publishes his landmark *De Motu Cordis*, giving first accurate description of human circulatory system.

c. 1630 Peruvian Indian remedy—cinchona bark—for malarial fever is introduced in Europe.

1633 Publication of first book on first aid.

1642 Beriberi is clinically described.

1650 Rickets is clinically described.

1656 Cataracts are found to be caused by clouding of eye's lens.

1658 Red corpuscles are discovered.

1660 Discovery is made that nasal secretion does not come from the pituitary gland, correcting long-standing belief (since Galen).

1660 English scientist Robert Boyle (1627–91) proves experimentally that air is necessary for life, important step toward understanding respiration.

1661 Italian Marcello Malpighi (1628–94) publishes his discovery of capillary circulation. Founder of histology and embryology, he also discovers skin layers, lymph nodes, and brain's cortical cells.

1663 Hospital on Long Island, New York becomes first in American colonies.

1667 Artificial respiration is successfully performed on animal.

1669 English physician Richard Lower (1631–91) reveals interaction of air and blood in lungs.

1674 Tourniquet is invented.

1683 Dutch scientist Anton van Leeuwenhoek (1632–1723) uses his primitive microscopes to discover bacteria, though he does not realize their connection with disease. He also observes striated muscle fibers (1679).

1721 Demonstration of obstetric forceps is made in France.

1726 First measurement of blood pressure is performed on horse.

1736 First successful appendectomy is performed.

1746 Compressive bandage is introduced to stop bleeding from wounds.

1768 Clinical description is made of angina pectoris.

1772 English chemist Joseph Priestley (1733–1804) discovers nitrous oxide.

▶

Winners of the Nobel Prize for Physiology or Medicine

1901 Emil A. von Behring (1854–1917, Germany)

1902 Sir Ronald Ross (1857–1932, Great Britain)

1903 Niels R. Finsen (1860–1904, Denmark)

1904 Ivan P. Pavlov (1849–1936, Russia)

1905 Robert Koch (1843–1910, Germany)

1906 Camillo Golgi (1844–1926, Italy) and Santiago Ramón y Cajal (1852–1934, Spain)

1907 Charles L. A. Laveran (1845–1922, France)

1908 Paul Ehrlich (1854–1915, Germany) and Élie Metchnikoff (1845–1916, France)

1909 Emil T. Kocher (1841–1917, Switzerland)

1910 Albrecht Kossel (1853–1927, Germany)

1911 Allvar Gullstrand (1862–1930, Sweden)

1912 Alexis Carrel (1873–1944, France)

1913 Charles R. Richet (1850–1935, France)

1914 Robert Bárány (1876–1936, Austria)

1915 No prize awarded

1916 No prize awarded

1917 No prize awarded

1918 No prize awarded

1919 Jules Bordet (1870–1961, Belgium)

1920 Schack A. S. Krogh (1874–1949, Denmark)

1921 No prize awarded

1922 Archibald V. Hill (1886–1977, Great Britain) and Otto F. Meyerhof (1884–1951, Germany)

1923 Frederick G. Banting (1891–1941, Canada) and John J. R. Macleod (1876–1935, Great Britain)

1924 Willem Einthoven (1860–1927, Netherlands)

1925 No prize awarded

1926 Johannes A. G. Fibiger (1867–1928, Denmark)

1927 Julius Wagner von Jauregg (1857–1940, Austria)

1928 Charles J. H. Nicolle (1866–1936, France)

1929 Christiaan Eijkman (1858–1930, Netherlands) and Sir Frederick G. Hopkins (1861–1947, Great Britain)

1930 Karl Landsteiner (1868–1943, United States)

1931 Otto H. Warburg (1883–1970, Germany)

1932 Edgar D. Adrian (1889–1977, Great Britain) and Sir Charles S. Sherrington (1857–1952, Great Britain)

1933 Thomas H. Morgan (1866–1945, United States)

1934 George R. Minot (1885–1950, United States), William P. Murphy (b. 1892, United States), and G. H. Whipple (1878–1976, United States)

1935 Hans Spemann (1869–1941, Germany)

1936 Sir Henry H. Dale (1875–1968, Great Britain) and Otto Loewi (1873–1961, United States)

1937 Albert Szent-Györgyi (1893–1986, United States)

1938 Corneille J. F. Heymans (1892–1968, Belgium)

1939 Gerhard Domagk (1895–1964, Germany)

1940 No prize awarded

1941 No prize awarded

1942 No prize awarded

1943 Henrik C. P. Dam (1895–1976, Denmark) and Edward A. Doisy (1893–1987, United States)

1944 Joseph Erlanger (1874–1965, United States) and Herbert S. Gasser (1888–1963, United States)

1945 Ernst B. Chain (1906–1979, Great Britain), Sir Alexander Fleming (1881–1955, Great Britain), and Sir Howard W. Florey (1898–1968, Great Britain)

1946 Hermann J. Muller (1890–1967, United States)

1947 Carl F. Cori (1896–1984, United States), Gerty T. Cori (1896–1957, United States), and Bernardo A. Houssay (1887–1971, Argentina)

1948 Paul H. Müller (1899–1965, Switzerland)

1949 Walter R. Hess (1881–1973, Switzerland) and António Moniz (1874–1975, Portugal)

1950 Philip S. Hench (1896–1965, United States), Edward C. Kendall (1886–1972, United States), and Tadeusz Reichstein (b. 1897, Switzerland)

1951 Max Theiler (1899–1972, United States)

1952 Selman A. Waksman (1888–1973, United States)

1953 Hans A. Krebs (1900–81, Great Britain) and Fritz A. Lipmann (b. 1899, United States)

1954 John F. Enders (1897–1985, United States), Frederick C. Robbins (b. 1916, United States), and Thomas H. Weller (b. 1915, United States)

1955 Axel H. T. Theorell (1903–82, Sweden)

1956 André F. Cournand (b. 1895, United States), Werner Forssman (1904–79, German-born), and Dickinson W. Richards (1895–1973, United States).

1957 Daniel Bovet (b. 1907, Italy)

1958 George W. Beadle (b. 1903, United States), Joshua Lederberg (b. 1925, United States), and Edward L. Tatum (1909–75, United States)

Winners of the Nobel Prize for Physiology or Medicine	
1959	Arthur Kornberg (b. 1918, United States) and Severo Ochoa (b. 1905, United States)
1960	Sir F. Macfarlane Burnet (1899–1985, Australia) and Peter B. Medawar (b. 1915, Great Britain)
1961	Georg von Békésy (1899–1972, United States)
1962	Francis H. C. Crick (b. 1916, Great Britain), James D. Watson (b. 1928, United States), and Maurice H. F. Wilkins (b. 1916, Great Britain)
1963	Sir John C. Eccles (b. 1903, Australia), Alan L. Hodgkin (b. 1914, Great Britain), and Andrew F. Huxley (b. 1917, Great Britain)
1964	Konrad E. Bloch (b. 1912, United States) and Feodor Lynen (b. 1911, German-born)
1965	François Jacob (b. 1920, France), André Lwoff (b. 1902, France), and Jacques Monod (1910–76, France)
1966	Charles B. Huggins (b. 1901, United States) and Francis Peyton Rous (1879–1970, United States)
1967	Ragnar A. Granit (b. 1900, Sweden), Haldan Keffer Hartline (1903–83, United States), and George Wald (b. 1906, United States)
1968	Robert W. Holley (b. 1922, United States), H. Gobind Khorana (b. 1922, United States), and Marshall W. Nirenberg (b. 1927, United States)
1969	Max Delbrück (1906–81, United States), Alfred D. Hershey (b. 1908, United States), and Salvador E. Luria (b. 1912, United States)
1970	Julius Axelrod (b. 1912, United States), Sir Bernard Katz (b. 1911, Great Britain), and Ulf von Euler (1905–83, Sweden)
1971	Earl W. Sutherland, Jr. (1915–74, United States)
1972	Gerald M. Edelman (b. 1929, United States) and Rodney R. Porter (1917–85, Great Britain)
1973	Karl von Frisch (1885–1982, German-born), Konrad Lorenz (b. 1903, German-born), and Nikolaas Tinbergen (b. 1907, Great Britain)
1974	Albert Claude (1898–1983, United States), George E. Palade (b. 1912, United States), and Christian René de Duve (b. 1917, Belgium)
1975	David Baltimore (b. 1938, United States), Howard M. Temin (b. 1934, United States), and Renato Dulbecco (b. 1914, United States)
1976	Baruch S. Blumberg (b. 1925, United States) and Daniel Carleton Gajdusek (b. 1923, United States)
1977	Rosalyn S. Yalow (b. 1921, United States), Roger C. L. Guillemin (b. 1924, United States), and Andrew V. Schally (b. 1927, United States)
1978	Daniel Nathans (b. 1929, United States), Hamilton O. Smith (b. 1931, United States), and Werner Arber (b. 1929, Switzerland)
1979	Allan M. Cormack (b. 1924, United States) and Godfrey N. Hounsfield (b. 1919, Great Britain)
1980	Baruj Benacerraf (b. 1920, United States), George Snell (b. 1903, United States), and Jean Dausett (b. 1916, France)
1981	Roger W. Sperry (b. 1913, United States), David H. Hubel (b. 1926, United States), and Torsten N. Wiesel (b. 1924, United States)
1982	Sune Bergstrom (b. 1916, Sweden), Bengt Samuelsson (b. 1934, Sweden), and John R. Vane (b. 1927, Great Britain)
1983	Barbara McClintock (b. 1902, United States)
1984	Cesar Milstein (b. 1926, Argentina), Georges J. F. Köhler (b. 1946, West Germany), and Niels K. Jerne (b. 1911, Denmark)
1985	Michael S. Brown (b. 1941, United States) and Joseph L. Goldstein (b. 1940, United States)
1986	Rita Levi-Montalcini (b. 1909, United States) and Stanley Cohen (b. 1922, United States)
1987	Susumu Tonegawa (b. 1939, Japan)
1988	Gertrude Elion (b. 1918, United States), George Hitchings (b. 1905, United States), and Sir James Black (b. 1924, Great Britain)
1989	J. Michael Bishop (b. 1936, United States) and Harold Varmus (b. 1939, United States)

1777 French chemist Anton L. Lavoisier (1743–94) describes role of oxygen (which he names in 1779) in respiration and combustion.

1784 Benjamin Franklin (1706–90) invents bifocal eyeglasses.

1786 Clinical description is made of alcoholism.

1792 French military surgeon invents "flying ambulances" to care for wounded soldiers at battlefront.

▶

1794 Clinical description is made of color blindness.

1795 British Navy eliminates scurvy among its sailors by supplying them with lemon juice.

1796 English physician Edward Jenner (1749–1823) introduces vaccination with relatively mild cowpox germs to immunize against deadly smallpox.

1799 English chemist Sir Humphry Davy (1778–1829) discovers anesthetic property of laughing gas (nitrous oxide).

1800 Chlorination is used to purify water.

1805 Morphine is isolated. Its manufacture for medical purposes begins in 1827.

1811 Scotsman Sir Charles Bell (1774–1842) publishes landmark book on brain anatomy. He distinguishes between sensory and motor nerves.

1811 Iodine is isolated for first time. It is not used as disinfectant until 1910, however.

1812 Cholesterol is discovered.

1816 R.T.H. Läennec (1781–1826), French physician, invents stethoscope.

1818 Hydrogen peroxide is isolated.

1822 First successful hysterectomy is performed.

1823 Medical journal *Lancet* is founded.

1825 First tracheotomy to prevent patient from suffocating is performed successfully.

1826 Bromine is discovered.

1827 Mammalian ovum is found.

1832 Codeine is discovered.

1833 English physician Marshall Hall (1790–1857) discovers reflex action.

1833 German Johannes Müller (1801–58) publishes his *Manual of Human Physiology*, which helps define physiology as separate discipline.

1836 Theodor Schwann (1810–82) pinpoints pepsin as stomach's digestive agent.

1839 Schwann identifies cell as bBsic structural unit of animals and humans. He also names cell's chemical processes *metabolism*.

1842 First use is made of ether for anesthetic in surgery, by American surgeon Crawford W. Long (1815–78).

1845 Introduction of hypodermic syringe for medical use.

1846 French physiologist Claude Bernard (1813–78) describes function of pancreas in digestion.

1846 American Medical Association is founded.

1847 Scottish physician Sir James Simpson (1811–70) uses chloroform as anesthetic for the first time.

1851 Hemoglobin in blood is identified.

1851 Ophthalmoscope is invented for examining eye's inner structure.

1852 Cell division is found responsible for all tissue growth.

1853–56 English nurse Florence Nightingale (1820–1910) founds modern nursing practice by her work with sick and wounded during Crimean War.

1858 German pathologist Rudolf Virchow shows that cell is center of disease processes. This finding displaces humoral theory of disease, held since ancient times.

1858 Cocaine is isolated.

1861 French surgeon Pierre-Paul Broca (1824–80) proves localization of brain functions by finding center for speech (Broca's convolution).

1862 French scientist Louis Pasteur (1822–95) disproves theory of spontaneous generation of organisms by proving airborne bacteria's role in fermentation. This contributes to eventual acceptance of germ theory of disease.

▶

1867 Modern antiseptic surgery begins with introduction of sterilization and antiseptic procedures by Joseph Lister (1827–1912).

1869 German surgeon J. Friedrich A. von Esmarch (1823–1908) demonstrates first-aid bandage for battlefield. He introduces antiseptic bandage in 1877.

1880 Parathyroid gland is described.

1881 Steam sterilization is introduced.

1881 Pasteur develops the first artificially produced vaccine.

1882 German bacteriologist Robert Koch (1843–1910) links germ with disease for first time by identifying tuberculosis bacterium.

1882 Blood platelets are discovered.

1885 Pasteur administers first inoculation for rabies.

1888 Pasteur Institute is founded.

1890 Rubber gloves are introduced for surgery.

1890 German bacteriologist Emil von Behring (1854–1917), founder of immunology, develops serum antitoxin. By 1892 serum antitoxin for diphtheria is marketed that produces passive immunity to this disease.

1892 Existence of virus is shown by Russian scientist Dmitri Ivanovsky (1864–1920).

1893 Aspirin is rediscovered by German chemist Felix Hoffman. Drug is first marketed in 1899.

1893 First open-heart surgery is performed, by American surgeon Daniel Williams (1858–1931).

1895 William Roentgen (1845–1923) uses X-rays to photograph bones and internal organs of living patients for first time.

1897 Artificial-respiration method is introduced.

1898 Heroin is used for medical purposes.

1899–1900 Cancer is treated with newly discovered X-rays.

1900 Spinal anesthesia (with cocaine) is introduced.

1900 American biologist Karl Landsteiner (1868–1943) discovers human blood types, making possible life-saving blood transfusions.

1901 Hormone adrenaline is isolated.

1902 First electrical hearing aid is invented.

1902 Human sex chromosome is identified.

1903 Precursor of electrocardiograph for diagnosing heart troubles is invented.

1905 Heart action is restored by adrenaline injection.

1905 Novocaine is discovered.

1905 George Crile (1864–1943) is first to successfully transfuse blood directly.

1905 First artificial joint—hip—is created.

1906 Wasserman serum test for syphilis is developed.

1907 German bacteriologist Paul Ehrlich (1854–1915) introduces chemotherapy treatment—attacking particular organism with specific chemical. He successfully treats syphilis for first time (1907) with arsphenamine.

1910 Lung disease is diagnosed with X-rays.

1911 Dr. Francis Rouss (1879–1970) shows virus can cause cancer.

1911 Name "vitamine" (later vitamin) is given to essential nutrients earlier identified by Dutch physician Christian Eijkman (1858–1930) and Frederick G. Hopkins (1861–1947).

1911–27 Russian psychologist Ivan Pavlov (1849–1936) investigates and demonstrates conditioned reflex.

1913 Mammography, X-ray technique for detecting breast cancer, is developed.

1917 Vitamin D is isolated from cod-liver oil.

1921 Canadians Frederick Banting (1891–1941) and Charles Best (1899–1978) discover insulin.

1927 First iron lung, mechanical breathing apparatus for humans, is developed.

▶

1928 Scottish bacteriologist Alexander Fleming (1881–1955) discovers penicillin, which comes into wide use in 1940s.

1932 German scientist Gerhard Domagk (1895–1964) discovers antibacterial action of sulfa drugs.

1937 First blood bank is established.

1942 Streptomycin, first antibiotic effective against tuberculosis, is found by U.S. scientist Selman Waksman (1888–1973). He coins term *antibiotic*.

1942 First kidney machine is developed.

1944 First eye bank is developed.

1947 Polio virus is isolated.

1948 Great Britain adopts national program of socialized medicine.

1948 Cortisone is shown to be effective in treating rheumatoid arthritis. Cortisone manufacture begins in 1949.

1950 Reserpine, drug for treating mental disorders, is found effective in low dosages for treating hypertension (high blood pressure).

1950 Manual heart-massage technique saves patient.

1950 Antabuse is used to treat alcoholics.

1951 First television broadcast of a human birth.

1952 Amniocentesis is introduced to diagnose genetic disorders in fetus.

1952 First patient is kept alive by mechanical heart.

1952 First artificial heart valve is implanted.

1952 Patient's heart is restarted by electric shock.

1952 First sex-change operation is performed on George Jorgenson, thereafter known as Christine Jorgenson.

1952 Successful test of polio vaccine is developed by Jonas E. Salk (b. 1914). Vaccine is used widely by 1954.

1953 Heart-lung machine (developed in 1950) is used on first human patient during heart surgery.

1953 Tests on mice reveal that tars from cigarette smoke may cause cancer.

1956 Albert Sabin (b. 1906) develops oral polio vaccine.

1956 Large-scale tests of birth-control pills are conducted.

1957 Interferons, body's virus fighters, are discovered.

1958 First use is made of ultrasound to examine unborn fetus.

1959–62 Babies deformed by thalidomide are born to mothers who have used this sedative.

1960 Pacemaker is developed.

1961 IUD birth-control device is introduced.

1962 First eye surgery is performed using laser.

1965 Soft contact lenses are developed.

1965 Medicare system begins in United States.

1967 Coronary bypass operation is introduced by American surgeon René Favalero.

1967 South African surgeon Christiaan Barnard (b. 1922) performs first heart transplant. Patient Louis Washkansky lives for 18 days. Barnard's second transplant recipient lives 74 days (1968).

1968 Researchers find birth-control pills cause blood clots in some women.

1969 First human eye transplant is performed.

1969 U.S. surgeons implant first artificial heart in human.

1970 First nerve transplant is performed.

1970 First synthesis of artificial gene is accomplished.

1971 Diamond-bladed scalpel is introduced.

1971 First heart and lung transplants are performed.

1972 CAT-scan imaging system is introduced for diagnosis and research work.

1973 British scientists develop MRI (magnetic resonance imager) for medical diagnosis.

▶

1974 Heimlich maneuver is introduced as first aid for choking.

1976 Parents win court fight to turn off respirator keeping their comatose daughter, Karen Quinlan, alive. Karen continues to live until 1985 with just feeding tube, however.

1976 Outbreak of Legionnaires' disease occurs at Philadelphia convention of American Legion.

1977 New procedure, balloon angioplasty, is developed for reopening diseased arteries.

1978 First "test-tube" baby is born, in Great Britain.

1980 Sound waves are used to break up kidney stones without surgery.

1981 Researchers identify acquired immunodeficiency syndrome (AIDS). First (unrecognized) cases date from late 1970s.

1982 First successful gene transplant in mammals is performed (rat gene to mouse).

1982 U.S. FDA approves bacteria-produced insulin for diabetics, first commercial application of genetic engineering.

1982 Jarvik 7 artificial heart is implanted. Patient, Dr. Barney Clark, survives 112 days. Second patient survives 620 days (1984).

1983 FDA approves immunosuppressant cyclosporine, which is used during organ transplants.

1983 First successful embryo transfer is performed.

1984 First successful surgery is performed on fetus.

1984 Baboon heart is implanted in human baby.

1986 FDA approves first vaccine (against hepatitis B) produced by genetic engineering.

1987 Successful treatment of Parkinson's disease is effected by implanting adrenal gland tissue in patient's brain.

1988 Researchers find smoking increases likelihood of stroke in women.

1988 Minoxidil is approved by FDA as antibaldness medication.

1989 Researchers inject cancer patient with genetically altered cells in hope of developing new cancer-fighting technique.

1989 Deaths from heart disease are reportedly on the decline, though scientists cannot explain why.

1989 Ointment containing natural hormone, epidermal growth factor, is shown to speed healing of wounds.

1989 U.S. researchers announce discovery of gene believed responsible for causing cystic fibrosis.

1990 Experimental vaccine reportedly protects chimpanzees from infection by AIDS virus.

Astronomy

B.C.

c. 6500 Earliest known calendar is made of bone by primitive man in Zaire. Markings on bone probably are used to record months and moon phases.

c. 3500 Primitive time-telling device, gnomon, is in use. It consists of upright pillar or stick. Length of its shadow tells time of day.

c. 2773 Egyptians develop first calendar with 365-day year. Roman (and hence modern) calendar is based on it.

2296 Chinese make earliest recorded sighting of a comet.

c. 2000 Stonehenge is built in its present form in England. This religious monument also

▶

is constructed as primitive astronomical observatory.

c. 1800 Beginning of Babylonian astronomy. Over next millennium, Babylonian astronomers compile earliest detailed scientific record of celestial events.

1361 Earliest confirmed record of lunar eclipse is made by Chinese, who also record earliest confirmed solar eclipse, in 1217 B.C. Possible total solar eclipse (1375 B.C.) is mentioned on clay tablet from Mesopotamia.

1300 Egyptians by this time know of all five planets visible in sky to naked eye (Mercury, Venus, Mars, Jupiter, and Saturn) and have identified 43 constellations.

763 Solar eclipse is reported by Babylonians.

c. 636–546 Greek philosopher Thales lives. He discovers why solar eclipses happen (moon passes in front of sun) but believes Earth is flat disk floating on great sea.

585 Famous eclipse predicted by Thales occurs during battle between Lydians and Medes. Believing eclipse an omen, combatants conclude peace treaty.

c. 582– c. 507 Greek philosopher and mathematician Pythagoras lives. He believes Earth is spherical and that heavenly bodies revolve around it.

c. 460–370 Greek philosopher Democritus lives. He theorizes that Milky Way is made up of many stars.

c. 400 Zodiac circle (zone of sun's annual path) is established. Horoscopes based on planetary positions and time of birth are available in Babylonia.

384–322 Greek philosopher Aristotle lives. His Earth-centered concept of universe (finite, spherical shell) lays basis for later incorrect geocentric theories that survive into Middle Ages.

352 First recorded supernova is reported, by Chinese.

c. 276–194 Eratosthenes, librarian at Alexandria, lives. He calculates Earth's circumference by measuring angle of sun's light at noon at two distant places and then measuring distance between these places.

c. 270 Greek mathematician Aristarchus of Samos correctly describes our solar system (Earth and other planets revolving around sun) centuries before Copernicus. Aristarchus also estimates relative sizes and distances of sun and moon.

240 Comet now known as Halley's comet is first sighted.

c. 200 Astrolabe, device for measuring position of celestial bodies, is developed in Greece.

165 Chinese astronomers observe and record sunspots. Imperial astronomers keep continuous record of sunspots from 28 B.C. to A.D. 1638.

c. 129 Bithnyian astronomer Hipparchus (c. 180–c. 127 B.C.) completes detailed star catalog (850 stars, rated by brightness). He also discovers precession of equinoxes.

46 Julius Caesar, acting on advice of astronomers, orders Roman calendar revised from 365-day to 365¼-day year (Julian calendar).

A.D.

c. 150 Ptolemy, a Hellenic Greek, completes his famous work *Megale syntaxis tes astronomius* (Arab translation: *Almagest*), summary of astronomical knowledge. In it he puts forward his own geocentric theory (sun and planets revolve around Earth), view that persists until Copernicus's time.

c. 940 Chinese astronomers develop Mercator-type projection to create Dunhuang star map. Mercator himself will not develop projection technique for another six centuries.

1066 Bright comet (later identified as Halley's comet) is sighted during William the Conqueror's invasion of England.

▶

c. 1075 Arab astronomer Arzachel suggests correctly that orbital paths of planets are elliptical rather than circular, as was generally believed.

1100s Latin translations of Arabic text on Greek astronomy, preserved for centuries by Arabs, become available in medieval Europe. Works of ancient astronomers and astrologers were thus rediscovered in Europe and for centuries went unchallenged.

c. 1100 Chinese demonstrate cause of solar and lunar eclipses.

c. 1260 Observatory at Peking, China is built.

1272 Alphonsine tables are compiled. These planetary tables are used until 16th century.

c. 1400 Length of solar year is recognized to be 365.25 days by Chinese astronomers.

1428 Famous observatory at Samarkand (now in south-central USSR) is built by Mongol astronomer Ulugh Beg (1394–1449). Finished structure has a 180-foot-high quadrant, device for measuring positions of stars.

1543 *De Revolutionibus Orbium Coelestium* is published by Polish astronomer Nicolaus Copernicus (1473–1543), who argues for sun-centered (heliocentric) system in opposition to incorrect prevailing view, Ptolemy's earth-centered theory. Copernicus's revolutionary work lays foundation for modern astronomy.

1546–1601 Danish astronomer Tycho Brahe lives. Using equipment of his own design, he compiles the first accurate tables of planetary positions to prove that comet he is observing is four times farther than moon.

1582 Gregorian (modern) calendar is proclaimed by Pope Gregory XIII. This reformed Julian calendar is not adopted in Great Britain and its colonies until 1752.

1609 Galileo Galilei (1564–1642), Italian scientist, constructs first of his many telescopes. (Telescope has been in use, secretly, for military purposes in Netherlands for 20 years.) Galileo turns his telescopes skyward and discovers Jupiter's moons, phases of Venus, Saturn's rings, mountains on moon, and sun's rotation.

1609–19 German astronomer Johannes Kepler (1571–1630) publishes his three laws of planetary motion, based on his discovery that planets travel in elliptical, not circular, orbits. Kepler's work helps win acceptance of Copernicus's heliocentric system.

1611 Four astronomers—Galileo, Thomas Harriot, Father Scheiner, and Johannes Fabricius—independently discover sunspots. Arguments arise over who truly discovered them first.

1612 Andromeda galaxy is discovered.

1633 Galileo is tried by Inquisition at Rome and forced to recant his belief in the Copernican system, as set forth in his *Dialogue Concerning the Two Chief World Systems* (published in 1632). Pope John Paul II effectively reverses the decision, admitting Galileo was right, in 1989.

1668 English physicist Isaac Newton (1642–1726) invents reflecting telescope, making possible telescopes of much greater magnification.

1671 First relatively accurate estimate of distance between sun and Earth is calculated by Giovanni Cassini (1625–1712).

1672 French scientist N. Cassegrain (1625–?) invents Cassegrain reflecting telescope, design widely used today.

1675 England's Royal Observatory at Greenwich is established.

▶

Discovering Our Solar System

Mercury

1855 French astronomer Urbain Leverrier (1811–77) postulates existence of another planet, Vulcan, within Mercury's orbit.

1881 Italian scientist Giovanni Schiaparelli (1835–1910) begins eight-year study that concludes Mercury's orbital and axial periods of revolution are identical.

1964 Radio telescope at Arecibo, Puerto Rico measures planet's rotation using Doppler radar. It detects presence of dark, igneous rocks that minimize sunlight reflection and discovers that rotation is not identical to revolution (same hemisphere does not always face sun).

1973 U.S. *Mariner 10* spacecraft completes first successful flyby of planet, taking over 10,000 pictures. They show that surface is marked with craters and covered with massive cliffs as high as 2 miles.

Venus

1610 Italian astronomer Galileo Galilei (1564–1642) discovers phases of Venus.

1761 Presence of atmosphere surrounding Venus is deduced.

1932 Spectroscopy detects large amounts of carbon dioxide in planet's atmosphere.

1956 Radio astronomers record microwaves emanating from planet, from which they calculate surface temperature to be as hot as 600 degrees F.

c. 1960 Radar astronomy reveals Venusian rotation to be retrograde (turning in direction opposite its revolution around sun).

1962 U.S. *Mariner 2* spacecraft completes first successful flyby, transmitting spectacular pictures to Earth. It finds no detectable magnetic field.

1967 Soviet *Venera 4* spacecraft makes first successful atmosphere entry. It drops capsule containing scientific instruments by parachute into Venusian atmosphere and finds atmosphere is 90 percent carbon dioxide. Spacecraft detects large amounts of water vapor in planet's upper atmosphere.

1970 Soviet spacecraft *Venera 7* lands safely on surface. It measures temperature, pressure, and gas densities.

1972 Soviet spacecraft *Venera 8* finds Venusian surface to be similar to Earth's granite rock. It measures solar radiation penetrating thick cloud cover.

1973 U.S. *Mariner 10* spacecraft is first dual planet probe (probing Venus and Mercury). It transmits first television pictures of Venus.

1975 Soviet *Venera 9* and *10* spacecraft provide panoramic pictures of surface. They discover prevalence of sharp-edged rocks, leading to speculation that Venus is young, "living" planet.

1978 U.S. *Pioneer-Venus 1* and *2* spacecraft conduct first radar survey of planet's cloud-covered surface. Spacecraft maps 93 percent of planet's surface and returns more than 1,000 pictures. Data indicate flat plains, great mountains and valleys, and possible surface fires.

1978 Soviet spacecraft *Venera 11* and *12* detect as many as 25 lightning impulses per second while descending to planet's surface.

1983 Soviet spacecraft *Venera 15* and *16* begin orbiting planet, using special imaging radar to secure finely detailed photographs. Radar altimeters measure geographical heights and detect hot spots and mountains.

1985 U.S. spacecraft *Vega 1* and *2* deposit landers in Venusian atmosphere while passing en route to Halley's comet. Each releases small balloon containing atmospheric and meteorological instruments. Data are transmitted until batteries fail 46 hours later.

1990 U.S. spaceprobe Galileo flys by Venus, using the planet's gravity to propel it on way to rendezvous with Jupiter in 1995.

Earth

1957 First artificial Earth satellite, *Sputnik I*, is launched by USSR (October 4). *Sputnik II*, carrying dog Laika, is launched (November).

1958 *Explorer I*, first U.S. Earth satellite, is launched (January 31). It discovers Van Allen radiation belt.

1958 U.S. Project Score launches first satellite communications experiment. It broadcasts first voice message from space.

1959 U.S. satellite *Vanguard II* is first satellite to transmit weather information to Earth.

Discovering Our Solar System

1959	U.S. satellite *Explorer 6* takes first television pictures of Earth's cloud cover.
1960	U.S. weather satellite *Tiros 1* sends back nearly 23,000 photographs of Earth's cloud cover.
1961	Soviet cosmonaut Yuri Gagarin (1934–68) becomes first man to orbit Earth, aboard satellite *Vostok I* (April 12).
1962	U.S. satellite *Telstar I* is first commercial communications satellite. It provides television and voice communications among United States, Great Britain, and France.
1965	Early Bird commercial communications satellite is launched.
1967	U.S. *Biosatellite 2* is launched to conduct biological experiments.
1972	First of U.S. Landsat satellites is launched. It studies mineral, agricultural, and other Earth resources. (Other Landsats are launched in 1975, 1978, and 1982.)
1973	U.S. space station *Skylab 3* orbits Earth.
1975	*GOES I*, geostationary satellite for Earth environmental studies, is launched by United States.
1977	International Sun–Earth Explorer (ISEE) is launched to study interaction of solar activity and Earth environment.
1978	U.S. *Sensat I* is launched to monitor oceanographic data.
1980	U.S. *Magsat* satellite completes mapping of Earth's magnetic field.
1987	Twenty-ton Soviet radar satellite is launched (July 25). It has applications in mapmaking, oceanography, crop predictions, ice monitoring, and prospecting for minerals.

Moon

1959	Soviet moon probe *Luna 1* makes first flyby of moon (January).
1959	Soviet spacecraft *Luna 2* makes crash landing on moon (September 12).
1959	Soviet moon probe *Luna 3* takes first photographs of dark side of moon (November).
1964	U.S. *Ranger 7* photographs moon surface (July 28).
1966	First soft landing on moon is made by Soviet spacecraft *Luna 9* (February 3). It transmits pictures of moon's surface to Earth.
1966	First moon orbit is accomplished by Soviet satellite *Luna 10* (April 3).
1966	U.S. *Surveyor 1* spacecraft lands on moon (June 2). For six weeks it sends back thousands of close-up photographs of lunar surface.
1968	Soviet spacecraft *Zond 5* is first to return to Earth after orbiting moon.
1968	First manned orbit of moon is accomplished by U.S. *Apollo 8* spacecraft (December 21–27). Astronauts Frank Borman (b. 1928), James Lovell, Jr. (b. 1928), and William Anders (b. 1933) orbit 10 times and are first to see dark side of moon.
1969	U.S. *Apollo 11* mission brings first men to moon (July 16). Astronaut Neil Armstrong (b. 1930) becomes first man to walk on moon (July 20).
1969	*Apollo 12* mission is second manned moon landing (November 14–24).
1971–72	U.S. manned moon landings continue with *Apollo 14* (January–February 1971), *Apollo 15 (July–August, 1971), Apollo 16* (April 1972), and *Apollo 17* (December 1972).
1989	United States announces renewed commitment to manned space exploration, including possibility of establishing colony on moon.

Mars

1784	British astronomer Sir William Herschel (1738–1822) observes mists or clouds around Mars that occasionally dim or distort its features.
1877	U.S. astronomer Asaph Hall (1829–1907) discovers two Martian satellites, Phobos and Deimos, whose small size and closeness to planet make them unique in solar system.
1877	Italian astronomer Giovanni Schiaparelli (1835–1910) observes network of straight lines on Mars, calling them canali ("channels"). American astronomers mistakenly term them canals and surmise that they are dug by highly civilized inhabitants to transport water.
1937	Studies indicate presence of "blue mist," opacity to blue, violet, and ultraviolet light, which clears on rare occasions.
1964	U.S. spacecraft *Mariner 4* completes first successful flyby, obtaining remarkable close-up pictures, rich in surface detail. It provides more information about Mars than is known about any other body in solar system except the moon and Earth.

Discovering Our Solar System

1969 U.S. *Mariner 6* and *7* spacecraft fly by, returning hundreds of detailed pictures before entering solar orbit.

1971 U.S. *Mariner 9* spacecraft orbits Mars, becoming first space probe to orbit another planet. It returns thousands of pictures.

1973 Soviet spacecraft *Mars 2* and *3*, first space probes to enter Martian atmosphere, send landing capsules to surface. Capsules contain television equipment, but signals end shortly after landing. Spacecraft continue orbiting and returning scientific data.

1976 U.S. spacecraft *Viking 1* and *2* complete first successful landings on planet. Spectacular pictures show planetary surface shaped by volcanic activity and water erosion and thin layers of frost in northern latitudes. Spacecraft transmit over 1,800 pictures, including first televised pictures of planet's surface.

1987 United States and Soviet Union work toward coordinating future Mars missions and exchanging data.

1989 U.S. space probe *Magellan*, launched from Earth-orbiting U.S. space shuttle *Atlantis*, will reach Venus some 15 months later.

Jupiter

1610 Italian astronomer Galileo Galilei (1564–1642) discovers Jupiter's four largest satellites: Callisto, Ganymede, Europa, and Io.

1630 Planet's parallel belts of color are first detected.

1857 Great Red Spot is first observed. Closer observations, after 1878, show changes in color, motion, and form.

1892 U.S. astronomer Edward Barnard (1857–1923) discovers Amalthea, small satellite in interior of Io's orbit.

1901 Southern tropical disturbance is first observed.

1904–51 Sixth through twelfth moons are discoverd by American and British astronomers.

1955 Radio astronomers in Washington, D.C. pick up strong radio waves from Jupiter.

1973–74 U.S. spacecraft *Pioneer 11* and *12* transmit first close-up color photos of planet back to Earth. Scientific data show Jupiter to be composed of liquid hydrogen. Weather is caused more by internal heat than by solar radiation (planet radiates twice the heat it receives from sun).

1974 American astronomer discovers Jupiter's 13th moon.

1979 U.S. spacecraft *Voyager 1* discovers new 18-mile-wide ring around planet while looking for additional moons. It detects turbulent atmosphere on planet, including frequent lightning, and finds three new moons (one is solar system's fastest-moving satellite). It returns close-up photographs of Jupiter's known satellites, including pitted surface of Io, desertlike and cracked surface of Ganymede, and giant crater on Callisto.

1989 United States launches space probe Galileo, which will rendezvous with Jupiter in 1995.

Saturn

1655 Dutch physicist Christiaan Huygens (1629–95) discovers planet's largest moon, Titan.

1671 Italian-born French astronomer Gian Cassini (1625–1712) locates second moon, Iapetus. In 1672 he discovers another, Rhea, and in 1684 two more, Tethys and Dione.

1789 British astronomer Sir William Herschel (1738–1822) discovers two moons, Mimas and Enceladus.

1898 American astronomer William Pickering (1858–1938) discovers ninth moon, Phoebe.

1933 Large white spot is detected on Saturn's equator. It is one of several temporary markings that appear occasionally on planet.

1979 U.S. spacecraft *Pioneer 11* discovers planet's 11th, 12th, and 13th moons as well as magnetic field.

1980 U.S. spacecraft *Voyager 1* completes first successful flyby of Saturn, detecting three new moons and returning close-up pictures of several of them.

1981 U.S. *Voyager 2* spacecraft finds evidence of six possible new moons and gives new details on Saturn's known moons. It shows Enceladus has varied geography, possibly from tidal heating and crustal expansion/contraction; Tethys is marked by large impact crater; Hyperion has strange shape, resembling a hamburger patty; and Iapetus is a dark satellite, possibly covered with thick, organic material. Planet's atmosphere is marked by swirls and spots from turbulence.

Discovering Our Solar System

Uranus

1781 First detected by British astronomer Sir William Herschel (1738–1822), Uranus is first planet discovered with modern telescope. Herschel at first believes he is viewing a comet.

1787 Herschel discovers two moons, Titania and Oberon.

1851 British astronomer William Lassell (1799–1880) discovers moons Ariel and Umbriel.

1932 Astronomers find that planet probably has large quantities of methane in its atmosphere. Methane gives Uranus its greenish color.

1948 American astronomer discovers fifth moon, Miranda.

1976 Astronomers detect approximately nine narrow rings of dark particles orbiting planet.

1986 U.S. spacecraft *Voyager 2* flies by planet, revealing that methane atmosphere may cover huge ocean of heated water. It sends back pictures of some 100 dark carbon rings and discovers and sends back pictures of 10 new moons.

Neptune

1846 German astronomer Johann Galle (1812–1910) discovers Neptune. He uses theoretical calculations made by French astronomer Urbain-Jean-Joseph Leverrier (1811–77), based on irregularities in orbit of Uranus.

1846 Neptune's moon Triton is discovered by British astronomer William Lassell (1799–1880).

1949 American astronomers discover another satellite, Nereid.

1984 Astronomers discover planet has partial ring.

1989 U.S. *Voyager 2* spacecraft makes first-ever flyby of planet. It discovers three new moons, partial rings, and evidence of volcanic activity on Neptune's moon Triton.

Pluto

1915 American astronomer Percival Lowell (1855–1916) postulates existence of planet beyond Neptune, based on irregularities in orbits of Neptune and Uranus.

1930 Planet is discovered by U.S. astronomer Clyde Tombaugh (b. 1906).

1976 Pluto is found to have covering of frozen methane.

1978 American astronomer James Christy (b. 1938) discovers Pluto's moon Charon through careful study of photographic images.

1980 Astronomers discover traces of atmosphere on Pluto. Finding is confirmed in 1988.

1985 New estimates put diameter of planet at 1,900 miles.

1687 Newton publishes his landmark *Philosophiae Naturalis Principia Mathematica*, setting forth his law of universal gravitation. His work in gravitational astronomy provides the basis for modern celestial mechanics.

1705 English astronomer Edmund Halley (1656–1742) predicts comet of 1682 will return 76 years after sighting. Halley's comet returns Christmas Day 1758, proving periodic comets orbit sun.

1743 Earth is shown to be flattened at poles, not a perfect sphere, as previously thought.

1754 Refracting telescope called heliometer is invented. It helps calculate angular distances between stars.

1755 German philosopher Immanuel Kant (1724–1804) theorizes that our solar system is part of larger system of stars (now called galaxy) and that other such "island universes" exist.

1762 British astronomer James Bradley (1693–1762) finishes his star catalog, listing over 60,000 stars.

1781 Planet Uranus is discovered by British astronomer Sir William Herschel (1738–1822).

▶

1785 William Herschel describes shape of Milky Way with fair accuracy.

1789 Herschel constructs then world's largest telescope (with 48-inch lens).

1794 Extraterrestrial origin of meteors is proved.

1796 French astronomer Pierre-Simon Laplace (1749–1827) theorizes that solar system is formed by condensation of gigantic gas cloud. This becomes basis of modern theory. Laplace also publishes important works on celestial mechanics.

1798 Black holes are first postulated by Laplace.

1801 First asteroid, Ceres, is observed by Giuseppe Piazzi (1746–1826). Over 1,000 asteroids are subsequently discovered.

1802 Herschel studies two brighest stars of Castor and discovers they are binary stars that revolve around common center. He also catalogs over 2,500 stars, galaxies, and star clusters.

1811 Star's origin from nebula is postulated by Herschel.

1814 Joseph von Fraunhofer (1787–1826) studies solar spectrum and catalogs over 500 of its lines.

1834 Magellanic clouds are found to be clusters of millions of individual stars by English astronomer John Herschel (1792–1871), son of Sir William Herschel. Clouds later are identified as galaxies closest to our own Milky Way.

1838 Distance to a star (61 Cygni, 6 light-years) is measured successfully for first time by Friedrich Bessel (1784–1846).

1843 Sunspot cycle, semiregular fluctuation in number of sunspots, is discovered.

1846 Planet Neptune is first sighted by German astronomer Johann Galle (1812–1910), who is aided by computations of French astronomer Urbain-Jean-Joseph-Leverrier (1811–77).

1850 Astronomers first use photographic plates to record images of stars, making study of star fields and spectra more objective and precise.

1855 Spiral structure of galaxies is observed by William Parsons (1800–67).

1861 German physicist Gustav Kirchoff (1824–87) determines composition of sun's atmosphere by spectral analysis of sunlight.

1877 "Canals" are observed on surface of Mars, giving rise to widespread belief that life exists or has existed there.

1888 Evolutionary cycle of star from birth to extinction is first outlined by Sir Joseph N. Lockyer (1836–1920).

1888 *New General Catalog of Nebulas and Clusters of Stars (NGC)* is published in England. It revises and updates John Herschel's 1864 catalog, which listed over 5,000 nebulas and star clusters. *NGC* includes over 8,000 nebulas.

1892 Edward Barnard presents first evidence that novas are actually exploding stars.

1897 World's largest refracting telescope at this time, with 40-inch lens, is in operation at Yerkes Observatory in Wisconsin.

1908 Sunspots are shown to be magnetic phenomena by American astronomer George E. Hale (1868–1938).

1911 Meteorite landing in Egypt strikes and kills dog in only recorded accidental death by meteorite.

1912 Evidence of existence of cosmic rays is discovered by Victor Franz Hess (1883–1964).

c. 1912 Danish astronomer Ejnar Hertzsprung (1873–1967) and American scientist Henry Russell (1877–1957) develop separate theories of stellar evolution. Hertzsprung-Russell diagram, still in use, graphs stars' magnitudes against their spectral types.

1915 Proxima Centauri, closest star to Earth other than sun, is located.

1916 German-born U.S. physicist Albert Einstein (1879–1955) devises his general

▶

theory of relativity, giving new theory of gravity. Einstein's calculations concerning bending of light by strong gravitational forces are proved by astronomical observations in 1919.

1916 Einstein publishes his *Foundation of the General Theory of Relativity*, which holds that in space–time continuum universe is curved as result of effects of gravitation.

1917 World's largest reflecting telescope at this time, 100-inch Hooker telescope, is in operation at Mount Wilson in California.

1918 Approximate size of Milky Way galaxy is determined (c. 100,000 light-years) by American astronomer Harlow Shapley (1885–1972), who also calculates location of galactic center and sun's position in galaxy.

1918–24 Henry Draper Catalogue, first major catalog of stars by spectral type, is compiled.

1921 Publication of ancient Chinese records revealing star explosion in A.D. 1054 confirms astronomers' theory of supernova, explosion of star, which releases huge amounts of energy and light.

1923 British astronomer Arthur Eddington (1882–1944) recognizes importance of star's mass in its evolution. He describes star's opposing tendencies to contract due to internal gravitational pull and to fly apart due to outward radiation of energy.

1925 Classification of galaxies by their structure (elliptical, spiral, or irregular) is introduced by Edwin P. Hubble (1889–1953). Milky Way is eventually shown to be spiral galaxy.

1927 Theory of rotation of Milky Way galaxy is proposed by Bertil Lindblad (1895–1965). It completes one revolution every 210 million years.

1927 "Big-bang" theory is published by Belgian astrophysicist Georges-Henri Lemaître (1894–1966). It holds that universe began with gigantic explosion and that it continues to expand outward today because of this. Studies in 1987 estimate big bang occurred 11 to 12 billion years ago.

1929 Hubble develops Hubble's law, regarding universe's uniform expansion, and Hubble's constant, which holds that the more distant a galaxy is from our own galaxy, the faster it is receding from us.

1930 Bernhard Schmidt (1879–1935) invents Schmidt telescope, important because it is free from spherical aberration coma.

1930 Planet Pluto is located by astronomer Clyde Tombaugh (b. 1906).

1933 Karl G. Jansky (1905–50) discovers that Milky Way center emits powerful electromagnetic radiation at radio frequencies, marking beginning of radio astronomy.

1933 German-American astronomer Walter Baade (1893–1960) and others suggest that supernova, after exploding, may leave small, extremely dense remnant called neutron star.

1937 First radio telescope (31-foot dish) is put into operation, in Illinois.

1938–39 Thermonuclear process inside sun is shown to be source of its energy.

1942 First radio maps of universe are produced by Grote Reber (b. 1911).

1946 First radio galaxy, Cygnus A, is identified.

1948 World's largest telescope is now the new 200-inch reflecting telescope at Mount Palomar, California.

1948 Steady-state theory of universe, which holds the universe is expanding at a constant rate, is put forward by three British astronomers who are opposed to big-bang theory.

1949 Comets are likened to "dirty snowballs" by American astronomer Fred L. Whipple (b. 1906), who believes them to be composed of ice and dust.

1951 Computer is used for the first time to calculate planetary orbits.

▶

c. 1955 Schwarzschild radius is identified as radius at which star in gravitational collapse becomes black hole. Named for German astronomer Karl Schwarzschild (1873–1916), it also is known as "event horizon."

1958 Solar wind is discovered by Eugene Parker (b. 1927).

1958 Earth's Van Allen radiation belt is discovered.

1963 World's largest radio telescope, with fixed spherical dish 1,000 feet across, is completed in Puerto Rico. Largest steerable dish (328 feet across) is built in Germany in 1970.

1963 First of the quasi-stellar radiation sources (quasars), most distant objects in known universe, are identified. Farthest quasar is about 12 billion light-years from Earth.

1965 Cosmic background (radio-frequency) radiation, predicted in big-bang theory, is discovered. It helps establish big bang as prevailing theory of cosmic origins.

1967 First pulsar (rotation neutron star) is discovered, by Cambridge student Jocelyn Bell.

1970 First X-ray satellite observatory (*Uhuru*) is launched.

c. 1970 Astronomers locate what they believe is first black hole (Cygnus X–1), a very dense body probably left over from the gravitational collapse of a star. Black hole has such powerful gravity that even light cannot escape its gravitational field. Second one is found in 1983.

1976 Soviets build 236-inch (6-meter) reflecting telescope in Caucasus Mountains, making it world's largest, but mechanical problems render it inoperative.

1979 Multiple-mirror telescope (MMT), with six 1.8-meter mirrors operating in unison (equal to 4.5-meter telescope), is installed at Mount Hopkins in Arizona.

1979 Radio astronomers studying interstellar matter identify over 40 kinds of organic molecules dispersed in rarefied gas that fills space between stars.

1980 Very Long Array (VLA) radio telescope is completed in New Mexico. It has 27 separate antennas that operate in unison.

1981 Biggest known star in universe, R136a, is located. It is 2,500 times more massive than our sun.

1983 Universe is mapped in infrared wavelength by orbiting infrared telescope of IRAS (Infrared Astronomy Satellite). IRAS finds first evidence of planetary material around star (Vega) outside our solar system.

1983 SETI program, search for extraterrestrial life, begins.

1983 Astronomers estimate that some 30 white dwarf stars (faint stars in final stages of stellar evolution) in our galaxy annually become novas. These stars suddenly become thousands of times brighter, then fade to their original magnitude.

1987 Planet-sized objects are found orbiting stars Gamma Cephei and Epsilon Eridani.

1987 Astronomers observe the supernova explosion SN1987A within Large Magellanic Cloud, galaxy closest to our own. It is first since 1604 visible to naked eye. Scientists discover massive streams of particles called neutrinos emitted during event.

1988 Astronomers find most distant (and oldest) galaxy at edge of known universe 12 billion light-years away.

1989 Astronomers conclude that Andromeda and its companion galaxy, M32, have black holes at their centers.

1989 Astronomers discover river of gas at center of Milky Way, indicating existence of a black hole at center of our galaxy.

1989 Theory on creation of pulsar from supernova explosion is documented by observation for first time.

▶

1989 Complete orbit by a planetlike body circling a star other than our sun is observed for first time at star HD 114762.

1990 Hubble Space Telescope is launched into orbit. Position above Earth's atmosphere allows images in far greater detail of distant objects.

Biology

See also Unraveling the Mysteries of the Cell; Medicine; and Environmentalism.

B.C.

c. 400 Greek physician Hippocrates (c. 460–c. 370 B.C.) proposes that all diseases have natural causes, basis of modern medical theory. He formulates Hippocratic oath.

384–322 Greek philosopher Aristotle's writings lay foundation for modern embryology and include early biological classification of organisms.

371–288 Greek scientist Theophrastus makes important advances in botany.

c. 350 Earliest description of human fetus is written, in Greece.

A.D.

c. 70 Roman naturalist Pliny the Elder (A.D. 23–79) publishes 37-volume *Natural History*, including information (and misinformation) about plants and animals.

c. 150 Greek physician and experimental physiologist Galen (A.D. 130–99) makes important advances in anatomy and physiology after treating injured gladiators and dissecting apes and pigs. (He discovers that arteries contain blood, not air.)

c. 1482 Italian artist and anatomist Leonardo da Vinci (1452–1519) publishes detailed sketches of human body.

1543 First accurate textbook on human anatomy, *De Humani Corporis Fabrica*, is published.

1551 Swiss naturalist Konrad von Gesner (1516–65) publishes his five-volume illustrated encyclopedia on animals, now considered the origin of modern zoology.

1580 Scientist Prospero Alpini (1553–1616) discovers that plants have two sexes.

1590 Compound microscope is invented.

1596–1650 French philosopher René Descartes speculates, mistakenly, that heart generates "animal spirits" that animate human body.

1628 British physician William Harvey (1578–1657) discovers that blood, pumped by heart's muscular contractions, circulates through body.

c. 1660 Italian anatomist Marcello Malpighi (1627–94) observes, through microscope, capillary network connecting veins and arteries.

1665 British scientist Robert Hooke (1635–1703) publishes first drawings of cells and cell walls.

c. 1670 Italian physician Francesco Redi (1626–97) performs controlled experiments with maggots, attempting to disprove theory of spontaneous generation; his findings are not universally accepted.

c. 1675 Dutch scientist Anton van Leeuwenhoek (1632–1723), working with crude microscopes, discovers microscopic life forms. He describes spermatozoa (1677) and bacteria (1683).

1682 English botantist Nehemiah Grew (1641–1712) describes sexual role of

▶

Unraveling the Mysteries of the Cell

1663	English scientist Robert Hooke (1635–1703) identifies and names *cell.*
1780	Invention of slicing machine, which can cut specimens as thin as .0005 inch, helps scientists prepare slides of specimens for study under microscope.
1824	French scientist René Dutrochet (1776–1847) recognizes that cells are responsible for growth.
1831	Cell nucleus is discovered by Scottish botanist Robert Brown (1773–1858) during study of orchids.
1835	French scientist Felix Dujardin (1801–62) describes material moving between food vacuoles in protozoa, calling it *sarcode.*
1837	German botanist Hugo von Mohl (1805–72) observes cell division in algae and describes steps in detail.
1838	German botanist Matthias Schleiden (1804–81) claims that cells are basic structural units of plant life and that cell nuclei are their essential internal parts. He is credited with establishing foundations of cell theory.
1839	German physiologist Theodor Schwann (1810–82) theorizes that cell also is basis of animal life and recognizes that egg is single cell. He coins term *cell theory.*
1839	Bohemian physiologist J. E. Purkyně (1787–1869) coins term *protoplasm.*
1841	Swiss embryologist Albrecht von Kolliker (1817–1905) notes that spermatozoa as well as ova are cellular. In 1844 he studies embryonic development as form of cellular division.
1845	German zoologist Karl von Siebold (1804–84) describes unicellular nature of protozoans and function of cilia.
1851	Hugo von Mohl (1805–72) advances cell theory for plants. He discovers that plant cell walls are fibrous.
1858	German pathologist Rudolf Virchow (1821–1902) theorizes that every cell in body tissue comes from another cell. His theory is initially criticized.
1863	Scientists describe protoplasm as "the physical basis of life."
1864	French microbiologist Louis Pasteur (1822–95) discovers that fermentation is caused by organisms in the air and not by spontaneous generation. Discovery ultimately leads to theory that diseases are caused by germs.
1879	Research into cell nucleus begins as scientists describe splitting of chromosomes and existence of equal number of chromosomes in every cell of body.
1882	Detailed and scientific terms for mitosis are complete by this time.
c. 1885	Individuality of chromosomes is noted and described.
1892	Complete stages of spermatogenesis and oogenesis are described by scientists.
1892	German biologist August Weismann (1834–1914) describes complex hierarchy of particles and particle groups in explaining heredity.
1898	Golgi apparatus, structures in the cytoplasm, are identified by Italian neurologist Camillo Golgi (1844–1926).
1903	Walter Sutton (1877–1916) shows that material determining heredity is carried within chromosomes.
1905	Scientists link single characteristic, sexual determination, to specific chromosome for first time.
1907	Individual nature of cells is shown in several scientific studies.
1907	Reliable technique of culturing cells outside body is introduced, leading to development of cytology.
1908	"Gene pool" theory of evolution is suggested in research by British mathematician G. H. Hardy (1877–1947).
1909	Russian-born American chemist Aaron Levene (1869–1940) discovers RNA (ribonucleic acid) in nucleus of cells.
1910	American biologist Thomas Morgan (1866–1945) explains that specific genes are assigned to specific chromosomes. Theory on genes earns him 1933 Nobel Prize for Physiology or Medicine.
1913	Studies prove that genetic factors follow linear arrangement in chromosomes.
1915–17	Simple procedure is developed for culturing and assaying viruses that attack bacteria, marking beginning of modern virology.
1924	Cytochrome, enzyme involved in cell respiration, is discovered.
1929	Levene discovers DNA (deoxyribonucleic acid) in cell nuclei.

Unraveling the Mysteries of the Cell

c. 1932 Electron microscope is developed, permitting direct observation of the minute working parts of cells.

1941–44 U.S. scientists discover that DNA, not protein, is main component in cell for determining hereditary characteristics.

1952 U.S. scientists show that virus that attacks bacteria can transmit genetic material to host cell, proving foreign genetic material can be introduced to host cell.

1953 U.S. biologist James Watson (b. 1928) and British physicist Francis Crick (b. 1916) describe their famous model of the DNA molecule, showing its structure to be double helix.

1956 DNA is first synthesized chemically. It is not biologically active, however.

1961 First breakthrough is made in search for keys to genetic code. RNA code that produces specific amino acid in cell is found.

1965 Molecular structure of transfer RNA molecule is described.

c. 1965 Experiments performed on connective tissue cells suggest "biological clock" causes cell death after predetermined period.

1966 Enzyme that RNA molecules use to duplicate themselves is found.

1967 Biologically active DNA is synthesized.

1968 Enzymes that can cut DNA strands at specific point are discovered. This later helps make genetic engineering possible.

1970 First complete synthesis of gene is done.

1970 Scientists discover process by which RNA code is transcribed onto DNA, reverse of normal cell process.

1973 First genetic engineering is accomplished by U.S. researchers Stanley Cohen (b. 1922) and Herbert Boyer. They successfully implant foreign genetic material into bacterium *Escherichia coli (E. coli)*.

1976 Functioning gene is completely synthesized.

1980 Gene is successfully transferred from one mouse to another by team of scientists.

1983 First chromosome is created artificially.

1984 Genetic fingerprinting, based on certain DNA codes unique to each individual, is discovered.

1989 Scientists have mapped over 2,000 human genes by this date, small fraction of estimated 50,000 to 100,000 genes in human chromosomes. Some of these genes have been found responsible for such diseases as muscular dystrophy and Down's syndrome.

flower parts and fertilization of seeds by pollen.

1691 Fossils are identified as animal and plant remains from earlier eras.

1728–93 Scottish anatomist John Hunter is influential in establishing natural-history museums. He dissects and describes more than 500 animal species.

1733–1804 English scientist Joseph Priestley proves that plants give off oxygen.

1735 Swedish biologist Carolus Linnaeus (1707–78), called father of taxonomy, classifies animals according to their structures. His two-name system (genus and species) is forerunner of modern scientific classification.

1759 Swiss nobleman Albrecht von Haller (1708–77) makes important advances in knowledge of central nervous system.

c. 1775 French chemist Antoine Lavoisier (1743–94) makes important advances in human physiology, specifically in understanding respiration and conversion of food to energy. He discovers that animals use oxygen and give off carbon dioxide.

1787–1869 Bohemian physiologist J. E. Purkyně lives. He coins term *protoplasm*.

1792 French anatomist George Cuvier (1769–1832), pioneer in comparative anatomy and paleontology, publishes studies on fossil structure. He is first to name "fossil species."

1796 First inoculation is performed in England by Edward Jenner (1749–1823).

▶

Evolving Plant and Animal Life

Million B.C.	
4,000–570	Precambrian time, spanning time from formation of Earth's crust to Cambrian period.
3,400–3,000	Oldest known living organisms, bacteria and nonphotosynthetic algae, are fossilized.
2,900	Photosynthetic organisms appear. They begin producing the oxygen now found in Earth's atmosphere.

Paleozoic Era

Million B.C.	
570–500	**Cambrian Period**. Appearance of marine shelled arthropod trilobite marks beginning of this period. Other animal life includes coelenterata, gastropods, and spongelike animals.
500–435	**Ordovician Period**. Corals appear. Many forms of sea life flourish, including first bivalve mollusks, echinoderms, and worms. First primitive jawless fish appear.
435–395	**Silurian Period**. First major coral reefs appear. Ostracoderms (fishlike vertebrates akin to lampreys with bony external armor plates) flourish.
395–345	**Devonian Period**. Many species of fish appear, including shark. First terrestrial life, including vascular plants, amphibians, and insects, appears. Terrestrial plant life includes fungi and first ferns and rushlike horsetails.
345–280	**Carboniferous Period**, divided into Pennsylvanian and Mississippian periods. Large, fern-filled swamps form. Greater part of Earth's coal deposits begin forming. Early conifers appear. Insect species, including giant dragonflies and 800 types of cockroaches, flourish. Reptiles appear toward end of this period.
280–225	**Permian Period**. Mass extinction of many marine invertebrates occurs late in this period. Reptiles flourish. Seed ferns appear.

Mesozoic Era

Million B.C.	
225–190	**Triassic Period**. Reptiles are dominant animals, and first dinosaurs, lizards, and turtles appear. Earliest mammals also appear. Conifers emerge as dominant tree as climate becomes drier.
190–136	**Jurassic Period**. Reptiles continue to dominate animal life. Dinosaurs flourish. Crocodiles and flying reptiles appear. Palmlike trees (cycads) dominate plant life. Birds appear toward end of this period.
136–65	**Cretaceous Period**. Among dinosaurs, *Tyrannosaurus* and *Brontosaurus* appear. Snakes appear, along with lobsters, crabs, and barnacles. Modern insects begin to appear. Flowering plants and many modern trees emerge and become dominant (fig, magnolia, birch, elm, and oak). Mass extinction of dinosaurs occurs at end of this period. Mammals are small and very primitive.

Cenozoic Era

Million B.C.	
65–54	**Paleocene Epoch**. Early mammal species dominate animal life. First horse, hare, monkey, ape, and whale appear, as well as rats, mice, and squirrels.
54–38	**Eocene Epoch**. First bats appear. Early horses and camels and first cats are among mammals of this period.
38–26	**Oligocene Epoch**. Ancestral dogs and pigs appear, along with early elephants, monkeys, and apes. Many older types of mammals become extinct.
26–5	**Miocene Epoch**. Forests decline as climate cools. Extensive grasslands form, giving rise to species of mammals that graze. Mastodons, weasels, raccoons, and early deer and antelopes appear, along with manlike apes.
5–2	**Pliocene Epoch**. Climate becomes cooler and drier. Some modern plants appear, and some species of mammals decline. First possible apelike humans appear.
2–.01	**Pleistocene Epoch**. Four ice ages occur (2.2 million to 10,000 years ago), interrupted by warmer periods that melt away ice sheets. Primitive men appear. Large Ice Age mammals (mastodon and saber-toothed cat) become extinct by end of this period.
.01–present	**Holocene or Recent**. During this period (roughly from 10,000 years ago) man and his rapidly developing civilization spread throughout world.

The Age of Dinosaurs

Million B.C.

c. 245 *Dimetrodon*, early dinosaur, flourishes. This carnivorous quadruped is 7 feet long with tall spiny back fin resembling sail. Fin possibly helps regulate body temperature.

c. 200 *Plesiosaurus*, marine reptile 15 feet long with long neck and short, paddlelike fins, flourishes. It has long, sharp teeth for eating fish.

c. 165 *Ichthyosaurus*, porpoiselike, fish-eating reptile 10 feet long, flourishes. It has large eyes and long jaws with numerous sharp teeth.

c. 165 *Ultrasaurus*, gigantic quadrupedal bronotosaurus-like herbivore, flourishes. It measures more than 100 feet long (with 40-foot neck) and stands 50 to 60 feet tall (according to modern estimates, based on partial fossil remains only).

c. 150 *Brontosaurus*, giant herbivore 70 feet long and weighing as much as 30 tons, flourishes. It has long skull and is at home in marshes and on dry land. Scientists in 1978 reject older short-skull representation (caused by mixup at fossil site) in favor of longer skull.

c. 150 *Stegosaurus*, quadrupedal herbivore 20 feet long, flourishes. It has very small skull and brain and short forelegs. Two staggered rows of triangular bony plates along back protect spinal cord. Four pointed spikes on end of tail are for self-defense.

c. 150 *Archeopteryx*, earliest known bird, flourishes. Crow-sized and very reptilelike, it has numerous teeth and long tail with feathers. In modern times, scientist speculate that *Archeopteryx* may have been warm-blooded.

c. 130 *Allosaurus*, bipedal carnivore weighing up to 2 tons and measuring 34 feet long, flourishes. It has massive hind legs; its small forelimbs are used only for grasping. It possesses very large skull with large pointed teeth.

c. 130 *Iquanodon*, bipedal herbivore, flourishes. It is 12 to 15 feet tall, weighs up to 3 tons, and has handlike appendages equipped with large spikes in place of thumbs for self-defense.

c. 130 *Deinonychus*, small bipedal carnivore, lives. About 5 feet tall and 160 pounds, it has powerful, slender hind legs enabling it to run like large bird. *Deinonychus* has serrated teeth and walks on two toes. Its third toe is 5-inch-long sickle claw for slashing prey and attackers.

c. 100 *Tyrannosaurus*, largest bipedal carnivore at 20 feet tall and 8 tons, flourishes. It has 4-foot-long skull with 6-inch serrated teeth. Though *Tyrannosaurus* has massive body, powerful tail, and strong hind legs, its forelimbs are tiny.

c. 100 *Triceratops*, massive quadrupedal herbivore 25 feet long and weighing 8 to 9 tons, flourishes. It has long skull with bony frill around its neck and is armed with small horns on its nose and above its eyes.

c. 100 *Ankylosaurus*, armored herbivore 15 feet long, flourishes. Its low, flat body is covered with bony plates (spikes along its flanks), and its tail has knob of bone used as club for self-defense.

c. 100 *Phobosuchus*, prehistoric crocodile up to 40 feet long, flourishes. It preys on larger dinosaurs.

c. 100 *Parasaurolophus*, bipedal duck-billed herbivore standing 16 feet tall and weighing up to 4 tons, flourishes. It has 5-foot-long crest curving backward from head. Crest is used for breathing, perhaps to improve sense of smell or perhaps to act as resonating chamber for mating calls.

c. 100 *Pteranodon*, large flying reptile with wingspan up to 25 feet, flourishes. Its body is no larger than that of turkey. It has toothless, pelicanlike jaws for eating fish. Long crest at back of skull may counterbalance long jaws.

c. 65 Mass extinction of dinosaurs as result of unknown cause.

1803–73 German chemist Justus von Liebig develops improved methods for analyzing organic materials.

1809 French naturalist Jean de Lamarck (1744–1829) develops first theory of organic evolution, believing mistakenly that an organ's size is proportional to its use.

1813–78 French physiologist Claude Bernard, considered founder of experimental medicine, makes important contributions to knowledge of liver, heart, brain, and placenta.

▶

1827 Karl Ernst von Baer (1792–1876) discovers eggs in reproductive system of female mammals, including humans.

1828 German chemist Friedrich Wohler (1800–1882) is first to synthesize an organic material from inorganic substances.

1831 Cell nucleus is first described by British botanist Robert Brown (1773–1858).

c. 1838 German botanist Matthias Schleiden (1804–81) and physiologist Theodor Schwann (1810–82) propose that cell is the basic unit of all plant and animal tissue.

1843 British scientist Richard Owen (1804–92), first director of the British Museum of Natural History, develops concepts of homology and analogy, basic distinctions in morphology.

1858 German pathologist Rudolf Virchow (1821–1902) proposes that all diseases are diseases of cell. (With the discoveries of Schleiden and Schwann, these ideas constitute cell theory.)

1859 French chemist Louis Pasteur (1822–59) effectively disproves theory of spontaneous generation. With German bacteriologist Robert Koch (1843–1910), Pasteur also concludes that bacteria cause some diseases.

1859 British naturalist Charles Darwin (1809–82) publishes *The Origin of Species*, his theory of organic evolution, which becomes basic axiom of biological science.

1866 Austrian abbot Gregor Mendel (1822–84) publishes his discoveries on basic principles of heredity, thus pioneering field of genetics.

1869 German biologist Ernst Haeckel (1834–1919) states that individual is product of heredity plus environment; he terms this relationship "oecology."

1876 German botanist Eduard Strasburger (1844–1912) first describes mitosis (cell division) in plants.

1880 Pasteur proposes germ theory of disease and reports his discovery on vaccines.

1881 Pasteur produces first successful vaccine.

c. 1890 Swiss, Danish, and U.S. scientists are at work in new field of biology, ecology.

1892 Russian scientist Dmitri Ivanovsky demonstrates existence of viral organisms.

1897 German chemist Eduard Buchner (1860–1917) proves that enzymes are inorganic, important distinction in biochemistry.

c. 1900 Term *biochemistry*, to describe the chemistry of living organisms, comes into use.

1901 First known essential amino acid, tryptophan, is discovered by Frederick Hopkins (1861–1947).

1901 Dutch botanist Hugo de Vries (1848–1935) coins term *mutation* to describe sudden genetic change.

1901 British mathematician Karl Pearson (1857–1936) coins term *biometry* to describe measurements of living phenomena.

1905 British physiologist Ernest Starling (1866–1927) coins term *hormone*.

1906 British biologist William Bateson (1861–1926) coins term *genetics*.

1910 U.S. biologist Thomas Morgan (1866–1945) finds that genes are located on chromosomes within cells.

1912 Polish-born American biochemist Casimir Funk (1884–1955) coins term *vitamin*. Frederick Hopkins had first suggested existence of these substances in 1906.

1928 Alexander Fleming discovers penicillin in molds.

1937 Krebs cycle, common pathway of oxidation of food for producing energy in living cells, is described by Hans Adolf Krebs (1900–1981).

▶

Timetable of Mammal Evolution

Million B.C.	
310–275	Synapsids, reptilian ancestors of mammals, evolve.
275–225	First warm-blooded creatures, known as therapids, evolve as branch of synapsids. They are carnivorous reptiles.
225–180	First mammals appear. They are characterized by warm-bloodedness, hair, and mammary glands.
180–66	Dinosaurs are predominant during this period, though primitive mammals continue to evolve.
100	Placentals and marsupials are distinguished. Placentals develop offspring within mother's uterus, while marsupials give birth prematurely and continue development of offspring in external pouch.
66	Cenozoic Era, or Age of Mammals, begins. Extinction of dinosaurs allows widespread expansion and evolution of mammals.
66–58	Paleocene epoch, during which primitive forms of modern mammals develop. They include opposumlike marsupials, early primates, and rodentlike herbivores. First known hare and first horse appear.
65	First monkeys and apes evolve.
60	Primitive whales evolve from carnivorous land mammals.
60	Rats, mice, squirrels, and other rodents develop.
58–37	Eocene epoch, during which all esssential groups of mammals evolve. First appearance of early rhinoceroses, deer, cattle, and sheep and development of whales, porpoises, and dolphins.
55	Rabbits and hares develop.
45	Bats evolve as only mammal with ability to fly.
40	Cat family evolves from *Miacis*, weasel-like mammal that also is ancestor of bears and dogs.
37–24	Oligocene epoch marks development of first pigs, mastodons, and primitive beavers. Bats become widespread. Primitive horses are extant in North America. Large cats and pigs evolve, as well as 17-foot rhinoceros.
24–5	Miocene epoch, during which half of modern mammal families are extant. First dogs, bears, foxes, and hyenas appear, along with primitive deer, antelope, walruses, seals, and saber-toothed cats.
5.5–3.5	Earliest known ancestor to man, *Australopithecus afarensis*, appears.
5–2	Pliocene epoch, during which most mammals increase greatly in size.
2–.01	Camels, cattle, zebrine horses, rhinoceroses, and certain elephants evolve during Pleistocene Epoch. Large mammals such as mammoth, mastodon, and saber-toothed cat become extinct.

1948 U.S. mathematician Norbert Wiener (1894–1964) coins term *cybernetics* to describe science of communication and control in animals and machines.

1950 Embryo transplants are first performed on cattle.

1953 U.S. biologist James Watson (b. 1928) and British physicist Francis Crick (b. 1916) successfully model structure of deoxyribonucleic acid (DNA), material in chromosomes controlling heredity.

1953 American chemist Harold Urey (1893–1981) demonstrates that amino acids can be formed from water vapor, methane, ammonia, and hydrogen. He concludes that life may have originated from chemical reactions of inorganic materials.

1954 Humans are found to have 46 rather than 48 chromosomes.

1960 U.S. scientist J. E. Steel coins term *bionics* to describe study of living organisms as models for man-made devices.

c. 1960 Reports of environmental effects of pollution spur research in ecology.

1965 Artificial sex attractants called pheremones are developed for certain insects.

▶

1970 Researchers at University of Wisconsin succeed in making first completely synthetic gene.

c. 1970 Biologists begin to question gradual evolutionary changes described by Darwin, suggesting instead that abrupt changes to species also may be part of process.

1973 Scientists experimentally remove gene from one bacterium and insert it into another, marking beginning of genetic engineering.

1978 First test-tube baby is born, in England.

1980 New theory holds that mass extinction of dinosaurs was caused by impact of large asteroid on earth about 65 million years ago.

1983 First successful human embryo transfers are completed.

1984 Sheep are successfully cloned by inserting sheep-embryo cell nuclei into specially prepared sheep ova. These are then implanted in female sheep.

1984 Fetus is successfully operated on.

1986 First license to market living organism, product of genetic engineering, is issued.

1989 U.S. scientists obtain first direct visual image of a DNA molecule.

Anthropology and Archaeology

c. 1400 Italian architect Filippo Brunelleschi (1377–1440) begins first archaeological excavations in Rome.

1506 Laocoön sculpture is discovered in vineyard near Santa Maria Maggiore, Italy. This first-century B.C. sculpture has profound influence on Michelangelo and other Renaissance sculptors.

1771 British philologist Sir William Jones (1746–94) discovers relationships among Sanskrit, Latin, and Greek, thus laying foundation for modern comparative philology and deciphering of ancient texts.

1790 Archaeologist John Frese (1740–1807) discovers flint tools and theorizes that they were fashioned by prehistoric peoples.

1799 Napoleon's troops discover Rosetta Stone while stationed in northern Egypt. Basalt slab is inscribed in hieroglyphic, demotic, and Greek. Inscriptions probably are work of priests of Ptolemy V, second-century B.C. Egyptian ruler.

1802 German scholar Georg F. Grotefend (1775–1853) deciphers Persian cuneiform script. Although he is first to translate cuneiform text, his discovery goes unnoticed.

1805 Science of physical anthropology is founded with work of Johann Friedrich Blumenbach (1752–1840).

c. 1820 Famous statue Venus de Milo is unearthed by farmer on island of Melos, Greece.

1822 Using Rosetta Stone as guide, Jean-François Champollion (1790–1832) deciphers Egyptian hieroglyphs.

c. 1839 Ruins of Mayan civilization are rediscovered in Central America.

1846 Working independently of Grotefend, British Orientalist Sir Henry Creswicke Rawlinson (1810–95) publishes his key to deciphering Mesopotamian cuneiform script.

1846 Mounds built by American Indians in Mississippi River Valley are explored.

1849–50 British archaeologist Sir Austen Henry Layard (1817–94), digging in Nineveh, unearths major portions of clay tablet library of King Assurbanipal of Assyria (reigned 669–663 B.C.).

1852–54 Further excavation of Nineveh, by Iraqi archaeologist Hormuzd Rassam (1826–

▶

1910), yields more of Assurbanipal's library, including parts of Gilgamesh epic.

1856 Discovery of first Neanderthal skeleton, in cave near Düsseldorf, Germany.

1868 First skeletal remains of Cro-Magnon man, who lived about 35,000 years ago, are found in France.

1870 German archaeologist Alexander Conze (1831–1914) determines chronological styles of pottery, thus providing key to comparative dating of stratified sites.

1871 German archaeologist Heinrich Schliemann (1822–90), seeking archaeological proof of existence of legendary city of Troy mentioned in Homer's *Iliad*, begins excavations at Hissarlik in Turkey. He later finds that two of four superimposed ruins are most likely walls of ancient Troy.

1871 English anthropologist Sir Edward B. Taylor (1832–1917) publishes *Primitive Culture*. He is first to quantify cultural phenomena for use in comparative study of social institutions.

1877 American anthropologist Louis Henry Morgan (1818–81) publishes *Ancient Society*, presenting stage theory of social evolution (that all human social evolution goes through progressive stages of savagery, barbarism, and civilization). *Origin of the Family, Private Property, and the State* (1884) by Friedrich Engels (1820–95) popularizes this theory.

1879 Cro-Magnon paintings are discovered in cave in Altamira, Spain.

1880 M. Kalokairinos, excavating at Knossos on Crete, discovers walls of legendary labyrinth at palace of King Minos.

1881 Emil Brugsch, while working on a dig headed by French Egyptologist Gaston Masperio (1846–1916), finds mummies of 40 Egyptian kings in cave at Dier el-Bahri, including mummy of Ramses the Great.

1890 Scottish anthropologist Sir James George Frazer (1854–1941) publishes *The Golden Bough*, influential but highly speculative comparative study of folklore and religion.

1893 French sociologist Émile Durkheim (1858–1917), a founder of sociology, publishes *The Division of Labor*, which becomes important in the development of anthropological theory. Durkheim interprets society as an organism made up of interdependent parts whose purpose is defined by common beliefs and values of whole.

1894 First skeletal remains of *Homo erectus* are discovered in Java by Marie Eugène Dubois (1858–1940). This early ancestor of man probably lived 500,000 to 1.6 million years ago.

1899 Boer War (1888–1902) in South Africa marks first time British anthropologists are employed by government to study non-Western kinship and political structures to aid in colonial administration.

1899 American anthropologist Franz Boas (1858–1942) becomes first professor of anthropology at Columbia University. Boas subsequently founds "American historical" school, which emphasizes extensive historical research into folklore and contemporary cultural beliefs as groundwork for theoretical formulations.

1899 German art historian Robert Koldewy (1855–1925) begins 18-year excavation of Babylon. Among other significant finds, he unearths remains of Nebuchadezzar's palace, Gate of Ishtar, and Tower of Babel.

1900 Palace of Knossos, center of Minoan civilization on Crete, is located by Arthur Evans.

1901 American anthropologist Alfred Lewis Kroeber (1876–1960) founds the Department of Anthropology at University of California at Berkeley. He stresses idea of historically recurring patterns of behavior as defining limits of cultural change.

1902 American Anthropological Association (AAA) is founded. It publishes journal *American Anthropologist*.

1902 Code of Hammurabi, first known set of laws, is discovered engraved on tablets.

1911 Amadeo Maiuri begins scientific excavations of Pompeii and Herculaneum in southern Italy. Cities lie buried beneath layer of volcanic ash from eruption of

▶

Mount Vesuvius (A.D. 79), which kept them in remarkably good condition.

1911–12 Fossil remains of "Piltdown man" are found in England. Not until 1953 is this discovery found to be clever hoax.

1912 American archaeologist Hiram Bingham (1875–1956) discovers Inca fortresses of Machu Picchu and Vitcos near Cuzco, Peru.

1912 Durkheim publishes *Elementary Forms of Religious Life*, which becomes seminal work in development of cultural anthropological theory.

1915–18 British anthropologist Bronislaw Malinowski (1884–1942) does his now-famous fieldwork with Trobriand Islanders. He later produces ethnographies *Argonauts of the Western Pacific* (1922) and *Coral Gardens and Their Magic* (1936). Malinowski founds theory of functionalism (systemic interdependence of cultural practices and beliefs).

1918 British archaeologist Leonard Woolley (1880–1960) begins excavations leading to discovery of Ur.

1919 Franz Boas is officially censured by American Anthropological Association for open letter accusing unnamed anthropologists of being employed by government as spies.

1920s Harvard University Department of Anthropology specializes in archaeology and physical anthropology. Columbia University continues to focus on ethnology and linguistics.

1922 Harward Carter (1873–1939) discovers tomb of King Tutankhamen. Excavation of tomb reveals one of richest and most significant finds in Egypt.

1924 Skull of *Australopithecus africanus* is identified by British anthropologist Louis Leakey (1903–72). It constitutes first remains of this ancestor of man who probably lived 2 million to 3 million years ago.

1925 *The Gift* is published by French anthropologist Marcel Mauss. It focuses on social bond of debt created through exchange and gift-giving.

1926 International Institute of African Languages is founded. It promotes anthropological and linguistic research in African territories.

1927 First Peking man fossil remains are found. Intact skulls are unearthed in 1937.

1928 American anthropologist Margaret Mead (1901–78) publishes *Coming of Age in Samoa*, which analyzes maturation and rites of passage in primitive society.

1932 Fragment of jaw of primate believed to be earliest human ancestor is discovered and named *Ramapithecus*. Later studies show *Ramapithecus* is probably an ancestor of orangutan, however.

1934 American anthropologist Ruth Benedict (1887–1948) publishes *Patterns of Culture*, describing whole cultures in terms of different personality types ("psychological sets"). This marks development of "cultural psychology" in United States during interwar years.

1935 Margaret Mead publishes *Sex and Temperament*, ethnography based on her fieldwork in New Guinea. This book looks at different social expectations of men and women in three different cultures.

1937 Human Relation Area Files (HRAF) are started at Yale University. They quantify ethnographic data for statistical comparison of cultures. HRAF modes of comparison and evaluation are widely criticized.

1940 Cave paintings 17,000 years old are discovered in Lascaux, France.

1940 British anthropologist Edward Evan Evans-Pritchard (1902–73) publishes *The Nuer*, first of noted Nuer trilogy that includes *Nuer Religion* (1956) and *Kinship and Marriage Among the Nuer* (1951).

1941 Society for Applied Anthropology is founded in United States to study effects of rapid industrial change on people.

1945 American anthropologist Ralph Linton (1893–1953) publishes *Cultural Background of Personality*. It attempts to systematize interdisciplinary approach in study of culture and personality.

▶

1946 Ruth Benedict (1887–1948) publishes *The Chrysanthemum and the Sword*, a study of Japanese culture.

1947 Dead Sea Scrolls are found by Bedouin in cave northwest of Dead Sea (in modern Israel). Stored in earthen jars, two of scrolls contain copies of Book of Isaiah almost 1,000 years older than any other Hebrew manuscript previously known. Over 10 other caves with scrolls are discovered in 1950s and 1960s.

1949 French anthropologist Claude Lévi-Strauss (b. 1908) publishes his first important work, *Elementary Structures of Kinship*. He becomes leading advocate of structuralism (how cultural practices and beliefs are structured in relation to one another). His other works include *Structural Anthropology* (1958), *Savage Mind* (1962), and *Totemism* (1962).

1952 British anthropologist Arnold Reginald Radcliffe-Brown (1881–1955) publishes *Structure and Function in Primitive Society*, outlining his influential theory of structural functionalism. He views society as whole system but considers it possible to study one aspect of society's structure apart from others.

1952 Linear B, ancient Cretan language, is deciphered.

1952 American chemist Willard F. Libby (1908–80) publishes his method of radiocarbon dating, in which amount of radioactive carbon-14 naturally occurring in rocks is measured against its half-life.

1955 In *Theory of Culture Change: The Methodology of Multilinear Evolution*, leading neo-evolutionist American anthropologist Julian Steward (1902–72) focuses on evolution as cultural adaptation.

1956 William and Lyle Boyd postulate existence of 13 distinct human races after study of blood groups.

1959 Louis Leakey discovers remains of ancestral humanoid (*Australopithecus robustus*) who lived about 1.75 million years ago.

1960 Leakey next finds remains of another species of ancestral man, *Homo habilis*, who probably lived 1.6 million to 1.9 million years ago.

1962 Essays by British anthropologist Victor Turner (b. 1920) on African ritual in Ndembu society are published as *Forest of Symbols*, setting tenor for symbolic studies in 1970s.

1967 Thirty-million-year-old skull of primate *Aegyptopithecus* is discovered by Elwyn Simons. Primate is oldest known belonging to hominid line.

1970 *Homo Hierarchicus* by French anthropologist Louis Dumont is published. It becomes definitive work on caste system in India.

1973 *Interpretation of Culture* by Clifford Geertz is published. It holds that symbols have value through their use in social interaction, as opposed to personal interpretation.

1974 Skeletal remains of new species of ancestral man, *Australopithecus afarensis*, are found by archaeological team led by Don Johnson and Maurice Taieb. Nicknamed Lucy, this ancestor of man belongs to species who probably lived 3.5 million to 5.5 million years ago.

1975 Farmers digging well in Shensi Province, China discover underground tomb of Chinese Emperor Ch'in Shih Huang Ti (d. 210 B.C.). Tomb contains army of 7,500 life-size terra-cotta human figures, placed there to guard emperor in death.

1981 Ice Age religious sanctuary is discovered in mountain cave in northern Spain. American and Spanish archaeologists determine that limestone altar and remains or burned offerings indicate that Neanderthals ritually worshiped a deity.

1982 Henry VIII's flagship *Mary Rose* is raised from bottom of Portsmouth Harbor, where she was sunk by French fleet on July 19, 1545. Hull, 120 feet long, is found to contain Tudor artifacts such as table settings, musical instruments, board games, leather boots, and jerkins.

1983 Fossil jawbone of *Sivapithecus*, apparently 16 million to 18 million years old, is found in Kenya.

1983 Roman baths in Bath, England, including Temple of Sulis Minerva, are excavated.

▶

1984 Underwater site known as Altit-Yam off coast of Israel is discovered. It is believed to be remains of 8,000-year-old settlement.

1985 Luxury liner *Titanic*, which sank in 1912 after colliding with iceberg, is located in deep waters of North Atlantic by French and American oceanographers. Use of remote-control camera enables scientists to take close-up photographs of wreckage.

1986 A stone stele dating from about A.D. 1 and inscribed with writing in an unknown language is found in Mexico.

1987 First undisturbed site occupied by first humans in America, the Clovis people, is discovered in Washington State.

1988 Newly discovered fossils show modern *Homo sapiens* first appeared twice as early as previously thought. Fossils found in Israel date back to 92,000 years ago.

1989 American archaeologists discover site of Babylonian city Mashkan-shapir, one of oldest cities in history. Built in c. 1840 B.C., city's remained undisturbed until this time.

The Development of Man

B.C.

5.5–3.5 million *Australopithecus afarensis*. Earliest known ancestor of man, its projecting canine teeth and long face give this extinct species an appearance much closer to that of ape family than to later species.

3–2 million *Australopithecus africanus*. Animal habitually walks on two feet and has a deep, heavy face with large molars—characteristics well suited to exploiting species' African savanna and woodland habitat.

2–1.4 million *Australopithecus robustus*. Robustus is about size of chimpanzee and is similar to *Australopithecus africanus*.

2–1.3 million *Australopithecus boisei*. Larger than *Australopithecus robustus*, this species also has greater brain size.

1.9–1.6 million *Homo habilis*. Brain size is larger than that of earlier species and changes in teeth indicate changes in diet.

1.6 million–500,000? *Homo erectus*. Characteristics are similar to those of *Homo habilis*. *Homo erectus* has thick, helmetlike skull, however, and apparently used crude stone tools and fire.

350,000–present *Homo sapiens*. All later stages of man's development belong to this species. In addition to differing physical characteristics, later stages show clear pattern of increasingly complex social organization.

100,000–40,000 *Neanderthal*. Barrel-chested and stocky-limbed subspecies with thick brow bone and flat-topped skull. Ritual burial practice of Neanderthals suggests some social organization.

40,000-30,000 *Cro-Magnon*. West European subspecies with body proportions close to those of modern man (*Homo sapiens sapiens*, who appears in various parts of the world in subsequent millennia). Among other differences, Cro-Magnon has thicker skull and somewhat larger teeth than modern man.

Chemistry

1627–91 English chemist Robert Boyle lives. He helps found modern science of chemistry (as distinct from alchemy). Boyle studies calcination of metals and develops definition of element.

1649 Arsenic is discovered by Johann Schröder (1600–1664).

1660–1734 German chemist George Stahl lives. He formulates incorrect phlogiston theory

▶

Discovering the Elements

Elements dated B.C. have been known since ancient times.

B.C. Carbon

B.C. Copper

B.C. Gold, called *juel* ("to shine") in ancient Sanskrit and *aurum* ("shining dawn") in Latin

B.C. Iron

B.C. Lead

B.C. Mercury

B.C. Silver; its ancient Assyrian name is *sarpu*

B.C. Sulfur

B.C. Tin

B.C. Zinc

A.D.

c. 1600 Antimony and bismuth, discovered by Basil Valentine (probable pseudonym for Johann Thölde)

1649 Arsenic, discovered by Johann Schröder (1600–1664)

1669 Phosphorus, discovered by Hennig Brand (d. c. 1692)

1735 Cobalt, discovered by Georg Brandt (1694–1768)

1735 Platinum, discovered by Antonio de Ulloa (1716–95)

1751 Nickel, discovered by Baron Axel F. Cronstedt (1722–65)

1766 Hydrogen, discovered by Henry Cavendish (1731–1810)

1772 Nitrogen, discovered by Daniel Rutherford (1749–1819)

1774 Chlorine and manganese, discovered by Carl W. Scheele (1742–86)

1774 Oxygen, discovered by Joseph Priestley (1733–1804)

1778 Molybdenum, discovered by Carl W. Scheele (1742–86)

1783 Tungsten, discovered by Juan Elhuyar (1754–96) and Fausto Elhuyar (1755–1833)

1784 Tellurium, discovered by Franz J. Müller, baron de Reichenstein (1740–1825)

1789 Uranium and zirconium, discovered by Martin H. Klaproth (1743–1817)

1791 Titanium, discovered by William Gregor (1761–1817)

1794 Yttrium (color TV's red phosphors), discovered by Johan Gadolin (1760–1852)

1797 Beryllium and chromium, discovered by Louis N. Vauquelin (1763–1829)

1801 Niobium, discovered by Charles Hatchett (c. 1765–1847)

1802 Tantalum, discovered by Anders G. Ekeberg (1767–1813)

1803 Cerium, discovered by Jöns Jakob Berzelius (1779–1848), Martin H. Klaproth, and Wilhelm Hisinger (1766–1852)

1803 Osmium (densest element occurring naturally), discovered by Smithson Tennant (1761–1815)

1803 Palladium and rhodium, discovered by William H. Wollaston (1766–1828)

1804 Iridium, discovered by Smithson Tennant

1807 Sodium and potassium, discovered by Sir Humphry Davy (1778–1829)

1808 Barium, calcium, strontium, and magnesium, discovered by Sir Humphry Davy

1808 Boron, discovered by Joseph-Louis Gay-Lussac (1778–1829), Louis-Jacques Thénard (1777–1857), and Sir Humphry Davy

1811 Iodine, discovered by Bernard Courtois (1777–1838)

1817 Cadmium, discovered by Friedrich Stromeyer (1776–1835)

1817 Lithium, discovered by Johan A. Arfwedson (1792–1841)

1818 Selenium, discovered by Jöns Jakob Berzelius

c. 1824 Silicon, discovered by Jöns Jakob Berzelius

1825 Aluminum, discovered by Hans C. Oersted (1777–1851)

1826 Bromine, discovered by Antoine J. Balard (1802–76)

1828 Thorium, discovered by Jöns Jakob Berzelius

1830 Vanadium, discovered by Nils G. Sefström (1787–1845)

1839 Lanthanum, discovered by Carl G. Mosander (1797–1858)

1843 Erbium and terbium, discovered by Carl G. Mosander

Discovering the Elements

1844	Ruthenium, discovered by Karl K. Klaus (1796–1864)
1860	Cesium and rubidium, discovered by Robert Bunsen (1811–99) and Gustav R. Kirchhoff (1824–87)
1861	Thallium, discovered by Sir William Crookes (1832–1919)
1863	Indium, discovered by Ferdinand Reich (1799–1882) and Theodor Richter (1824–98)
1868	Helium, discovered by Pierre Janssen (1824–1907) and Joseph N. Lockyer (1836–1920)
1875	Gallium, discovered by P. Lecoq de Boisbaudran (c. 1838–1912)
1878	Holmium, discovered by Jacques L. Soret (1827–90) and Marc Delafontaine (1838–8?)
1879	Samarium, discovered by P. Lecoq de Boisbaudran
1879	Scandium, discovered by Lars F. Nilson (1840–99)
1879	Thulium, discovered by Per T. Cleve (1840–1905)
1885	Neodymium and praseodymium, discovered by Carl von Welsbach (1858–1929)
1886	Dysprosium, discovered by P. Lecoq de Boisbaudran
1886	Fluorine, discovered by Henri Moissan (1852–1907)
1886	Gadolinium, discovered by P. Lecoq de Boisbaudran and Jean C. de Marignac (1817–94)
1886	Germanium, discovered by Clemens Winkler (1838–1904)
1894	Argon, discovered by John Strutt, Baron Rayleigh (1842–1919), and Sir William Ramsay (1852–1916)
1896	Europium, discovered by Eugène Demarçay (1852–1903)
1898	Krypton, neon, and xenon, discovered by Sir William Ramsay (1852–1916) and Morris W. Travers (1872–1961)
1898	Radium and polonium (first element discovered by radiochemical analysis), discovered by Pierre Curie (1859–1906) and Marie Curie (1867–1934)
1899	Actinium, discovered by André-Louis Debierne (1874–1949)
1900	Radon, discovered by Friedrich E. Dorn (1848–1916)
1906	Lutetium, discovered by Carl von Welsbach and Georges Urbain (1872–1938)
1907	Ytterbium, discovered by Georges Urbain and Carl von Welsbach
1917	Protactinium, discovered by Otto Hahn (1879–1968) and Lise Meitner (1878–1968)
1923	Hafnium, discovered by Dirk Coster (1889–1950) and George C. de Hevesy (1885–1966)
1925	Rhenium, discovered by Walter Noddack (1893–1960), Ida Noddack, and Otto C. Berg
1937	Technetium, discovered by Emilio Segré (b. 1905) and Carlo Perrier
1939	Francium, discovered by Marguerite Perey (b. 1909)
1940	Astatine, discovered by Emilio Segré, Dale Corson, and K. R. Mackenzie
1940	Neptunium (first transuranium element produced), discovered by Edwin M. McMillan (b. 1907) and Philip H. Abelson (b. 1913)
1940	Plutonium, discovered by Glenn T. Seaborg (b. 1912) et al.
1944	Americium, discovered by Glenn T. Seaborg et al.
1944	Curium, discovered by Glenn T. Seaborg et al.
1947	Promethium (last rare earth element found), discovered by J. A. Marinsky, L. E. Glendenin, and C. D. Coryell
1949	Berkelium, discovered by Glenn T. Seaborg et al.
1950	Californium, discovered by Glenn T. Seaborg et al.
1952	Einsteinium, discovered by Albert Ghiorso (b. 1915) et al.
1953	Fermium, discovered by Albert Ghiorso et al.
1955	Mendelevium, discovered by Albert Ghiorso et al.
1958	Nobelium, discovered by Albert Ghiorso et al.
1961	Lawrencium, discovered by Albert Ghiorso et al.
1969	Hahnium, discovered by Albert Ghiorso et al.
1969	Rutherfordium, discovered by Albert Ghiorso et al.
1974	Element 106, discovered by scientists in United States and USSR
1981	Element 107, discovered by scientists in USSR
1982	Element 109, discovered by West German scientists

of combustion, which holds that phlogiston escapes from a substance when it is burned.

1662 Boyle formulates his law on relationship of pressure and volume of gas at given temperature (Boyle's law).

1669 Phosphorus is discovered by Hennig Brand (1630–92).

1674 English chemist and physiologist John Mayow (1640–79) discovers action of oxygen in burning of metals. He also studies respiration and identifies role of oxygen in it.

1728–99 British scientist Joseph Black studies thermal properties of materials, discovering latent heat, heat required to make ice melt from solid to liquid state (or required to make boiling water change to steam).

1735 Cobalt is discovered by Georg Brandt (1694–1768).

1735 Platinum is discovered by Antonio de Ulloa (1716–95).

1746–1823 French scientist Jacques Charles lives. He formulates Charles's law on relationship of temperature and volume of gases.

1751 Nickel is discovered by Baron Axel F. Cronstedt (1722–65).

1766 English chemist and physicist Henry Cavendish (1731–1810) isolates hydrogen. He shows that water is made of oxygen and hydrogen.

1772 French scientist Antoine Lavoisier (1743–1794), often regarded as a father of modern chemistry, disproves phlogiston theory. He identifies gain or loss of oxygen as explanation for increase or decrease in weight resulting from burning substance.

1772 Nitrogen is discovered by Daniel Rutherford (1749–1819).

1774 Oxygen is discovered by Joseph Priestley (1733–1804). He calls it "dephlogisticated air." Previous discovery by Karl W. Scheele (1742–86) is not published until 1777.

1779 Lavoisier proposes name *oxygen* for Priestley's dephlogisticated air.

1779–1848 Swedish chemist Jöns Jacob Berzelius, a founder of modern chemistry, lives. Among his many contributions to chemistry are introduction of modern chemical symbols, classification of organic and inorganic chemicals, and introduction of such basic analytic equipment as filter paper, rubber tubing, and the dessicator.

c. 1783 Cavendish discovers that water is made up of hydrogen and oxygen.

1789 Uranium and zirconium are discovered by Martin H. Klaproth (1743–1817).

1797 French chemist Joseph Proust (1754–1826) develops law of constant proportions of compounds, though his theory is challenged.

1800 Italian physicist Alessandro Volta (1745–1827) invents first battery, proving that electricity can be generated by chemical action. His "voltaic pile" uses disks of silver and zinc.

1802 French chemist Joseph Gay-Lussac (1778–1829) introduces his law on gases, stating relationship of temperature and pressure of gas.

1803 Osmium (densest element occurring naturally) is discovered by Smithson Tennant (1761–1815).

1807 Sodium and potassium are discovered by British chemist Sir Humphry Davy (1778–1829). He discovers these particular elements by electrolysis and holds that this is best technique for breaking down compounds into their constituent parts.

1808 Barium, calcium, strontium, and magnesium are discovered by Davy.

1808–10 English chemist John Dalton (1766–1844) publishes his revolutionary atomic theory of matter in his *New System of Chemical Philosophy*. He holds

▶

that all elements are made of tiny atoms, each of same weight. His work further confirms Proust's theory of constant proportions.

1811 Italian physicist Amedeo Avogadro (1776–1856) proposes Avogadro's law, which holds that equal volumes of different gases contain same number of molecules, provided temperature and pressure of each is same.

1814 *Theory of Chemical Proportions and the Chemical Action of Electricity* is published by Berzelius. It suggests positively charged and negatively charged components make up compound.

1818 Berzelius publishes his table of combining proportions and atomic weights.

1825 Berzelius discovers titanium. Danish physicist Hans C. Oersted (1777–1851) discovers aluminum.

1826 Bromine is discovered by Antoine J. Balard (1802–76).

1828 German chemist Friedrich Wöhler (1800–1882) synthesizes the first organic compound, urea, from inorganic substances. His work contributes to development of field of organic chemistry.

1830 Berzelius names chemical phenomenon of isomerism (same atoms arranged in different structure produce different substances).

1833 British chemist and physicist Michael Faraday (1791–1867) formulates his law of electrolysis.

1835 Berzelius introduces name *catalysis* to describe action of catalytic agents, which aid chemical reaction but are unchanged by it.

1852 System of chemical valence for elements is proposed.

1855 Bunsen burner is popularized by German chemist Robert Bunsen (1811–99).

1858 Italian scientist Stanislao Cannizzaro (1826–1910) applies Avogadro's law to calculate atomic weights of atoms in molecules of volatile compounds.

1859 Technique of spectrum analysis of substances is developed by Robert Bunsen and Gustav Kirchoff (1824–87).

1862 English chemist develops early pyroxylin plastic substance, Parkesine.

1865 German chemist Friedrich Kekulé (1829–96) discovers ring structure of benzene, major contribution to understanding of molecular structure in organic chemistry.

1868 Celluloid is developed by U.S. inventor John Hyatt (1837–1920).

1868 Helium is discovered by Pierre Janssen (1824–1907) and Joseph N. Lockyer (1836–1920).

1869 Dmitri Mendeléev (1834–1907) first publishes his periodic table of elements.

1871 Concept of chemical chain reaction is developed.

1880s German chemist Emil Fischer (1852–1919) begins his extensive researches into sugars, purines, proteins, and enzymes. He helps lay foundation for modern biochemistry.

1884 French chemist invents first manufactured fiber, viscose rayon.

1886 Fluorine is discovered by Henri Moissan (1852–1907).

1887 Swedish chemist Svante Arrhenius (1859–1927), founder of modern physical chemistry, develops his theory of solutions of acids and bases. He holds that acids and bases in water form ions, electrically charged molecules or atoms.

1890 Discovery of acetylene chemistry.

1895 Coal-tar chemistry is developed, leading to a variety of new products.

1898 Krypton, neon, and xenon are discovered by Sir William Ramsay (1852–1916) and Morris W. Travers (1872–1961).

▶

Winners of the Nobel Prize for Chemistry

1901 Jacobs H. Van't Hoff (1852–1911, Netherlands)

1902 Emil Fischer (1852–1919, Germany)

1903 Svante A. Arrhenius (1859–1927, Sweden)

1904 Sir William Ramsay (1852–1916, Great Britain)

1905 Adolf von Baeyer (1835–1917, Germany)

1906 Henri Moissán (1852–1907, France)

1907 Eduard Buchner (1860–1917, Germany)

1908 Ernest Rutherford (1871–1937, Great Britain)

1909 Wilhelm Ostwald (1853–1932, Germany)

1910 Otto Wallach (1847–1931, Germany)

1911 Marie Curie (1867–1934, France)

1912 Victor Grignard (1871–1935, France) and Paul Sabatier (1854–1941, France)

1913 Alfred Werner (1866–1919, Switzerland)

1914 Theodore W. Richards (1868–1928, United States)

1915 Richard M. Willstätter (1872–1942, Germany)

1916 No prize awarded

1917 No prize awarded

1918 Fritz Haber (1868–1934, Germany)

1919 No prize awarded

1920 Walther H. Nernst (1864–1941, Germany)

1921 Frederick Soddy (1877–1956, Great Britain)

1922 Francis W. Aston (1877–1945, Great Britain)

1923 Fritz Pregl (1869–1930, Austria)

1924 No prize awarded

1925 Richard A. Zsigmondy (1865–1929, Germany)

1926 Theodor Svedberg (1884–1971, Sweden)

1927 Heinrich O. Wieland (1877–1957, Germany)

1928 Adolf O. R. Windaus (1876–1959, Germany)

1929 Sir Arthur Harden (1865–1940, Great Britain) and Hans von Euler-Chelpin (1873–1964, Sweden)

1930 Hans Fischer (1881–1945, Germany)

1931 Friedrich Bergius (1884–1949, Germany) and Karl Bosch (1874–1940, Germany)

1932 Irving Langmuir (1881–1957, United States)

1933 No prize awarded

1934 Harold C. Urey (1893–1981, United States)

1935 Frédéric Joliot-Curie (1900–1958, France) and Irène Joliot-Curie (1897–1956, France)

1936 Peter J. W. Debye (1884–1966, Netherlands)

1937 Walter N. Haworth (1883–1950, Great Britain) and Paul Karrer (1889–1971, Switzerland)

1938 Richard Kuhn (1900–67, Germany)

1939 Adolf F. J. Butenandt (b. 1903, Germany; declined award) and Leopold Ruzicka (1887–1976, Switzerland)

1940 No prize awarded

1941 No prize awarded

1942 No prize awarded

1943 George de Hevesy (1885–1966, Hungary)

1944 Otto Hahn (1879–1968, Germany)

1945 Artturi I. Virtanen (1895–1973, Finland)

1946 James B. Sumner (1887–1955, United States), John H. Northrop (b. 1891, United States), and Wendell M. Stanley (1904–71, United States)

1947 Sir Robert Robinson (1886–1975, Great Britain)

1948 Arne W. K. Tiselius (1902–71, Sweden)

1949 William F. Giauque (1895–1982, United States)

1950 Kurt Alder (1902–58, Germany) and Otto P. H. Diels (1876–1954, Germany)

1951 Edwin M. McMillan (b. 1907, United States) and Glenn T. Seaborg (b. 1912, United States)

1952 Archer J. P. Martin (b. 1910, Great Britain) and Richard L. M. Synge (b. 1914, Great Britain)

1953 Hermann Staudinger (1881–1965, Germany)

1954 Linus C. Pauling (b. 1901, United States)

1955 Vincent Du Vigneaud (1901–78, United States)

1956 Sir Cyril N. Hinshelwood (1897–1967, Great Britain) and Nikolai N. Semyonov (1896–1986, USSR)

1957 Sir Alexander R. Todd (b. 1907, Great Britain)

1958 Frederick Sanger (b. 1918, Great Britain)

1959 Jaroslav Heyrovsky (1890–1967, Czechoslovakia)

1960 Willard F. Libby (1908–80, United States)

1961 Melvin Calvin (b. 1911, United States)

1962 John C. Kendrew (b. 1917, Great Britain) and Max F. Perutz (b. 1914, Great Britain)

1963 Giulio Natta (1903–79, Italy) and Karl W. Ziegler (1898–1973, West Germany)

1964 Dorothy C. Hodgkin (b. 1910, Great Britain)

1965 Robert B. Woodward (1917–79, United States)

1966 Robert S. Mulliken (1896–1987, United States)

Winners of the Nobel Prize for Chemistry

1967 Manfred Eigen (b. 1927, West Germany), Ronald G. W. Norrish (1897–1978, Great Britain), and George Porter (b. 1920, Great Britain)

1968 Lars Onsager (1903–76, United States)

1969 Derek H. R. Barton (b. 1918, Great Britain) and Odd Hassel (1897–1981, Norway)

1970 Luis F. Leloir (b. 1906, Argentina)

1971 Gerhard Herzberg (b. 1904, Canada)

1972 Christian B. Anfinsen (b. 1916, United States), Stanford Moore (1913–82, United States), and William H. Stein (b. 1911, United States)

1973 Ernst Otto Fischer (b. 1918, West Germany) and Geoffrey Wilkinson (b. 1921, Great Britain)

1974 Paul J. Flory (1910–85, United States)

1975 John Cornforth (b. 1917, Great Britain) and Vladimir Prelog (b. 1906, Switzerland)

1976 William N. Lipscomb (b. 1919, United States)

1977 Ilya Prigogine (b. 1917, United States)

1978 Peter Mitchell (b. 1920, Great Britain)

1979 Herbert C. Brown (b. 1912, United States) and Georg Wittig (b. 1897, West Germany)

1980 Paul Berg (b. 1926, United States), Walter Gilbert (b. 1932, United States), and Frederick Sanger (b. 1918, Great Britain)

1981 Fukui Kenichi (b. 1918, Japan) and Roald Hoffmann (b. 1937, United States)

1982 Aaron Klug (b. 1926, Great Britain)

1983 Henry Taube (b. 1915, United States)

1984 Bruce Merrifield (b. 1921, United States)

1985 Herbert A. Hauptman (b. 1917, United States) and Jerome Karle (b. 1918, United States)

1986 Dudley R. Herschbach (b. 1932, United States), Yuan T. Lee (b. 1936, United States), and John C. Polanyi (b. 1929, Canada)

1987 Donald J. Cram (b. 1919, United States), Charles J. Pederson (b. 1904, United States), and Jean-Marie Lehn (b. 1939, France)

1988 Johann Deisenhofer (b. 1948, West Germany), Robert Huber (b. 1937, West Germany), and Hartmut Michel (b. 1948, West Germany)

1989 Sidney Altman (b. 1939, United States) and Thomas R. Cech (b. 1947, United States)

1898 Radium and polonium are discovered by Pierre Curie (1859–1906) and Marie Curie (1867–1934).

1901 First synthetic vat dye is manufactured.

1904 Silicones are synthesized.

1905 Chemical structure of chlorophyll is discovered.

1906 Technique of paper chromatography, later important for study of organic molecules, is introduced.

1908 Swiss chemist discovers cellophane.

1909 First totally synthetic plastic, Bakelite, is invented by Belgian-born American chemist Leo Baekeland (1863–1944).

1909 System for measuring acidity and alkalinity using pH numbers is developed.

1910 Process of refining oil by thermal cracking is developed.

1916 Theory of shared electrons is developed to explain chemical bonding.

1920 German scientist Hermann Staudinger (1881–1965) accurately describes molecular makeup of polymers (chemical compounds made up of chainlike molecules), leading to greater understanding and development of plastics.

1923 Ultracentrifuge, used for analysis in organic chemistry, is invented.

1927 Cellulose acetate is developed.

1927 Polyvinyl chloride is developed.

1928 Diels-Alder reaction is introduced. It links atoms into molecules and finds applications in making synthetics.

1930 Freon is discovered. It becomes widely used as refrigerant.

1930 Electrophoresis, technique for studying proteins, is developed.

1931 Synthetic rubber called neoprene is discovered.

▶

1931 U.S. chemist Linus Pauling (b. 1901) introduces chemical theory of resonance to explain bonding of atoms in certain molecules, notably benzene.

1933 Heavy water (deuterium oxide) is discovered.

1934 Invention of the pH meter for electronic measurement of acidity and akalinity.

1936 Introduction of catalytic cracking technique for refining oil.

1937 Polyurethane is patented, in Germany.

1938 Nylon, discovered in 1934 by U.S. chemist Wallace Carothers (1896–1937), is first manufactured.

1938 Discovery of Teflon (polytetrafluoroethylene) by U.S. scientists at Du Pont.

1939 DDT is discovered to be effective insecticide.

1939 Polyvinyl chloride is developed.

1940 Radioactive tracer carbon-14 is discovered.

c. 1955 Italian scientist Giulio Natta (1903–79) discovers that Karl W. Ziegler's (1898–1973) earlier use of catalysts to produce strong polyethylene works with many other plastic materials. (Ziegler and Natta share 1963 Nobel Prize for work on polymers.)

1957 Polypropylene, lightweight plastic, is developed.

c. 1964 Simplified technique for synthesizing proteins is introduced by American researcher Bruce Merrifield (b. 1921). It soon is adapted to automated machines and becomes important in gene synthesis in 1980s.

1970 Human growth hormone is synthesized.

1981 "Glassy" metal alloys are created by rapidly cooling molten metal. Technique prevents normal crystallization that takes place during cooling and produces new characteristics, such as high strength combined with lightness.

1983 American Chemical Society reports the number of chemicals it has recorded to date has reached 6 million. Millions of others are believed known but not formally recorded.

1983 Group Transfer Polymerization (GTP) process is developed. It becomes first polymerization process introduced in decades.

1984 MEEP, polymer that conducts electricity, is discovered. It makes possible efficient, lightweight batteries.

1985 Discovery of lanxides, new material with characteristics of both metal and ceramic, is made public in United States.

Physics

B.C.

c. 624–c. 548 Greek philosopher Thales of Miletus lives. He postulates incorrectly that water is basis of all matter in universe.

c. 460–370 Greek philosopher Democritus lives. He proposes that matter is made up of atoms, but this theory is not widely accepted.

427–347 Plato lives. He postulates that there are four elements of all matter (earth, water, air, and fire).

384–322 Aristotle, perhaps most influential of all early Greek philosopher-scientists, lives.

▶

By his voluminous observations of natural world, he lays foundations for modern natural science. He believes incorrectly that heavy objects fall faster than light ones, however.

287–212 Archimedes, Greek inventor and mathematician, lives. He invents Archimedes screw and discovers Archimedes principle of buoyancy and principle of lever. Of the latter, he is traditionally believed to have said, "Give me a place to stand, and I will move the world."

A.D.

85–165 Ptolemy, Hellenistic Greek astronomer and mathematician, lives. He advances theory of earth-centered universe, which is accepted for centuries.

c. 1010–30 Parabolic mirrors are developed by Arabian physicist Alhazen (965–1039), who also provides correct explanation of how lenses work.

c. 1220–30 Jordanus Nemorarius formulates early law of lever and law of composition of movements.

1304 Theodonic of Fribourg (c. 1250–c. 1310) correctly explains many aspects of rainbow formation.

1490 Capillary action is first described by Leonardo da Vinci (1452–1519).

1543 Polish astronomer Nicolaus Copernicus (1473–1543) publishes his *De Revolutionibus Orbium Coelestium,* refuting Ptolemy's Earth-centered system and proposing sun-centered system. It is not widely accepted for some time.

1564–1642 Italian astronomer and physicist Galileo Galilei lives. He disproves Aristotle's theory that objects of different weights fall at different speeds (1590), discovers the principle of the pendulum, and finds the path of a projectile is a parabola. Galileo's work provides the basis for modern mechanics.

1586 Dutch mathematician Simon Stevinus (1548–1620) anticipates Galileo's later finding by publishing results of experiment with falling bodies. He drops two lead spheres of different weights and observes that they hit ground at same time.

1593 Galileo constructs crude thermometer that uses air rather than liquid.

1600 Basic principles of magnetism and Earth's magnetic field are described in pioneering work *Concerning Magnetism, Magnetic Bodies, and the Great Magnet Earth* by English physician William Gilbert (1544–1603).

1604 Galileo learns that falling body increases its distance as square of time.

1609–19 German astronomer Johannes Kepler (1571–1630) publishes his three laws of planetary motion, based on his discovery that planets travel in elliptical, not circular, orbits. Kepler's findings support Copernicus's theory that planets revolve around sun.

1632 Galileo completes his *Dialogue Concerning Two World Systems.* In it he backs Copernicus's theory of a heliocentric system.

1640 Evangelista Torricelli (1608–47) becomes father of hydrodynamics by applying laws of motion to liquids.

1644 French philosopher René Descartes (1596–1650) publishes *Principles of Philosophy,* in which he describes all natural phenomena of world and universe in mechanistic terms. Descartes's earlier work on philosophy, *Discourse on Method* (1637), also contains discoveries on law of refraction of light and on cause of rainbows.

1661 English scientist Robert Boyle (1627–91) advances theory that all matter is made up of substance that clusters together to form corpuscles, building blocks of matter.

1662 Boyle publishes his findings on fundamental relationship of pressure and volume of gas (Boyle's law).

1668 Famed English mathematician and physicist Isaac Newton (1642–1727) invents reflecting telescope. (Today's big telescopes are all reflecting telescopes.)

▶

Winners of the Nobel Prize for Physics

1901 Wilhelm C. Roentgen (1845–1923, Germany)

1902 Hendrik A. Lorentz (1853–1928, Netherlands) and Pieter Zeeman (1865–1943, Netherlands)

1903 Pierre Curie (1859–1906, France), Marie Curie (1867–1934, France), and Antoine H. Becquerel (1852–1908, France)

1904 John W. Strutt (1842–1919, Great Britain)

1905 Philipp E. A. von Lenard (1862–1947, Germany)

1906 Sir Joseph J. Thomson (1856–1940, Great Britain)

1907 Albert A. Michelson (1852–1931, United States)

1908 Gabriel Lippmann (1845–1921, France)

1909 Karl F. Braun (1850–1918, Germany) and Guglielmo Marconi (1874–1937, Italy)

1910 Johannes van der Waals (1837–1923, Netherlands)

1911 Wilhelm Wien (1864–1928, Germany)

1912 Nils G. Dalén (1869–1937, Sweden)

1913 Hedike K. Onnes (1853–1926, Netherlands)

1914 Max von Laue (1879–1960, Germany)

1915 Sir W. Lawrence Bragg (1890–1971, Great Britain) and Sir William H. Bragg (1862–1942, Great Britain)

1916 No prize awarded

1917 Charles G. Barkla (1877–1944, Great Britain)

1918 Max Planck (1858–1947, Germany)

1919 Johannes Stark (1874–1957, Germany)

1920 Charles E. Guillaume (1861–1938, Switzerland)

1921 Albert Einstein (1879–1955, Germany)

1922 Niels Bohr (1885–1962, Denmark)

1923 Robert A. Millikan (1868–1953, United States)

1924 Karl Siegbahn (1886–1978, Sweden)

1925 James Franck (1882–1964, Germany) and Gustav Hertz (1887–1975, Germany)

1926 Jean B. Perrin (1870–1942, France)

1927 Arthur H. Compton (1892–1962, United States) and Charles T. R. Wilson (1869–1959, Great Britain)

1928 Sir Owen W. Richardson (1879–1959, Great Britain)

1929 Louis V. de Broglie (1892–1987, France)

1930 Sir Chandrasekhara V. Raman (1888–1970, India)

1931 No prize awarded

1932 Werner Heisenberg (1901–76, Germany)

1933 Paul A. M. Dirac (1902–84, Great Britain) and Erwin Schrödinger (1887–1961, Austria)

1934 No prize awarded

1935 Sir James Chadwick (1891–1974, Great Britain)

1936 Victor F. Hess (1883–1964, Austria) and Carl D. Anderson (b. 1905, United States)

1937 Clinton J. Davisson (1881–1958, United States) and Sir George P. Thomson (b. 1892, Great Britain)

1938 Enrico Fermi (1901–54, Italy)

1939 Ernest O. Lawrence (1901–58, United States)

1940 No prize awarded

1941 No prize awarded

1942 No prize awarded

1943 Otto Stern (1888–1969, United States)

1944 Isidor I. Rabi (1898–1988, United States)

1945 Wolfgang Pauli (1900–58, United States)

1946 Percy W. Bridgman (1882–1961, United States)

1947 Sir Edward V. Appleton (1892–1965, Great Britain)

1948 Patrick M. S. Blackett (1897–1974, Great Britain)

1949 Hideki Yukawa (1907–81, Japan)

1950 Cecil F. Powell (1903–69, Great Britain)

1951 Sir John D. Cockcroft (1897–1967, Great Britain) and Ernest T. S. Walton (b. 1903, Ireland)

1952 Felix Bloch (1905–83, United States) and Edward M. Purcell (b. 1912, United States)

1953 Frits Zernike (1888–1966, Netherlands)

1954 Max Born (1882–1970, Great Britain) and Walther Bothe (1891–1957, German-born)

1955 Willis E. Lamb, Jr. (b. 1913, United States) and Polykarp Kusch (b. 1911, United States)

1956 William B. Shockley (b. 1910, United States), John Bardeen (b. 1908, United States), and Walter H. Brattain (b. 1902, United States)

1957 Tsung-dao Lee (b. 1926, China) and Chen Ning Yang (b. 1922, China)

1958 Pavel A. Cherenkov (b. 1904, USSR), Ilya M. Frank (b. 1908, USSR), and Igor Y. Tamm (1895–1971, USSR)

1959 Emilio G. Segrè (1905–89, United States) and Owen Chamberlain (b. 1920, United States)

1960 Donald A. Glaser (b. 1926, United States)

Winners of the Nobel Prize for Physics

1961 Rudolf L. Mössbauer (b. 1929, German-born) and Robert Hofstadter (b. 1915, United States)

1962 Lev D. Landau (1908–68, USSR)

1963 J. Hans D. Jensen (1907–73, Germany), Maria Mayer (1906–72, United States), and Eugene P. Wigner (b. 1902, United States)

1964 Nikolai G. Basov (b. 1922, USSR), Aleksandr M. Prokhorov (b. 1916, USSR), and Charles H. Townes (b. 1915, United States)

1965 Richard Feynman (1918–88, United States), Julian S. Schwinger (b. 1918, United States), and Shinichiro Tomonaga (1906–79, Japan)

1966 Alfred Kastler (1902–84, France)

1967 Hans A. Bethe (b. 1906, United States)

1968 Luis W. Alvarez (b. 1911, United States)

1969 Murray Gell-Mann (b. 1929, United States)

1970 Louis E. F. Néel (b. 1904, France) and Hannes O. G. Alfvén (b. 1908, Sweden)

1971 Dennis Gabor (1900–79, Great Britain)

1972 John Bardeen (b. 1908, United States), John R. Schrieffer (b. 1931, United States), and Leon N. Cooper (b. 1930, United States)

1973 Leo Esaki (b. 1925, Japan) and Ivar Giaever (b. 1929, United States)

1974 Sir Martin Ryle (1918–84, Great Britain) and Anthony Hewish (b. 1924, Great Britain)

1975 James Rainwater (1917–86, United States), Ben Mottelson (b. 1926, Denmark), and Aage Bohr (b. 1922, Denmark)

1976 Burton Richter (b. 1931, United States) and Samuel Chao Ching Ting (b. 1936, United States)

1977 John H. Van Vleck (1899–1980, United States), Nevill F. Mott (b. 1905, Great Britain), and Philip W. Anderson (b. 1922, United States)

1978 Arno A. Penzias (b. 1933, United States), Robert W. Wilson (b. 1936, United States), and Pëtr Kapitsa (1894–1984, USSR)

1979 Sheldon L. Glashow (b. 1933, United States), Steven Weinberg (b. 1933, United States), and Abdus Salam (b. 1926, Pakistan)

1980 James W. Cronin (b. 1931, United States) and Val L. Fitch (b. 1923, United States)

1981 Kai M. Siegbahn (b. 1918, Sweden), Arthur L. Schawlow (b. 1929, United States), and Nicolaas Bloembergen (b. 1920, United States)

1982 Kenneth G. Wilson (b. 1936, United States)

1983 William A. Fowler (b. 1911, United States) and Subrahmanyan Chandrasekhar (b. 1910, United States)

1984 Carlo Rubbia (b. 1934, Italy) and Simon van der Meer (b. 1925, Netherlands)

1985 Klaus von Klitzing (b. 1943, West Germany)

1986 Ernst Ruska (b. 1906, West Germany), Heinrich Rohrer (b. 1933, Switzerland), and Gerd Binnig (b. 1947, Germany)

1987 Karl A. Müller (b. 1927, Switzerland) and J. Georg Bednorz (b. 1950, Germany)

1988 Leon Lederman (b. 1922, United States), Melvin Schwartz (b. 1932, United States), and Jack Steinberger (b. 1921, United States)

1989 Norman Ramsey (b. 1915, United States) and Hans Dehmelt (b. 1922, United States)

1676 Olaf Römer (1644–1710) is first to calculate value for speed of light, based on observations of eclipses of moons of Jupiter.

1687 Newton publishes his landmark work *Principia,* which introduces his three laws of motion and concept of universal gravitation. He explains motion of heavenly bodies (celestial mechanics) and of objects on earth. As part of this work, Newton develops new branch of mathematics, calculus.

1690 Wave theory of light, anticipating modern quantum theory of light, is published by Huygens in his *Treatise on Light.*

1704 *Optics* is published by Newton. He discusses his earlier finding that white light is made up of the colors of the spectrum and theorizes that light is made up of particles called corpuscles.

▶

1714 German physicist Gabriel Fahrenheit (1686–1736) constructs mercury thermometer and develops Fahrenheit temperature scale.

1731–1810 English physicist and chemist Henry Cavendish's work on electricity (unpublished until 1879) anticipates discoveries of Coulomb, Faraday, and Ohm.

1733 French scientist discovers that materials charged with similar kinds of electricity repel one another and that those charged with opposite kinds of charge attract each other.

1738 Swiss scientist Daniel Bernoulli (1700–82) publishes *Hydrodynamica*, putting forward now famous Bernoulli principle concerning velocity of flow of fluids and gases.

1742 Celsius scale is devised by Swedish astronomer Anders Celsius (1701–44).

1745 Leyden jar, device in which electrical charge can be collected and briefly stored, is developed.

1752 American inventor Benjamin Franklin (1706–90) experiments with kite in thunderstorm to prove that lightning is electricity. He gives terms *positive* and *negative* to different electrical charges.

1762 Scottish chemist and physicist Joseph Black (1728–99) develops theory of "latent heat." He later develops theory of specific heat. His work provides basis for caloric theory of heat (heat as kind of fluid).

1775 Italian physicist Alessandro Volta (1745–1827) invents device that can produce and store charges of static electricity.

1785 French physicist Charles Coulomb (1736–1806) describes mathematically how positive and negative electrical charges attract and repel one another. His theory is known as Coulomb's law.

1786 Italian physician Luigi Galvani (1737–98) observes contractions in severed frog legs and that twitches also are caused by certain metals. He mistakenly theorizes animal tissues produce electricity, but his research leads to study of electric currents.

1788 French mathematician Joseph-Louis Lagrange (1736–1813) publishes *Analytical Mechanics*, expanding on Newton's mechanics.

1791 Pierre Prevost (1751–?) demonstrates that cold is absence of heat and that all bodies continuously give off heat.

1797–98 First calculation of gravitational constant (G in Newton's theory) results from Cavendish experiment, performed by Henry Cavendish (1731–1810). This makes possible calculation of Earth's mass.

1798 Theory of heat as increased motion of particles of heated object is demonstrated experimentally by British physicist Benjamin Thompson (1753–1814). He disproves earlier theory of heat as caloric, type of fluid.

1800 Volta disproves Galvani's animal-electricity theory, showing that chemical action of moisture and two different metals can generate flow of electricity. He makes world's first battery, "voltaic pile," using disks of silver and zinc. Electric volt is named for Volta.

1801 English scientist Thomas Young (1773–1829) reintroduces wave theory of light developed earlier by Dutch physicist Christiaan Huygens (1629–95).

1802 French chemist Joseph Gay-Lussac (1778–1850) publishes his theory on relationship of temperature and expansion of gases.

1807 English chemist Humphry Davy (1778–1829) makes important strides in new field of electrochemistry, following Volta's discovery that electric current can decompose water.

1807 Young develops coefficients of elasticity of materials.

1808–10 Modern theory of atomic structure of matter is introduced by British physicist and chemist John Dalton (1766–1844). He makes early effort at establishing atomic weights and sets weight of hydrogen at 1.

▶

1819 French physicists prove experimentally relationship of specific heat of substance and its atomic weight.

1820 Danish physicist Hans Christian Oersted (1777–1851) demonstrates important relationship between electricity and magnetism (that compass needle is deflected when placed near electric current), beginning study of electromagnetism.

c. 1820 French physicist André-Marie Ampère (1775–1836) continues Oersted's work, demonstrating that two parallel wires carrying electrical current in same direction will attract each other, whereas if currents travel in opposite directions, wires repel one another. He invents the static needle. (Electric current unit ampere is named after him.)

1821 Ampère and others have developed galvanometers for measuring current, its direction, and its strength.

1821 English scientist Michael Faraday (1791–1867) invents instruments to demonstrate effects of magnetic "field of force" around current and magnet; this becomes basic principle of electric motor.

1821 Work by Augustin Fresnel (1788–1827) on polarized light provides further support for wave theory of light.

1823 British invent first electromagnet, iron bar with current-carrying wire wrapped around it.

1824 *Reflections on the Motive Power of Fire* is published by French scientist Nicholas (Sadi) Carnot (1796–1832). This study of physical processes at work in steam engines later provides basis for first and second laws of thermodynamics.

1826 German physicist Georg Simon Ohm (1787–1854) describes electrical conduction in solids, from which he formulates Ohm's law (for measuring electrical current). Ohm, unit of electrical resistance, is named for him.

1827 Brownian motion of particles in fluids is discovered by botanist Robert Brown (1773–1858).

c. 1830 Swedish chemist Jöns Jakob Berzelius (1779–1848) determines systematic group of atomic weights. He introduces H_20 as formula for water.

1831 Faraday discovers electromagnetic induction, phenomenon by which moving magnet can generate electric current. It is also discovered independently by American physicist Joseph Henry (1797–1878). Faraday builds the first crude electric generators.

1840 British physicist James Joule (1818–98) publishes Joule's law, concerning relationship of both current flow and resistance in wire to heat produced.

1842 Doppler effect is described by Austrian physicist Christian Doppler (1803–53).

1843 German physicist Julius Mayer (1814–78) discovers direct relationship between heat and mechanical work, showing they are different forms of energy. Joule later also discovers mechanical equivalent of heat.

1848 Red shift, in which light moving away from observer shifts to red side of spectrum, is observed.

1848 Lord Kelvin (William Thomson, 1824–1907) introduces his absolute temperature scale, today main temperature scale used by scientists.

1850 French physicist Jean Foucault (1819–68) establishes accurate (to 1 percent) value for speed of light.

1850 German physicist Rudolf Clausius (1822–88) formulates first and second laws of thermodynamics, thus pioneering field of thermodynamics.

1851 Kelvin publishes paper synthesizing earlier work of Joule and Carnot, introducing another version of second law of thermodynamics.

1852 Joule and Kelvin discover that gas under pressure cools if allowed to expand, which later makes modern refrigeration equipment possible.

1859 Spectrum analysis is developed by German chemist Robert Bunsen (1811–99)

▶

and German physicist Gustav Kirchhoff (1824–87).

1864 Scottish physicist James Maxwell (1831–79) unifies known facts about electricity into concise, four-equation mathematical theory, providing framework for understanding all electric and magnetic phenomena in nature.

1869 Periodic law of elements is introduced by Russian chemist Dmitri Mendeleev (1834–1907). He groups elements by atomic weight to create early periodic table of elements.

1881 English physicist Joseph John Thompson (1856–1940) discovers electromagnetic mass.

1886 German physicist Heinrich Hertz (1857–94) confirms Maxwell's theories by generating and detecting electromagnetic waves, leading to development of radio, television, and other media.

1895 X-rays are discovered by German physicist Wilhelm C. Roentgen (1845–1923).

1896 French scientist Henri Becquerel (1852–1909) discovers radioactivity.

1897 J. J. Thomson discovers electron, first known particle smaller than atom.

1898 French scientists Pierre Curie (1859–1906) and Marie Curie (1867–1934) discover radioactive element radium.

1900 Alpha, beta, and gamma radiation are recognized.

1900 German physicist Max Planck (1858–1947) formulates his quantum theory to explain discrepancies in classical theories of physics. With Einstein's theories, this revolutionizes physics.

1903 Radioactivity is explained as disintegration of atom.

1905 German physicist Albert Einstein (1879–1955) proposes his special theory of relativity to explain discrepancies in classical theories of physics. His second paper on subject contains famous equation $E = mc^2$. He also postulates light quantum, theory that light may behave like particles.

1910 J. J. Thomson confirms existence of isotopes.

1911 Ernest Rutherford (1871–1937) suggests model of atom consisting of nucleus (positive charge) surrounded by electrons (negative charge).

1911 Dutch scientist discovers that mercury loses all electrical resistance at temperatures close to absolute zero, beginning study of phenomenon of superconductivity.

1912 Wilson cloud chamber is devised, providing visual evidence of alpha particle emissions.

1913 Danish physicist Niels Bohr (1885–1962) proposes new model of atom that incorporates elements of quantum theory. It stipulates specific orbits for electrons and relationship of change in orbit to energy loss or gain.

1915 Einstein completes his general theory of relativity, in which he holds that gravity creates curved universe in space-time continuum.

1918 Third law of thermodynamics is developed by German chemist Walther Nernst (1864–1941).

1919 Rutherford discovers proton, which he proves is part of nucleus of atom. His experimental bombardment of nitrogen with alpha particles, which produces oxygen isotope, points way for later research into fission.

1925 Austrian physicist Wolfgang Pauli (1900–1958) develops his exclusion principle, which further clarifies Bohr's atomic model.

1926 British physicist Paul Dirac (1902–84) simplifies and further extends field of quantum mechanics. In 1929 he develops fusion of quantum mechanics and relativity, which predicts existence of what are later called positrons and antiprotons.

1930 Superfluidity of helium is discovered at −456 degrees F.

▶

1930 Italian physicist Enrico Fermi (1901–54) develops well-defined theory of nuclear magnetic moment.

1931 Cyclotron is invented by American physicist Ernest Lawrence (b. 1901).

1931 First electron microscope is invented.

1932 Carl Anderson discovers first type of antimatter, positively charged electron called positron. Existence of antimatter had first been postulated by Paul Dirac in 1931.

1932 British scientists artificially split atom for first time by bombardment with accelerated protons.

1932 Neutrons are identified for first time, although their existence was first proposed in 1920.

1933 Meissner effect is discovered. Superconductoring materials are found to lose a magnetic field in their interiors when cooled to superconducting temperatures.

1933 Improved electron microscope is built.

1934 Cherenkov effect is discovered: when particle passing through liquid or transparent solid reaches speed greater than light in medium, it emits light.

1935 U–235 isotope of uranium is discovered. It becomes important in nuclear fission.

1935 Existence of *pi* meson (pion) is postulated by Hideki Yukawa.

1935 Helium is cooled to just 0.1° above absolute zero.

1937 Muon, electronlike particle over 200 times more massive than electron, is discovered.

1937 First artificial element is produced (technetium).

1938 Uranium atom is split. This step leads to possibility of chain reaction and to development of atomic bombs.

1942 Enrico Fermi and colleagues produce first controlled fission chain reaction.

1948 Renormalizable quantum electrodynamics (QED) is developed by American physicist Richard Feynman (b. 1918) to resolve problems with aspects of earlier quantum theory.

1951 Synchrotron particle accelerator is built in United States.

1952 H-bomb is built by physicist Edward Teller (b. 1908) and team of scientists.

1953 Maser is developed in United States.

1954 Bevatron particle accelerator is built at University of California at Berkeley.

1955 Antiproton is first observed. Antineutron is observed in 1956.

1955 First images of individual atoms are produced by newly invented field ion microscope.

1956 Neutrino, particle without mass or charge that was postulated theoretically in 1930, is discovered.

1957 Theoretical explanation for phenomenon of superconductivity is advanced (BCS theory).

1960 First laser is developed by American scientist Theodore Maimen (b. 1927).

1961 Inner structure of protons and neutrons is discovered. Two meson shells (with aggregate negative charge) surround central core (positively charged).

1961 Beginnings of quark theory emerge with method of classifying heavy subatomic particles.

1967 Electroweak theory is developed, unifying electromagnetic and electroweak forces by postulating existences of three new subatomic particles (discovered in 1983).

1973 American scientists build continuous-wave laser that can be tuned.

1974 First theoretical unification of electromagnetic, weak, and strong forces is attempted in the first grand unified theory (GUT).

1975 Tauon, very large muon, is discovered.

1978 Wolfgang Pauli discovers that average lifetime of neutron is 15 minutes.

1979 Gluon, subatomic particle that carries strong force, is discovered.

▶

1980 Scanning tunneling microscope, which produces images of individual atoms on material's surface, is built.

1983 Subatomic particles predicted by electroweak theory (two W particles and Z particle) are discovered.

1986 Quantum jumps in single atoms are observed for first time.

1987 First of "warm-temperature superconductors" is discovered. Material becomes superconductor at −321 degrees F, temperature of liquid nitrogen.

1988 Positron transmission microscope is built.

1988 Electric motor that employs new "warm-temperature superconductors" is demonstrated.

1989 Cold-fusion experiment at University of Utah becomes center of controversy over feasibility of producing fusion (and thus energy) at room temperature. Scientists at other universities later discredit cold fusion after attempting to duplicate the experiment.

1990 IBM researchers use scanning tunneling microscope to move individual atoms on a surface for first time.

Mathematics

B.C.

c. 3500–c. 3000 Needs of agriculture and business in Egypt, Mesopotamia, India, and China give rise to primitive numerical systems.

2400 Sumerians develop mathematics based on number 60 (sexagesimal).

c. 1800 Babylonian multiplication tables are written, among earliest known mathematical tables.

1750 Egyptian Moscow papyrus is written. It includes text on Egyptian system of geometry.

c. 1700 Babylonians make use of squares and square roots, cubes and cube roots, and quadratic equations and have calculated approximate value for *pi* by this time.

c. 1700 Egyptian Rhind papyrus is written. It contains record of Egyptian ciphered numeral system and early mathematical table of values of fractions.

c. 1350 Chinese incorporate decimals into their numerical system.

c. 1000 Chinese develop counting boards, forerunner of abacus.

876 Earliest use of symbol for 0, in India.

c. 525 Greek mathematician Pythagoras (c. 582–507 B.C.) founds Pythagorean school. Pythagoreans make important contributions to mathematics. Pythagorean theorem of right triangles is popularly attributed to them.

c. 400 Greek mathematician Eudoxus of Cnidus (c. 408–355 B.C.) develops geometric theory of irrational numbers.

323 Greek mathematician Euclid (fl. fourth century B.C.) completes his landmark work *Elements* on what is today known as Euclidean geometry. He establishes such basic concepts in geometry as point and line.

287–212 Greek mathemetician Archimedes lives. He formulates displacement principle and calculates *pi*.

c. 260 Roman numeral system, dominant number system in West into medieval times, has reached advanced stage. Romans use it to represent numbers into millions.

c. 100 Chinese mathematicians use negative numbers.

▶

A.D.

c. 140 Alexandrian astronomer and mathematician Ptolemy completes his 13-volume *Almagest*, which includes writings on trigonometry and table of chords.

c. 190 Chinese calculate *pi* to 3.14159.

c. 250 Diophantus of Alexandria, often called father of algebra, produces first book on subject.

271 Chinese use first simple compass for drawing circles.

c. 400 Hindus develop number system based on the number 10, modern system.

c. 680 Symbol "0" is first used for zero in Sumatra and Cambodia.

780–c. 850 Indian mathematician Al-Khawarizmi lives. A latin translation of his work *Kitab al jabr wa'l muqabala* on mathematics gives rise to term *algebra*. Translation of another work (c. 1150) introduces Hindu-Arabic numeral system (in use today) to Europe, then still using Roman numerals.

1202 Italian mathematician Leonardo Fibonacci (1170–1230) writes *Liber Abaci*, which introduces Islamic algebra and modern "0" symbol to Europe.

c. 1300 Decimal system is introduced in Europe.

1514 First use of plus (+) and minus (−) signs for addition and subtraction in equations.

1525 Square-root symbol is introduced by Christoff Rudolf (c. 1500–45) in *Die Cass*, important also for its use of decimal fractions.

1545 First solutions to cubic equations are published by Geronimo Cardano (1501–76).

1557 First known use of the modern equals (=) sign, by English mathematician Robert Recorde.

1564–1642 Galileo Galilei, Italian astronomer and mathematician, lives. He develops mathematical formulations for many natural phenomena.

1572 Complex numbers are first used to solve equations.

1585 Simon Stevin (1548–1620) of Holland establishes use of decimal fractions and discovers rules for locating roots of equations.

1591 French mathematician François Vieta (1540–1603) introduces algebraic sign language. He uses consonants for known quantities and vowels for unknowns.

1614 Tables of logarithms are published by Scottish mathematician John Napier (1550–1617).

1617 First table of logarithms to base 10 is published by English mathematician Henry Briggs (1561–1631).

1621 First known slide rule is constructed by British mathematician William Oughtred (1574–1660). He uses two sliding logarithm scales.

1631 "×" symbol for multiplication is introduced.

1637 French philosopher and mathematician René Descartes (1596–1650) publishes his new analytic geometry in *Discours de la méthode*. He originates Cartesian coordinates and curves and system of exponential notation.

1640 Modern number theory is developed by French mathematician Pierre de Fermat (1601–65).

1642 Adding machine is invented by French mathematician and scientist Blaise Pascal (1623–62). He also contributes to development of differential calculus.

1654 Theory of probability is founded by Pascal and Fermat.

1659 Modern division sign (÷) is introduced.

1671 Machine to multiply and divide is invented by German mathematician and philosopher Gottfried W. Leibniz (1646–1716).

1679 Leibniz devises system that represents every number with combinations of 1 and 0, thereby introducing binary arithmetic.

1687 Newton publishes his work on calculus (which he developed in 1666) in

▶

Philosophiae naturalis principia mathematica.

1690 Mathematische Gesellschaft, oldest mathematical society still in existence, is founded in Germany.

1706 First use of Greek letter *pi* as symbol for ratio of circle's circumference to its diameter.

1731 Bernoulli numbers are introduced by posthumous publication of work of Swiss mathematician Jakob Bernoulli (1654–1705).

1733 Swiss mathematician Leonhard Euler (1707–83) publishes work that begins modern mathematical analysis. He also devises such mathematical symbols as f(x) and Σ (summation).

1736–1813 Noted French mathematician Joseph Lagrange lives.

1801 Noted German mathematician Carl F. Gauss (1777–1855) publishes his work on law of quadratic reciprocity and theory of congruences.

1812 Pierre Laplace (1749–1827), French scientist and mathematician, publishes his work on analytic probability theory.

c. 1820 English mathematician Charles Babbage (1792–1871) begins work on his mechanical computing machines.

1822 Fourier series of sines and cosines are defined in *Théorie analytique* by Frenchman J.B.J. Fourier (1768–1830).

1829 Russian mathematician Nikolai Lobachevski (1793–1856) publishes first account of non-Euclidean geometry.

1846 Paper by Evariste Galois (1811–32) on group theory is published posthumously—after having been lost or rejected numerous times by the French Academy.

c. 1850 Boolean algebra, which later becomes useful in computer science, is developed by Englishman George Boole (1815–64).

1853 Calculation of *pi* to 707 decimal places is made by William Shanks (1821–82). Only value to 527 places is believed correct.

1862 German Bernhard Riemann (1826–66) invents a non-Euclidean geometry that later proves useful to the relativity theory.

1869 Algebraic geometry is founded by British mathematician Arthur Cayley (1821–95).

1880–84 American mathematician Josiah W. Gibbs (1839-1903) evolves system of vector analysis.

1895 French mathematician Jules-Henri Poincaré (1854-1912) publishes founding work in topology.

1908 The basis of modern mathematics is established by Ernst Zermelo with his axiomatic treatment of set theory.

1913 British mathematicians and philosophers Bertrand Russell (1872–1970) and Alfred North Whitehead (1861–1947) publish *Principia Mathematica*, which influences fields of symbolic logic and set theory.

1915 American immigrant Albert Einstein (1879–1955) announces his general theory of relativity. His mathematical formulations on nature of universe revolutionize modern physics and astronomy.

1931 American immigrant Kurt Gödel (1906–78) develops Gödel's Proof, which holds that any mathematical system contains propositions that cannot be proved or disproved using given axioms.

c. 1943 Precursor of the modern electronic computer is developed in United States. In coming years, computers make it possible to do enormously complex mathematical computations in seconds.

1944 *Theory of Games and Economic Behavior* is published by American immigrants John von Neumann (1903–57) and Oskar Morgenstern (b. 1902). Their work is important in development of theory of games.

▶

1948 American mathematician Norbert Wiener (1894–1964) publishes *Cybernetics*, on his theory of computer-based artificial intelligence.

c. 1960 "New math" is introduced in U.S. public schools.

1961 U.S. scientist Edward Lorenz (b. 1917) begins work on what will become the chaos theory, which leads to new branch of mathematics.

c. 1970 Industry and science make extensive use of computer-based mathematical models in research, providing new incentives for mathematical research.

1975 Term *fractals* is coined by American mathematician Benoit Mandelbrot to describe irregular mathematical patterns and structures in nature.

1980 University-based American mathematicians complete exhaustive classification of finite simple groups, basic building blocks of modern algebra.

Earth Sciences

B.C.

780 Chinese are recording major earthquakes and inventing instruments to detect them.

c. 600 Greek philosopher Thales (624–546 B.C.) theorizes that Earth's surface is curved and postulates existence of equator. He suggests making geographic measurements in terms of circles parallel to equator (parallels) and lines perpendicular to them (meridians).

c. 575 Greek philosopher Anaximander (611–547 B.C.) states that fossil fish are remains of early life. He invents device for measuring latitude and sundial for telling time.

530 Greek philosopher Pythagoras (c. 582–507 B.C.) theorizes that Earth is sphere.

c. 450 Greek historian Herodotus (c. 485–428 B.C.) observes that land is shaped by water that once covered Earth; he is first to designate Europe, Africa, and Asia as separate continents.

c. 400 Greek philosopher Plato (c. 427–347 B.C.) incorrectly believes origin of rivers and streams to be huge underground reservoir within Earth.

c. 350 Greek philosopher Aristotle (c. 384–322 B.C.) defines torrid, temperate, and frigid climate zones. He makes important advances in knowledge of earthquakes and water erosion. (He incorrectly believes that volcanoes are caused by hot winds moving underground.)

c. 330 Greek explorer Pytheas identifies moon as cause of tides in Atlantic.

c. 300 First written study of rocks and minerals is made, by Greek philosopher Theophrastus (c. 373–287 B.C.).

c. 200 Greek mathematician Eratosthenes (c. 275–195 B.C.) calculates tilt of Earth's axis and size of Earth's circumference (to within 50 miles of true measurement).

A.D.

c. 20 Greek geographer Strabo (c. 63 B.C.–A.D. 21) writes 17-volume geography of known world.

79 Roman encyclopedist Pliny the Elder (b. A.D. 23) is killed while observing volcanic eruption of Mount Vesuvius. His nephew writes famous account of eruption and accompanying earthquake.

c. 100 Greco-Egyptian geographer Ptolemy flourishes. He is famed for his knowledge of mapping and geography, but his mistaken belief that *terra incognita* joins Africa and Asia discourages exploration around Africa for centuries.

▶

c. 132 Chinese mathematician Zhang Heng (78–139) invents first seismograph.

c. 800 Petrified wood is described by Chinese scholars.

c. 1080 Chinese scientists maintain that fossilized plants prove historical climate changes.

1086 Principles of erosion, uplift, and sedimentation are described.

1492 Italian explorer Christopher Columbus (1451–1506) voyages westward from Spain in search of China. His voyage is first of many at about this time that lead to new knowledge about world geography.

1512 Polish astronomer Nicholas Copernicus (1473–1543) suggests that Earth rotates on its axis every 24 hours and revolves around sun.

1520 Portuguese navigator Ferdinand Magellan (c. 1480–1521) travels around world, proving continuity of oceans as well as fact that Earth is round.

1600 British physician William Gilbert (1544–1603) describes Earth's magnetic properties, comparing it to large spherical magnet.

▶

Major Geologic Epochs

Million B.C.	
5,000–4,500	Earth is formed along with rest of the solar system.
4,500–2,500	Early Precambrian period. Earth's core, mantle, and crust form. Formation of crystalline rock on Earth's surface probably begins.
2,500	Five protocontinents exist, including Gondwanaland (which later breaks up to form Antarctica, Africa, South America, India, and Australia), China, Siberia, Europe, and North America. Continents are not stationary, however, and "continental drift" eventually leads to formation of modern landmasses.
2,500–570	Late Precambrian period, characterized by metamorphosed sedimentary rocks, lava flows, and granite. Oldest known rocks date from this time or earlier. Oldest known living organisms, bacteria and non-photosynthetic algae, are fossilized. Photosynthetic organisms appear. They begin producing oxygen now found in Earth's atmosphere.

Paleozoic Era

Million B.C.	
570–500	Cambrian Period. Appearance of marine shelled arthropod trilobite marks beginning of this period. Shallow seas cover large areas of continents.
570–400	Protocontinents of China, Siberia, Europe, and North America drift toward each other until they form giant continent of Laurasia. Gondwanaland is in Southern Hemisphere and Laurasia is in Northern Hemisphere.
500–435	Ordovician Period. Inland seas of North America reach greatest extent.
435–395	Silurian Period. Shallow seas again flood much of continental landmasses, except in Africa.
400–250	Supercontinents of Laurasia and Gondwanaland join, forming massive continent called Pangea.
395–345	Devonian Period. First terrestrial life appears. Large areas of North America, Europe, and parts of other continents are again flooded for time.
345–280	Carboniferous Period. Large, fern-filled swamps form. Greater part of Earth's coal deposits begin forming. Ural Mountains form.
280–225	Permian Period. Climate is cooler and drier. Glaciers form in Southern Hemisphere. Appalachian Mountains are formed.

Mesozoic Era

Million B.C.	
225–190	Triassic Period, marked by uplifting of western North America. Climate warms.
245–65	Gradual breakup of supercontinent Pangea. Modern continents of Eurasia,

Major Geologic Epochs

	North and South America, and Africa begin to form and move toward present positions.
190–136	Jurassic Period. Inland sea covering large area of modern-day western United States retreats. Dinosaurs flourish. Mountains rise in western North America.
136–65	Cretaceous Period. South America and Africa break apart as modern continents continue moving toward present positions. Large North American coalbeds begin to form. Mass extinction of dinosaurs occurs.
	Cenozoic Era
Million B.C.	
65–54	Paleocene Period. Mammal species flourish. Greenland and North America break apart.
54–38	Eocene Period. Climate of North America is subtropical, and forests are widespread. Atlantic and Gulf coastal regions are submerged. Rockies, Andes, Alps, Himalayas, and other mountains rise.
54–2	Formation of modern continents is completed. India breaks away from Africa and joins with Asia. Himalaya Mountains form. Mediterranean Sea is formed. Antarctica moves to its present location.
38–26	Oligocene Period. Mountain-building in Alps reaches greatest intensity. Large sea covers part of northern Europe.
26–5	Miocene Period, marked by renewed mountain-building in Andes, Himalayas, and Alps. Extensive grasslands form, giving rise to species of mammals that graze. Sea level becomes so low that Mediterranean is cut off and temporarily evaporates to salt lake 6.5 million years ago.
5–2.2	Pliocene Period. First of four Ice Ages in recent times begins at end of this period. Isthmus joining North and South America is formed.
2.2–.01	Pleistocene Period. Four most recent Ice Ages occur (2.2 million to 10,000 years ago), interrupted by warmer periods that melt ice sheets. Glaciers create Great Lakes. Primitive men and large Ice Age mammals appear.
.01–present	Holocene or Recent. Climate warms, melting glaciers and creating deserts. During this period (roughly from 10,000 years ago) man and his rapidly developing civilization spread throughout world.

1607 Italian physicist Galileo Galilei (1564–1642) develops thermometer for measuring air temperature.

1664 Jesuit philosopher Athanasius Kircher (1602–80) incorrectly maintains that tides pump seawater to mountaintops, where they reemerge as rivers, springs, and streams.

1669 Danish physician Nicolaus Steno (1638–86) shows that rock strata are deposited according to age, with the oldest rock layers on bottom (law of superposition).

c. 1672 British scientist Isaac Newton (1642–1727) theorizes that centrifugal force causes bulge at Earth's equator. He develops theory of universal gravitation in 1687.

1701 British astronomer Edmund Halley (1656–1742) publishes first magnetic chart. He makes important advances in studies of ocean evaporation and salinity.

c. 1730 Sextant, for measuring latitude, is invented.

1774 Mineral classification by physical characteristics is introduced.

c. 1775 Accurate clocks enable sailors to determine nautical longitude.

1779 Horace de Saussure (1740–99) coins term *geology*.

▶

Meteorology

B.C.

c. 350 Greek philosopher Aristotle (384–322 B.C.) writes of weather and climate in *Meteorologica* (and possibly coins term *meteorology*).

c. 300 Crude rain gauge is in use in India.

c. 100 Greeks have installed wind vanes on Acropolis in Athens.

A.D.

1607 Italian physicist Galileo Galilei (1564–1642) develops air thermometer.

1643 Mercury barometer is invented.

c. 1660 Simple hygrometer is in use.

1667 British physicist Robert Hooke (1635–1703) invents anemometer.

1747 U.S. statesman Benjamin Franklin (1706–90) experiments with kite during thunderstorm and discovers presence of electricity in lightning.

1783 Hot-air balloon is invented.

1803 Cloud names (cirrus, cumulus, stratus, and nimbus) are coined by British meteorologist Luke Howard.

1817 German naturalist Alexander von Humboldt (1769–1859) publishes first map using isothermal lines in mapping temperature.

1828 Prussian meteorologist Heinrich Dove (1803–79) discovers that tropical storm winds move counterclockwise in Northern Hemisphere, clockwise in Southern Hemisphere.

1835 U.S. meteorologist James Espy (1785–1860) studies energy sources of storms, recognizing and describing frontal surfaces. He becomes first official meteorologist to the U.S. government, in 1842.

1835 French physicist Gustave Coriolis (1792–1843) accurately explains apparent sideways drift of winds and other objects due to Earth's rotation and speed; phenomenon is called Coriolis effect.

1844 First U.S. telegraph line is strung, between Washington, D.C. and Baltimore, making daily weather data available for first time.

1854 Unexpected storm wrecks British and French warships, leading to establishment in 1856 of national storm-warning service in France.

1855 Capt. Henry Piddington studies revolving storms in Bay of Bengal and Arabian Sea, naming them cyclones.

1860 U.S. telegraph companies, equipped with meteorological instruments, provide daily weather information to Smithsonian Institution.

1861 Region in northeastern India is drenched by heaviest recorded rainfall over twelve-month period, 1,041.78 inches.

1861 Heaviest rainfall in single month (366.14 inches) is recorded in northeastern India.

1863 U.S. meteorologist Francis Capen attempts to persuade President Abraham Lincoln (1809–65) of wartime applications of weather forecasting but is dismissed.

1863 Paris Observatory begins publishing first modern weather maps.

1868 British meteorologist Alexander Buchan (1829–1907) publishes map showing the movement of cyclonic depression across parts of America and Europe, possibly first example of modern meteorology.

1868 U.S. meteorologist Cleveland Abbe (1838–1916) publishes maps to forecast storms dangerous to Great Lakes commerce. He later is important figure in founding of U.S. Weather Bureau.

1870 First headquarters are established for U.S. Weather Bureau, originally Division of Telegrams and Reports for the Benefit of Commerce.

1876 Daily weather bulletins appear in over 7,000 U.S. post offices.

1878 First daily weather maps are published in United States.

1882 First world map of annual precipitation is published.

c. 1895 Kites with meteorgraphs are in use as atmospheric probes.

1901 French develop "antihail cannon" designed to change damaging hailstones to rain; device fails.

1913 Highest temperature in United States is recorded in Death Valley, California. Thermometer reaches 134 degrees F.

1916 Browning, Montana experiences widest range of temperatures recorded in 24-hour period when mercury plummets from 44 degrees F to −56 degrees F (January 23–24).

1918 Norwegian meteorologist Vilhelm Bjerknes (1862–1951) refers to "polar front," borrowing term from combat language of World War I.

Meteorology

1921	Silver Lake, Colorado is blanketed by the heaviest recorded snowfall in 24 hours. Total of 6 feet, 4 inches falls April 14–15.
1922	El Azizia in Libya is baked by highest recorded temperature in the shade, 136 degrees F (September 13).
1922	Norwegian Weather Service begins issuing first five-day forecasts.
c. 1925	Radio has replaced telegraph as means for transmitting weather information.
1934	Mount Washington, New Hampshire experiences highest surface winds recorded anywhere except in tornado. Wind speeds reach 231 mph (April 12).
c. 1935	Airplanes are used for weather reporting.
1936	Radio meteorgraph (radiosonde), to transmit weather information from unmanned balloons, is developed.
c. 1941	World War II planes discover "jet streams," air currents that slow their flight as they fly westward in Pacific.
1944	U.S. military flight crews hunt hurricanes in Atlantic and Pacific, part of joint Weather Bureau and armed services project.
1945	Lowest barometric pressure on record to date, 25.55 inches, is recorded in eye of typhoon near Okinawa, Japan.
c. 1945	Cloud-seeding experiments are conducted in United States in attempt to produce rain.
1946	American mathematician John von Neumann (1903–57) works on electronic device to simplify complicated weather computations. (By 1950 his computer forecasts weather with great accuracy.)
1946	U.S. scientist creates first artificial snowstorm by seeding a cloud with dry ice.
1947	U.S. scientists attempt to diminish hurricane's strength by seeding its clouds; they are unsuccessful.
1952	Small island of Réunion in Indian Ocean experiences heaviest one-day downpour on record, when 75 inches of rain fall (March 15–16).
1953	Meteorologists in United States begin giving women's names to hurricanes. (Beginning in 1979, men's names are used as well.)
1955	National Hurricane Center is established in Florida, following destruction caused by hurricanes Connie and Diane.
1958	U.S. satellite *Explorer I* is launched to measure cosmic radiation in atmosphere.
11958	Scientists warn of global temperature warming; phenomenon comes to be known as "greenhouse effect."
1958	Record surface wind speed of 280 mph is reached during a tornado at Wichita Falls, Texas (April 2).
1959	U.S. Weather Bureau begins using temperature-humidity index (THI).
1960	*TIROS I*, first all-weather satellite, is launched. (Polar-orbiting satellite, it produces picture every 24 hours.)
1962	U.S. Weather Bureau and U.S. Navy create Project Stormfury, attempting to control hurricanes.
1966	First geostationary satellites, unmoving in relation to Earth's surface, are launched.
1966	Death Valley, California, which already holds record for highest temperature in United States, hits high THI of 98.2.
1968	Highest barometric pressure to date, 32 inches, is recorded in Siberia.
1969	St. Petersburg, Florida has enjoyed unbroken streak of 768 sunny days (March 17).
1970	First system for dissipating fog is installed at Orly Airport, Paris.
1971	Doppler radar is used to study storm systems. (Radio signals bounce off raindrops within storms as far away as 140 miles.)
1971	Winter at Prospect Creek, Alaska brings record cold temperature for United States. Mercury drops to −80 degrees F (January 23).
1971–72	Record snowfall during 1-year period piles up 1,224.5 inches at Mount Rainier, Washington.
c. 1975	United States tries unsuccessfully to defuse thunderstorms by dropping aluminum-coated thread fibers through clouds.
1981	Canton, Ohio newspaper reports hailstorm dropped hailstones weighing up to 30 pounds.
1983	Vostok station in Antarctica experiences record low temperature of −128.6 degrees F.
1986	Over 90 people are reported killed by hailstorm in Bangladash (April 14). Hailstones weighing over 2 pounds fall during storm.

1783 Hot-air balloon is invented. It is later used for atmospheric study.

1785 Scottish geologist James Hutton (1726–97) develops principle of uniformitarianism, theory that Earth is gradually changing and will continue to do so.

1791 English engineer William Smith (1769–1839) uses fossils to demonstrate age of rock strata.

1801 First of two studies by French scientist René-Just Haue establishes science of crystallography.

1827 Polarizing microscope is developed; it is widely used to view rock and mineral samples in thin sections.

c. 1830 Swiss naturalist Louis Agassiz (1807–73) studies European glaciers, theorizing that glaciers once covered much of Europe and changed land's surface.

1831 British naturalist Charles Darwin (1809–92) begins five-year voyage during which he develops various theories, including those for origins of coral reefs and explanations for species' survival.

1834 Swedish scientist Carl Gauss (1777–1855) proves origin of magnetic field to be deep within Earth.

c. 1840 First use of diving gear for scientific underwater exploration.

1847 U.S. naval officer Matthew Maury (1806–73) compiles first charts of North Atlantic ocean winds and currents. He writes what is considered first oceanographic textbook (1855).

1848 First known attempt to cross continent of Australia. Explorer perishes en route.

1868 British meteorologist Alexander Buchan (1829–1907) publishes map showing movement of cyclonic depression, possibly first example of modern scientific meteorology.

1872 Three-year ocean expedition of vessel *Challenger* begins in England, marking beginning of modern oceanography.

1880 Seismological Society of Japan is formed to study earthquakes. Director, English scientist John Milne (1850–1913), develops first accurate seismographs.

1882 Scottish scientist Balfour Stewart (1828–87) proposes existence of the ionosphere.

1883 American geologist James Hall (1811–98) observes that mountainous portions of Earth's crust are affected by downward warping over time, now known as geosynclines.

1884 Mechanical analog tide-predicting device is invented in United States.

1895 Discovery of X-rays enables scientists to study internal structures of minerals.

1896 Relationship between amount of carbon dioxide in atmosphere and global temperature is postulated by Svante Arrhenius (1859–1927).

1899 American scientist Clarence Dutton (1841–1912) coins term *isostasy* to describe equilibrium of Earth's crust through vertical movements of its outer layers.

1899 *National Atlas* of Finland, first to include glacial forms, is published.

1902 Leon Teisserenc de Bort (1855–1913) discovers layers in Earth's atmosphere. He names them troposphere and stratosphere.

1905 English physicist Ernest Rutherford (1871–1937) theorizes that half-life of radioactive minerals can be used to date other minerals and rocks (radioactive dating).

1906 English seismologist Richard Oldham shows that Earth has core.

1909 International map of world is adopted by International Geographic Union. (Full world coverage becomes available only with aerial photographs taken during World War II.)

1909 Croatian geophysicist Andrija Mohorovičić discovers gap between Earth's crust and mantle below, now known as Moho.

▶

Exploring Earth's Upper Atmosphere

1783	Hot-air balloon is invented. It becomes important tool to study upper atmosphere.
1804	French chemist Joseph Gay-Lussac (1778–1850) ascends to about 21,000 feet to take air samples. He determines that rarefied air at that level contains same percentage of oxygen (21.49 percent) as air at ground level.
1874	Austrian meteorologist Julius von Hann (1839–1921) determines that some 90 percent of atmospheric water vapor exists below 18,000 feet.
c. 1890	U.S. scientists equip kites with meteorgraphs to record conditions such as temperature and barometric pressure automatically.
1898	French scientist discovers troposphere and stratosphere by using weather balloons to study temperature variations at high altitudes. He discovers that above about 7 miles, temperature stops declining with increasing height.
c. 1930	High-altitude planes are used for atmospheric research.
1936	Radiosonde is developed. It transmits information on temperature, humidity, and barometric pressure from unmanned balloon.
c. 1940	Radiosonde/balloon experiments reveal that columns of warm air rise to more than 1 mile above Earth. Lower atmosphere is discovered to have separate "layers" of wind activity, sometimes blowing in different directions.
c. 1945	B-29 Superfortress is world's first large aircraft to cruise regularly at high altitudes. Pilots discover strong high-altitude wind systems, later called jet streams.
1957	U.S. Air Force major ascends to 101,486 feet in plastic balloon. He sets record for highest balloon ascent and longest time spent at this high altitude.
1958	First U.S. artificial satellite, *Explorer I*, is launched into orbit. Satellite sensors discover Van Allen radiation belt, zone of radiation encircling Earth 400 to 40,000 miles up.
1958	American scientist proposes using radar to study ionosphere at levels higher than 60 miles. Huge antennas that suffer little from ionospheric refraction are eventually used.
1960	United States launches world's first all-weather satellite, *TIROS I* (Television Infra-Red Observation Satellite).
1960	National Center for Atmospheric Research is established in Boulder, Colorado.
1961	Two U.S. Navy scientists set new balloon altitude record, 113,500 feet, aboard balloon *Stratolab High No. 5*.
1966	U.S. ESSA (Environmental Science Service Administration) launches first of its sophisticated weather satellites.
1974	International research team launches GARP (Global Atmospheric Research Program), detailed study of one section of atmosphere. It helps develop mathematical model of Earth's atmosphere for weather forecasting. Worldwide atmospheric experiments take place in 1978.
1976	American Panel on Atmospheric Chemistry warns that release of chlorofluorocarbons from spray cans and refrigeration systems may destroy ozone in atmosphere.
1976–79	International Magnetosphere Study observes magnetosphere and its effects on lower stratosphere (including magnetic storms that can disrupt communications).
1977	Soviet pilot sets world record for jet aircraft altitude, 123,523 feet.
1977	Scientists warn that global warming effects may increase eightfold in next 200 years. They believe cause is increase in atmospheric levels of carbon dioxide from pollution.
1979	Thirteen-year global meteorological study is completed by National Academy of Sciences. It improves understanding of atmospheric circulation.
1982	Four balloons carrying several tons of scientific equipment are released to measure vertical distribution of stratospheric chemicals and solar radiation.
1985	Scientists inject barium into Earth's magnetosphere to study interaction of magnetosphere with solar winds.
1987	U.S. scientists aboard Sabreliner jet observe levels of carbon monoxide and nitric acid at high elevations around thunderstorm clouds, proving that thunderstorm systems can push pollutants as high as lower stratosphere.
1988	Nike-Orion sounding rocket launches scientific payload to measure concentration and distribution of positive and negative ions in ionosphere.

1912 German geologist Alfred Wegener (1880–1930) theorizes that original continental mass drifted apart, causing eventual creation of continents and oceans (continental drift).

c. 1920 Invention of mass spectrometer.

1925 Sonar is used to discover Mid-Atlantic Ridge.

1930 First use of bathysphere, marking beginning of exploration of oceans at great depths.

1935 Charles Richter (b. 1900) develops the Richter scale to measure strength of earthquakes.

1946 Following fatal Hawaii tsunami (tidal wave), U.S. Coast and Geodetic Survey establishes warning system to locate and predict underwater earthquakes.

1957–58 In International Geophysical Year (IGY), various nations focus scientific efforts on study of Earth. Soviets launch *Sputnik I*, first artificial Earth satellite.

1958 U.S. satellite *Explorer I* is launched to measure atmospheric radiation. Van Allen radiation belts are discovered.

1958 Scientists warn of possible global temperature warming trend, "greenhouse effect."

1960 *TIROS 1* weather satellite is launched by United States.

1960 Geothermal power is first produced in United States.

1960 Theory that new sea floor is formed at midocean ridges and spreads toward deep-sea trenches, where it reenters mantle, is proposed by Harry Hess (1906–69).

c. 1965 Scientists are using electronic computers and satellite data to make precise geodesic measurements.

1966 Borehole is cut in Greenland glacier to date layers of ice. (Bottom is found to be more than 150,000 years old.)

1967 Brittany dam and electrical generating facility becomes first major power project to harness tidal motion.

1968 American Earth scientists theorize that Earth's outer shell consists of rigid plates in constant motion, the theory of plate tectonics.

1969 Joint Oceanographic Institutions Deep Earth Sampling Project tests theory of sea floor spreading following separation of continents.

1972 U.S. Earth Resources Technology Satellite (ERTS) photographs and studies Earth's surface for oil and mineral exploration, forestry, water management, and other purposes using remote-sensing equipment. (It is renamed Landsat in 1975.)

1977 Deep-sea vents are discovered near Galápagos Islands. Unusual forms of sea life exist in hot, sulfurous water found surrounding vents.

1978 U.S. National Research Council confirms that fluorocarbons from refrigerants and aerosol sprays damage ozone layer.

1980 Thin layer of clay rich in iridium is found around world in soil laid down about time of dinosaurs' extinction. Some scientists theorize that large meteor collided with Earth, spreading layer of debris worldwide and somehow killing off dinosaurs.

1983 Studies from *Lageos* satellite (launched in 1976) show that Earth's gravitational field is changing.

1985 Scientists discover that hole in Earth's protective ozone layer periodically forms in upper atmosphere over Antarctica.

1986 U.S. satellite is deployed to measure solar radiation absorbed by Earth.

1988 U.S. National Aeronautics and Space Administration, using radio noise from quasars and long baseline interferometry, detects motion of Earth's crust.

1990 Scientists use drill to retrieve sample of what is believed to be oldest segment of ocean bottom anywhere. They drill into Pacific ocean bottom submerged under 18,600 feet of water near Saipan. Sample is about 170 million years old.

Environmentalism

B.C.

c. 1000 Phoenicians terrace hillside land to prevent soil erosion.

c. 300 Greek farmers use crop rotation to maintain soil quality.

A.D.

1626 Plymouth colony ordinance restricts cutting and sale of timber.

1639 Newport, Rhode Island law allows deer hunting during six months of year only.

1681 English Quaker leader William Penn (1644–1718) requires that one of every five settled acres be left in forested state.

1849 U.S. Department of Interior is established. In c. 1880, agency recommends scientific care of forests and creation of forest reservations.

c. 1855 Term *acid rain* is first used to describe precipitation containing relatively high amounts of acid-producing chemicals.

1858 British naturalist Charles Darwin (1809–82) publishes his findings on evolution and promotes interest in man's relation to environment.

1869 German naturalist Ernst Haeckel (1834–1919) coins term *ecology*.

1872 Beginning of environmentalism as a U.S. political issue. Government creates Yellowstone National Park, to be maintained in its natural condition.

1872 April 10 is officially named Arbor Day.

1875 First society emphasizing conservation, American Forestry Association, is created in United States.

1891 U.S. Forest Reserve Act becomes law, designating 13 million acres as reservations. (Twenty-one million additional acres are added during succeeding administration).

1899 First U.S. pollution-control law forbids liquid-waste dumping, other than from sewers, in navigable waters.

1905 National Audubon Society meets for first time.

1907 Head of U.S. Division of Forestry coins term *conservation*.

1908 President Theodore Roosevelt (1859–1919) holds conference to assess conservation of wildlife and natural resources.

1918 Creation of U.S. Save-the-Redwoods League.

1933 U.S. Civilian Conservation Corps is organized; workers plant trees, build dams, and make other environmental improvements.

1933 Tennessee Valley Authority (TVA) is created to preserve area resources.

c. 1935 Dust Bowl area of the U.S. Midwest suffers huge soil losses in windstorms because of drought and poor farming practices. (Soil Conservation Service is created as result.)

1937 Federal Aid to Wildlife Restoration Act taxes hunting weapons and ammunition.

1958 Higher-than-predicted radioactivity in soil due to test explosions of nuclear weapons arouses concern about radioactive fallout. It leads to Nuclear Test-Ban Treaty (1963), ending above-ground tests.

1962 U.S. marine biologist Rachel Carson (1907–64) publishes *Silent Spring*, book about damage to environment caused by pesticides.

1963 U.S. Clean Air Act is passed; it uses federal funds to attack air pollution.

1964 U.S. Wilderness Act is passed to preserve wilderness areas.

c. 1965 Studies in environmental science are introduced in U.S. schools.

1966 First of several U.S. Endangered Species acts restricts trade of endangered animals and their products.

▶

List of Endangered Animals

This selective list provides examples of endangered species from a wide range of major animal groups. A complete and up-to-date list is available in the U.S. government's *Code of Federal Regulations* (see the index volume, Endangered Species List).

Year	Animal	Scientific name / Range
1967	bacterian camel	(*Camelus bactrianus*) Mongolia, China
	grizzly bear	(*Ursus arctos horribilis*) northern United States, Canada
	bald eagle	(*Haliacetus leucocephalus*) North America, northern Mexico
	black-footed ferret	(*Mustela nigripes*) western United States, western Canada
	Florida manatee	(*Trichechus manatus*) southeastern United States, Caribbean Sea, South America
	red wolf	(*Canis rufus*) southern United States
	ivory-billed woodpecker	(*Campephilus principalis*) United States, Cuba
1970	cheetah	(*Acinonyx jubatus*) Africa to India
	gibbons	(*Hylobates* spp.) China, India, Southeast Asia
	gorilla	(*Gorilla gorilla*) central and western Africa
	orangutan	(*Pongo pygmaeus*) Borneo, Sumatra
	tiger	(*Panthera tigris*) temperate and tropical Asia
	brown hyena	(*Hyaena brunnea*) southern Africa
	lemurs	(*Lemuridae*) (all species) Madagascar
	gray whale	(*Eschrichtius robustus*) northern Pacific, Bering Sea
	blue whale	(*Balaenoptera musculus*) oceanic
	great Indian rhinoceros	(*Rhinoceros unicornis*) India, Nepal
	Kemp's ridley sea turtle	(*Lepidochelys kempii*) tropical and temperate Atlantic Ocean
	brown pelican	(*Pelecanus occidentalis*) southern United States, West Indies, Central and South America
1972	tiger cat	(*Felis tigrinus*) Costa Rica to northern Argentina
	snow leopard	(*Panthera uncia*) Central Asia
	ocelot	(*Felis Pardalis*) southwestern United States, Central and South America
1973	Puerto Rican nightjar	(*Caprimulgus noctotherus*) Puerto Rico
1975	American crocodile	(*Crocodylus acutus*) Florida, Mexico, Caribbean, Central and South America
1975	Asian elephant	(*Elephas maximus*) South-central and Southeast Asia
1976	Przewalski's horse	(*Equus przewalskii*) Mongolia, China
	howler monkey	(*Alouatta palliata*) Mexico to South America
	resplendent quetzel	(*Pharomachrus mocinno*) Mexico to Panama
1979	Arabian gazelle	(*Gazella gazella*) Arabian Peninsula, Israel, Jordan, Sinai
	red-necked parrot	(*Amazona arausiaca*) Dominica
1980	black rhinoceros	(*Diceros bicornis*) sub-Saharan Africa
1984	African wild dog	(*Lycaon pictus*) sub-Saharan Africa
	giant panda	(*Alluropoda melanoleuca*) China
	peregrine falcon	(*Falco peregrinus*) almost worldwide
1985	least tern	(*Sterna antillarum*) southern United States, West Indies, Mexico, Central and South America
1987	Alabama red-bellied turtle	(*Pseudemys alabamensis*) Alabama
	black-capped vireo	(*Vireo atricapillus*) Kansas, Oklahoma, Texas, Mexico
1988	Visayan deer	(*Cervus alfredi*) Philippines

1969 National Environmental Policy Act is created, requiring public agencies to publish environmental impact statements prior to proposed projects or activities.

1970 Passage of U.S. Water Quality Act, which aims to clean up federal waterways.

1970 Environmental Protection Agency (EPA) is created to set and enforce U.S. pollution standards.

1972 Almost all use of pesticide DDT is banned in United States.

1972 U.S. Ports and Waterways Safety Act is enacted to prevent oil spills at sea.

1972 U.N. Conference on Human Environment meets in Stockholm to address ocean dumping, commercial whaling, and pollution control, among other issues.

1973 International conference in Lebanon warns that sewage and industrial waste dumping in Mediterranean will turn it into "dead sea."

1973 Report by U.S. National League of Cities warns of dangerous levels of garbage accumulation and recommends widespread recycling.

1973 Australia bans sale of live kangaroos and kangaroo hides to other countries, fearing extinction of species.

1974 Construction begins on U.S. Alaska pipeline for transporting oil. It ends long delay of project to ensure protection of area's environment and wildlife.

c. 1975 Major fuel shortages result in gas rationing and emphasize need for energy conservation.

1977 United States creates Department of Energy to promote new sources of energy and conserve existing supplies.

1978 Residents in Love Canal area of New York State are forced to evacuate their homes because of contamination from industrial wastes in canal.

1979 Three-Mile Island near-meltdown and some radioactive leakage arouse strong opposition to nuclear power plants in United States.

1980 EPA "superfund" is created by U.S. government to clean up toxic-waste sites and prosecute cases of illegal dumping.

1981 U.S. Committee on Atmosphere and Biosphere finds evidence linking acid rain to power-plant emissions.

1982 International Whaling Commission votes for 1985 ban on commercial whaling; USSR and Japan object.

c. 1982 Environmentalists oppose U.S. government efforts to open wilderness areas to commercial development.

1984 Poisonous gas leak from pesticide plant in Bhopal, India kills more than 2,000 people.

1984 EPA identifies 538 abandoned hazardous-waste dump sites in United States that threaten public health. Number of sites identified grows considerably in subsequent years.

1985 U.S. National Wildlife Federation criticizes federal regulatory policies on strip mining, citing major environmental damage to streams and hillsides.

1985 United States and Canada sign pact to examine causes of acid rain.

1985 Scientists discover a hole develops periodically in ozone layer over Antarctica.

1986 Explosion and fire at nuclear plant in Chernobyl, USSR release large amounts of radiation into atmosphere.

1987 Clean Water Bill is passed to deal with pollution of estuaries and rainwater in United States.

1987 International treaty is signed to limit production of chemicals responsible for destruction of ozone layer.

1988 USSR announces end to all commercial whaling by Soviet fishermen.

1989 Exxon oil tanker runs aground off pristine Alaska coastline, causing worst U.S. oil spill.

1989 U.S. scientists find Arctic air pollution as severe as in any U.S. city.

10 Necessities to Notoriety: An Omnium Gatherum

Foods and Beverages

B.C.

c. 7000 Earliest-known cultivation of foods begins in Middle East. Includes growing of barley, peas, lentils, bitter vetch, and wheat.

7000 Red peppers, gourds, avocados, and squash are under cultivation in Central America.

5200 Wild corn is known in Central America by or before this time.

4000 Grapes and olives are cultivated in Middle East.

c. 3500 Wine is being made in Turkestan. Mesopotamians have discovered how to make beer.

c. 3500 Rice, millet, sorghum, and soybeans are farmed in Asia.

c. 3000 Popped corn is a native American Indian dish.

2600 Egyptian bakers create more than 50 varieties of bread, including whole-wheat and sourdough.

c. 2600 Pancakes are prepared by Egyptian bakers.

2500 Potatoes and sweet potatoes are cultivated by South American farmers.

2500 Cabbages are grown in Middle East.

c. 2000 Bananas and tea are cultivated in India.

c. 2000 Ice cream is invented in China when soft milk-and-rice mixture is further solidified by packing it in snow.

2000 Watermelons are cultivated by African farmers.

2000 Apples are cultivated in Middle East.

1600 Spices are being traded between China and Baghdad.

c. 1500 Babylonians stuff animal intestines with spiced meats, creating first sausages.

1500 South Americans cultivate peanuts.

1500 Oranges are grown on Asian farms.

c. 1000 Pasta is first prepared in China.

c. 500 Artichokes are grown in Middle East.

c. 300 Cookies first appear in Rome.

c. 300 Ketchup is first prepared in Roman kitchens as a sauce originally made of vinegar, oil, pepper, and anchovies.

100 Cocoa plants are cultivated in South America by this date.

c. 100 First wedding cake is developed as fertility symbol, in Rome. (Instead of being served to bride, it is thrown at her by guests.)

100 Sugarcane is cultivated in Far East.

A.D.

610 Pretzels are invented by Italian monk (and awarded to children who learn their prayers).

700 Central American farmers are now growing tomatoes.

850 Coffee is cultivated in Middle East.

1200s Birthday cakes are baked in Germany to celebrate children's birthdays. Candles on cake are kept lit throughout day to symbolize life.

c. 1514 Pineapples first appear on European tables.

1517 Coffee is introduced in Europe.

1520 Chocolate is first brought to Spain by conquistador Hernán Cortés (1485–1547), who learned of it from Aztecs. Kept a secret in Spain for nearly a century, use of chocolate spread throughout Europe in 1600s.

1565 Sweet potatoes and tobacco are introduced in England.

c. 1565 Ice cream in its modern form, frozen hard and made of sweetened cream, is made in Florence.

c. 1600 Doughnut originates in Holland (but does not have a hole in its center until c. 1850).

1609 Tea is first shipped to Europe, by Dutch East India Company.

1632 First London coffee shop opens for business.

1657 Hot chocolate is introduced in England.

1666 First cheddar cheese is made, in English village named Cheddar.

1724 Gin becomes popular drink in England.

c. 1727 First Brazilian coffee is under cultivation.

▶

1760 Modern sandwich is created in England when notorious fourth earl of Sandwich (John Montagu, 1718–92) refuses to leave his gaming tables to eat. He insists that meat and cheese be brought to him between pieces of bread.

c. 1800 Mayonnaise is introduced in United States and originally is considered delicate and exotic creation.

1820 Tomato, previously thought poisonous, is first accepted as food in United States.

c. 1830 Graham crackers, named after health-food advocate the Reverend Sylvester Graham (1794–1851), originate in New England.

c. 1850 Worcestershire sauce is first prepared by British nobleman who learns its recipe during military service in India.

1852 German sausages come to be known as frankfurters, after Frankfurt butchers develop special, streamlined shape with thin casing.

1853 Potato chips are developed in Saratoga Springs, New York.

1860 San Francisco bartender invents martini.

1872 Square-bottomed grocery bag (and machine to make it) are patented in United States.

1880 First appearance of canned fruits and meats on store shelves.

1885 Morton Salt is first marketed in United States.

1886 New soft drinks, Coca-Cola and Dr. Pepper, are introduced.

1889 First pre-mixed, self-rising pancake mix is marketed.

1892 Melba toast is first prepared in London and is named after opera star Nellie Melba.

1893 Kellogg Company in America develops a new cereal, shredded wheat. (It introduces corn flakes in 1906.)

1893 Cracker Jack, a combination of candied popcorn and peanuts, is first sold at small popcorn stand in Chicago, Illinois.

1894 First Hershey bar is sold after being developed by Pennsylvania caramel maker.

1895 First pizzeria opens in New York City.

1898 Pepsi-Cola is introduced.

1900 New Haven, Connecticut restaurant owner invents hamburger, first served between two slices of toast.

1900 U.S. vendor sells frankfurters from his cart, calling them "hot dachshund sausages," whence comes "hot dogs."

1902 Animal crackers are first sold in United States, having originated as "animal cookies" in England in 1890s.

1903 Sanka, instant decaffeinated coffee, is introduced.

1904 First ice-cream cone is sold, at Louisiana Purchase Exposition in St. Louis, when vendor serves ice cream in rolled waffle.

1916 First U.S. supermarkets open in Tennessee. They feature self-service and checkout counter.

1921 Eskimo Pie ice-cream bars are developed.

1923 Minnesota confectioner invents the Milky Way. (His profits are nearly $800,000 in two years.)

1924 Popsicles, brand of frozen ice on stick, are first sold.

1932 New candy bar, 3 Musketeers, is marketed.

1933 New soft drink, 7-Up, is introduced.

1933 Ritz crackers are first sold in the United States.

1933 Spam, an economical meat product, is developed.

1935 Krueger Beer, brewed in New Jersey, is first beer sold in cans in United States.

1937 Store owner in Oklahoma invents first grocery cart.

1940 Mars, Inc. develops M&Ms, sugar-coated chocolate candy. (Intended for GIs, M&Ms are designed to resist melting in field.)

1948 First self-service McDonald's hamburgers restaurant opens, in California.

1954 First "TV dinners" are sold, by Swanson Company in United States.

1955 Kentucky restaurant owner begins selling his fried-chicken-recipe products under name Kentucky Fried Chicken.

1958 First Pizza Hut opens, in Kansas City. It becomes largest U.S. pizza chain.

1960 Soft-drink manufacturers introduce aluminum cans. Pop-top can follows in 1963.

1961 Mocha Mix is introduced as refrigerated nondairy cream subsitute.

c. 1962 Freeze-dried foods are introduced.

▶

1962 New diet soda, Diet-Rite, is marketed. Tab appears a year later; Diet Pepsi appears in 1965.

1965 Variety of frozen dinner entrées increases markedly at about this time with growing popularity of convenience foods.

c. 1970 Organic foods, grown without chemical fertilizers and pesticides, become available. Market for such "natural foods" increases sharply throughout decade.

1970s Yogurt becomes popular food among dieters and the health conscious.

c. 1972 Meat substitutes, made from textured vegetable protein, are introduced.

1975 First low-calorie ("lite") beers are marketed.

1980s Host of new low-calorie, low-cholesterol foods are developed for dieters and the health conscious. New substitutes for conventional foods are also introduced, notably the popular tofu (soybean curd).

1983 NutraSweet is introduced as a synthetic sugar substitute.

1985 Coca-Cola introduces new-formula Coke. Public outcry results in return of old formula as Classic Coke in 1986.

1986 Massachusetts hospital develops line of low-fat, low-salt frozen dinners named Just What the Doctor Ordered.

c. 1987 Fluffy cellulose, high-fiber no-calorie substitute for flour, is developed. Many other synthetic substitute foods are developed in late 1980s, including new fat substitutes, Olestra and Simplesse.

1988 Egg Watchers, pourable liquid egg substitute made from soybean curd, is introduced.

1989 Carnation introduces low-cholesterol, liquid form of Coffeemate.

1990 Food researchers invent ice-cream substitute made from oat bran.

Clothing: From Egyptian Robes to French Bikinis

B.C.

c. 3000 Egyptian women are wearing ankle-length robes. Men wear either long robes or short kilts. Sandals are commonly worn outdoors.

c. 3000 Male and female Egyptian monarchs adorn themselves with false beards, made of metal and tied in place, as sign of royalty.

c. 3000 In Crete, men wear short loincloths, while noblewomen wear bell-skirted dresses with corseted bodices to emphasize their exposed breasts.

c. 1600 Babylonians introduce moccasinlike shoes.

c. 1100 Assyrian soldiers develop first boot, a calf-high, laced leather boot with metal-reinforced sole.

c. 1000 Mesopotamians wear long rectangular pieces of fabric draped around their bodies. Men wrap their garments counterclockwise, women clockwise.

c. 750 Greek men and women wear *chiton*, two long rectangular pieces of cloth sewn up the sides, with holes left for head and arms. This is generally belted and sometimes accompanied by *himation*, or rectangular mantle, draped around torso.

c. 600 By this time metal corsets are commonly worn by Minoan women. Although whalebone later becomes most common corset material in Europe, Marie de Médici (1573–1642) briefly revives metal corset in 1600s.

c. 500 Ancient Persians first begin wearing breeches instead of robes.

300s Roman men and women wear togas.

200s Roman shoemakers begin to fashion shoes specifically designed for left and right feet.

▶

A.D.

1200s Buttons and buttonholes come into use in Europe.

1200s Pattens, or wooden undershoes, are strapped to leather shoes to protect them from muddy streets.

1200s European men begin wearing two-piece hose.

1370s Men's hose evolve into waist-high, one-piece tights. This necessitates introduction of codpiece, bag sewn to front of tights to cover crotch. Codpiece additionally serves as pocket.

1380s Extended toes on shoes become vogue in Europe, some reaching lengths of up to two feet.

c. 1460 Women's steeple hats become popular in France.

c. 1465 Chopines, or platform shoes, become popular among European women. Venetian women wear chopines up to 2½ feet high.

c. 1470 Spanish women begin wearing *verdugados*, wide skirts reinforced with strips of wood. This is precursor of English farthingale.

1500s Men's shirts, previously worn long and belted over hose, are now shortened to allow for tucking into breeches.

1550s Ruff first appears in Spain. Frill of folded linen worn around neck, it is originally shaped by wires and later stiffened by starch.

1580s Pockets in trousers are introduced.

1600s French king Louis XIV (1638–1715) has his boots equipped with high heels to compensate for his short height. The fashion quickly sweeps through Western Europe.

1650s Louis XIV begins another fashion trend as he begins wearing long, curly wigs to disguise his approaching baldness.

1666 King Charles II of England (1630–85) introduces fashion of wearing waistcoat with open coat. This style eventually evolves into the modern three-piece suit.

1700s Englishmen are first to begin wearing suspenders to hold up their pants.

1730s Whalebone panniers, hoop skirts that extend sideways instead of in circular fashion, are popular in England.

1750s Clothing designed specifically for health and comfort of children debuts about this time, replacing previous fashion of dressing children as miniature adults.

1790s European men give up wigs and resume wearing their own hair.

1795 Napoleon Bonaparte (1769–1821) attempts to revive styles of Roman Empire, and women's fashion follows suit: Bustles are abandoned, lighter fabrics are used, and waistlines rise to just below bust.

1797 London haberdasher John Etherinton introduces his latest creation, the top hat.

1840s Men's bathing trunks are made of non-elasticized worsted wool that becomes weighted down when wet. Women's bathing suits consist of flannel knickers and knee-length dress.

1850s Crinoline, stiff underskirt made of horsehair and linen, becomes popular. Hoops appear shortly thereafter.

1850s Bowler hat makes its first appearance.

1851 Amelia Bloomer (1818–94) begins publicly wearing baggy pants gathered at ankles. They are named bloomers, after her.

1860s Bustle begins to replace hoops in women's fashion.

1860s Levi Strauss markets his denim jeans in San Francisco as heavy, durable pants for miners.

1860s Ladies' blouse–skirt combination debuts.

1860s John B. Stetson (1830–1906) designs his famous "10-gallon" hat.

1882 Soft felt hat with center crease is worn by Sarah Bernhardt (1844–1923) in title role of play *Fedora*.

1886 Tailless dinner jackets debut in Tuxedo Park, New York.

1890s Women begin wearing knickerbockers for bicycling.

1890s Button-down collar is designed by British polo players to prevent their collars from flapping during game.

1893 "Clasplocker," an early form of zipper, is patented (August 29). Zipper is perfected in 1913.

c. 1910 Eastern pajamas are introduced into Western Europe and quickly begin re-

▶

placing nightshirts as dominant sleepwear.

1910s V-neck first appears and is branded dangerous to health and morals.

1913 First modern bra is designed by New York socialite Mary Phelps Jacobs out of handkerchiefs, ribbon, and cord. She patents her design in 1914.

1917 U.S Rubber introduces Keds, first tennis shoe.

1920 Women's fashion abandons corsets and waistline, resulting in tubular look.

1927 Women's hemlines rise sharply, to just below knee, after slight rise in 1925.

1930s Women begin to wear trousers to play golf and ride horses.

1933 René Lacoste (b. 1905), "Le Crocodile" of the tennis courts, launches his world-famous cotton tennis shirt.

1938 Nylon stockings are invented by DuPont chemicals company. They become available to public in 1940, and their popularity is such that silk stockings, already scarce, become obsolete overnight.

1946 Daring, two-piece bikini bathing suit debuts in Paris only four days after atomic testing in Bikini Atoll (July 5). French designer Louis Réard reportedly finds bikini name appropriate for explosive impact of his creation on fashion world.

1947 Christian Dior (1905–57) introduces the "New Look" in women's fashions, featuring skin-tight bodices, nipped-in waists, and full, above-the-ankle skirts.

1950s Dior and Cristobal Balenciaga (1895–1972) dominate Paris and New York fashion scenes, emphasizing cut, fit, and line in their ultrafeminine designs.

1950s Men favor severe "military" styles typified by the Eisenhower jacket.

1951 Hemlines, measured in inches from the ground, are fashionable as 13 to 15 inches. In 1952, they would drop to 11¼, only to rise again to a high of 16 inches in 1953.

Mid-1950s Bustlines are accentuated in figure-hugging evening wear. Suits and dresses are the fashion in women's daywear.

1957 The sack dress is introduced by Balenciaga, featuring an unfitted waistline and a short hemline (1 inch below the knee). Dior popularizes the style.

c. 1958 Antiestablishment "beat" style for men and women appears. Female "beatniks" don oversized, dark-colored sweaters and jeans, loosely arranged long hair, and stiletto-heeled shoes or long boots in black leather. Male beatnics favor dark clothes, jeans, and black leather jackets or waistcoats.

1960 Norman Norell (1900–72), innovator behind the sheath dress (1944), opens his own fashion company, featuring in his first collection harem pants and low-cut evening dresses.

c. 1960 Movie stars become fashion symbols, popularizing the creations of Hollywood costume designers such as Edith Head (1899–1981).

1960s "Fun furs," imitation fur coats often dyed bold, "mod" colors, are made from mixture of acrylic and polyester.

1960s Coco Chanel (1883–1971) introduces the Chanel suit, perhaps her biggest success since the "little black dress" of the 1940s.

1961 Jacqueline Kennedy popularizes the short, tailored two-piece suit and pillbox hat created by Oleg Cassini (b. 1913).

1963 The characteristic "'60s" clothes emerge, featuring miniskirts, stretch pants, hip-hugger bell-bottomed trousers, and collarless jackets.

1964 André Courrèges (b. 1923) introduces his "futuristic" designs, including straight-cut minidresses, helmet-shaped hats, and calf-length boots.

1964 Rudi Gernreich (1922–85), a youth-oriented sportswear designer, introduces the scandalous topless bathing suit. Few women wear it.

c. 1965 Mary Quant (b. 1934) and a new generation of British designers set the new fashion trends of the late 1960s. Carnaby Street becomes synonynous with *haute couture*.

c.1965–75 The Afro hairstyle is favored among America's black youth and long, straight hair among whites.

c. 1966 Women's fashion is geared toward youth and features the "boy look" popularized

▶

by Twiggy (b. 1949) and other models of the time.

1967 Oscar de la Renta (b. 1932) introduces his gypsy collection and later is recognized as a leading designer of evening wear and cocktail dresses.

1968 Ralph Lauren (b. 1939) creates Polo, featuring the "Ivy League" style. In 1978 he introduces the "prairie" look.

1969 The "maxi" is introduced, featuring the longest hemlines seen in women's fashion since c. 1915.

1969 "Wet-look" raincoats and crocheted bikinis are big fashion sellers.

1970 Bill Blass (b. 1922) renames Maurice Renther Ltd, a company he'd bought in 1967, Bill Blass Inc.

1970s Flare trousers and "hot pants," short shorts, are popular, as well as platform shoes and cloche hats. Unisex dressing is the rage among youth.

1972 Japanese designer Kenzo (b. 1940), owner of Jungle Jap, popularizes boldly colored oriental-styled tunics, wide-legged pants, and in 1973, the samurai dress.

1973 John Bates (b. 1935) introduces the controversial backless evening dress.

1975 Giorgio Armani (b. 1935), famous for his simple and elegant men's suits, establishes himself as a consultant. He has worked since 1961 for several leading designers, including Nino Cerruti (b. 1930) and Emanuel Ungaro (b. 1933).

c. 1975 Jeans, a veritable wardrobe essential, are adopted and created anew by fashion designers such as Calvin Klein (b. 1942), ushering in a decade or more of popular, higher-priced "designer" jeans.

1976 Peasant-style, or "ethnic," clothes feature the "layered" look.

1976 Diane von Fürstenberg (b. 1946) creates the jersey wrap dress, now a fashion classic.

1976 Mary McFadden (b. 1938) leaves *Vogue* to form her own company. She becomes famous for her oriental-inspired designs and her hallmark pleated fabrics.

late 1970s Punk designs, once the style of a lower-class cult market, are adapted for the mainstream.

1980s Man-tailored clothing is made popular for women by designers such as Perry Ellis (1940–1986).

1981 Norma Kamali (b. 1945) introduces her "sweats" collection of fashions styled from soft fleece-backed sweatshirt material.

1984 The T-shaped jacket is introduced by American designer Geoffry Beene (b. 1927), who in the 1960s was a popularizer of the Empire-style dress.

Fashion Fads Since 1950

1951 "Crazy beach hats" are popular this year. Straw hats are topped with bizarre adornments such as figures of little animals, chairs, and athletes.

1951 Teenage girls wear poodle skirts and saddle shoes.

1952 Over 30 million American children wear propeller-topped beanies.

1954 Short-lived fashion fads for young men include jungle jackets, neon-blue suits, and huge cuff links.

1955 Garrison belt, 2½-inch-wide black leather belt with oversize metal buckle, gives young men tough look.

1955 Davy Crockett craze results in $100 million industry, which includes sales of Crockett caps, raccoon coats, and even ladies' panties.

c. 1955 Clothing fashion for sophisticated college girls this year is straight skirt with matching pastel sweater (decorated with a circle pin).

c. 1955 Some American college men favor button-down Ivy League look this year. Others choose padded shoulders, pegged pants, and open-neck Hawaiian shirts.

1956 U.S. businessmen have adopted the "Madison Avenue look," consisting of dark suit, button-down shirt, bow tie, and narrow-brimmed hat.

1956 Hairstyles for America's youth include long ponytails for girls, crew cuts and flat tops for boys.

1956 Young American men are wearing blazers and dirty-white buck shoes.

1956 Bermuda shorts become a necessity for stylish dressers of all ages.

▶

1957 Ankle bracelets increase in popularity among young women.

1957 European designers introduce "sack" or "bag" look, shapeless dresses that quickly replace the form-fitting elegant fashion previously in style.

1957 Some teenage boys depart from fashion customs and begin wearing ducktail haircuts and black leather jackets.

1957 Bouffant hairdos begin a six-year period of popularity as American women seek elegant, regal appearance.

1958 British bowler hats are extremely popular among young adults.

1958 New York businessman discovers surpus oilskin bobbies' capes in London warehouse, leading to profitable but brief craze.

1959 With appearance of new "cowboy" stars on television, Americans purchase $283 million worth of childrens' toy spurs, guns, boots, and lassos in four years.

1961 New American First Lady Jacqueline Kennedy (b. 1929) inspires fad for pillbox hats this year.

1962 The twist dance inspires special shoes, shirts, pajamas, fringed dresses, cuff links, and fur-trimmed caps.

1964 The Beatles appear on the music scene and spawn an annual $50 million industry, including wigs, buttons, and T-shirts. Teens adopt their longer, shaggier hairstyle as well.

1965 Cotton handmade ankle-length "granny" dress becomes fad among young women.

1965 Teenagers are cutting holes in their shoes and sneakers, seeking "air-conditioned" effect.

1965 Girls in many states adopt "knee designs," painted or pasted pictures of butterflies and flowers, in lieu of stockings. (It leads to overall "body painting" within a year.)

1965 Teenagers in Florida make rings by filing down necks of Coca-Cola bottles.

1965 Glassless glasses (without lenses) are style among young people, especially in California.

1965 Miniskirt's midthigh hemline becomes fashionable.

c. 1965 "Flower child" image is popular around this time. It eventually includes such accouterments as blue jeans, T-shirts, beads, wire-frame glasses, and sandals.

1966 Ear-piercing becomes part of American fashion scene.

1966 Batman craze sweeps country. Fad items that include costumes and clothing sell for $600 million. (In 1989, new Batman movie and craze appear.)

1967 Small, wire-rimmed "granny" glasses become popular.

1967 Nehru jacket, collarless and many-buttoned, is fashionable briefly, then disappears.

1967 Antiwar protest buttons are worn by millions of young people.

1967 Mickey Mouse watches are worn by millions throughout United States.

1968 "Radically chic" fashion trends this year include cashmere turtlenecks and French-cut jeans.

c. 1968 Flamboyant, wide ties for men become extremely popular.

c. 1968 British rock stars popularize "mod" fashion look. Young men wear Dutch-boy caps, silk shirts, and bell-bottom trousers. Women wear high boots, white stockings, and microminiskirts.

1969 Spiro Agnew watches are popular among his critics.

1970 American women begin wearing chokers, close-fitting necklaces made of fabric, jewels, or metal.

1970 Midiskirt, midcalf length, replaces mini.

1971 Hot pants, women's short shorts appropriate for almost any occasion, become fashionable and controversial.

1972 American men are wearing flamboyant bow ties in larger numbers.

1973 "Gatsby look," consisting of 1920s clothing such as baggy trousers for men and delicate evening dresses for women, is very popular.

1973 American Indian culture inspires widespread popularity of Indian squash-blossom necklaces, handmade Indian silver, and turquoise jewelry of all kinds.

1973 Tattoos become rage as tattoo parlors spring up nationwide.

1973 Puka-shell necklaces and other jewelry become brief fad.

1973 New York City women don inexpensive nightgowns to wear as evening dresses.

▶

1974 String bikinis are sensation. They are first worn in Brazil, where government tries to ban them.

1980 "Designer" jeans manufactured by top fashion houses and conspicuously labeled are popular among young people nationwide.

1980 Preppy fashions, including alligator shirts, topsider shoes, blazers, and madras jackets, are growing trend in United States.

1985 American women begin wearing, again, clothing with shoulder pads, previously popularized by Joan Crawford.

1988 Denim jeans, ragged and torn at knees and elsewhere, are fad among U.S. teenagers.

1989 "Chanel" look, first introduced in 1920s, is popular, as women wear gold buttons, gold-chain belts, and gold jewelry and trim of various types.

Holidays and Festivals

B.C.

c. 400 Festival of Samhain (October 31) is first observed by ancient Celts in Ireland. Bonfires celebrate summer's harvest and frighten away evil spirits. It becomes known as Halloween.

165 Hanukkah is first celebrated, in Palestine, to honor Jewish victory in Maccabean revolt against Syrians.

A.D.

61 Romans outlaw human sacrifices as part of Halloween celebrations.

313 Christians who come into power in Rome adopt several pagan holidays (to encourage Romans to convert), among them Feast of St. Valentine (February 14). Young men draw names of eligible women from box as part of traditional festivities.

325 Emperor Constantine proclaims that Easter will be celebrated on first Sunday after first full moon on or after vernal equinox. Originally a Saxon celebration of Eastre (goddess of spring), day now is a Christian holiday celebrating Christ's resurrection.

336 Christmas is first celebrated on December 25 as Christian holiday. It falls on same date as first-century B.C. Roman holiday for the sun god Mithra.

342 Death of Turkish bishop St. Nicholas (December 6), whose generosity and fondness for children make him symbol of Christmas gift-giving. December 6 becomes children's holiday. (English in New York adopt the Dutch "Sinterklass," calling him Santa Claus from the 1600s.)

432 St. Patrick's Day (March 17), a religious holiday honoring arrival of St. Patrick, is first celebrated in Ireland.

c. 750 Undecorated Christmas tree originates in Germany.

c. 850 European Christians practice "souling," walking through villages to beg for "soul cakes" (November 1), precursor of Halloween tradition of trick-or-treating.

1415 Earliest known Valentine's Day card, a love letter, is sent by Charles, duke of Orleans, to his wife during his imprisonment in Tower of London.

c. 1550 Groundhog Day begins as German farming legend about badger forecasting coming of spring by seeing his shadow. Nineteenth-century German immigrants to United States switch to groundhog, and February 2 becomes known as Groundhog Day.

1582 Christians move New Year holiday to January 1 on adopting Gregorian calendar. It replaces medieval celebration that lasted from March 25 to April 1.

1582 New Gregorian calendar displaces April 1 as French New Year's Day. April 1 eventually becomes known as a "fool's holiday" (after those who refused to change) and is a time for giving mock gifts and playing tricks.

1621 Colonists in Plymouth, Massachusetts join with American Indians in November to celebrate their first successful harvest, later called Thanksgiving.

▶

c. 1700 Mardi Gras celebrations, days of feasting and festivities preceding Lent, are celebrated in many cities by this time.

1723 Earliest commercial valentines—decorated papers for writing personal messages—appear. Printed valentine cards originate in United States in 1840.

1737 St. Patrick's Day (March 17) is first celebrated in United States by Protestant group organized to help homeless and unemployed Irishmen.

1777 U.S. Independence Day (July 4) is first celebrated, with church bells, fireworks, bonfires, and music. Quakers complain about excessive noise and broken windows.

1789 Parisian mob storms the Bastille fortress to free prisoners and seize ammunition during French Revolution (July 14). Day becomes known as Bastille Day, important French holiday.

1789 Newly elected U.S. President George Washington (1732–99) first proclaims Thanksgiving Day (then November 26) a national holiday, to give thanks for the Constitution. But political opposition blocks its acceptance until 1863, when President Abraham Lincoln (1809–65) makes Thanksgiving a national holiday.

1822 American professor Clement Clarke Moore (1779–1863) writes a poem for Christmas Eve. Its unauthorized distribution makes "A Visit from St. Nicholas," also known as "'Twas the Night Before Christmas," a Christmas tradition.

1843 London artist prints first Christmas cards.

1872 Arbor Day is first celebrated, in Nebraska, as a day for planting trees.

1908 Mother's Day (May 10), known earlier in Europe, is first celebrated in United States, in West Virginia. By 1911 all U.S. states recognize the holiday.

1910 First celebration of Father's Day, in Spokane, Washington.

1954 Veterans' Day (November 11) is proclaimed a holiday to honor all who have fought for the nation. It replaces Armistice Day, which marked end of World War I.

1972 President Richard M. Nixon (b. 1913) proclaims Father's Day an official holiday.

1983 Martin Luther King Day becomes a U.S. national holiday, in memory of the slain civil-rights leader.

Fads and Crazes Since 1900

c. 1900 Baseball cards are packaged as bonus with cigarettes. Later such cards reappear with bubble gum and become collector's items.

1901 Americans discover Ping-Pong. Craze sweeps country for two years, then is gone. Game reappears later.

1902 Brooklyn store owner develops and sells first "teddy bear," named after U.S. President Theodore Roosevelt (1858–1919).

c. 1909 Jigsaw puzzles become enormously popular with children and adults.

1913 First Kewpie dolls go into production and earn their creator $1.5 million.

1918 Raggedy Ann dolls first go on sale in United States.

1919 Ouija boards go on sale. They prove extremely popular during both world wars and Korean War. Sales reach all-time high during Vietnam War.

1922 Mah-jongg is introduced in United States from China. Within a year it becomes a hot new game fad.

1923 Americans flock to stores to buy King Tut memorabilia, including hats, rings, and even home furnishings.

1923 Charleston begins catching on in United States. By 1925 the dance craze reaches New York City Harlem dance clubs. Later it becomes symbol of Jazz Age.

1923 First recorded dance marathon.

▶

1924 First crossword-puzzle book (with pencil attached) is published in United States. Interest in crossword puzzles is blamed for downfall of mah-jongg in United States.

1924 Marathons become popular. Ex-sailor sits atop a flagpole for 13 hours, 13 minutes. Schoolgirl stays on a backyard perch for 10 days.

1926 Two businessmen market game of miniature golf on roof of New York skyscraper.

1929 U.S. businessman introduces new toy, yo-yo. Sales are poor until it becomes fad in 1961, making him a millionaire overnight.

1933 U.S. engineer creates Monopoly and has trouble finding buyer for game.

1935 First chain letters, promising instant wealth, appear. Post Office Department is flooded with the mail.

1939 Goldfish swallowing becomes fad on college campuses. Animal-preservationist societies protest.

1939 Roller-skate dancing attracts hundreds of thousands as rinks are hurriedly refurbished. "Skatarina," full skirt with matching bloomers, becomes popular.

c. 1940 "Zoot suit," with shoulder padding and baggy trousers, becomes fad.

1940 Faddish dances such as conga line, lindy hop, and kangaroo jump sweep through American dance halls.

1941 Teenagers popularize slumber parties by this time.

1943 Scrabble is introduced but does not catch on right away.

1945 Silly Putty is developed as manufacturers seek wartime substitute for synthetic rubber. Substance, which bounces and lifts print from newspaper, is marketed as toy in 1949.

1945 Massachusetts inventor Earl S. Tupper develops Tupperware. He markets his polyethylene containers through in-home sales parties.

1946 Bikini bathing suit appears on European beaches, shocking even modern-thinking sunbathers of this time.

1947 Bubble-gum-blowing contests become popular.

c. 1950 Fallout shelters, including prefabricated models with wall-to-wall carpeting, are growing in popularity. Couples compete by trying to stay inside the longest.

1951 Women and girls wear saddle shoes and crinolines.

1952 Over 30 million children are wearing propeller-topped beanies.

1955 Popularity of TV's Davy Crockett leads to coonskin cap fad.

1955 Pizza has become a favorite food by this time.

1955 College students stuff themselves into automobiles, competing to see how many will fit.

1957 First plastic Frisbee is introduced, in California. Frisbee is developed after one of its creators watched college students tossing metal pie plates to one another. Fad is now recognized sport.

1958 U.S. toy manufacturer markets first "Barbie doll."

1958 Biggest toy fad in history, hula hoop, is introduced by California toy manufacturers. Twenty million plastic hoops sell in less than a year. (Hula hoops are outlawed in Japan.)

1959 "Twist" dance is introduced. It spawns its own industry in accessories and records.

1959 Students on college campuses pile into telephone booths.

1964 The Beatles take over music scene, creating Beatlemania. Related products bring in $50 million in first year alone.

c. 1965 Nonconformist dress, including blue jeans, T-shirts, and long hair, becomes fad among the young. It evolves into hippie life-style.

1973 Tattooing is, briefly, a teenagers' fad.

1974 Streaking—running naked—at parties, ceremonies, sports events, and the like becomes a short-lived fad.

1975 Pet Rocks go on sale. Creator cites obedience, loyalty, and low maintenance costs.

1975 Skateboarding has swept across United States. Tournaments are popular.

1975 Disco becomes current dance craze.

1983 One-of-a-kind Cabbage Patch dolls become overnight sensation.

1989 Teenage Mutant Ninja Turtles, characters from a TV cartoon series, spawn a new craze among the young.

Basic Household Items

c. 1490 Toothbrush is invented in China. (It becomes common in Europe c. 1600. Nylon-bristle toothbrushes are sold in 1938.)

1690 French physicist Denis Papin (1647–1712) invents pressure cooker. He calls it "steam digester."

1778 Forerunner of modern flush toilet is patented by English inventor Joseph Brahma.

1806 Carbon paper is invented in England. It consists of paper soaked in ink, then dried.

1810 Tin can is patented in England but, because it is handmade, proves expensive. Machine-stamped cans are introduced in 1847. Thinner cans, allowing use of first can openers, are available in 1865.

1829 First sewing machine is invented in France. Commercial home model is later invented and marketed by U.S. inventor Isaac Singer (1811–75) in 1851.

1836 Early gas stove is manufactured commercially.

1840 First adhesive postage stamp, "penny black," goes on sale in England. First postage meter is manufactured in 1920.

1845 Earliest rubber bands are patented in Great Britain.

1848 Yale lock is invented by American Linus Yale.

1872 First square-bottom grocery bag is invented in United States.

1877 Telephone (local service) is introduced.

1879 American inventor Thomas Alva Edison (1847–1931) demonstrates first practical electric light bulb, in New Jersey.

1879 Well-to-do Illinois housewife Josephine Cochrane invents first mechanical dishwasher, after becoming annoyed at breakage of her fine china by kitchen servants.

1887 Earliest contact lenses are developed. First plastic lenses are manufactured in 1939.

1888 U.S. inventor John H. Loud patents first ballpoint pen.

1888 Kodak markets its first home camera.

1889 Earliest electric oven is installed in Swiss hotel. It is available in United States in 1891.

1889 First aspirin is introduced, in Germany. Aspirin tablets become available in United States in 1915.

1890 Aluminum saucepan is invented in Ohio.

1893 Early form of zipper is introduced. It is demonstrated at World's Columbian Exposition in Chicago.

1896 Earliest trading stamps, S&H stamps, are introduced in United States for redemption of merchandise.

1900 Paper clip is patented.

1902 First "teddy bear," named after U.S. President Theodore Roosevelt (1858–1919), is marketed.

1903 Safety razors go on sale.

1907 Chicago company develops "Thor," first self-contained electric clothes washer.

1908 Disposable paper cups are introduced.

1913 Brillo pads are commercially available.

1913 First home refrigerator is manufactured in Chicago.

1915 Lipstick is marketed for first time.

1918 First pop-up toaster is patented in United States. It is marketed in 1930.

c. 1920 Home radios are introduced.

1920 New Jersey company, Johnson & Johnson, introduces Band-Aids.

1924 First disposable handkerchiefs are manufactured in United States. Originally called Celluwipes, they are eventually renamed Kleenex.

1927 Wall-mounted can openers become available.

1928 Introduction of adhesive tape.

1932 Flexible rubber ice tray is invented in United States.

1936 American band leader Fred Waring successfully markets electric blender invented by Polish American Stephen Poplawski.

1937 First cellophane tape is commercially available, known as "Sellotape."

▶

c. 1939 Television receivers are commercially available. Sales of the new electronic device mushroom after World War II.

1945 American chemist manufactures airtight kitchen storage bowls and boxes, calling them Tupperware.

1947 U.S. scientist Percy Spencer accidentally discovers microwave cooking when microwave signals melt candy bar in his pocket. First commercial microwave oven becomes available this year. (Compact microwave ovens are introduced in 1967.)

1947 Polaroid camera is developed.

1953 Low-cost aerosol cans are available, following invention of simple plastic valve device.

c. 1954 Color television is marketed.

1955 Velcro fastener is marketed. It contains hundreds of tiny hooks that hold fast when pressed together.

1955 Inexpensive transistor radio goes on sale.

1956 British firm, Wilkinson Sword, introduces stainless-steel razor blades.

1960 First felt-tip pens go on sale.

1960 Introduction of aluminum can. Tab-top can follows in 1963.

1961 Self-wringing mop is invented and marketed in United States.

1961 Electric toothbrush is introduced.

1963 Home video recorder becomes available.

1963 Introduction of push-button telephone.

1964 First electric carving knives are commercially available.

1966 Small, disposable photographic four-shot flashcube is introduced for sale.

1970 Phillips develops first videocassette recorder (VCR) especially designed for home use.

1970s Compact-disc (CD) technology becomes available.

1972 Kodak introduces pocket Instamatic cameras with cartridges.

1973 Push-through tabs are invented for use on cans, in answer to environmental and safety concerns.

1974 Electronic pocket calculators are first marketed.

1975 First desktop microcomputer is marketed. MITS' Altair computer sells rapidly.

c. 1976 Culsinart home food processor is introduced.

1978 First automatic-focus camera is marketed by Konica.

1979 Sony Walkman, portable tape player with headphones, is introduced.

1981 First talking wristwatch is introduced.

1982 Wristwatch-sized television is introduced by Japanese, with a 1.2-inch screen.

1982 Kodak markets 15-picture film disk.

1985 Sony introduces world's largest television: 80-by-150-foot Jumbo Tron.

1989 Household plastic wrap in various colors is marketed.

Furniture and Furniture Styles

B.C.

c. 1320 Furniture in Egypt by this time is collapsible for easy transport. Beds are simple wooden frames with woven cords lashed across. Wooden stools, footstools, chairs, and headrests are in use. Tables are virtually unknown.

c. 700 Egyptians are using tables made of wood by this time. Bronze tables become popular in Greece.

c. 500 Bed legs begin to project above the frame, evolving into headboards and footboards.

A.D.

300 *Klismos* chair, with front legs curving forward and rear legs curving backward, has by now become most popular chair in Greece. It later influences Roman and European design.

▶

c. 800 Charlemagne (c. 742–814), ruler of medieval European empire, is said to have had three special tables, one made of gold and two of silver.

1300 Craftsmen develop frame-and-panel method of construction for furniture. Thin wood panel set inside frame of thicker wood makes for lighter-weight but strong cabinetry.

1300s Cupboards, desks, and boxes with compartments are first built in England.

1400s High-backed settle makes first appearance, in England. English oak is primary furniture material. Gothic style is dominant.

1400s Fixed writing desk is developed in Italy. Furniture is frequently decorated with painted stucco.

1500s Credenza and armoire are developed in France.

1500s *Vargueno*, tall chest with fall front and small drawers inside, is invented in Spain.

1500s In England, beds are now heavily draped for privacy. Gothic style is replaced by Italianate Renaissance ornament. Chairs are more widespread, although stools remain the dominant form of seating. Gateleg tables are introduced.

1570 "The Great Bed of Ware," measuring 12 feet square, is built in England for third earl of Huntingdon.

1600s Classic wing chair appears in England. Heavily upholstered chair has wings, or lugs, at either side to block drafts.

1620 Simple chest of drawers is introduced in England.

1620s Baroque style sweeps through Europe.

c. 1650 Furniture manufacture begins in colonial America. Alcove bed, high-backed settle chair, and settle table (dining table that converts into high-backed seat) are popular items.

1660 Louis XIV (1638–1715) organizes royal furniture workshop, which develops the influential Louis XIV style in cabinetry and furnishings. There is extensive use of veneers, inlays, and gilding with metals.

1680 Banister-back chair makes first appearance, in New England.

c. 1685 Pennsylvania walnut stretcher table, with drawers along top and braced by stretcher at bottom, is introduced.

1690 Highboy is invented in America by adding short legs to chest of drawers.

1700s In France, highly crafted rococo Louis XV style cabinetry comes into vogue. Chiffonier, étagère, and bookcase are developed.

c. 1700 Slat-back chair appears simultaneously in New York, Pennsylvania, and New England.

c. 1700 Elegantly curving Queen Anne style of furniture, marked by extensive use of cabriole leg, is developed in England and remains popular through early 1700s. Dropleaf table, corner chair, corner cupboard, and china cabinet have recently been developed in England.

1730 Wallpaper becomes popular among wealthy colonial merchants in United States.

1740s Resurgence of Gothic style in England.

1750s French develop neoclassical style, also called Louis XVI style.

c. 1750s Spindle-backed Windsor chair begins to become popular.

1754 Thomas Chippendale (1718–79), world-famous furniture maker, publishes *The Gentleman and Cabinet-Maker's Director*, stressing flamboyant rococo style.

1788 *Cabinet-Maker and Upholsterer's Guide*, by English furniture maker George Hepplewhite (d. 1786), is published. He moves away from rococo to simpler styles and includes first designs for three-part dining table and sideboard.

1790 Federal style, variation of European neoclassical, first appears in United States.

1790 Shakers begin selling their distinctive slat-back chairs in New Lebanon, New York.

1791 English furniture maker Thomas Sheraton (1751–1806) publishes his *Cabinet-Maker and Upholsterers' Drawing Book*. He promotes modified neoclassical style.

1791 American carpet industry begins with manufacture of Turkish and Axminster carpets.

1792 Duncan Phyfe (1768–1854), one of America's greatest furniture makers, opens his cabinetry shop in New York City. A master of neoclassical style, he successfully employs the (then) new factory system in cabinetmaking, eventually employing over 100 craftsmen.

1804 French emperor Napoleon orders creation of new style to reflect his great achievements. The resulting Empire style, featuring

▶

grandiose furniture designs, borrows heavily from Roman styles.

1820 Noted chair factory is founded in Hitchcockville, Connecticut, by Lambert Hitchcock (1795–1852). Hitchcock chair is produced in parts sent to South Carolina for assembly.

1820s Gothic style again comes into vogue in England, partly due to influence of Sir Walter Scott's medieval novels.

1830 Biedermeier style, stressing classic Roman lines with decorative detail, is developed in Europe.

1848 Power loom to weave Brussels, Wilton, velvet, and tapestry carpets is invented by American Erastus Bigelow (1814–79).

1850 By this time, metal springs are used in chair and sofa construction in England. Mass production of machine-made furniture begins late in 1800s.

1893 Art Nouveau style comes into vogue in Belgium and soon spreads throughout Europe.

1917 New art movement in Netherlands, De Stijl, emphasizes rectangular forms in primary colors.

1919 Bauhaus school of design is established. It has marked influence on interior design.

1925 First use of tubular steel, in Wassily chair of Marcel Breuer (1902–81).

1940 New York Museum of Modern Art establishes a Department of Industrial Art for sponsoring design competitions and furniture exhibitions.

1940 Union workers oppose innovation of foam upholstery.

1957 Tulip chair, fiberglass chair mounted on single aluminum support, is designed by Eero Saarinen (1910–61).

Customs and Traditions

B.C.

3500 Sumerians by this time customarily throw salt over their shoulders after accidental spills of the valued seasoning, to prevent bad luck.

c. 3500 Sumerian noblemen and noblewomen adorn themselves with jewelry made of precious metals and gems by this time, earliest known instance of what becomes long-standing tradition among powerful and rich.

c. 3000 Egyptians establish annual celebration of New Year. They celebrate at autumnal equinox, and it is not until Roman times that January 1 is celebrated as New Year's Day (after 153 B.C.).

3000 Birthday celebrations begin as Egyptian pharaohs preside over lavish feasts in honor of themselves.

c. 2800 Egyptians begin wearing wedding rings, believing the circular shape to symbolize eternity.

c. 1800 Babylonian kings initiate use of handshake. They grasp the hand of god Marduk's statue to receive his power symbolically. This gesture later comes to symbolize good faith.

c. 600 Priests in India, believing that repeating a prayer makes it more powerful, begin using knotted strings as rosaries to help them keep track of number of times a prayer is repeated. Monks of eastern Christian Church adopt rosary in third century A.D.

100s Druids in British Isles celebrate winter by hanging mistletoe in their homes, a tradition that later becomes associated with Christmas.

c. 100 Wedding cakes debut in Rome as small individual wheat cakes for wedding guests. Since wheat is symbol of fertility, guests often fling their cake at bride to bless her.

44 Roman Senate declares Julius Caesar's birthday a holiday, first in long tradition of birthday-holidays.

A.D.

c. 200 Germanic Goths initiate tradition of "best man" by taking their best fighter with them to abduct brides from neighboring villages.

▶

c. 750 Tradition of Christmas tree begins in Germany, though tree is at first undecorated. By 1500s, practice of decorating tree is well established in Germany.

1200s Birthday cakes with candles are first used to celebrate children's birthdays at German *Kinderfests*. Candles, symbolizing life, are kept lit all day.

1500s Brides in England and France begin wearing white wedding dresses as symbol of virginity. Many priests find such advertisement inappropriate and voice their preference for customary yellow dresses.

1600s Italian noblemen begin practice of crossing knife and fork on their plate to form cross as symbol of thanksgiving.

1858 Queen Victoria's daughter Princess Victoria establishes new tradition in wedding music by selecting for her wedding Richard Wagner's "Bridal Chorus" from *Lohengrin* (processional) and Felix Mendelssohn's "Wedding March" (recessional). These two pieces continue to be popular at weddings today.

1893 Robert H. Coleman publishes song called "Good Morning to All" by Mildred Hill of Louisville, Kentucky, adding second stanza of his own that starts with "Happy birthday to you." Song becomes nationwide hit, and today no birthday is complete without it.

Villains: Infamous Criminals and Crimes Since 1900

See also Historic Assassinations chronology.

1906 Enraged husband Harry K. Thaw shoots and kills noted American architect Stanford White (1853–1906) at Madison Square Garden (June 25). Murder is prompted by White's affair with Thaw's wife, actress Evelyn Nesbit. Thaw successfully pleads insanity and is institutionalized.

1922 Henry Desire Landru (1869–1922), known as "modern Bluebeard," is executed (February 22) in France for murdering 12 women to whom he had been engaged.

1927 Italian-Americans Nicola Sacco (1891–1927) and Bartolomeo Vanzetti (1888–1927) are executed (August 23) for murdering two factory officials in South Braintree, Massachusetts (1920). Many believe the conviction unfair, citing ethnic origin and anarchist political beliefs of accused.

1929 Valentine's Day Massacre in Chicago (February 14). This brutal gangland killing of seven members of "Bugs" Moran gang is part of struggle for control of bootleg liquor traffic. Mobster Al "Scarface" Capone (1899–1947) is thought responsible for massacre.

1931 Capone is sentenced to 11 years in prison and fined $50,000 (October 24) for tax evasion. Sentence ends his control over the mob.

1932 Charles Lindbergh, Jr., infant son of American aviator Charles Lindbergh (1902–74), is kidnapped (March 1) and later found dead (May 12). German-American carpenter Bruno Hauptmann (1899–1936) is executed in 1936 for the murder. Some believe Hauptmann was framed.

1934 Bonnie Parker (1911–34) and Clyde Barrow (1909–34) are shot dead by Texas Rangers near Shreveport, Louisiana (May 23), ending two-year string of bank robberies and 12 murders.

1934 John Dillinger (1902–34), American bank robber and "public enemy number one," dies (July 22) in shootout with FBI agents outside Chicago movie theater. Dillinger is responsible for string of bank robberies and 16 murders.

1934 American gangster Charles "Pretty Boy" Floyd (1904–34) is killed by federal agents during manhunt in Ohio (October 22). Floyd is called "most dangerous man alive" for committing series of bank robberies armed with machine gun.

▶

1936 American mobster Charles "Lucky" Luciano (1897–1962) is convicted for running ring of over 1,000 prostitutes in New York City. Luciano is known as boss of all Mafia bosses.

1947 Al Capone dies.

1950 Brinks, Inc. armored-car company is robbed of $2.7 million in Boston by masked robbers (January 17). The robbers are later caught and convicted (1956).

1952 Bank robber Willie Sutton is caught by authorities after escaping from jail in 1947. His second jailbreak over (after a 1932 escape from Sing Sing), Sutton remained behind bars until 1969.

1958 Nineteen-year-old Charles Starkweather and 15-year-old girlfriend Caril Fugate are arrested near Douglas, Wyoming following spree of 11 murders, including those of Fugate's mother, stepfather, and stepsister. Starkweather is sentenced to execution and Fugate to life imprisonment.

1963 Robbers steal $5 million worth of jewelry and cash from British train outside London (August 14) in one of greatest train robberies in history.

1964 Kitty Genovese is murdered in Queens, New York while 37 witnesses do nothing to prevent the crime (March 13). Case become notorious indictment of city residents' fear of becoming "involved" in such attacks.

1966 Richard Speck massacres eight student nurses in Chicago by stabbing and strangulation. He is convicted in 1967.

1966 Architecture student Charles Whitman shoots 12 people to death from tower at University of Texas in Austin (August 1). He finally is killed by off-duty policeman.

1967 "Boston Strangler" Albert DeSalvo, confessed murderer of 13 women in Boston between 1962 and 1964, is sentenced to life on lesser charges (January 18). He is never charged with the murders because of "insufficient" evidence.

1969 Actress Sharon Tate and four others are brutally murdered in Los Angeles (August 9) by members of cult led by Charles Manson. Manson is convicted of the murders in 1971.

1971 Alleged Mafia boss Joseph Colombo, Jr. (1923–71), is shot to death in New York City (June 28).

1972 Reputed Mafia leader Joseph Gallo is shot and killed in New York City while celebrating his birthday at a Little Italy restaurant (April 7).

1974 American newspaper heiress Patricia Hearst is kidnapped by terrorist Symbionese Libration Army (February 4) and later assists her captors in San Francisco bank robbery (April 15, 1975). She is convicted of aiding in the robbery (1976); William and Emily Harris are convicted of the kidnapping (1976).

1974 Zebra serial murders in San Francisco involve random killings of 12 whites. Four black Muslims are indicted.

1975 Former Teamster president Jimmy Hoffa (1914–75) disappears and is presumed murdered (July 30).

1977 Gary Gilmore, convicted of 1976 murders of two Brigham Young University students, asks for and receives death penalty. He becomes first person executed in United States in 10 years.

1978 David Berkowitz, "Son of Sam" killer who terrorized New York City, pleads guilty (May 8) to six murders and seven assaults.

1978 Former Italian premier Aldo Moro (1916–78) is kidnapped (March 16) and murdered (May 8) by Red Brigade terrorists.

1978 Rev. Jim Jones reportedly orders mass suicide by his fanatical followers at Jonestown, Guyana, retreat. Jones and over 900 followers are later found dead, from drinking poisoned Kool-Aid.

1980 Herman Tarnower, author of the best-selling *The Complete Scarsdale Medical Diet*, is murdered by his mistress Jean Harris (March 11).

1981 Jack Ratcliffe, "Yorkshire Ripper" reportedly driven by "divine commission" to kill prostitutes, is convicted of murders and beatings of 13 women in England.

1982 Wayne B. Williams is convicted for murdering two of 28 black children and teenagers missing or murdered in Atlanta during past two years.

1982 Tylenol capsules are removed from the market after eight people who took them are poisoned by strychnine contamination.

1982 Thieves escape with $9.8 million in cash stolen from New York City armored-truck company.

1983 Mass murderer Henry Lee Lucas is sentenced to life imprisonment (November 9) for murder of his 15-year-old common-law wife and his mother. Lucas claims he has killed more than 100 people.

▶

1984 Passenger Bernhard Goetz shoots and seriously wounds four black youths after they reportedly menace him on New York City subway. Ongoing problem of subway crime and racial overtones make this a highly sensational case. Goetz is convicted only on illegal-weapons charge.

1987 Italian court convicts 338 members of Sicilian Mafia in largest of such cases to date (December).

1989 Serial killer Theodore Bundy is executed (January 24) in Florida for 1978 murder of 12-year-old Kimberly Leach. Bundy, also convicted for two other murders, confesses to at least 16 additional murders before being executed.

1989 California ''Nightstalker'' Richard Ramirez is convicted of 13 murders and 30 other crimes, including rape and other sex offenses committed during 1984 and 1985. He is sentenced to death.

11 Wars and Military History

Wars, from Ancient Greece to the Modern World

See also national and regional chronologies.

B.C.

c. 735–461 Messenian wars, between Sparta and Messenia. Spartan expansionism results in first war (c. 735–c. 715) and conquest of Messenia. The second (c. 650–c. 600) and third (464–461) wars begin with Messenian revolts and end in Spartan victories.

c. 1203–c. 1193 Trojan War. Greek warriors under leadership of Agamemnon, Achilles, and Ulysses lay siege to city of Troy, which ultimately falls c. 1193.

499–c. 449 Persian wars between Persia and Greek city-states: 492, invading Persians turn back after storm mauls invasion fleet; 490, Persians withdraw after outnumbered Athenians win at Marathon; 480, Persians are defeated despite early victories (Thermopylae and the capture, in 480, of Athens). Greeks crush Persian fleet at Salamis (480) and defeat army at Plataea and Mycale (both in 479). Final peace c. 449.

431–404 Peloponnesian War between Athens and Sparta, ending Athenian dominance over Greece. Spartans besiege Athens (431), which holds out but is ravaged by plague (430–428). Later Athens is weakened by disastrous expedition against Syracuse on Sicily (415–413). Following naval battles at Notium, Arginusae, and Aeogospotami (405, Athenian Navy destroyed), Sparta captures Athens but does not destroy it.

395–387 Corinthian War between Sparta and Athens and allies (including Persia). Sparta wins at Nemea (394) and Coronea (394) but loses decisive naval battle at Cnidus (394). Spartan control of Athens is ended.

355–338 Sacred wars (third and fourth) among Greek city-states. The two wars (355–346 and 339–338) end in the Macedonians gaining control over all of Greece.

343–290 Samnite wars between Rome and Samnite tribes in central Italy. Rome engages in three wars (343–341, 316–304, and 298–290) before subjugating the Samnites.

340–338 Latin Wars. Latin cities unsuccessfully rebel against control by Rome.

334–324 Alexander's wars of conquest. Alexander the Great (356–323), king of Macedonia and Greek peninsula, rapidly conquers vast empire including Persia, Egypt, Media, Scythia, and part of India. Notable battles include Issus (333), Tyre (332), Gaugamela (331), and Hydaspes (326).

323–322 Lamian War between Macedonia and Athens and allies. Athens, after Alexander the Great's death, rebels against Macedonia's rule and is defeated at Crannon (322).

323–218 Wars of Diadochi between rival generals for control of Alexander's empire after his death. After Battle of Corupedion (281), empire is permanently divided into Macedonia, Seleucid Asia Minor, and Ptolemaic Egypt.

282–272 War between Rome and Tarentum. Tarentum unsuccessfully contests expansion of Roman control of Italian peninsula. Battle of Asculum (279), a costly victory for forces led by King Pyrrhus of Epirus over the Romans, gives rise to expression "Pyrrhic victory."

276–195 Syrian Wars, series of five conflicts (276–272, 260–255, 245–241, 221–217, and 201–195) between Ptolemaic Egypt and Seleucid Empire for control of eastern Mediterranean region.

266–262 Chremondian War. War between Sparta and Athens ends with Spartan victory.

264–241 First Punic War between Rome and its rival Carthage. War begins when Rome and Carthage take sides in local dispute in Sicily. Following victory at Mylae (260), a failed invasion of Carthage, and

▶

a naval victory at Aegates (241), Rome wins war and gains Sicily.

218–201 Second Punic War between Rome and Carthage. Carthaginian conquests in Spain lead to war. Great Carthaginian general Hannibal (247–182?) crosses Alps from Spain to Italy and conquers much of Italy before being halted by lack of supplies. Roman general Scipio Africanus Major (234–183?) in turn invades Carthage and defeats Hannibal at Zama (202). Carthage falls and never regains its former power.

214–148 Macedonian wars between Macedonia and Rome. In the first (214–205), Macedonia wins advantageous terms after siding with Carthage during Second Punic War. Rome wins second war (200–196) and increases its control over Greece. By third (171–168) and fourth (149–148) wars, Rome reduces Macedonia to province and thereafter dominates region.

149–146 Third Punic War between Rome and Carthage. Commercial rivalry and popular hatred for Carthage in Rome lead to war. Rome invades Carthage and, after capturing it in bloody fight, razes it and sells survivors into slavery.

88–64 Mithridatic Wars. Unsuccessful wars against Rome by king of Pontus (in Asia Minor) in 88–44, 83–81, and 74–63 result in Roman mastery of Asia Minor.

58–51 Gallic Wars, in which Julius Caesar (c. 102–44) leads Roman troops in conquest of Gaul. Between 58 and 51, Caesar's armies conquer Gaul, invade British Isles twice (in 55 and 54), and put down revolt in Gaul. Caesar gains great prestige by these victories.

49–46 Roman Civil War. Julius Caesar defies Senate order to disband his armies, makes his famous crossing of Rubicon (January 10–11), and marches into Italy against his rival Pompey (106–48). Victorious, Caesar enters Rome and is made consul. Thereafter, by victories at Pharsalus (48), Zela (47), and Thapsus (46), he defeats his enemies.

A.D.

167–75 Wars of the Marcomanni and Quadi, series of campaigns by Roman emperor Marcus Aurelius (121–180) against invading German tribes.

406–19 Visigoths invade Roman Empire, resulting in sacking of Rome by Alaric (410) and establishment of Visigoths in Gaul (412) and Spain (416).

441–53 Attila the Hun (406?–53) invades the Roman Empire. He mounts devastating attack on Roman Empire, advancing deep into Eastern Empire (441, 447), Gaul (451), and Italy (452). Famine and pestilence in Italy force Attila to withdraw. He dies (453) before mounting new attack.

533–54 Byzantines reconquer parts of Western Roman empire. Emperor Justinian (527–65) attempts to restore Roman Empire, conquering Africa (533–34) and Italy (535–54).

711–1492 Muslims invade Europe. Crossing from North Africa to Iberian Peninsula, they drive northward until Franks turn them back at Poitiers (732). Eastern conquests include Syria (634–41), Persia (635–42), and Egypt (639–42). By 800s, Christian kingdoms in northern Spain begin pushing Muslims southward, finally driving them out of Spain (1492).

c. 771–804 Charlemagne's wars of conquest. After securing his control over Frankish domains, Charlemagne (c. 742–-814) conducts successful campaigns in Italy and against Germanic tribes to east, creating vast empire that briefly unites much of Europe.

787–c. 1000 Norsemen invade Europe. Danes (Norsemen from Denmark) begin attacks on England in 787 and by 800s control much territory in British Isles. Danes eventually rule England for a time in 11th century. Norsemen invade France c. 845, settling what becomes duchy of Normandy. Norsemen settle at Novgorod in 800s, establishing ruling Rurik dynasty of early Russia.

1066 Norman Conquest. William I the Conqueror (1027–87), claiming English throne, invades England and defeats his rival at Battle of Hastings. William subsequently puts down revolts and imposes Norman rule and customs on English.

1096–1291 Crusades, by European Christians against Muslim Seljuk Turks for control of Holy Land. On First—and only successful—Crusade (1096–99), crusaders take

▶

Jerusalem (1099) and set up Crusader States. Second Crusade (1147–49) fails completely. In Third Crusade (1189–92), crusaders capture Acre (1191) but fail to take Jerusalem, now back in Muslim hands. Fourth Crusade (1202–4) never reaches Holy Land, crusaders diverting to sack Constantinople instead. They set up Latin Empire of Constantinople. Fifth (1218–21) and Seventh (1248–54) crusades involve unsuccessful invasions of Egypt. Sixth (1228–29) and ill-fated Eighth (1270–72) crusades wind up in truce negotiations rather than in fighting. Muslims take Christians' last outpost in the Holy Land, Acre, in 1291.

1207– Mongol invasions begin. Genghis Khan (1167?–1227) leads Mongol hordes in invasions of territories from China (from c. 1207) and Persia (1218) to Russia (1223). He and his successors create vast empire that for a time includes much of Asia and part of Eastern Europe.

1264–67 Barons' War. English barons led by Simon de Montfort (1208–65) rebel against inept rule of Henry III (1207–72), and by victory at Lewes (1264) force reforms.

1337–1453 Hundred Years' War between England and France over English-held territories on Continent (in modern France) and English claims to French throne. During intermittent warfare, England gains much of France. Joan of Arc (c. 1412–31) lifts siege of Orléans (1429), however, and then French drive English forces off Continent. Notable battles include Crécy (1346), Poitiers (1356), Agincourt (1415), Orléans (1429), and Patay (1429).

1455–85 Wars of the Roses. Dynastic wars in England fought by House of York to gain throne from ruling House of Lancaster. Warfare ends with crowning of Henry VII (1457–1509), founder of Tudor line. Notable battles include St. Albans (1455), Towton (1461), and Bosworth Field (1485).

1494–1559 Italian wars, between Spain and France for control of Italy. Unsuccessful French invasion of Italy (1494) draws Spain into fighting. After further intermittent fighting, Spain gains control of Italy.

1519–1522 Conquest of Mexico. Hernando Cortés (1485–1547) conquers Aztec Empire in what is now central and southern Mexico.

1557–82 Livonian War. War among Russia, Sweden, and Poland over Baltic territories.

1562–98 Wars of Religion, series of civil wars between Catholics and Protestants (Huguenots) in France. Wars end with accession of Protestant King Henry IV (1553–1610), who converts to Catholicism and then issues Edict of Nantes (1598), ordering toleration of Protestantism.

1618–48 Thirty Years' War, complex war involving, variously, Denmark, Sweden, and France against Holy Roman Empire and Hapsburg dynasty, which dominate Europe. Ultimately France replaces Hapsburg-dominated Spain as Europe's leading power, and Holy Roman Empire is devastated by the war. Notable battles include Breitenfeld, Lützen, Nordlingen, and Rocroi.

1642–49 English Civil War. (*See page 453.*)

1652–78 Dutch wars. Two trade wars between England and United Provinces (Netherlands), in 1652–54 and 1664–67, and general European war resulting from French expansionism, 1672–78.

1655–60 First Northern War. Succession to Swedish throne sparks this war between Sweden and Poland and its allies. Poland loses its claim and Sweden gains some Danish territory.

1667–68 War of Devolution. French claims to Spanish Netherlands (by right of marriage) result in invasion that Netherlands, Sweden, and England successfully oppose.

1676–1878 Russo-Turkish wars, long series of wars between Russia and the Ottoman Empire in which Russia greatly expands its borders at expense of weakening Ottoman Empire. Among wars are those of 1676–81, 1695–96, 1710–11, 1736–39, 1768–74, 1787–92, 1806–12, 1828–29, 1853–56, and 1877–78.

1689–97 War of the Grand Alliance. Expansionist aims of French king Louis XIV (1638–1715) result in French invasion of Rhineland and Grand Alliance of Holy Roman Empire, England, and other states to oppose French. War ends inconclusively.

▶

Notable battles include Fleurus (1690) and La Hougue (1692).

1700–21 Second Northern War, in which Swedish hegemony in Baltic region is successfully challenged by coalition including Russia, Denmark, and Poland. Russia gains Baltic territories and emerges as a leading power in region.

1701–14 War of Spanish Succession. Great Britain, Netherlands, Holy Roman Empire, Poland, and others war against Spain and France after Philip V (1683–1746), grandson of French king Louis XIV (1638–1715), is named king of Spain. Though Philip remains in power, British are generally successful against France (see Queen Anne's War). Notable battles include Blenheim (1704), Ramillies (1706), and Malplaquet (1709).

1702–13 Queen Anne's War between France and Great Britain for control of colonial territories in North America; part of larger War of Spanish Succession. Despite bloody attacks on frontier settlements by allied French and Indian raiders, Great Britain gains control of French Acadia (modern Nova Scotia), Newfoundland, and the region around Hudson Bay.

1733–38 War of the Polish Succession, struggle for Polish throne between candidates supported by France on one side and Austria and Russia on other.

1740–48 War of Austrian Succession. Complex war involving rivalries over succession to considerable Hapsburg domains in Europe, succession to throne of Holy Roman Empire, and the continuing struggle between France and Britain over colonial territories. France, Spain, Prussia, and others war against Great Britain and Austria. Nevertheless, Maria Theresa (1717–80) succeeds to Hapsburg domains in Austria; her husband, Francis I (1708–65), becomes Holy Roman emperor; and Prussia gains Silesia at expense of Austria.

1744–48 King George's War. Inconclusive war for control of North American colonial territories, it is part of larger War of Austrian Succession. War ends without significant territorial changes.

1754–63 French and Indian War, between Great Britain and France over colonial territories in North America; part of larger Seven Years War. Fighting begins over control of Ohio Valley, then extends into Canada. British victories at Plains of Abraham (September 13, 1759) and Montreal (September 8, 1760) lead to conquest of all French Canada. French later concede control of North America to Great Britain.

1756–63 Seven Years War. France, Austria, Russia, Saxony, Sweden, and Spain oppose Great Britain and Prussia. In Europe, Prussian king Frederick the Great succeeds in establishing Prussia as important European power. By victories in North America (see French and Indian War, above) and India, Great Britain wins most of France's colonial territories and becomes world's leading colonial power.

1767–99 Mysore wars, series of wars between British colonial forces and kingdom of Mysore in India, ending with British conquest of Mysore.

1775–83 American Revolution. (*See page 454.*)

1775–1818 Maratha wars, series of wars in India (1775–82, 1803–5, and 1817–18) in which British gain control of Maratha confederacy.

1789–1802 French Revolution. (*See page 456.*)

1801–5 Tripolitan War, conflict between United States and pasha of Tripoli over his demands for tribute as protection from Barbary Coast pirates. U.S. naval blockade of Tripoli fails, but invasion by U.S. troops (1805) quickly produces peace agreement.

1803–15 Napoleonic wars. (*See page 458.*)

1809–25 Latin American wars for independence. Among states that secured independence are Argentina (1816), Chile (1818), Peru (1824), Bolivia (1825), Venezuela (1821), and Colombia (1819).

1812–15 War of 1812 between United States and Great Britain, sparked by British harassment of American shipping during British embargo of Napoleonic France. War is declared June 18, 1812. Americans invade Canada and burn Toronto (1813); Great Britain captures and burns Washington, D.C. (1814) in retaliation. British bombardment of Baltimore's Ft. McHenry (September 13–14) is recounted

▶

in "The Star-Spangled Banner" by Francis Scott Key (1779–1843). Andrew Jackson (1767–1845) wins Battle of New Orleans (January 8, 1815), which takes place before he receives word the war has ended (by the Treaty of Ghent, December 24, 1814).

1821–32 War of Greek Independence, successful Greek rebellion against Ottoman rule. With aid of Egyptian fleet, Ottoman Turks invade Greece and by 1827 win near-complete victory. Intervention by European powers beginning at Battle of Navarino (1827) results in destruction of Egyptian fleet and eventual Greek victory.

1838–1919 Anglo-Afghan wars. Series of wars (1838–42, 1878–80, and 1919) in which Afghanistan established its independence from control by British authorities in India.

1846–48 Mexican War, between United States and Mexico, resulting in U.S. annexation of Mexican territory, comprising U.S. Southwest and California. Sparked by border dispute and U.S. annexation of Texas, war effectively ends with capture of Mexico City (1847) by U.S. invasion force.

1846–49 Sikh Wars, in which British colonial forces defeat the Sikhs and annex Punjab region of northern India.

1853–56 Crimean War. Russia sparks war by invading domains of the declining Ottoman Empire. Great Britain, France, and others aid Turks, launching attack in Crimea at Sevastopol. After long siege, Russia evacuates city (1855). Notable battles include Alma (1854) and Balaklava (1854).

1861–65 American Civil War. (*See page 460.*)

1865–70 War of the Triple Alliance, bloody war between Paraguay and allied Brazil, Argentina, and Uruguay. Paraguay initially invades Brazil but is ultimately invaded and conquered by allied forces. It loses over half its prewar population of some 525,000 to various causes.

1866 Seven Weeks War, between Prussia and Austria, which Prussian chancellor Otto von Bismarck (1815–98) provokes. By subsequent treaty, victorious Prussia excludes Austria from confederation of northern German states, precursor of modern Germany.

1870–71 Franco-German War, initiated by Otto von Bismark (1815–1898) to promote German unity. French suffer humilating defeat, lose province of Alsace-Lorraine, and must pay war reparations.

1879–84 War of the Pacific, over mineral-rich Atcama Desert, between Chile and Bolivia, with Peru declaring war on both nations. Chile successfully invades both nations and by subsequent treaties gains all of Bolivia's coastal territory and mineral-rich Peruvian lands as well.

1894–95 Sino-Japanese War (First), between Japan and China for control of Korea, long a vassal-state of China. Japan successfully invades Chinese territory, forcing a peace treaty in which China gives up Taiwan and other territories and recognizes Korea's independence.

1895–96 Ethiopian War. Italy's attempt to expand beyond its colony of Eritrea results in failure.

1898 Spanish-American War. Brief war between United States and Spain, which is sparked by Spain's harsh treatment of independence fighters in Cuba and by the sinking of *U.S.S. Maine* in Havana harbor (February 15). United States defeats Spanish colonial forces in Philippines and Cuba, gaining control of both territories. Major battles include Manila Bay (1898) and Santiago (1898).

1899–1902 Boer War, bitter war between British and Dutch colonists (Boers) in Transvaal and Orange in South Africa. After hard fighting, Boers are forced to accept British rule.

1904–05 Russo-Japanese War between Russia and Japan over control of Korea and Manchuria. Japanese deliver Russians a humiliating defeat at Battle of Mukden and off Tsushima Islands, where the Russian fleet is destroyed.

1911–12 Italo-Turkish War. Italy takes advantage of weakening Ottoman empire and invades North Africa, taking from the Turks what is Libya today. Ottoman Turks accede by Treaty of Lausanne (1912).

1912–13 Balkan Wars, two wars that arouse nationalistic fervor in Eastern Europe,

which in turn contributes to outbreak of World War I. First war (1912) is fought by Serbia, Montenegro, Greece, and Bulgaria against Ottoman Turks, who are driven out of Eastern Europe. Second (1913), over Serbia's demand for territory from Bulgaria, ends in Serbia's loss of territory to Bulgaria, Greece, Rumania, and Turkey.

1914–18 World War I. *(See page 462.)*

1917–22 Russian Revolution and Civil War. *(See page 465.)*

1918–20 Baltic War of Liberation. Estonia, Latvia, and Lithuania, having declared independence from Russia during Bolshevik revolution of 1917, successfully fight off invasions by both Bolsheviks and Germans at end of World War I.

1919–20 Russo-Polish War. Poland joins Ukrainian nationalists in attempting to gain Ukraine's independence from Communist Russia. Poland gains some territory in Ukraine and Byelorussia by the Treaty of Riga (1921).

1921–22 Greco-Turkish War. Greeks unsuccessfully invade Turkey and thereby lose eastern Thrace (by Treaty of Lausanne, 1923).

1927–49 Chinese Civil War. *(See Revolution and Civil Wars, page 466).*

1932–35 Chaco War, between Bolivia and Paraguay over disputed territory. Paraguay is awarded the bulk of the land.

1935–36 Italo-Ethiopian War, invasion and conquest of Ethiopia by Italy. It reveals weakness of the League of Nations as a peacekeeping force and contributes to rising international tensions on eve of World War II.

1936–39 Spanish Civil War. Conservative nationalist faction in Spain (aristocracy, military, fascists, and the church) defeats Loyalist faction (liberals, socialists, communists, and anarchists) in bloody war that also serves as first combat test for weaponry that later appears in World War II. It results in long Fascist dictatorship of Gen. Francisco Franco (1892–1975) and contributes to outbreak of World War II.

1937–45 Sino-Japanese War (Second). Chinese nationalist and communist factions, warring for control of China, unite (1936) in unsuccessful bid to prevent Japan from expanding further into Chinese territory from Manchuria. Japan quickly captures Peking and other major cities after outbreak of war (1937). After 1941, this war merges with World War II.

1938–45 World War II. *(See page 468.)*

1939–40 Russo-Finnish War. Soviet Union fights to acquire Finnish territory.

1944–49 Greek Civil War, for control of Greece, fought between rightist and communist factions after end of German occupation in World War II. Rightists win with British and U.S. aid.

1945–54 Indochina War. Communist nationalists oppose France's attempt to reestablish its colonial rule in Indochina after Japan's surrender in World War II. Communists under Ho Chi Minh (1890–1969) gain control of North Vietnam by 1950 and break French resistance at Dien Bien Phu (1954).

1948–49 First Arab–Israeli War. Israel successfully repulses invasions by Egypt, Iraq, Jordan, Syria, and Lebanon, all opposing creation of Israeli state. Separate armistice agreements end the conflict. Israel loses no territory, but some 400,000 Palestinian refugees flee from Israeli territories to neighboring Arab countries.

1950–53 Korean War. *(See page 471.)*

1956 Second Arab–Israeli War. Israel invades Egyptian Sinai amid Suez Canal crisis and captures Egyptian territory up to Suez east bank (October 29–November 6). Gaining guarantees of access to the Gulf of Aqaba, Israel turns captured territory over to the U.N. in 1957.

1967 Third Arab–Israeli War (Six-Day War, June 5–10). Israelis quickly establish air superiority and emerge victorious from ground engagements on various fronts. They capture Sinai Peninsula, Old City of Jerusalem and West Bank, and Golan Heights.

1960–63 Civil war in Kinshasa (Congo). Chaos erupts with independence from Belgium. Katanga province secedes but returns.

1960–75 Vietnam War. *(See page 472.)*

1967–70 Nigerian Civil War, unsuccessful attempt of Biafra to secede.

▶

1969–88 Ethiopian War. Sparked by attempted secession of Eritrea, war involves neighboring Somalia and other nations until signing of accord (1988).

1971 Pakistani Civil War. Separatist movement in East Pakistan revolts after West Pakistani government bans separatist Awami League. East Pakistan, aided by India's invasion of West Pakistan, finally established itself as independent Bangladesh.

1973 Fourth Arab–Israeli War (Yom Kippur War). Egypt, Syria, and Iraq attack Israel on Jewish Holy Day, but Israelis beat back Arab invaders before fighting ends. Israelis agree to limited withdrawal in Sinai as part of peace accords.

1975– Lebanese Civil War. Muslim factions rebel against the Christian government (April 1975), prompting invasion by Syria, which temporarily halts fighting. Israel later invades Lebanon (1982) and forces evacuation of PLO militiamen. A peacekeeping force of U.S. Marines is stationed in Beirut by 1983 but is withdrawn (1984) after truck bomb kills 239 in 1983.

1975–89 Angolan Civil War between U.S.- and communist-backed factions. Cuba sends troops. War ends in cease-fire.

1978–79 Cambodian (Kampuchean) War. Communist Vietnam invades Cambodia following border clashes and captures capital of Phnom Penh (1979). Vietnam installs new government, though supporters of various factions continue guerrilla warfare.

1979–89 Afghan Civil War. Soviet troops invade Afghanistan as part of Soviet-backed coup, in which pro-Soviet ruler Hafizullah Amin is put into power. Afghans rebel, however, mounting a guerrilla war that finally forces Soviets to withdraw troops (1989). Fighting continues after Soviet departure.

1982 Falkland Islands War. Argentina seizes these British-held islands off Argentine coast (April), but a British invasion quickly forces Argentine garrison to surrender (June).

1980– Iran–Iraq War. Border clashes erupt into full-scale war between Iran and Iraq soon after Muslim fundamentalist revolution in Iran. Bloody war destabilizes Persian Gulf region and leads to U.S. naval presence in area until a cease-fire agreement is reached.

English Civil War

War between proroyalist and proconstitutional factions, in which religion also plays a part. Generally, royalists are Anglicans and Catholics, while parliamentarians are Puritans. War briefly ends the English monarchy.

1641 Parliament issues Grand Remonstrance (December), listing grievances against King Charles I (1600–49).

1642 Parliament issues 19 Propositions (June 2) in attempt to limit king's powers. Charles resists, and war is declared (August 22).

1642 First battle of war, at Edgehill (October 23), proves indecisive, as do subsequent skirmishes in 1642.

1643 Battle of Newbury (September 20); victory for Parliamentarians.

1643 Parliament enters alliance with Scotland. Scots get in return Solemn League and Covenant (September 25), calling for religious reforms.

1644 Battle of Marston Moor (July 2), major victory for Parliamentarians led by Oliver Cromwell (1599–1658). They take control in northern England, formerly royalist stronghold.

1645 Charles rejects Uxbridge Propositions, which contain earlier demands and require he approve Solemn Leaue and Covenant (January–February).

▶

1645 Cromwell and independent faction gain control in Parliament. Parliamentary forces are reorganized as New Model Army.

1645 Battle of Naseby (June 14); victory for Parliament and decisive battle of war.

1645 General Leslie, dispatched by Parliament to Philipaugh (September 13), defeats royalist partisans in Scotland.

1646 Battle of Stowe-on-the-Wold (March 26), Parliamentary victory ending this phase of war.

1646 King Charles flees to Scotland (May 5).

1647 Scots turn king over to Parliament (January 30).

1647 King finally consents to Newcastle Propositions (May 31), giving Parliament control of the militia for 10 years and requiring him to accept Presbyterianism for three years.

1647 New Model Army seizes Charles (June 4) in split between Army, dominated by Independent faction, and Parliament, dominated by Presbyterians, over back pay and other issues. Army refuses to disband.

1647 Charles rejects new peace proposal (August), alternative to Newcastle Propositions (nullified by his seizure).

1647 Agreement of the People is issued by Agitators (Army faction). It contains basic concepts concerning sovereignty of people and their rights that later appear in U.S. Constitution.

1647 King escapes (November) and concludes alliance (December 26) with Scots, who oppose the increasing radicalism of Parliament.

1648 Battle of Preston (August 17–19). Parliamentary forces under Oliver Cromwell defeat Scots.

1648 Pride's Purge. Colonel Thomas Pride (d. 1658) arrests some 140 members of Parliament (December 6–7) to prevent negotiated peace between king and Parliament. Rump Parliament moves to put king on trial.

1649 King Charles is tried and executed at Whitehall (January 30), ending English monarchy for a time. Cromwell becomes leader of newly proclaimed Commonwealth (republican government).

American Revolution

1775 British Parliament proclaims Massachusetts to be in revolt against crown (February 9).

1775 Patrick Henry (1736–99) delivers famous "Give me liberty or give me death" speech before provincial Congress in Massachusetts (March 23).

1775 Paul Revere (1735–1818) sets out on historic ride (April 18) to warn Minutemen of coming British raid on Concord.

1775 Lexington and Concord (April 19) are sites of first skirmishes between colonial Minutemen and British regulars. Colonial siege of Boston begins soon after.

1775 Colonials seize Fort Ticonderoga and Crown Point (May 10–12).

1775 Continental Army is organized from colonial troops in Boston vicinty (May 31).

1775 George Washington (1732–99) is named commander of newly created Continental Army (June 15).

1775 Battle of Bunker Hill (June 17). Outnumbered Americans withstand several British charges before retiring in this first major battle of war.

1775 American expedition in Canada. Americans capture Montreal (November 12). They unsuccessfully attack Quebec (December 31) and withdraw in spring 1776.

1775 Congress creates American Navy (October).

1776 Pamphlet *Common Sense* by Thomas Paine (1737–1809) is issued in January, helping many Americans decide in favor of revolt.

1776 British attempts to rally support in Carolinas meet defeat in battle at Moore's Creek Bridge (February).

1776 British evacuate Boston (March 17) in face of American siege.

▶

1776 Congress proclaims Declaration of Independence (July 4).

1776 British occupy New York City (September 15). They win Battle of Long Island (August 27), forcing eventual American withdrawal from New York, across New Jersey, to Pennsylvania.

1776 Americans under Washington retreat after Battle of White Plains (October 28). Forts Washington and Lee fall to British (November 16–20).

1776 Washington crosses Delaware River (December 25), capturing over 900 Hessians in raid at Trenton. He routs British relief column at Princeton (January 3) and returns to New Jersey.

1777 French soldier Marquis de Lafayette (1757–1834) is commissioned major general in Continental Army. He helps Americans get French aid.

1777 Americans defeat British at Bennington, Vermont (August 16), weakening British and boosting American morale.

1777 British occupy Philadelphia (September 26) following victory at Brandywine Creek (September 11).

1777 Washington's attack at Germantown fails (October 4). Americans make winter quarters at Valley Forge.

1777 British campaign in New England fails (October 17). British attackers initially succeed at Fort Ticonderoga (July), then fail at Fort Stanwix (August), Freeman's Farm (September 19), and Bemis Heights (October 7). British Gen. John Burgoyne (1722–92) surrender 5,700 troops (October 17). Tide of war turns in Americans' favor.

1777 Congress adopts Articles of Confederation (November 15; ratified 1781).

1778 Treaty establishes formal alliance with France (February 6).

1778 Prussian general Baron Friedrich von Steuben is commissioned inspector general in the Continental Army (May). He organizes American Army into disciplined fighting unit.

1778 British, under new commander General Henry Clinton (1738?–95), abandon Philadelphia (June) as French fleet approaches American waters. They defeat Americans at Monmouth (June 28) and withdraw to New York City.

1778–79 British advance in South, taking Savannah (December 28, 1778) and Augusta (January 29, 1779) and holding them against American counterattacks.

1779 Spain joins war on American side (June), hoping to regain Florida and Gibraltar.

1779 American captain John Paul Jones (1747–92), commanding *Bonhomme Richard*, is victorious over *Serapis* in British waters (September 23).

1780 British take Charleston (May 12), capturing 5,400 American troops. They drive off attacking Americans at Camden, South Carolina (August 16).

1780 American general Benedict Arnold (1741–1801) joins British (September 25) after his treasonous activities are exposed by capture of British go-between, Maj. John André (1751–80).

1780 Kings Mountain (October 7) is first victory in South. It gives colonists moral support and time to reorganize.

1781 Mutinies by American troops in Pennsylvania and New Jersey (January) are quashed.

1781 Americans win Battle of Cowpens in South (January 17). By fall of 1781 they have British in Carolinas bottled up at Charleston.

1781 Americans win strategic victory at Battle of Guilford Court House and force Gen. Charles Cornwallis (1738–1805) to withdraw to Wilmington (March 15).

1781 British army of 7,500 conducts raids in Virginia until Americans force retreat to Yorktown (August 1).

1781 Siege of Yorktown begins (August). French fleet drives off British Navy, isolating British at Yorktown. Allied American and French army (including troops from North under Washington) surrounds and forces surrender (October 19) of British troops under Cornwallis, last battle of war.

1782 British Parliament orders end to further fighting (February 27). Prime Minister Lord Frederick North (1732–92) is ousted, and British troops in America withdraw to New York City.

1783 Treaty of Paris is signed (September 3). British troops are withdrawn (November 25).

French Revolution and Revolutionary Wars

1787 French economy is in sharp decline.

1788 Government suspends French *parlements* (law courts) on May 8 to end resistance to taxes needed to avert bankruptcy of government. French nobles retaliate by instigating popular revolt against government.

1788 King Louis XVI (1774–92) reestablishes *parlements* to end unrest (August 8). He agrees to convene Estates-General (with representatives from the three estates: clergy, nobility, and commons).

1788 Bread prices rise following poor harvest, causing hardship and unrest among poor.

1789 Estates-General meets (May 5) and, after reconstituting itself as National Assembly (June 17), announces it has constitutional powers.

1789 King Louis XVI bars National Assembly from its assembly hall (June 20). Assembly takes famous "Tennis Court Oath," vowing to write constitution (June 25).

1789 Storming of Bastille (July 14) by Paris mob seeking arms stored there.

1789 Peasants revolt (late July) against nobility throughout France.

1789 Assembly enacts legislation to end feudal privileges of *ancien régime* (August 4).

1789 Declaration of Rights of Man and of Citizen is adopted by Assembly (August 26).

1789 March on Versailles (October 5–6). King is forced to move to Paris.

1791 King Louis XVI fails in attempt to escape to Austria (June 20–25); he is kept under arrest at Tuileries Palace.

1791 Declaration of Pillnitz (August 27) by Austria and Prussia threatens military intervention to reestablish monarchy.

1791 King accepts Constitution of 1791 (September 30).

1791 Legislative Assembly convenes (October 1). Girondists, Jacobins, and others soon vie for power.

1792 Austria and Prussia become allies against France (February 7).

1792 Revolutionary wars begin (April). Austria and Prussia invade France.

1792 French organize voluntary army to help defend country (July).

1792 Mob storms Tuileries Palace (August 10). Assembly suspends monarchy. September massacres follow; mobs kill royalists.

1792 Battle of Valmy (September 20). French halt invasion by Austro-Prussians, forcing Prussians to retreat across Rhine (September). Mainz, Frankfurt, Nice, and Savoy are then taken.

1792 New legislative body, National Convention, abolishes monarchy (September 21) and proclaims republic (September 22).

1792 Tide of war turns in French favor. By Battle of Jemappes (November 6), France occupies Netherlands.

1793 King Louis XVI is executed for treason (January 21).

1793 Britain joins war against France (February).

1793 Belgium is annexed (February).

1793 Battle of Neerwinden. Austrians defeat French (March 18) and regain Brussels.

1793 Wars of the Vendée, counterrevolutionary fighting in western France, begin (March) following new government conscription program. They drag on until 1796.

1793 Arrest of moderate Girondist deputies (May 31). Committee of Public Safety gains dictatorial powers (July) and becomes ruling body of government.

1793 Radical leader Jean Paul Marat (1743–93) is stabbed to death in his bath by Charlotte Corday (July 13).

1793 Mainz is back under Prussian control. Toulon is under siege by British.

1793 Massive conscription program (August 23) begins. France creates army of 1 million, enough to field fourteen armies.

1793–94 Reign of Terror begins (September 5). Some 40,000 royalists and others are executed before it ends (July 1794).

▶

1793 Battle of Wattignies (October 15–16) and lifting of siege of Toulon (December 19). French victories turn back new allied invasion, paving way for French invasion of Germany.

1793 De-Christianization movement is begun by radical Hebértistes (November).

1794 Radical Georges Danton (1759–94) is executed April 6 after opposing excesses of Reign of Terror.

1794 Battle of Fleurus (June 26). Victory gains France southern Netherlands (modern Belgium), turning tide of war again in French favor.

1794 Coup of 9 Thermidor (July 27). Convention delegates, threatened by Committee of Public Safety leader Maximilien Robespierre (1758–94), arrest and execute Robespierre and supporters. Subsequently terror and other radical measures are relaxed.

1795 Unrest over food shortages and other problems leads to rioting and attacks on National Convention, beginning in spring.

1795 Treaty of Basel (March 5), by which Prussia agrees to cease fighting.

1795 British and émigré force lands in Brittany (June 27) but is overwhelmed (July 16–21).

1795 Constitution of 1795 is in force (August 22), ruling by five-man Directory.

1796 Napoleon becomes national hero by his victories in command of French forces in Italy.

1796 Battle of Lodi (May 10). Austrian defeat here leads to occupation of Milan (May 15).

1796–97 Napoleon's successful siege of Mantua. He forces surrender of Austrian forces there, effectively gaining control of Italy.

1797 Treaty of Tolentino. France acquires Bologna, Ferrara, and Romagna (February 19).

1797 Napoleon, confronted by new Austrian army and revolts in Italy, negotiates with Austria. He divides northern Italy with Austria (April 18) and gains Belgian provinces.

1797 Napoleon creates Cisalpine Republic (July) to rule French territories in northern Italy.

1797 Coup of 18 Fructidor (September 4). Republicans, with help of troops sent by Napoleon, seize power and set up dictatorship.

1797 War with Austria ends by Treaty of Campo Formio (October 17). Only Great Britain is still at war with France.

1798 Napoleon occupies Rome (February) and declares Roman Republic. Switzerland is invaded in April.

1798 French under Napoleon invade Egypt to cut British trade with India (July). French are victorious on land (Battle of Pyramids, July 21), but British isolate them in Egypt by defeating its navy (Battle of Nile, August 1–2).

1798 New coalition (including Austria, Russia, and Ottoman Empire) joins Britain in war against France (December).

1799 Napoleon invades Syria and takes Jaffa (February), but outbreak of plague forces retreat to Egypt.

1799 French reverses in Italy at Magnano (April 5), Cassano (April 27), and Turin (May 27).

1799 Napoleon is victorious at Aboukir, Egypt (July 25). French position in Egypt is hopeless, however, and he returns to France (August 24).

1799 French victory at Zurich (September 26) knocks Russia out of war.

1799 Coup of 18 Brumaire (November 9). Napoleon seizes power, becoming first consul of Consulate.

1800–1 Napoleon returns to wars, defeating Austrians (Marengo, Italy, June 14; Hohenlinden, December 3), and, with Spain, conquers Portugal (September 1801). He loses Egypt to Great Britain (1801).

1802 Treaty of Amiens (March 27) ends fighting with France's last remaining opponent, Great Britain, ending French Revolution wars.

Napoleonic Wars

See also French Revolution.

1803 Mutual distrust leads to renewed war between France and Great Britain. Great Britain institutes naval blockade of Napoleon's empire.

1803–5 Napoleon (1769–1821) masses his Grand Army and prepares fleet to invade Great Britain. He later abandons plan.

1805 Russia, Austria, and Sweden join Great Britain in opposing Napoleon.

1805 Capitulation of Ulm (October 17). Napoleon gains impressive strategic victory by encircling and forcing surrender of 30,000 Austrian troops.

1805 Battle of Trafalgar (October 21), decisive British naval victory over allied French and Spanish fleet (33 ships). It establishes British naval supremacy, but British commander, Admiral Horatio Nelson (1758–1805), is killed.

1805 Battle of Austerlitz (December 2), notable strategic victory for Napoleon's 68,000 troops, who rout some 90,000 Austrians and Russians. Austria subsequently surrenders and Russians withdraw.

1805 Prussia signs peace treaty with France (December 15). Treaty of Pressburg (December 26) between Austria and France leaves France in control of Italy.

1806 Confederation of Rhine is established by Napoleon (July 12), leaving much of Germany under French control.

1806 Battle of Jena-Auerstadt (October 14). Some 122,000 French under Napoleon crush somewhat smaller Prussian force in Saxony. Napoleon conquers Prussia soon after and advances into Poland.

1806 Napoleon declares blockade of Great Britain (November 21) and inaugurates Continental System. Berlin is occupied.

1807 Battle of Eylau (February 7–8). Napoleon's army is battled to draw for first time in major engagement. Allied Russian and Prussian forces successfully fight French in blinding snowstorm. Losses are heavy on both sides.

1807 Battle of Friedland (June 14). Napoleon crushes smaller Russian force, forcing Russia and Prussia to sue for peace. By Treaty of Tilsit (July 7–9), Prussia is forced to cede territory and is occupied by France. Russia agrees to alliance with France. Only Great Britain now opposes Napoleon.

1807 French invade Portugal (November) and capture Lisbon (December 1), closing Great Britain's avenue for commercial trade with continental Europe.

1808 French invade Spain (March), depose King Charles IV (1788–1808), and install Napoleon's brother Joseph Bonaparte (1768–1844). This sparks resistance by Spanish and Portuguese, however.

1808 Battle of Baylen (July 19, first surrender of Napoleonic army). Some 35,000 Spanish troops force surrender of 20,000 French troops, who subsequently are massacred.

1808 French army in Portugal is forced to surrender (August 30), by popular uprising and by British expeditionary force under Arthur Wellesley (1769–1852), later duke of Wellington. British forces then join Spanish in pushing back French in Spain.

1808 Napoleon takes command in Spain and recaptures Madrid (December 4). He reestablishes French control in Spain and forces British to withdraw from Continent (1809).

1809 Austria rejoins war against Napoleon, who leaves Spain to take up the fight (January).

1809 Battle of Abensberg-Eckmühl (April 19–23). Napoleon defeats larger Austrian force of 190,000 by rapid maneuvers and hard fighting.

1809 Battle of Aspern-Essling (May 21–22). Napoleon's first defeat comes in failed attempt to cross Danube River against stiff Austrian resistance.

1809 British send Wellesley to Lisbon to protect Portugal (spring). They again invade northern Spain. The inconclusive Battle of Talavera (July 28) saves Portugal from invasion.

▶

1809 Battle of Wagram (July 5–6). Napoleon's decisive victory, near Vienna, hinges on massed artillery fire (greatest up to that time). Battle ends Austrian resistance. By Treaty of Schönbrunn, Austria cedes territory and becomes French ally. Great Britain alone opposes Napoleon.

1810 Russia occupies Bessarabia, Moldavia, and Wallachia.

1810 Napoleon marries Marie-Louise, daughter of Austrian emperor (April). Marriage is arranged as ploy to give Austria time to recover.

1811 British forces in Spain score victories at Fuentes de Onoro (May 5) and Albuera (May 16).

1812 War preparations begin after friction between France and Russia. Napoleon forms 450,000-man army to invade Russia (by June), but only about 200,000 are French.

1812 Napoleon's ill-fated invasion of Russia begins (June 24). Russian forces facing Napoleon number only about 220,000, but they put up stiff resistance as they retreat toward Moscow.

1812 French in Spain are decisively beaten by Wellington at Salamanca (July 22). Joseph Bonaparte flees Madrid.

1812 Battle of Borodino (September 7). Outside Moscow, Russian general Mikhail Kutuzov (1745–1813), with army of 120,000, opposes a somewhat larger French force. Napoleon suffers first of his seizures in midbattle and puts subordinates in charge. Russians thereby escape with remaining forces intact.

1812 French occupy Moscow (September 14). City is set afire—probably by Russians—with French in it. Lacking supplies, Napoleon abandons Moscow (October 19). Bitter Russian winter plus harassing Russians turn withdrawal into disorganized retreat to Poland.

1812 Crossing Beresina River (November 26–28). Napoleon, commanding only 37,000 combat-ready troops, fights 144,000 Russians deployed on both sides of Beresina to make crossing. He reaches Poland soon after, his invasion having cost 300,000 casualties.

1813 Napoleon is defeated at battles of Laon (March 9–10) and Arcis-sur-Aube (March 20–21).

1813 Russian troops occupy Hamburg (March 18) and Dresden (March 27). There is massive military buildup by both sides.

1813 Russia, Sweden, and Prussia rejoin Great Britain to oppose Napoleon, who is quickly raising new army (200,000 fresh recruits by April).

1813 Allied victory in southern campaign is complete with the Battle of Toulouse (April 10).

1813 Counterattacking French troops gain victories at Lützen (May 2), Bautzen (May 20), and Wurschen (May 21). Hamburg is taken by French (May 30).

1813 Battle of Vittoria (June 21). Major victory by Wellington forces French to abandon Spain.

1813 Battle of Dresden (August 26–27). Napoleon, on verge of defeating army (Austrian, Russian, and Prussian troops) twice as large as his, suffers new seizure. Allied troops narrowly escape nearly complete encirclement as result.

1813 Battle of Leipzig (October 16–19), decisive victory for some 320,000 Austrians and allies over Napoleon's army (now about 185,000 men). Nearly encircled at Leipzig, Napoleon escapes after losses of 60,000.

1813 Napoleon retreats across Rhine (November) and is followed by first of Allied armies (December 21–25).

1813 Wellington invades southern France and besieges Bayonne (December).

1814 Paris surrenders (March 31). Hopelessly outnumbered, Napoleon cannot stop allied advance on Paris, and city surrenders.

1814 Napoleon abdicates unconditionally (April 11). Louis XVIII (1755–1824) becomes French king. French borders are restored to those of 1792 by Treaty of Paris.

1815 Napoleon, banished to the island of Elba, returns (March 1), retakes power (March 20), and raises army (by June).

1815 Napoleon's troops are victorious at Ligny and Quatre-Bras (June 16) but fail to follow up, giving allies time to regroup for Waterloo.

1815 Waterloo (June 18). Napoleon, with 72,000 troops, attacks somewhat smaller British force commanded by Wellesley, duke of Wellington. British stall Napoleon's attack, however, and Prussian troops arrive under Gebhard von Blücher (1742–1819). French retreat in disarray.

1815 Napoleon surrenders to British after abdicating (June 21) and is exiled to St. Helena, where he dies (1821).

American Civil War

1860 South Carolina votes to secede from Union (December 20) following election of Abraham Lincoln (1809–65) as U.S. President on antislavery platform. Mississippi, Florida, Alabama, Georgia, Louisiana, and Texas follow suit (January 9–February 1).

1861 Confederacy is established at Montgomery Convention (February 4). Jefferson Davis (1808–89) becomes Confederate president (February 9).

1861 Confederates bombard Fort Sumter, to force its surrender (April 13). Civil War begins.

1861 Virginia, Arkansas, North Carolina, and Tennessee secede (April–June). Richmond, Virginia becomes capital of Confederacy.

1861 Union naval blockade of Confederate coastline is ordered (April 19).

1861 First Battle of Bull Run (July 21). Confederates under Gen. Pierre Beauregard (1818–93) turn back invading Union army at Manassas Junction.

1861 Union forces capture forts Clark and Hatteras (August 28–29), first steps in enforcing naval blockade of Confederacy.

1861 Gen. George B. McClellan (1826–85) is in command of Union army (November) but is removed in 1862.

1862 Federal troops capture Fort Henry on Tennessee River (February 6) and Fort Donelson on Cumberland River (February 16), forcing Confederate withdrawal to Memphis–Chattanooga line.

1862 Union navy takes Roanoke Island and Elizabeth City (February 8 and 9) and Amelia Island and Jacksonville, Florida (March 4 and 12).

1862 Historic first battle of ironclad warships, the Union *Monitor* and Confederate *Merrimac*, ends in draw (March 9).

1862 Union troops advance to outskirts of Richmond (March–May 31), where Gen. McClellan halts to await reinforcements.

1862 Confederate military conscription begins (April).

1862 Battle of Shiloh (April 6–7). In Tennessee, advancing Union troops under Gen. Ulysses S. Grant (1822–85) turn back Confederate counterattack.

1862 New Orleans falls (May 1) to combined Union forces of Adm. David Farragut (1801–70) and Gen. Benjamin F. Butler (1728–96).

1862 Battle of Fair Oaks (May 31) is first in series of engagements near Richmond.

1862 Robert E. Lee (1807–70) is named commander in chief of the Army of Northern Virginia (June 1).

1862 Memphis, Tennessee falls to Union troops (June 6).

1862 Seven Days' Battles (June 26–July 2). In Virginia, Confederates under Gen. Robert E. Lee unsuccessfully attack Gen. McClellan's positions and then withdraw. Confederate losses are over 20,500; Union, nearly 16,000.

1862 Gen. Henry W. Halleck (1815–72) becomes commander of all Union armies (July).

1862 Second Battle of Bull Run (August 29–30) is successful Confederate attack that forestalls major Union offensive on Richmond.

1862 Battle of Antietam (September 17). McClellan forces Lee to withdraw from Maryland to Virginia, but heavy casualties on both sides are highest for any one day of fighting in this war (11,729 for Confederacy; 11,657 for Union).

1862 Battle of Perryville (October 8). Confederate troops fail to gain local support in Kentucky and are forced to retreat south.

1862 Gen. Ambrose E. Burnside replaces McClellan as head of Union troops (November 7).

1862 Battle of Fredericksburg (December 13). Gen. Burnside's defeat here leads him to be replaced by Gen. Joseph Hooker (January 25, 1863).

1862–63 Battle of Murfreesboro (December 31–January 2). In Tennessee, Confederates under

▶

Gen. Braxton Bragg (1817–76) withdraw after bloody battle in which each side suffers over 9,000 casualties.

1863 President Lincoln issues Emancipation Proclamation (January 1).

1863 Military-conscription law is enacted in the North (March 3), leading to draft riots by poor in New York City (July).

1863 Battle of Chancellorsville (May 2–4). Some 60,000 Confederates under Gen. Lee send Union army over twice as large into retreat. Gen. Thomas (Stonewall) Jackson (1824–63) is accidentally shot and killed in battle.

1863 Siege of Vicksburg (May 22–July 4). Gen. Grant's 20,000 Union troops force surrender of 40,000 Confederates and gain control of Mississippi River.

1863 Gettysburg (July 1–4). Confederates under Gen. Lee invade Pennsylvania, engaging larger Union force under Gen. George G. Meade (1815–72) at Gettysburg. Successive Confederate charges fail to dislodge Union troops, forcing Confederate withdrawal. Battle marks turning point of war.

1863 Battle of Chickamauga Creek (September 19–20), bloodiest single battle of this war (losses: Union, 16,000; Confederacy, 18,000).

1863 Battle of Chattanooga (November 23–25). Union troops gain control of Tennessee, opening way into Georgia.

1864 Gen. Grant becomes commander of all Union armies (March 9).

1864 Grant's Army of Potomac suffers heavy losses in Battles of Wilderness (May 5–6), Spotsylvania (May 8–12), Cold Harbor (June 1–3), and Petersburg (June 15–18) but is able to withstand them. Grant lays siege to Petersburg (June), outside Richmond.

1864 Gen. Jubal A. Early (1816–94) leads raiding party and threatens Washington, D.C. (July). He is defeated by Gen. Philip Sheridan (1831–88), who lays waste to Shenandoah Valley.

1864 Battle of Atlanta (July 22). Gen. John B. Hood, who replaced Gen. Joseph E. Johnston, fails to halt Union advance on Atlanta. Confederate army abandons Atlanta (September 2).

1864 Battle of Kenesaw Mountain (July 27). Retreating Confederate troops under Gen. Johnston engage pursuing Union troops, whom they defeat. Union general McCook is killed.

1864 Battle of Mobile Bay (August 5). Union ships commanded by Adm. David Farragut (1801–70) engage Confederate fleet and successfully cut off Mobile's link to sea. Farragut makes his famous exhortation "Damn the torpedoes" during this battle.

1864 Capture of Atlanta, Georgia (September 2) by some 100,000 Union troops under Gen. William T. Sherman (1820–91), who have marched from Tennessee.

1864 Sherman's March to Sea (November 14–December 22). Countryside is devastated along way to capture of Savannah (December 22).

1864 Lincoln is reelected President (November), in part because of Sherman's victories in South.

1865 Gen. Lee becomes commander of all Confederate armies (February).

1865 Desertions and shortages of food and supplies weaken war-weary Confederacy. President Davis approves arming of slaves (February).

1865 Sherman marches north and takes Columbia, South Carolina (February 17) and Goldsboro (March 19). Union fleet captures Charleston (February 18).

1865 Battle of Five Forks (April 1). Confederate attack on Union siege force at Petersburg fails. Lee withdraws from Petersburg and Richmond.

1865 Surrender at Appomattox Courthouse (April 9). Surrounded by Grant's army, Lee agrees to surrender terms. His 30,000 soldiers are allowed to return to their homes.

1865 Lincoln is assassinated in Washington theater by actor John Wilkes Booth (April 14).

1865 Last Confederate army surrenders (May 26). Jefferson Davis is captured in Georgia (May 10). Total war losses: Union, almost 360,000 dead; Confederacy, about 258,000 dead.

World War I

1914 Archduke Francis Ferdinand (1863–1914) is assassinated in Bosnia by Serbian nationalist (June 28). Austria declares war on Serbia (July 28), and World War I begins. By subsequent war declarations, two opposing sides emerge: Allies, including France, Great Britain, Russia, and Italy; and Central Powers, including Germany, Austria-Hungary, and Ottoman Empire.

1914 Germany invades Luxembourg (August 1). Russia invades Prussia (August 13).

1914 Battle of Frontiers (August 14–25). French offensive is broken with heavy losses, and they are forced to evacuate Lorraine.

1914 Germans take Liège and Brussels in August.

1914 British troops land in France to join fight against Germans (August).

1914 Battle of Mons, first engagement between British and German troops. Germans send British into retreat from Belgium (August 23–24).

1914 Japan declares war on Germany (August 23). Shortly thereafter Japan occupies German islands in Pacific, including Marshall Islands, Marianas, and Carolines.

1914 German troops advance into France, taking Namur (August 25). They continue successful drive westward to Rheims (September 3), and Maubeuge (September 7).

1914 Several German colonial possessions are taken over by Allied forces, including Togoland (August 26), Samoa (August 30), and New Guinea (September 21).

1914 In East, Germans rout Russians at Battles of Tannenberg and Mazurian Lakes (August–September). Total of 225,000 prisoners are taken, demoralizing Russians.

1914 First Battle of Marne. German attack in France falters 25 miles short of Paris (September 5–9).

1914 Gen. Erich von Falkenhayn (1861–1922) replaces Helmuth von Moltke (1848–1916) as commander of German forces (September).

1914 Germans advance in North in Belgium (October), taking Antwerp (October 10), Ghent, and Bruges.

1914 Battle of Coronel. German admiral Maximilian von Spee (1861–1914) is victorious over British naval squadron (November 1).

1914 First Battle of Ypres (October 19–November 22). German offensive in Belgium (near French border) is halted by British.

1914 Germans take Lodz (December 6) as they advance eastward through Poland.

1914 Battle of Falkland Islands, British naval victory over von Spee (December 8).

1914 Great Britain declares protectorate over Egypt (December 18) and rushes forces to protect vital Suez Canal.

1914 Allied offensive on Western Front fails (December). Subsequent fighting along entire battlefront settles into trench warfare.

1914 Ottoman Turks lose entire army during failed invasion of southern Russia (December–January). British meanwhile occupy parts of Persian Gulf region.

1915 German zeppelin raids begin on England (January). Raids peak with attack of 14 ships on September 2, 1916.

1915 Second Battle of Mazurian Lakes. Germans again rout Russians (February).

1915 Disastrous Gallipoli campaign in Turkey begins with landings by Allies (February 19). Withdrawal begins in 1916 after Allies suffer heavy losses.

1915 Germans begin submarine blockade of Great Britain (February). British impose blockade of Germany (March).

1915 Second Battle of Ypres (April 22–May 25), marking first combat use of poison gas (chlorine, by Germans).

1915 Turkish troops begin their infamous massacre and deportation in Caucasus of Armenians, whom they accuse of aiding Russia (April).

▶

1915 German and Austrian forces send Russians into headlong retreat in East (May–October).

1915 Italy, after gaining territorial concession by secret Treaty of London, enters war as one of Allies.

1915 German submarine sinks liner *Lusitania* off Irish coast (May 7); 1,195 of passengers and crew are killed.

1915 Battle of Artois (May 9–June 18). French forces under Gen. Henri Pétain (1856–1951) break through German lines.

1915 German South-West Africa falls to British forces (July 9).

1915 Second major German offensive begins on Eastern Front. Warsaw (August 4–7) and Brest-Litovsk (August 25) fall. With capture of Vilnius (September 19), nearly all of Poland, Lithuania, and Kurland are in Axis hands.

1915 Third Battle of Artois (September 25–October 15). British troops employ poison gas and gain initial victories, but battle ends in virtual stalemate.

1915 Allied troops land at Salonica in Greece (October 5).

1915 Germans introduce Fokker airplane, equipped with machine gun synchronized to shoot through the propeller (October), giving them air superiority for nearly a year.

1915–17 Battles of Isonzo. Series of 11 battles over two years between Italy and Austria fail to establish clear-cut victory.

1916 Battle of Verdun (February 21–December 18). German attack ultimately fails, with both sides suffering heavy losses.

1916 Early Austrian gains in the Trentino are reversed by costly Italian counteroffensive (May 15–June 3).

1916 Battle of Jutland (May 31–June 1). British naval forces suffer losses but retain command of seas.

1916 In East, new Russian offensive breaks through Austrian lines but at cost of 1 million casualties (June 4–September 20).

1916 Arab revolt against Turks begins with attack on the garrison in Medina (June 5).

1916 British war secretary Horatio Kitchener (1850–1916) dies when his battleship *Hampshire* is sunk by submarine or mine (June 5).

1916 Battle of Somme, Allied offensive to relieve Verdun (July 1–November 18). British use tanks in combat for first time. Germans replace Falkenhayn with Gen. Paul von Hindenburg (1847–1934).

1916 German airplane raids on Great Britain begin (November 28).

1916 Bucharest falls into German hands (December 6). Within a month most of Romania is in Axis hands.

1916 Gen. Robert Nivelle (1856–1924) replaces Gen. Joseph Joffre (1852–1931) as commander of French forces, in December.

1917 Germans declare submarines will attack unarmed merchant ships at will (February), prompting United States to enter war against Germany (April 6).

1917 Germans begin to pull back to Hindenburg Line of defensive positions in February.

1917 Russian Revolution begins (March), further weakening Russia's ability to fight invading Central Powers. Later, Bolsheviks seize power (November) and begin negotiating separate peace settlement with Central Powers.

1917 British capture Baghdad (March 11).

1917 Allied offensive fails to make major gains. French troops, after heavy losses at Aisne (April–May), mutiny. Gen. Pétain replaces Nivelle as French commander.

1917 United States declares war on Germany (April 6).

1917 Battle of Arras (April 9–May 4). British troops under Gen. Edmund Allenby (1861–1936) gain 4 miles after heavy use of poison gas.

1917 New Allied offensive stalls (June–November).

1917 First U.S. troops arrive in France (June 26), under command of Gen. John J. Pershing (1860–1948).

1917 Col. T. E. Lawrence (1888–1935, Lawrence of Arabia) takes Aqaba (July 6). He inspires Arab revolt against rule by Turks.

1917 Germans advance in East (September-October) after Russian offensive fails.

1917 Dutch dancer Mata Hari (1876–1917) is executed in France on charges of spying (October 15).

▶

1917 First Americans in combat (October) on Western Front.

1917 Battle of Cambrai. British make first use of massed tanks in battle (November 20–December 4).

1917 British, advancing from Egypt, take Jerusalem (December 9).

1918 U.S. President Wilson (1856–1924) announces his 14 Points in peace plan (January 8).

1918 Bolsheviks sign Treaty of Brest-Litovsk, ending war with Central Powers (March 3). By treaty, Ukraine, Poland, Baltic provinces, Finland, and Transcaucasia are abandoned by Russians.

1918 German forces invade Ukraine, taking Kiev (March 3) and pushing to Sevastopol, in Crimea (May 1).

1918 Battle of Somme. Germans unsuccessfully attempt to split Allies (March 21–April 6).

1918 Gen. Ferdinand Foch (1851–1929) becomes overall commander of Allied forces (March 26).

1918 Red Baron, German air ace Baron Manfred von Richthofen (1892–1918), is killed (April 21).

1918 Germany invades Finland (April) and intervenes on side of anti-Bolsheviks, ending civil war (May 7).

1918 Third Battle of Aisne. Germans advance to Marne in surprise attack (May 27–June 3).

1918 Battle of Marne. French turn back German advance across Marne River, inflicting heavy losses (July 15–August 6). Battle proves decisive victory of war. Allied counteroffensive (August) pushes Germans back to Hindenburg Line.

1918 Battle of St. Mihiel, American victory on Western Front (September 12–13).

1918 Final British offensive in Palestine begins (September 18). British and their Arab allies capture Damascus (October 1–2), Beirut (October 7), and Aleppo (October 26).

1918 Battle of Argonne Forest, Allies' last offensive in West, begins (September 26–October 31).

1918 War in Bulgaria is ended by armistice (September 30).

1918 British break through Hindenburg Line (September 27).

1918 Americans capture St. Étienne (October 6).

1918 Allies take Cambrai, Roncroy, Osten, and Bruges in quick succession (October 17).

1918 American sergeant Alvin York (1887–1964) becomes hero by single-handedly capturing 132 Germans (October 18).

1918 Battle of Vittorio Veneto (October 24–November 4). Austrian resistance collapses, and Italian troops take Trieste (November 3) and Fiume (November 5).

1918 Armistice in fighting with Ottoman Turks (October 30).

1918 Poland, Hungary, and other nations proclaim independence as Austro-Hungarian Empire collapses (October).

1918 Austria surrenders (November 3).

1918 Kiel mutiny by German sailors presages collapse of German government (November 3).

1918 German kaiser Wilhelm II (1859–1941) abdicates (November 9).

1918 Armistice ends fighting in World War I (November 11).

1918 American troops cross over Rhine (December 13).

1919 Germans scuttle warships at Scapa Flow rather than turn them over to Allies (June 21).

1919 Versailles Peace Conference begins in France (January 18). Final treaty is signed June 28, setting harsh terms and fixing blame for war on Germany.

Russian Revolution and Civil War

1917 Russian workers in Petrograd demonstrate against involvement in World War I and famine in Russia (March 8), latest in months of such protests.

1917 Czar Nicholas II (1868–1918) orders dissolution of legislative Duma (March 11), but members refuse.

1917 Strikers organize Petrograd Workers' and Soldiers' Soviet (March 12). They soon vie for power with newly formed provisional government.

1917 Czar Nicholas abdicates as unrest mounts (March 15). Provisional government subsequently enacts liberal reforms.

1917 Czar's troops in capital mutiny (March 10).

1917 Special "sealed train" (provided by Germans) carries Communist leader Vladimir Ilyich Lenin (1870–1924) back to Russia (April) after long exile. He advocates Bolshevik-dominated government.

1917 Ukraine, important agricultural region, is in revolt for autonomy (July). Bolshevik attempt to seize power in Petrograd fails (July 16–18). Lenin flees to Finland.

1917 Non-Communist leader Alexander Kerensky (1881–1970) becomes head of provisional government (July 20).

1917 Conservative attack on government is led by Gen. Kornilov (1870–1918) and causes Kerensky to appeal to Bolsheviks for help. Trotsky and others are freed (September 9–14).

1917 Lenin secretly returns to Russia (October 21). Soon after, Bolsheviks vote for armed revolt against provisional government (October 25).

1917 Bolsheviks overthrow Kerensky's provisional government at Petrograd (November 7). Leon Trotsky (1879–1940) is named head of state, though Lenin is in control.

1917 Kerensky unsuccessfully attacks Petrograd with Eastern Front troops (November 11), then flees (October).

1917 Social Revolutionaries control Duma, with 420 seats to Bolsheviks' 225, following elections (November 25).

1917 Japanese forces land at Vladivostok (December), which they occupy until October 25, 1922.

1917 Bolsheviks imprison Czar Nicholas and family (December).

1917 Gen. Kornilov and Gen. Kaledin lead Don Cossacks in open revolt against Bolshevik government (December 9).

1917 Lithuania and Moldavia declare independence (December 11 and 15).

1917 Russia signs armistice with Germany, ending Russia's disastrous involvement in World War I (December 15).

1917 Bolsheviks organize Cheka (December 20), group to crush anti-Bolshevik opposition.

1918 Bolsheviks dissolve Duma by force (January) after deputies refuse to vote for Lenin's government by workers' councils (soviets).

1918 Cossacks suffer defeats (February 13 and April 13) that result in deaths of Kornilov and Kaledin.

1918 Nationalization of all land in Russia occurs (February 19).

1918 Ukraine, Russian Georgia, and Armenia declare independence early in year. All are eventually restored to USSR.

1918 Moscow becomes Soviet capital (March).

1918 Counterrevolutionaries take over in Petrograd (April), naming Grand Duke Alexis as ruler.

1918 Russian civil war between Bolsheviks and various counterrevolutionary groups begins (May).

1918 British troops land in Murmansk (June 23) to fight Germans. By August they are backing puppet government of northern Russia.

1918 Nationalization of Russian industry is complete (June 28).

1918 Bolsheviks are victorious against rebel Cossacks (June 30).

1918 Allied British and American expeditionary force gains control of Russia's Murmansk region (July). British and French troops land in Archangel (August 2). Allied troops are withdrawn at end of World War I.

▶

1918 Bolsheviks execute czar and family (July 16).

1918 White Russians lose Kazan to Red Army (August 8).

1918 Socialist revolutionary shoots and wounds Lenin (August 30). Red Terror begins, in which tens of thousands of opponents and suspected opponents are killed.

1918 Conservative coup (November 18) places Adm. Kolchak (1874–1920) at head of government in Omsk, in western Russia. His White Army advances into eastern Russia and captures Perm (December 24).

1919 Bolshevik counteroffense against Kolchak results in capture of Orenburg and Ekaterinburg (January).

1919 Central Russia erupts in peasant rebellion against Bolsheviks (January 15).

1919 Bolsheviks retake Kiev (February 3) and Odessa (April 8).

1919 Lenin forms Third Communist International to foment world revolution (March).

1919 Communist First Army surrenders in Ukraine (April).

1919 White Russians at Archangel defect to Red Army (July).

1919 Gen. Anton Denikin's (1872–1947) White armies take Kiev (August). They are expelled by Bolsheviks in December.

1919 Allied forces abandon Archangel (September 30) and Murmansk (October 12) to Bolsheviks.

1919 White Russians advance to just outside Moscow by early October. Red Army soon forces them into retreat, however, at Moscow and also at Petrograd. Red Army also advances into Urals by late October.

1919 Omsk is retaken by Bolsheviks (November 14).

1920 Soviets recognize independence of Baltic states of Estonia (February 2), Lithuania (July 12), and Latvia (August 11).

1920 Red Army launches offensives on Polish Front and into Finland (March). After advancing to vicinity of Warsaw, Red Army suffers disastrous defeat (October 3), and Reds sign armistice. They also recognize Finnish independence (October) after March offensive.

1920 Bolshevik forces advance south and capture Baku (April 28). Meanwhile, White forces under Gen. Pyotr Wrangel (1878–1928) invade southern Russia from Crimea (June–November).

1920 White Army resistance is broken by Red Army victory in Crimea (November 16). Wrangel flees to Constantinople. Some 30,000 White Army soldiers are killed and 40,000 captured.

1921 Lenin announces his New Economic Policy (March). He appeals for world aid to feed 18 million starving Russians (August).

1921 Unsuccessful Kronstadt Revolt against Bolsheviks is crushed (March).

1922 Lenin has stroke (June).

1922 Remnants of White Army are driven from Siberia into Manchuria (October).

1922 Union of Soviet Socialist Republics is formally adopted as new national name (December 30).

Chinese Civil War

1921 Canton proclaims independence as provincial warlords continue to undermine central government power. Sun Yat-sen (1866–1925), Nationalist leader, becomes president of Canton (April). Sun asks for U.S. aid in reunifying China but is rebuffed.

1921 Chinese Communist Party (CCP) is founded by Ch'en Tu'hsiu (1879–1942) at Shanghai (December).

1923 Sun asks for Soviet aid and agrees to alliance with CCP to reunify country.

1925 Sun dies, leaving no clear successor. Communist influence in Kuomintang is considerable by this time.

1926 Chiang Kai-shek (1886–1975), commander of Nationalist Whampoa Military Academy, leads anti-Communist coup. Soon after, however, he agrees to restore Kuomintang–CCP alliance.

▶

1926–27 Kuomintang launches successful expedition against warlords in North, but rift opens between Nationalists and Communists in Kuomintang.

1927 Communists try to oust Chiang from Kuomintang (March), but his support is too strong.

1927 Chiang arrests and executes Communists at Shanghai (April).

1927 Chiang and his conservative allies split with radicals in Hankow government and establish new government, at Nanking (April 18).

1927 Autumn Harvest Uprising, organized by Communist leader Mao Tse-tung (1893–1973), fails (September). Nevertheless, Mao is impressed by potential of peasant revolt.

1927 Mao leads Chinese Communists to mountains in Southeast to escape attacking Nationalist armies.

1928 Kuomintang, now in control of most of China, becomes national government of China. Chiang Kai-shek becomes president.

1931 Nationalist Constitution is enacted. Chiang steps down as president.

1931 Communists establish Chinese Soviet Republic in Kiangsi Province under Mao's leadership (November).

1931 Japanese invasion of Manchuria temporarily ends warfare between Nationalists and Communists.

1933 Nationalists surround Communists in Kiangsi after signing armistice with Japanese.

1934–35 In Long March, Mao leads Communist army on 6,000-mile trek to escape Nationalists. CCP elects him party chairman (January 1935) during march, which ends in Shensi Province in northwestern China (October).

1936 Chiang is kidnapped (December 12) by CCP to force him to accept Nationalist–CCP alliance against invading Japanese. Alliance is implemented (1937) after his release.

1937 Territory under CCP control is only about 35,000 square miles. Army numbers 100,000 troops.

1941 Confrontation between Nationalist and CCP forces (January). Nationalists blockade territory under CCP domination.

1943 CCP begins rapid expansion into Japanese-controlled countryside in North. It exploits hatred of Japanese and institutes reforms. By 1945, CCP controls about 300,000 square miles and 95 million Chinese.

1944 Nationalists and CCP hold unsuccessful talks on postwar China government.

1945 CCP Congress. Mao urges creation of coalition government in China after defeat of Japanese, though goal remains Communist government. CCP army now numbers 900,000.

1945 Japan surrenders (August), ending World War II. Nationalists and CCP resume their warfare.

1945 CCP troops clash with Nationalists (August). Nationalist forces take Nanking and Shanghai.

1945 United States sends Gen. George C. Marshall to mediate Communist–Nationalist conflict (December 14).

1946 Truce is signed between Nationalist and Communist forces (January 10). Nationalists ignore Communists' demand for joint control in Manchuria (February 17).

1946 CCP troops in North attack Peking–Mukden railway (April). Second truce is declared (May 12–June 30). CCP begins offensive in Yangtze region (July). Mao orders total offensive (August).

1946 Chiang Kai-shek is reelected president by Kuomintang (October 10).

1947 United States gives up its mediation efforts in China (January 29).

1947 Nationalists start offensive in Communist-held Shantung region (February). They capture Yenan, but Communists have evacuated their former capital (March).

1947 CCP conquest of Manchuria is complete (October). Nationalists suffer heavy losses.

1948 Formation of the North China People's Government is announced by Communists (September 1).

1948 United States sharply cuts aid because of widespread corruption in Nationalist Army.

1949 CCP takes Peking (January) after month-long siege. Surrender talks open. Communists demand punishment of so-called war criminals (including Chiang) and want Mao Tse-tung to lead coalition cabinet.

▶

1949 CCP army's sweep of country begins after negotiations with Nationalists break down (April).

1949 People's Republic of China is proclaimed (October 1).

1949 Nationalists flee mainland China for Formosa (December).

World War II

See also Battles from Ancient to Modern Times.

1939 Nazi Germany, led by Adolf Hitler (1889–1945), invades Poland and starts World War II (September 1). German *Blitzkrieg* tactics include heavy use of Panzer tanks, mechanized infantry, and air support, which overwhelm Polish forces.

1939 British announce naval blockade of Germany (September 3).

1939 Great Britain and France declare war on Germany (September 3).

1939 Japan and USSR sign nonaggression pact (September 15).

1939 Soviets, allied with Germany at this point, invade Poland from East (September 17). Nazis capture Warsaw (September 27) and later divide up Poland with USSR.

1939 Soviet Union gains access to military bases in Baltic states in series of agreements, with Estonia (September 29), Latvia (October 5), and Lithuania (October 10).

1939 United States begins supplying arms to Britain and France on cash-and-carry basis (November).

1939 Soviets invade Finland (November 30). Soviet advance stalls, however, and they sign treaty early in 1940.

1939 German pocket battleship *Admiral Graf Spee* is bottled up by British warships in Montevideo Harbor. Germans scuttle her (December 17).

1940 German *Blitzkrieg* continues. Nazis occupy Denmark and Norway (April–June).

1940 Winston Churchill (1874–1965) becomes British prime minister (May), replacing Neville Chamberlain (1869–1940), who has been discredited by his appeasement policy toward Hitler.

1940 Germans roll through Luxembourg, Netherlands, and Belgium (May).

1940 Germans break through Allied defenses on Meuse (May 12). They push toward English Channel ports to trap British and French troops.

1940 Netherlands surrender (May 14), and Queen Wilhelmina flees to England.

1940 Dunkirk. British make emergency evacuation of troops from France across English Channel (May 29–June 4). British and other Allied troops are rescued.

1940 Italy joins Axis (June 10). Italian forces attack southern France.

1940 Germans take Paris (June 14). France surrenders (June 22). Mar. Henri Pétain (1856–1951) becomes Vichy government head.

1940 British sink French fleet at Oran to prevent it from falling into German hands (July 3).

1940 Gen. Charles de Gaulle (1890–1970), head of French National Committee in London, vows continued resistance.

1940 France and Italy sign armistice (June 24).

1940 Battle of Britain begins after fall of France (June). Massive German bomber offensive begins against British air fields and key industrial facilities (August). Bombing of London begins (September 7). British fighters ultimately defeat German Luftwaffe's offensive, forcing Hitler to call off his planned invasion of Great Britain.

1940 Italy invades British Somaliland (August 6). Conquest is complete August 19. British retake all of East Africa by 1941.

1940 British withdraw their troops from northern China and Shanghai (August 9).

▶

1940 Germany announces total naval blockade of Great Britain (August 17).

1940 British, in desperate need of destroyers to protect transatlantic supply convoys from German submarines, get 50 aging U.S. destroyers. German U-boats inflict heavy losses on Allied shipping in Atlantic until 1943.

1940 Japanese occupy French Indochina (September).

1940 Italy invades Egypt from Libya (September 13–15), but British turn them back.

1940 U.S. Congress enacts Selective Service Act (September 16).

1940 Germany occupies Romania (October 7).

1940 Italian troops invade Greece (October 28). British and Soviets aid resistance.

1940 Hungary and Romania join Axis Powers.

1940 Greek forces score victories over Italians in Albania (December 3). Germans send reinforcements.

1941 Bulgaria joins Axis.

1941 Lend-Lease Act is signed (March 11), providing critically needed supplies to Great Britain.

1941 Battle of Cape Matapan (March 30). Five Italian warships are sunk by British off Crete.

1941 Germany invades and conquers Greece and Yugoslavia (March–April).

1941 Nazi Rudolf Hess (1894–1987) makes mysterious flight to Great Britain, landing in Scotland by parachute (May 10).

1941 German battleship *Bismarck* is sunk by British (May 27).

1941 U.S. President Franklin D. Roosevelt (1882–1945) declares national emergency (May 27). United States, still officially neutral, steps up preparations for war.

1941 Operation Barbarossa begins (June 22). Nazi troops invade USSR on broad front. Soviets adopt scorched-earth tactic as they retreat.

1941 British and Free French forces wrest control of Syria from Vichy government (July).

1941 German successes in Russian invasion are rapid. Most of Ukraine is conquered August 19; Kiev and Poltava are taken September 19; Odessa is taken October 16. Crimea is reached at month's end.

1941–44 Siege of Leningrad. City is badly damaged, and over 1 million die in heroic stand against Nazis (September 8, 1941–January 27, 1944).

1941 Allies agree to aid former enemy, Soviet Union. United States, Great Britain, and Soviet Union sign aid pact (October 1).

1941 Germans approach Moscow (November 15), but Russian winter forces them to delay final attack. Soviets counterattack (December) and retake Kalinin (December 16).

1941 German siege of Sevastopol begins (November 15); it falls during new summer offensive (July 2, 1942).

1941 Battle of Tobruk (April 10–November 29). British push back German Afrika Korps of Gen. Erwin Rommel (1891–1944) in Libya.

1941 Pearl Harbor attack. Japanese sneak attack on U.S. naval base destroys much of U.S. Pacific fleet (December 7) but brings United States into war (December 8). Attacks are also made on Hong Kong, Malaysia, and Philippines, opening Japan's Pacific offensive that results in capture of Indonesia (January), Philippines (May), other islands, and much of Southeast Asia.

1942 Gen. Douglas MacArthur (1880–1964), evacuated from Philippines but vowing to return, arrives in Australia (March).

1942 Lt. Col. James H. Doolittle (b. 1896) leads U.S. bombing raid on Tokyo (April 18).

1942 Battle of Coral Sea (May 7–8). U.S. Navy is victorious in first naval battle fought entirely by attack planes from opposing aircraft carriers.

1942 USSR, invaded by its erstwhile German ally, now joins Allies (May 26).

1942 Battle of Midway (June 4–7). U.S. Navy turns back Japanese invasion fleet.

1942 Japanese troops land in Aleutian Islands (June 12) and take control of Attu and Kiska.

1942 In North Africa, Rommel drives British back into Egypt (June).

1942 Marines land on Guadalcanal (August 7), finally taking island in February 1943.

1942 Dieppe raid (August 19) by Allies on French coast fails.

▶

1942 Germans enter Stalingrad in Soviet Union (September).

1942 Battle of El Alamein (October 23–November 4). British under Gen. Bernard Montgomery (1887–1976) turn back Rommel's invasion of Egypt.

1942 Operation TORCH. U.S. troops under Gen. Dwight D. Eisenhower (1890–1969) land in North Africa and occupy Morocco and Algiers (November 8).

1942 Despite German victory at Kasserine Pass (February 22), Allied invasion of Tunisia begins (November). Allies push eastward and secure North Africa by capturing Tunis and Bizerte (May 8–12). North Africa is secure by May 1943.

1942–43 Battle of Stalingrad (November 19–February 2). Soviets kill or capture troops of 22 Nazi divisions and turn tide of war in East.

1943 German spring campaign briefly pushes back Soviet lines (March). By summer Soviets regain initiative and advance to Smolensk (September 25) and Kiev (November 6).

1943 Warsaw ghetto uprising. This revolt against Nazis by tens of thousands of Jews fails (May).

1943 Allied invasion of Sicily (July–August). Messina is captured (August 18).

1943 Italian Fascist leader Benito Mussolini (1883–1945) resigns July 25.

1943 Allied invasion of Italian mainland begins in South (September 3); Italy surrenders immediately. United States lands at Salerno (September 9). German troops seize control of northern Italy.

1943 U.S. invasion of Gilbert Islands (November 24). Allied Pacific offensive is under way.

1944 Anzio landings. Allies make successful landings at Anzio (January 22) but are bottled up by German resistance. Rome is finally captured (June 4).

1944 Advancing Soviets push Germans back across border with Poland (February).

1944 United States invades Marshall Islands (February 1–22).

1944 Massive Normandy invasion (June 6) begins Allied effort to retake continental Europe and totally defeat Nazi Germany.

1944 First Battle of Philippine Sea (June 19). U.S. Navy wins major victory in what is primarily battle between carrier-based aircraft.

1944 Assassination attempt on Hitler fails (July 20). He is injured, but not killed, by bomb.

1944 St. Lo breakout. U.S. troops, stalled on Normandy beachhead by German resistance, finally break through (July 25–27). Allies' rapid eastward advance toward Germany begins.

1944 Second Allied amphibious landings occur on French coast between Nice and Marseilles (August 15).

1944 Soviet forces sweep into Poland and Romania. Romania surrenders (August 24).

1944 Allies liberate Paris (August 25) and Brussels (September 2).

1944 American forces cross German border (near Eupen) for first time (September 12).

1944 Battle of Arnhem (September 17–26), airborne attack in Netherlands by British and Polish troops, fails.

1944 Second Battle of Philippine Sea (October 21– 22). Massive Japanese losses occur in failed attempt to halt American invasion. Japanese fleet withdraws (October 25).

1944 U.S. troops under Gen. MacArthur land on Leyte (October 20). Manila is finally retaken February 23, 1945.

1944 Nazis start V-2 rocket attacks against Great Britain (October 7).

1944 B-29 bombing raids on Japan begin (November 29).

1944–45 Battle of Bulge. Germans punch through overextended American lines in southern Belgium (December 16) in last major German offensive of war; they withdraw by mid-January.

1945 Soviets, continuing westward drive, take Warsaw, Poland (January 17).

1945 Yalta Conference (February 7). Allies reaffirm demand Germany must surrender unconditionally and plan shape of postwar Europe.

1945 Allies firebomb Dresden, Germany (February 13–14).

1945 Battle of Iwo Jima (February 19–March 17) is hard-fought victory for U.S. Marines on this Pacific island.

▶

1945 U.S. and British forces, driving eastward into Germany, cross Rhine after capturing bridge at Remagen intact (March 7).

1945 Battle of Okinawa (April 1–June 21) is bloody U.S. victory for this strategic island on Japan's doorstep.

1945 President Franklin D. Roosevelt dies (April 12).

1945 In Battle of Berlin, Soviet army crushes German resistance in massive attack (April 16–May 2).

1945 U.S. and Soviet troops link up at Torgau on Elbe River (April 25).

1945 German resistance in Italy is overcome, and Italy falls (April 28).

1945 Hitler commits suicide in his Berlin bunker (April 30). Adm. Karl Doenitz (1891–1980) becomes head of state.

1945 Germany signs surrender at Berlin (May 8, V-E Day).

1945 Massive air offensive against Japan (May–August) shatters Japanese industry and puts remnants of Japanese fleet out of action.

1945 Secret test explosion of atomic bomb at Los Alamos, New Mexico (July 16).

1945 Atomic bombs are dropped on Hiroshima (August 5) and Nagasaki (August 8) after Japanese steadfastly refuse unconditional surrender.

1945 USSR belatedly declares war on Japan (August 8) and invades Manchuria.

1945 Unconditional surrender by Japan (August 14, V-J Day) ends fighting. Formal surrender is signed aboard *USS Missouri* in Tokyo Bay (September 2).

Korean War

1948 U.S.-backed Republic of Korea is proclaimed (August 15), with Syngman Rhee as president. People's Democratic Republic of Korea is proclaimed in the North (September 9), led by Communist Kim Il-sung.

1948 United Nations declares South Korean government lawful and establishes commission to promote unification (December 12).

1949 U.N. effort to promote peaceful reunification of Korea fails, and commission warns of civil war (September 2).

1950 Surprise invasion of South Korea (June 25) by North Korean tank and infantry units under Marshal Choe Yong-gun. Ill-equipped South Korean army mounts little resistance as Communists drive toward Seoul.

1950 U.N. emergency session begins (June 25). United Nations calls for withdrawal of North Korean troops and for aid from member nations (June 27).

1950 U.S. President Harry S. Truman (1884–1972) orders use of U.S. combat forces (June 30) after fall of Seoul. Transfer of ill-equipped U.S. units from Japan begins (June 30). Troops from other nations begin arriving in subsequent weeks.

1950 Gen. Douglas MacArthur (1880–1964) becomes U.N. forces' commander in South Korea (July 8).

1950 Battlefront stabilizes. U.S. and Allied forces hold only small area on southeastern coast of Korean Peninsula (August) but have much-needed port of Pusan.

1950 United States launches surprise amphibious landing at Inchon, well behind enemy lines (September 15–25), coordinated with breakout from perimeter in South. It sends Communists into disorganized retreat. Seoul is retaken (September 26).

1950 U.S. and South Korean forces launch counteroffensive (September 13).

1950 U.S. and other U.N. forces push northward to Yalu River on the border with Manchuria (October 1–November 24). MacArthur orders troops across 38th parallel (October 9). Rapid advance occurs north, toward Manchurian border. North Korean capital, Pyongyang, is taken (October 20).

▶

1950 Gen. MacArthur meets with President Truman at Wake Island (October 15). He reassures Truman Chinese Communists will not cross Yalu River from Manchuria.

1950 South Korean and U.S. forward units report fighting with Chinese Communists south of Yalu (November 1). U.S. Eighth Army pulls back from Yalu.

1950 Chinese Communists 300,000 strong advance southward (November 25) and nearly encircle Eighth Army as it again advances toward Yalu. U.N. forces withdraw to 38th parallel.

1951 Chinese offensive (400,000 Chinese, 100,000 North Koreans) push Eighth Army (200,000) back to just south of Seoul (January 1–15).

1951 U.N. forces slowly push Chinese northward in series of limited counterattacks (January–February).

1951 Operation Ripper. Renewed U.N. attacks force Chinese northward to about 38th parallel (March 7–31). Seoul is retaken (March 14).

1951 MacArthur's offer to talk with Communists on ways to end fighting (March 24) is rejected (March 29). U.N. forces counterattack across 38th parallel (April 3).

1951 President Truman removes Gen. MacArthur from command (April 11) for publicly criticizing policy of limiting war to Korean Peninsula. Gen. Matthew B. Ridgway (b. 1895) replaces MacArthur.

1951 U.N. forces defeat Communists' spring offensive (April 22–May 20), inflicting about 160,000 casualties in fierce fighting.

1951 U.N. counteroffensive (May) drives Chinese northward, above 38th parallel, but is halted as threats from Soviets mount (May 31).

1951 Cease-fire negotiations at Kaesong (July–August) prove unsuccessful as Communists use delaying tactics.

1951 U.N. limited offensives resume (August), slowly pushing Communists farther northward.

1951 Capture of "Heartbreak Ridge" (September 23) caps 37-day effort to secure strategic point near Yanggu.

1951–52 Peace talks are held at Panmunjon (from October). Prisoners of war, particularly 50,000 Communist POWs refusing repatriation, become central issue.

1952 Gen. Mark W. Clark (1886–1984) succeeds Gen. Ridgway as U.N. Far Eastern commander (April 28).

1952 Peace talks break off (October). Fighting is resumed.

1953 Chinese propose renewed peace talks (March 28), just days after death of Soviet dictator Joseph Stalin (1879–1953). They agree to limited POW exchange (April).

1953 Chinese attack positions held by South Koreans (June) after South Korean President Syngman Rhee (1875–1965) refuses to accept divided Korea. Chinese suffer some 70,000 casualties.

1953 Renewed peace talks result in end to war on signing of armistice (July 27). Boundary of divided Korea is drawn along battle lines.

Vietnam War

1957 North Vietnamese guerrilla activity directed against South Vietnamese government begins.

1959 First U.S. military advisers die in Vietcong attack (July 8).

1960 Communists form National Liberation Front (NLF) (December).

1961 United States agrees to increase its 685-member military advisory group and to arm and supply 20,000 South Vietnamese troops (June 16).

1961 U.S. aircraft carrier arrives off Vietnam with armed helicopters to aid South Vietnamese (December 11).

▶

1961–62 U.S. military advisers are in Laos.

1962 U.S. President John F. Kennedy (1917–63) sends more military advisers as Vietcong guerrilla force increases. U.S. noncombatant troops in Vietnam number 12,000 in 1962.

1963 Some 200 Vietcong defeat 2,000 attacking South Vietnamese army (ARVN) regulars at Ap Bac in Mekong Delta (January 2).

1963 South Vietnamese dictator Ngo Dinh Diem (1901–63) is overthrown and killed in military coup (November 1). Vietcong make major gains during subsequent period of political instability. Military dictator Nguyen Cao Ky (b. 1930) finally takes control of South Vietnamese government in 1965.

1964 Gen. Nguyen Khanh comes to power in bloodless coup (January 30) and is elected President of South Vietnam (August 19).

1964 Gen. William Westmoreland (b. 1914) is now in command of U.S. troops in South Vietnam.

1964 Gulf of Tonkin Incident (August 2–4). North Vietnamese patrol boats attack U.S. destroyers in Gulf of Tonkin. U.S. Congress passes resolution (August 7) that President Lyndon Johnson (1908–73) uses as basis for later troop buildup in Vietnam.

1964 Tran Van Huong succeeds Khanh (November 4), but Buddhist agitation results in Khanh's return to power January 26, 1965.

1964 United States announces massive aid increase to counter Hanoi's support of Vietcong (December 11).

1965 Military coup brings Gen. Nguyen Cao Ky and Gen. Nguyen Van Thieu to power (February 20).

1965 United States begins air attacks against selected military targets in North Vietnam (February 24) in retaliation for Communist attacks on U.S. bases in South Vietnam.

1965 "Rolling Thunder" bombing campaign takes place against North (March 2).

1965 First U.S. ground combat troops (3,500 Marines) land, at Danang (March 7–9).

1965 Vietcong bomb U.S. embassy in Saigon (March 30).

1965 Australian and New Zealand troops arrive (July).

1965 First major U.S. victory comes in operation Starlight (August 18–21) against Vietcong near Chu Lai.

1966 Operation Masher-White Wing. Some 20,000 U.S., ARVN, and South Korean troops sweep coastal province of Binh Dinh (January 24–March 6).

1966 First bombing raid by B-52s based on Guam (April 12) occurs, attacking infiltration routes into South Vietnam.

1966 Buddhist uprisings against South Vietnamese government result in armed intervention throughout country (May 15–June 23).

1966 U.S. bombers attack oil installations near Haiphong and Hanoi (June 29).

1966–67 Operation Prairie. U.S. 3rd Marine Division is in hard fighting against North Vietnamese just south of DMZ (August–January).

1966 Operation Attleboro. U.S. force of 22,000 sweeps area northwest of Saigon, near Cambodian border (September 14–November 24).

1966 U.S. troops in Vietnam number 389,000 by year's end.

1967 Operation Cedar Falls. U.S.–ARVN forces attack Communist strongholds in Iron Triangle, just 25 miles from Saigon (January 8–26).

1967 Operation Junction City. Twenty-six U.S. and ARVN battalions sweep area along Cambodian border.

1967 First U.S. mining of rivers in North Vietnam.

1967 First of B-52s to be stationed in Thailand arrive there (April 10).

1967 Vietcong siege (August–September) of U.S. base at Con Thieu (in Quang Tri Province) fails.

1967 U.S. troops in Vietnam number 480,000 by year's end.

1968 Unsuccessful siege of U.S. base at Khesanh in Quang Tri Province by about 20,000 Communist troops (January 21–April 8). United States mounts relief operation Pegasus (April 1–15) as Communists retreat into Laos.

1968 Communist Tet offensive. Saigon, Hue, and many other South Vietnamese cities are attacked simultaneously by some 50,000 North Vietnamese and Vietcong

▶

troops (January 30–February 29). Communists are driven back but gain major psychological victory.

1968 My Lai massacre (March 16) of about 200 Vietnamese villagers by U.S. troops is kept secret for over a year.

1968 President Johnson halts air and naval attacks on North Vietnam north of 20th parallel (March 31).

1968 Operation Complete Victory. U.S.–ARVN force of 100,000 troops sweep area around Saigon (April).

1968 U.S. Navy Riverines begin combat role in Mekong Delta waterways (April).

1968 Spring offensive by Communists involves unsuccessful attacks of Saigon and over 100 other targets in South Vietnam (May).

1968 Start of Paris peace talks (May 10).

1968 Communists bombard Saigon with rockets on nightly basis (May 19–June 21).

1968 U.S. general Creighton Abrams (1914–74) replaces Gen. Westmoreland as military commander in South Vietnam.

1968 Battleship *New Jersey* starts combat duty off Vietnam coast.

1969 "Vietnamization" of war, turning over greater combat responsibility to ARVN, begins (February).

1969 Renewed Communist offensive occurs against Saigon and over 100 other targets (February 23–March 29).

1969 Massive B-52 bombing attacks occur against Communist positions northwest of Saigon and near Cambodian border.

1969 Total U.S. troop number in South Vietnam reaches highest level of war, at about 543,000 (April).

1969 Battle of Hamburger Hill (Hill 937, Ap Bia Mountain) in Quang Tri Province. U.S. troops seize hill after 10 unsuccessful attempts (May 20). Communists retake it June 17.

1969 U.S. troop withdrawal from Vietnam, ordered by President Richard M. Nixon (b. 1913) on May 14, begins (July 8). Nixon announces further withdrawal of 60,000 soldiers (November 3).

1969 North Vietnamese Communist leader Ho Chi Minh (1890–1969) dies in September.

1969 Cambodia's Sihanouk permits entry of 40,000 North Vietnamese troops (October 8).

1970 U.S.–ARVN force sweeps area near Cambodia (February–March), uncovering Communist supply depots.

1970 United States admits American civilian and military personnel have been involved in fighting in Laos (April).

1970 President Nixon announces his intention to withdraw additional 150,000 U.S. troops by year's end (April 30).

1970 Some 40,000 U.S. and ARVN troops briefly invade Cambodia's Parrot's Beak and Fish Hook border region to attack Communist bases there (April 30– June 30).

1970 U.S. Senate repeals Gulf of Tonkin resolution (June 24) and passes amendment barring use of troops in Cambodia (June 30).

1970 U.S. orders heavy bombing of targets in North Vietnam in retaliation for attacks on reconnaissance flights (November 21).

1970 Son Tay raid. U.S. commandos unsuccessfully raid prison camp (already abandoned) near Hanoi to free U.S. POWs (November 21).

1971 South Vietnamese troops invade Laos in Operation Lam Son 719 (February 8–April 9), capturing major Communist supply depot.

1971 Publication of classified Pentagon Papers in United States (June) reveals how country became so heavily involved in Vietnam War, further increasing U.S. antiwar sentiment.

1971 U.S. security adviser Henry A. Kissinger (b. 1923) begins secret negotiations with North Vietnamese (June).

1971 United States turns over all ground-combat responsibilities to South Vietnamese (August 11).

1971 U.S. troops in Vietnam number only 158,000 by year's end.

1972 Easter offensive by Communists begins (March 30). Some 20,000 North Vietnamese cross DMZ to take Quang Tri City. About 50,000 Communists invade South Vietnam from Laos and Cambodia (April 5), while two other major Communist

▶

offensives are launched in central regions (April).

1972 President Nixon orders mining of Haiphong and other North Vietnamese ports (May 8).

1972 Gen. Frederick C. Weyand replaces Gen. Abrams as commander of U.S. forces in Vietnam (June 28).

1972 Some 20,000 South Vietnamese troops, with heavy air support, retake Quang Tri City (September 15).

1972 Kissinger announces tentative U.S. and North Vietnamese agreement on peace plan (October 26). Talks are suspended in December.

1972–73 Massive bombing raids on Hanoi and Haiphong are resumed (December 18–January 15) after North Vietnamese stall peace negotiations.

1972 U.S. troops in Vietnam number only 24,200 by year's end.

1973 Cease-fire agreement goes into effect (January 28).

1973 Last U.S. troops are withdrawn from Vietnam (March 29). Some 8,500 American civilian technicians remain.

1973 Last U.S. prisoners of war are released (April 1).

1974 Communists and South Vietnamese resume fighting in central South Vietnam (March).

1975 South Vietnamese withdrawal of troops from northern and central regions becomes retreat (March 16).

1975 North Vietnamese begin final offensive in South (January). Last Americans are evacuated (April 29). South Vietnam surrenders to Communists (April 30).

Battles from Ancient to Modern Times

B.C.

c. 1479 Meggido, site in northern Israel. Egyptian armies under Pharaoh Tuthmosis III defeat confederation of Palestinian states by taking fortress of Meggido.

c. 1296 Kadesh. Egyptian armies under Pharaoh Ramses II battle Hittites for mastery of Syria. At Kadesh Egyptians and Hittites strike first recorded peace treaty in history.

c. 1193 Troy. City of Troy in Asia Minor falls to confederation of Greek warriors, an event immortalized by Homer in *The Iliad.*

c. 1190 Egyptian Delta. Pharaoh Ramses III defeats armada of invading "Peoples of the Sea," thus preserving Egypt's independence.

605 Carchemish, battle fought in Mesopotamia. Babylonian king Nebuchadnezzar (c. 630–562) defeats armies of Egyptian pharaoh Necho (c. 609–593) and thereby gains control of Assyrian Empire.

586 Jerusalem. City is destroyed by Babylonian king Nebuchadnezzar, thus beginning exile and Babylonian captivity of Jews.

539 Babylon. Persian ruler Cyrus the Great conquers city and frees Jews.

490 Marathon (Persian wars). Army of Miltiades (d. 489)—11,000 Athenians and Plataeans—defeats much larger Persian force under Darius I (c. 549–486), turning back Persia's second invasion of Greece.

480 Himera. Armies of city-state Syracuse led by Gelon (d. 478) defeat Carthaginians led by Hamilcar (who dies in battle). Syracuse gains control of Sicily.

480 Thermopylae (Persian wars). Small force of Spartans led by King Leonidas (d. 480) bravely holds strategic pass against invading Persian army under King Xerxes I (c. 519–465).

480 Salamis (Persian wars). Celebrated Greek naval victory in which Themistocles's (c. 525–c. 460) fleet sinks about 200–300 Persian ships after bottling up 600-ship Persian fleet in the narrow straits off island of Salamis.

▶

479 Plataea (Persian wars). Spartan general Pausanius (fl. 5th century B.C.) wins decisive victory over Persian forces led by Mardonius (who dies in battle).

479 Mycale (Persian wars). Greeks win decisive naval victory against Persian fleet, which with Greek victory at Plataea ends any further threat of invasion by Persia.

466 Eurymedon River (Persian wars). Athenian statesman-general Cimon (d. 449 B.C.) defeats Persia's land and naval forces in Asia Minor.

422 Amphibpolis (Peloponnesian War). Decisive victory for Sparta over Athenians, who afterward arrange temporary peace. Both Spartan general, Brasidas, and Athenian leader, Cleon, die in battle.

415–413 Syracuse (Peloponnesian War). Athens lays siege to Syracuse in Sicily, committing 40,000 troops and 200 ships. With reinforcements from Sparta, Syracuse destroys entire Athenian force, loss from which Athens never recovers.

407 Notium (Peloponnesian War). Spartan fleet commanded by Lysander (d. 395) defeats Athenian naval forces.

406 Arginusae (Peloponnesian War). Athenian commander Conon (d. c. 390) wins this naval battle with Spartans.

405 Aegospotami (Peloponnesian War). Spartan commander Lysander catches Athenians unaware and destroys their fleet, leading to capture of Athens.

401 Cunaxa, fought near Babylon. Cyrus the Younger (d. 401), with 13,000 Greek mercenaries, here loses his attempt to overthrow Persian king Artaxerxes II (d. c. 359). The mercenaries afterward begin their famous "Retreat of the 10,000" to safety.

394 Nemea, in Greece. After Peloponnesian War, number of city-states challenge Sparta's domination of Greece. In this battle, Spartan troops surround and defeat larger Athenian force.

371 Leuctra, in Greece. Theban commander Epaminondas (d. 363), with his new battle formation to attack Spartan phalanx, leads allied Greeks to victory over Sparta, breaking Spartan power in Greece.

338 Chaeronea, in Greece. Macedonian king Philip II (382–336), with army of 40,000, defeats allied Athenian and Theban army and thereby wins control of Greece. Philip's conquest lays foundation for the great empire of his son, Alexander the Great.

333 Issus (Alexander the Great's wars of conquest). Invading Persia, Macedonian king Alexander the Great (356–323) routs far larger force led by Darius III (d. 330), losing only some 450 soldiers.

332 Tyre (Alexander's wars of conquest). Alexander lays seven-month siege to this fortified city (in modern Lebanon) before capturing it. Macedonians kill 10,000 survivors and sell 30,000 others into slavery. Jerusalem also is taken.

331 Gaugamela (Alexander's wars of conquest), decisive victory for Alexander's 47,000 Macedonian troops over some 250,000 Persian soldiers led by Darius III. Alexander's conquest of Persia is all but complete.

330 Babylon. Alexander the Great conquers city before invading India.

326 Hydaspes (Alexander's wars of conquest), Alexander's last major battle. Macedonians rout 34,000-man army of kingdom in northern India, adding this territory to Alexander's empire.

301 Ipsus (Wars of the Diodochi), great battle between Alexander's successors for control of Asia Minor. The 80,000-man army of Antigonus I (382?–301) is defeated by allied force under Seleucus I Nicator (d. 280) and Lysimachus (c. 355–281).

281 Corupedion (Wars of the Diodochi). Seleucus I defeats Lysimachus for control of Asia Minor.

279 Asculum (war between Rome and Tarentum). Pyrrhus, king of Epirus and ally of Tarentum, defeats Romans at Asculum, but costly win originates concept of "Pyrrhic victory."

260 Mylae (First Punic War), Rome's first naval victory. Roman consul Gaius Duilius (fl. 3d century) uses boarding ramps he developed in successful battle against Carthaginian fleet off Sicily.

241 Aegates (First Punic War). Romans destroy Carthaginian fleet, bringing this war to a close.

216 Cannae (Second Punic War). Carthaginian army led by Hannibal (247–182?) routs 85,000-man Roman army under consuls

▶

Gaius Terentius Varro and Lucius Aemilius Paulus (both fl. third century). Romans lose 50,000 in one of their worst defeats.

207 Metaurus (Second Punic War). Romans evade certain defeat in war by winning this battle against Carthaginians led by Hasdrubal, Hannibal's brother (he dies in battle).

202 Zama (Second Punic War). Romans under Scipio Africanus Major (234–183?) crush Hannibal's Carthaginian army, making its last stand near Carthage. This battle ends war.

191 Thermopylae. Seleucid King Antiochus III is defeated by Romans. Battle marks onset of Rome's domination of Western world.

146 Carthage (Third Punic War). Rome destroys its longtime rival and razes city; 450,000 are killed and 50,000 taken as slaves, thus ending Third Punic War.

102 Aquae Sextiae (in Gaul). Romans under Gaius Marius (c. 155–86) rout invading Teutones.

53 Carrhae (in Asia Minor). Parthians rout Roman army led by Marcus Crassus. Death of Crassus in battle leaves Pompey (106–48) effective ruler of Rome.

48 Pharsalus (Roman civil war). Julius Caesar's 22,000-man army defeats Pompey's 45,000-man army in Greece. Victory marks turning point in Caesar's battle for control of empire.

46 Thapsus (Roman civil war). Caesar defeats (in North Africa) last remnants of his opponents' armies.

42 Philippi (in Macedonia). After Caesar's assassination, forces under Octavian (63 B.C.–A.D. 14) and Marc Antony (c. 83–30) defeat army led by Marcus Brutus (85–42) and Crassius (d. 42), who were among assassins.

31 Actium (off Greek coast). Decisive naval battle between forces of Octavian and coruler of Roman Empire Marc Antony (September 2). By his victory Octavian becomes sole ruler of empire.

A.D.

9 Teutoburg Forest (in modern Germany). German tribes led by Arminius (d. A.D. c. 19) ambush and slaughter three Roman legions under Publius Quintilius Varus (d. A.D. 9).

70 Jerusalem. Conquest and destruction of Jerusalem ends revolt of Jews against Rome.

151 Lyons (in modern France). Albinus revolts in Great Britain, proclaiming himself Roman emperor, but is killed in Battle of Lyons.

258 Milan. Invading Alemanni and Suevi are defeated at Milan, preserving Roman rule in northern Italy.

323 Adrianople (in modern Turkey). Constantine I (280–337), emperor of Eastern Roman Empire, defeats Lucinius (d. 323), emperor of Western Roman Empire. Constantine thereby becomes sole ruler of reunited empire.

351 Mursa (in modern Yugoslavia). Bloody battle between forces of Roman emperor Constantius II (317–61) and a usurper, Magnentius (d. 353). Losses on both sides are over 50,000.

378 Adrianople (in modern Turkey). Battle marks start of major invasions by Germanic tribes, in which 20,000 Visigoths and allies rout Roman army under Emperor Valens (c. 328–378), who dies in battle.

410 Rome. Visigoths under the leadership of Alaric sack city for the first time since 390 B.C.; 45 years later Vandals also will sack city.

451 Chalons (in modern France). Only defeat of Attila the Hun (A.D. 406–53). Romans under Aëtius (396?–454) and allied Visigoths defeat Huns, thereby turning back Attila's invasion of Western Europe.

476 Pavia (in northern Italy). Here German leader Odoacer (c. 435–93) defeats Romulus Augustus (fl. 5th century), last emperor of Western Roman Empire. Thereafter Odoacer deposes emperor, marking end of Western Roman Empire.

507 Campus Vogladensis. Frankish King Clovis kills Alaric II and annexes kingdom of Toulouse to his realm.

537 Camlan. Semilegendary King Arthur is killed in this battle in Great Britain.

552 Busta Gallorum. Byzantines win major victory over Ostrogoths led by Totila (d. 552 in battle) in Italy, thereby gaining much of Italy.

624 Badr (March). First victory for prophet Muhammad over Meccans in Arabia, a key event in establishment of Islam.

▶

627 Nineveh. Byzantine troops under Emperor Heraclius decisively defeat Persians at Nineveh, ending Persian wars next year.

641 Alexandria. Arab conquest reaches Byzantine Egypt, where invading Muslims burn the famous Alexandrian library.

697 Carthage. Muslim invaders conquer and destroy Byzantine Carthage. Shortly thereafter they will have complete mastery of North Africa.

711 Xeres de la Frontera (in modern Spain). Arab forces defeat King Roderic, ensuring Muslim domination of Spain. Seville is conquered next year.

718 Constantinople. Emperor Leo III defeats reinforced Muslim attack and decimates their fleet, saving Byzantine Empire once more.

732 Poitiers. Decisive victory for Franks, led by Charles Martel (688–741), over invading Muslims, marking end of Muslim advance up Iberian Peninsula.

751 Samarkand. Arab victory signals loss of Chinese dominance in western Asia.

773–74 Pavia. Frankish king Charlemagne (c. 742–814) lays siege to Lombard capital of Pavia in Italy after Lombards threaten Rome, forcing Lombard king Desiderius to surrender and making Charlemagne king of Lombards.

778 Roncesvalles (August 15). Basques ambush and wipe out rear guard of Charlemagne's army in Pyrenees. Battle is basis for *La Chanson de Roland.*

878 Edington. Anglo-Saxon king of Wessex, Alfred the Great (849–99), defeats invading Danes, preventing complete takeover of Anglo-Saxon kingdoms in Great Britain.

955 Lechfeld (August 10). German king Otto I (912–73) by this battle ends Magyar raids of German territories and strengthens his rule over the territories.

1014 Clontarf (April 23). Irish king Brian Boru (940?–1014) turns back Norse invasion near Dublin but is killed in battle.

1041 Montemaggiore. Byzantines are defeated by alliance of Normans and Lombards, marking decline of Byzantine power in Italy.

1054 Dunsinane. Usurper Macbeth is overthrown by Malcolm and Sinard of Northumbria.

1058 Varaville (in modern France). Geoffrey of Anjou is defeated by William of Normandy.

1066 Hastings (October 14). William I the Conqueror (1027–87) invades England and defeats his rival claimant to English throne, Harold II (1022?–66), who dies in battle. Key victory in Norman Conquest.

1071 Manzikert. Seljuks, led by Sultan Alp Arslan (1029–72), destroy Byzantine army, led by Emperor Romanus IV (d. 1071), in Armenia and win control of most of Asia Minor.

1097 Dorylaeum (present Eskisehir, Turkey, First Crusade). Crusaders defeat Turks here and soon conquer Nicaea on their way to Holy Land.

1098 Antioch (First Crusade). Crusaders take city from Turks.

1099 Jerusalem (First Crusade). Victorious Christian forces take control of city from the Muslims, first step toward the establishment of Latin kingdom next year.

1138 Standards. Defeat of David I of Scotland in invasion of England on behalf of beleaguered Queen Matilda.

1176 Legnano. Lombard League defeats Emperor Frederick I, slowing German designs on Italian Peninsula.

1176 Myriocephalon (September). Seljuk victory over Byzantines (Emperor Manuel I Comnenus, 1120–80) confirms Seljuk conquest of Asia Minor.

1191 Acre (July 12, Third Crusade). Crusaders, after long siege (1189–91), take city and massacre some 3,000 Muslims.

1204 Constantinople (Fourth Crusade). Crusaders turn their attack on Byzantine capital and force emperor to flee. They found Latin empire that lasts until 1261.

1212 Las Navas de Tolosa (July 16). Key victory for allied Spanish Christian forces in the reconquest of Muslim-held territories on Iberian Peninsula.

1214 Bouvines (July 27). French victory over allied force led by England. English lose domains on Continent to France.

1223 Kalka River. Mongols invade Russia. They complete their conquest by taking Moscow in 1240.

1263 Largs. Scots defeat Norwegian king Haakon, forcing him to cede Hebrides Islands.

▶

1264 Lewes (May 14, Barons' War). Decisive battle in which rebellious English barons capture King Henry III (1207–72).

1291 Acre (May 18, after Crusades). Muslim victory after brief siege of Christian-held city of Acre by 160,000 Muslims. It effectively ends Christian presence in Holy Land.

1302 Golden Spurs (July 11). Ill-trained Flemish militia defeats professional French force in Flanders, preventing French annexation of region.

1316 Faughart, Ireland. This battle near Dundalk results in death of Irish king and Lancastrian pretender to English throne Edward Bruce.

1340 Sluis (June 24, Hundred Years War). English naval victory in which archers aboard English ships decimate crews of massed French fleet, which is destroyed.

1346 Crécy (August 26, Hundred Years War). English archers prove crucial in this decisive victory over French forces in France.

1354 Gallipoli. Ottoman Turks occupy town and gain their first permanent foothold on continent.

1356 Poitiers (September 19, Hundred Years War). Major victory for invading English troops, who maul larger French army, killing some 2,000 and capturing King John.

1358 Zurich. Two defeats by Hapsburg forces result in their treaty with and recognition of Swiss League.

1371 Maritsa River (September 26). Ottoman Turks slaughter larger 70,000-man Serbian army in Balkans, paving way for westward expansion of Ottoman Empire.

1380 Battle of Chioggia. Decisive sea battle between Genoa and Venice marks triumph of Venice.

1380 Kulikovo (September 8, Tartar invasions). Russians defeat Tartars in battle that marks turning point in struggle against them.

1385 Aljubarrota (August 14). Battle in which Portuguese turn back Spanish invasion and confirm Portugal's independence.

1396 Nicopolis (September 25). Ottoman Turks decimate European Christian army sent to counter Ottoman siege of Constantinople, ending effective European resistance to Ottoman expansion into Eastern Europe.

1402 Ankara (July). Mongol leader Tamerlane (c. 1336–1405) defeats and captures Ottoman Sultan Bajazet I (1347–1403).

1415 Agincourt (October 25, Hundred Years War). Major victory for English over larger French force. French suffer some 6,000 deaths.

1428–29 Siege of Orléans (Hundred Years War). Joan of Arc (c. 1412–31) successfully breaches English siege with fresh troops and supplies to relieve city (April 30, 1429), and by May 8 French forces lift siege. This proves to be turning point in war.

1429 Patay (June 18, Hundred Years War). French rout English forces in decisive battle after Orléans.

1444 Varna. Vladislav III, King of Poland and Hungary, is defeated and killed by Turks. Disaster results in breakup of anti-Turk Balkan coalition.

1453 Constantinople (April 6–May 29). Ottoman Turks succeed in breaching city's defenses. Conquest of city marks end of Byzantine Empire.

1455 Saint Albans (May 22, Wars of the Roses). First battle of these English wars; Yorkist victory.

1461 Towton (March 29, Wars of the Roses). Yorkists and Lancastrians each field about 20,000 troops in biggest battle of the wars. Yorkists win decisively (in a snowstorm), putting Yorkist king Edward IV (1442–83) in power.

1485 Bosworth Field (August 22, Wars of the Roses). Decisive victory for Lancastrian claimant Henry VII (1457–1509) over Yorkist forces, effectively closing battle of the wars.

1492 Granada. Spanish troops capture this last Muslim stronghold in Spain, ending 800 years of rule.

1495 Fornovo. Charles VIII of France defeats Holy League, coalition of states allied against French involvement in Italy. Defeat marks end of Holy League.

1513 Flodden. King James IV of Scotland is killed. His son becomes James V under regency of mother, Margaret Tudor.

1525 Pavia (February 24, Italian wars). Armies of Hapsburg Holy Roman emperor Charles V (1500–58) fighting in Lombardy nearly wipe out 28,000-man French army and capture King Francis I (1494–1547).

▶

1526 Mohács (August 29). Ottoman armies invading Hungary crush small Hungarian army at Mohács and break Hungarian resistance. They eventually take control of much of Hungary.

1526 Panipat (April 21). In India, decisive victory for Moguls led by Baber (1483–1530) over the sultan of Delhi. It marks start of Mogul Empire.

1540 Formigny (April 15, Hundred Years War). Major French victory over English, leading to the French conquest of all Normandy. It marks first use of field artillery by French.

1558 Gravelines (July 13, in France). Final victory for Spanish Hapsburgs in wars with France. It marks early use of offshore naval bombardment in support of land attack.

1571 Lepanto (October 7). Famous naval battle, first major defeat of Ottoman forces by Christians. Only about 40 of 200 Ottoman ships escape.

1620 White Hill (November 8, Thirty Years War). Johannes Tserklaes, count of Tilly (1559–1632), leads German Catholic forces to victory in this first battle of war.

1631 Breitenfeld (September 17, Thirty Years War). Swedes win first major Protestant victory over Catholic forces of Holy Roman Empire. They introduce tactics of mobility, which soon ends long-standing use of massive infantry squares on battlefield.

1632 Lützen (November 16, Thirty Years War). Swedish king Gustavus II Adolphus (1594–1632) is killed in this battle in Saxony.

1634 Nördlingen (September 5–6, Thirty Years War). Allied army of Holy Roman Empire and Spain routs Swedish forces in Bavaria. France is forced to enter war.

1636 Wittstock (October 4, Thirty Years War). Swedish forces win bloody victory over imperial army near Berlin.

1643 Rocroi (May 19, Thirty Years War). French victory over Spanish forces in northern France in which French destroy 28,000-man Spanish army.

1643 Newbury (September 20, English Civil War). First victory for Parliamentarians.

1644 Marston Moore (July 2, English Civil War). Major victory for Parliamentary forces over Royalists; 4,000 Royalist soldiers are killed.

1645 Naseby (June 14, English Civil War). Decisive victory for Parliament's New Model Army, in which force of 14,000 Royalists is routed.

1646 Stowe-on-the-Wold (March 26, English Civil War). Parliamentary victory ends first phase of war.

1648 Preston (August 17–20, English Civil War). Closing battle of war, victory for Parliamentarian leader Oliver Cromwell (1599–1658) over last Royalist supporters.

1653 Downs (off Folkestone). English defeat Dutch in this opening salvo of Anglo-Dutch War.

1665 Montes Claros and Villa Viciosa. Spanish army is defeated by alliance of Portuguese and British forces, ensuring independence of Portugal.

1675 Sasbach and Altenheim (July 27 and August 1, respectively, Dutch War). Decisive victory for army of Holy Roman Empire fighting French in German territory. French commander Henri de Turenne (1611–75) is killed at Sasbach.

1683 Vienna (September 12). Polish forces under Jan Sobieski join with troops of Charles of Lorraine and rout besieging Turks.

1685 Sedgemoore (July 16, in England). Royalist troops crush Duke of Monmouth's revolt by massacring peasants and other rebels supporting James Scott, duke of Monmouth (1649–85).

1690 Fleurus (July 1, War of the Grand Alliance). French victory in southern Belgium against forces of Grand Alliance.

1690 Boyne (July 12, Ireland). Royalist army of English king William III (1650–1702) defeats allied Jacobite force supporting restoration of James II as king.

1692 La Hogue (May 29–June 2, War of the Grand Alliance). Decisive victory for allied English and Dutch fleet off French coast, in which French navy is destroyed.

1697 Zenta (September 11, in modern Yugoslavia). Major Austrian victory over Ottoman army, establishing Austrian power in Europe.

1700 Narva. An 8,000-man Swedish force under Charles XII crosses Baltic and routs force of 60,000 Russians under Peter the Great.

▶

1704 Blenheim (August 13, War of Spanish Succession). Fighting in German territory, 52,000 allied English and Austrian troops defeat 60,000-man French and Bavarian army, inflicting some 18,000 casualties. This marks decline of French military power.

1706 Ramillies (May 23, War of Spanish Succession). In Belgium, English and allied troops defeat French forces, inflicting heavy losses and gaining much of Spanish Netherlands.

1709 Poltava (June). Peter the Great defeats Swedish forces under Charles XII, paving way for Russian dominance of eastern Baltic.

1709 Malplaquet (September 11, War of Spanish Succession). Bloody victory for army of 100,000 allied English and Dutch troops over 90,000-man French force; last major battle of war.

1717 Belgrade (August 16, War of Austrian Succession). Prince Eugène leads Austrian troops and shatters Turkish army, forcing surrender of city next day. Turks are permanently evicted from Hungary.

1741 Mollwitz (April). Frederick II establishes reputation of Prussian army by shattering Austrian troops.

1757 Kolin (June 18, Seven Years War). In Bohemia, first Austrian defeat of Prussian forces, forcing Prussia out of Bohemia.

1759 Kunersdorf (August 12, Seven Years War). Major victory in Poland for allied Austrian and Russian army over army of Frederick the Great (1712–86).

1759 Quiberon Bay (November 20, Seven Years War). Decisive victory for British fleet over French off French coast, ending threat of French invasion of Great Britain.

1760 Torgau (November 3, Seven Years War). Frederick the Great's Prussian army defeats larger Austrian force, but with some 13,000 casualties.

1760 Wandiwash (January 22, India). British victory here against larger French colonial force ends rivalry for control of India.

1775 Lexington and Concord (April 19, American Revolution). First engagement of Revolution; American colonial Minutemen skirmish with British regulars. Losses are 95 Americans to 273 British.

1775 Ticonderoga (May 10, American Revolution). Colonists seize this important fort from British in early phase of their revolt.

1775 Bunker Hill (June 17, American Revolution). Famous victory for American colonists against British regulars early in war. Battle costs British about 1,000 casualties and builds Americans' confidence in fighting ability.

1775–76 Colonial siege of Boston (American Revolution). Battle begins soon after Lexington and ends (March 17, 1776) when British evacuate.

1776 Long Island (August 27, American Revolution). British victory (in Brooklyn) in their drive to seize control of New York City.

1776 White Plains (October 28, American Revolution). British victory over Americans forces them to continue retreating.

1776 Trenton (December 25, American Revolution). American colonials secretly cross Hudson River and capture garrison of Hessian troops. Later, at Princeton, Americans force British retreat (January 2, 1777).

1777 Bennington (August 16, American Revolution). Early American victory weakens the British and boosts morale.

1777 Fort Stanwix (August, American Revolution). First victory for Americans in series that prevents British from splitting colonies in two.

1777 Brandywine Creek (September 11, American Revolution). British victory here leads to occupation of Philadelphia (September 26).

1777 Freeman's Farm (September 19–October 7, American Revolution). Important victory near Saratoga, in which American colonials halt a British advance toward Albany.

1777 Bemis Heights (October 2, American Revolution). Second American victory near Saratoga. Soon after, British army in area is forced to surrender.

1777 Germantown (October 4, American Revolution). Bold attack by Americans on British stronghold near Philadelphia fails.

1780 Charleston (May 12, American Revolution). British victory in which siege forces surrender of 5,000 Americans at Charleston and makes possible British invasion of Carolinas.

▶

1780 Camden (August 16, American Revolution). British victory in Carolinas in which American force of 3,400 suffers over 2,000 casualties. It is American army's worst defeat of war.

1780 Kings Mountain (October 7, American Revolution). Battle proves a turning point in southern and final phase of war.

1781 Siege of Yorktown (August 10–October 19, American Revolution). Allied American and French land forces and French naval detachment bottle up 8,000 British troops and force their surrender, assuring American victory in Revolution.

1792 Valmy (September 20, French Revolution wars). First major battle of wars ends in victory for French revolutionary army over attacking force of Prussians and Austrians.

1792 Jemappes (November 6, French Revolution wars). French victory over Austrians. France occupies Netherlands.

1793 Siege of Toulon (August 28–December 19, French Revolution wars). French artillery under Napoleon Bonaparte (1769–1821) forces British to evacuate city after long siege, marking beginning of Napoleon's rise to power.

1793 Wattignies (October 15–16, French Revolution wars). Inexperienced French force turns back Austrian invasion force.

1794 Fleurus (June 26, French Revolution wars). French defeat with heavy casualties smaller Austrian and Dutch force, thereby gaining southern Netherlands (now Belgium).

1796 Lodi (May 10, French Revolution wars). Decisive battle gives Napoleon control of Lombardy.

1796–97 Siege of Mantua (ending February 2, French Revolution wars). Victory in Italy for Napoleon, who turns back several Austrian relief forces with adroit tactical maneuvers, thereby gaining northern Italy.

1797 Cape Saint Vincent (February 14, French Revolution wars). Small British fleet defeats much larger Spanish fleet preparing to join in French invasion of England.

1798 The Pyramids (July 21, French Revolution wars). Victory for Napoleon in Egypt over much larger Egyptian force. Napoleon first uses his massive "divisional squares."

1798 Nile (August 1–2, French Revolution wars). Great naval victory for British admiral Horatio Nelson (1758–1805), in which French lose all but two of 13 ships supporting Napoleon's invasion of Egypt.

1799 Aboukir (July 25, French Revolution wars). Victory over Egyptians for Napoleon's much smaller French force. Some 12,000 Egyptians die and 3,000 are captured.

1800 Marengo (June 14, French Revolution wars). Important victory for Napoleon over more numerous Austrians in Italy, by which Napoleon gains much of Lombardy.

1800 Hohenlinden (December 3, French Revolution wars). Major French victory over larger Austrian force, costing Austria some 23,000 casualties and captured to about 1,800 for France. Austria signs armistice soon after.

1805 Trafalgar (October 21, Napoleonic wars). Decisive British naval victory over allied French and Spanish fleet (33 ships). It ends threatened invasion of Great Britain, but British commander, Admiral Horatio Nelson (1758–1805), is killed.

1805 Austerlitz (December 2, Napoleonic wars). Victory for 68,000 French under Napoleon in battle against about 90,000 Austrians and Russians. French suffer 9,000 casualties; Austrians and Russians, 26,000 casualties and prisoners.

1806 Jena-Auerstadt (October 14, Napoleonic wars). Some 122,000 French under Napoleon crush somewhat smaller Prussian force in Saxony. Napoleon conquers Prussia soon after.

1807 Eylau (February 7–8, Napoleonic wars). Napoleon's army is battled even for first time in major engagement, by allied Russian and Prussian forces. Losses number about 20,000 on each side.

1807 Friedland (June 14, Napoleonic wars). Napoleon crushes smaller Russian force, leading to Treaty of Tilsit between France and Russia.

1809 Abensberg-Eckmühl (April 23, Napoleonic wars). Victories for Napoleon over larger Austrian force of 190,000.

1809 Wagram (July 5–6, Napoleonic wars). Victory for Napoleon near Vienna. It ends Austrian attempts to push French out of German territories.

▶

1810 Celaya (September 28). Mexican revolutionaries win first victory over ruling Spanish.

1811 Tippecanoe (November 7, U.S. Indian wars). Famous victory for General William H. Harrison (1773–1841) over Shawnee Indians in Indiana. Harrison uses "Tippecanoe and Tyler, too" as presidential campaign slogan in 1840.

1812 Borodino (September 7, Napoleonic wars). Napoleon's army of 130,000 mauls somewhat smaller Russian force and occupies Moscow. Russia suffers 45,000 casualties; France, 30,000.

1813 Dresden (August 26–27, Napoleonic wars). Victory for Napoleon over larger Austrian, Russian, and Prussian army.

1813 Thames (October 5, War of 1812). American victory over combined British and Indian force.

1813 Lake Erie (September 10, War of 1812). With only nine small boats, Admiral Perry captures British squadron.

1813 Leipzig (October 16–19, Napoleonic wars). Decisive victory for some 320,000 Austrians and allies over Napoleon's 185,000-man army.

1813 Chrysler's Farm (November 11, War of 1812). British victory against American invasion force in Canada, halting planned attack on Montreal.

1814 Lundy's Lane (July 25, War of 1812). British halt American force invading Canada in bloodiest battle of war.

1814 Lake Champlain (September 11, War of 1812). American destruction of British ships prevents British invasion from Canada.

1814 Fort McHenry (September 13–14, War of 1812). British fleet bombards fort, defending Baltimore Harbor. American forces resist siege, inspiring Francis Scott Key to write "The Star-Spangled Banner."

1815 New Orleans (January 8, War of 1812). Decisive American victory over British. Battle is fought before news of peace treaty reaches New Orleans.

1815 Waterloo (June 18, Napoleonic wars). Final defeat of Napoleon, who fielded 72,000 troops, by allied forces totaling 129,000 under the duke of Wellington (1769–1852) and Gebhard von Blücher (1742–1819). French suffer 25,000 casualties; allied forces, 23,000.

1818 Maipu (April 5). Victory for independence fighters over Spanish forces, securing Chilean independence.

1819 Boyaca (August 7). New Granada (modern Colombia and Venezuela) gains independence from Spanish.

1821 Carabobo (June 24). Venezuela gains independence from Spanish.

1824 Ayachucho (December 9). Peru and Bolivia are freed of Spanish.

1827 Navarino (October 20, Greek War of Independence). In last major naval battle by wooden ships, 24 British, French, and Russian ships aiding Greeks sink 50 Turkish and Egyptian ships.

1836 Siege of Alamo (February 24–March 6). Some 200 Texas independence fighters unsuccessfully try to hold fort against overwhelming Spanish forces.

1836 San Jacinto (April 21). Texas independence fighters defeat larger Mexican force, assuring Texas independence.

1847 Buena Vista (February 22–23, Mexican War). Invading U.S. forces under Gen. Zachary Taylor (1784–1850) defeat Mexicans.

1847 Cerro Gordo (April 18, Mexican War). Invading U.S. forces rout Mexicans under Gen. Antonio López de Santa Anna (1794–1876).

1854 Alma (September 20, Crimean War), initial allied British and French victory over Russians.

1854 Balaklava (October 25, Crimean War), indecisive battle made famous by poem "Charge of the Light Brigade" by Tennyson.

1859 Solferino (June 24, wars of Italian unification). Some 120,000 allied French and Italian troops rout equal Austrian army and gain much of Lombardy.

1859 Harpers Ferry raid (October 16–18). John Brown (1800–59) and his abolitionist raiders seize and hold federal arsenal in Virginia for two days.

▶

1861 Firing on Fort Sumter, South Carolina (April 13, American Civil War). Confederates bombard fort to force its surrender, thus beginning war.

1861 First Battle of Bull Run (July 21, American Civil War). Some 35,000 Confederates send 37,000-man Union force into retreat, ending threatened attack on Richmond, Virginia.

1861–65 Siege of Charleston, South Carolina (American Civil War). Union blockade begins soon after Confederate capture of Fort Sumter (April 1861), but city resists capture until February 18, 1865.

1862 Shiloh (April 6–7, American Civil War), bloody Union victory in Tennessee, with about 10,000 casualties on both sides.

1862 Seven Days' Battles (June 25–July 1, American Civil War), series of Confederate attacks on Union forces threatening Richmond. Union army retreats (July 11).

1862 Second Battle of Bull Run (August 29–30, American Civil War), successful Confederate attack that forestalls major Union offensive on Richmond.

1862 Antietam (September 17, American Civil War). Bloody battle in which Union army thwarts Lee's attempt to capture Washington by way of Maryland. Losses are war's highest for any one day of fighting (casualties: 11,729 for Confederacy; 11,657 for Union).

1862 Fredericksburg (December 13, American Civil War). Some 75,000 Confederates turn back attacks by 113,000-man Union army, inflicting heavy casualties.

1862–63 Murfreesboro (December 31–January 2, American Civil War). In Tennessee, Confederates withdraw after bloody battle. Each side suffers over 9,000 casualties.

1863 Chancellorsville (May 2–4, American Civil War). Some 60,000 Confederates under Gen. Lee send Union army over twice as large into retreat. Gen. Thomas (Stonewall) Jackson (1824–63) is accidentally killed.

1863 Siege of Vicksburg (May 22–July 4). Gen. Grant's 20,000 Union troops force surrender of 40,000 Confederates and gain control of Mississippi River.

1863 Chattanooga (November 23–25, American Civil War). Decisive Union victory for control of strategic Confederate rail junction.

1863 Gettysburg (July 1–4, American Civil War). Some 75,000 Confederates under Gen. Lee invade Pennsylvania and engage 88,000 Union troops under Gen. Meade. After successive Confederate charges fail, Lee withdraws. Battle marks turning point of war. Union suffers 23,000 casualties; Confederacy, over 20,000.

1863 Chickamauga Creek (September 19–20, American Civil War). Bloodiest single battle of war, fought in Tennessee. Union suffers 16,000 casualties; Confederacy, 18,000.

1864 Cold Harbor (June 3–12, American Civil War). Confederates repulse attacks by much larger Union army, inflicting one of worst Union defeats of war.

1864 Wilderness (May 5–6, American Civil War), bloody but indecisive battle fought during Union army approach toward Richmond.

1864–65 Siege of Petersburg and Richmond. Grant begins siege in June 1864. Last Confederate attack on Union siege force at Five Forks fails (April 1, 1865), and Lee withdraws from Petersburg and Richmond.

1866 Königgratz (July 3, Seven Weeks War), decisive Prussian victory over Austrians, leading to Prussian victory in war.

1870 Sedan (September 1, Franco-Prussian War), decisive German victory resulting in capture of strategic fortress of Sedan, surrender of 83,000 French troops, and toppling of French government.

1876 Little Bighorn (June 25, Wyoming). Gen. George A. Custer (1839–76) and all 264 troops of his 7th Cavalry are killed by Indians led by Sitting Bull (c. 1831–1890). Only a horse named Comanche survives attack.

1885 Khartoum (January 26, Sudan), victory for Sudanese rebels over British colonial governor-general Charles G. Gordon (1833–85). Rebels kill Gordon and his entire garrison at Khartoum.

1896 Adua (March 1, Ethiopian War). Italian troops meet complete defeat in their attempt to conquer Ethiopia.

1898 Manila Bay (May 1, Spanish-American War), U.S. Navy victory over Spain's Pacific Fleet, leading to capture of Manila and U.S. victory in war.

▶

1898 Omdurman (September 2, Sudan), victory over Sudanese rebels by better-equipped British force led by Maj. Gen. Herbert Kitchener (1850–1916). British destroy rebel army.

1898 Santiago (July 1–17, Spanish-American War), final hostilities of war. U.S. forces destroy Spanish fleet and capture Santiago, Cuba. Attacks by U.S. land forces on Santiago include Battle of San Juan Hill (July 1), fought by Theodore Roosevelt (1858–1919) and Rough Riders.

1904 Battle of Port Arthur (February 8–May 2, Russo-Japanese War). Japanese bottle up 16-ship Russian fleet at Port Arthur and then force surrender of city by land assault. 60,000 Japanese are killed.

1905 Battle of Mukden (February 21–March 10, Russo-Japanese War). Bloody battle between Japanese and Russian forces of about 300,000 each. Japan wins. Russians suffer 100,000 casualties; Japanese, 50,000.

1905 Battle of Tsushima Strait (May 27–28, Russo-Japanese War), naval battle in which Japanese humiliatingly defeat Russian fleet. Japanese sink 28 ships, losing only three torpedo boats.

1914 Mons (August 23–24, World War I). Germans send British into retreat from Belgium. British suffer 1,600 casualties; Germans, about 3,000.

1914 Tannenberg and Mazurian Lakes (August–September, World War I). In East, Germans under Paul von Hindenburg (1847–1934) envelop Russians at Tannenburg and successfully counter attempt to envelop them at Mazurian Lakes.

1914 First Battle of Marne (September 5–9, World War I). German attack in France falters. Germans suffer 200,000 casualties; French, 250,000.

1914 Coronel (November 1, World War I). German admiral Maximilian von Spee (1861–1914) is victorious over British naval squadron.

1914 First Battle of Ypres (October 19–November 22, World War I). German offensive in Belgium (near French border) is halted by British. Germans suffer 130,000 casualties; French, 50,000; British, 58,000; Belgians, 32,000.

1914 Falkland Islands (December 8, World War I). British naval victory over von Spee. Four of five ships in von Spee's squadron are sunk. Some 2,300 German sailors, including von Spee, are killed.

1915 Second Battle of Mazurian Lakes (February, World War I). Germans again rout Russians after attacking in snowstorm. Russian losses are 200,000 casualties and prisoners.

1915 Second Battle of Ypres (April 22–May 25, World War I). Battle marks first combat use of poison gas (chlorine, by Germans). Germans suffer 35,000 casualties; Allies, 60,000.

1916 Verdun (February 21–December 18, World War I). German attack ultimately fails, with both sides suffering heavy losses. Some 40 million shells are fired during fighting. French suffer 543,000 casualties; Germans, 434,000.

1916 Jutland (May 31, World War I). British naval forces suffer heavy losses but retain command of seas. Germans lose two battleships, four cruisers, and eight destroyers; British, five cruisers and eight destroyers.

1916 Somme (July 1–November 18, World War I), Allied offensive to relieve Verdun. British use tanks in combat for first time.

1917 Aisne (April–May, World War I). French suffer heavy losses, leading to mutiny.

1917 Third Battle of Ypres (June 21–November 4, World War I). British attack bogs down in mud after heavy rains. British suffer 400,000 casualties; Germans, 65,000.

1917 Cambrai (November 20–December 4, World War I). British make first use of massed tanks in battle.

1918 Somme (March 21–April 6, World War I). Germans attempt to split French and British force, advancing some 40 miles at Somme before being halted. Germans suffer 150,000 casualties; Allies, 160,000 casualties and 70,000 prisoners.

1918 Second Battle of Aisne (May 27–June 6, World War I). Germans, mounting 4,600-gun artillery barrage, advance 36 miles to Marne in surprise attack.

1918 Second Battle of Marne (July 15–August 6, World War I). French turn back German advance across Marne River, inflicting

▶

heavy losses (100,000 German; 60,000 Allied). Allied win proves decisive victory of war. Allied counteroffensive (August) pushes Germans back to Hindenburg Line.

1918 St. Mihiel (September 12–13, World War I). Americans under Gen. John J. Pershing (1860–1948) are victorious on Western Front. They suffer 7,000 casualties; Germans, 5,000 casualties and 15,000 prisoners.

1918 Argonne Forest (September 26–October 31, World War I). Allies' last offensive in West begins here with 500,000 troops. Germans suffer 100,000 casualties; American and French, 117,000.

1918 Americans capture St. Etienne (October 6, World War I).

1922 Sakarya (August 24–September 16). Turks battle Greeks in successful defense of Ankara. Greece had invaded along with Italy in 1919 to gain territory.

1937 Bombing of Guérnica by German planes (April 26, Spanish Civil War). Bombing raid on behalf of Spanish Fascists causes heavy damage and arouses public outrage in democratic nations.

1938 Ebro River (July 24–November 18, Spanish Civil War). Republican army offensive with 100,000 men advances across Ebro River, stalls, and is driven back. Republicans suffer 50,000 deaths and 20,000 taken prisoner; Nationalists, 33,000 casualties.

1938–39 Barcelona (December 23–January 26, Spanish Civil War). Massive Nationalist attack drives Republican defenders out of Barcelona and into headlong retreat.

1939 Madrid (March 26–31, Spanish Civil War). Final Nationalist offensive of war ends in surrender by Republican garrison in Madrid.

1940 Dunkirk (May 29–June 4, World War II). British make emergency evacuation of 340,000 British, French, and Belgian troops by ferrying them from France across English Channel to Great Britain. They leave behind huge amounts of needed war materiel, however.

1940-41 Britain (June–June, World War II). Intensive bombing campaign begins after fall of France and continues until opening of Eastern Front.

1941 German battleship *Bismarck* is sunk by British planes and naval fire (May 27, World War II). British lose battleship in engagement.

1941 Operation Barbarossa begins (June 22, World War II). Nazi troops (138 divisions) invade USSR on a broad front in greatest ground offensive ever. Soviets, who field 148 divisions at start, adopt scorched-earth tactic as they retreat. Germans advance 550 miles along 1,500-mile front before being stalled by the onset of winter. Russian losses to this point are 2.5 million casualties and 1 million prisoners; German casualties, about 1 million.

1941 Tobruk (April 10–November 29, World War II). German Afrika Korps of Gen. Erwin Rommel (1891–1944) attacks and besieges Tobruk for 240 days before being forced to withdraw.

1941 Pearl Harbor attack (December 7, World War II). Japanese sneak attack on U.S. naval base destroys much of U.S. Pacific Fleet but brings United States into war (December 8). Attack opens Japan's Pacific offensive.

1941-43 Moscow (World War II). German forces reach outskirts of Moscow (November 1941), but Russian winter forces them to delay final attack. Soviets counterattack (December) with 100 fresh divisions but do not force German withdrawal until March 1943. This is first major German defeat of the war. Soviet casualties in Moscow's defense are about 500,000.

1941–44 Siege of Leningrad (September 8, 1941–January 27, 1944, World War II). City is badly damaged, and over 1 million die in heroic stand against Nazis.

1942 Bataan (January 2–April 9, World War II). Allied forces hold out here after fall of Manila. Corregidor doesn't fall until May 6.

1942 Doolittle's raid (April 18, World War II). Lt. Col. James Doolittle (b. 1896) leads U.S. bombing raid on Tokyo.

1942 Coral Sea (May 7–8, World War II). U.S. Navy is victorious in first carrier-to-carrier battle and first naval battle in which opposing ships do not directly fire at one another. United States and Japan each lose one carrier.

▶

1942 Midway (June 4–7, World War II). U.S. Navy turns back Japanese invasion fleet and by this battle turns tide of Pacific war. United States loses carrier *Yorktown*; Japan, four carriers, 275 planes, and about 5,000 men.

1942 Guadalcanal (August 7, World War II). Marines invade island, finally taking it in February 1943. United States suffers 5,800 casualties and 12,000 stricken by tropical disease; Japan, 23,000 dead by battle and disease and 1,000 prisoners.

1942 Dieppe Raid (August 19, World War II). Allied raid on French coast fails. Canadians lose 1,000, with another 2,000 taken prisoner.

1942 Stalingrad (September 22, World War II). Germans take city, only to find themselves enveloped by Russians (November). Of some 300,000 Germans trapped in Stalingrad, only 132,000 are alive at German surrender there (January 1, 1943).

1942 El Alamein (October 23–November 4, World War II). British under Gen. Bernard Montgomery (1887–1976) turn back Rommel's invasion of Egypt. British suffer 13,500 casualties; Germans, about 40,000.

1942 Operation Torch (November 8, World War II). U.S. troops under Gen. Dwight D. Eisenhower (1890–1969) land in North Africa and occupy Morocco and Algiers.

1943 Salerno (September 9, World War II). Allied invasion of Italian mainland begins in south (September 3), and Italy surrenders. United States then lands at Salerno. German troops seize control of northern Italy.

1944 Anzio landings (January 22, World War II). Allies, hoping to outflank Germans stalling their drive to Rome, make successful landing but are bottled up by German resistance. Rome is not finally captured until June 4. Germans suffer about 11,000 casualties; Allied troops, 21,000.

1944 Battle of Java Sea (February 27–March 1, World War II). Japanese defeat here paves way for Allied conquest of East Indies.

1944 Normandy invasion (June 6, World War II). This is largest amphibious assault in history, involving over 5,000 ships, 11,000 planes, and 1 million men (put ashore by June 20). It begins Allied effort to retake continental Europe and defeat Nazi Germany. During first two weeks, Allies suffer about 140,000 casualties; Germans, 94,000 casualties plus 60,000 taken prisoner.

1944 First Battle of Philippine Sea (June 19, World War II), also called "Marianas Turkey Shoot." U.S. Navy wins major victory in what is primarily battle between carrier-based aircraft. Japanese lose three carriers, over 400 planes, and over 4,000 men.

1944 St. Lo breakout (July 25–27, World War II). U.S. troops, stalled on Normandy beachhead by German resistance, finally break through. Allies' rapid eastward advance toward Germany begins.

1944 Arnhem (September 17–26, World War II). Unsuccessful airborne attack in Netherlands by 10,000 British and Polish troops. Some 6,000 are taken prisoner.

1944 Leyte Gulf (October 23–26, World War II). Japanese fleet of some 50 ships unsuccessfully attempts to block the American invasion of Philippines by engaging American fleet of over 200 ships. Japanese lose four carriers, three battleships, ten cruisers, and nine destroyers; United States loses three carriers and three destroyers.

1944 Leyte landing (October 20, World War II). U.S. troops under Gen. MacArthur land on Leyte in Philippines. Fighting continues until December 25. Japanese suffer 70,000 casualties and troops taken prisoner; Americans lose about 16,000.

1944 Luzon (December 15, World War II). United States lands 68,000 troops, taking island and recapturing Manila. However, some 50,000 Japanese on island hold out until war's end.

1944–45 Battle of Bulge (December 16, World War II). Germans, with 1,000 tanks and assault guns and 250,000 men, punch through overextended American lines in southern Belgium in the last major German offensive of war; they withdraw by mid-January. Germans lose 100,000 troops; Allies, 76,000.

1945 Iwo Jima (February 19–March 17, World War II). U.S. Marines get hard-fought victory against some 1,500 fortifications on this tiny Pacific island. Almost 21,000 Japanese die; 6,821 Americans die, and 18,200 are injured.

1945 Okinawa (April 1–June 21, World War II), bloody U.S. victory for this strategic island

▶

on Japan's doorstep that is defended by 110,000 troops. Japanese suffer 100,000 deaths; Americans, 7,500.

1945 Berlin (April 16–May 2, World War II). Soviet army crushes German resistance in massive attack. German *Führer* Adolf Hitler (1889–1945) commits suicide (April 30) as Soviets near central Berlin.

1949 Siege of Peking (January, Chinese Civil War). Communist forces take Peking after month-long siege of Nationalist forces there, marking virtual end of long and bloody battle between Nationalists and Communists for control of China.

1950 Inchon landing (September 15–25, Korean War). United States launches surprise amphibious landing at Inchon, well behind enemy lines, coordinated with breakout from perimeter in South, sending Communists into disorganized retreat. Seoul is retaken (September 26). U.S. loses 6,000 troops.

1951 Operation Ripper (March 7–31, Korean War). Renewed U.N. attacks force Chinese northward to about 38th parallel. Seoul is retaken (March 14).

1954 Dien Bien Phu (March–May 7, French Indochina War). Vietnamese general Vo Nguyen Giap (b. 1912) masses 50,000 men against this fortified garrison of 10,000 French troops and eventually forces its surrender by siege, ending French colonial rule in Indochina. Vietnamese suffer 22,000 casualties; French, 2,600.

1961 Bay of Pigs invasion (April 17–19, CIA-backed Cuban takeover attempt). U.S.-trained force of 1,500 Cuban exiles unsuccessfully attempts to overthrow Communist government of Fidel Castro (b. 1926) by military force. United States is humiliated by fiasco; 300 exiles die.

1963 Ap Bac (January 2, Vietnam War). Some 200 Vietcong defeat 2,000 attacking South Vietnamese army (ARVN) regulars at Ap Bac in Mekong Delta.

1964 Gulf of Tonkin incident (August 2, Vietnam War). Alleged attack on U.S. naval forces in gulf by North Vietnamese gunboats spurs increased U.S. involvement in war.

1965 Operation Starlight (August 18–21, Vietnam War). First major U.S. victory against Vietcong, near Chu Lai. Some 5,000 Americans trap and defeat 2,200 Vietcong. Vietnamese suffer 1,000 deaths; United States, 50.

1966 A Shau (Vietnam War). U.S. Special Forces camp in northern South Vietnam, defended by 17 Americans and 360 Vietnamese, is attacked by 3,000 North Vietnamese regulars. Helicopters rescue 200 defenders after 30 hours of fighting. Some 500 Communists are killed.

1966 Operation Attleboro (September 14–November 24, Vietnam War). U.S. force of 22,000 sweeps area northwest of Saigon, near Cambodian border.

1967 Operation Cedar Falls (January 8–26, Vietnam War). U.S.–ARVN forces attack Communist strongholds in Iron Triangle, just 25 miles from Saigon.

1967 Siege of Con Thieu (August–September, Vietnam War). Vietcong siege of U.S. position fails.

1968 Siege of Khesanh (January 21–April 8, Vietnam War). About 20,000 Communist troops lay unsuccessful siege of this U.S. base in Quang Tri Province. United States mounts relief operation Pegasus (April 1–15) as Communists retreat into Laos.

1968 Communist Tet offensive (January 30–February 29, Vietnam War). Saigon, Hue, and many other South Vietnamese cities are attacked simultaneously by some 50,000 Communist troops. Communists are driven back but gain major psychological victory.

1969 Hamburger Hill (Vietnam War). U.S. troops seize this hill (Hill 937) after 10 unsuccessful attempts. Communists retake it June 17.

1970 Son Tay raid (November 21, Vietnam War). U.S. commandos unsuccessfully raid prison camp (already abandoned) near Hanoi to free U.S. POWs.

1971 Operation Lam Son 719 (February 8–April 9, Vietnam War). South Vietnamese troops invade Laos and capture major Communist supply depot.

1972 Quang Tri City (September 15, Vietnam War). Some 20,000 South Vietnamese troops, with heavy air support, retake city.

1975 Fall of Phnom Penh (April 17, Cambodian Civil War). Extremist Communist Khmer Rouge army takes this Cambodian capital city, ending their long war for control of Cambodia. They forcibly relocate entire city to countryside, where estimated 1 million to 3 million die.

▶

1978 Phnom Penh (December, Vietnamese–Cambodian War). Vietnamese troops take capital in their invasion of Cambodia to oust extremist Khmer Rouge.

1982 Operation Ramadan (July, Iran–Iraq War). Iran launches this major offensive, using successive waves of Revolutionary Guardsmen. It develops into possibly largest battle to date since 1945, though Iranian advance eventually stalls.

1982 Operation Peace for Galilee (June 6, Israeli invasion of Lebanon). Israelis launch 60,000-man invasion of Lebanon to clear out PLO terrorists. They encircle PLO forces in Beirut by June 14, forcing them to withdraw from Lebanon by boat. About 3,000 PLO and Syrian troops die; 368 Israelis die.

1983 Ammara (February 6, Iran–Iraq War). Some 200,000 Iranians attack Iraqi positions in desert southeast of Baghdad. Iranians suffer 6,000 deaths.

1984 Operation Dawn V (February 22, Iran–Iraq War). Iranians launch major offensive by 500,000 troops into Iraq with little gain in territory. Iranians suffer 19,000 deaths.

1984 Battle of Majnoon (April, Iran–Iraq War). Fanatical Iranians attack Iraqi positions in marshland by sending army of thousands of children, roped together, against Iraqis.

1985 (March, Iran–Iraq War). Major Iranian offensive is crushed with estimated loss of 10,000 Iraqis and 30,000 Iranians.

Military Techniques and Technology

B.C.

c. 23,000 Primitive bow and arrow are invented.

c. 15,000 Spear-throwing device is invented.

c. 3500 Two-wheeled war chariots are in use in Mesopotamia.

c. 3000 Sumerians have devised helmet as head protection.

c. 2550 Art of ancient Ur depicts soldiers with square shields in formation.

c. 1500 Fabric armor (linen in 14 layers) and bronze armor are in use in Mycenae. Chinese use armor of layered rhinoceros skin by c. 1000 B.C.

c. 1200 Iron is in use for making weapons.

c. 1000 Wheeled battering rams are used by Assyrians.

c. 1000 Mounted cavalry soldiers figure in combat.

c. 700 Catapult-type device for hurling large missiles at enemy is constructed during reign of King Uzziah of Judah, according to reference in Bible.

c. 700 Greek phalanx, formation of eight rows of infantrymen, is in use throughout Greece. Soldiers (hoplites) each carry shield, pike, and 2-foot sword.

c. 500 Early form of chain mail is in use near modern-day Kiev. Romans adopt shirt made of mail for soldiers.

481 Greek engineer constructs pontoon bridge over Hellespont for invading Persian armies.

424 Greeks use sulfur fumes as gas attack in their siege of Delium.

c. 400 Ballista, giant crossbow, is used in defense of Syracuse against Sicily.

371 Theban commander Epaminondas (d. 362) introduces warfare tactics during Battle of Leuctra. Departing from the usual frontal assault of opposing phalanxes, he puts bulk of his forces on one flank. This turns enemy's opposing flank, resulting in Theban victory.

357 Macedonian king Philip II (382–336 B.C.) develops highly successful battle formation. He doubles Greek phalanx size and creates light infantry force of archers and javelin throwers. Heavy and light cavalry and portable catapult-type artillery also are used.

▶

350 Romans develop basic battle formation they use to conquer ancient world. It is based on legion, about 4,500 men arranged in three rows. Soldiers, mainly heavily armed and on foot, wear metal helmets, carry round shields, and fight with javelins and 20-inch broadswords.

c. 325 Earliest known book on strategy and war, *The Art of War*, is written in China by Sun-tzu.

c. 300 Hellenistic engineers construct fighting towers to scale defensive walls of cities. Battering rams up to 180 feet long also are in use.

216 Battle of Cannae. Carthaginian general Hannibal (247–182?) uses cavalry to annihilate 85,000-man Roman army. In subsequent years Roman legion is enlarged and cavalry and light infantry are added.

c. 200 Roman ballista are in use. Romans use this engine extensively in field operations for throwing javelins or stones.

A.D.

c. 300 Romans develop onager, catapultlike device for siege and field use.

527–65 Byzantines develop formidable army of mounted archers. Archers shoot while at full gallop in cavalry charge and also carry spears and swords.

c. 575 Byzantine military theory is codified in *Strategicon* of Emperor Maurice (539–602) and later in *Tactica* of Emperor Leo the Wise (886–911). War is considered practical, not heroic effort.

c. 600 Chinese invent stirrup, which vastly improves fighting ability of men on horseback.

673 Greek fire, secret flammable compound, helps save Constantinople from being overrun by attacking Muslims.

c. 1000 Crossbow, possibly developed as early as 500 B.C. in China, appears in Italy. Weapon is considered so terrible that Lateran Council (1139) bans its use by Christian soldiers. Crossbow with steel bow is developed c. 1325.

1066 In Battle of Hastings, mounted soldiers under William the Conqueror (1027–87) defeat English army of foot soldiers. Soldier on horseback replaces foot soldier as backbone of battle formation in Europe for a time.

c. 1100 Trebuchet, device for throwing heavy missiles, comes into general use. Its long throwing arm is driven upward by releasing heavy counterweight, propelling missile toward target.

c. 1200 Chinese develop first rocket-propelled weapons, arrows propelled by black-powder rockets, by this time.

1200s Development of armor as protection in combat begins in Europe. Armor plate begins to replace chain mail.

c. 1213 Mongols under Genghis Khan (1167?–1227) develop effective cavalry system. Mounted soldiers carry scimitar, bow, and lance.

c. 1300 Construction of fortifications (city walls and castles) reaches peak. Defense is now superior to offense.

c. 1330s Longbow, developed as hunting tool in Wales, is adapted by King Edward III for war.

1346 Battle of Crécy. Superiority of rapid-fire English longbow over crossbow is demonstrated in victory over French. English also use early cannon fired by gunpowder.

c. 1400 Primitive breech-loading gun is developed. Most cannon are muzzle-loading over next centuries, however.

c. 1400 Complete suit of armor as worn in medieval Europe appears. Its use declines rapidly after 1650 as improved firearms render it ineffective.

c. 1400 Swiss armies introduce halberd, 8-foot spear that also has ax blade and hook (for pulling armored soldiers off their horses). Swiss restore foot soldier as backbone of their armies, discarding almost all armor in favor of mobility.

c. 1420 Siege cannon mounted on wheels is introduced during Hussite wars.

1450 French field artillery proves superiority of mobile artillery at Battle of Formigny. Ranks of English longbowmen are decimated by French fire.

c. 1450 Spanish invent harquebus, riflelike weapon that is first gun to be fired from shoulder.

▶

c. 1450 Cast-iron cannonballs replace stones, greatly increasing power of the weapons.

1453 Turks use 70 cannons to breach massive walls of Constantinople, which fall after centuries of successful resistance to Muslim invaders. Largest of bombards could hurl 1,600-pound ball more than 1 mile. Artillery thus renders ineffective medieval defensive walled fortifications.

1494–95 French king Charles VIII introduces mobile horse-drawn cannon and uses iron cannonballs in his invasion of Italy. Ability to move cannon rapidly changes conduct of battle.

c. 1500 Spanish develop battle formation combining soldiers armed with harquebuses and units of pikemen to protect harquebusiers during reloading. Spanish armies are victorious in both Europe and New World during this century.

1520 *Art of War* is written by Niccolò Machiavelli (1496–1527).

1540s English produce cast-iron cannon using techniques described in Vannocio Biringuccio's *De re Pirotechnia* of 1540.

c. 1550s Italian military engineer Niccolò Tartaglia attempts to compute range of cannon and invents gunner's quadrant.

1571 In Battle of Lepanto, naval guns spell end of oared galleys in combat. Galleons carrying many cannons subsequently rule seas.

1611–32 King Gustavus II Adolphus, sometimes called father of modern tactics, reigns in Sweden. He organizes Swedish infantry into brigades and regiments and introduces lighter field artillery, lighter muskets, and paper cartridges (to speed reloading). Gustavus's innovations prove highly successful in Thirty Years' War. His methods of careful planning and transportation allow winter campaigns to be fought successfully.

c. 1650 Flintlock gun, invented years earlier, now is in general use.

mid-1600s Maurice of Nassau develops professional standing army, more stable and disciplined fighting force.

1655 French engineer Sebastion Vauban (1633–1707) devises new systems of defensive fortifications to better withstand artillery attack. His writings become final word.

c. 1670s French establish naval training academies to systematize training of naval personnel.

1674 Publication of Robert Anderson's *Genuine Use and Effects of the Gunne* introduces Galileo's parabolic theory into study of ballistics.

1688 French introduce socket bayonet on muskets. It eliminates need for pikemen and makes possible linear warfare (advancing line of exposed troops, pausing to fire platoon volleys before final bayonet charge).

c. 1700 Grenades, originally used in sieges, are adopted by infantry, giving rise to grenadier.

1742 Benjamin Robins's *New Principles of Gunnery* greatly advances science of ballistics.

1756–63 Prussian king Frederick the Great (1712–86) develops strategy of attack and of interior lines to take advantage of Prussia's central position during Seven Years' War. It successfully counters prevailing strategy of siege warfare. He also introduces limited conscription, superior method of raising troops and achieving uniformity and discipline.

1775 Battle of Bunker Hill. Americans exploit weakness of linear warfare with more accurate gunfire and by firing from behind cover.

1784 Shrapnel shell is invented.

1793 French revolutionary government adopts conscription as means of building army. Citizen-soldier replaces paid mercenary.

1796–1815 Napoleon I (1769–1821), first great strategist of modern times, develops his basic principles of surprise, mobility, and concentration of forces against enemy at crucial point. He makes especially effective use of artillery in battle.

c. 1800 Eli Whitney revolutionizes weapons production with mass-production techniques, first used to produce muskets.

1805 French general Antoine-Henri Jomini (1779–1869) publishes his *Treatise on Grand Military Operations*, which codifies Napoleon's methods. His work later influences conduct of U.S. Civil War.

▶

1806 First Congreve military rockets are used by British in combat against French port of Boulogne during Napoleonic wars.

1815 Percussion cap is invented in United States. It makes possible breech-loading firearms (1820–65), which are easier to reload and which increase soldiers' firepower.

c. 1818 Prussian theorist on military strategy Karl von Clausewitz (1780–1831) begins work on his famous book *On War*. He holds success in battle is key to strategy. Book remains basic text even today.

1853–56 Crimean War. Use of explosive shells in naval combat renders wooden warships useless.

1858 Rifled artillery is in use in France. It becomes widespread in 1860s.

1859 Superior breech-loading rifle is developed by American Christian Sharps. It gives rise to term "sharpshooter."

1861–65 U.S. Civil War. Railroads add new element of speed and mobility in strategy. First, partial mobilization of industry to produce war materiel is effected, foreshadowing complete mobilizations of next century. Long-range infantry rifle and quick-loading one-piece metal cartridges are introduced. Improved rifles now effectively counter Napoleon's tactic of rapidly massed attack. Land mines and flame projectors also are introduced.

1862 First battle between ironclad ships, during U.S. Civil War. Naval mines and torpedoes also are introduced.

1862 Hand-cranked Gatling gun is invented, forerunner of modern machine guns.

1870–71 Franco-Prussian War culminates strategic and organizational innovations of Prussian field marshal Helmuth von Moltke (1800–1891). He introduces system of subordinate commanders operating under broad directives, not detailed orders, to better control movements of massive troop formations. Superior Krupp cast-steel cannon is used.

1884 French invent first smokeless gunpowder.

1884 Maxim machine gun, first successful automatic machine gun, is invented.

c. 1890s Improved gun carriage and recoil mechanism are developed in Europe, allowing artillery to be fired frequently and accurately.

1890–92 American naval scholar Thayer Mahan (1840–1914) publishes his works on history and theory of sea power. He sees navy as tool of national policy and notes importance of sea power in economic warfare.

1899 First automatic cannon is introduced, Maxim pom-pom.

1900–27 German military historian Hans Delbrück (1848–1929) publishes works advancing theory that there are two ways to conduct war, by strategy of annihilation and by strategy of exhaustion. Latter includes blockade, destruction of crops, interruption of commerce, and occupation of territory.

1905 Japanese win first major battle of steam-powered warships mounting long-range guns, during Russo-Japanese War.

1905 Famous Schlieffen Plan is finalized by German chief of staff Alfred von Schlieffen (1833–1913). It aims at victory by annihilation of enemy in battle. Schlieffen calls for victorious attack against French forces (by envelopment), then for drive into Russia.

1906 British launch *Dreadnought*, battleship heavily armed with long-range guns. Until World War II, big battleships with heavy guns remain centerpiece of navies of major powers.

1914–18 World War I, planned as offensive, mobile war, quickly becomes stalemated conflict fought from defensive trenches. Artillery and, more importantly, machine gun give defensive forces strong new advantage over offensive ones. Radio and telephone increase speed of communication and, along with rail and motor transportation, make it possible to marshal armies of millions of men.

1914 Hand grenades, known since 1500s, are widely used by soldiers from this time forward.

1914 Airplanes, at first used for reconnaissance, are armed with machine guns, bombs, and sometimes even rockets for attack missions.

1914–18 Flamethrowers are introduced.

▶

1915 Anthony H. G. Fokker (1890–1939) invents synchronizing gear that allows machine gun to fire through airplane's propellers.

1915 Germans make first use of poison gas in combat, in Second Battle of Ypres.

1916 British make first use of tanks in battle. Tank offers means to overcome defensive advantage of machine gun. Trucks for transporting troops quickly to front also are used extensively from this time.

1916 Standardized guns and "director" system of fire greatly increase accuracy of naval firepower. Long-range fighting is now possible.

1917 Germany begins unrestricted submarine warfare on Allied merchant shipping in Atlantic. U-boats prove effective at attacking vital seagoing supply lines. Submarines also force changes in naval tactics for defending surface warships.

1918 United States develops idea of airborne parachute divisions.

1918 Germans use huge "Paris gun" to shell Paris from up to 74.5 miles away. Gun fires small shell into upper atmosphere to achieve long range.

1918 Germans introduce submachine gun in combat.

1918 British build first aircraft carrier. Aircraft radically change nature of naval warfare during World War II.

1919 Browning air-cooled machine gun is developed.

c. 1929–39 French construct Maginot Line, string of defensive fortifications designed to prevent new German invasion of France. Such defensive strategy fails utterly before speed and mobility of Germans in World War II.

1930s-40s Various nations develop powerful chemical and biological weapons. Although not used, their development continues into postwar era.

1935 British construct first radar installations. This new invention proves vital military defensive weapon after 1939.

1935 German general and noted strategist Erich Ludendorff (1865–1937) publishes book expounding his theory of "total war," involving complete mobilization of society for war effort. It presages World War II mobilizations in Axis and Allied nations.

1939 Germans introduce *Blitzkrieg* tactics, using aircraft and tanks to spearhead rapid offensives in Poland.

1941 Germans successfully test V-1, jet-powered flying bomb. V-1s are precursors of today's cruise missiles.

1942 Development of proximity fuse vastly increases effectiveness of artillery and airborne bombing.

1942 Battle of Midway is first naval battle fought solely by planes from opposing aircraft carriers. It heralds new age in naval warfare.

1943 Improved sonar and depth charges turn tide against German U-boats.

1943 Strategic bombing comes into its own as U.S. and British planners organize day and night bombing offensives against military and economic targets in Nazi Germany, preparing way for eventual invasion.

1944 Normandy invasion, largest amphibious invasion in history, marks culmination of developing amphibious tactics in World War II.

1944 Germans make first combat use of V-2 rockets, precursors of modern guided missiles. Rocket travels 120 miles.

1945 Development of napalm enhances effectiveness of incendiary bombing, which is used effectively against Japan.

1945 Dropping of atomic bombs on Hiroshima and Nagasaki opens age of nuclear warfare. Until mid-1950s, United States would enjoy nuclear superiority. Strategic bombing is now of penultimate importance.

1945–52 Indochina War. French colonial forces are defeated by Communists employing guerrilla tactics designed to overcome superior conventional military forces with small, lightly equipped groups of soldiers.

1950–53 Korean War sees first extensive aerial combat between jet fighters, first combat use of helicopters (tactical and logistic), and introduction of synthetic-fiber bulletproof vest for infantry soldiers.

▶

1952 United States tests first hydrogen bomb, many times more powerful than atomic bomb.

1953 Soviets explode their first hydrogen bomb. By mid-1950s Soviets become credible nuclear power, although United States retains nuclear superiority.

1953 United States successfully tests its first large missile, intermediate-range ballistic missile (IRBM) called Redstone.

1955 United States, to counter Soviet nuclear threat, establishes extensive early-warning system. It keeps some Strategic Air Command (SAC) bombers, armed with nuclear bombs, in air at all times to shorten response time in event of Soviet attack.

1957 First intercontinental ballistic missile (ICBM) is tested by Soviets. Atlas, first U.S. ICBM, is tested in 1958. Advent of ICBMs, capable of delivering nuclear warheads accurately and in minutes, makes massive nuclear war real possibility. It also forces strategic bombing (with nuclear bombs) into secondary strategic role.

1960 First underwater firing of U.S. IRBM, solid-fuel Polaris missile with 1,400-mile range. Nuclear submarines, introduced in 1954, are difficult to find and track and thus become excellent mobile launch platforms for nuclear-tipped missiles. Submarines now serve as key element in both first-strike and massive retaliation strategies.

1960s–70s United States uses chemical defoliants in its battle against Communists in Vietnam.

1962–63 Cuban missile crisis. Soviets nearly provoke nuclear war by installing nuclear missiles in Communist Cuba. Thereafter, superpowers make efforts to avoid such confrontations and to avoid accidental nuclear war. "Arms race" proceeds, to build bigger and more advanced arsenals.

1964 First test of U.S. Minuteman II missile.

1965–73 U.S. involvement in Vietnam War. Communists again use guerrilla tactics successfully against conventional military forces. While relying heavily on conventional forces, United States also develops some Special Forces units trained to operate in guerrilla fashion.

1970s Introduction of MIRV missiles with multiple, independently targeted warheads.

c. 1972 New guidance systems based on lasers and electro-optical systems used in highly accurate guided bombs are used in Vietnam.

1978 United States and Soviet Union both delay building neutron bomb.

1979–89 Soviet conventional forces fail in protracted war against guerrilla forces in Afghanistan. As in case of Vietnam, guerrilla-style war of attrition effectively counters large conventional military force.

c. 1980 Cruise missiles are developed. These jet-powered missiles are guided by extremely accurate navigational systems and fly very low to ground, thereby escaping detection by radar.

1984 United States achieves first midair direct hit of one missile by another, demonstrating significant advance in antiballistic missile (ABM) technology.

1989 U.S. Delta Star, "Star Wars" satellite is launched into space. It successfully detects and tracks test missiles immediately after they are launched.

Military Small Arms Development

1300s Hand cannons, cumbersome miniature cannons, are in use.

c. 1400 Slow-burning match, fiber cord soaked in combustible chemicals, is developed for igniting powder in hand cannons.

c. 1425 Invention of muskets with matchlock, which holds slow-burning match and touches it to gun's powder priming pan when trigger is pulled. It allows soldier to concentrate on aiming.

c. 1500 Spanish introduce first infantry tactics making effective use of small firearms.

1500s First guns with rifled barrels are produced. Rifling produces spin to bullet, which increases accuracy.

Military Small Arms Development

c. 1515	Invention of wheel-lock musket, in Germany. Less reliable than matchlock, wheel-lock muskets are not widely used. Cavalry uses wheel locks, however.
c. 1540	Pistols come into use.
c. 1550	Flintlock musket is introduced. Musket, which ignites powder by striking flint against steel to produce sparks, is not widely used in military for about another century. It is perfected in France in 1630.
c. 1500–c. 1675	Infantrymen in Europe carry either musket or pike (pikemen provide protection for riflemen during reloading).
c. 1650	French infantry develops first bayonet, of type that fits into the musket muzzle, eliminating need for pikemen.
c. 1660s	French equip army with flintlock guns, first such widespread adaptation.
1668	Socket bayonet, which fits around gun muzzle to allow firing of weapon, is introduced by French.
1700s	"Brown Bess" musket becomes standard issue in British infantry, remaining British infantry rifle well into 1800s.
c. 1795	Austrian sharpshooters use Girandoni air rifle, only air-powered weapon ever used in battle.
c. 1798	American Eli Whitney applies principles of interchangeable parts and mass production to rifle-making.
1815	Percussion cap for detonating power charge is invented by Philadelphian, Joshua Shaw. Muskets using percussion caps, simpler and more reliable than flintlocks, are in wide use after mid-1800s.
1820–65	Percussion caps make possible development of many types of breech-loading guns (loaded at back rather than through muzzle).
1835	Samuel Colt patents revolver, which can be fired several times without reloading.
1841	Invention of needle gun, a breech-loading rifle. It can be loaded in prone position and fired rapidly.
1850s	Introduction of specially shaped Minié bullets, accurate to 1,000 yards, alters battlefield tactics. Infantry using muskets with these bullets are able to fire on approaching cavalry much earlier and to field artillery units to positions far in rear.
1859	Christian Sharp's breech-loading gun is perfected. It becomes most popular single-shot rifle of American Civil War.
1860s	U.S. Spencer rimfire, used in Civil War, is first all-in-one metal cartridge (case, bullet, primer, and powder) to be used successfully in war. After 1865, suitable brass and copper cases make practical center-fire cartridge, with primer cap embedded in center of back of cartridge. These become military standard issue.
1862	Gatling gun is patented by American inventor Richard J. Gatling (1818–1903).
1866	Winchester rifle, repeating rifle, is introduced. Its effectiveness is demonstrated in Russo-Turkish War of 1877.
1866	Chassepot rifle is adopted by French military. It has an increased range over needle rifle.
1880s	Repeating rifles, with internal magazines to hold extra bullets, are widely used.
1884	American inventor Hiram Maxim (1840–1916) invents recoil-operated Maxim machine gun, first successful automatic machine gun.
1885	French use smokeless powder for military cartridges for first time. It is more powerful propellant than standard black powder.
1887–88	Alfred B. Nobel (1853–96) patents smokeless propellants.
1889	Bolt-action Mauser with box magazine is manufactured for Belgians. This rifle becomes prototype for infantry rifles around world by early 1900s.
1895	Browning-Colt machine gun is introduced. It is used by United States early in World War I.
1900	First Browning revolvers are manufactured.
1903	Springfield 1903A rifle is developed. U.S. military uses it in World War I.
c. 1914	Introduction of recoil-operated semiautomatic pistols in most armies. German Luger and Mauser are among these pistols. Colt .45 semiautomatic, developed in 1911, becomes U.S. military standard-issue sidearm.
1918	Germans develop first successful submachine gun, 1918-model Schmeisser used in battle.
1919	Development of Browning air-cooled machine gun. U.S. military uses it into 1950s.
1920	Browning Automatic Rifle (BAR) becomes U.S. standard-issue light infantry machine gun.

Military Small Arms Development

1928 Schmeisser completes his redesigned submachine gun with long, straight magazine out to one side. Germans use it widely in World War II.

1928 Thompson 1928-model submachine gun, tommy gun, is produced.

1930s Development of .50-caliber Browning heavy machine gun. With its high-powered ammunition, it becomes successful weapon for use against armored vehicles.

1936 U.S. Army adopts Garand rifle, semiautomatic weapon capable of firing 40 shots a minute.

1942 U.S. Army begins issuing M-1 semiautomatic rifles.

1943 Germans develop their intermediate-powered 7.92 "Kurtz" round. It makes possible development of their World War II "assault rifle," MP-44, with greater range and accuracy than submachine gun.

1947 Soviets borrow from MP-44 design in creating their AK-47.

1948 Israeli major Uziel Gal begins work on now-famous Uzi submachine gun.

1950s U.S. military adopts M-60 machine gun.

1952 NATO countries adopt standard 7.62-millimeter, full-power rifle round.

1967 U.S. Army adopts M-14, .22-caliber assault rifle.

1970s U.S. Army introduces M-16 rifle, with full or semiautomatic fire, for use in Vietnam War.

1970s U.S. M-79 breech-loading grenade launcher is used in Vietnam.

Tanks and Troop Carriers in the 20th Century

1904 Turreted armored cars are built in France for Russians.

1911 German army turns down proposal to build armored vehicle with tracks.

1914 Col. Ernest Swinton conceives of tank and wins backing of Winston Churchill.

1915 British construct "Little Willie," first tank with tracks.

1916 British army orders 100 "Big Willie" tanks, large version of their 1915 "Little Willie" prototype.

1916 First use of tanks in battle (September 15), by British at Battle of Somme.

1917 Massed tanks prove effective shock weapon in battle (November 20), by British at Cambrai, but have limited range and not enough speed (4 mph).

1918 Mass tank assault by British near Amiens opens up 8-mile break in German defenses.

1918 French Renault, 6-ton tank, is introduced. Light tank proves effective for infantry support. French help to popularize theory of tanks as only weapons of infantry support.

1919 Medium D, early amphibious tank, is built.

1920s British develop fast, mobile Vickers medium tank capable of moving 20 mph.

1928 U.S. tank designer builds fastest tank of this time, experimental version moving over 42 mph.

1934 Germans produce Panzer IV tank design with 75mm gun to meet battlefield need for support from heavy tanks.

1939 Soviets, recognizing need for heavier gun on medium tank, are at work on T-34 tank with 76mm gun.

1939 Soviet tank-building program produces world's largest tank corps, with some 20,000 tanks of various types. Germans have 3,195; French, over 2,600.

1939–40 German successes in battle prove effectiveness of concentrating tanks into Panzer corps and of transporting Panzer infantry in new half-track personnel carriers with light armor. Other nations quickly imitate these innovations.

1940 Work begins on producing jeep, all-purpose four-wheel vehicle used widely in World War II and after.

1941 Russians' new T-34 tanks, with heavier armor and more firepower, help turn back German assault on Moscow.

1942 United States begins producing M-4 Sherman medium tank with medium-velocity 75mm gun. Nearly 50,000 tanks are eventually built,

Tanks and Troop Carriers in the 20th Century

	though by 1944 they are no match for newer enemy tanks.
1943	Germans introduce Panther medium tank with high-velocity 75mm gun.
1944	Soviets introduce their new heavy tank, 46-ton Stalin, with 122mm gun.
1944	First tank (British) is flown into battle by glider, on D-Day at Normandy. Specially equipped floating tanks also take part in Normandy invasion, though many are lost in rough seas.
1944	Germans build their 68-ton Tiger II heavy tank with 88mm gun. It is heaviest tank of World War II.
1945	U.S. M-44 armored personnel carrier is built. Full-tracked and fully enclosed, it carries 27 men.
1945	United States introduces its new Pershing M-26 heavy tank with 90mm gun.
1948	British upgrade their Centurion heavy tank from 76mm to 83.4mm; it fires new high-velocity armor-piercing ammunition.
c. 1949	Soviets introduce T-54 heavy tank with 100mm gun. It becomes Communist-bloc standard heavy tank.
1951	United States introduces the M-48 tank with 90mm gun. It becomes main U.S. battle tank.
1953	United States adopts M-75 enclosed and full-tracked troop carrier, scaled-down version of earlier M-44.
1955	Tests show troops inside enclosed armored troop carriers can withstand nuclear blast. Tank crews are similarly protected.
1959	British again upgrade their Centurion heavy tank, to 105mm.
1960	U.S. M-60 tank (47 tons; 105mm gun) succeeds M-48 as main U.S. battle tank.
1960	M-113 troop carrier is introduced, first vehicle built in large numbers to have new lightweight aluminum armor.
1962	Soviets develop new heavy tank, T-62, with 115mm gun.
1966	British put 51-ton Chieftain tank in service. It carries 120mm gun and full armor.
c. 1975	Soviets put their new main battle tank, T-72, in service. It carries 120mm gun and moves 35 mph.

Warships

B.C.	
c. 1400	Mediterranean pirates use early form of war galley to attack shipping.
c. 800	Galleys designed for ramming appear. Ramming replaces use of javelin and bow and arrow as main method of attacking enemy vessel.
c. 700	Two-banked galley called penteconter is introduced by Greeks and Phoenicians. It has 24 rowers on each side.
c. 550	Greek trireme, with three banks of oars on each side, is introduced. Over 100 feet long with 170 oarsmen, it also is designed for ramming and boarding.
c. 450	Quinquireme, with five banks of oars (and more than one man at oar), appears. This vessel becomes workhorse of ancient navies.
c. 300	Galleys reach their maximum size, using over 2,000 oarsmen (several men at each oar). They carry catapults and other armament.
c. 260	Romans introduce corvus, spiked boarding ramp that, when dropped on enemy galley, locks two vessels together and allows speedy boarding of enemy vessel by Roman infantrymen.
31	Liburnian, small, fast galley developed for chasing pirates and escorting merchant ships, is used successfully by Roman commander Agrippa at Battle of Actium.
A.D.	
673	Byzantine galleys are equipped with hollow tubes that squirt Greek fire onto enemy ships.
700	Vikings develop their double-ended warship with square sail and some 60 oarsmen. It has open hull made of overlapping planks and proves exceptionally seaworthy.
c. 800	Galleys using lateen sails appear in Greece, and in Italy by 900.

Warships

c. 800	Vikings' "longship" (Gokstadt ship) is in use. It has true keel and single steering oar with tiller handle.
c. 1200	Sternpost rudder appears on Italian galleys. It makes tacking easier and greatly improves accurate navigation.
c. 1350	Guns are mounted aboard ships for first time. Early guns are designed to kill soldiers instead of damaging enemy ship.
1410	British arm their ship *Christopher of the Tower* with small guns, French mount cannon in 1494.
c. 1450	Three-masted caravels are built by Portuguese. They are best and most maneuverable ships of their day.
c. 1500	British navy switches to one-piece brass, muzzle-loading cannon, replacing inferior cannon barrels made of welded wrought-iron bars.
1514	Famous English man-of-war *Henry Grace à Dieu*, carrying 186 small and large guns, is built by Henry VIII (1491–1547). He also introduces gunport in English ships.
c. 1540	Galleons are introduced, establishing basic configuration of warships for next centuries. Powered by sail alone, galleon is slimmer and has lower sterncastle and forecastle.
1550	Brass muzzle-loading cannon capable of firing 60-pound cannonball is introduced.
1571	Naval cannons are used in combat in Battle of Lepanto, last major battle between warships powered by oars.
1580s	English navy man Sir John Hawkins (1532–95) effects redesign of English warships, arming them with longer-range cannons and making them faster than enemy vessels. This enables English ships to attack and sink enemy ships instead of battling infantry aboard them.
1588	Spanish Armada is defeated by cannons of redesigned English warships, setting style of naval battle for next centuries: ships under sail fight close-range cannon duels.
1637	British launch famous warship *Sovereign of the Seas*.
1673	British navy's *Fighting Instructions* are reissued. They formalize tactic of line-ahead formation for battle (one ship after another), giving rise to expression "ships of the line."
1765	British launch 100-gun warship *Victory*, flagship of Admiral Horatio Nelson (1758–1805), at Battle of Trafalgar (1805).
1797	United States launches *Constitution*, fast 44-gun frigate. Able to outrun other warships, she also outgunned typical 38-gun frigate of this period.
1798	British warship *Ajax* is launched. It typifies class of warship that is workhorse of navies of this time. *Ajax* carries 74 guns and offers best compromise of seaworthiness and combat effectiveness.
1812	*USS Constitution* defeats British frigate *Guerrière* in famous War of 1812 naval battle (August 19). Ship is said to be nicknamed "Old Ironsides" when Guerrière's cannonballs bounce off its sturdy oak hull.
1814	First steam-powered warship, called *Demologos Fulton*, is built in United States. Designed by Robert Fulton (1765–1815), it carries 32 guns but never sees action. Steam power does not become widespread until 1850s.
1830s	Naval guns firing explosive shells are introduced.
1836	Screw propeller is patented.
c. 1840s	Breech-loading naval guns begin to replace muzzle-loaders. Rifling of barrels to increase accuracy also is introduced.
1842	*Phlegethon* becomes first iron ship used in battle, during war in China.
1843	United States launches *Princeton*, world's first ship driven by screw propeller. Propeller is less vulnerable to cannon fire than paddle wheels of earlier steamships.
1853–56	Crimean War. Use of explosive shells in naval combat renders wooden ships useless. Steam power also ends reliance of warships on winds and tides.
1859	*La Gloire* is first iron-armored wooden ship, built by France. Great Britain then builds all-iron, armored *Warrior*.
1862	First battle between ironclad ships, Union *Monitor* and Confederate *Merrimack*, during American Civil War. The two ships fight to a draw and withdraw (March 9).
1869–73	British launch *HMS Devastation*, early example of steam-powered, heavily armored turret warship (having two turrets with two 12-inch guns each).
1880	Armor plate on battleships reaches 24-inch thickness.

Warships

1880s United States and other powers begin building big capital ships mounting large- and small-caliber guns. Smaller, cruiser-class warship also is developed at about this time; it carries less armor and has medium-caliber guns.

1880s Steel armor-piercing shells are introduced. Explosive armor-piercing shells appear in 1890s.

1890–92 American naval scholar Thayer Mahan (1840–1914) publishes his works on history and theory of sea power. He sees navy as tool of national policy and notes importance of sea power in economic warfare.

1893 British develop first torpedo-boat destroyer, which evolves into destroyer-class warship.

1895 Krupp develops superior "new process" armor plate. It is used by most navies.

1898 *Holland VI*, first successful military submarine, is built by American inventor John P. Holland (1840–1914). United States and other nations buy the new subs for their navies.

c. 1900 Steam turbine is introduced, sharply increasing speed of warships.

1905 Battle of Tsushima Strait (May 27–28). Big guns of Japanese battleships (British-made) soundly defeat Russian Baltic fleet during this battle of Russo-Japanese War. It is first major battle involving new battleships.

1906 British launch first of new all-big-gun battleships, *Dreadnought*. It carries 10 12–inch guns and can move at 21 knots. During the early 1900s major powers compete to build biggest battleships. Battleships remain most powerful warships until World War II.

1907–9 "Great White Fleet" of 16 U.S. battleships makes round-the-world cruise to demonstrate U.S. naval power.

1914–18 World War I. Major naval battle of war occurs at Jutland (May 31–June 1, 1916) between British and German battleships; Great Britain retains naval superiority. During this war, Germans develop tactic of submarine attacks on Allied merchant shipping, and Allies respond by building hundreds of new destroyers for antisubmarine duty.

1918 British build first aircraft carrier, *HMS Argus*, converted passenger liner with 560-foot deck. It carries 20 planes.

1921 U.S. Army pilot Gen. William Mitchell (1879–1936) demonstrates vulnerability of naval vessels to air attack by sinking target ships in test. By 1930s newer ships have improved antiaircraft defenses.

1930s Major powers develop aircraft carriers into effective new type of warship.

1936 Major powers resume building big battleships, suspended since 1922 by international agreement. Among them are German *Bismarck* (sunk by the British in May 1941) and Japanese *Yamato* and *Mushashi* (largest ever built; sunk by U.S. bombs).

1939–45 World War II sees revolution in naval warfare. Submarine again proves effective weapon against merchant shipping, nearly crippling Allied war effort. Aircraft carriers replace battleships as leading class of warships. Tactics of amphibious invasion are first used by Japanese during this war and later are fully developed by United States in Pacific and European theaters.

1940s Germans launch first experimental rockets from submarines.

1941 Pearl Harbor. Carrier-based Japanese planes make sneak attack on U.S. naval base in Hawaii (December 7). They damage or sink all eight U.S. battleships and 11 other vessels of the Pacific Fleet.

1942 United States launches its first amphibious invasion of World War II, at Guadalcanal (August 7).

1942 Battle of Coral Sea (May 7–8) is first naval battle fought entirely by attack planes from opposing aircraft carriers. No U.S. or Japanese ships make contact during engagement.

1942 United States launches battleship *New Jersey*.

1944 Normandy Invasion (June 6). Allies mount largest-ever amphibious invasion, along French coast. It involves over 5,000 ships and craft engaged in landing tens of thousands of Allied troops on D-Day.

1947 United States fires its first guided missile from submarine (surfaced).

1950s Soviets put short-range ballistic missiles aboard submarines.

1954 United States launches 319-foot *Nautilus*, world's first nuclear submarine (January 21). Nuclear submarines revolutionize naval warfare. Addition of nuclear-missile capabilities in next years makes nuclear submarines primary weapons in event of strategic nuclear war.

Warships

1955 Big guns of U.S. cruiser *Boston* are replaced with mounts for firing new Terrier guided missile. Thereafter, missiles steadily replace heavy guns as main armament of warships.

1955 First of big new U.S. carriers of Forrestal class is built.

1958 U.S. Navy successfully launches Polaris missile from submerged sub.

1958 Decommissioning of last active battleship, *USS Wisconsin*, marks end of battleship era.

c. 1959 Soviets build their first nuclear sub (November class). They maintain large nuclear and conventional submarine fleet thereafter.

1959 United States launches *George Washington*, first of its fleet of nuclear-powered subs carrying Polaris ballistic missiles. Sub launches its first Polaris while submerged (July 20, 1960).

1959 U.S. Navy launches first nuclear-powered surface warship, cruiser *Long Beach*, which is armed with guided missiles.

1960 U.S. Navy sub *Triton* makes first round-the-world voyage without resurfacing (36,000 miles, 2½ months), from February 24 to May 10.

1960 First nuclear-powered aircraft carrier, *Enterprise*, is launched by U.S. Navy. It becomes largest warship afloat to date.

1961 First ship designed as helicopter carrier, *Iwo Jima*, is commissioned.

1962 *Bainbridge*, first nuclear-powered destroyer, is built by U.S. Navy.

1967 Soviets build their first carrier for helicopters.

1968–69 Battleship *New Jersey* is put back into service for coastal bombardment during Vietnam War.

c. 1970 Soviets deploy cruise missiles aboard their nuclear and conventional submarines for attacking surface ships.

1975 Nuclear-powered aircraft carrier *Nimitz* is launched. It becomes largest warship afloat.

Warplanes

1908 U.S. Army buys its first warplane, biplane designed and built by Wright brothers.

1911 First successful takeoffs and landings by airplane on U.S. Navy ship are made, marking beginning of development of aircraft-carrier operations.

1911–12 Airplanes are used for reconnaissance and bombing in Italians' Tripoli campaign.

1914 World War I begins with France having largest air force, some 1,500 planes. Germany has 1,000, Britain only a few hundred, and United States less than 100.

1914–18 Airplane design advances rapidly during World War I. By 1918 planes make top speeds of 145 mph, reach altitudes of 30,000 feet, and can carry over 3,000-pound payloads.

1914–18 Combat role of airplanes emerges during World War I. Warplanes, at first used for scouting, soon are armed. Among notable warplanes that see combat are German triwing Fokker D-7 fighter; Sopwith Camel, British fighter; and French Spad and de Havilland bomber.

1915 Germans develop mechanism to synchronize fire of machine gun with rotation of propeller, creating first effective airplane armament.

1915 British forces employ torpedo-carrying seaplanes in attack on Axis ships in Dardanelles.

1916 Oswald Boelcke (1891–1916), German fighter pilot who developed first effective tactics for deploying fighters in combat, is killed in noncombat crash.

1918 British build first aircraft carrier, *HMS Argus*, converted passenger liner with 560-foot deck. It carries 20 planes.

1918 German air ace Baron Manfred von Richthofen (1892–1918) is killed in action (April 21) after downing 80 planes, most of any fighter pilot. American air ace Edward V. Rickenbacker (1890–1973) ends war with 21 kills.

1921 Italian army general Giulio Douhet (1869–1930) writes *The Command of the Air*, concerning his pioneering work on strategic use of air power, particularly use of massed bombers to strike deep in enemy territory.

1921 U.S. Army pilot Gen. William Mitchell (1879–1936) gives dramatic demonstration of vulnerability of naval vessels to air power. He sinks series of captured German warships.

1923 First midair refueling of plane is accomplished by U.S. Army pilots.

Warplanes

1925 Gen. Mitchell is court-martialed as result of his unbending crusade for air-power and for U.S. Air Force separate from Army. He is convicted, but he nevertheless calls attention to military policies that ignore importance of warplanes in future combat.

1926 U.S. Army Air Corps is formed under overall control of Army.

1930s U.S. Navy develops aircraft-carrier capabilities. Army Air Corps begins developing long-range bombers.

1931 B-9, first twin-engine, modern all-metal bomber with retractable landing gear, is produced in United States.

1932 Soviets are first to develop techniques of mass paratroop assaults.

1935 Now-famous B-17 Flying Fortress bomber crashes on maiden test flight for U.S. Army.

1935 Radar detection system is developed by British, who use it early in World War II to detect German bombers attacking Great Britain. It remains key antiaircraft detection and tracking system in modern times.

1935 Germans produce Heinkel He-111. It is fastest plane for its time and is used extensively in Spanish Civil War and later in invasion of Poland and France.

1936 First test flight of British Spitfire fighter prototype.

1936 Douglas DC-3 goes into service. It becomes workhorse military transport.

1939 Junkers Ju-87 Stuka, tested during Spanish Civil War, becomes Germany's prime aircraft in early successes of World War II. Poorly armed, it is outclassed by superior aircraft later developed by Allies.

1939 B-29 bomber prototype completes successful test flights.

1939 German Messerschmitt Bf-109 fighter sets new world speed record of 469 mph. Plane is armed with four 7.7mm machine guns plus a 23mm fully automatic model that fires explosive shells.

1939 German Luftwaffe is Europe's most powerful air force. Japan, meanwhile, has focused on building up strong carrier-based air arm.

1939–45 In World War II, role of airplane in combat increases dramatically. Airplane becomes important as basic military transport, as combat support weapon (ground and naval), and as strategic weapon in its own right.

1940 Great Britain's Spitfire fighters are victorious in war of attrition against German Luftwaffe during Battle of Britain.

1941 Japanese air attack on Pearl Harbor demonstrates importance of air power.

1941 First B-17 Flying Fortress bombers go into active service during World War II. United States introduces additional bombers during war years, including B-24 Liberator and B-29 Superfortress.

1943 Grumman Hellcat and P-38 Lightning are put into service. They prove effective counterweapon to Zero, Japanese Navy fighter.

c. 1943 American P-51 Mustang fighter is introduced. Its long range makes it possible to provide fighter escort for bombers on raids deep inside Germany.

c. 1944 Pilot's Universal Sighting System (PUSS) is developed. It allows pilot to direct his weapons—guns, rockets, bombs, and torpedoes—automatically.

c. 1945 Germans produce some 1,200 Messerschmitt Me-262 jet fighters by this time but do not use them extensively in combat.

1945 B-29 Superfortress drops first atomic bomb on Hiroshima (August 6).

1947 U.S. Air Force is established as distant branch of military service, separate from Army.

1947 First U.S. jet fighter, F-86, is tested.

1947 B-47 Stratojet has its first test flight. Five models in all are produced, including one that sets 1949 speed record of 607 mph.

1948 U.S. Convair B-36, powered by four jet and six piston engines, becomes first operational intercontinental bomber. It has range of 10,000 miles.

1950–53 Korean War sees first extensive use of jet fighters in combat, including U.S. F-86 Sabre and Soviet MiG-15. Military helicopters also are widely used for first time for troop transportation.

1950s Development of Mach 2 jet fighters begins. United States introduces Convair F-106 and Lockheed F-104G Starfighter. Soviets build MiG-21; British, Lightning P-1; and French, Mirage III.

Warplanes

1956	Supersonic B-58 medium bomber flies round trip from Los Angeles to New York in 4¾ hours.
1957	United States builds new B-52 long-range bomber. With range of 12,000 miles, it replaces B-47, introduced in early 1950s, as SAC bomber.
1960	High-flying American spy plane, U-2, is shot down over USSR, revealing role of spy planes in international intelligence-gathering.
1960	First American supersonic bomber, Convair B-58, goes in service. Plane can cruise at Mach 2.
1960s	Interest in building new long-range bombers declines in U.S. military as intercontinental missiles become key weapons in nuclear arsenal.
1961	Soviets display two supersonic jet bombers, called Bounder and Blinder, at air show.
1964–65	U.S. F-111 is introduced with variable sweep wing for subsonic and supersonic flight. Fighter-bomber makes Mach 2.5.
1965	U.S.-made Lockheed A-11 reconnaissance jet becomes first warplane to make Mach 3 speed. Soviet MiG-23 also reaches Mach 3.
1970	First flight of F-14 Tomcat, carrier-based fighter that is missile-armed fleet defense craft.
c. 1970	British introduce Harrier fighter plane, with vertical takeoff and landing capabilities.
1972	U.S. Air Force introduces F-15 fighter.
1976	Defecting Soviet pilot flies advanced Soviet fighter, MiG-25 Foxbat, to Japan. Foxbat reportedly has closed-circuit speed record of 1,800 mph and climbs to record high altitude.
1976	U.S. F-16 Falcon, proficient multipurpose fighter, is introduced.
1978	U.S. F/A-18 Hornet, single-seat, carrier-based, all-weather fighter, is introduced.
1980s	United States develops Stealth bomber technology. Specially designed surfaces help aircraft escape detection by enemy radar.
1984	New American B-1 long-range bomber crashes during test flight.
1989	New "vortex flap" is tested in United States. It greatly improves maneuverability of swept-wing supersonic fighters.
1989	First flight of Stealth bomber (July).

12 Accidents and Disasters

Accidents in the Air

1907 Lt. Thomas E. Selfridge is first airplane fatality when plane he is flying with Orville Wright crashes in Fort Meyer, Virginia (September 17). Wright is badly injured.

1912 First U.S. dirigible, *Akron*, explodes over Atlantic City, New Jersey (July 2). Five perish.

1919 Dirigible crashes into skylight of Chicago bank, killing 13 (July 21).

1933 U.S. dirigible plunges into Atlantic Ocean (April 14); 73 die.

1937 German dirigible *Hindenburg* explodes while mooring in Lakehurst, New Jersey (May 6); 36 perish.

1938 Colombian stunt plane crashes into grandstand in Bogotá, Colombia (July 24); 53 are killed.

1944 U.S. bomber crashes into school in Freckelton, England (August 23); 76 are killed, including 51 children.

1945 U.S. B-25 bomber smashes into Empire State Building, New York City (July 28); 14 die, including 10 in building.

1947 Eastern Airlines DC-4 crashes outside Fort Deposit, Maryland (May 30); 53 are killed.

1952 U.S. Air Force plane burns and crashes near Moses Lake, Washington (December 20); 87 are killed.

1953 Canadian jet crashes in Karachi, Pakistan (March 3), first commercial jet to crash; 11 perish.

1953 U.S. Air Force plane crashes near Tokyo, Japan (June 18); 129 perish.

1955 United Air Lines DC-6B explodes and crashes after passenger's son plants bomb to collect insurance money (November 1); 44 are killed.

1956 Bomb accidently released by Thai Air Force jet falls into crowd in Prachuabkhiri, Thailand (April 15); 30 die.

1956 DC-7 and TWA jet collide over Grand Canyon (June 30); 128 are killed.

1957 British airliner overshoots runway and crashes into housing development (March 14); 22 are killed.

1957 Canadian DC-4 crashes near Quebec City, Canada (August 11); 79 perish.

1958 Dutch plane crashes into ocean off Ireland (August 14); 99 perish.

1959 American airliner crashes in New York City's East River (February 3); 66 perish.

1960 Midair explosion of Northwest Airlines plane over Tell City, Indiana (March 17); 63 die.

1960 Midair collision of United DC-8 and TWA jet over Staten Island (December 16); 136 are killed.

1961 Sabena Boeing 707 crashes outside Brussels, Belgium (February 15); 73 are killed.

1961 Small plane crashes into crowd in Seville, Spain, after hitting high-tension wire (December 19); 30 are killed.

1962 American Airlines Boeing 707 crashes into Jamaica Bay in New York City shortly after takeoff (March 1); 95 die.

1962 British DC-7 crashes in jungle in Cameroon (March 4); 111 perish.

1962 Two Air France Boeing 707s crash within days: 130 die in Paris, France crash (June 3); 113 in crash in West Indies (June 22).

1963 Chartered DC-7 goes down off Alaska (June 3); 101 perish.

1963 TWA airliner crashes at Montreal, Canada (November 29); 118 perish.

1963 Jetliner crashes near Elkton, Maryland, after being struck by lightning (December 8); all 81 on board die.

1965 Eastern Airlines DC-7B crashes into Atlantic after leaving Kennedy Airport in New York (February 8); 84 die.

1965 Pakistani Boeing 707 crashes in Cairo, Egypt (May 20); 124 die.

1966 Indian Boeing 707 crashes into Mont Blanc in French Alps (January 24); 117 are killed.

1966 British Boeing 707 catches fire and crashes into base of Mount Fuji in Japan (March 5); 124 die.

1966 U.S. military transport plane crashes near Binh Thai in South Vietnam (December 24); 129 die.

▶

1967 South Korean Air Force jet smashes into church dome in Seoul (April 8); 55 are killed.

1967 Swiss airliner crashes at Nicosia, Cyprus (April 20); 126 perish

1968 First manslaughter charges are lodged against commercial airline pilots in connection with crash in Taiwan (February 16); 22 are killed in crash.

1968 Boeing 707 crashes at Windhoek, South-West Africa (April 20); 123 perish.

1969 DC-9 crashes in suburb of La Coruba, Venezuela, after pilot is shot by gunman (March 16); 155 perish.

1969 DC-9 and plane piloted by student collide near Indianapolis, Indiana (August 9); 83 are killed.

1970 Charter airliner crashes at Barcelona, Spain (July 3); 112 perish.

1970 Air Canada DC-8 crashes near Toronto, Canada, after two engines and part of wing fall off (July 5); 109 are killed.

1971 Midair collision of Japanese Boeing 727 and Japanese fighter over Morioka, Japan (July 30); 162 perish.

1971 Alaskan Airlines Boeing 727 crashes in Tongass National Forest east of Juneau, Alaska (September 4); 109 die.

1972 Crash of Iberian Airlines Caravelle on island of Ibiza, Spain (January 7); 104 are killed.

1972 Mountainside crash of Danish airliner in the Persian Gulf country of Oman (March 15); 112 perish.

1972 Alitalia DC-8 crashes west of Palermo, Sicily (May 5); 115 die.

1972 British airliner crashes shortly after takeoff from Heathrow Airport in London, England (June 18); 118 are killed.

1972 East German-chartered Ilyushin jet crashes outside of Berlin just after takeoff (August 14); 156 die.

1972 Soviet airliner crashes while landing in Kranaya Polyana, USSR (October 13); 176 perish.

1972 Spanish-chartered Convair 990A crashes just after taking off from Tenerife in Canary Islands (December 3); 155 are killed.

1972 United Air Lines Boeing 737 plows into houses near Chicago's Midway Airport (December 8); 45 die.

1972 Eastern Airlines Lockheed Tri-Star jet crashes into Everglades in Florida (December 29); 101 die.

1973 Boeing 707 crashes in thick fog in Nigeria while carrying pilgrims returning from Mecca (January 22); 176 perish.

1973 British European Airways jet crashes during attempted landing at Basel, Switzerland (April 10); 104 die.

1974 Turkish DC-10 crashes after takeoff near Paris, France (March 3); 346 are killed in worst crash to date.

1974 Pan Am Boeing 707 crashes into mountain on the island of Bali, Indonesia (April 27); 107 perish.

1974 DC-8 en route to Mecca, Saudi Arabia crashes in Sri Lanka during rainstorm (December 4); 191 die.

1975 U.S. Air Force jet crashes on takeoff from Saigon during airlift of 2,000 Vietnamese orphans to United States (April 4); 172 are killed.

1975 Eastern Airlines Boeing 727 crashes on approach to New York's Kennedy International Airport during electrical storm (June 24); 113 perish.

1975 Charter Boeing 727 crashes into Atlas Mountains near Azadir, Morocco (August 3); 188 die.

1975 Czechoslovakian Airlines jetliner crashes south of Damascus, Syria (August 20); 126 perish.

1976 Turkish airliner crashes into mountainside during premature landing near Isparta in southwestern Turkey (September 19); 155 die.

1976 Midair collision between British Airways jet and Yugoslavian chartered DC-9 (September 10); 176 die.

1977 American and Dutch jets collide on runway in Canary Islands (March 27); 583 are killed in worst crash to date.

1977 Portuguese airliner overshoots runway at Funchal, Madeira and bursts into flames (November 19); 130 die.

1978 Indian airliner explodes and falls into sea near Bombay (January 1); 213 perish.

1978 Midair collision between Pacific Southwest jetliner and private plane over San Diego, California (September 25); 144 die.

1978 Chartered Icelandic jet crashes just short of runway at Sri Lanka's Colombo Airport (November 15); 183 die.

1978 Alitalia jet crashes into sea near Palermo, Sicily (December 23); 109 perish.

▶

1979 American Airlines plane crashes at Chicago's O'Hare Airport (May 25); 275 die in worst U.S. air crash to date.

1979 Pakistan International jet crashes into mountains north of Jidda, Saudi Arabia, shortly after takeoff (November 26); 156 perish.

1979 New Zealand jetliner crashes into Mount Erebus in Antarctica (November 28); 257 are killed.

1980 Iranian Airlines Boeing 727 crashes in mountains outside Teheran (January 21); all 128 people aboard perish.

1980 Saudi Arabian jetliner explodes in flames on the runway in Riyadh shortly after returning (August 19); 301 die.

1981 Midair explosion of Far Eastern Air jet over Sanyi, Taiwan (August 22); 110 die.

1981 Chartered Yugoslavian DC-9 smashes into mountains en route to Ajaccio, Morocco (December 1); 180 perish.

1982 U.S. Boeing 737 goes down into Potomac River in downtown Washington, D.C. (January 13); 78 perish.

1982 Pan Am jetliner crashes after takeoff in Kenner, Louisiana (July 9); 153 die.

1983 Soviet fighter plane shoots down South Korean airliner, which is off course in Soviet airspace (September 1); 269 are killed.

1983 Colombian jetliner crashes near Spain's Madrid airport (November 27); 183 die.

1985 Indian jetliner crashes into sea off Ireland (June 23); 329 perish.

1985 Japanese jetliner crashes into mountain near Tokyo (August 12); 520 perish in worst crash to date involving single plane.

1985 Plane returning U.S. soldiers from Mideast crashes at Gander, Newfoundland (December 12); 256 are killed.

1986 Mexican jet crashes in mountains near Mexico City (March 31); 166 perish.

1987 Soviet jetliner crashes just after takeoff from Warsaw, Poland (May 9); 183 die.

1987 Northwest Airlines jetliner disintegrates in fireball near Detroit, Michigan airport (August 16); 156 perish.

1988 Italian jets collide during air show and crash into spectators in Ramstein, West Germany; 50 are killed and over 500 are wounded.

1988 Iranian airbus is shot down after flying too close to U.S. Navy ship then engaged in combat with Iranian gunboats in Persian Gulf (July 3); 290 Iranians die.

1988 Terrorist bomb explodes aboard Pan Am Boeing 747 near Lockerbie, Scotland (December 21); 270 are killed, including 11 on ground.

1989 Suriname DC-8 jetliner crashes near Suriname's Paramaribo Airport (June 7); 168 die.

Train Wrecks and Auto Accidents

1842 Rail carriages catch fire at Versailles, France (May 8); estimated casualties are as high as 100.

1864 Collision of Erie and prisoner trains at Shohla, Pennsylvania (July 15); 148 die.

1879 Military train collapses bridge at Phillippopolis, Turkey (January 11); 200 die.

1881 Train plunges into San Antonio River after bridge collapses in Cuartla, Mexico (June 24); 216 die.

1882 Train derails on rail line to Moscow, Russia (July 13); 178 perish.

1896 Holiday streetcar wreck (May 26) in Victoria, British Columbia, Canada; 54 die.

1915 Train derails into gorge (January 18) in Guadalajara, Mexico; 600 perish.

1915 Head-on collision of two trains in Gretna Green, Scotland (May 22); 227 perish.

1915 Wreck of Veracruz-bound train (November 19); 150 die.

1917 Train wreck in Tshura, Rumania (January 28); 500 killed.

▶

1917 Troop train derails in Modane, France (December 12); 543 die.

1920 Train wreck in Petrograd, Russia (December 22); 212 die.

1926 Train derails in Hanover, Germany (August 19); 248 are killed.

1930 Trolley plunges into river in Buenos Aires, Argentina (July 12); 60 die.

1933 Rear-end collision between local and express trains in Evreux, France (December 23); 150 are killed.

1935 Troop train wreck in Loyang, China (September 24); 200 die.

1937 Calcutta–Delhi train derailment in Patna, India (July 16); 160 are killed.

1940 Head-on train collision and fire in Osaka, Japan (January 28); 200 are killed.

1944 Victims trapped in train in tunnel in Salerno, Italy suffocate (March 2); 426 die.

1944 Collision and fire in train tunnel in Aguadilla, Spain (November 7); 500 die.

1946 Train derailment in Aracaju, Brazil (March 20); 185 are killed.

1947 Train plunges into river in Canton, China (July 10); 200 die.

1949 Train derailment in Nowy Dwor, Poland (October 22); 200 die.

1952 Train wreck near Pzepin, Poland (July 9); 160 die.

1953 Train derails in Waiouri, New Zealand following volcanic eruption (December 24); 155 die.

1955 Train plunges into canyon in Guadalajara, Mexico (April 3); 300 die.

1955 Two cars collide and plunge into spectators at Grand Prix auto race in Le Mans, France (June 11); 83 die.

1957 Train plunges into ravine in Kendal, Jamaica (September 1); 175 are killed.

1957 Passenger train collides with fuel carrier in Gambar, West Pakistan (September 29); 300 die.

1960 Bus crash in Ahmedabad, India (May 30); 69 die.

1960 Bus plunges into Turvo River, Brazil (August 25); 59 are killed.

1962 Three-train collision in Tokyo, Japan (May 3); 163 die.

1963 Bus crash in Tecunaman, Guatemala (March 1); 64 are killed.

1963 Three-train wreck in Yokohama, Japan (November 9); 162 die.

1965 Electric trolley plunges into Nile in Cairo, Egypt (November 1); 74 die.

1965 Two trucks veer out of control into crowd in Sotouboua, Togo (December 6); more than 125 are killed.

1966 Bus crash in Moscow, Soviet Union (June 17); 64 die.

1967 Two-bus crash south of Manila, Philippines (January 5); 83 die.

1968 Three trucks are swept away on flooded road in Bolivia (December 6); 50 die.

1970 Injured train-wreck passengers are in truck accident on way to hospital in Kafanchan, Nigeria (February 21); 52 die.

1970 Bus plunges into river in Quito, Ecuador (March 21); 56 die.

1971 Bus plunges into reservoir in Kapyong, South Korea (May 10); 77 are killed.

1972 Train derailment in Saltillo, Mexico (October 9); 204 die.

1973 Bus wreck in Alwar, India (July 6); 78 are killed.

1974 Truck is struck by train in Belém, Brazil (July 28); 69 are killed.

1974 Train crashes into station in Zagreb, Yugoslavia (August 30); 150 die.

1975 Bus and truck collision in New Delhi, India (May 19); 66 are killed.

1978 City bus plunges into Nile in Cairo, Egypt (July 17); at least 55 die.

1979 School bus plunges into river in Zamora Province, Spain (April); 52 die.

1979 Bus and gasoline-truck collision in Phang Nga Province, Thailand (June 2); 52 die.

1980 Bus plunges off bridge into river in Bilaspur, India (June 25); at least 100 die.

1981 Train is swept off bridge into river in Mansi, India (June 7); at least 268, possibly 500, die.

1981 Truck crossing dam plunges into Nile at Owens Falls, Uganda (September 25); unconfirmed reports claim 80 dead.

1989 Two passenger trains are engulfed by gas-line explosion near Asha, Soviet Union (June 5); 650 or more die.

Explosions and Mining Disasters Since 1900

1900 Coal-mine explosion in Scofield, Utah (May 1); 200 die.

1904 Coal-mine explosion in Cheswick, Pennsylvania (January 25); 179 die.

1906 Coal-mine explosion in Courripères, France (March 10); 1,060 die.

1906 Coal-mine explosion in Monongah, West Virginia (December 6); 361 die.

1907 Jacobs Creek, Pennsylvania coal-mine explosion (December 19); 239 die.

1908 Radbod, Germany coal-dust explosion; 360 die.

1909 Coal-mine fire at Cherry, Illinois (November 13); 259 die.

1910 *Los Angeles Times* Building explosion (October 1); 21 die.

1913 Coal-mine disaster in Senghenydd, Wales (October 14); 439 die.

1913 Coal-mine disaster in Dawson, New Mexico (October 22); 263 die.

1914 Eccles, West Virginia coal-mine disaster (April 28); 181 die.

1915 Havre, Belgium munitions-plant explosion (December 11); 110 die.

1916 Munitions explosion at Skoda, Austria (February 6); 195 die.

1916 Munitions-factory explosion at Kent, England (April 4); 170 die.

1916 Munitions explosion at La Satannaya, Russia (November 16); 1,000 die.

1917 Silvertown, England munitions-plant explosion (January 19); 300 die.

1917 Ammunitions-ship explosion during unloading at Archangel, Russia; 1,500 die.

1917 Munitions-plant explosion at Bloeweg, Bohemia (June 23); 1,000 die.

1917 Henningsdorf, Germany munitions-plant explosion (August 6); 300 die.

1917 Ammunition ship explodes at Halifax, Nova Scotia, Canada (December 6), destroying 75-acre area; over 1,500 die.

1918 TNT explodes at Oakdale, Pennsylvania chemical plant (May 18); about 200 are killed.

1918 Trains loaded with German munitions at Hamont Station, Belgium explode (August 3); 1,750 die.

1918 German munitions factory at Woellersdorf, Austria explodes (September 22); 382 die.

1921 Munitions explosion at Hiroshima, Japan (August 10); 100 are killed.

1921 Chemicals plant at Oppau, Germany explodes, destroying the town (September 21); 565 are killed and 4,000 injured.

1921 Oil tank explodes at Nobel dynamite plant in Prussia (December 6); 100 die.

1922 Lightning detonates stored munitions at La Spezia, Italy (September 28); 174 die.

1924 Coal-mine explosion at Castle Gate, Utah (March 8); 171 die.

1924 Dynamite explosion at Otaru Harbor, Japan (December 27); 120 are killed.

1925 Kharput, Turkey munitions-factory explosion (March 1); 160 die.

1925 Peking, China arsenal explosion (May 25); 300 die.

1928 Coal-mine explosion at Mather, Pennsylvania (May 19); 195 die.

1929 Hospital explosion in Cleveland, Ohio (May 15); 121 are killed.

1931 Naval munitions-plant explosion at Nictheroy, Brazil; 100 die.

1931 Munitions explosion at Yuchu Fortress, China (May 5); 100 die.

1932 Nanking, China munitions-dump explosion (July 10); 100 are killed.

1933 Gas-tank explosion at Neunkirchen, Germany factory (February 10); 100 die.

1934 Stored dynamite explodes at La Libertad, Salvador (March 14); 250 die.

1935 Munitions dump explodes at Shanghai, China (October 23); 190 die.

1935 Arsenal explodes at Lanchow, China (October 30); 2,000 are killed.

1937 Natural gas explodes at New London, Texas school (March 18); 294 die.

▶

1937 Munitions-factory explosion at Chunking, China (July 17); 110 die.

1938 Munitions stored in subway in Madrid, Spain explode (January 11); over 100 die.

1941 Stored munitions explode at Fort Smederovo, Yugoslavia (June 9); 1,500 die.

1942 Worst mine explosion to date takes at least 1,550 lives in Honkeiko, China (April 26).

1942 Limbourg, Belgium chemicals-factory explosion (July 21); 200 die.

1944 Ship explosion in Bombay, India port (April 14); 1,376 are killed.

1944 TNT and cordite aboard ships at Port Chicago, California explode (July 17); over 320 die.

1944 Gas-company explosion at Cleveland, Ohio sets 50 city blocks afire (October 20); over 100 die.

1944 Stored bombs explode at Burton-on-Trent, England (November 28); 175 die.

1945 U.S. Liberty ship explodes in Bari, Italy (April 9); 360 die.

1946 Grimberg Monopol mine explosion, West Germany; 439 die.

1947 French ship *Grandcamp* explodes in Texas City, Texas harbor (April 16), destroying most of city; 552 die.

1947 Naval munitions-plant explosion at Cadiz, Spain (August 18); 149 are killed.

1948 I. G. Farben chemicals-plant explosion at Ludwigshafen, Germany (July 28); 184 are killed and 6,600 injured.

1948 Stored chemicals explode in Hong Kong (September 22); 135 die.

1956 Dynamite explodes in trucks at Cali, Colombia (August 7); over 1,000 die.

1958 Buried nuclear wastes explode in Blagoveschensk, USSR, killing hundreds.

1958 Stored fireworks explode at Rio de Janeiro, Brazil (June 23); 110 die.

1959 Bomb salvaged from sunk World War II ship explodes in Philippine fishing village (April 10); 38 are killed.

1960 Coal-mine explosion at Coalbrook, South Africa (January 21); 437 die.

1962 Coal-mine explosion at Saarland, West Germany (February 7); 298 die.

1963 Gas explosion at Indianapolis State Fair Grounds Coliseum (October 31); over 65 die.

1963 Coal-mine explosion at Omuta, Japan (November 9); 447 are killed.

1965 Nation's worst mining disaster claims about 400 lives in Dhanbad, India (May 28). Another mining disaster 10 years later takes at least 250–350 lives.

1965 Coal-mine explosion at Fukuoka, Japan (June 1); 237 die.

1965 Titan 2 missile-silo explosion at Searcy, Arkansas (August 9); 53 die.

1967 Explosion aboard *USS Forrestal*, anchored off South Vietnam (July 29); 134 die.

1970 Explosion at subway construction site in Osaka, Japan (April 9); 73 are killed.

1972 Coal-mine explosion at Wankee, Rhodesia (June 6); 427 die.

1973 Staten Island, New York liquefied-gas storage tank—world's largest—explodes (February 10); 40 die.

1975 Mine explosion at Chasnala, India (December 27); 431 die.

1978 Tanker truck explodes along coast of Spain (July 11); 150 are killed.

1980 Munitions-dump explosion in Bangkok, Thailand (November 16); at least 60 are killed.

1982 Salang Tunnel explosion in Afghanistan (November 2); over 1,000 die.

1984 Oil-pipeline explosion at Cubatão, Brazil (February 25); 508 die.

1984 Soviet naval-supply-dump explosion at Severomorsk (June 21); over 200 die.

1984 Gas-facility explosion at Mexico City (November 19); 334 are killed.

1986 North Sea oil-rig explosion (July 6); 166 die.

1988 Train loaded with chemicals explodes at Arzamas, USSR (June 4); 83 die.

Fires

B.C.

538 Fire destroys most of city of Babylon and kills many inhabitants.

146 Victorious Romans burn city of Carthage at end of Third Punic War (149–146 B.C.).

86 Athens is partly destroyed by fires set during attack by Romans in First Mithradatic War (88–84 B.C.).

47 Main library at Alexandria, Egypt, containing hundreds of thousands of papyrus scrolls, burns. Subsidiary library survives until it, too, is burned in A.D. 391.

A.D.

64 Three-quarters of Rome burns when accidental fire at Circus Maximus spreads out of control. Afterward Nero is wrongly blamed for fire, though he does use opportunity to rebuild Rome in grandiose fashion.

798 Nearly all of London is destroyed by fire; it is nearly destroyed again in 982.

1106 Venice is largely destroyed by fire.

1212 Most of London burns after fire starts at London Bridge.

1405 Berlin is devastated by fire.

1491 Dresden (in modern Germany) fire leaves city in ruins.

1570 Moscow burns in great fire; some 200,000 persons are killed.

1577 Fire ravages Venice.

1614 Destructive fire at Stratford-upon-Avon, England, Shakespeare's birthplace.

1666 Great Fire of London (September 2–6), city's worst, burns about 80 percent of city, including over 13,000 houses. Buildings subsequently are rebuilt of brick, wood construction being outlawed.

1676 Historic Jamestown, Virginia burns.

1728 Much of Copenhagen, Denmark is lost in fire.

1729 Constantinople, in modern Turkey, burns, including over 12,000 houses. Some 15,000 houses burn in great fire of 1756, and 10,000 in 1782 fire.

1752 Moscow again burns in great fire. Some 18,000 buildings are destroyed.

1788 In New Orleans fire over 800 buildings are burned.

1812 Moscow is set afire by Russians after Napoleon's invading army enters city; over 30,000 structures burn (September 15–19).

1822 Most of Canton, China burns in great fire.

1845 New York City fire burns over 1,000 buildings.

1851 Almost three-quarters of San Francisco is destroyed by fire.

1857 Tokyo burns after severe earthquake sets city afire (March 21); over 100,000 persons die.

1866 Quebec City, Canada fire burns some 2,500 structures.

1871 Peshtigo, Wisconsin is hit by massive forest fire (October 8) that burns 1.28 million acres; 1,500 die.

1871 Great Chicago Fire occurs, one of worst fires in U.S. history (October 8–11). It starts in barn on DeKoven Street on Chicago's West Side and spreads out of control (Mrs. O'Leary's cow did not start it by knocking over a lantern, as is popularly believed.) Fire ultimately kills about 250 persons and destroys almost 18,000 buildings.

1872 Boston is hit by great fire that burns almost 1,000 buildings.

1876 Fire destroys New York's Brooklyn Theater (December 5); 285 die.

1881 Ring Theater in Vienna burns (December 8), taking estimated 850 lives.

1883 In circus fire in Berditschott, Russian Poland (January 13), 430 die.

1887 Fire during performance at *Opéra Comique* (May 25) kills about 200 in Paris.

1894 Twelve towns in northern Minnesota are destroyed by forest fire that claims 600 lives (September 1).

1900 Pier fire in Hoboken, New Jersey (June 30) kills 326.

▶

1903 Chicago's Iroquois Theater burns, killing over 600 (December 30); it is worst theater fire in U.S. history.

1906 Three-day San Francisco fire begins (April 18). It is started by powerful earthquake and destroys about three-quarters of city (some 25,000 buildings burn). Over 500 are killed.

1908 Elementary-school fire in Collinwood, Ohio (March 4) kills 176.

1909 Theater fire in Acapulco, Mexico (February 14) kills 300.

1910 Fire sweeps Hungarian town of Okorito (March 28), killing 320.

1911 Spectacular fire at Triangle Shirtwaist Factory in New York (March 25) claims 145 victims, most of them girls.

1918 Cloquet, Minnesota, a sawmill town, is completely destroyed by forest fire (October 12); 559 die.

1923 Yokohama and much of Tokyo, Japan are destroyed by fire set by major earthquake (measuring 8.2); over 140,000 die.

1924 Fire in Shantung Province, China (March 24); 300 die.

1930 In Columbus, Ohio, 320 convicts are killed as fire sweeps severely overcrowded Ohio State Penitentiary (April 21).

1930 Fire in teahouse district of Wuchow, China (October 19); 650 die.

1934 Massive fire in Hakodate, Japan (March 21) kills 1,500.

1934 U.S. luxury liner *Morro Castle* burns off New Jersey coast (September 7–8); 134 die.

1936 Theater fire in Tuliuchen, China (March 17) kills 221.

1937 Theater fire in Antung, China (February 13) kills 685.

1939 Fire in oil refinery spreads to town of Lagunillas, Venezuela, killing 500 (November 14).

1940 Natchez, Mississippi dance-hall fire claims 198 lives (April 23).

1942 Boston, Massachusetts nightclub fire, one of the most famous in American history. 491 people die at Cocoanut Grove (November 28).

1944 Residential area of Bombay, India burns after steamship explodes nearby (April 14); almost 1,400 die.

1944 Fire destroys Ringling Brothers and Barnum and Bailey Circus tent at Hartford, Connecticut; 168 die in minutes (July 9).

1945 Dresden is hit by Allied firebombing raid (February 13–14). Resulting firestorm virtually destroys city.

1945 Copper-mine fire in Santiago, Chile (June 19) kills 383.

1946 Atlanta's Wincroft Hotel burns (December 7), in worst hotel fire in U.S. history. 119 die.

1947 Fire ignited by explosion of French liner *Grandcamp* destroys most of Texas City; 552 die (April 16–18).

1947 Hankow, China harbor-area fire (December 28) kills 400.

1949 Hospital fire in Effingham, Illinois (April 5) kills 77.

1949 Over 10,000 buildings burn in Chunking, China (September 2); 1,700 die.

1952 Over 5,000 houses burn in Tottori, Japan (April 18).

1957 Nursing-home fire in Warrenton, Missouri (February 17) kills 72.

1958 Chicago parochial-school fire (December 1) kills 93.

1958 Store fire in Bogotá, Colombia (December 16) kills 84.

1960 Mental-hospital fire in Guatemala City (July 14) kills 225.

1960 In movie-theater fire at Amude, Syria (November 13), 152 die.

1960 In Brooklyn, New York, fire hits aircraft carrier *Constellation* as it is being built at naval shipyard; 50 workers die (December 16).

1961 Circus fire in Niterói, Brazil (December 17) kills 323.

1962 Widespread fires in Brazilian state of Paraná leave 300,000 homeless (September 7); 250 die.

1967 Store fire in Brussels, Belgium (May 22) kills 322.

1970 Dance-hall in St. Laurent du Pont, France (November 1) kills 144.

1971 Hotel fire in Seoul, South Korea (December 25) kills 162.

1972 Nightclub fire in Osaka, Japan (May 13) kills 116.

1973 In department store in Kumamoto, Japan (November 29), 107 die.

▶

1974 Bank fire in São Paulo, Brazil (February 1) kills 189.

1974 Discothèque in Port Chester, New York (June 30) kills 24.

1975 In tent city at Mina, Saudi Arabia (December 12), 138 die.

1977 Nightclub fire in Southgate, Kentucky (May 28) kills 164.

1978 In Abadan, Iran, arsonists touch off conflagration in crowded movie house, killing 377 (August 19).

1980 Nursing home in Kingston, Jamaica burns (May 20), killing 157.

1980 MGM Grand Hotel fire in Las Vegas, Nevada (November 21) kills 84.

1980 Stouffer Inn fire in Harrison, New York (December 4) kills 26.

1981 In Bangalore, India, circus fire kills more than 100 (February 7).

1981 Discothèque fire in Dublin, Ireland (February 14) kills 44.

1981 In Mandalay, Burma, 35,000 are left homeless as fires burn out of control for two days (May 10–12).

1983 Movie-theater fire in Turin, Italy (February 13) takes 64 lives.

1983 Discothèque fire in Madrid, Spain (December 17) kills 83.

1985 Hospital fire in Buenos Aires, Argentina (April 26) kills 79.

1985 In soccer-stadium fire in Bradford, England (May 11), 53 die.

1986 Dupont Plaza Hotel fire in Puerto Rico (December 31) kills 96.

1987 Forest fires sweep northern China (May 6–June 2); 193 die.

1988 Some 2,000 homes burn in Lashio, Burma (March 20); 113 die.

1988 Street-market fire in Mexico City (December 11) kills over 60.

Ship Sinkings and Other Accidents at Sea Since 1900

1900 Three ships of North German Lloyd Steamship Company docked in Hoboken, New Jersey catch fire (June 30); 326 crew members die.

1902 British steamer *Camorta* sinks off Rangoon (May 6); 739 die.

1904 Excursion steamer *General Slocum* burns off New York (June 15); 1,021 die.

1904 Danish steamer *Norge* sinks off Scottish coast (June 28); 620 die.

1905 Japanese battleship *Mikasa* explodes at Sasebo (September 10); 599 die.

1912 British liner *Titanic* collides with iceberg in North Atlantic (April 14–15); 1,513 die. Loss of "unsinkable" *Titanic* on maiden voyage is among most famous marine accidents of all time.

1912 Japanese steamer *KicheMaru* sinks off Japanese coast (September 28); 1,000 die.

1914 British steamer *Empress of Ireland* collides with Norwegian freighter in St. Lawrence River (May 29); 1,027 die.

1914 British battleship *Bulwark* explodes in Sheerness Harbor (November 26); 788 die.

1915 British liner *Lusitania* sinks by torpedo from German U-boat off Irish coast (May 7); 1,198 die.

1915 Steamer *Eastland* sinks in Chicago River (July 24); 852 die.

1916 French cruiser *Provence* goes down in Mediterranean after German sub attack (February 26); 3,100 are killed.

1916 Chinese steamer *Hsin Yu* sinks off Chinese coast (August 29); 1,000 die.

1917 Belgian steamer *Imo* collides with French ammunition ship *Mont Blanc* in Halifax Harbor,

▶

Nova Scotia, Canada (December 30). Explosion causes widespread damage; 1,600 die.

1917 British warship *Vanguard* explodes dockside at Scapa Flow (July 9); 800 die.

1918 U.S. naval vessel *Cyclops* disappears in Atlantic en route from Rio de Janeiro to Baltimore, Maryland (March); 324 aboard are presumed lost.

1918 Chinese steamer *Kiang-Kwan* sinks in Chinese coastal harbor (April 25); 500 die.

1918 Japanese battleship *Kawachi* explodes in Japanese waters (July 12); 500 die.

1918 Canadian steamer *Princess Sophia* sinks off Alaska coast (October 25); 398 die.

1919 French steamer *Chaouia* sinks off Greek coast (January 17); 460 die.

1919 Spanish steamer *Valbanera* sinks off Florida coast (September 9); 500 die.

1921 Steamer *Hong Kong* sinks off Swatlow Harbor (March 18); 1,000 die.

1926 Chinese troop transport blows up in Chinese waters (October 16); 1,200 perish.

1928 Chinese steamer *Hsin Hsu-Tung* sinks in Yangtze River; 500 die.

1929 Chinese steamer *Hsin Wah* sinks off Wagan Island (January 16); 401 die.

1931 French ship *St. Philibert* sinks off St. Nazaire, France (June 14); 450 die.

1934 U.S. luxury liner *Morro Castle* burns off New Jersey coast (September 7–8); 134 die.

1935 Fire aboard luxury liner *Morro Castle* en route to New York takes 133 lives.

1939 Submarine *Thetis* is lost in Liverpool Bay (May 23); 99 die.

1939 Russian steamer *Indigirka* sinks off Hokkaido Island (December 12); 750 perish.

1942 British cruiser *Curaçao* collides with *Queen Mary* in British waters (October 2); 338 die.

1944 Three ships of U.S. Third Fleet sink in typhoon in western Pacific (December 17–18); 769 die.

1945 German passenger ship *Wilhelm Gustoff* sinks after Soviet submarine attack in Baltic Sea (January 30); 7,700 die.

1945 U.S. Liberty ship explodes in Bari, Italy (April 9); 360 die.

1945 Chinese riverboat sinks off Hong Kong (November 8); 1,550 perish.

1946 Japanese ship *Ebisu Maru* sinks off Korea (October 29); 497 die.

1947 Greek steamer *Himara* sinks in Saronic Gulf after hitting relic mine (January 10); 392 die.

1947 Chinese river steamer sinks in Yangtze River near Woosung, China (January 18); more than 400 die.

1947 French freighter *Grandcamp* explodes and sets fire to Texas City, Texas (April 16–18); 552 die.

1948 Chinese steamer *Kiangya* explodes off Shanghai (December 3); over 2,750 perish.

1949 Chinese steamer *Tai Ping* collides with another ship in Chinese waters (January 27); 1,500 die.

1949 Great Lakes vessel *Noronic* burns dockside at Toronto, Canada (September 17); 128 die.

1949 Chinese army troop transport explodes (November); 6,000 die.

1951 Soviet steamer *Eshghabad* sinks in Caspian Sea (July 14); 270 perish.

1952 Midatlantic collision between U.S. naval vessels *Hobson* and *Wasp* (April 26); 176 servicemen die.

1953 South Korean ferryboat sinks off coast near Pusan (January 9); 249 die.

1954 Japanese ferryboat *Toya Maru* capsizes in Hakodate Bay off Japanese coast (September 26); over 700 perish.

1955 Soviet warship *Novorossiisk* sinks while on maneuvers in Black Sea (October); 1,500 lives are lost.

1956 Italian liner *Andrea Doria* sinks after collision with Swedish liner *Stockholm* off Nantucket (July 25); 52 die.

1959 Egyptian Nile boat *Dandarah* capsizes (May 8); 200 die.

1961 Portuguese steamer *Save* sinks off coast of Mozambique (July 7); 227 perish.

1962 British liner *Dara* is blown up by time bomb in Persian Gulf (April 8); 236 die.

1963 U.S. Navy nuclear submarine *Thresher* is lost in North Atlantic (April 10); 129 die.

1963 Japanese ferryboat *Midori Maru* sinks near Okinawa (August 17); 128 die.

1964 Iranian motor launch sinks in Persian Gulf en route to Kuwait (April 10); 113 perish.

1965 River ferry capsizes in Malawi (May 23); more than 150 die.

▶

1966 Greek ferryboat *Iraklin* sinks in Aegean Sea (December 8); 241 die.

1967 U.S. Navy carrier *Forrestal* catches fire off Vietnam coast (July 29); 134 die in worst maritime disaster for U.S. Navy since World War II.

1968 U.S. Navy nuclear submarine *Scorpion* is lost off Azores (May 21); 99 die.

1969 U.S. destroyer *Evans* collides with Australian aircraft carrier *Melbourne* (June 2); 74 die.

1970 South Korean ferryboat *Namhong-Ho* sinks off Korean coast (December 15); over 250 perish.

1973 Burmese ferryboat collides with Japanese freighter in Rangoon Harbor (February 1); more than 200 die.

1974 Soviet destroyer sinks after shipboard fire in Black Sea (September 26); over 200 die.

1975 Two Chinese excursion boats collide off Canton (August 3); 400 perish.

1975 Ore carrier *Edmund Fitzgerald* sinks in storm on Lake Superior (November 10); 29 die.

1978 Fleet of cargo boats sinks in Bay of Bengal during storm (April 4); estimates of dead exceed 1,000.

1978 Fishing boat sinks off coast of Malaysia (November 22); 200 Vietnamese refugees die.

1978 Offshore-oil service platform capsizes at Staranger, Norway (March 27); 123 workers drown.

1980 Bangladesh ferryboat sinks near Dacca (April 20); at least 230 perish.

1981 Brazilian river steamer *Novo Amapa* sinks in Amazon River near Macapá (January 6); 260 die.

1981 Indonesian ship *Tamponas II* sinks after fire in Java Sea (January 27); 580 perish.

1981 Riverboat *Sobral Santos* sinks in Amazon River near Óbidos (September 19); at least 300 die.

1983 Egyptian riverboat *10th of Ramadan* sinks in Lake Nasser (May 25); 357 die.

1986 Soviet liner *Admiral Nakhimov* collides with Soviet freighter *Pyotr Vasev* in Black Sea (August 31); 398 die.

1987 British ferryboat sinks off Belgian coast (March 6); 134 die.

1987 Philippine boat *Doña Paz* sinks after collision in Philippine waters (December 20); over 3,000 die.

1988 Explosion and fire on Occidental Petroleum oil rig off coast of Scotland (July 6); 166 die.

1988 Indian ferryboat sinks in Ganges River (August 6); 400 die.

1989 Explosion in gun turret aboard *USS Iowa* (April 19); 47 die.

Natural Disasters

Droughts

484 Drought in Africa causes widespread famine.

592 England suffers drought and famine. Drought strikes again in 605.

772 Ireland suffers drought and famine.

968 Drought in Africa brings low Nile in Egypt, resulting in as many as 600,000 deaths.

1022 Drought in India kills thousands.

1064–72 Drought in Africa causes seven-year failure of Nile flooding, resulting in widespread famine.

1135–37 England suffers famine brought on by drought.

1199–1202 Severe drought in Africa causes several years of low Nile; upward of 100,000 die.

1200–1300 Prolonged drought conditions over southwestern United States brings advanced Indian agricultural societies to end.

1412–13 India is again hit by drought.

1631 Drought conditions across much of Asia cause widespread famine.

▶

1661	Punjab region of India suffers drought; thousands die.
1769–78	Drought causes famine and 3 to 10 million deaths.
1790–92	Severe drought in India leads to "Skull Famine," in which cannibalism becomes widespread in Bombay and other cities.
1803–04	Drought compounded by war and locust plagues causes thousands of deaths in India.
1837–38	India's northwestern region is hit by severe drought; some 800,000 perish.
1860–61	Drought in India's New Delhi region.
1866	About 1.5 million die in drought-related famine in Bengal and nearby provinces in India.
1876–78	Drought in India kills some 5 million.
1876–79	Prolonged drought in northern China kills 9 to 13 million.
1877	Northern Brazil suffers severe drought; famine kills thousands.
1892–94	Drought causes severe famine in China; about 1 million perish.
1896–97	India again suffers severe drought. Resulting famine and pestilence kill 5 million.
1899–1900	Severe drought and famine in India; 1.25 million die.
1920–21	Drought in China's northern regions causes famine that kills 500,000.
1921–22	Drought strikes Soviet Union, then reeling under upheavals of Communist takeover. United States sends aid, but millions still die of famine.
1928–29	Some 3 million in northwestern China die during drought.
1933–39	Severe drought in U.S. Midwestern farming region results in huge dust storms as winds blow dry topsoil up into thick clouds. About 60 percent of area's population eventually migrates to other parts of country.
1936	Severe drought in China's western regions; some 5 million die in resulting famine.
1950	Drought followed by flooding kills 10 million in northern China.
1961–66	Great Northeast Drought in the United States, most prolonged and severe drought in U.S. history.
1965	Severe drought in India; relief efforts keep death toll in thousands.
1968–74	Drought-prone Sahel region of Africa is hit; 500,000 perish.
1972	India is hit by drought; it causes severe crop damage but only 800 perish.
1973–74	Severe drought in Africa and India; hundreds of thousands perish in resulting famine.
1976–77	California is parched by severe drought.
1980	Severe summer drought and heat wave roast U.S. Midwest, Southwest, and South; over 1,200 perish, and crops and livestock are heavily damaged.
1983–85	Drought throughout much of Africa. Tens of thousands die.

Earthquakes

The measure of earthquake magnitude, the Richter Scale, did not come into official use until the 1930s. Estimates for some earlier earthquakes have been made.

B.C.	
464	Severe earthquake kills tens of thousands in Sparta.
217	Quake destroys over 100 cities in North Africa; over 50,000 perish.

A.D.	
19	Over 100,000 perish in Syria.
79	Pompeii is hit by earthquake as Mount Vesuvius erupts. Inhabitants and ruined city are buried beneath volcanic ash.
365	Alexandria, Egypt is destroyed by quake (July 21); the 600-foot lighthouse, one of Seven Wonders of Ancient World, is toppled; some 50,000 perish.
526	Antioch, Syria is devastated by earthquake (May 26); estimated 250,000 perish.
557	Much of Constantinople is destroyed.
856	Some 45,000 persons are killed in Corinth, Greece (December).
893	Wide area of India is struck; estimated 180,000 perish.
936	Constantinople is again devastated.
1036	Shansi, China quake kills 23,000.

▶

1040 Tabriz, Persia is left in ruins by quake that kills 50,000.

1170 Sicily is struck; 15,000 are killed.

1268 City of Cieilia in Asia Minor is destroyed; 60,000 perish.

1290 Chihli (Hopeh), China is hit (September 27); 100,000 perish.

1456 Naples, Italy is struck (December 5); 35,000 perish.

1531 Lisbon, Portugal is hit by large quake (January 26); 30,000 die.

1556 Shaanxi (Shensi), China earthquake (February 2) is one of deadliest of all; estimated 830,000 perish.

1626 Naples, Italy again is hit by massive quake (July 30); 70,000 perish.

1667 Shemakha in Caucasia is struck by quake (November); 80,000 perish.

1692 Port Royal, Jamaica, home of pirates and known as "Wickedest City in the World," is struck by violent quake and tsunami (June 7); 2,000 are killed outright.

1693 Naples, Italy and Catania, Sicily are struck by separate quakes, killing estimated 93,000 in Naples and 60,000 in Catania.

1703 Tokyo, Japan is hit (December 30); 200,000 perish.

1730 Hokkaido, Japan is hit (December 30); 137,000 perish.

1731 Peking, China quake devastates city (November 30); 100,000 perish.

1737 Estimated 300,000 are killed in Calcutta, India quake (October 11).

1754 Quake destroys half the dwellings in Cairo, Egypt; 40,000 die.

1755 Lisbon, Portugal is demolished and estimated 60,000 are killed in one of most violent earthquakes in recorded history (November 1). Estimated at 8.75 on Richter scale at its center, quake is felt from southern France to North Africa.

1773 Santiago, Guatemala is totally destroyed by quake (June 7); 58,000 die.

1783 Massive shock of series of quakes levels more than 180 towns in Calabria, Italy (February 4–5); 30,000 are killed outright.

1793 Quake hits Unsen, Japan (April 1); 53,000 die.

1797 Quito, Ecuador is leveled; 41,000 die.

1797 Calabria, Italy is severely shaken; 50,000 die.

1811–12 Quake in New Madrid, Missouri causes surface distortions of 30 feet and more (December 16–February 7).

1828 About 30,000 are killed in Echigo, Japan quake (December 28).

1847 Zenkoji, Japan is leveled by quake; 34,000 perish.

1857 Tokyo is struck by earthquake and fires whipped by 60-mph winds (March 21); 107,000 perish.

1863 Three-day series of quakes rocks Peru and Ecuador (August 13–15); 25,000 perish.

1868 Southern Peru is hit by series of earthquakes that kill over 25,000 and cause $300 million in damages (August 13).

1872 One of largest quakes to strike California fractures earth for 100 miles and kills 60. Quake is felt from San Diego to Mount Shasta (March 26).

1875 Colombia and Venezuela are struck by quake (May 16) that kills 16,000.

1876 Quake in northeastern Japan causes widespread coastal damage from 100-foot-high tsunamis (June 15); 28,321 are killed.

1891 Mino-Owari, Japan is shaken by quake felt throughout Japanese archipelago (October 28); 7,300 lives are lost.

1896 Japanese city of Sanriku is leveled by quake and tsunamis (June 15); 28,000 die.

1897 Most powerful earthquake in history strikes Assam, India but results in few deaths (June 12).

1905 Several villages in central India are destroyed by quake (April 4) that kills more than 20,000 people.

1906 Total of 4 square miles of downtown San Francisco, California are destroyed by earthquake and subsequent fire (beginning April 18). Quake measures 8.3 on Richter scale; over 500 perish.

1908 Messina, Italy is leveled by earthquake measuring 7.5 (December 28); 160,000 perish throughout Sicily and southern mainland Italy.

1915 Earthquake strikes Avezzano, Italy (January 13); 30,000 die.

▶

1920 Estimated 80,000 are killed in Kansu, China during earthquake measuring 8.6 (December 16).

1923 Quake in Japan measuring 8.2 destroys Yokohama and much of Tokyo (September 1); over 140,000 perish in quake and subsequent fire.

1927 Kansu, China is hit (May 22); 100,000 perish.

1932 Kansu, China is hit by quake measuring 7.6 that kills 70,000 (December 25).

1935 Quetta, India (now in Pakistan) is devastated (May 31); earthquake measures 7.5 and causes estimated 50,000 deaths.

1939 Concepción, Chile is hit by earthquake measuring 8.3 (January 24); some 30,000 perish.

1939 About 50,000 are killed in Erzincan, Turkey. City is leveled by quake measuring 7.9 (December 27).

1949 Pelileo, Ecuador earthquake measures 6.8 (August 5); some 6,000 perish.

1950 Assam, India is again struck by massive quake, measuring 8.6 on Richter scale. About 1,500 die (August 15).

1954 Orleansville, Algeria is shaken by 12-second quake (September 10) that kills some 1,460.

1960 Agadir, Morocco is largely demolished by earthquake measuring 5.8 and subsequent fire and tidal waves (February 29); about 12,000 perish.

1960 Quake hits Concepción, shaking coastal Chile in series of shocks that kills 5,700; tsunamis generated by quake kill 199 in Hawaii and Japan (May 21–30).

1962 Northwestern Iran is hit by quake (September 2); 10,000 perish.

1964 Anchorage, Alaska is rocked by strongest recorded earthquake in North America (March 27), measuring 8.5; 118 perish.

1966 Eastern Turkey is struck by quake measuring 7 (August 19–23); 2,500 perish.

1967 Moderate 6.5 quake shakes Caracas, Venezuela (July 29), killing more than 250.

1968 Some 6,000 perish in earthquake (August 31) in Khurasan, Iran.

1970 Total of 254 villages in wide area of western Turkey are destroyed by 7.1 quake (March 28); more than 1,300 die.

1970 Northern Peru is hit by earthquake measuring 7.7 (May 30); death toll is estimated at 50,000 to 70,000.

1971 Quake measuring 6.6 shakes California's San Fernando Valley (February 9), causing $500 million in damage and taking 64 lives.

1972 Managua, Nicaragua is devastated by earthquake measuring 6.2 (December 23); about 10,000 perish.

1974 Remote region of northern Pakistan is struck by quake (December 28) that kills over 5,200 people.

1975 Town of Lice in eastern Turkey suffers 6.8 quake (September 6) that kills over 2,000.

1976 Guatemala City area is hit by quake measuring 7.5 (February 4); about 23,000 perish.

1976 More than 900 die in a quake in northern Italy (May 7–8) measuring 6.5.

1976 A 7.2 quake (June 26) kills over 9,000 in West Irian Province, Indonesia.

1976 Tangshan, China earthquake (July 28) measuring 8.2 levels about 20 square miles of city and kills some 242,000 persons.

1976 Offshore quake measuring 7.8 hits Mindanao, Philippines; quake and resulting tsunami (August 17) kill over 5,000.

1976 Muradiye, Turkey, near Mount Ararat, suffers 7.9 quake (November 24) that takes lives of at least 5,000.

1977 Bucharest, Romania is destroyed by quake (March 5) felt in Rome and Moscow; more than 1,100 are killed here and in neighboring Bulgaria.

1978 Tabas, Iran is razed by earthquake measuring 7.7 (September 16); about 25,000 perish.

1980 Al Asnam, Algeria is hit by two quakes measuring 7.5 and 6.5 (October 10); at least 2,600 die.

1980 Southern Italy is shaken by earthquake measuring 7.2 (November 23); almost 5,000 perish.

1981 Kerman Province in southeastern Iran suffers 6.6 quake (June 11) that kills over 2,000.

1982 Northern Yemen earthquake measuring 6.0 (December 13) results in almost 3,000 deaths.

▶

1985 Earthquakes up to 8.1 (September 19–21) damage much of Mexico City and kill 7,000 or more there and in surrounding region.

1987 Northeastern Ecuador is hit by quake measuring 7.3 (March 5–6); some 4,000 or more perish.

1988 Soviet Armenia is devastated by quake measuring 6.9 (December 8); some 60,000 perish.

1989 Major quake hits Santa Cruz and northern California (October 17); scores are killed.

1990 Quake strikes Peru (May 29), leaving at least 161 dead.

Floods, Tsunamis, and Tidal Waves Since 1200

1211 Tay and Anan rivers overflow in Perth, Scotland. King manages to escape by boat, though most of his court drowns; hundreds die.

1219 St. Lawrence Lake in Norway bursts; 36,000 perish.

1228 Sea floods Holland; 100,000 die.

1273 Seismic sea wave ravages coast of Japan; 30,000 are killed.

1287 Collapse of seawall results in flood of Zuider Zee in Holland, taking 50,000 lives (December 14).

1421 Sea floods Holland (April 17). 72 counties are submerged, 20 of which never resurface; 100,000 die.

1530 Dikes burst in Holland and flood country (November 1); 400,000 are killed.

1570 Dikes fail in Holland, leading to widespread flooding (November 1); 50,000 die.

1574 Flood washes away Spanish troops besieging Leyden, Holland (October 1–2); 20,000 perish.

1617 Catalonian region of Spain is inundated; 35,000 die.

1642 Flooding in China kills some 300,000.

1646 Massive flooding throughout Holland; 110,000 die.

1755 Earthquake centered in Lisbon spawns seismic waves 15–20 feet high that ravage Spanish and Portuguese coasts (November 1); 31,000 are killed.

1824 Ice jam on Neva River causes severe flooding in St. Petersburg (modern Leningrad) and Kronstadt; 10,000 die.

1851–66 Sunken area between Peking, Shanghai, and Hankow is repeatedly flooded during 15-year period; 40 to 50 million perish.

1854 Tidal wave hits coast of Japan and levels 5,000 buildings in Yamato, Iga, and Ise (July 9); 2,400 are killed.

1864 Ganges River overflows in India (October 5); 45,000 die.

1876 Cyclonic flooding in Bengal region of India takes 200,000 lives.

1882 Bombay, India is overwhelmed by storm and tidal wave (June 5); 100,000 perish.

1883 Eruption of Krakatoa generates 100-foot tsunami that inundates Java and Sumatra (August 27); 295 towns and villages are submerged; 36,000 perish. (*See also Volcanic Eruptions.*)

1887 Yellow River in China overflows throughout spring months, submerging 50,000 square miles; 1.5 million die.

1889 Collapse of South Fork Dam causes flash flood to hit Johnstown, Pennsylvania, killing 2,209 (May 31).

1893 Tidal wave generated by hurricane hits U.S. southeastern coast, submerging islands between Charleston, South Carolina and Savannah, Georgia (August 27); 1,000 die.

1900 Hurricane and accompanying tidal wave inundate Galveston, Texas (September 8); 6,000 die.

1906 Great flood at Hong Kong kills 10,000.

1911 Yangtze River overflows in China, forming lake 80 miles long and 35 miles wide (September); 100,000 drown, and another 100,000 die of starvation in following weeks.

1915 Canton and neighboring cities in China are hit by severe floods (June 12); 100,000 perish.

1916 Sea floods Holland (January 14); 10,000 perish.

1927 Widespread flooding of Mississippi River Valley takes 313 lives and causes $285 million in damage (April).

▶

1928 St. Francis Dam in California bursts and floods valley below (March 12–13); 350 perish.

1928 Lake Okeechobee, swollen by hurricane, inundates southern Florida (September 10–16); 2,000 are killed.

1931 Hwang-ho River overflows in China (August), causing one of worst floods in history; 3.7 million perish. Flooding along Yangtze River causes another 140,000 deaths.

1937 Widespread flooding of Ohio and Mississippi rivers (January–February) takes more than 250 lives and causes $300 million in damage.

1938 Extensive flooding in New York and New England causes $350 million in damage (September 23); 500 perish.

1939 Flooding in northern provinces of China kills 200,000.

1941 Landslide causes flood that inundates Huaráz, Peru (December 14); 3,000 are killed.

1946 Tsunami hits Hilo, Hawaii with such force that railroad tracks are wrapped around trees (April 1); 179 perish.

1948 Flooding along seacoast of Fukien Province in China (June) kills upward of 3,500.

1949 Floods in China leave 20 million homeless.

1951 Flooding of Kansas River at Kansas City, Missouri and Topeka and Lawrence, Kansas (July) causes first $1 billion flood loss in U.S. history; only 41 people die.

1951 Typhoon-related flooding in Manchuria, China takes estimated 4,800 lives (August 6–7).

1953 Catastrophic flooding in Holland takes 2,000 lives, with another 300 deaths reported in Great Britain (January 31).

1953 Extensive flooding in Kyushu, Japan (June 27) leaves 1 million homeless and kills 684; renewed flooding (July 17) kills another 638.

1954 Yangtze overflows in China, forcing 10 million people to evacuate the flooded area (August 1); 40,000 are killed.

1954 Flash flood in Farahzad, Iran sweeps away worshipers at Muslim shrine (September 17); 2,000 perish.

1955 Extensive summer flooding at mouth of Ganges River in India and East Pakistan (now Bangaladesh) kills 2,000.

1955 Severe flooding in New England resulting from rains of Hurricane Diane (August 17–19) kills 190 people. Damage is estimated at $1.8 billion.

1960 Cyclone and tidal wave ravage coast of East Pakistan, destroying 900,000 homes (October 31); 10,000 are killed.

1962 Dikes break near Hamburg, Germany, leaving 500,000 homeless (February 16); 343 perish.

1964 Mekong River Delta in South Vietnam is inundated by rains from typhoons Iris and Joan (November–December); 5,000 lives are lost.

1966 Severe floods inundate Italian cities of Florence, Rome, Naples, and Venice (November). Millions of dollars' worth of fine art is severely damaged; 113 die.

1967 Heavy rains flood Rio de Janeiro and São Paulo in Brazil (January 23); 620 are killed.

1969 Storm-generated flood tides strike Shantung Province, China (April 23). Japanese sources estimate death toll in hundreds of thousands.

1970 Cyclone and tidal wave devastate Ganges River Delta in East Pakistan (now Bangladesh) (November 12); 300,000 to 500,000 perish, 100,000 are missing.

1971 Severe floods inundate North Vietnam (August 30); more than 100,000 die.

1973 Flooding in Indus River system in Pakistan (August) kills estimated 300 people.

1974 Unusual monsoon rains cause widespread flooding in Bangladesh (August); at least 2,500 die. Property damage is over $2 billion.

1978 Northern India is hit by summer monsoon flooding; more than 1,000 lives are lost. Estimated 5 percent of nation's population is affected.

1980 Heavy autumn monsoon rains cause extensive flooding in West Bengal, India; almost 1,500 die.

1985 Five Chinese provinces are heavily inundated (June 6–August 8); 527 die.

1988 Severe flooding in Bangladesh leaves 25 million homeless (August 19–September 6); over 1,000 perish.

▶

Hurricanes, Typhoons, and Cyclones Since 1900

1900 Hurricane and accompanying tidal wave strike Galveston, Texas (September 8), destroying half city's buildings and killing 6,000.

1906 Hong Kong hurricane batters coast (September 18), killing over 10,000.

1906 Some 100 railroad laborers living on houseboats in Florida Keys are killed when hurricane strikes area (October 18).

1909 Hurricane lashes Mexico's northeastern coast (August 27); 1,500 die and property damage is extensive.

1911 Trieste, Italy suffers major property damage and 100 die when hurricane hits area (June 15).

1915 Worst hurricane of century to date ravages Galveston, Texas (August 16); 275 perish.

1919 Spanish steamer *Valbanera* disappears with 488 on board when hurricane strikes Straits of Florida (September 9–10).

1919 Major hurricane ravages Gulf Coast from Texas to Florida (September 14–17), taking over 300 lives.

1926 Miami and eastern coast of Florida are raked by hurricane (September 18) that takes 250 lives and causes $150 million in damage.

1926 Fierce hurricane strikes Havana, Cuba (October 20), leveling 325 buildings in hour.

1928 Hurricane attacks Puerto Rico and Lake Okeechobee area of Florida, killing 5,000 (September 10–16).

1930 Deadly hurricane strikes Dominican Republic, killing some 2,000 (September 3).

1932 Hurricane storm surge inundates Santa Cruz del Sur, Cuba, killing at least 2,500 (November 9).

1933 Second hurricane in 10 days leaves Tampico, Mexico in ruins (September 10); hundreds perish.

1934 Honshu Island, Japan is struck by typhoon that takes over 4,000 lives (September 21).

1935 Newfoundland, Canada is lashed by severe hurricane that destroys many villages (August 25).

1935 Florida is struck by hurricane that kills some 400 (September 2). Barometer falls to North American record low of 26.35.

1935 Hurricane-induced flooding strikes Haiti (October 25); more than 2,000 die.

1938 Worst hurricane to strike Long Island and New England in over 100 years derails railroad cars, topples buildings and church steeples, and alters coastline from New Jersey to Massachusetts (September 21–22).

1942 Bengal, India is struck by deadly hurricane system (October 16); 40,000 perish.

1944 Deadly hurricane strikes Cuban and eastern U.S. coasts (October 13–21); hundreds lose their lives and $20 million citrus crop is ruined.

1953 Typhoon destroys one-third of Japan's industrial city of Nagoya.

1954 Hurricane Carol attacks New England coastline, causing nearly $500 million in damages (August 30).

1954 Hurricane Edna drops 5 inches of rain on New York City in 14 hours before moving north into Canada, causing widespread destruction (September 10–13).

1954 Typhoon striking Hokkaido, Japan kills more than 1,600 people, most when Japanese ferryboat *Toya Maru* capsizes in Hakodate Bay (September 26).

1954 Hurricane Hazel, one of strongest storms to strike North America, destroys three towns in Haiti and coastal cities and towns in South Carolina, then rages north through New York and Canada (October 5–18).

1955 Hurricanes Connie and Diane rage from North Carolina to New England (August 12–21), causing floods and highest property damage to date, $1.75 billion.

1957 Coastal areas of Louisiana, Mississippi, and Texas are hit by Hurricane Audrey's 100-mph winds (June 27). Cameron, Louisiana is leveled, and some 550 die.

1959 Japan and South Korea are battered by Typhoon Sarah (September 17–19); over 2,000 die.

1959 Typhoon Vera, worst in Japan's history, kills 5,000 (September 26–27); it leaves 1.5 million people homeless on island of Honshu.

1959 Mexico's Pacific coast is hit by rare hurricane; 2,000 perish.

1960 Hurricane Donna, most destructive in U.S. history to date, lashes entire Eastern Seaboard (September 4–12).

1961 Hurricane Carla kills 46 in Texas and causes hundreds of millions of dollars in damage (Sep-

▶

tember 11–14); it is fiercest hurricane of century to date.

1961 Hurricane Hattie destroys most of Belize, British Honduras (October 31); over 300 die.

1961 Typhoon Muroto II kills 32 in Osaka, Japan and floods city (September).

1963 At least 22,000 die in East Pakistan (now Bangladesh) when cyclone sweeps in from Bay of Bengal (May 28–29).

1963 Hurricane Flora rages through Haiti, Cuba, and Dominican Republic (September 30–October 9), killing over 6,000 and destroying 90 percent of Cuban coffee crop.

1964 Typhoon Gloria hits Taiwan (September 11–12); some 330 die.

1965 Two successive cyclonic storms ravage East Pakistan (now Bangladesh), killing 35,000 to 40,000 in Ganges River Delta (May 11–12, June 1–2).

1965 Over $1.4 billion in property damage is caused by Hurricane Betsy's devastation in Bahamas, southern Florida, and Louisiana (September 7–12).

1966 Haiti is called "valley of death" following destruction of Hurricane Inez, which also rages through Cuba and Guadeloupe (September 24–30); some 3,600 die.

1967 Caribbean islands, Mexico, and Texas are hit hard by Hurricane Beulah (September 20), which spawns 60 tornados and destroys 328 coastal towns; 58 die, and damage exceeds $1 billion.

1968 Hurricane Gladys strikes Florida (October 18).

1969 Hurricane Camille rages through seven states, from Louisiana to Virginia, with winds of 200 mph (August 17–18); widespread damage totals more than $1.5 billion.

1970 Texas coast near Corpus Christi is devastated by Hurricane Cecilia (August 3); property damage totals $400 million.

1970 Cyclone and accompanying tidal wave devastate East Pakistan (now Bangladesh) (November 12); 300,000 to 500,000 die, 100,000 are missing.

1972 Hurricane Agnes causes destruction from Florida to New York exceeding $2 billion (June 21–26); some 122 die.

1974 Casualties number 8,000 in Honduras when Hurricane Fifi batters coast (September 19–20).

1974 Hurricane Tracy runs ashore at Darwin, Australia, decimating city and forcing evacuation of all residents (December 24–25).

1977 Hurricane Frederick strikes Alabama. Despite storm's strength and force, early forecasting and warnings lead to extensive evacuation (September 12); property damage exceeds $2.3 billion (most expensive U.S. hurricane to date), but only five people die.

1977 Andhra Pradesh State in India is struck by huge cyclonic storm (November 19–20); resultant surge takes at least 10,000 lives.

1979 Dominican Republic, Puerto Rico, and southeastern United States are battered by Hurricane David (August 30–September 7); 2,000 killed.

1979 Hurricane Frederick, following in the wake of David, slams into Florida–Alabama coast and causes $2 billion damage in southeastern United States (September 12–15).

1980 Hurricane Allen strikes several Caribbean islands (August 4–11), killing 272 and causing widespread devastation.

1984 Southern Philippines are devastated by Typhoon Ike (September 2); 1,363 die.

1985 Cyclonic storm causes estimated 10,000 deaths in Bangladesh (May 25).

1985 Southeastern U.S. coastal areas are struck by Hurricane Juan (October 26–November 6); 97 die.

1988 Hurricane Gilbert causes estimated $5 billion in damages from Caribbean to Mexico (September 10–18); record low pressure (26.22 inches of mercury) makes it most powerful hurricane recorded.

1989 Hurricane Hugo rips up U.S. east coast, causing extensive property damage and virtually leveling Charlestown, South Carolina.

Snowstorms and Blizzards

1719 Severe snowstorm in Sweden strikes caravan en route to Drontheim; 7,000 perish.

1798 Hundreds die as New England is hit by blizzard that completely buries houses in massive drifts (November 17–21).

1846 Group of 90 settlers known as Donner Party is trapped in record-deep snow at Truckee Pass, California, en route to Sacramento (April).

▶

Several of them turn to cannibalism to survive; 42 perish.

1886 Kansas loses 80 percent of its cattle when blizzard ravages five midsouthwestern states (January 6–13); at least 80 lose their lives.

1888 "The Blizzard of '88" buries New York City and New England in snowdrifts up to 15 feet high (March 11–14); 800 people die, more than 200 of those in New York City alone.

1891 Iowa, Nebraska, and South Dakota are wracked by severe snowstorm with accompanying winds of up to 80 mph (February 8); 23 die.

1896 Blizzard claims 29 lives in Minnesota and Dakotas (November 25–28).

1898 Storm and heavy snowfall destroy 142 ships and fishing boats off coast of New England (November 26–27); 455 perish.

1909 Christmas blizzard ravages eastern United States, causing more than $20 million in damages (December 25–26); 28 die.

1913 Snowstorms dump 35 inches in Great Lakes area of United States, wrecking eight ships on Lake Huron (November 7–11); 230 perish.

1922 Exceptionally heavy snowfall on Eastern Seaboard between South Carolina and Massachusetts causes roof of Knickerbocker Theater in Washington, D.C., to collapse, killing 98 (January 27–29); total of 140 die in two-day storm.

1923 Blizzard with accompanying winds of 63 mph hits Michigan, Wisconsin, and Dakotas (February 13–14); 28 die.

1925 More than 200 die when exceptionally heavy snowfall buries Primorsk region of Korea (April 25).

1927 Niigata Prefecture of Japan is struck by blizzard (February 13); 91 die.

1928 Blizzard hits caravan of pilgrims in India (September 13), killing 60.

1930 Hundreds die when severe snowstorm hits China's Yangtze Valley (January 9).

1931 Blizzard derails train and tears roofs off more than 1,000 houses in Japan (January 12); 30 are killed.

1933 Heavy snowfall in New York and Pennsylvania causes 68 deaths (December 26–31).

1936 Bulgaria is ravaged by blizzard while temperature plummets to −49° F (February 11); 100 die.

1936 Nevada and California are blanketed by 32 inches of snow in under 24 hours (February 23–24); 26 die.

1938 Record-breaking snowfall causes 37 deaths on eastern coast between New York and Virginia (November 24–25).

1940 Blizzard sweeps through South Dakota to Michigan, killing 157 (November 11–12).

1941 Minnesota to South Dakota is swept by blizzard and gales (March 15–16); 151 die.

1947 New York, New Jersey, and New England receive 30 inches of snow in 24 hours (December 26–27); 55 die.

1948 Mississippi, Arkansas, and South Carolina are ravaged by ice storm that does $20 million in damage and claims 38 lives (January 24–31).

1949 Severe blizzard conditions spread from Great Basin to northwestern Great Plains (January 1–14); 121 die.

1949 Blizzard leaves 73,000 homeless in Teheran, Iran (January 1); 60 die.

1950 Severe snowstorm and winds over 72 mph sweep through Oregon, killing 75 (January 12–14).

1952 Worst blizzard in 50 years hits California (January 13), stranding 196 passengers on stalled train for three days; 26 die.

1952 Snowstorm dumps 31 inches on New England in under 24 hours (February 18–19); 47 die.

1953 Record-breaking snowfall leaves 16-foot drifts in parts of Colorado, Iowa, and Minnesota (February 18–21); 24 die.

1955 Blizzard with accompanying winds of 65 mph ravages Nebraska, Wyoming, and Dakotas (February 18–20), killing 23.

1955 Severe blizzard injures more than 200 in Colorado, Idaho, and Montana (March 20–25); 47 are killed.

1956 Ice storm in New England causes $2 million in damage and is responsible for 40 deaths (January 8–10).

1956 Snowstorms and bitter cold sweep from New England to Siberia (February 1–8); 907 die.

1956 Continuous snow for 92 hours causes millions of dollars of damage in Texas and New Mexico (February 1–8); 23 die.

1956 Severe snowstorms in New England, New York, and Pennsylvania claim 162 lives (March 16–17).

▶

1957 Four-day blizzard sweeps Kansas, New Mexico, and Oklahoma, killing 40 (March 22–25).

1958 Four-day blizzard drops 37 inches of snow on Syracuse, New York (February 7–8); 21 die.

1958 Severe snowstorm blasts Eastern Seaboard from Delaware to North Carolina, causing $520 million in damage and killing over 500 (February 15–16).

1958 Blizzard stretching from New York to Virginia does $20 million in damages and claims 63 lives (March 19–22).

1959 Blizzard conditions hit New York and New England (March 11–13); 93 die.

1960 Four-day blizzard causes extensive damage from Virginia to New England (December 10–13); 137 die.

1966 Blizzard with accompanying 60 mph winds dump 3 feet of snow on Eastern Seaboard between Virginia and New York (January 29–31); 201 die.

1971 Snowstorm in Wyoming and Utah claims 24 lives (October 30).

1972 Thousands are missing and another thousand die when week-long snowfall leaves 26-foot drifts in Iran (February 3–10).

1973 Colorado blizzard claims lives of 30 people and 10,000 cattle (April 4–7).

1975 Blizzard with accompanying 50 mph winds hits the American Midwest, causing 50 deaths (January 14).

1986 Most of Europe is paralyzed by heavy snows and temperatures as low as −13°F (February 11–March 7); 25 deaths are reported in Great Britain, Austria, and Netherlands.

1986 Severe blizzard with winds reaching 100 mph causes six deaths in Great Britain (March 24).

Tornadoes Since 1900

1900 Damage of $500,000 is caused by six tornadoes that hit Alabama, Arkansas, and Mississippi (November 20).

1905 Tornado touches down in Marquette, Kansas, destroying town (May 9).

1905 Twenty buildings are leveled by deadly tornado in Texas (July 5).

1909 Hundreds of buildings are destroyed and 64 die as tornado strikes Arkansas (March 8).

1911 Tornado strikes Kansas and Oklahoma, leaving four towns in ruins (April 13).

1913 Deadly tornado levels Omaha, Nebraska and causes severe damage in neighboring areas (March 23); 115 perish.

1913 Freetown, Sierra Leone is struck by devastating tornado (December 11); some 250 perish.

1915 Savage tornado strikes Kansas, Nebraska, and North Dakota, creating swath of destruction 16 miles wide (November 10).

1917 Tornado in Indiana kills 45 and leaves 2,500 homeless (March 23).

1917 Large area from Illinois to Alabama and Arkansas to Tennessee is hit by many tornadoes (May 26–27); 249 die.

1920 Eleven tornadoes devastate five U.S. Midwestern states, doing millions of dollars in damage. In Chicago, over 100 buildings are destroyed (March 28).

1920 Series of tornadoes sweeps Mississippi, Alabama, and Tennessee (April 20); 220 are killed.

1923 Tornado in U.S. Midwest devastates three states, killing 900 people and injuring 2,500 (March 18).

1924 About 22 tornadoes rip through South and Midwest, causing over $10 million in damages (April 29–30).

1924 Ohio and Pennsylvania are struck by four tornadoes (June 28); 96 are killed.

1925 Worst U.S. killer tornado to date cuts 220-mile path (in some places 1 mile wide) through Missouri, Illinois, and Indiana (March 18); 689 die.

1926 Bengal, India is struck by tornado that levels seven villages.

1927 Large area from Texas to Michigan is devastated by series of tornadoes (May 8–9); 227 die.

1927 Fierce tornado rages through St. Louis, Missouri, wrecking city in five minutes (September 29).

1930 Turkey's worst tornado in a century strikes Adrianople, destroying most of its homes (July 28).

1931 Tornado rips through Moorhead, Minnesota, picking up five 70-ton railroad cars and carrying them 80 feet.

1932 Four-state area of the U.S. South is hit by 27 tornadoes (March 21); 321 die.

▶

1932 Tornado sweeps through Mymensingh, India (May 5), burying 70 victims in ruins.

1936 Killer tornado flattens Tupelo, Mississippi (April 5); 216 die.

1936 Two tornadoes converge on Gainesville, Georgia, destroying shopping district and taking 203 lives (April 6).

1936 Fierce tornado strikes India, leveling tobacco depot and killing all within; 15,000 people are left homeless (October 30).

1941 Missouri tornado destroys over 200 houses, causing $4 million in damages in five minutes (May 21).

1942 Pryor, Oklahoma is annihilated by tornado that takes 100 of 400 lives there (April 27).

1944 Four tornadoes cause large path of destruction through Pennsylvania, Maryland, and West Virginia (June 23); 153 die.

1947 Three U.S. Southwestern states are devastated by series of eight tornadoes. Woodward, Oklahoma is left in ruins (April 9).

1951 Hombo, Madagascar is struck by tornado that kills 500 (January 4–5).

1952 Series of 31 tornadoes sweeps from Missouri to Alabama, killing 343 and causing over $15 million in property damage (March 21–22).

1952 Tornado flattens area of South Africa, killing 35 and injuring over 500 (November 31).

1953 Severe tornado takes 114 lives in Waco, Texas, where it causes some $39 million in property losses (May 11).

1953 Southeastern Michigan is raked by tornadoes that destroy section of Flint (June 8); 113 die.

1953 Worst tornado to strike New England in 75 years claims 90 lives and leaves thousands homeless (June 9).

1955 Kansas and Oklahoma are hit with series of 19 tornadoes (May 25); more than 100 die.

1959 Damages of $12 million result from tornado in St. Louis, Missouri (February 10); it leaves 5,000 homeless.

1961 Colla, East Pakistan is left in ruins by savage tornado (March 19); over 250 perish.

1965 Tornadoes declared worst in Indiana history devastate Midwest (April 11); almost 300 are killed.

1966 Topeka, Kansas is struck by tornado that causes over $100 million in damages (June 8); 4,000 lose their homes.

1970 Violent tornado ravages Coffeyville, Kansas, dropping record-size hailstones, including one measuring 6 inches in diameter and weighing 1.7 pounds (September 3).

1971 Mississippi and Louisiana are ravaged by series of tornadoes that kill 117 (February 21).

1972 Deadly tornado strikes Bangladesh (April 2); 200 are killed and 25,000 are left homeless.

1973 Tornado strikes San Justo, Argentina, leveling every building in city (January 10).

1974 At least 148 tornadoes touch down in 13 U.S. states, destroying entire towns and killing 315 (April 3–4). In Madison, Illinois, tornado destroys entire house—except for closet where family had sought safety.

1975 Omaha, Nebraska suffers $100 million in property damage from single tornado (May 6). Weather Service alert limits fatalities to just three.

1979 Tornado with six vortices devastates Wichita Falls, Texas (April 10), destroying about 8,000 homes; more than 60 die.

1984 Fierce tornadoes strike USSR, killing over 400 (June 9).

1985 Ohio, Pennsylvania, and New York are hit by series of 30 tornadoes (May 31); 76 die.

1987 Saragosa, Texas is heavily damaged when tornadoes rip through town (May 22).

Volcanic Eruptions

B.C.

c. 5000 Mazama, 9,900-foot volcano in modern-day Oregon, erupts with great force, spreading ash over large area. Volcano's cone collapses to form 6-mile-wide Crater Lake.

c. 1628 Volcano on island of Thera, near Crete, explodes with massive force (probably greatest known explosion on Earth). Resulting hail of debris and tidal wave devastate Minoan cities on Crete and cause the civilization to decline. Disaster probably also gives rise to legend of Atlantis.

1226 Earliest known eruption of Mount Etna (Greek for "I burn"), on Sicily. It erupts often in ancient times, including 1170, 1149, 525, 477, and 122 B.C..

396 Eruption of Mount Etna forces invading Carthaginians to divert around area,

▶

preventing them from reaching city of Catania.

A.D.

79 Great eruption of Mount Vesuvius (near modern Naples) kills thousands and buries cities of Pompeii and Heraculaneum (August 24). Pompeii is rediscovered in 1595 and later excavated by archaeologists.

130 Massive eruption of Taupo, volcano in modern New Zealand, levels some 6,000 square miles of surrounding countryside.

186 Taupo, New Zealand explodes with unprecedented force. 666,667 cubic feet of material representing 80 percent of mountain is blown as high as 137 miles.

203 Vesuvius erupts again. New eruption in 472 spews ash across Europe.

1169 First eruption of Mount Etna in centuries destroys about 50 nearby cities and kills some 15,000.

1198 Violent eruption of Vesuvius. It erupts again in 1302.

1536 Mount Etna eruption kills thousands.

1591 Thousands die in eruption of volcano Taal on Luzon, Philippines.

1616 Volcano Mayon in Philippines erupts for first time; thousands perish.

1631 Vesuvius, near Naples, erupts and spews streams of lava (December 13); some 3,000 die.

1638 Peak, volcano almost 10,500 feet high on island of Timor (in modern Indonesia), blows up; remnants become a lake.

1669 Italian city of Catania is partly destroyed by lava flow from eruption of Mount Etna (March 11–July 15); 20,000 perish in this, volcano's most violent known eruption. Citizens dig trench in first known attempt to divert lava flow. City again is hit by Etna eruption on January 11, 1693.

1741 Major eruption of world's highest volcano, Cotopaxi (19,500 feet), in Ecuador, sends avalanche of lava and ice crashing down on villages below. Volcano erupts again in 1744.

1755 Massive eruption of Mount Etna occurs in conjunction with major earthquake and kills some 36,000.

1766 Eruption of Philippine volcano Mayon (October 23–30) is accompanied by tornadoes and floods in incredible disaster; some 2,000 perish.

1772 Eruption of volcano Papandayan on Java blows away top 3,700 feet of 8,700-foot cone; 3,000 perish.

1783–84 Massive eruption of Skaptar in Iceland kills 9,000 persons, about fifth of total population (December–January). Over half of all livestock there is also killed.

1793 Javanese volcano Unsen explodes (April 1); 50,000 die.

1793 Javanese volcano Miyi-Yama erupts, killing 53,000.

1794 Tunquraohua, volcano in Ecuador, erupts; 40,000 perish.

1814 Philippine volcano Mayon erupts and showers nearby villages with hot stones and lava; 2,200 perish.

1815 Indonesian volcano Tambora erupts (April 5–7), spewing estimated 36 cubic miles of ejecta, greatest for a known eruption; blast rips away over 4,000 feet of volcano's cone; nearly 12,000 perish.

1822 Javanese volcano Galung Gung, long inactive, erupts (October); about 4,000 perish and over 1,000 villages are destroyed.

1883 Massive eruption of Indonesian volcano Krakatoa (August 27), believed second-largest in history after Thera (c. 1628 B.C.). Eruption throws rocks upward 34 miles and spreads dust over 3,300 miles away. Blast, with estimated force of 26 hydrogen bombs, is heard some 3,000 miles away. Resulting tidal wave kills over 36,000.

1888 Japanese volcano Bandaisan erupts, killing over 400 (July 15). It spreads rock and debris over 10 square miles.

1902 LaSoufriere, on island of St. Vincent in West Indies, explodes (May 7); more than 1,500 killed.

1902 Pelée on Martinique erupts, spewing lava and destroying port of St. Pierre (May 8); over 30,000 perish.

1902 Guatemalan volcano Santa Maria erupts, killing 6,000.

1911 Philippine volcano Taal erupts, destroying 13 villages and killing 1,335 (January 30).

1912 Massive and violent eruption of Mount Katmai in Alaska results in creation of Valley of 10,000 Smokes (June).

▶

1919 Kelud in Java erupts after lying dormant for 18 years, releasing avalanche of water and mud from crater lake into valley below; 5,500 perish.

1926 Mauna Loa volcano in Hawaii erupts (April 17), spewing 40-foot-deep lava flow.

1928 Etna erupts (November 7), spewing lava flow.

1929 Vesuvius eruption destroys nearby villages with lava flow.

1944 Mexican volcano Parícutin, growing since 1943, erupts (June 10), inundating nearby towns with lava; 3,500 perish.

1951 Mount Lamington, volcano in New Guinea, erupts (January 18–21); 3,000 perish.

1951 Mount Catarman in Philippines explodes in its first major eruption since 1875 (December 4); more than 500 die instantly.

1952 Scientific research boat investigating a submarine volcano southeast of Tokyo disappears in an underwater explosion (September 24); 29 perish.

1956 Following its return to activity in 1955, Bezymianny on Kamchatka Peninsula in eastern USSR explodes, ejecting 0.5 cubic mile of material (March 20).

1960 Chilean volcanoes erupt (May 21–30), causing earthquakes and tsunamis that kill 5,700.

1963 Mount Agung in Bali erupts (March 17–21). Lava flow inundates nearby towns, killing some 1,500 and leaving 200,000 homeless.

1968 Explosion of Mount Arenal in Costa Rica claims 80 lives when *nurée ardente* races down its western slope (July 29).

1971 Etna erupts, sending out new lava flows (April 5).

1973 Eruption of Helgafell volcano on island of Heimaey, Iceland after 7,000 years of inactivity forces evacuation of all 5,500 of island's inhabitants (January 23).

1975 Soviet scientists report birth of new land volcano on the Kamchatka Peninsula (July), first since Parícutin in 1943.

1975 Mauna Loa in Hawaii erupts.

1979 Poison gases from eruption of Mount Sinila on Java, Indonesia claim more than 175 lives (February 20).

1980 Mount St. Helens in Washington State erupts explosively (May 18) for first time in 123 years. Blast, 500 times more powerful than that of atomic bomb dropped on Hiroshima, blows out one side of volcano and kills dozens of onlookers. Eruption is largest in U.S. history.

1983 Hawaiian volcano Kilauea erupts (January), spewing lava flows for months.

1984 Mauna Loa, world's largest active volcano, erupts (March). Event is broadcast live on television throughout Hawaii.

1985 Some 25,000 are killed by eruption of Nevado del Ruiz in Colombia (November 13).

1986 Toxic gas cloud created by volcanic activity under Lake Nios in Cameroon kills 1,734 persons living near lake (August 21).

Plagues and Epidemics

B.C.

767 First recorded worldwide plague.

430 Famous plague in Athens takes life of Pericles.

187 Plague occurs throughout Egypt, Syria, and Greece. Classical author Pliny notes 2,000 deaths daily.

A.D.

79–88 Pestilence occurs throughout Italy; one-day death total in Rome is said to be 10,000.

167–89 Roman Empire is swept by three separate waves of plague.

▶

250–65 Bubonic plague decimates Roman Empire; 5,000 are said to have died daily in Rome.

410 Rome is hit with plague; thousands die.

430 England is struck with plague; barely enough survivors are left to bury dead.

434 Plague occurs throughout Italy; thousands die.

444 Bubonic plague epidemic in England; thousands die.

542–94 Waves of bubonic plague wash over Europe, Asia, and Africa; estimated half of population dies.

746–49 Plague devastates Constantinople and spreads westward into Greece and Italy; estimated 200,000 die.

863 Scotland is hit with epidemic disease; thousands die.

1021 Epidemics of St. Vitus's dance strike Europe.

1094–95 Plague sweeps England and Ireland, where thousands die.

1097 Egypt and Palestine are afflicted with severe outbreak of plague (September–December); 100,000 die.

1111 Springtime outbreak of plague in London; thousands die.

1123–24 France and Germany are swept by plague; thousands die.

1172–73 First major influenza epidemic devastates British Isles; thousands die, including Henry II.

1204 Plague in Ireland; thousands die.

1218 Severe plague in Egypt; 67,000 die.

1235 Plague in England; thousands die, 20,000 in London alone.

1262 Plague in Ireland, recurring in 1271 and 1295; deaths are in thousands.

1299 Pestilence sweeps Persia; thousands die.

1316 England and Ireland are devastated by pestilence; mortality is high.

1332 Bubonic plague breaks out in India; it also is reported in China, where millions are said to die.

1347–51 Greatest known outbreak of bubonic plague sweeps into Europe from Orient. Black Death becomes history's greatest disaster; estimated 75 million die.

1358–70 Repeated outbreaks of plague decimate British Isles and France; thousands die.

1382–85 Pestilence called "Fourth" grips Ireland; thousands die.

1408 London is hit with pestilence; 30,000 die.

1485 First English outbreak of "sweating sickness" kills thousands; repeated outbreaks in 1506, 1517, 1528, and after ultimately claim several million victims.

1495 Epidemic of syphilis sweeps Europe.

1499–1500 Plague in England; 30,000 die.

1507 First outbreak of smallpox in New World. By 1520 it spreads into Mexico; several million die.

1528–30 Typhus epidemic in Italy; in Tuscany alone 100,000 lives are lost.

1545 Cuba is decimated by typhus outbreak; 250,000 are said to perish.

1560 Smallpox epidemic sweeps Brazil; deaths reportedly are in millions.

1600 Pestilence and famine strike Russia; 500,000 perish.

1601–4 Plague ravages Ireland and England; deaths exceed 50,000.

1611 Constantinople suffers great attack of disease; 200,000 reportedly die.

1630 Venice and surrounding Italy are devastated by plague; 500,000 die.

1632 Plague strikes France; at least 80,000 die.

1656 Plague spreads from Sardinia to Naples; 400,000 die.

1664 Great plague in London; 100,000 die.

1672 Naples again is struck by bubonic plague; 400,000 die.

1711 Austria and Germany are devastated by plague; 500,000 deaths are estimated.

1760 Plague kills at least 100,000 in Syria.

1792 Egypt suffers great outbreak of plague; 800,000 die.

1799 North Africa is struck by plague; 300,000 die.

1826–37 Repeated outbreaks of cholera ravage Europe; millions die, 900,000 in 1831 alone.

▶

1840–62 Cholera spreads worldwide; fatalities are in millions.

1851–55 Tuberculosis ravages England; 250,000 die.

1863–75 Cholera continues global attack. In 1866 300,000 die in Eastern Europe; total deaths are in millions.

1876–77 Great crop failure in India leads to outbreak of cholera, responsible for 3 million deaths.

1889–90 Great influenza epidemic afflicts 40 percent of world; deaths are in millions.

1893–94 Renewed worldwide outbreak of cholera takes millions of lives.

1898–1923 Repeated outbreaks of plague continue to ravage India; death total may be as high as 12 million.

1910–13 Bubonic plague causes great devastation in China; death toll may be in millions.

1915 Serbia is decimated with typhus epidemic that takes 150,000 lives (June–August).

1917–19 Influenza pandemic; estimates of dead range as high as 50 million.

1917–21 Savage and persistent invasion of typhus claims as many as 3 million lives in Russia.

1921 Cholera strikes India anew; 500,000 die. Another outbreak, two years later, claims additional 300,000.

1926–30 Smallpox epidemic in India kills 500,000.

1933 Unidentified disease kills 50,000 in China (October 1).

1958 Outbreaks of cholera and smallpox in India and East Pakistan (now Bangladesh) kill more than 75,000 (January–July).

1974 Smallpox epidemic takes 30,000 lives in India (January–June).

1981– AIDS is officially recognized in United States, where by June 1989 more than 58,000 fatalities occurred. In much of central Africa, AIDS has reached epidemic proportions.

13 Media, Entertainment, and Contemporary Music

Newpapers and Magazines

See also Printing and Reproduction Technology in the Technology Section.

B.C.

59 Roman ruler Julius Caesar (c. 102–44) creates first public news bulletin, *Acta diurna,* which is posted in public places in Rome.

A.D.

c. 1450 Invention of printing with movable type makes coming proliferation of newspapers possible.

1594 *Mercurius gallobelgicus,* believed to be world's first magazine, is published in Cologne.

1643 English Parliament enacts laws for censorship of publications.

1660s First regularly published newspapers appear in Germany and England. Academic journals published in Europe lay groundwork for later magazine development.

1690 *Publick Occurrences,* first newspaper in American colonies, is printed in Boston (September 25). Operation is shut down days later because publisher neglected to get official permission.

1702 *The Daily Courant,* England's first daily, also is probably world's first daily newspaper.

1712 England enacts stamp duty for newspapers.

1733 John Peter Zenger (1697–1745) launches *The New York Weekly Journal.* Two years later he is brought to trial for libel but is acquitted. His is first victory for freedom of press in America.

1741 *American Magazine* appears in Philadelphia, leading the way for other magazines, many of which develop in mid—1800s.

1783 *Pennsylvania Evening Post,* first American daily, is published.

1791 U.S. Bill of Rights is adopted. It contains guarantee of freedom of press.

1798 Alien and Sedition Acts are passed in America. Press is restricted for a time.

1800s American newspapers are voice of partisan politics during first quarter of century. Later they establish independent voice.

1814 *The Times* of London, started by John Walter (1739–1812) in 1785, begins using steam-driven press.

1826 *Le Figaro* is founded in France.

1830s Women's magazines and inexpensive pulp periodicals appear.

1833 Charter Institute of Journalists forms in England as journalism's first professional association.

1833 Penny press begins with establishment of *New York Sun.* Papers are small, featuring social exposés and human-interest topics. Timeliness becomes reporting virtue.

1841 Famed newspaper publisher Horace Greeley (1811–72) founds *New York Tribune.*

1847 High-speed rotary press is developed.

1848 New York Associated Press news agency is founded. Rapid expansion of newly invented telegraph over next decades creates fast communications-system ideal for gathering news.

1850 *Harper's New Monthly Magazine* uses woodcut illustrations and articles from abroad. It establishes reputation as a quality, informative periodical.

1851 Reuters news agency is started by Great Britain's Paul J. Reuter (1821–99).

1854 *The Times* of London reaches circulation of 50,000.

1855 *Daily News* is founded in New York City.

1855 *Daily Telegraph* is founded in London.

1860s Censorship is imposed on American newspapers during Civil War.

1877 *Washington Post* is started by Melville Stone (1848–1929).

▶

1878 *Philadelphia Times* successfully publishes Sunday edition for first time.

1879 University of Missouri offers first journalism course.

1880s Halftones are used for newspaper photo reproduction.

1880s Joseph Pulitzer (1847–1911) founds St. Louis *Post-Dispatch* and purchases *New York World.*

1881 *Los Angeles Times* is founded.

1884 Linotype is used for first time.

1889 *Wall Street Journal* is founded.

1895 William Randolph Hearst (1863–1951) buys *New York Journal* and converts it to a competitive, sensationalistic broadsheet. Others soon imitate its format, which includes heavy use of photos, gossip columns, and sensationalistic reporting.

1896 First advice-to-lovelorn column, written under pseudonym Dorothy Dix, appears in *New Orleans Picayune.*

1896 Southern newspaper publisher Adolph S. Ochs (1858–1935) buys failing *New York Times* and soon turns it into a successful and highly respected newspaper. Slogan "All the news that's fit to print" first appears October 25, 1896. Ochs later starts first book-review supplement.

1897 Monotype typesetting machine is developed. It casts individual characters and is driven by punched tape.

1897 First regular comic strip ("Katzenjammer Kids") appears in *New York Journal.*

1898 Sensational reporting in Hearst papers is generally considered important factor in outbreak of Spanish-American War.

c. 1900 Four-column, tabloid-style paper appears.

1900 *Daily Express* is founded in England by C. Arthur Pearson (1866–1921).

1908 *Christian Science Monitor* is founded by Mary Baker Eddy (1821–1910)

1914 Newspapers published in United States reach high of 15,000. Number declines thereafter.

1918 Existing Russian papers become propaganda arm of revolutionary government as TASS (Telegraphnoye Agentstvo Sovyetskova Soyuza) news agency is formed.

1922 Scripps-Howard becomes first newspaper chain in America.

1922 *Reader's Digest* is founded in Pleasantville, New York by DeWitt Wallace (1889–1981).

1923 Henry Luce (1898–1967) founds *Time* magazine. In 1936 he creates photojournalism magazine *Life.*

1925 *The New Yorker* is founded by Harold Ross (1892–1951).

1939 First offset-printed newspaper is produced. By 1980 almost every American newspaper is offset.

1955 *Playboy*, the first large-circulation magazine in America featuring photos of scantily clad women, is launched.

1955 Alternative newspaper, *Village Voice*, appears.

1966 *Pravda* ("Truth," founded 1912) has 6 million readers in Soviet Union. *Izvestia* ("Spark") has over 8 million.

1969 *Saturday Evening Post* publishes final issue. Published since 1821, it is just one of several major U.S. magazines to succumb to competition from television and other media and to increasing employee, paper, and postage costs, among them *Life* (1972), *Look* (1972), and *Saturday Review* (1982).

c. 1970 Circulation of *The Times* of London is about 375,000.

1970s Electronic age overtakes newspaper and magazine publishing. Computer terminals for writing, editing, page layout, and typesetting speed up the editorial process. Satellite communications aid newsgathering and even make possible typesetting and printing at remote locations.

1972–74 Watergate scandal. Investigated by rookie *Washington Post* reporters Woodward and Bernstein, it results in resignation of U.S. President Richard M. Nixon (b. 1913).

1977 Australian publishing magnate Rupert Murdoch (b. 1931) makes headlines by buying failing American newspapers, including *New York Post* (1977) and *Chicago Sun-Times* (1983).

▶

Pulitzer Prize for Reporting

Until 1953, awards include all categories of reporting. From 1953, this award is for local general reporting.

1917 Herbert B. Swope (*New York World*)
1918 Harold A. Littledale (*New York Evening Post*)
1919 No prize awarded
1920 John J. Leary, Jr. (*New York World*)
1921 Louis Seibold (*New York World*)
1922 Kirke L. Simpson (Associated Press)
1923 Alva Johnston (*New York Times*)
1924 Magner White (*San Diego Sun*)
1925 James W. Mulroy and Alvin H. Goldstein (*Chicago Daily News*)
1926 William B. Miller (*Louisville Courier-Journal*)
1927 John T. Rogers (*St. Louis Post-Dispatch*)
1928 No prize awarded
1929 Paul Y. Anderson (*St. Louis Post-Dispatch*)
1930 Russell D. Owen (*New York Times*)
1931 A. B. MacDonald (*Kansas City Star*)
1932 W. C. Richards, D. D. Martin, J. S. Pooler, F. D. Webb, and J.N.W. Sloan (*Detroit Free Press*)
1933 Francis A. Jameson (Associated Press)
1934 Royce Brier (*San Francisco Chronicle*)
1935 William H. Taylor (*New York Herald Tribune*)
1936 Lauren D. Lyman (*New York Times*)
1937 John J. O'Neill (*New York Herald Tribune*), William L. Laurence (*New York Times*), Howard W. Blakeslee (Associated Press), Gobind Behair Lai (Universal Service), and David Dietz (Scripps-Howard Newspapers)
1938 Raymond Sprigle (*Pittsburgh Post-Gazette*)
1939 Thomas L. Stokes (Scripps-Howard Newspapers)
1940 S. Burton Heath (*New York World-Telegram*)
1941 Westbrook Pegler (*New York World-Telegram*)
1942 Stanton Delaplane (*San Francisco Chronicle*)
1943 George Weller (*Chicago Daily News*)
1944 Paul Schoenstein (*New York Journal-American*)
1945 Jack S. McDowell (*San Francisco Call-Bulletin*)
1946 William L. Laurence (*New York Times*)
1947 Frederick Woltman (*New York World-Telegram*)
1948 George E. Goodwin (*Atlanta Journal*)
1949 Malcolm Johnson (*New York Sun*)
1950 Meyer Berger (*New York Times*)
1951 Edward S. Montgomery (*San Francisco Examiner*)
1952 George de Carvalho (*San Francisco Chronicle*)
1953 *Providence Journal-Bulletin*
1954 *Vicksburg Sunday Post-Herald*
1955 Mrs. Caro Brown (*Alice Daily Echo*)
1956 Lee Hills (*Detroit Free Press*)
1957 *Salt Lake Tribune*
1958 *Fargo Forum*
1959 Mary Lou Werner (*Washington Evening Star*)
1960 Jack Nelson (*Atlanta Constitution*)
1961 Sanche de Gramont (*New York Herald Tribune*)
1962 Robert D. Mullins (*Deseret News*)
1963 Sylvan Fox, William Longgood, and Anthony Shannon (*New York World-Telegram & Sun*)
1964 Norman C. Miller (*Wall Street Journal*)
1965 Melvin H. Ruder (*Hungry Horse News*)
1966 *Los Angeles Times*
1967 Robert V. Cox (*Chambersburg Public Opinion*)
1968 *Detroit Free Press*
1969 John Fetterman (*Louisville Courier-Journal & Times*)
1970 Thomas Fitzpatrick (*Chicago Sun-Times*)
1971 *Akron Beacon Journal*
1972 Richard Cooper and John Machacek (*Rochester Times-Union*)
1973 *Chicago Tribune*
1974 Hugh F. Hough and Arthur M. Petacque (*Chicago Sun-Times*)
1975 *Xenia Daily Gazette*
1976 Gene Miller (*Miami Herald*)
1977 Margo Huston (*Milwaukee Journal*)
1978 Richard Whitt (*Louisville Courier-Journal*)
1979 *San Diego Evening Tribune*
1980 *Philadelphia Inquirer*
1981 *Longview Daily News*
1982 *Kansas City Star* and *Kansas City Times*
1983 *Fort Wayne News-Sentinel*
1984 *Newsday*
1985 Thomas Turcol (*Virginian-Pilot and Ledger-Star*)
1986 Edna Buchanan (*Miami Herald*)
1987 *Akron Beacon Journal*
1988 *Alabama Journal* and *Lawrence Eagle-Tribune*
1989 *Louisville Courier-Journal*

1980 Two highly respected U.S. magazines, *Atlantic* and *Harper's*, are rescued from financial failure.

1980s Sixty-five million Americans read daily newspapers.

1982 *USA Today,* first national newspaper, is launched by Gannett.

1984 CBS, Inc. makes largest-ever magazine acquisition, paying $362,500,000 for 12 periodicals.

1989 Two media conglomerates, Warner Communications and Time, Inc., agree to merge. Resulting Time Warner, Inc. will have $10 billion yearly revenue from media and entertainment, including magazines and books.

1989 *New York Times* buys *McCall's* magazine, reportedly for $80 million.

Radio and Television Broadcasting

See also Radio and Television chronology in Technology section.

1897 Italian scientist Guglielmo Marconi (1874–1937) achieves first radio transmission of code signals over long distances by this time, using basic components of transmitter, receiver, and antenna.

1906 American Lee De Forest (1873–1961) builds first vacuum tube, making it possible to transmit voice and even music over the air.

1906 American inventor Reginald Aubrey Fessenden (1866–1932) broadcasts first voice and music program via his wireless at Brant Rock, Massachusetts, on Christmas Eve.

1910 First daily radio broadcasts begin. Station belongs to Charles Herrold School of Radio Broadcasting in San Jose, California. It is still operating in 1990, as KCBS, San Francisco—the oldest continuously run station in the world.

1916 David Sarnoff (1891–1971) outlines basic plan for radio broadcasting industry. He recommends building stations to broadcast voice and music and specifies manufacture of a "radio music box" to be bought by potential listeners.

1919 Radio Corporation of America (RCA) is formed by General Electric Company and Westinghouse. Radios are gaining in popularity. Soon they will change from a novelty to a necessary household appliance.

1920 First commercial radio station, KDKA in Pittsburgh, goes on the air (November 2).

1920 KDKA makes first broadcast of presidential election results.

1922 United States has 500 radio stations in operation by this time. Radio advertising also begins in 1922, ushering in era of commerical broadcasting.

1926 First public demonstration of television (January 27), by British inventor J. L. Baird (1888–1946). System uses mechanical scanner to convert image for electronic transmission.

1926 NBC is founded by David Sarnoff. NBC buys New York station WEAF to feed programs to subsidiary stations via telephone lines, thereby establishing first permanent radio network.

1927 British Broadcasting Corporation (BBC) is chartered for radio broadcasting in Great Britain.

1927 Radio broadcast of Dempsey–Tunney fight causes such excitement among listeners that deaths of 10 fans are attributed to it.

1927 CBS is formed. In following decades it becomes a leading network under direction of President William S. Paley (b. 1901).

▶

1928 *The Voice of Firestone* program debuts. It broadcasts live musical performances.

1928 Color television is demonstrated for first time by Baird. He uses mechanical scanner.

1928 *Amos 'n' Andy* debuts. Show is so popular by early 1930s that movie theaters pipe in program.

1930 About 13.7 million radios are operating in United States by this time. Popular programs such as *The Shadow,* featuring elusive detective Lamont Cranston, attract large listening audiences nationwide.

1930 United States brings antitrust suit against RCA because of its patents control.

1930 Philo T. Farnsworth (b. 1909) develops cathode-ray tube, basis of modern video viewing screens.

1930 Vladimir Zworykin (1889–1982) develops image orthicon tube, basis of television camera.

1930 RCA opens experimental TV station WZRBX, New York on July 30.

1932 RCA demonstrates all-electronic television system, operating on 120-line scan.

1933 President Franklin D. Roosevelt (1882–1945) uses radio to broadcast first of his famous "fireside chats" to American public (March 12).

1934 Communications Act establishes U.S. Federal Communications Commission (June) to regulate broadcast industry.

1936 FM (frequency modulation) broadcast is developed by Edwin H. Armstrong (1890–1954).

1936 BBC begins first public high-definition television broadcasting service.

1937 News coverage of *Hindenburg* disaster becomes first radio program broadcast coast-to-coast (May 6).

1938 *CBS World News Roundup* debuts. Notable for live, on-the-spot reports from European correspondents (such as Edward R. Murrow) as Continent heats up, it is still on the air.

1938 H. G. Wells's *The War of the Worlds* is broadcast (October 30), as Halloween prank by Orson Welles's *Mercury Theatre On the Air.* Done as series of on-the-spot news broadcasts, show panics 1.5 million Americans.

1938 NBC televises news event live, for first time. Mobile unit in Queens, New York chances on raging fire and switches on cameras.

1939 NBC begins first regular television broadcast service in United States.

1939 Roosevelt become first President to appear on TV, during NBC telecast of his speech.

1939 CBS and Dumont networks begin programming, but wartime restrictions impede growth of industry.

1941 FCC grants first commercial TV-station licenses to NBC and CBS (July 1) in New York.

1943 NBC, which has two radio networks, is forced to divest itself of one as result of antitrust action. Divested radio network becomes ABC.

1945 FCC allots 13 channels for commercial broadcasting in United States.

1946 Lifting of U.S. wartime restrictions on manufacture of television sets. Boom in television-receiver sales begins, and television soon replaces radio as leading U.S. communications industry.

1946 First regular TV network, NBC, feeds New York programs to Philadelphia and Schenectady.

1947 First TV broadcast of World Series. New York Yankees beat Brooklyn Dodgers in seven games.

1948 TV debuts of Ted Mack's *Original Amateur Hour*, Milton Berle's *Texaco Star Theater*, and Ed Sullivan's *Toast of the Town.*

1948 Voice of America begins foreign radio broadcasts.

1949 National Academy of Television Arts and Sciences presents first Emmy Awards for television programs.

1950 CATV (Community Antenna Television) system is introduced. By late 1980s rapidly expanding cable television systems compete successfully with regular broadcast networks.

1951 Fifteen million television sets are now operating in United States, up from 1.5 million in 1950.

1951 First transcontinental TV broadcast.

1951 *I Love Lucy* debuts on CBS. For rest of decade it will be among TV's most popular shows.

1952 Congress launches first investigation of possible behavioral effects caused by violent television programming.

1952 *The Today Show* airs on NBC. Talk shows become enduring popular TV genre.

1953 *You Are There*, weekly newsmagazine hosted by Walter Cronkite (b. 1916), debuts on CBS. Cronkite goes on to become a leading figure in television news.

1954 NBC broadcasts Festival of Roses parade in color, first network color broadcast (January 1).

1954 RCA markets first electronic color television meeting FCC standards.

1954 Eurovision network begins broadcast.

1955 TV income surpasses that of radio this year.

1955 First televised (filmed) presidential news conference is held by Dwight D. Eisenhower (1890–1969) on January 19.

1955 Commercial television broadcasting begins in Great Britain.

1956 *The Huntley-Brinkley Report* begins airing on NBC.

1957 Sony introduces pocket-size transistor radio.

1958 NBC pays $200,000 for rights to broadcast 1958 championship football game. Football and other sports become a staple of network television broadcasting.

1958 Sony sells first transistorized FM radio.

1959 Philco markets Safari, portable television that runs for four hours on batteries.

1959 Sony introduces first transistorized television set.

1960 First televised presidential election debate, between John F. Kennedy (1917–63) and Richard M. Nixon (b. 1913). It later is considered important factor in Kennedy's election victory.

1960 Americans own 85 million TV sets by this time. Television now is the leading form of mass communication.

1961 FCC chairman Newton Minow (b. 1926) declares that American television programming is "vast wasteland."

1962 Legislation is passed in the United States establishing communications satellite corporation to handle global satellite communication. Satellite is in operation three years later.

1963 Live television broadcasts of assassination of President John F. Kennedy and killing of Lee Harvey Oswald, his assassin, shock nation. In later years audiences regularly view graphic news coverage of Vietnam War, protests, and riots.

1963 Roper poll finds that television has become the main news source for Americans.

1964 First public demonstration of satellite television feed via "stationary" (geosynchronous) satellite (October 10). Tokyo Olympic Games are relayed to North America.

1965 Some 240 million radios and 61.8 million television sets are in use in the United States.

1965 News cameras are banned from U.S. courtrooms as possible negative influence on jurors.

1966 FCC requires separate programming on FM stations. FM had been developed before World War II but had languished. It takes off in 1960s and 1970s, because better sound is ideal for rock 'n' roll.

1968 CBS investigative news program *60 Minutes* debuts, popularizing aggressive, muckraking reporting style.

1969 Landing on Moon is carried worldwide by satellite to some 100 million television viewers.

1969 Accuracy in Media is founded to study accuracy of news coverage.

1970 Approximately 231 million television sets are in use in world by this time.

1970 Cigarette ads are banned from television and radio, eliminating major source of advertising revenue for networks.

1980 About 4,225 cable-television channels, 750 commercial television stations, and over 7,000 radio stations are now operating in United States.

1982 National Institute of Mental Health announces violent TV programming is detrimental to viewers.

▶

1985 First buyout of a major U.S. television network, ABC, by Capital Cities Communications Inc, for $3.5 billion.

1985 Live Aid, internationally broadcast rock telethon, reaches worldwide audience of 1.5 billion viewers and raises $70 million for starving Africans.

1986 New commercial TV network, Fox, begins operations as alternative to big three, NBC, ABC, and CBS.

1989 Japanese begin limited broadcasts of new high-definition television (HDTV), though only a few hundred special HDTV receivers are in use in Japan.

Long-Running Television Shows

1947–58 "Kraft Television Theatre," weekly dramatic anthology. Ed Herlihy and Charles Stark host.

1947–60 "Howdy Doody," children's show. Buffalo Bob Smith hosts.

1947–present "Meet the Press," live interview program with moderator, panel of journalists, and guests.

1948–54 "The Author Meets the Critics," book talk program. Hosts include John K. M. McCaffery (1948–51), Faye Emerson (1952), and Virgilia Peterson (1952–54).

1948–55 "Who Said That?", quiz game show, featuring panelists John Cameron Swayze, June Lockhart, Morey Amsterdam, and John Mason Brown. Hosts include Robert Trout (1948–51), Walter Kiernan (1951–54), and John Daly (1955).

1948–55 "Philco TV Playhouse," dramatic anthology program. Bert Lytell, host.

1948–57 "Break the Bank," game show. Bert Parks, host.

1948–57 "Kukla, Fran & Ollie," children's program with puppets, starring Fran Allison.

1948–58 "Studio One," dramatic anthology program with Betty Furness as spokesperson.

1948–58 "Arthur Godfrey's Talent Scouts," talent variety program. Arthur Godfrey hosts.

1948–63 "The Perry Como Show," musical variety show, featuring Perry Como and regulars including Fontane Sisters, Ray Charles Singers, Peter Gennaro Dancers, Louis Da Pron Dancers, and Mitchell Ayres Orchestra.

1948–67 "Candid Camera," humorous predicament anthology program. Allen Funt hosts.

1948–67 "The Milton Berle Show," comedy variety show, with long run interrupted in 1957 and between 1960 and 1966.

1948–71 "The Ed Sullivan Show" (originally "Toast of the Town,") variety show hosted by Sullivan.

1948–71 "The Original Amateur Hour," variety show featuring amateur performers. Ted Mack hosts.

1949–55 "Captain Video and His Video Rangers," children's program. Richard Coogan is first "Captain Video," followed by Al Hodge (1951–55).

1949–55 "Twenty Questions," quiz show. Hosts include Bill Slater (1949–52) and Jay Jackson (1953–55).

1949–56 "Mama," comedy starring Peggy Wood, Judson Laire, Dick Van Patten, and Rosemary Rice.

1949–56 "Man Against Crime," detective drama, starring Ralph Bellamy and, later, Frank Lovejoy (1956).

1949–56 "Stop the Music," quiz show. Bert Parks hosts.

1949–56 "This Is Show Business," variety program and panel. Panelists include George S. Kaufman, Abe Burrows, Sam Levenson, and Walter Slezak. Clifton Fadiman hosts

1949–57 "The Big Story," dramatic anthology program. Bob Sloane narrates.

1949–57 "Ford Theatre," dramatic anthology program.

1949–58 "The Life of Riley," situation comedy, starring Jackie Gleason and Rosemary DeCamp (1949–50) and later William Bendix and Marjorie Reynolds (1953–58).

▶

1949–58 "Red Barber's Corner," sports commentary program with Barber reporting.

1949–59 "Arthur Godfrey and His Friends," musical variety program, featuring host Godfrey and regulars Tony Marvin, the McGuire Sisters, Pat Boone, and Archie Bleyer Orchestra.

1949–61 "The Lone Ranger," weekly western series with Clayton Moore (1949–52, 1954–57) and John Hart (1952–54) as Lone Ranger.

1949–63 "Fireside Theatre," dramatic anthology. Hosts include Frank Wisbar, Gene Raymond, and Jane Wyman.

1949–63 "The Voice of Firestone," musical variety program, featuring narrator John Daly and Firestone Concert Orchestra.

1950–56 "Big Town," journalism drama, starring Patrick McVey and, later, Mark Stevens (1954–56).

1950–56 "Gene Autry Show," western, starring Autry and Pat Buttram.

1950–56 "Life Begins at Eighty," panel discussion show. Jack Barry hosts.

1950–57 "Lux Video Theatre," dramatic anthology. Hosts over the years include James Mason, Otto Kruger, Gordon MacRae, and Ken Carpenter.

1950–57 "Robert Montgomery Presents," dramatic anthology program. Montgomery hosts.

1950–57 "The Web," dramatic anthology. Jonathan Blake and, later, William Bryant (1957), host.

1950–58 "The George Burns and Gracie Allen Show," situation comedy.

1950–58 "Beat the Clock," game show. Bud Collyer hosts.

1950–59 "You Asked for It," viewer participation "dream-come-true" program. Art Baker and, later, Jack Smith (1958–59) host.

1950–59 "Your Hit Parade," musical variety show. Andre Baruch and Del Sharbutt host.

1950–60 "The Arthur Murray Party," musical variety show, featuring Arthur and Kathryn Murray and the Arthur Murray Dancers.

1950–61 "You Bet Your Life," quiz game show. Groucho Marx hosts and George Fenneman announces.

1950–63 "Armstrong Circle Theatre," dramatic anthology. Over the years, Nelson Case, Joe Ripley, Bob Sherry, Sandy Becker, John Cameron Swayze, Douglas Edwards, Ron Cochran, and Henry Hamilton host.

1950–63 "The Pantomime Quiz," game show, featuring regulars Hans Conried, Vincent Price, Jackie Coogan, Rocky Graziano, Dorothy Hart, Carol Burnett, and Stubby Kaye. Series is interrupted between 1959 and 1962. Mike Stokey, and later Pat Harrington, Jr., host.

1950–67 "What's My Line," nighttime game show. John Daly hosts. Celebrity panel members (including Dorothy Kilgallen, Arlene Francis, and Bennett Cerf) attempt to guess occupations of contestants.

1950–77 "The Jack Benny Show," comedy/variety show, featuring Benny and Eddie "Rochester" Anderson in a variety of skits. Regulars include Don Wilson, Dennis Day, Mary Livingstone, and Mel Blanc.

1951–57 "The Roy Rogers Show," western, starring Roy Rogers, Dale Evans, and Pat Brady.

1951–59 "Midwestern Hayride," musical variety show. Hosts include Bill Thall (1951–54), Bob Shrede (1951), Hugh Cherry (1955–56), Paul Dixon (1957–58), and Dean Richards (1959).

1951–59 "Schlitz Playhouse of Stars," dramatic anthology program. Irene Dunne hosts.

1951–60 "Goodyear TV Playhouse," dramatic anthology program.

1951–60 "Mark Saber," detective drama, starring Tom Conway and, later, Donald Gray (1955–60).

1951–61 "I Love Lucy," situation comedy, starring Lucille Ball, Desi Arnaz, Vivian Vance, and William Frawley.

1951–65 "Watch Mr. Wizard," children's science show. Don Herbert hosts.

▶

1951–72 "The Red Skelton Show," comedy/variety show, featuring Skelton in such roles as Clem Kadiddlehopper, Sheriff Deadeye, and Freddie the Freeloader.

1951–present "Search for Tomorrow," soap opera, starring Mary Stuart.

1951–present "Hallmark Hall of Fame," specials.

1952–59 "Dragnet," crime show, starring Jack Webb and Ben Alexander. It is revived in 1967–70, with Webb and Harry Morgan.

1952–60 "Masquerade Party," game show. Moderators over the years include Bud Collyer, Douglas Edwards, Peter Donald, Eddie Bracken, Robert Q. Lewis, and Bert Parks.

1952–61 "This Is Your Life," testimonial program. Ralph Edwards hosts.

1952–66 "The Adventures of Ozzie and Harriet," situation comedy, featuring the Nelson family—Ozzie, Harriet, David, and Ricky.

1952–67 "I've Got a Secret," game show. Hosts are Gary Moore (1952–64) and Steve Allen (1964–67). It is briefly revived in 1976, with host Bill Cullen.

1952–69 "Art Linkletter's House Party," daytime variety show hosted by Linkletter. It is notable for interviews with children.

1952–present "The Today Show," morning news program. Among past hosts are Dave Garroway, Hugh Downs, Barbara Walters, Tom Brokaw, and Jane Pauley.

1953–59 "Name That Tune," game show. Hosts include Rod Benson (1953–54), Bill Cullen (1954–55), and George De Witt (1955–59).

1953–61 "The Loretta Young Show," dramatic anthology program. Loretta Young hosts.

1953–61 "Person to Person," talk show. Edward R. Murrow and, later, Charles Collingwood (1959–61) host.

1953–62 "General Electric Theatre," dramatic anthology program. Ronald Reagan hosts.

1953–71 "The Danny Thomas Show" (originally "Make Room for Daddy"), situation comedy, starring Danny Thomas, Jean Hagen (1953–56), Marjorie Lord (1957–71), Rusty Hamer, Angela Cartwright, and Hans Conried.

1954–60 "The George Gobel Show," comedy/variety program. Regulars include Frank DeVol Orchestra, Eddie Fisher, and Anita Bryant.

1954–60 "The Lineup," police drama, starring Warner Anderson.

1954–61 "December Bride," situation comedy, starring Spring Byington, Frances Rafferty, Dean Miller, Verna Felton, and Harry Morgan.

1954–61 "People Are Funny," game show. Art Linkletter hosts.

1954–63 "Father Knows Best," situation comedy, starring Robert Young, Jane Wyatt, Elinor Donahue, Billy Gray, and Lauren Chapin.

1954–71 "Lassie," children's adventure show. Six different dogs descended from the same line portray Lassie during its 17-year run. Show is syndicated after network run.

1954–present "The Tonight Show," late-night talk show with Johnny Carson. Past hosts include Steve Allen, Ernie Kovacs, and Jack Paar.

1954–present "Walt Disney Show," called variously "Disneyland" (1954–58), "Walt Disney Presents" (1958–61), "Walt Disney's Wonderful World of Color," and other titles. It is currently aired as "The Disney Family Movie."

1955–61 "The Life and Legend of Wyatt Earp," western, starring Hugh O'Brian.

1955–63 "Cheyenne," western drama, starring Clint Walker and L.Q. Jones.

1955–65 "Alfred Hitchcock Presents," dramatic anthology, featuring Alfred Hitchcock as host. Series is revived briefly (1985–86) for network and has continued to run since 1987 on cable.

1955–71 "The Honeymooners," situation comedy, starring Jackie Gleason, Art Carney, Audrey Meadows, and Joyce Randolph. In 1971, Sheila MacRae and Jane Kean replaced Meadows and Randolph.

1955–71 "The Lawrence Welk Show," musical variety show hosted by Welk. Show features

▶

"traditional popular music." It is syndicated after network run.

1955–75 "Gunsmoke," western set in Dodge City, starring James Arness, Amanda Blake, and Milburn Stone.

1955–83 "Captain Kangaroo," entertainment program for preschool children. Bob Keeshan hosts.

1956–62 "Dick Powell's Zane Grey Theatre," western drama anthology program. Powell hosts.

1956–63 "The Dinah Shore Chevy Show," musical variety program, starring Shore and regulars the Skylarks, the Even Dozen, the Tony Charmoli Dancers, the Nick Castle Dancers, the Harry Zimmerman Orchestra, and Frank DeVol and His Orchestra.

1956–67 "To Tell the Truth," game show. Bud Collyer hosts. Panel members (including Kitty Carlisle, Orson Bean, and Tom Poston) try to guess which of three guests is telling the truth. Syndicated after network run, 1969–77.

1957–63 "Have Gun Will Travel," western, starring Richard Boone, Kam Tong, and Lisa Lu.

1957–63 "Leave It To Beaver," situation comedy, starring Barbara Billingsley, Hugh Beaumont, Jerry Mathers, and Tony Dow.

1957–63 "The Real McCoys," situation comedy, starring Walter Brennan, Richard Crenna, Kathy Nolan, Lydia Reed, and Michael Winkleman.

1957–64 "The Price Is Right," game show. Bill Cullen hosts. Show is revived in 1986 with Bob Barker, host.

1957–65 "Wagon Train," western, starring Ward Bond, Robert Horton, Terry Wilson, and Frank McGrath.

1957–66 "Perry Mason," courtroom drama series, featuring Raymond Burr, Barbara Hale, William Hopper, and William Talman. It is revived in 1973–74 with a different cast. Beginning in 1985, Burr and Hale returned in occasional "Perry Mason" TV movies.

1957–70 "The 20th Century," documentary. Walter Cronkite hosts.

1957–87 "American Bandstand," afternoon musical entertainment program for teenagers. Dick Clark hosts. Show is syndicated after network run.

1958–66 "The Donna Reed Show," situation comedy, starring Donna Reed, Carl Betz, Shelley Fabares, and Paul Peterson.

1958–71 "The Andy Williams Show," musical variety program, featuring host Williams and regulars the New Christy Minstrels, the Osmond Brothers, the Goodtime Singers, and the Nick Castle Dancers.

1959–65 "The Twilight Zone," science-fiction anthology program. Serling hosts. Show is revived 1985–87.

1959–66 "Rawhide," western, starring Clint Eastwood, Paul Brinegar, and Steve Raines.

1959–68 "The Bell Telephone Hour," musical variety program, featuring the Bell Telephone Orchestra, Donald Voorhees, conductor.

1959–73 "Bonanza," western series, featuring Lorne Green, Pernell Roberts, Dan Blocker, and Michael Landon.

1960–66 "The Flintstones," animated children's program.

1960–70 "The Andy Griffith Show," situation comedy set in Mayberry, North Carolina, featuring Andy Griffith, Don Knotts, Frances Bavier, and Ron Howard. Network reruns air daytime, 1964–70.

1960–72 "My Three Sons," situation comedy, starring Fred MacMurray, Tim Considine, Don Grady, Stanley Livingston, William Frawley, and William Demarest.

1961–69 "The Dick Van Dyke Show," situation comedy, featuring Dick Van Dyke and Mary Tyler Moore. Daytime network reruns air 1967–69.

1961–present "Wide World of Sports," regular sports program. Announcers include Jim McKay, Howard Cosell, and others.

1962–71 "The Virginian," TV's first 90-minute Western series, featuring James Drury, Doug McClure, and Lee J. Cobb.

▶

1962–71 "The Beverly Hillbillies," situation comedy, starring Buddy Ebsen, Irene Ryan, Max Baer, Jr., and Donna Douglas.

1962–74 "The Lucy Show," situation comedy, starring Lucille Ball, Vivian Vance, Gale Gordon, Lucie Arnaz, and Desi Arnaz, Jr.

1963–70 "Petticoat Junction," situation comedy, starring Bea Benaderet and Edgar Buchanan.

1964–70 "Daniel Boone," western, starring Fess Parker.

1964–70 "Gomer Pyle, U.S.M.C.," situation comedy, starring Jim Nabors, Frank Sutton, and Ronnie Schell.

1964–70 "The Hollywood Palace," variety program, featuring Mitchell Ayres Band.

1964–72 "Bewitched," situation comedy, starring Elizabeth Montgomery, Dick York (1964–69), Dick Sargent (1969–72), Agnes Moorehead, and David White.

1965–71 "Green Acres," situation comedy, starring Eddie Albert and Eva Gabor.

1965–71 "Hogan's Heroes," situation comedy, starring Bob Crane, Werner Klemperer, John Banner, Robert Clay, Richard Dawson, and Ivan Dixon.

1965–74 "The Dean Martin Show," comedy variety program, featuring host Dean Martin and regulars the Golddiggers, the Ding-a-Ling Sisters, Marian Mercer, Tom Bosley, Dom DeLuise, Nipsey Russell, Rodney Dangerfield, and Les Brown and His Band of Renown.

1965–74 "The F.B.I.," police drama, starring Efrem Zimbalist, Jr., Philip Abbott, Lynn Loring, Stephen Brooks, William Reynolds, and Shelly Novack.

1966–73 "Mission: Impossible," spy drama, starring Peter Graves, Greg Morris, Peter Lupis, Leonard Nimoy, Barbara Bain, Martin Landau, and Lesley Ann Warren.

1967–75 "Mannix," detective series, starring Mike Connors as Mannix and Gail Fisher as his secretary.

1967–75 "Ironside," police drama, starring Raymond Burr, Don Galloway, Don Mitchell, Barbara Anderson, and Elizabeth Baur.

1967–78 "The Carol Burnett Show," comedy/variety show, featuring Carol Burnett, Harvey Korman, Vicki Lawrence, Lyle Waggoner, and Tim Conway.

1968–75 "Adam 12," police drama, starring Martin Milner, Kent McCord, William Boyett, and Gary Crosby.

1968–80 "Hawaii Five-O," crime series, featuring Steve McGarrett (Jack Lord).

1968–present "60 Minutes," newsmagazine program. Reporters over the years include Mike Wallace, Harry Reasoner, Morley Safer, Dan Rather, and Ed Bradley. Since 1978, Andy Rooney has offered humorous social commentary.

1969–76 "Marcus Welby, M.D.," medical drama, starring Robert Young, James Brolin, and Elena Verdugo.

1969–76 "Medical Center," medical drama, starring James Daly and Chad Everett.

1969–present "Sesame Street," educational preschool-children's show on Public Broadcasting Service. Original cast includes Loretta Long, Matt Robinson, Will Lee, and Jim Henson's Muppets: Big Bird, Oscar the Grouch, Bert and Ernie, and Cookie Monster.

1970–77 "The Mary Tyler Moore Show," situation comedy, featuring Mary Tyler Moore, Edward Asner, Valerie Harper, Ted Knight, and Gavin McLeod.

1970–77 "McCloud," police drama, starring Dennis Weaver, J.D. Cannon, Terry Carter, Diana Muldaur, and Ken Lynch.

1970–present "Monday Night Football," weekly football special. Broadcasters include Howard Cosell (1970–83), Don Meredith (1970–73, 1977–84), Frank Gifford (1971–present), and Al Michaels (1986–present).

1971–77 "Columbo," police drama starring Peter Falk.

1971–77 "McMillan and Wife," police drama, starring Rock Hudson and Susan Saint James.

1971–77 "The Sonny and Cher Comedy Hour," musical comedy/variety program, starring Sonny Bono and Cher.

▶

1971–83 "All in the Family," situation comedy, featuring Carroll O'Connor, Jean Stapleton, Sally Struthers, and Rob Reiner. From 1979 to 1983 O'Connor played Archie in a new series, "Archie Bunker's Place," with a different supporting cast.

1971–present "Masterpiece Theater," PBS anthology of miniseries, usually British adaptations of traditional English novels. Alistair Cooke hosts.

1972–81 "The Waltons," drama, starring Ralph Waite, Michael Learned, Will Geer, and Richard Thomas.

1972–78 "The Bob Newhart Show," situation comedy, starring Bob Newhart, Suzanne Pleshette, Bill Daily, Peter Bonerz, and Marcia Wallace.

1972–78 "Maude," situation comedy, spinoff of "All in the Family," starring Beatrice Arthur, Bill Macy, and Adrienne Barbeau.

1972–83 "M*A*S*H," comedy/drama series set during Korean War, featuring Alan Alda, Loretta Swit, Gary Burghoff, Larry Linville, and Harry Morgan.

1973–80 "Barnaby Jones," detective drama, starring Buddy Ebsen and Lee Meriwether.

1973–81 "The Midnight Special," musical variety program. Helen Reddy hosts.

1973–82 "The Tomorrow Show," late-night talk show with Tom Snyder and Rona Barrett (1980–81).

1974–80 "The Rockford Files," detective drama, starring James Garner and Noah Beery, Jr.

1974–83 "Little House on the Prairie," dramatic series, starring Michael Landon, Karen Grassle, and Melissa Gilbert.

1975–82 "Barney Miller," situation comedy, starring Hal Linden, Abe Vigoda, Maxwell Gail, Jack Soo, Ron Glass, Steve Landesberg, and Ron Carey.

1975–84 "Happy Days," situation comedy, featuring Tom Bosley, Marion Ross, Ron Howard, and Henry Winkler.

1975–84 "One Day at a Time," situation comedy, starring Bonnie Franklin, Mackenzie Phillips, Valerie Bertinelli, and Pat Harrington, Jr.

1975–85 "The Jeffersons," situation comedy and spinoff of "All in the Family," starring Sherman Hemsley and Isabel Sanford.

1975–present "Saturday Night Live," late-night comedy show, featuring celebrity guest host and regular cast of comedians. Original cast includes Chevy Chase, Dan Aykroyd, John Belushi, Jane Curtin, and Gilda Radner. Over the years, regulars have included Eddie Murphy, Bill Murray, Joe Piscopo, Billy Crystal, Nora Dunne, and Dana Carvey.

1976–83 "Laverne and Shirley," situation comedy, starring Penny Marshall and Cindy Williams.

1976–85 "Alice," situation comedy, starring Linda Lavin, Philip McKeon, Polly Holliday, Beth Howland, and Vic Tayback.

1977–83 "CHiPS," police drama, starring Larry Wilcox and Erik Estrada.

1977–84 "Three's Company," situation comedy, starring John Ritter, Joyce DeWitt, and Suzanne Somers.

1977–86 "The Love Boat," situation comedy, starring Gavin McLeod, Bernie Kopell, Fred Grandy, Lauren Tewes, and Ted Lange.

1978–84 "Fantasy Island," fantasy drama, starring Ricardo Montalban and Herve Villechaize.

1978–86 "Diff'rent Strokes," situation comedy, starring Conrad Bain, Gary Coleman, and Todd Bridges.

1978–88 "Dallas," serial drama, starring Larry Hagman, Barbara Bel Geddes, Patrick Duffy, Victoria Principal, Linda Gray, Charlene Tilton, and an ensemble cast.

1978–present "20/20," news magazine program. Hugh Downs and Barbara Walters host.

1979–85 "The Dukes of Hazzard," comedy drama, starring Tom Wopat, John Schneider, Catherine Bach, and Denver Pyle.

1979–86 "Benson," situation comedy, starring Robert Guillaume, James Noble, Missy Gold, Inga Swenson, and Ethan Phillips.

1979–86 "Trapper John, M.D.," medical drama, starring Pernell Roberts and Gregory Harrison.

▶

1979–89 "The Facts of Life," situation comedy, starring Charlotte Rae, Lisa Whelchel, Kim Fields, Mindy Cohn, Nancy McKeon, and (after 1986) Cloris Leachman.

1979–89 "Knot's Landing," serial drama, starring Ted Shackelford, Joan Van Ark, Michele Lee, Pat Peterson, and Donna Mills.

1980–88 "Magnum, P.I.," detective series, featuring Thomas Magnum (Tom Selleck) and Jonathan Higgins (John Hillerman).

1981–87 "Hill Street Blues," serial police drama, starring Daniel J. Travanti, Veronica Hamel, James B. Sikking, and a large and distinguished ensemble cast.

1981–87 "Gimme a Break," situation comedy, starring Nell Carter.

1981–88 "Falcon Crest," serial drama, starring Jane Wyman, Susan Sullivan, Lorenzo Lamas, and an ensemble cast.

1981–88 "Simon & Simon," detective drama, starring Jameson Parker, Gerald McRaney, and Mary Carver.

1981–89 "Dynasty," serial drama, starring John Forsythe, Linda Evans, Joan Collins, and an ensemble cast.

1982–88 "Cagney & Lacey," police drama, starring Tyne Daly and Sharon Gless.

1982–89 "Family Ties," situation comedy, starring Meredith Baxter-Birney, Michael Gross, Michael J. Fox, Justine Bateman, and Tina Yothers.

1982–90 "Newhart," situation comedy, starring Bob Newhart, Mary Frann, Tom Poston, Julia Duffy, and Peter Scolari.

1982–present "Cheers," situation comedy, starring Ted Danson, Shelley Long (1982–87), Rhea Perlman, Nicholas Colasanto (1982–85), George Wendt, John Ratzenberger, Kelsey Grammer (1984–present), Woody Harrelson (1985), and Kirstie Alley (1987-present).

1982–present "Late Night with David Letterman," late-night talk show. David Letterman is host and Paul Shaffer, bandleader.

Book Publishing

See also Printing and Reproduction Technology in Technology section.

c. 1438 German goldsmith and printer Johannes Gutenberg (c. 1398–c. 1468) invents the printing press, thus beginning modern book publishing. (He publishes his famous Mazarin Bible in 1455.)

1461 Woodcuts are in use to illustrate books by or before this date.

c. 1466 Handbills or broadsheets advertising books are circulated.

c. 1474 Englishman William Caxton (c. 1421–91) is first to print in English instead of Latin.

1478 Copper engravings appear in published works; they are especially suited to reproducing detailed illustrations such as maps.

c. 1480 Printer's colophon (date and place of printing; printer's name and advertisement) begins to appear on title page with title and author.

1490–1597 Aldine Press in Venice prints its Aldine editions. In addition to printing quality editions, it makes first use of italic type.

1500 Latin, language for at least 75 percent of books until this date, begins to decline. Vernacular languages become more common in publishing.

1538 British government opposes "naughty printed books" and requires official licensing prior to publication of each book in English.

▶

1543 Catholic Church decrees that no book may be published without its permission. It publishes list of banned books.

1564 First book trade catalog appears, at Frankfurt Book Fair.

1586 British controls on publishing are tightened when printing is confined to London and to one press each at Oxford and Cambridge universities.

1593 Elzevir publishing company is founded in Netherlands.

1594 First Shakespearean play, *Titus Andronicus*, is published by publishing "pirate" who later prints faulty version of *Romeo and Juliet*. Pirate works from notes taken during a performance.

1640 National press of France is established by monarchy.

1640 Cambridge Press of Massachusetts begins publishing. First book published in colonial America is *Whole Booke of Psalmes*.

c. 1647 Boston shopkeeper becomes first American bookseller when he adds books to his merchandise.

1709 England's Copyright Act of 1709 is first such act to be passed anywhere. It protects creative property of author.

1724 Founding of Longmans publishing company in Great Britain.

c. 1735 Decline in practice of "patronage," whereby authors are supported by wealthy patrons (to whom they dedicate their work). Henry Fielding dedicates his 1737 *Historical Register* to "the public at large."

1750 English book publishers produce about 100 new titles a year at this time.

1766 Swedish government forbids book censorship. Other governments soon follow.

1790 New United States of America adopts first copyright legislation, to protect U.S. authors from unwarranted publication of their works.

1792 Founding of U.S. publishing company J. B. Lippincott & Co.

1798 Thomas Nelson & Sons is founded in Great Britain.

1800s Technological advances in printing presses during this century revolutionize book publishing industry.

1817 Harper & Brothers publishing company is founded in United States

1819 Founding of William Collins, Sons & Co. in Great Britain.

c. 1820 Cloth is replacing leather for book casings. Publishers begin to issue hardbound books.

1820 Founding of British publisher W. & R. Chambers Ltd.

c. 1825 Subscription book selling is profitable business in United States by this time.

1828 American publisher John Wiley & Sons is founded.

1833 French publisher Garnier is founded.

1836 G. P. Putnam's Sons is founded in United States.

1844 Macmillan & Co., a bookshop, enters publishing business by successfully publishing textbooks, then popular novels.

1846 Charles Scribner's Sons is founded in United States.

c. 1850 Dime novels (in the United States) and "penny dreadfuls" (in Great Britain)—small, cheap sensational paperback novels—are published successfully.

▶

Bestsellers 1945–1989

1945 **Fiction**: *Forever Amber* by Kathleen Winsor; *The Robe* by Lloyd Douglas; *The Black Rose* by Thomas Costain; *The White Tower* by James Ullman; *Cass Timberlane* by Sinclair Lewis
Nonfiction: *Brave Men* by Ernie Pyle; *Dear Sir* by Juliet Lowell; *Up Front* by Bill Mauldin; *Black Boy* by Richard Wright; *Try and Stop Me* by Bennett Cerf

1946 **Fiction**: *The King's General* by Daphne du Maurier; *This Side of Innocence* by Taylor Caldwell; *The River Road* by Frances Keyes; *The Miracle of the Bells* by Russell Janney; *The Hucksters* by Frederic Wakeman
Nonfiction: *The Egg and I* by Betty MacDonald; *Peace of Mind* by Joshua Liebman; *As He Saw It* by Elliott Roosevelt; *The Roosevelt I Knew* by Frances Perkins; *Last Chapter* by Ernie Pyle

1947 **Fiction**: *The Miracle of the Bells* by Russell Janney; *The Moneyman* by Thomas Costain; *Gentlemen's Agreement* by Laura Hobson; *Lydia Bailey* by Kenneth Roberts; *The Vixens* by Frank Yerby
Nonfiction: *Peace of Mind* by Joshua Liebman; *Information Please Almanac,* 1947 edition by John Kieran; *Inside USA* by John Gunther; *A Study of History* by Arnold Toynbee; *Speaking Frankly* by James Byrnes

1948 **Fiction**: *The Big Fisherman* by Lloyd Douglas; *The Naked and the Dead* by Norman Mailer; *Dinner at Antoine's* by Frances Keyes; *The Bishop's Mantle* by Agnes Turnbull; *Tomorrow Will Be Better* by Betty Smith
Nonfiction: *Crusade in Europe* by Dwight D. Eisenhower; *How to Stop Worrying and Start Living* by Dale Carnegie; *Peace of Mind* by Joshua Liebman; *Sexual Behavior in the Human Male* by Alfred Kinsey; *Wine, Women, and Words* by Billy Rose

1949 **Fiction**: *The Egyptian* by Mika Waltari; *The Big Fisherman* by Lloyd Douglas; *Mary* by Sholem Asch; *A Rage to Live* by John O'Hara; *Point of No Return* by John Marquand
Nonfiction: *White Collar Zoo* by Clare Barnes, Jr. *How to Win at Canasta* by Oswald Jacoby; *The Seven Storey Mountain* by Thomas Merton; *Home Sweet Zoo* by Clare Barnes, Jr.; *Cheaper by the Dozen* by Frank Gilbreth, Jr. and Ernestine Gilbreth Carey

1950 **Fiction**: *The Cardinal* by Henry Robinson; *Joy Street* by Frances Keyes; *Across the River and into the Trees* by Ernest Hemingway; *The Wall* by John Hersey; *Star Money* by Kathleen Winsor
Nonfiction: *Betty Crocker's Picture Cook Book*; *Look Younger, Live Longer* by Gayelord Hauser; *How I Raised Myself from Failure to Success in Selling* by Frank Bettger; *Kon-Tiki* by Thor Heyerdahl

1951 **Fiction**: *From Here to Eternity* by James Jones; *The Caine Mutiny* by Herman Wouk; *Moses* by Sholem Asch; *The Cardinal* by Henry Robinson; *A Woman Called Fancy* by Frank Yerby
Nonfiction: *Look Younger, Live Longer* by Gayelord Hauser; *Betty Crocker's Picture Cook Book*; *Washington Confidential* by Jack Lait and Lee Mortimer; *Better Homes and Gardens Garden Book*; *Better Homes and Gardens Handyman's Book*

1952 **Fiction**: *The Silver Chalice* by Thomas Costain; *The Caine Mutiny* by Herman Wouk; *East of Eden* by John Steinbeck; *My Counsin Rachel* by Daphne du Maurier; *Steamboat Gothic* by Frances Keyes
Nonfiction: *The Holy Bible: Revised Standard Version; A Man Called Peter* by Catherine Marshall; *U.S.A. Confidential* by Jack Lait and Lee Mortimer; *The Sea Around Us* by Rachel Carson; *Tallulah* by Tallulah Bankhead

1953 **Fiction**: *The Robe* by Lloyd Douglas; *The Silver Chalice* by Thomas Costain; *Désirée* by Annemarie Selinko; *Battle Cry* by Leon Uris; *From Here to Eternity* by James Jones
Nonfiction: *The Holy Bible: Revised Standard Edition; The Power of Positive Thinking* by Norman Vincent Peale; *Sexual Behavior in the Human Female* by Alfred Kinsey; *Angel Unaware* by Dale Evans Rogers; *Life Is Worth Living* by Fulton J. Sheen

1954 **Fiction**: *Not As a Stranger* by Morton Thompson; *Mary Anne* by Daphne du Maurier; *Love Is Eternal* by Irving Stone; *The Royal Box* by Frances Keyes; *The Egyptian* by Mika Waltari
Nonfiction: *The Holy Bible: Revised Standard Edition*; *The Power of Positive Thinking* by Norman Vincent Peale; *Better Homes and Gardens New Cook Book*; *Betty Crocker's Good and Easy Cook Book*; *The Tumult and the Shouting* by Grantland Rice

1955 **Fiction**: *Marjorie Moringstar* by Herman Wouk; *Auntie Mame* by Patrick Dennis; *Andersonville* by MacKinlay Kantor; *Bonjour Tristesse* by Françoise Sagan; *The Man in the Gray Flannel Suit* by Sloan Wilson
Nonfiction: *Gift from the Sea* by Anne Morrow Lindberg; *The Power of Positive Thinking* by Norman Vincent Peale; *The Family of Man* by Edward Steichen; *A Man Called Peter* by Catherine Marshall; *How to Live 365 Days a Year* by John Schindler

1956 **Fiction**: *Don't Go Near The Water* by William Brinkley; *The Last Hurrah* by Edwin O'Connor;

Bestsellers 1945–1989

Peyton Place by Grace Metalious; *Auntie Mame* by Patrick Dennis; *Eloise* by Kay Thompson
Nonfiction: *Arthritis and Common Sense* by Dan Alexander; *Webster's New World Dictionary of the English Language* edited by David Guralnik; *Betty Crocker's Picture Cook Book*; *Etiquette* by Frances Benton; *Better Homes and Gardens Barbecue Book*

1957 **Fiction**: *By Love Possessed* by James Cozzens; *Peyton Place* by Grace Metalious; *Compulsion* by Meyer Levin; *Rally Round the Flag, Boys!* by Max Shulman; *Blue Camellia* by Frances Keyes
Nonfiction: *Kids Say the Darndest Things!* by Art Linkletter; *The FBI Story* by Don Whitehead; *Stay Alive All Your Life* by Norman Vincent Peale; *To Live Again* by Catherine Marshall; *Better Homes and Gardens Flower Arranging*

1958 **Fiction**: *Doctor Zhivago* by Boris Pasternak; *Anatomy of a Murder* by Robert Traver; *Lolita* by Vladimir Nabokov; *Around the World with Auntie Mame* by Patrick Dennis; *From the Terrace* by John O'Hara
Nonfiction: *Kids Say the Darndest Things!* by Art Linkletter; *'Twixt Twelve and Twenty* by Pat Boone; *Only in America* by Harry Golden; *Masters of Deceit* by J. Edgar Hoover; *Please Don't Eat the Daisies* by Jean Kerr

1959 **Fiction**: *Exodus* by Leon Uris; *Doctor Zhivago* by Boris Pasternak; *Hawaii* by James Michener; *Advise and Consent* by Allen Drury; *Lady Chatterley's Lover* by D.H. Lawrence
Nonfiction: *'Twixt Twelve and Twenty* by Pat Boone; *Folk Medicine* by D. C. Jarvis; *For 2 Cents Plain* by Harry Golden; *The Status Seekers* by Vance Packard; *Act One* by Moss Hart

1960 **Fiction**: *Advise and Consent* by Allen Drury; *Hawaii* by James Michener; *The Leopard* by Giuseppe de Lampedusa; *The Chapman Report* by Irving Wallace; *Ourselves to Know* by John O'Hara
Nonfiction: *Folk Medicine* by D. C. Jarvis; *Better Homes and Gardens First Aid for Your Family*; *The General Foods Cook Book*; *May This House Be Safe From Tigers* by Alexander King; *Better Homes and Gardens Dessert Book*

1961 **Fiction**: *The Agony and the Ecstasy* by Irving Stone; *Franny and Zooey* by J. D. Salinger; *To Kill a Mockingbird* by Harper Lee; *Mila 18* by Leon Uris; *The Carpetbaggers* by Harold Robbins
Nonfiction: *The New English Bible: The New Testament*; *The Rise and Fall of the Third Reich* by William Shirer; *Better Homes and Gardens Sewing Book*; *Casserole Cook Book*; *A Nation of Sheep* by William Lederer

1962 **Fiction**: *Ship of Fools* by Katherine Anne Porter; *Dearly Beloved* by Anne Morrow Lindbergh; *A Shade of Difference* by Allen Drury; *Youngblood Hawke* by Herman Wouk; *Franny and Zooey* by J. D. Salinger
Nonfiction: *Calories Don't Count* by Dr. Herman Taller; *The New English Bible: The New Testament*; *Better Homes and Gardens Cook Book*; *O Ye Jigs and Juleps!* by Virginia Hudson; *Happiness Is a Warm Puppy* by Charles M. Schulz

1963 **Fiction**: *The Shoes of the Fisherman* by Morris West; *The Group* by Mary McCarthy; *Raise High the Roof Beam, Carpenters, and Seymour* by J. D. Salinger; *Caravans* by James Michener; *Elizabeth Appleton* by John O'Hara
Nonfiction: *Happiness Is a Warm Puppy* by Charles M. Schulz; *Security Is a Thumb and a Blanket* by Charles M. Schulz; *JFK: The Man and the Myth* by Victor Lasky; *Profiles in Courage* by John F. Kennedy; *O Ye Jigs and Juleps!* by Virginia Hudson

1964 **Fiction**: *The Spy Who Came in from the Cold* by John Le Carré; *Candy* by Terry Southern; *Herzog* by Saul Bellow; *Armageddon* by Leon Uris; *The Man* by Irving Wallace
Nonfiction: *Four Days* by American Heritage and United Press International; *I Need All the Friends I Can Get* by Charles M. Schulz; *Profiles in Courage* by John F. Kennedy; *In His Own Write* by John Lennon; *Christmas Is Together-time* by Charles M. Schulz.

1965 **Fiction**: *The Source* by James Michener; *Up the Down Staircase* by Bel Kaufman; *Herzog* by Saul Bellow; *The Looking Glass War* by John Le Carré; *The Green Berets* by Robin Moore
Nonfiction: *How to Be a Jewish Mother* by Dan Greenburg; *A Gift of Prophecy* by Ruth Montgomery; *Games People Play* by Eric Berne; *World Aflame* by Billy Graham; *Happiness Is a Dry Martini* by Johnny Carson

1966 **Fiction**: *Valley of the Dolls* by Jacqueline Susann; *The Adventurers* by Harold Robbins; *The Secret of Santa Vittoria* by Robert Crichton; *Capable of Honor* by Allen Drury; *The Double Image* by Helen MacInnes
Nonfiction: *How to Avoid Probate* by Norman Dacey; *Human Sexual Response* by Masters and Johnson; *In Cold Blood* by Truman Capote; *Games People Play* by Eric Berne; *A Thousand Days* by Arthur Schlesinger, Jr.

1967 **Fiction**: *The Arrangement* by Elia Kazan; *The Confessions of Nat Turner* by William Styron; *The Chosen* by Chaim Potok; *Topaz* by Leon Uris; *Christy*

Bestsellers 1945–1989

by Catherine Marshall
Nonfiction: *Death of A President* by William Manchester; *Misery Is a Blind Date* by Johnny Carson; *Games People Play* by Eric Berne; *Stanyan Street and Other Sorrows* by Rod McKuen; *A Modern Priest Looks at His Outdated Church* by Father James Kavanaugh

1968 **Fiction**: *Airport* by Arthur Hailey; *Couples* by John Updike; *The Salzburg Connection* by Helen MacInnes; *A Small Town in Germany* by John Le Carré; *Testimony of Two Men* by Taylor Caldwell
Nonfiction: *Better Homes and Gardens New Cook Book*; *The Random House Dictionary of the English Language: College Edition*; *Listen to the Warm* by Rod McKuen; *Between Parent and Child* by Haim Ginott; *Lonesome Cities* by Rod McKuen

1969 **Fiction**: *Portnoy's Complaint* by Philip Roth; *The Godfather* by Mario Puzo; *The Love Machine* by Jacqueline Susann; *The Inheritors* by Harold Robbins; *The Andromeda Strain* by Michael Crichton
Nonfiction: *American Heritage Dictionary of the English Language*; *In Someone's Shadow* by Rod McKuen; *The Peter Principle* by Laurence Peter and Raymond Hull; *Between Parent and Teenager* by Haim Ginott; *The Graham Kerr Cook Book* by Graham Kerr, the Galloping Gourmet

1970 **Fiction**: *Love Story* by Erich Segal; *The French Lieutenant's Woman* by John Fowles; *Islands in the Stream* by Ernest Hemingway; *The Crystal Cave* by Mary Stewart; *Great Lion of God* by Taylor Caldwell
Nonfiction: *Everything You Wanted to Know About Sex but Were Afraid to Ask* by David Reuben, M.D.; *The New English Bible*; *The Sensuous Woman* by "J"; *Better Homes and Gardens Fondue and Tabletop Cooking*; *Up the Organization* by Robert Townsend

1971 **Fiction**: *Wheels* by Arthur Hailey; *The Exorcist* by William Blatty; *The Passions of the Mind* by Irving Stone; *The Day of the Jackal* by Frederick Forsyth; *The Betsy* by Harold Robbins
Nonfiction: *The Sensous Man* by "M"; *Bury My Heart at Wounded Knee* by Dee Brown; *Better Homes and Gardens Blender Cook Book*; *I'm OK, You're OK* by Thomas Harris; *Any Woman Can!* by David Reuben, M.D.

1972 **Fiction**: *Jonathan Livingston Seagull* by Richard Bach; *August 1914* by Alexander Solzhenitsyn; *The Odessa File* by Frederick Forsyth; *The Day of the Jackal* by Frederick Forsyth; *The Word* by Irving Wallace
Nonfiction: *The Living Bible* by Kenneth Taylor; *I'm OK, You're OK* by Thomas Harris; *Open Marriage* by Nena and George O'Neill; *Harry S Truman* by Margaret Truman; *Dr. Atkins' Diet Revolution* by Robert Atkins

1973 **Fiction**: *Jonathan Livingston Seagull* by Richard Bach; *Once Is Not Enough* by Jacqueline Susann; *Breakfast of Champions* by Kurt Vonnegut, Jr.; *The Odessa File* by Frederick Forsyth; *Burr* by Gore Vidal
Nonfiction: *The Living Bible* by Kenneth Taylor; *Dr. Atkins' Diet Revolution* by Richard Atkins; *I'm OK, You're OK* by Thomas Harris; *The Joy of Sex* by Alex Comfort; *Weight Watchers Program Cookbook* by Jean Nidetch

1974 **Fiction**: *Centennial* by James Michener; *Watership Down* by Richard Adams; *Jaws* by Peter Benchley; *Tinker, Tailor, Soldier, Spy* by John Le Carré; *Something Happened* by Joseph Heller
Nonfiction: *The Total Woman* by Marabel Morgan; *All the President's Men* by Carl Bernstein and Bob Woodward; *Plain Speaking* by Merle Miller; *More Joy* by Alex Comfort; *Alistair Cooke's America* by Alistair Cooke

1975 **Fiction**: *Ragtime* by E. L. Doctorow; *The Moneychangers* by Arthur Hailey; *Curtain* by Agatha Christie; *Looking for Mister Goodbar* by Judith Rossner; *The Choirboys* by Joseph Wambaugh
Nonfiction: *Angels: God's Secret Agents* by Billy Graham; *Winning Through Intimidation* by Robert Ringer; *TM: Discovering Energy and Overcoming Stress* by Harold Bloomfield; *The Ascent of Man* by Jacob Bronowski; *Sylvia Porter's Total Money Book* by Sylvia Porter

1976 **Fiction**: *Trinity* by Leon Uris; *Sleeping Murder* by Agatha Christie; *Dolores* by Jacqueline Susann; *The Deep* by Peter Benchley; *1876* by Gore Vidal
Nonfiction: *The Final Days* by Bob Woodward and Carl Bernstein; *Roots* by Alex Haley; *Your Erroneous Zones* by Dr. Wayne Dyer; *Passages* by Gail Sheehy; *Born Again* by Charles Colson

1977 **Fiction**: *The Silmarillion* by J.R.R. Tolkien; *The Thorn Birds* by Colleen McCullough; *Illusions* by Richard Bach; *The Honourable Schoolboy* by John Le Carré; *Oliver's Story* by Erich Segal
Nonfiction: *Roots* by Alex Haley; *Looking Out for #1* by Robert Ringer; *All Things Wise and Wonderful* by James Herriot; *Your Erroneous Zones* by Dr. Wayne Dyer; *The Book of Lists* by David Wallechinsky

1978 **Fiction**: *Chesapeake* by James Michener; *War and Remembrance* by Herman Wouk; *Fools Die* by Mario Puzo; *Bloodlines* by Sidney Sheldon; *Scruples* by Judith Krantz
Nonfiction: *If Life Is a Bowl of Cherries—What Am I*

Bestsellers 1945–1989

Doing in the Pits? by Erma Bombeck; *Gnomes* by Wil Huygen; *The Complete Book of Running* by James Fixx; *Mommie Dearest* by Christina Crawford; *Memoirs of Richard Nixon* by Richard Nixon

1979 **Fiction**: *The Matarese Circle* by Robert Ludlum; *Sophie's Choice* by William Styron; *Overload* by Arthur Hailey; *Memories of Another Day* by Harold Robbins; *Jailbird* by Kurt Vonnegut, Jr.
Nonfiction: *Aunt Erma's Cope Book* by Erma Bombeck; *The Complete Scarsdale Medical Diet* by Herman Tarnower; *How to Prosper During the Coming Bad Years* by Howard Ruff; *Cruel Shoes* by Steve Martin; *The Pritiken Program for Diet and Exercise* by Nathan Pritiken

1980 **Fiction**: *The Covenant* by James Michener; *The Bourne Identity* by Robert Ludlum; *Rage of Angels* by Sidney Sheldon; *Princess Daisy* by Judith Krantz; *Firestarter* by Stephen King
Nonfiction: *Crisis Investing* by Douglas Casey; *Cosmos* by Carl Sagan; *Free to Choose* by Milton and Rose Friedman; *Anatomy of an Illness* by Norman Cousins; *Thy Neighbor's Wife* by Gay Talese

1981 **Fiction**: *Noble House* by James Clavell; *The Hotel New Hampshire* by John Irving; *Cujo* by Stephen King; *An Indecent Obsession* by Colleen McCullough; *Gorky Park* by Martin Cruz Smith
Nonfiction: *Beverly Hills Diet* by Judy Mazel; *The Lord God Made Them All* by James Herriot; *"Never Say Diet" Book* by Richard Simmons; *A Light in the Attic* by Shel Silverstein; *Cosmos* by Carl Sagan

1982 **Fiction**: *ET: The Extraterrestrial Storybook* by William Kotzwinkle; *Space* by James Michener; *The Parsifal Mosaic* by Robert Ludlum; *Master of the Game* by Sidney Sheldon; *Mistral's Daughter* by Judith Krantz
Nonfiction: *Jane Fonda's Workout Book* by Jane Fonda; *Living, Loving, and Learning* by Leo Buscaglia; *And More by* Andy Rooney; *Better Homes and Gardens New Cook Book; Life Extension* by Durk Pearson

1983 **Fiction**: *The Return of the Jedi Storybook* by Joan Vinge; *Poland* by James Michener; *Pet Sematary* by Stephen King; *The Little Drummer Girl* by John le Carré; *Christine* by Stephen King
Nonfiction: *In Search of Excellence* by Tom Peters; *Megatrends* by John Naisbitt; *Motherhood: The Second Oldest Profession* by Erma Bombeck; *The One-Minute Manager* by Kenneth Blanchard; *Jane Fonda's Workout Book* by Jane Fonda

1984 **Fiction**: *The Talisman* by Stephen King; *The Aquitaine Progression* by Robert Ludlum; *The Sicilian* by Mario Puzo; *Love and War* by John Jakes; *The Butter Battle Book* by Dr. Seuss
Nonfiction: *Iacocca: An Autobiography* by Lee Iacocca; *Loving Each Other* by Leo Buscaglia; *Eat to Win* by Robert Haas, M.D.; *Pieces of My Mind* by Andrew Rooney; *Weight Watchers' Fast and Fabulous Cookbook*

1985 **Fiction**: *The Mammoth Hunters* by Jean Auel; *Texas* by James Michener; *Lake Wobegon Days* by Garrison Keillor; *If Tomorrow Comes* by Sidney Sheldon; *Skeleton Crew* by Stephen King
Nonfiction: *Iacocca: An Autobiography* by Lee Iacocca; *Yeager: An Autobiography* by Chuck Yeager; *Elvis and Me* by Priscilla Presley; *Fit for Life* by Harvey and Marilyn Diamond; *The Be-Happy Attitudes* by Robert Schuller

1986 **Fiction**: *It* by Stephen King; *Red Storm Rising* by Tom Clancy; *Whirlwind* by James Clavell; *The Bourne Supremacy* by Robert Ludlum; *Hollywood Husbands* by Jackie Collins
Nonfiction: *Fatherhood* by Bill Cosby; *Fit for Life* by Harvey and Marilyn Diamond; *His Way* by Kitty Kelly; *The Rotation Diet* by Martin Katahn; *You're Only Old Once* by Dr. Seuss

1987 **Fiction**: *The Tommyknockers* by Stephen King; *Patriot Games* by Tom Clancy; *Kaleidoscope* by Danielle Steele; *Misery* by Stephen King; *Leaving Home* by Garrison Keillor
Nonfiction: *Time Flies* by Bill Cosby; *Spy Catcher* by Peter Wright; *Family: The Ties That Bind and Gag* by Erma Bombeck; *Veil: Secret Wars of the CIA* by Bob Woodward; *A Day in the Life of America* by Rick Smolan

1988 **Fiction**: *The Cardinal of the Kremlin* by Tom Clancy; *The Sands of Time* by Sidney Sheldon; *Zoya* by Danielle Steel; *The Icarus Agenda* by Robert Ludlum; *Alaska* by James Michener
Nonfiction: *The Eight-Week Cholesterol Diet* by Robert Kowalski; *Talking Straight* by Lee Iacocca; *A Brief History of Time* by Stephen Hawking; *Trump* by Donald Trump; *Gracie: A Love Story* by George Burns

1989 **Fiction**: *The Joy Luck Club* by Amy Tan; *The Satanic Verses* by Salman Rushdie; *While My Pretty One Sleeps* by Mary Higgins Clark; *The Russia House* by John Le Carré; *Star* by Danielle Steel
Nonfiction: *A Brief History of Time* by Stephen Hawking; *All I Really Needed to Know I Learned in Kindergarten* by Robert Fulghum; *Wealth Without Risk* by Charles Givens; *The 8-Week Cholesterol Cure*, Revised Edition by Robert Kowalski; *The T-Factor Diet* by Martin Katahn

1852 Immediately following U.S. publication of *Uncle Tom's Cabin* by American author Harriet Beecher Stowe (1811–96), 1.5 million "pirated" copies appear in England, selling for as little as a sixpence (with no royalties going to author).

1860 U.S. Government Printing Office is founded.

1865 Harper's turns down manuscript of *Alice in Wonderland* on advice of two readers.

c. 1875 First literary agencies begin.

1884 Society of Authors is founded in England to address fair contract procedures and royalty payments.

1885 Berne Convention approves international system of copyright providing protection for the author's lifetime plus fifty years. (United States and Russia do not accept the convention, so copyright registration lasts only 28 years with another 28-year renewal possible in United States.)

1890 First pictorial dust jacket appears, on book by U.S. author William Dean Howells (1837–1920), *The Shadow of a Dream*.

1896 New York Supreme Court rules that authors may have access to sales records for their books, thus ending publishers' alleged juggling of books to reduce royalty payments.

1897 American publisher Frank Nelson Doubleday (1862–1934) begins first of several publishing companies: Doubleday, McClure and Co. His writers eventually include such authors as Rudyard Kipling (1865–1936), Joseph Conrad (1857–1924), and Sinclair Lewis (1885–1951).

1900 British book publishers are producing some 6,000 new titles a year by this time.

1900 American Booksellers is organized to increase promotion and distribution of books.

1909 Introduction in United States of "on-approval" method of selling books by mail.

1912 U.S. Authors' League is founded.

1915 Alfred A. Knopf publishing company is founded. By the time Random House acquires it in 1960, Knopf is publisher of quality books.

1919 Harcourt, Brace and Howe is founded in New York City. It begins operations in a secretary's apartment while founders Alfred Harcourt (1881–1954) and Donald Brace (1881–1955) look for offices.

1920 National Association of Book Publishers (NABP) is organized in United States.

1922 "Bookleggers" do brisk, profitable business in banned books as censorship in United States market becomes stiffer.

1922 *Reader's Digest* magazine is founded. It eventually publishes series of condensed books.

1924 American publishers Richard Simon (1900–61) and Max Schuster (b. 1897) found Simon and Schuster. Their first published work is a crossword puzzle book.

1925 Random House is founded by U.S. publishers Bennett Cerf (1898–1971) and Donald Klopfer (b. 1902), originally as Modern Library, Inc.

1926 Book-of-the-Month Club and Literary Guild are founded.

1939 Founding of Pocket Books, U.S. paperback publisher.

1939–45 American book publishers undergo rapid expansion during years of World War II. Sharp increase in demand for books spreads to other free-world countries after the war.

1942 British government ignores nationwide wartime shortage and releases additional paper for reprinting 'important" books.

1946 Bantam Books is founded as paperback publisher.

▶

c. 1950 Literary scouts, who alert publishers to promising new books, appear.

1950s Popularity of paperback editions spreads worldwide.

1952 Signet-Mentor paperbacks are launched.

1955 United States accepts terms of Universal Copyright Convention as originally proposed by UNESCO in 1952. USSR signs in 1973, China in 1984.

c. 1968 Bookstore chains, such as B. Dalton, begin to grow in importance in the book trade.

1976 U.S. Senate passes major revisions in copyright law to deal with proliferation of electronic copying and storing devices. Reproduction permissions and royalty payments are strictly defined.

1980 *Thor Power Tool Company* tax ruling alters reporting of value of publishers' inventories of unsold books. In reaction, publishing houses prepare to destroy or sell at discount millions of warehoused books.

1980 American book publishers explore possible industry involvement in new electronic innovations such as videodiscs, audiocassettes, and computer programs.

1980s American book industry undergoes period of uncertainty and widespread corporate takeovers. For much of decade, larger publishing houses buy up smaller ones, which often are ailing from such industrywide problems as rapidly increasing costs, tight money, competition from television and other entertainment media, and reduced demand. By end of decade, however, even largest of U.S. publishers become targets of successful corporate takeovers—this time by foreign publishing concerns.

1984 Total annual receipts by U.S. book publishers reach $9.12 billion. Approximately 2.168 billion books are sold.

1985 Bookstores in United States and Canada number 23,749.

1989 Second edition of venerable *Oxford English Dictionary* is published, replacing original 1928 edition and supplements.

Motion Pictures Industry

1887 Celluloid roll film is developed for still photography. It later makes motion-picture photography possible.

1889 William Dickson (1860–1937), employed at Thomas A. Edison laboratories, develops kinetograph (patented 1893), a viable motion-picture camera. His kinetoscope (marketed 1894) plays short film to single viewer (peep show).

1893 "Black Maria," early film studio, is built at Edison's New Jersey laboratory. Covered with tar paper, building swivels to move with sun.

1895 French inventors Auguste Lumière (1862–1954) and Louis Lumière (1864–1948) demonstrate first commercially successful projector, the *cinématographe*. It gives rise to term *cinema*.

1895–1902 Permanent film theaters are established in Paris, New Orleans, and Los Angeles.

1896 Founding of French Pathé film company, which later becomes a leading producer of news film.

▶

Academy Awards

1927–28 Emil Jannings, best actor, *The Way of All Flesh*; Janet Gaynor, best actress, *Seventh Heaven*; *Wings*.

1928–29 Warner Baxter, best actor, *In Old Arizona*; Mary Pickford, best actress, *Coquette*; *Broadway Melody*, best picture.

1929–30 George Arliss, best actor, *Disraeli*; Norma Shearer, best actress, *The Divorcée*; best picture, *All Quiet on the Western Front*.

1930–31 Lionel Barrymore, best actor, *A Free Soul*; Marie Dressler, best actress, *Min and Bill*; best picture, *Cimarron*.

1931–32 Fredric March, best actor, *Dr. Jekyll and Mr. Hyde*, and Wallace Beery, best actor, *The Champ*; Helen Hayes, best actress, *Sin of Madelon Claudet*; best picture, *Grand Hotel*.

1932–33 Charles Laughton, best actor, *The Private Life of Henry VIII*; Katharine Hepburn, best actress, *Morning Glory*; best picture, *Cavalcade*.

1934 Clark Gable, best actor, *It Happened One Night*; Claudette Colbert, best actress, *It Happened One Night*; best picture, *It Happened One Night*.

1935 Victor McLaglen, best actor, *The Informer*; Bette Davis, best actress, *Dangerous*; best picture, *Mutiny on the Bounty*.

1936 Paul Muni, best actor, *The Story of Louis Pasteur*; Luise Rainer, best actress, *The Great Ziegfeld*; best picture, *The Great Ziegfeld*.

1937 Spencer Tracy, best actor, *Captains Courageous*; Luise Rainer, best actress, *The Good Earth*; best picture, *The Life of Emile Zola*.

1938 Spencer Tracy, best actor, *Boys Town*; Bette Davis, best actress, *Jezebel*; best picture, *You Can't Take It With You*.

1939 Robert Donat, best actor, *Goodbye, Mr. Chips*; Vivien Leigh, best actress, *Gone with the Wind*; best picture, *Gone with the Wind*.

1940 James Stewart, best actor, *The Philadelphia Story*; Ginger Rogers, best actress, *Kitty Foyle*; best picture, *Rebecca*.

1941 Gary Cooper, best actor, *Sergeant York*; Joan Fontaine, best actress, *Suspicion*; best picture, *How Green Was My Valley*.

1942 James Cagney, best actor, *Yankee Doodle Dandy*; Greer Garson, best actress, *Mrs. Miniver*; best picture, *Mrs. Miniver*.

1943 Paul Lukas, best actor, *Watch on the Rhine*; Jennifer Jones, best actress, *The Song of Bernadette*; best picture, *Casablanca*.

1944 Bing Crosby, best actor, *Going My Way*; Ingrid Bergman, best actress, *Gaslight*; best picture, *Going My Way*.

1945 Ray Milland, best actor, *The Lost Weekend*; Joan Crawford, best actress, *Mildred Pierce*; best picture, *The Lost Weekend*.

1946 Fredric March, best actor, *The Best Years of Our Lives*; Olivia DeHavilland, best actress, *To Each His Own*; best picture, *The Best Years of Our Lives*.

1947 Ronald Colman, best actor, *A Double Life*; Loretta Young, best actress, *The Farmer's Daughter*; best picture, *Gentleman's Agreement*.

1948 Laurence Olivier, best actor, *Hamlet*; Jane Wyman, best actress, *Johnny Belinda*; best picture, *Hamlet*.

1949 Broderick Crawford, best actor, *All the King's Men*; Olivia DeHavilland, best actress, *The Heiress*; best picture, *All the King's Men*.

1950 José Ferrer, best actor, *Cyrano de Bergerac*; Judy Holliday, best actress, *Born Yesterday*; best picture, *All About Eve*.

1951 Humphrey Bogart, best actor, *The African Queen*; Vivien Leigh, best actress, *A Streetcar Named Desire*; best picture, *An American in Paris*.

1952 Gary Cooper, best actor, *High Noon*; Shirley Booth, best actress, *Come Back, Little Sheba*; best picture, *The Greatest Show on Earth*.

1953 William Holden, best actor, *Stalag 17*; Audrey Hepburn, best actress, *Roman Holiday*; best picture, *From Here to Eternity*.

1954 Marlon Brando, best actor, *On the Waterfront*; Grace Kelly, best actress, *The Country Girl*; best picture, *On the Waterfront*.

1955 Ernest Borgnine, best actor, *Marty*; Anna Magnani, best actress, *The Rose Tattoo*; best picture, *Marty*.

1956 Yul Brynner, best actor, *The King and I*; Ingrid Bergman, best actress, *Anastasia*; best picture, *Around the World in 80 Days*.

1957 Alec Guinness, best actor, *The Bridge on the River Kwai*; Joanne Woodward, best actress, *The Three Faces of Eve*; best picture, *The Bridge on the River Kwai*.

1958 David Niven, best actor, *Separate Tables*; Susan Hayward, best actress, *I Want To Live*; best picture, *Gigi*.

1959 Charlton Heston, best actor, *Ben-Hur*; Simone Signoret, best actress, *Room at the Top*; best picture, *Ben-Hur*.

1960 Burt Lancaster, best actor, *Elmer Gantry*; Elizabeth Taylor, best actress, *Butterfield 8*; best picture, *The Apartment*.

Academy Awards

1961 Maximilian Schell, best actor, *Judgment at Nuremburg*; Sophia Loren, best actress, *Two Women*; best picture, *West Side Story*.

1962 Gregory Peck, best actor, *To Kill a Mockingbird*; Anne Bancroft, best actress, *The Miracle Worker*; best picture, *Lawrence of Arabia*.

1963 Sidney Poitier, best actor, *Lilies of the Field*; Patricia Neal, best actress, *Hud*; best picture, *Tom Jones*.

1964 Rex Harrison, best actor, *My Fair Lady*; Julie Andrews, best actress, *Mary Poppins*; best picture, *My Fair Lady*.

1965 Lee Marvin, best actor, *Cat Ballou*; Julie Christie, best actress, *Darling*; best picture, *The Sound of Music*.

1966 Paul Scofield, best actor, *A Man for All Seasons*; Elizabeth Taylor, best actress, *Who's Afraid of Virginia Woolf?*; best picture, *A Man for All Seasons*.

1967 Rod Steiger, best actor, *In the Heat of the Night*; Katharine Hepburn, best actress, *Guess Who's Coming to Dinner?*; best picture, *In the Heat of the Night*.

1968 Cliff Robertson, best actor, *Charly*; Katharine Hepburn, best actress, *The Lion in Winter*, and Barbra Streisand, best actress, *Funny Girl*; best picture, *Oliver!*

1969 John Wayne, best actor, *True Grit*; Maggie Smith, best actress, *The Prime of Miss Jean Brodie*; best picture, *Midnight Cowboy*.

1970 George C. Scott, best actor, *Patton* (refused); Glenda Jackson, best actress, *Women in Love*; best picture, *Patton*.

1971 Gene Hackman, best actor, *The French Connection*; Jane Fonda, best actress, *Klute*; best picture, *The French Connection*.

1972 Marlon Brando, best actor, *The Godfather* (refused); Liza Minnelli, best actress, *Cabaret*; best picture, *The Godfather*.

1973 Jack Lemmon, best actor, *Save the Tiger*; Glenda Jackson, best actress, *A Touch of Class*; best picture, *The Sting*.

1974 Art Carney, best actor, *Harry and Tonto*; Ellen Burstyn, best actress, *Alice Doesn't Live Here Anymore*; best picture, *The Godfather, Part II*.

1975 Jack Nicholson, best actor, *One Flew Over the Cuckoo's Nest*; Louise Fletcher, best actress, *One Flew Over the Cuckoo's Nest*; best picture, *One Flew Over the Cuckoo's Nest*.

1976 Peter Finch, best actor, *Network*; Faye Dunaway, best actress, *Network*; best picture, *Rocky*.

1977 Richard Dreyfuss, best actor, *The Goodbye Girl*; Diane Keaton, best actress, *Annie Hall*; best picture, *Annie Hall*.

1978 Jon Voight, best actor, *Coming Home*; Jane Fonda, best actress, *Coming Home*; best picture, *The Deer Hunter*.

1979 Dustin Hoffman, best actor, *Kramer vs. Kramer*; Sally Field, best actress, *Norma Rae*; best picture, *Kramer vs. Kramer*.

1980 Robert DeNiro, best actor, *Raging Bull*; Sissy Spacek, best actress, *Coal Miner's Daughter*; best picture, *Ordinary People*.

1981 Henry Fonda, best actor, *On Golden Pond*; Katharine Hepburn, best actress, *On Golden Pond*; best picture, *Chariots of Fire*.

1982 Ben Kingsley, best actor, *Gandhi*; Meryl Streep, best actress, *Sophie's Choice*; best picture, *Gandhi*.

1983 Robert Duvall, best actor, *Tender Mercies*; Shirley MacLaine, best actress, *Terms of Endearment*; best picture, *Terms of Endearment*.

1984 F. Murray Abraham, best actor, *Amadeus*; Sally Field, best actress, *Places in the Heart*; best picture, *Amadeus*.

1985 William Hurt, best actor, *Kiss of the Spider Woman*; Geraldine Page, best actress, *The Trip to Bountiful*; best picture, *Out of Africa*.

1986 Paul Newman, best actor, *The Color of Money*; Marlee Matlin, best actress, *Children of a Lesser God*; best picture, *Platoon*.

1987 Michael Douglas, best actor, *Wall Street*; Cher, best actress, *Moonstruck*; best picture, *The Last Emperor*.

1988 Dustin Hoffman, best actor, *Rain Man*; Jodie Foster, best actress, *The Accused*; best picture, *Rain Man*.

1989 Daniel Day-Lewis, best actor, *My Left Foot*; Jessica Tandy, best actress, *Driving Miss Daisy*; best picture, *Driving Miss Daisy*.

1899 French filmmaker George Mèliès (1861–1938), who pioneered stop-action camera technique, expands on original peep-show format to film fictional narratives. He produces short movies that tell simple stories, as in *A Trip to the Moon* (1902).

1903 *The Life of an American Fireman* and *The Great Train Robbery* by American Edwin S. Porter (1869–1941) are first narrative films successfully to portray events occurring simultaneously at different locations (parallel editing).

1905 First "nickelodeon" (nickel theater) opens in Pittsburgh. Within four years 5,000 nickelodeons open across United States.

1906 British filmmaker Cecil Hepworth (1874–1953) produces *Rescued by Rover*, noted for its fluid and innovative style of film editing.

1908 National Board of Censorship (later National Board of Review), industry-supported voluntary censorship body, is formed. It tries to establish uniform censorship guidelines for state and local censors.

1908 D. W. Griffith (1875–1948) directs his first film, *The Adventures of Dollie*.

1909 Motion Picture Patents Company, film-company trust, pools patents on motion-picture equipment and attempts to freeze out competitors. Trust is rendered ineffective by 1912.

1910 Star system for promoting studios' leading actors and actresses is adopted. Previously, film actors were little-advertised or even anonymous. By this time U.S. movie audiences have mushroomed to millions a week.

1911 Mack Sennett (1880–1960), veteran of D. W. Griffith's troupe, begins making comedies at Keystone Studio. His stock company at Keystone and, after 1917, his own studio would include Mabel Normand, "Fatty" Arbuckle, Charlie Chaplin, and the Keystone Kops at various times.

1912 French-made *Queen Elizabeth* becomes first popular full-length (multiple-reel) feature film shown in the United States. It proves that audiences are ready for longer, more complex films.

1912 American producer-director Thomas H. Ince (1882–1924) devises production-budgeting formula later used by all studios.

1912 Universal Pictures is formed by mergers.

1914 Paramount Pictures is founded as film-distribution outlet for a number of film production companies.

1914 Luxury movie theater, Strand (3,300 seats), opens on Broadway. It accommodates growing middle-class movie audience. Over 20,000 movie "palaces" are in United States by 1916.

1915 D. W. Griffith releases his masterpiece *Birth of a Nation*, portraying life in U.S. South during Civil War and Reconstruction. Despite controversy over its pro-Southern bias and unremorseful portrayal of racist attitudes, the film establishes many fundamentals of narrative style for movies. It is huge box-office and critical success.

1915 Hollywood becomes the center of American movie-making by about this time.

1919 United Artists is formed by Charlie Chaplin (1889–1977), D. W. Griffith, Douglas Fairbanks (1883–1939), and Mary Pickford (1893–1979). The artists control production and distribution of their films.

1919 Germany's *The Cabinet of Dr. Caligari*, directed by Robert Wiene, is released. Considered the first horror film, its stark, angled sets give rise to German Expressionism.

1920s Fox and Paramount acquire U.S. theaters, while continuing to produce and distribute films.

1921 Scandal rocks Hollywood, confirming its freewheeling reputation. Actor-film-

▶

	maker Roscoe "Fatty" Arbuckle (1887–1933) is accused (and later cleared) of bizarre killing at Hollywood party.
1922	American producers' and directors' association adopts Hays Code for self-censorship of films.
1922	Rin Tin Tin makes Hollywood debut.
1923	Release of *The Covered Wagon*, first of "big" Westerns.
1923	Experimental sound movie, using sound-on-film system, is demonstrated by inventor Lee De Forest (1873–1961) in New York City (March 13).
1923	Warner Bros. Pictures is founded.
1924	French moviemakers pioneer avant-garde film style with *Entr'acte*.
1924	Metro-Goldwyn-Mayer (MGM) is formed by merger.
1924	Columbia Pictures is founded.
1925	Russian Sergei Eisenstein (1898–1948) releases his masterpiece *Battleship Potemkin*.
1927	Release of *The Jazz Singer*, early sound-on-film movie with four singing and speaking sequences. Warner Bros. releases the first all-talking film in 1928.
1927	*King of Kings*, super spectacle by noted director Cecil B. DeMille (1881–1959), is released. He makes career of big-budget spectacles.
1928	Walt Disney (1901–66) releases first animated Mickey Mouse cartoon. Disney eventually builds an entertainment empire based on his animated cartoons.
1929	First Academy Awards are given for outstanding film performances.
1929–30	Technicolor is introduced. Color film, first demonstrated in United States in 1928, eventually replaces black and white.
1931	Release of *Public Enemy*, one of many gangster movies popular in 1930s.
1933	First U.S. drive-in theater opens in New Jersey. United States has 2,200 by 1950.
1935	20th Century-Fox is formed.
1939	Release of *Gone with the Wind*, one of the most successful movies of all time.
1946	First Cannes Film Festival.
1947	U.S. congressional committee claims there are Communists in movie industry, leading to blacklisting of suspect writers and actors.
1948	Antitrust ruling against major studios forces divestment of studio-owned theaters.
1948	Television begins its phenomenal rise in popularity. By 1950s movie audiences are declining as a result.
1953	Wide-screen projection (CinemaScope) in theaters is introduced.
1953	Industry introduces 3-D movies in attempt to compete with growing popularity of TV.
1960	Release of *Psycho*, one of best known of many suspense dramas directed by Alfred Hitchcock (1899–1980).
1960s	Takeovers of financially strapped movie studios by multinational companies.
1968	Film rating system—using ratings G, PG, R, and X—is adopted in United States. It replaces self-censorship. Number of U.S. films with explicit sex increases sharply.
1969	*Bonnie and Clyde*, one of first films targeted at American youth market, is released.
1974	*Earthquake*, one of spate of disaster films, introduces special-effects system Sensurround.
1978	Release of first psycho-slasher film, *Halloween*. Genre remains popular into 1980s.
1978	Release of *Superman*, believed to be most expensive movie to date. Cost is estimated at $55 million.
1980s	Cable networks and videocassette recorders force movie industry to produce more films for home market.

▶

1984 New rating category, PG–13, is adopted by the industry.

1984 70mm photography and Dolby sound are introduced.

1987 Case of accidental deaths during filming of *Twilight Zone* in 1982 is settled. Director and others are acquitted of criminal charges.

1989 U.S. Copyright Office, in report to Congress, recommends drafting of guidelines to control colorization of classic black-and-white films.

Memorable Films from the Archives

The list includes box-office hits, critical successes, movie classics, and all-time favorites.

1902 *Voyage to the Moon*, starring Georges Méliès. Georges Méliès, director.

1903 *The Great Train Robbery*, starring George Barns and A. C. Abadie. Edwin S. Porter, director.

1906 *Rescued by Rover*, starring Cecil M. Hepworth, Mrs. Hepworth, and Barbara Hepworth. Cecil M. Hepworth, director.

1912 Keystone Kops films. Mack Sennett, director.

1912 *Queen Elizabeth*, starring Sarah Bernhardt. Louis Mercanton and Henri Desfontaines, directors.

1914 *Tillie's Punctured Romance*, starring Marie Dressler and Charlie Chaplin. Mack Sennett, director.

1915 *The Birth of a Nation*, starring Henry B. Walthall, Lillian Gish, and Mae Marsh. D. W. Griffith, director.

1916 *Intolerance*, starring Mae Marsh and Robert Harron. D. W. Griffith, director.

1917 *The Little Princess*, starring Mary Pickford.

1919 *The Cabinet of Dr. Caligari*, starring Werner Krauss and Conrad Veidt. Robert Wiene, director.

1920 *The Mark of Zorro*, starring Douglas Fairbanks and Noah Beery. Fred Niblo, director.

1921 *The Kid*, starring Charlie Chaplin and Jackie Coogan. Charlie Chaplin, director.

1921 *The Sheik*, starring Rudolph Valentino and Diane Mayo. George Melford, director.

1922 *Nosferatu*, starring Max Schreck and Alexander Granach. Frederich Wilhelm Marnau, director.

1922 *Nanook of the North*. Robert Flaherty, director.

1922 *Robin Hood*, starring Douglas Fairbanks and Wallace Beery. Allan Dwan, director.

1924 *The Thief of Baghdad*, starring Douglas Fairbanks. Raoul Walsh, director.

1924 *The Iron Horse*, starring George O'Brien and Madge Bellamy. John Ford, director.

1924 *Greed*, starring Gibson Gowland, ZaSu Pitts, and Jean Hersholt. Erich von Stroheim, director.

1925 *The Big Parade*, starring John Gilbert and Renée Adoree. King Vidor, director.

1925 *The Battleship Potemkin*, starring Alexander Antonov and Vladimir Barsky. Sergi Eisenstein, director.

1925 *The Gold Rush*, starring Charlie Chaplin and Mack Swain. Charlie Chaplin, director.

1925 *The Phantom of the Opera*, starring Lon Chaney and Mary Philbin. Rupert Julian, director.

1926 *Metropolis*, starring Brigitte Helm and Alfred Abel. Fritz Lang, director.

1927 *The King of Kings*, starring H. B. Warner and Dorothy Cumming. Cecil B. DeMille, director.

1927 *The Jazz Singer*, starring Al Jolson and Warner Oland. Alan Crosland, director.

1928 *The Passion of Joan of Arc*, starring Marie Falconetti. Carl Theodor Dreyer, director.

1929 *Coconuts*, starring the Marx Brothers. Joseph Santley, director.

1930 *All Quiet on the Western Front*, starring Lew Ayres, and Louis Wolheim. Lewis Milestone, director.

1930 *Hell's Angels*, starring Ben Lyon and James Hall. Howard Hughes, director.

1930 *The Blue Angel*, starring Marlene Dietrich and Emil Jannings. Josef von Sternberg, director.

1931 *Frankenstein*, starring Colin Clive and Mae Clark. James Whale, director.

▶

1931 *Dracula*, starring Bela Lugosi and Helen Chandler. Tod Browning, director.

1931 *The Public Enemy*, starring James Cagney and Jean Harlow. William A. Wellman, director.

1931 *City Lights*, starring Charlie Chaplin and Virginia Cherrill. Charlie Chaplin, director.

1931 *Platinum Blond*, starring Loretta Young, Robert Williams, and Jean Harlow. Frank Capra, director.

1931 *Cimarron*, starring Richard Dix and Irene Dunne. Wesley Ruggles, director.

1931 *The Front Page*, starring Adolphe Menjou and Pat O'Brien. James Whale, director.

1931 *Monkey Business*, starring the Marx Brothers. Norman Z. McLeod, director.

1932 *Tarzan, The Ape Man*, starring Johnny Weissmuller. W. S. Van Dyke, director.

1932 *Scarface*, starring Paul Muni. Howard Hawks, director.

1932 *Grand Hotel*, starring Greta Garbo, John Barrymore, Joan Crawford, Lionel Barrymore, and Wallace Beery. Edmund Goulding, director.

1932 *Dr. Jekyll and Mr. Hyde*, starring Fredric March and Miriam Hopkins. Rouben Mamoulian, director.

1932 *I Am a Fugitive from a Chain Gang*, starring Paul Muni and Glenda Farrell. Mervyn LeRoy, director.

1932 *42nd Street*, starring Warner Baxter, Ruby Keeler, and Dick Powell. Lloyd Bacon, director; Busby Berkeley, choreographer.

1933 *Zero de Conduit*, starring Jean Dasté and Louis Lefebvre. Jean Vigo, director.

1933 *King Kong*, starring Fay Wray and Bruce Cabot. Merian Cooper and Ernest Schoedsack, directors.

1933 *The Invisible Man*, starring Claude Rains and Gloria Stuart. James Whale, director.

1933 *She Done Him Wrong*, starring Mae West and Cary Grant. Lowell Sherman, director.

1933 *Duck Soup*, starring the Marx Brothers. Leo McCarey, director.

1933 *The Power and the Glory*, starring Spencer Tracy and Colleen Moore. William K. Howard, director.

1933 *Queen Christina*, starring Greta Garbo and John Gilbert. Rouben Mamoulian, director.

1934 *Death Takes a Holiday*, starring Fredric March and Evelyn Venable. Mitchell Leisen, director.

1934 *Triumph of the Will*. Leni Riefenstahl, director.

1934 *It Happened One Night*, starring Clark Gable and Claudette Colbert. Frank Capra, director.

1934 *The Thin Man*, starring William Powell and Myrna Loy. W. S. Van Dyke, director.

1934 *Treasure Island*, starring Wallace Beery.

1934 *Cleopatra*, starring Claudette Colbert and Warren William. Cecil B. DeMille, director.

1934 *The Gay Divorcée*, starring Fred Astaire and Ginger Rogers. Mark Sandrich, director.

1934 *Of Human Bondage*, starring Leslie Howard and Bette Davis. John Cromwell, director.

1934 *The Scarlet Pimpernel*, starring Leslie Howard and Merle Oberon. Harold Young, director.

1935 *Top Hat*, starring Fred Astaire and Ginger Rogers. Mark Sandrich, director.

1935 *Mutiny on the Bounty*, starring Clark Gable and Charles Laughton. Frank Lloyd, director.

1935 *The Bride of Frankenstein*, starring Boris Karloff and Colin Clive. James Whale, director.

1935 *Captain Blood*, starring Errol Flynn and Olivia DeHavilland. Michael Curtiz, director.

1935 *David Copperfield*, starring Freddie Bartholomew, W. C. Fields, and Roland Young. George Cukor, director.

1935 *A Night at the Opera*, starring the Marx Brothers. Sam Wood, director.

1935 *The Littlest Rebel*, starring Shirley Temple, John Boles, and Bill Robinson. David Butler, director.

1935 *Les Misérables*, starring Fredric March and Charles Laughton. Richard Boleslawski, director.

1935 *Naughty Marietta*, starring Jeanette MacDonald and Nelson Eddy. W. S.VanDyke, director.

▶

1935 *A Tale of Two Cities*, starring Ronald Colman and Elizabeth Allan. Jack Conway, director.

1935 *The 39 Steps*, starring Madeleine Carroll and Robert Donat. Alfred Hitchcock, director.

1936 *Modern Times*, starring Charlie Chaplin and Paulette Goddard. Charlie Chaplin, director.

1936 *My Man Godfrey*, starring William Powell and Carole Lombard. Gregory LaCava, director.

1936 *Show Boat*, starring Irene Dunne and Allan Jones. James Whale, director.

1937 *Snow White and the Seven Dwarfs*. A Walt Disney production.

1937 *Grand Illusion*, starring Erich von Stroheim and Jean Gabin. Jean Renoir, director.

1937 *Captains Courageous*, starring Freddie Bartholomew, Spencer Tracey, Lionel Barrymore, and Mickey Rooney. Victor Fleming, director.

1937 *100 Men and a Girl*, starring Deanna Durbin and Adolphe Menjou. Henry Koster, director.

1937 *A Star is Born*, starring Janet Gaynor and Fredric March. William A. Wellman, director.

1937 *Topper*, starring Constance Bennett and Cary Grant. Norman Z. McLeod, director.

1938 *Boys Town*, starring Spencer Tracy and Mickey Rooney. Norman Taurog, director.

1938 *Angels with Dirty Faces*, starring James Cagney and Humphrey Bogart. Michael Curtiz, director.

1938 *The Adventures of Robin Hood*, starring Errol Flynn and Olivia DeHavilland. Michael Curtiz and William Keighley, directors.

1938 *The Adventures of Tom Sawyer*, starring Tommy Kelly and Jackie Moran. Norman Taurog, director.

1938 *Bringing Up Baby*, starring Cary Grant and Katharine Hepburn. Howard Hawks, director.

1938 *Dawn Patrol*, starring Errol Flynn and David Niven. Edmund Goulding, director.

1938 *Holiday*, starring Katharine Hepburn and Cary Grant. George Cukor, director.

1938 *Jezebel*, starring Bette Davis, Henry Fonda, and Fay Bainter. William Wyler, director.

1938 *The Lady Vanishes*, starring Margaret Lockwood and Michael Redgrave. Alfred Hitchcock, director.

1938 *Love Finds Andy Hardy*, starring Lewis Stone and Mickey Rooney. George B. Seitz, director.

1938 *Pygmalion*, starring Leslie Howard and Wendy Hiller. Anthony Asquith and Leslie Howard, directors.

1939 *The Wizard of Oz*, starring Judy Garland and Frank Morgan. Victor Fleming, director.

1939 *Gone with the Wind*, starring Clark Gable and Vivien Leigh. Victor Fleming, director.

1939 *Rules of the Game*, starring Marcel Dalio and Nora Gregor. Jean Renoir, director.

1939 *Stagecoach*, starring John Wayne and Thomas Mitchell. John Ford, director.

1939 *Goodbye, Mr. Chips*, starring Greer Garson and Robert Donat. Sam Wood, director.

1939 *Mr. Smith Goes to Washington*, starring James Stewart and Jean Arthur. Frank Capra, director.

1939 *Beau Geste*, starring Gary Cooper and Donald O'Connor. William A. Wellman, director.

1939 *Dark Victory*, starring Bette Davis, George Brent, Geraldine Fitzgerald, and Humphrey Bogart. Edmund Goulding, director.

1939 *Destry Rides Again*, starring Marlene Dietrich and James Stewart. George Marshall, director.

1939 *Drums Along the Mohawk*, starring Henry Fonda and Claudette Colbert. John Ford, director.

1939 *The Hound of the Baskervilles*, starring Basil Rathbone and Nigel Bruce. Sidney Lanfield, director.

1939 *The Hunchback of Notre Dame*, starring Charles Laughton and Maureen O'Hara. William Dieterle, director.

1939 *Ninotchka*, starring Greta Garbo and Melvyn Douglas. Ernst Lubitsch, director.

1939 *Of Mice and Men*, starring Burgess Meredith and Lon Chaney, Jr. Lewis Milestone, director.

1939 *Only Angels Have Wings*, starring Cary Grant and Jean Arthur. Howard Hawks, director.

▶

1939 *Young Mr. Lincoln*, starring Henry Fonda and Alice Brady. John Ford, director.

1940 *The Grapes of Wrath*, starring Henry Fonda and Jane Darwell. John Ford, director.

1940 *Fantasia*. A Walt Disney production.

1940 *My Little Chickadee*, starring W. C. Fields. Eddie Cline, director.

1940 *Knute Rockne*, starring Ronald Reagan and Pat O'Brien. Lloyd Bacon, director.

1940 *The Sea Hawk*, starring Errol Flynn. Michael Curtiz, director.

1940 *The Great Dictator*, starring Charlie Chaplin and Grace Hayle. Charlie Chaplin, director.

1940 *The Bank Dick*, starring W. C. Fields. Eddie Cline, director.

1940 *Dr. Ehrlich's Magic Bullet*, starring Edward G. Robinson, Otto Kruger, and Ruth Gordon. William Dieterle, director.

1940 *The Great McGinty*, starring Brian Donlevy and Akim Tamiroff. Preston Sturges, director.

1940 *His Girl Friday*, starring Cary Grant and Rosalind Russell. Howard Hawks, director.

1940 *The Mark of Zorro*, starring Tyrone Power and Linda Darnell. Rouben Mamoulian, director.

1940 *The Philadelphia Story*, starring Cary Grant, Katharine Hepburn, and James Stewart. George Cukor, director.

1940 *Rebecca*, starring Joan Fontaine and Laurence Olivier. Alfred Hitchcock, director.

1940 *The Thief of Bagdad*, starring Sabu and John Justin. Ludwig Berger, Tim Whelan, and Michael Powell, directors.

1941 *Citizen Kane*, starring Orson Welles and Joseph Cotten. Orson Welles, director.

1941 *Road to Zanzibar*, starring Bing Crosby and Bob Hope. Victor Schertzinger, director.

1941 *The Maltese Falcon*, starring Humphrey Bogart and Mary Astor. John Huston, director.

1941 *Dr. Jekyll and Mr. Hyde*, starring Spencer Tracy, Ingrid Bergman, and Lana Turner. Victor Fleming, director.

1941 *High Sierra*, starring Ida Lupino and Humphrey Bogart. Raoul Walsh, director.

1941 *How Green Was My Valley*, starring Walter Pidgeon, Maureen O'Hara, and Donald Crisp. John Ford, director.

1941 *The Little Foxes*, starring Bette Davis and Herbert Marshall. William Wyler, director.

1941 *Sergeant York*, starring Gary Cooper and Walter Brennan. Howard Hawks, director.

1941 *Suspicion*, starring Cary Grant and Joan Fontaine. Alfred Hitchcock, director.

1942 *The Magnificent Ambersons*, starring Joseph Cotten and Tim Holt. Orson Welles, director.

1942 *Casablanca*, starring Humphrey Bogart and Ingrid Bergman. Michael Curtiz, director.

1942 *Bambi*, A Walt Disney production.

1942 *Yankee Doodle Dandy*, starring James Cagney and Joan Leslie. Michael Curtiz, director.

1942 *Cat People*, starring Simone Simon and Kent Smith. Jacques Tourneur, director.

1942 *Mrs. Miniver*, starring Greer Garson, Walter Pidgeon, and Teresa Wright. William Wyler, director.

1942 *Road to Morocco*, starring Bing Crosby, Bob Hope, and Dorothy Lamour. David Butler, director.

1942 *To Be or Not to Be*, starring Carole Lombard and Jack Benny. Ernst Lubitsch, director.

1943 *Cabin in the Sky*, starring Ethel Waters, Eddie "Rochester" Anderson, and Lena Horne. Vincente Minnelli, director.

1943 *For Whom the Bell Tolls*, starring Gary Cooper, Ingrid Bergman, and Katina Paxinou. Sam Wood, director.

1943 *Girl Crazy*, starring Mickey Rooney and Judy Garland. Norman Taurog, director.

1943 *Heaven Can Wait*, starring Gene Tierney and Don Ameche. Ernst Lubitsch, director.

1943 *Madame Curie*, starring Greer Garson and Walter Pidgeon. Mervyn LeRoy, director.

1943 *Phantom of the Opera*, starring Nelson Eddy and Susanna Foster. Arthur Lubin, director.

1943 *Watch on the Rhine*, starring Paul Lukas and Bette Davis. Herman Shumlin, director.

▶

1944 *Arsenic and Old Lace*, starring Cary Grant and Josephine Hull. Frank Capra, director.

1944 *Gaslight*, starring Ingrid Bergman and Charles Boyer. George Cukor, director.

1944 *Going My Way*, starring Bing Crosby and Barry Fitzgerald. Leo McCarey, director.

1944 *Jane Eyre*, starring Joan Fontaine, Orson Welles, and Elizabeth Taylor. Robert Stevenson, director.

1944 *Laura*, starring Gene Tierney and Dana Andrews. Otto Preminger, director.

1944 *Lifeboat*, starring Tallulah Bankhead and William Bendix. Alfred Hitchcock, director.

1944 *Murder, My Sweet*, starring Dick Powell and Claire Trevor. Edward Dmytryk, director.

1944 *To Have and Have Not*, starring Humphrey Bogart and Lauren Bacall. Howard Hawks, director.

1944 *Children of Paradise*, starring Jean-Louis Barrault and Arletty. Marcel Carne, director.

1944 *Meet Me in St. Louis*, starring Judy Garland and Margaret O'Brien. Vincente Minnelli, director.

1945 *And Then There Were None*, starring Barry Fitzgerald and Walter Huston. René Clair, director.

1945 *The Body Snatcher*, starring Boris Karloff and Henry Daniell. Robert Wise, director.

1945 *The Corn is Green*, starring Bette Davis and John Dall. Irving Rapper, director.

1945 *The Picture of Dorian Gray*, starring Hurd Hatfield and George Sanders. Albert Lewin, director.

1945 *They Were Expendable*, starring Robert Montgomery, John Wayne, and Donna Reed. John Ford, director.

1945 *A Walk in the Sun*, starring Dana Andrews and Richard Conte. Lewis Milestone, director.

1945 *Rome, Open City*, starring Aldo Fabrizi and Anna Magnani. Roberto Rossellini, director.

1946 *It's a Wonderful Life*, starring James Stewart and Donna Reed. Frank Capra, director.

1946 *The Best Years of Our Lives*, starring Fredric March and Myrna Loy. William Wyler, director.

1946 *The Big Sleep*, starring Humphrey Bogart and Lauren Bacall. Howard Hawks, director.

1946 *The Postman Always Rings Twice*, starring Lana Turner and John Garfield. Tay Garnett, director.

1946 *The Razor's Edge*, starring Tyrone Power, Gene Tierney, and Anne Baxter. Edmund Goulding, director.

1947 *Gentleman's Agreement*, starring Gregory Peck, Dorothy McGuire, and Celeste Holm. Elia Kazan, director.

1947 *Great Expectations*, starring John Mills and Valerie Hobson. David Lean, director.

1948 *Key Largo*, starring Humphrey Bogart, Edward G. Robinson, Lauren Bacall, and Claire Trevor. John Huston, director.

1948 *The Treasure of the Sierra Madre*, starring Humphrey Bogart and Walter Huston. John Huston, director.

1948 *Hamlet*, starring Laurence Olivier and Eileen Herlie. Laurence Olivier, director.

1949 *The Third Man*, starring Joseph Cotten and Orson Welles. Carol Reed, director.

1949 *All The King's Men*, starring Broderick Crawford and Joanne Dru. Robert Rossen, director.

1949 *Adam's Rib*, starring Spencer Tracy, Katharine Hepburn, and Judy Holliday. George Cukor, director.

1950 *All About Eve*, starring Bette Davis, Anne Baxter, and George Sanders. Joseph L. Mankiewicz, director.

1950 *Cyrano de Bergerac*, starring Jose Ferrer and Mala Powers. Michael Gordon, director.

1950 *Sunset Boulevard*, starring Gloria Swanson and William Holden. Billy Wilder, director.

1951 *A Christmas Carol*, starring Alastair Sim and Mervyn Johns. Brian Desmond Hurst, director.

1951 *Oliver Twist*, starring Robert Newton and Alec Guinness. David Lean, director.

1951 *A Place in the Sun*, starring Montgomery Clift and Elizabeth Taylor. George Stevens, director.

▶

1951 *Quo Vadis*, starring Robert Taylor and Deborah Kerr. Mervyn LeRoy, director.

1951 *The Thing*, starring Kenneth Tobey and Robert Coruthwaite. Christian Nyby, director.

1951 *A Streetcar Named Desire*, starring Marlon Brando and Vivian Leigh. Elia Kazan, director.

1951 *The Desert Fox*, starring James Mason. Henry Hathaway, director.

1951 *An American in Paris*, starring Gene Kelly and Leslie Caron. Vincente Minnelli, director.

1951 *Rashomon*, starring Toshiro Mifune and Machiko Kyo. Akira Kurosawa, director.

1951 *The African Queen*, starring Humphrey Bogart and Katharine Hepburn. John Huston, director.

1951 *The Death of a Salesman*, starring Fredric March and Mildred Dunnock. Laslo Benedek, director.

1952 *High Noon*, starring Gary Cooper and Lloyd Bridges. Fred Zinnemann, director.

1952 *Singin' in the Rain*, starring Gene Kelly and Debbie Reynolds. Gene Kelly and Stanley Donen, directors.

1952 *The Lavender Hill Mob*, starring Alec Guinness and Stanley Holloway. Charles Crichton, director.

1952 *Monkey Business*, starring Cary Grant, Ginger Rogers, Charles Coburn, and Marilyn Monroe. Howard Hawks, director.

1953 *From Here to Eternity*, starring Burt Lancaster, Montgomery Clift, Frank Sinatra, and Donna Reed. Fred Zinnemann, director.

1953 *Gentlemen Prefer Blondes*, starring Jane Russell and Marilyn Monroe. Howard Hawks, director.

1953 *Julius Caesar*, starring Louis Calhern and Marlon Brando. Joseph L. Mankiewicz, director.

1953 *Shane*, starring Alan Ladd and Jean Arthur. George Stevens, director.

1953 *Peter Pan*, starring Mary Martin.

1954 *La Strada*, starring Giulietta Masina and Anthony Quinn. Federico Fellini, director.

1954 *On the Waterfront*, starring Marlon Brando and Eva Marie Saint. Elia Kazan, director.

1954 *Dial M for Murder*, starring Grace Kelly and Robert Cummings. Alfred Hitchcock, director.

1954 *The High and the Mighty*, starring John Wayne and Robert Newton. William A. Wellman, director.

1954 *20,000 Leagues Under the Sea*, starring Kirk Douglas and James Mason. Richard Fleischer, director.

1954 *Bad Day at Black Rock*, starring Spencer Tracy and Robert Ryan. John Sturges, director.

1954 *The Caine Mutiny*, starring Humphrey Bogart and Van Johnson. Edward Dmytryk, director.

1954 *Seven Brides for Seven Brothers*, starring Howard Keel and Jeff Richards. Stanley Donen, director.

1955 *Mister Roberts*, starring Henry Fonda and Jack Lemmon. John Ford, director.

1955 *Oklahoma!*, starring Gordon MacRae and Shirley Jones. Fred Zinnemann, director.

1955 *To Catch a Thief*, starring Cary Grant and Grace Kelly. Alfred Hitchcock, director.

1955 *The Trouble With Harry*, starring Edmund Gwenn, John Forsythe, and Shirley MacLaine. Alfred Hitchcock, director.

1955 *The Blackboard Jungle*, starring Sidney Poitier and Anne Francis. Richard Brooks, director.

1955 *East of Eden*, starring James Dean and Julie Harris. Elia Kazan, director.

1955 *The Seven-Year Itch*, starring Marilyn Monroe and Tom Ewell. Billy Wilder, director.

1955 *Rebel Without a Cause*, starring James Dean and Natalie Wood. Nicholas Ray, director.

1956 *The Searchers*, starring John Wayne and Jeffrey Hunter. John Ford, director.

1956 *The King and I*, starring Yul Brynner and Deborah Kerr. Walter Lang, director.

1956 *The Ten Commandments*, starring Charlton Heston and Yul Brynner. Cecil B. DeMille, director.

1956 *The Seventh Seal*, starring Bengt Ekerot and Nils Poppe. Ingmar Bergman, director.

1956 *Around the World in 80 Days*, starring David Niven and Shirley MacLaine. Michael Anderson, director.

▶

1956 *Bus Stop*, starring Marilyn Monroe and Don Murray. Joshua Logan, director.

1956 *Carousel*, starring Gordon MacRae and Shirley Jones. Henry King, director.

1956 *Giant*, starring Elizabeth Taylor, Rock Hudson, and James Dean. George Stevens, director.

1956 *Invasion of the Body Snatchers*, starring Kevin McCarthy and Dana Wynter. Don Siegel, director.

1957 *The Incredible Shrinking Man*, starring Grant Williams and Randy Stuart. Jack Arnold, director.

1957 *Peyton Place*, starring Lana Turner and Hope Lange. Mark Robson, director.

1957 *Silk Stockings*, starring Fred Astaire and Cyd Charisse. Rouben Mamoulian, director.

1957 *The Wings of Eagles*, starring John Wayne and Dan Dailey. John Ford, director.

1957 *Witness for the Prosecution*, starring Tyrone Power and Marlene Dietrich. Billy Wilder, director.

1957 *Wild Strawberries*, starring Bibi Andersson and Victor Sjostrom. Ingmar Bergman, director.

1957 *Twelve Angry Men*, starring Henry Fonda and Martin Balsam. Sidney Lumet, director.

1957 *Gunfight at the O.K. Corral*, starring Kirk Douglas and Jo Van Fleet. John Sturges, director.

1957 *Bridge on the River Kwai*, starring William Holden and Alec Guinness. David Lean, director.

1958 *Black Orpheus*, starring Brenno Melio and Marapessa Dawn. Marcel Camus, director.

1958 *Vertigo*, starring James Stewart and Kim Novak. Alfred Hitchcock, director.

1958 *Touch of Evil*, starring Charlton Heston and Orson Welles. Orson Welles, director.

1958 *Cat on a Hot Tin Roof*, starring Elizabeth Taylor and Paul Newman. Richard Brooks, director.

1958 *Gigi*, starring Leslie Caron and Maurice Chevalier. Vincente Minnelli, director.

1958 *Horror of Dracula*, starring Christopher Lee and Peter Cushing. Terence Fisher, director.

1958 *South Pacific*, starring Rossano Brazzi and Mitzi Gaynor. Joshua Logan, director.

1959 *Breathless*, starring Jean Seberg and Jean-Paul Belmondo. Jean-Luc Godard, director.

1959 *400 Blows*, starring Jean-Pierre Leaud and Claire Maurier. François Truffaut, director.

1959 *North by Northwest*, starring Cary Grant and Eva Marie Saint. Alfred Hitchcock, director.

1959 *Anatomy of a Murder*, starring James Stewart and Lee Remick. Otto Preminger, director.

1959 *La Dolce Vita*, starring Marcello Mastroianni and Anita Ekberg. Federico Fellini, director.

1959 *Ben-Hur*, starring Charlton Heston and Stephen Boyd. William Wyler, director.

1959 *Compulsion*, starring Dean Stockwell and Bradford Dillman. Richard Fleischer, director.

1959 *The Mouse That Roared*, starring Peter Sellers and Jean Seberg. Jack Arnold, director.

1959 *Pillow Talk*, starring Doris Day, Rock Hudson, and Tony Randall. Michael Gordon, director.

1959 *Room at the Top*, starring Laurence Harvey and Simone Signoret. Jack Clayton, director.

1959 *Some Like It Hot*, starring Marilyn Monroe, Tony Curtis, and Jack Lemmon. Billy Wilder, director.

1960 *The Apartment*, starring Jack Lemmon and Shirley MacLaine. Billy Wilder, director.

1960 *The Magnificent Seven*, starring Yul Brynner and Steve McQueen. John Sturges, director.

1960 *Psycho*, starring Anthony Perkins and Janet Leigh. Alfred Hitchcock, director.

1960 *The Alamo*, starring John Wayne and Richard Widmark. John Wayne, director.

1960 *Exodus*, starring Paul Newman and Eva Marie Saint. Otto Preminger, director.

1961 *A Raisin in the Sun*, starring Sidney Poitier and Claudia McNeil. Daniel Petrie, director.

1961 *Breakfast at Tiffany's*, starring Audrey Hepburn and George Peppard. Blake Edwards, director.

1961 *The Guns of Navarone*, starring Gregory Peck and David Niven. J. Lee Thompson, director.

1961 *West Side Story*, starring Natalie Wood and Richard Beymer. Robert Wise and Jerome Robbins, directors.

1961 *The Hustler*, starring Paul Newman and Jackie Gleason. Robert Rossen, director.

1961 *Jules and Jim*, starring Jeanne Moreau and Oskar Werner. François Truffaut, director.

1961 *King of Kings*, starring Jeffrey Hunter and Siobhan McKenna. Nicholas Ray, director.

1961 *The Parent Trap*, starring Hayley Mills and Maureen O'Hara. David Swift, director.

1962 *Advise and Consent*, starring Henry Fonda and Walter Pidgeon. Otto Preminger, director.

1962 *Birdman of Alcatraz*, starring Burt Lancaster and Karl Malden. John Frankenheimer, director.

1962 *The Children's Hour*, starring Audrey Hepburn, Shirley MacLaine, and James Garner. William Wyler, director.

1962 *Days of Wine and Roses*, starring Jack Lemmon and Lee Remick. Blake Edwards, director.

1962 *Lawrence of Arabia*, starring Peter O'Toole and Alec Guinness. David Lean, director.

1962 *The Loneliness of the Long Distance Runner*, starring Tom Courtenay and Avis Bunnage. Tony Richardson, director.

1962 *Long Day's Journey Into Night*, starring Katharine Hepburn and Ralph Richardson. Sidney Lumet, director.

1962 *To Kill a Mockingbird*, starring Gregory Peck and Mary Badham. Robert Mulligan, director.

1962 *Lolita*, starring James Mason, Sue Lyon, Shelley Winters, and Peter Sellers. Stanley Kubrick, director.

1963 *Scorpio Rising*, starring Bruce Byron and Johnny Sapienza. Kenneth Anger, director.

1963 *8½*, starring Marcello Mastroianni and Anouk Aimée. Federico Fellini, director.

1963 *The Birds*, starring Tippi Hedren and Rod Taylor. Alfred Hitchcock, director.

1963 *Dr. No*, starring Sean Connery and Joseph Wiseman. Terence Young, director.

1963 *The Great Escape*, starring Steve McQueen and James Garner. John Sturges, director.

1963 *The Nutty Professor*, starring Jerry Lewis and Stella Stevens. Jerry Lewis, director.

1963 *Tom Jones*, starring Albert Finney and Susannah York. Tony Richardson, director.

1964 *Zorba the Greek*, starring Anthony Quinn and Alan Bates. Michael Cacoyannis, director.

1964 *Dr. Strangelove, or: How I Learned to Stop Worrying and Love The Bomb*, starring George C. Scott, Peter Sellers, and Sterling Hayden. Stanley Kubrick, director.

1964 *The Pink Panther*, starring Peter Sellers and David Niven. Blake Edwards, director.

1964 *A Hard Day's Night*, starring the Beatles. Richard Lester, director.

1964 *Mary Poppins*, starring Julie Andrews and Dick Van Dyke. Robert Stevenson, director.

1964 *My Fair Lady*, starring Rex Harrison and Audrey Hepburn. George Cukor, director.

1964 *Goldfinger*, starring Sean Connery and Gert Frobe. Guy Hamilton, director.

1965 *Doctor Zhivago*, starring Omar Sharif and Julie Christie. David Lean, director.

1965 *The Sound of Music*, starring Julie Andrews and Christopher Plummer. Robert Wise, director.

1965 *The Spy Who Came in from the Cold*, starring Richard Burton and Claire Bloom. Martin Ritt, director.

1965 *A Thousand Clowns*, starring Jason Robards, Barry Gordon, and Martin Balsam. Fred Coe, director.

1966 *Alfie*, starring Michael Caine and Shelley Winters. Lewis Gilbert, director.

1966 *Blow-Up*, starring David Hemmings and Vanessa Redgrave. Michelangelo Antonioni, director.

1966 *A Man for All Seasons*, starring Paul Scofield and Wendy Hiller. Fred Zinnemann, director.

1966 *Fahrenheit 451*. François Truffaut, director.

1967 *The Graduate*, starring Dustin Hoffman, Anne Bancroft, and Katharine Ross. Mike Nichols, director.

1967 *Bonnie and Clyde*, starring Warren Beatty and Faye Dunaway. Arthur Penn, director.

▶

1967 *Cool Hand Luke*, starring Paul Newman and George Kennedy. Stuart Rosenberg, director.

1967 *The Dirty Dozen*, starring Lee Marvin and Ernest Borgnine. Robert Aldrich, director.

1967 *In the Heat of the Night*, starring Sidney Poitier and Rod Steiger. Norman Jewison, director.

1967 *King of Hearts*, starring Alan Bates and Pierre Brasseur. Philippe de Broca, director.

1968 *Petulia*, starring Julie Christie and George C. Scott. Richard Lester, director.

1968 *2001: A Space Odyssey*, starring Keir Dullea and Gary Lockwood. Stanley Kubrick, director.

1968 *The Lion in Winter*, starring Katharine Hepburn and Peter O'Toole.

1968 *Funny Girl*, starring Barbra Streisand and Omar Sharif. William Wyler, director.

1968 *Bullitt*, starring Steve McQueen and Robert Vaughn. Peter Yates, director.

1968 *The Odd Couple*, starring Jack Lemmon and Walter Matthau. Gene Saks, director.

1968 *Planet of the Apes*, starring Charlton Heston and Roddy McDowall. Franklin J. Schaffner, director.

1968 *The Producers*, starring Zero Mostel and Gene Wilder. Mel Brooks, director.

1968 *Rosemary's Baby*, starring Mia Farrow, John Cassavetes, and Ruth Gordon. Roman Polanski, director.

1969 *The Prime of Miss Jean Brodie*, starring Maggie Smith and Robert Stephens. Ronald Neame, director.

1969 *True Grit*, starring John Wayne and Kim Darby. Henry Hathaway, director.

1969 *Midnight Cowboy*, starring Jon Voight and Dustin Hoffman. John Schlesinger, director.

1969 *Hello, Dolly!*, starring Barbra Streisand and Walter Matthau. Gene Kelly, director.

1969 *Easy Rider*, starring Dennis Hopper, Peter Fonda, and Jack Nicholson. Dennis Hopper, director.

1969 *Butch Cassidy and the Sundance Kid*, starring Robert Redford and Paul Newman. George Roy Hill, director.

1969 *Satyricon*. Federico Fellini, director.

1970 *Patton*, starring George C. Scott and Karl Malden. Franklin Schaffner, director.

1970 *M*A*S*H*, starring Donald Sutherland and Elliott Gould. Robert Altman, director.

1970 *Catch-22*, starring Alan Arkin and Richard Benjamin. Mike Nichols, director.

1970 *Five Easy Pieces*, starring Jack Nicholson and Karen Black. Bob Rafelson, director.

1971 *Dirty Harry*, starring Clint Eastwood and Harry Guardino. Don Siegel, director.

1971 *Fiddler on the Roof*, starring Chaim Topol and Norma Crane. Norman Jewison, director.

1971 *The Go-Between*, starring Julie Christie and Dominic Guard. Joseph Losey, director.

1971 *Klute*, starring Jane Fonda and Donald Sutherland. Alan J. Pakula, director.

1971 *The French Connection*, starring Gene Hackman and Fernando Rey. William Friedkin, director.

1971 *A Clockwork Orange*, starring Malcolm McDowell and Patrick Magee. Stanley Kubrick, director.

1971 *The Sorrow and the Pity*. Marcel Ophuls, director.

1972 *The Godfather*, starring Marlon Brando and Al Pacino. Francis Ford Coppola, director.

1972 *Harold and Maude*, starring Bud Cort and Ruth Gordon. Hal Ashby, director.

1972 *Play It Again Sam*, starring Woody Allen and Diane Keaton. Herbert Rossi, director.

1972 *Cabaret*, starring Liza Minnelli, Joel Grey, and Michael York. Bob Fosse, director.

1972 *Deliverance*, starring Jon Voight and Burt Reynolds. John Boorman, director.

1972 *The Poseidon Adventure*, starring Gene Hackman and Ernest Borgnine. Ronald Neame, director.

1972 *Sleuth*, starring Laurence Olivier and Michael Caine. Joseph L. Mankiewicz, director.

1973 *Serpico*, starring Al Pacino and John Randolph. Sidney Lumet, director.

1973 *Last Tango in Paris*, starring Marlon Brando and Maria Schneider. Bernardo Bertolucci, director.

▶

1973 *American Graffiti*, starring Richard Dreyfuss and Ron Howard. George Lucas, director.

1973 *Sleeper*, starring Woody Allen and Diane Keaton. Woody Allen, director.

1973 *The Sting*, starring Paul Newman and Robert Redford. Sydney Pollack, director.

1973 *The Exorcist*, starring Linda Blair, Ellen Burstyn, and Max Von Sydow. William Friedkin, director.

1974 *The Godfather, Part II*, starring Al Pacino and Robert DeNiro. Francis Ford Coppola, director.

1974 *Chinatown*, starring Jack Nicholson and Faye Dunaway. Roman Polanski, director.

1974 *Blazing Saddles*, starring Cleavon Little and Gene Wilder. Mel Brooks, director.

1975 *The Man Who Would Be King*, starring Sean Connery and Michael Caine. John Huston, director.

1975 *Nashville*, starring Henry Gibson and Karen Black. Robert Altman, director.

1975 *One Flew over the Cuckoo's Nest*, starring Jack Nicholson and Louise Fletcher. Milos Forman, director.

1975 *Jaws*, starring Richard Dreyfuss and Roy Scheider. Steven Spielberg, director.

1975 *The Sunshine Boys*, starring George Burns and Walter Matthau. Herbert Ross, director.

1976 *Rocky*, starring Sylvester Stallone and Talia Shire. John G. Avildsen, director.

1976 *Taxi Driver*, starring Robert DeNiro, Cybill Shepard, and Jody Foster. Martin Scorsese, director.

1976 *All the President's Men*, starring Robert Redford, Dustin Hoffman, and Jason Robards. Alan J. Pakula, director.

1976 *Network*, starring Faye Dunaway, Peter Finch, and Beatrice Straight. Sidney Lumet, director.

1976 *The Shootist*, starring John Wayne and Lauren Bacall. Don Siegel, director.

1977 *The Spy Who Loved Me*, starring Roger Moore and Barbara Bach. Lewis Gilbert, director.

1977 *Annie Hall*, starring Diane Keaton and Woody Allen. Woody Allen, director.

1977 *Close Encounters of the Third Kind*, starring Richard Dreyfuss and Melinda Dillon. Steven Spielberg, director.

1977 *Star Wars* (trilogy: *Star Wars*, 1977; *The Empire Strikes Back*, 1980; *Return of the Jedi*, 1983), starring Mark Hamill and Harrison Ford. George Lucas, director.

1978 *Halloween*, starring Donald Pleasence and Jamie Lee Curtis. John Carpenter, director.

1978 *The Marriage of Maria Braun*, starring Hanna Schygulla and Ivan Desny. Rainer Werner Fassbinder, director.

1978 *Superman*, starring Christopher Reeve, Margot Kidder, and Gene Hackman. Richard Donnor, director.

1978 *The Deer Hunter*, starring Robert DeNiro, Christopher Walken, and Meryl Streep. Michael Cimino, director.

1978 *Grease*, starring John Travolta and Olivia Newton-John. Randal Kleiser, director.

1978 *Interiors*, starring Diane Keaton, Marybeth Hurt, and Maureen Stapleton. Woody Allen, director.

1979 *Alien*, starring Tom Skerritt and Sigourney Weaver. Ridley Scott, director.

1979 *Being There*, starring Peter Sellers, Shirley MacLaine, and Melvyn Douglas. Hal Ashby, director.

1979 *Breaking Away*, starring Dennis Christopher and Dennis Quaid. Peter Yates, director.

1979 *Kramer vs. Kramer*, starring Dustin Hoffman and Meryl Streep. Robert Benton, director.

1979 *Manhattan*, starring Woody Allen and Diane Keaton. Woody Allen, director.

1979 *Apocalypse Now*, starring Marlon Brando, Martin Sheen, and Robert Duvall. Francis Ford Coppola, director.

1979 *The China Syndrome*, starring Jane Fonda and Jack Lemmon. James Bridges, director.

1979 *10*, starring Bo Derek, Dudley Moore, and Julie Andrews. Blake Edwards, director.

1980 *The Green Room*. Francois Truffaut, director.

1980 *Raging Bull*, starring Robert DeNiro and Cathy Moriarty. Martin Scorsese, director.

▶

1980 *Coal Miner's Daughter*, starring Sissy Spacek. Michael Apted, director.

1981 *Raiders of the Lost Ark*, starring Harrison Ford and Karen Allen. Steven Spielberg, director.

1981 *The French Lieutenant's Woman*, starring Meryl Streep and Jeremy Irons. Karel Reisz, director.

1981 *On Golden Pond*, starring Henry Fonda, Katharine Hepburn, and Jane Fonda. Mark Rydell, director.

1982 *E.T., The Extraterrestrial*, starring Henry Thomas and Dee Wallace. Steven Spielberg, director.

1982 *Gandhi*, starring Ben Kingsley. Richard Attenborough, director.

1983 *Tender Mercies*, starring Robert Duvall and Tess Harper.

1984 *Ghostbusters*, starring Bill Murray, Harold Ramis, and Dan Aykroyd. Ivan Reitman, director.

1984 *Indiana Jones and the Temple of Doom*, starring Harrison Ford. Steven Spielberg, director.

1984 *Amadeus*, starring Tom Hulce and F. Murray Abraham. Milos Forman, director.

1985 *Ran*, starring Tatsuya Nakadai and So Terao. Akira Kurosawa, director.

1985 *Kiss of the Spider Woman*, starring William Hurt and Raul Julia. Hector Babenco, director.

1985 *Out of Africa*, starring Meryl Streep and Robert Redford. Sydney Pollack, director.

1986 *Platoon*, starring Tom Berenger and Willem Dafoe. Oliver Stone, director.

1987 *Wings of Desire*, starring Bruno Ganz and Otto Sander. Wim Wenders, director.

1987 *Good Morning Vietnam*, starring Robin Williams. Barry Levinson, director.

1987 *The Last Emperor*, starring John Lone and Joan Chen. Bernardo Bertolucci, director.

1987 *Full Metal Jacket*. Stanley Kubrick, director.

1988 *Rain Man*, starring Dustin Hoffman and Tom Cruise. Barry Levinson, director.

1988 *Dangerous Liaisons*, starring Glenn Close and John Malkovich. Stephen Frears, director.

1988 *Who Framed Roger Rabbit?*, starring Bob Hoskins and Christopher Lloyd. Roger Zemeckis, director.

1989 *Batman*, starring Michael Keaton, Jack Nicholson, and Kim Basinger. Tim Burton, director.

1989 *Do the Right Thing*, starring Danny Aiello and Spike Lee. Spike Lee, director.

1989 *Sex, Lies and Videotape*, starring James Spader and Andie McDowell. Steven Soderbergh, director.

Recording Industry

1877 American inventor Thomas A. Edison (1847–1931) invents the phonograph. He successfully records *Mary Had a Little Lamb* (November 20) on a tinfoil cylinder.

1877–79 Several hundred tinfoil phonographs are sold as novelty items.

1886 American inventor Alexander Graham Bell (1847–1922) receives patent on Graphophone, an improvement on Edison's invention that requires use of headphones. Graphophones are used as office dictating machines.

1888 German-born American inventor Emile Berliner (1851–1929) invents Gramophone, recording on a flat disc. Patent is purchased by Victor Talking Machine Co.

1888 Edison records symphonic music program at Metropolitan Opera House.

▶

Grammy Awards

1959 *Mack the Knife* by Bobby Darin, best record; *Come Dance with Me* by Frank Sinatra, best album.

1960 *Theme from a Summer Place* by Percy Faith, best record; *Button-Down Mind* by Bob Newhart, best album.

1961 *Moon River* by Henry Mancini, best record; *Judy at Carnegie Hall* by Judy Garland, best album.

1962 *I Left My Heart in San Francisco* by Tony Bennett, best record; *The First Family* by Vaughn Meader, best album.

1963 *The Days of Wine and Roses* by Henry Mancini, best record; *The Barbra Streisand Album*, best album.

1964 *The Girl From Ipanema* by Stan Getz and Astrud Gilberto, best record; *Getz/Gilberto*, best album.

1965 *A Taste of Honey* by Herb Alpert, best record; *September of My Years* by Frank Sinatra, best album.

1966 *Strangers in the Night* by Frank Sinatra, best record; *A Man and His Music* by Frank Sinatra, best album.

1967 *Up, Up, and Away* by the Fifth Dimension, best record; *Sgt. Pepper's Lonely Hearts Club Band* by the Beatles, best album.

1968 *Mrs. Robinson* by Simon and Garfunkel, best record; *By the Time I Get to Phoenix* by Glen Campbell, best album.

1969 *Aquarius/Let the Sunshine In* by the Fifth Dimension, best record; *Blood, Sweat, and Tears* by Blood, Sweat, and Tears, best album.

1970 *Bridge over Troubled Water* by Simon and Garfunkel, best record; *Bridge over Troubled Water* by Simon and Garfunkel,best album.

1971 *It's Too Late* by Carole King, best record; *Tapestry* by Carole King, best album.

1972 *The First Time Ever I Saw Your Face* by Roberta Flack, best record; *The Concert for Bangladesh*, best album.

1973 *Killing Me Softly with His Song* by Roberta Flack, best record; *Innervisions* by Stevie Wonder, best album.

1974 *I Honestly Love You* by Olivia Newton-John, best record; *Fulfullingness' First Finale* by Stevie Wonder, best album.

1975 *Love Will Keep Us Together* by The Captain and Tennille, best record; *Still Crazy After All These Years* by Paul Simon, best album.

1976 *This Masquerade* by George Benson, best record; *Songs in the Key of Life* by Stevie Wonder, best album.

1977 *Hotel California* by the Eagles, best record; *Rumours* by Fleetwood Mac, best album.

1978 *Just the Way You Are* by Billy Joel, best record; *Saturday Night Fever* by the Bee Gees, best album.

1979 *What a Fool Believes* by the Doobie Brothers, best record; *52nd Street* by Billy Joel, best album.

1980 *Sailing* by Christopher Cross, best record; *Christopher Cross* by Christopher Cross, best album.

1981 *Bette Davis Eyes* by Kim Carnes, best record; *Double Fantasy* by John Lennon and Yoko Ono, best album.

1982 *Rosanna* by Toto, best record; *Toto IV* by Toto, best album.

1983 *Beat It* by Michael Jackson, best record; *Thriller* by Michael Jackson, best album.

1984 *What's Love Got to Do with It* by Tina Turner, best record; *Can't Slow Down* by Lionel Richie, best album.

1985 *We Are the World* by U.S.A. for Africa, best record; *No Jacket Required* by Phil Collins, best album.

1986 *Higher Love* by Steve Winwood, best record; *Graceland* by Paul Simon, best album.

1987 *Graceland* by Paul Simon, best record; *The Joshua Tree* by U2, best album.

1988 *Don't Worry, Be Happy* by Bobby McFerrin, best record; *Faith* by George Michael, best album.

c. 1890 Gramophones are sold as toys in Germany.

1892 First "Berliner" records are sold, marking first mass production of music on records.

1898 Danish inventor Valdemar Poulsen (1869–1942) invents magnetic recording.

1903 Italian tenor Enrico Caruso (1873–1921) records *Vesti la Giubba*, first record to sell 1 million copies.

1914 American Society of Composers, Authors, and Publishers (ASCAP) is established.

▶

1915 78-rotation-per-minute (RPM) records largely replace cylinders. Their playing time is 4½ minutes per side.

1919 Radio Corporation of America (RCA) is founded.

1925 Bell Laboratories researchers J. P. Maxfield and H. C. Harrison invent first successful electric phonograph.

c. 1930 Coin-operated jukeboxes come into widespread use.

1931 U.S. record sales fall from high of 100 million per year to about 10 million during the Great Depression.

1934 Muzak Company is founded, offering easy-listening background music to offices and businesses.

1940 *Billboard* magazine begins publishing weekly chart of hit records, with *I'll Never Smile Again* by Tommy Dorsey (1905–56) at number one.

1940 Walt Disney (1901–66) film *Fantasia* is released, using stereophonic sound.

1942 Bing Crosby (1904–77) records *White Christmas* by Irving Berlin (1888–1989), the best-selling record of all time. By 1987, North American sales of recording reach 170 million copies.

1942 RCA Victor sprays its first golden disc, indicating sales of over 1 million, for *Chattanooga Choo Choo* by Glenn Miller (1904–44).

1945 Bing Crosby records are at top of charts, with *White Christmas*, *Silent Night*, and *Don't Fence Me In* all passing 1 million in sales.

1948 CBS Laboratories produces the long-playing (LP) record, expanding playing time for each record side from 5 minutes to more than 20 minutes.

1949 RCA-Victor markets 45 rpm records while Columbia issues 33⅓ rpm's. There are 16 million record-player owners by now.

1950 *Your Hit Parade* premieres on television.

1951 Deutsche Gramophone markets first 33 rpm LP record.

1951 About 190 million records are sold this year in United States.

1952 "American Bandstand" (originally called "Philadelphia Bandstand") premieres on television.

1956 Elvis Presley (1935–77) releases *Heartbreak Hotel*, first of his string of more than 170 hit singles and 80 hit albums.

1957 Stereo technique for recording two-channel sound in single groove is perfected.

1958 Stereo discs are first marketed, inaugurating era of two-track stereo recordings.

1959 Grammy Awards are first presented by National Academy of Recording Arts and Sciences.

1960 Payola scandal erupts. Congress holds hearings to investigate radio disc jockeys who reportedly take bribes in return for playing certain records. Congress later passes law making the practice a criminal offense.

1963 Beatles, best-selling musical group of all time, release *She Loves You*, their first U.S. hit single. Total record sales by Beatles are estimated at more than 1 billion as of 1985.

1969 Woodstock Music Festival in upstate New York attracts 500,000 people and spawns 2-million-selling LP.

1970s American electrical engineer Thomas Stockholm, Jr. invents compact disc (CD), on which sound waves are coded and stored on tape. CDs account for 10 percent of all U.S. album sales by 1986.

1981 Music Television (MTV) premieres on cable network, featuring video renditions of popular music.

1983 Michael Jackson (b. 1958) releases LP *Thriller*, which breaks all sales records. Sales for this LP reach 40 million in 1988.

1989 Digital audio tape recorders, which can compete with quality of compact discs, are marketed in the United States.

▶

Hit Singles—Rock, and More

Selected hit records of the 1950s, 1960s, 1970s, and 1980s

1955 *Mr. Sandman* by the Chordettes; *Sincerely* by the McGuire Sisters; *Rock Around the Clock* by Bill Haley and the Comets; *The Yellow Rose of Texas* by Mitch Miller; *Sixteen Tons* by Tennessee Ernie Ford

1956 *The Great Pretender* by the Platters; *Heartbreak Hotel* by Elvis Presley; *The Wayward Wind* by Gogi Grant; *Don't Be Cruel* by Elvis Presley; *Love Me Tender* by Elvis Presley

1957 *Young Love* by Tab Hunter; *All Shook Up* by Elvis Presley; *Love Letters in the Sand* by Pat Boone; *Teddy Bear* by Elvis Presley; *Tammy* by Debbie Reynolds; *Wake Up, Little Susie* by the Everly Brothers; *Jailhouse Rock* by Elvis Presley; *April Love* by Pat Boone

1958 *At The Hop* by Danny and the Juniors; *Get a Job* by the Silhouettes; *All I Have to Do Is Dream* by the Everly Brothers; *Yakety Yak* by the Coasters; *Volare* by Domenico Modugno; *Tom Dooley* by the Kingston Trio; *To Know Him Is to Love Him* by the Teddy Bears

1959 *Smoke Gets in Your Eyes* by the Platters; *Stagger Lee* by Lloyd Price; *Venus* by Frankie Avalon; *Lonely Boy* by Paul Anka; *Mack the Knife* by Bobby Darin; *Mr. Blue* by the Fleetwoods

1960 *Teen Angel* by Mark Dinning; *A Summer Place* by Percy Faith; *Cathy's Clown* by the Everly Brothers; *Everybody's Somebody's Fool* by Connie Francis; *It's Now or Never* by Elvis Presley; *The Twist* by Chubby Checker; *Save the Last Dance for Me* by the Drifters; *Georgia on My Mind* by Ray Charles; *Stay* by Maurice Williams and the Zodiacs; *Are You Lonesome Tonight?* by Elvis Presley

1961 *Will You Love Me Tomorrow?* by the Shirelles; *Blue Moon* by the Marcels; *Runaway* by Del Shannon; *Travelin' Man* by Ricky Nelson; *Quarter to Three* by Gary "U. S." Bonds; *Tossin' and Turnin'* by Bobby Lewis; *Hit the Road, Jack* by Ray Charles; *Big Bad John* by Jimmy Dean; *The Lion Sleeps Tonight* by the Tokens

1962 *Duke of Earl* by Gene Chandler; *I Can't Stop Loving You* by Ray Charles; *Roses Are Red* by Bobby Vinton; *Sheila* by Tommy Roe; *Sherry* by the Four Seasons; *Big Girls Don't Cry* by the Four Seasons; *Telstar* by the Tornadoes

1963 *Go Away, Little Girl* by Steve Lawrence; *Hey, Paula* by Paul and Paula; *He's So Fine* by the Chiffons; *It's My Party* by Lesley Gore; *Fingertips, Part II* by Little Stevie Wonder; *Sugar Shack* by Jimmy Gilmer and the Fireballs; *Deep Purple* by Nino Tempo and April Stevens

1964 *There! I've Said It Again* by Bobby Vinton; *I Want to Hold Your Hand* by the Beatles; *She Loves You* by the Beatles; *Can't Buy Me Love* by the Beatles; *A World Without Love* by Peter and Gordon; *Rag Doll* by the Four Seasons; *A Hard Day's Night* by the Beatles; *Oh, Pretty Woman* by Roy Orbison; *Baby Love* by the Supremes; *Mr. Lonely* by Bobby Vinton

1965 *Downtown* by Petula Clark; *You've Lost That Lovin' Feeling* by the Righteous Brothers; *Eight Days a Week* by the Beatles; *Stop! In the Name of Love* by the Supremes; *Mrs. Brown, You've Got a Lovely Daughter* by Herman's Hermits; *Help Me, Rhonda* by the Beach Boys; *Satisfaction* by the Rolling Stones; *I Got You, Babe* by Sonny and Cher; *Help!* by the Beatles; *Yesterday* by the Beatles; *Turn, Turn, Turn* by the Byrds

1966 *The Sounds of Silence* by Simon and Garfunkel; *These Boots Are Made for Walkin'* by Nancy Sinatra; *The Ballad of the Green Berets* by Sgt. Barry Sadler; *(You're My) Soul and Inspiration* by the Righteous Brothers; *Monday, Monday* by the Mamas and the Papas; *When a Man Loves a Woman* by Percy Sledge; *Paperback Writer* by the Beatles; *Summer in the City* by the Lovin' Spoonful; *Cherish* by the Association; *96 Tears* by Question Mark and the Mysterians; *Winchester Cathedral* by the New Vaudeville Band; *Good Vibrations* by the Beach Boys

1967 *I'm a Believer* by the Monkees; *Ruby Tuesday* by the Rolling Stones; *Penny Lane* by the Beatles; *Somethin' Stupid* by Nancy and Frank Sinatra; *Respect* by Aretha Franklin; *Windy* by the Association; *Light My Fire* by the Doors; *All You Need Is Love* by the Beatles; *Ode to Billie Joe* by Bobbie Gentry; *To Sir,*

▶

with Love by Lulu; *Daydream Believer* by the Monkees; *Hello, Goodbye* by the Beatles

1968 *Judy in Disguise* by John Fred and His Playboy Band; *(Sittin' On) The Dock of the Bay* by Otis Redding; *Honey* by Bobby Goldsboro; *Mrs. Robinson* by Simon and Garfunkel; *This Guy's in Love With You* by Herb Alpert; *People Got to Be Free* by the Rascals; *Harper Valley PTA* by Jeannie C. Riley; *Hey, Jude* by the Beatles; *I Heard It Through the Grapevine* by Marvin Gaye

1969 *Everyday People* by Sly and the Family Stone; *Aquarius/Let the Sunshine In* by the Fifth Dimension; *Get Back* by the Beatles; *Love Theme from Romeo and Juliet* by Henry Mancini; *In The Year 2525* by Zager and Evans; *Honky-tonk Woman* by the Rolling Stones; *Sugar Sugar* by the Archies; *Suspicious Minds* by Elvis Presley; *Wedding Bell Blues* by the Fifth Dimension; *Leaving on a Jet Plane* by Peter, Paul and Mary

1970 *Raindrops Keep Fallin' on My Head* by B. J. Thomas; *Bridge over Troubled Water* by Simon and Garfunkel; *Let It Be* by the Beatles; *ABC* by the Jackson Five; *American Woman* by the Guess Who; *The Long and Winding Road* by the Beatles; *Close to You* by the Carpenters; *Ain't No Mountain High Enough* by Diana Ross; *I'll Be There* by the Jackson Five

1971 *Knock Three Times* by Dawn; *One Bad Apple* by the Osmonds; *Joy to the World* by Three Dog Night; *It's Too Late* by Carole King; *How Can You Mend a Broken Heart?* by the Bee Gees; *Go Away, Little Girl* by Donny Osmond; *Maggie May* by Rod Stewart; *Theme from Shaft* by Isaac Hayes; *Family Affair* by Sly and the Family Stone

1972 *American Pie* by Don McLean; *A Horse with No Name* by America; *The First Time Ever I Saw Your Face* by Roberta Flack; *The Candy Man* by Sammy Davis, Jr.; *Song Sung Blue* by Neil Diamond; *Alone Again (Naturally)* by Gilbert O'Sullivan; *I Can See Clearly Now* by Johnny Nash; *Papa Was a Rolling Stone* by the Temptations; *I Am Woman* by Helen Reddy; *Me and Mrs. Jones* by Billy Paul

1973 *You're So Vain* by Carly Simon; *Killing Me Softly With His Song* by Roberta Flack; *The Night the Lights Went Out in Georgia* by Vicki Lawrence; *Tie a Yellow Ribbon* by Dawn with Tony Orlando; *You Are the Sunshine of My Life* by Stevie Wonder; *My Love* by Paul McCartney and Wings; *Bad, Bad Leroy Brown* by Jim Croce; *Touch Me in the Morning* by Diana Ross; *Angie* by the Rolling Stones; *Midnight Train to Georgia* by Gladys Knight and the Pips; *Keep On Truckin'* by Eddie Kendricks; *Time in a Bottle* by Jim Croce

1974 *You're Sixteen* by Ringo Starr; *The Way We Were* by Barbra Streisand; *Love's Theme* by Love Unlimited Orchestra; *Seasons in the Sun* by Terry Jacks; *Sunshine on My Shoulder* by John Denver; *Hooked on a Feeling* by Blue Swede; *Band on the Run* by Paul McCartney and Wings; *Rock the Boat* by Hues Corporation; *Annie's Song* by John Denver; *You're Having My Baby* by Paul Anka; *I Shot the Sheriff* by Eric Clapton; *Can't Get Enough of Your Love, Babe* by Barry White; *I Honestly Love You* by Olivia Newton-John; *Then Came You* by Dionne Warwick and the Spinners; *You Ain't Seen Nothing Yet* by Backman Turner Overdrive

1975 *Lucy in the Sky With Diamonds* by Elton John; *Mandy* by Barry Manilow; *Have You Never Been Mellow?* by Olivia Newton-John; *My Eyes Adored You* by Frankie Valli; *Philadelphia Freedom* by Elton John; *He Don't Love You* by Tony Orlando and Dawn; *Thank God I'm a Country Boy* by John Denver; *Love Will Keep Us Together* by The Captain and Tennille; *Listen to What the Man Said* by Paul McCartney and Wings; *The Hustle* by Van McCoy; *Jive Talkin'* by The Bee Gees; *Rhinestone Cowboy* by Glen Campbell; *Fame* by David Bowie; *Bad Blood* by Neil Sedaka; *Island Girl* by Elton John

1976 *I Write the Songs* by Barry Manilow; *Fifty Ways to Leave Your Lover* by Paul Simon; *Theme From S.W.A.T.* by Rhythm Heritage; *Disco Lady* by Johnny Taylor; *Silly Love Songs* by Wings; *Afternoon Delight* by the Starland Vocal Band; *Don't Go Breaking My Heart* by Elton John and Kiki Dee; *You Should Be Dancing* by the Bee Gees; *Play That Funky Music* by Wild Cherry; *If You Leave Me Now* by Chicago; *Tonight's the Night* by Rod Stewart

▶

Jazz Greats

1868–1917 Scott Joplin, pianist, cornetist, and pioneer ragtime composer, lives. Works include "Maple Leaf Rag" (1899).

1868–1931 Charles "Buddy" Bolden, cornetist, lives. First true jazz musician, he pioneeers New Orleans jazz style.

1873–1958 William Christopher "W. C." Handy, cornetist, composer, and "father of blues," lives. Works include "St. Louis Blues" (1914) and "Beale Street Blues" (1914).

1885–1938 Joseph "King" Oliver, cornetist, bandleader, and composer, lives. He writes important standard jazz pieces, including "Sugarfoot Stomp."

1885–1941 Ferdinand Joseph "Jelly Roll" Morton, pianist, combo leader, and composer in ragtime and blues styles, lives. Works include "Jelly Roll Blues" (1917) and "King Porter Stomp." Morton records history of jazz for Library of Congress in 1938.

1888–1949 Huddle (Leadbelly) Ledbetter, guitarist and blues singer, lives.

1895–1937 Bessie Smith, vocalist, lives. She is very influential in blues singing.

1898–1952 Fletcher "Smack" Henderson, influential bandleader in swing and big-band era, lives. Arranger for Benny Goodman band, his arrangements include "Jingle Bells," "Sandman," and "Honeysuckle Rose."

1899–1974 Edward Kennedy "Duke" Ellington, pianist, bandleader, composer, and arranger, lives. His illustrious career begins in 1918 and continues into early 1970s. Works include "Sophisticated Lady" (1933) and "In a Sentimental Mood" (1936).

1900–64 Don Redman, sideman, arranger, bandleader, and alto saxophonist, lives. He is a major influence in big-band jazz arrangements.

1900–71 Louis Armstrong (nicknames: Satchelmouth, Satchmo), trumpeter, vocalist, and combo leader, lives. Star pioneer of big-band style, he makes many classical jazz recordings in late 1920s with own combos Hot Five and Hot Seven.

1903–83 Earl "Father" Hines, pianist and bandleader, lives. He is originator of "trumpet" piano style inspired by Louis Armstrong and an influential pioneer of swing piano style.

1904–43 Thomas "Fats" Waller, pianist, vocalist, and composer, lives. Originator of stride piano style, his works include "Honeysuckle Rose" and "Ain't Misbehavin'."

1904–44 Glenn Miller, trombonist and big-band leader, lives.

1904–69 Coleman Hawkins (nicknames: Bean, The Hawk), first star tenor saxophonist, lives. He makes saxophone important in jazz, especially big-band jazz.

1904–84 William "Count" Basie, pianist and bandleader, lives. Leader of top swing band in 1930s–70s, he receives Kennedy Center honors for achievement in performing arts in 1981. Works include theme song, "One O'Clock Jump."

1905–58 Tommy Dorsey, trombonist and leader of popular dance band, lives.

1906–70 Johnny Hodges, alto saxophonist and composer, lives. This admired musician is featured soloist in Duke Ellington band for nearly 40 years.

1909–86 Benjamin David "Benny" Goodman, clarinetist and bandleader, lives. "King of swing," he pioneers in rise of swing jazz in mid-1930s.

1909– Lionel Hampton, bandleader, vibist, drummer, and pianist, lives. Pioneer in integrating vibraphone into jazz, he develops "trigger finger" method of piano playing.

1910–56 Art Tatum, pianist and combo leader, lives. Much-admired musician, he is almost totally blind.

1911–85 Jonathan "Jo" Jones, drummer, lives. Influential swing musician, he is member of Count Basie band rhythm section.

1911–89 David Roy Eldridge (nickname: Little Jazz), trumpeter, lives. Influential in swing jazz, he performs with Fletcher Henderson and Benny Goodman bands. Eldridge becomes member of *Down Beat* magazine Hall of Fame in 1971.

1912–79 Stan Kenton, pianist, composer, and orchestra leader, lives.

1912–86 Theodore "Teddy" Wilson, pianist and combo leader, lives. He performs with Benny Goodman band and teaches music at Juilliard School of Music.

1913–78 Woodrow Charles "Woody" Herman, bandleader, clarinetist, vocalist, and alto saxophonist, lives. A leader in development of big-band progressive jazz, he has illustrious bandleading career from 1936 to 1970s.

Jazz Greats

1914–85 Kenny "Klook" Clarke, drummer, lives. He is leader in development of bop drumming in early to mid-1940s.

1914– William Clarence "Billy" Eckstine (nickname: Mr. B), vocalist and bandleader, lives. Notable cool-jazz vocalist, he is leader of a bebop band.

1915–59 Eleanor Gough "Billie" Holiday (nickname: Lady Day), vocalist, lives.

1916–42 Charlie Christian, guitarist, lives. He is responsible for introducing jazz guitar as solo instrument and is leader in employing electric amplification.

1917–87 Buddy Rich, drummer and bandleader, lives.

1917– John Birks "Dizzy" Gillespie, trumpeter, lives. One of star trumpeters in jazz history, he is pioneer of bop trumpet and invents trumpet with bell extending upward from middle of horn. He wins Grammy Award in 1975 and 1980.

1918–82 Thelonius Monk, pianist, lives. He is giant in development of bop piano. Works include theme song "Straight, No Chaser."

1918– Ella Fitzgerald, vocalist, lives. Known for scat singing and vocal improvisation in swing and bebop jazz, she is winner of eight Grammy awards.

1919– Art Blakey, drummer who developed hand-swing bop drumming style, lives. He is modern jazz pioneer.

1920–55 Charlie "Bird" Parker, alto saxophonist, lives. He is originator of bop or progressive jazz. Works include "Embraceable You."

1920– Dave Brubeck, pianist and quartet leader, lives. His signature song is "Take Five."

1923– Dexter Gordon, tenor saxophonist, lives. He is influential in early development of bop tenor saxophone.

1924–63 Dinah Washington, jazz singer, lives.

1924– James Louis "J. J." Johnson, trombonist, lives. He develops bop trombone and becomes involved in emergence of modern jazz.

1925–85 Zoot Sims, tenor and alto saxophonist and clarinetist, lives.

1925– Oscar Peterson, pianist and composer, lives.

1925– Maxwell "Max" Roach, drummer, lives. He performs with Charlie Parker, Dizzy Gillespie, and Miles Davis and is professor of music at University of Massachusetts at Amherst.

1926– Miles Davis, trumpeter, lives. Star modern jazz trumpeter in 1950s to 60s, he introduces cool jazz and is pioneer of fusion jazz.

1926–67 John William Coltrane, tenor saxophonist, lives. He is originator of "new black music," incorporating African and Asian music.

1927– Stan Getz, tenor saxophonist, lives. He is a popular performer in modern jazz style.

1928–75 Julian "Cannonball" Adderley, alto saxophonist, lives.

1928– Horace Silver, pianist and combo leader, lives.

1930– Herbie Mann, jazz flutist, lives.

1930– Theodore Walter "Sonny" Rollins, tenor saxophonist, lives. He becomes leading master of instrument in modern jazz and is member of *Down Beat* magazine Hall of Fame.

1933– Quincy Jones, arranger, lives.

1940– Herbert Jeffrey "Herbie" Hancock, pianist and combo leader, lives. He develops use of electric piano in modern jazz and fusion.

1941– Chick Corea, pianist and composer, lives.

1945– Keith Jarrett, pianist known for inprovisatory modern jazz style, lives. He is also accomplished classical musician.

1961– Wynton Marsalis, trumpeter and combo leader, lives. Popular modern jazz musician, he receives unprecedented popular acclaim by winning Grammy awards for both jazz and classical performances.

1977 *You Make Me Feel Like Dancing* by Leo Sayer; *Torn Between Two Lovers* by Mary MacGregor; *Evergreen* by Barbra Streisand; *Rich Girl* by Daryl Hall and John Oates; *Dancing Queen* by Abba; *Southern Nights* by Glen Campbell; *Dreams* by Fleetwood Mac; *Looks Like We Made It* by Barry Manilow; *I Just Want to Be Your Everything* by Andy Gibb; *Best of My Love* by Emotions; *You Light Up My Life* by Debby Boone; *How Deep Is Your Love?* by the Bee Gees

1978 *Baby, Come Back* by Player; *Stayin' Alive* by the Bee Gees; *Saturday Night Fever* by the

▶

Bee Gees; *If I Can't Have You* by Yvonne Elliman; *With a Little Luck* by Wings; *Too Much Too Little Too Late* by Johnny Mathis and Deniece Williams; *Shadow Dancing* by Andy Gibb; *Three Times a Lady* by the Commodores; *Grease* by Frankie Valli; *Hot Child in the City by Nick Gilder; You Don't Bring Me Flowers* by Barbra Streisand and Neil Diamond

1979 *Too Much Heaven* by the Bee Gees; *I Will Survive* by Gloria Gaynor; *Reunited* by Peaches and Herb; *Bad Girls* by Donna Summer; *My Sharona* by the Knack; *Heartache Tonight* by the Eagles

1980 *Rock with You* by Michael Jackson; *Crazy Little Thing Called Love* by Queen; *Call Me* by Blondie; *It's Still Rock 'n' Roll to Me* by Billy Joel; *Magic* by Olivia Newton-John; *Upside Down* by Diana Ross; *Another One Bites the Dust* by Queen; *Women in Love* by Barbra Streisand; *Lady* by Kenny Rogers

1981 *(Just Like) Starting Over* by John Lennon; *The Tide Is High* by Blondie; *9 to 5* by Dolly Parton; *Morning Train (9 to 5)* by Sheena Easton; *Angel of the Morning* by Juice Newton; *Bette Davis Eyes* by Kim Carnes; *Jessie's Girl* by Rick Springfield; *Endless Love* by Diana Ross and Lionel Richie; *Private Eyes* by Daryl Hall and John Oates; *Physical* by Olivia Newton-John

1982 *Centerfold* by J. Geils Band; *I Love Rock 'n' Roll* by Joan Jett and the Blackhearts; *Eye of the Tiger* by Survivor; *Up Where We Belong* by Joe Cocker and Jennifer Warnes; *Maneater* by Daryl Hall and John Oates

1983 *Down Under* by Men at Work; *Billie Jean* by Michael Jackson; *Beat It* by Michael Jackson; *Flashdance (What a Feeling)* by Irene Cara; *Every Breath You Take* by the Police; *Total Eclipse of the Heart* by Bonnie Tyler; *Island in the Stream* by Kenny Rogers and Dolly Parton; *All Night Long* by Lionel Richie; *Say Say Say* by Paul McCartney and Michael Jackson

1984 *Jump* by Van Halen; *Footloose* by Kenny Loggins; *Against All Odds* by Phil Collins; *Hello* by Lionel Richie; *Let's Hear It for the Boy* by Deniece Williams; *Time After Time* by Cyndi Lauper; *What's Love Got to Do with It?* by Tina Turner; *Missing You* by John Waite; *I Just Called to Say I Love You* by Stevie Wonder; *Wake Me Up Before You Go Go* by Wham!; *Out of Touch* by Daryl Hall and John Oates

1985 *Say You, Say Me* by Lionel Richie; *We Are the World* by U.S.A. for Africa; *Careless Whisper* by Wham! featuring George Michael; *The Power of Love* by Huey Lewis and the News; *One More Night* by Phil Collins; *Crazy for You* by Madonna; *Part-time Lover* by Stevie Wonder; *Saving All My Love for You* by Whitney Houston; *You Belong to the City* by Glenn Frey; *Material Girl* by Madonna; *The Heat Is On* by Glenn Frey

1986 *That's What Friends Are For* by Dionne Warwick and Friends; *Greatest Love of All* by Whitney Houston; *Kiss* by Prince and the Revolution; *How Will I Know?* by Whitney Houston; *Addicted to Love* by Robert Palmer; *Live to Tell* by Madonna; *Manic Monday* by the Bangles; *Living in America* by James Brown; *My Hometown* by Bruce Springsteen; *Take Me Home* by Phil Collins

1987 *(I Just) Died in Your Arms* by Cutting Crew; *Alone* by Heart; *I Wanna Dance with Somebody* by Whitney Houston; *I Want Your Sex* by George Michael; *La Bamba* by Los Lobos; *I Think We're Alone Now* by Tiffany; *Lookin' for a New Love* by Jody Watley; *Brilliant Disguise* by Bruce Springsteen; *Faith* by George Michael; *Sign o' the Times* by Prince

1988 *Could've Been* by Tiffany; *Father Figure* by George Michael; *One More Try* by George Michael; *Sweet Child o' Mine* by Guns 'n' Roses; *Simply Irresistible* by Robert Palmer; *Don't Worry, Be Happy* by Bobby McFerrin; *Love Bites* by Def Leppard

1989 *Look Away* by Chicago; *My Prerogative* by Bobby Brown; *Every Rose Has Its Thorn* by Poison; *Straight Up* by Paula Abdul; *Miss You Much* by Janet Jackson; *Wind Beneath My Wings* by Bette Midler; *Love Shack* by B-52s; *She Drives Me Crazy* by Fine Young Cannibals; *Bust a Move* by Young M.C.

14 Sports

Football

1800s Football gradually evolves as game separate from soccer and rugby.

1862 Oneida Football Club is founded in Boston.

1869 First intercollegiate football game is played between Princeton and Rutgers in New Brunswick, New Jersey. Game is similar to soccer, with 25 players per team and a rule prohibiting running with ball.

1873 Representatives from Princeton, Rutgers, Yale, and Columbia universities meet to standardize rules. They agree on soccerlike form of game. Harvard University refuses to participate, playing "Boston game," in which running with ball is allowed.

1874 Harvard defeats McGill University in Montreal in one of two games, 3–0. Neither side scores in second game. Harvard adopts McGill's rugby-style rules, which include using 15-man teams, dropkicks, and oblate ball.

National College Football Champions

The *National Collegiate A: A Football Guide* recognizes as unofficial national champion the team selected each year by press association polls. Where the Associated Press poll (of writers) does not agree with the United Press International poll (of coaches), the guide lists both teams selected.

1936	Minnesota
1937	Pittsburgh
1938	Texas Christian
1939	Texas A&M
1940	Minnesota
1941	Minnesota
1942	Ohio State
1943	Notre Dame
1944	Army
1945	Army
1946	Notre Dame
1947	Notre Dame
1948	Michigan
1949	Notre Dame
1950	Oklahoma
1951	Tennessee
1952	Michigan State
1953	Maryland
1954	Ohio State and UCLA
1955	Oklahoma
1956	Oklahoma
1957	Auburn and Ohio State
1958	Louisiana State
1959	Syracuse
1960	Minnesota
1961	Alabama
1962	USC
1963	Texas
1964	Alabama
1965	Alabama and Michigan State
1966	Notre Dame
1967	USC
1968	Ohio State
1969	Texas
1970	Texas and Nebraska
1971	Nebraska
1972	USC
1973	Notre Dame and Alabama
1974	Oklahoma and USC
1975	Oklahoma
1976	Pittsburgh
1977	Notre Dame
1978	Alabama and USC
1979	Alabama
1980	Georgia
1981	Clemson
1982	Penn State
1983	Miami
1984	Brigham Young
1985	Oklahoma
1986	Penn State
1987	Miami
1988	Notre Dame
1989	Miami

Heisman Trophy Winners

Year	Winner	Year	Winner
1935	Jay Berwanger, halfback, Chicago	1962	Terry Baker, quarterback, Oregon State
1936	Larry Kelley, end, Yale	1963	Roger Staubach, quarterback, Navy
1937	Clinton Frank, halfback, Yale	1964	John Huarte, quarterback, Notre Dame
1938	David O'Brien, quarterback, Texas Christian	1965	Mike Garrett, halfback, USC
1939	Nile Kinnick, quarterback, Iowa	1966	Steve Spurrier, quarterback, Florida
1940	Tom Harmon, halfback, Michigan	1967	Gary Beban, quarterback, UCLA
1941	Bruce Smith, halfback, Minnesota	1968	O. J. Simpson, halfback, USC
1942	Frank Sinkwich, halfback, Georgia	1969	Steven Owens, fullback, Oklahoma
1943	Angelo Bertelli, quarterback, Notre Dame	1970	Jim Plunkett, quarterback, Stanford
1944	Leslie Horvath, quarterback, Ohio State	1971	Pat Sullivan, quarterback, Auburn
1945	Felix (Doc) Blanchard, fullback, Army	1972	Johnny Rogers, fullback, Nebraska
1946	Glenn Davis, halfback, Army	1973	John Cappelletti, halfback, Penn State
1947	John Lujack, quarterback, Notre Dame	1974	Archie Griffin, running back, Ohio State
1948	Doak Walker, halfback, SMU	1975	Archie Griffin, running back, Ohio State
1949	Leon Hart, end, Notre Dame	1976	Tony Dorsett, running back, Pittsburgh
1950	Vic Janowicz, halfback, Ohio State	1977	Earl Campbell, running back, Texas
1951	Richard Kazmaier, halfback, Princeton	1978	Billy Sims, running back, Oklahoma
1952	Billy Vessels, halfback, Oklahoma	1979	Charles White, running back, USC
1953	John Lattner, halfback, Notre Dame	1980	George Rogers, running back, South Carolina
1954	Alan Ameche, fullback, Wisconsin	1981	Marcus Allen, running back, USC
1955	Howard Cassady, halfback, Ohio State	1982	Herschel Walker, running back, Georgia
1956	Paul Hornung, quarterback, Notre Dame	1983	Mike Rozier, running back, Nebraska
1957	John Crow, halfback, Texas A&M	1984	Doug Flutie, quarterback, Boston College
1958	Pete Dawkins, halfback, Army	1985	Bo Jackson, running back, Auburn
1959	Billy Cannon, halfback, Louisiana State	1986	Vinny Testaverde, quarterback, Miami of Florida
1960	Joe Bellino, halfback, Navy	1987	Tim Brown, wide receiver, Notre Dame
1961	Ernie Davis, halfback, Syracuse	1988	Barry Sanders, tailback, Oklahoma State

1876 Intercollegiate Football Association is formed by Harvard, Yale, Princeton, Rutgers, and Columbia. Rugby style of football is standardized, including rushing, carrying ball, and kicking.

1880 Walter Camp (1859–1925), known as "father of football," initiates first pivotal rules changes. Scrimmage line is first used, teams are reduced to 11 members, and playing field is shortened from 140 yards to 100. Use of scrimmage line brings about position of quarterback.

1882 Camp initiates rules requiring loss of ball if team fails to advance five yards in three downs. Rules speed up game by discouraging stalling tactics and decrease use of kicking in offensive play.

1887 Football reaches South with game between Virginia Military Institute and College of William and Mary.

1888 Tackling below waist is first allowed.

1890 Navy defeats Army, 24–0, at West Point in first Army–Navy game.

▶

NFL and AFL Championships

1933 Chicago Bears defeat New York Giants, 23–21, in first NFL championship playoff.

1934 New York Giants defeat Chicago Bears, 30–13, in NFL championship.

1935 Detroit Lions defeat New York Giants, 26–7, in NFL championship.

1936 Green Bay Packers defeat Boston Redskins, 21–6, in NFL championship.

1937 Washington Redskins defeat Chicago Bears, 28–21, in NFL championship.

1938 New York Giants defeat Green Bay Packers, 23–17, in NFL championship.

1939 Green Bay Packers defeat New York Giants, 27–0, in NFL championship.

1940 Chicago Bears defeat Washington Redskins, 73–0, in NFL championship.

1941 Chicago Bears defeat New York Giants, 37–9, in NFL championship.

1942 Washington Redskins defeat Chicago Bears, 14–6, in NFL championship.

1943 Chicago Bears defeat Washington Redskins, 41–21, in NFL championship.

1944 Green Bay Packers defeat New York Giants, 14–7, in NFL championship.

1945 Cleveland Rams defeat Washington Redskins, 15–14, in NFL championship.

1946 Chicago Bears defeat New York Giants, 24–14, in NFL championship.

1947 Chicago Cardinals defeat Philadelphia Eagles, 28–21, in NFL championship.

1948 Philadelphia Eagles defeat Chicago Cardinals, 7–0, in NFL championship.

1949 Philadelphia Eagles defeat Los Angeles Rams, 14–0, in NFL championship.

1950 Cleveland Browns defeat Los Angeles Rams, 30–28, in NFL championship.

1951 Los Angeles Rams defeat Cleveland Browns, 24–17, in NFL championship.

1952 Detroit Lions defeat Cleveland Browns, 17–7, in NFL championship.

1953 Detroit Lions defeat Cleveland Browns, 17–16, in NFL championship.

1954 Cleveland Browns defeat Detroit Lions, 56–10, in NFL championship.

1955 Cleveland Browns defeat Los Angeles Rams, 38–14, in NFL championship.

1956 New York Giants defeat Chicago Bears, 47–7, in NFL championship.

1957 Detroit Lions defeat Cleveland Browns, 59–14, in NFL championship.

1958 Baltimore Colts defeat New York Giants, 23–17, in NFL championship.

1959 Baltimore Colts defeat New York Giants, 31–16, in NFL championship.

1960 Philadelphia Eagles defeat Green Bay Packers, 17–13, in NFL championship.

1961 Houston Oilers defeat Los Angeles Chargers in the AFL championship for 1960 (January 2). Oilers defeat San Diego Chargers, 10–3, for 1961 championship. Green Bay Packers defeat New York Giants, 37–0, in NFL championship.

1962 Dallas Texans defeat Houston Oilers, 20–17, in AFL championship. Green Bay Packers defeat New York Giants, 16–7, in NFL championship.

1963 Chicago Bears defeat New York Giants, 14–0, in NFL championship.

1964 San Diego Chargers defeat Boston Patriots, 51–10, in 1963 AFL championship (January 5). Buffalo Bills defeat San Diego Chargers, 20–7, for 1964 title. Cleveland Browns defeat Baltimore Colts, 27–0, in NFL championship.

1965 Buffalo Bills defeat San Diego Chargers in AFL championship. Green Bay Packers defeat Cleveland Browns, 23–12, in 1965 NFL championship (January 2).

1966 Kansas City Chiefs defeat Buffalo Bills, 31–7, in AFL championship. Green Bay Packers defeat Dallas Cowboys, 34–27, in NFL championship.

1967 Oakland Raiders defeat Houston Oilers, 40–7, in AFL championship. Green Bay Packers defeat Dallas Cowboys, 21–17, in NFL championship.

1968 New York Jets defeat Oakland Raiders, 27–23, in AFL championship. Baltimore Colts defeat Cleveland Browns, 34–0, in NFL championship.

1970 Kansas City Chiefs defeat Oakland Raiders, 17–7, in 1969 AFL championship. Minnesota Vikings defeat Cleveland Browns, 27–7, in 1969 NFL championship.

1971 Baltimore Colts defeat Oakland Raiders, 27–17, in 1970 AFC championship. Dallas Cowboys defeat San Francisco 49ers, 17–10, in 1970 NFC championship.

1972 Miami Dolphins defeat Baltimore Colts, 21–0, in AFC championship. Dallas Cowboys defeat San Francisco 49ers, 14–13, in NFC championship.

NFL and AFL Championships

1973 Miami Dolphins defeat Pittsburgh Steelers, 21–17, in AFC championship. Washington Redskins defeat Dallas Cowboys, 26–3, in NFC championship.

1974 Miami Dolphins defeat Oakland Raiders, 27–10, in AFC championship. Minnesota Vikings defeat Dallas Cowboys, 27–10, in NFC championship.

1975 Pittsburgh Steelers defeat Oakland Raiders, 24–13, in AFC championship. Minnesota Vikings defeat Los Angeles Rams, 14–10, in NFC championship.

1976 Pittsburgh Steelers defeat Oakland Raiders, 16–10, in AFC championship. Dallas Cowboys defeat Los Angeles Rams, 37–7, in NFC championship.

1977 Oakland Raiders defeat Pittsburgh Steelers, 24–7, in AFC championship. Minnesota Vikings defeat Los Angeles Rams, 14–7, in NFC championship.

1978 Denver Broncos defeat Oakland Raiders, 20–17, in AFC championship. Dallas Cowboys defeat Minnesota Vikings, 23–6, in NFC championship.

1979 Pittsburgh Steelers defeat Houston Oilers, 34–5, in AFC championship. Dallas Cowboys defeat Los Angeles Rams, 28–0, in NFC championship.

1980 Pittsburgh Steelers defeat Houston Oilers, 27–13, in AFC championship. Los Angeles Rams defeat Tampa Bay Buccaneers, 9–0, in NFC championship.

1981 Oakland Raiders defeat San Diego Chargers, 34–27, in AFC championship. Philadelphia Eagles defeat Dallas Cowboys, 20–7, in NFC championship.

1982 Cincinnati Bengals defeat San Diego Chargers, 27–7, in AFC championship. San Francisco 49ers defeat Dallas Cowboys, 28–27, in NFC championship.

1983 Miami Dolphins defeat New York Jets, 14–0, in AFC championship. Washington Redskins defeat Dallas Cowboys, 31–17, in NFC championship.

1984 Los Angeles Raiders defeat Seattle Seahawks, 30–14, in AFC championship. Washington Redskins defeat San Francisco 49ers, 24–21, in NFC championship.

1985 Miami Dolphins defeat Pittsburgh Steelers, 45–28, in AFC championship. San Francisco 49ers defeat Chicago Bears, 23–0, in NFC championship.

1986 New England Patriots defeat Miami Dolphins, 31–14, in AFC championship. Chicago Bears defeat Los Angeles Rams, 24–0, in NFC championship.

1987 Denver Broncos defeat Cleveland Browns, 23–20, in AFC championship. New York Giants defeat Washington Redskins, 17–0, in NFC championship.

1988 Denver Broncos defeat Cleveland Browns, 38–33, in AFC championship. Washington Redskins defeat Minnesota Vikings, 17–10, in NFC championship.

1989 Denver Broncos defeat Cleveland Browns, 37–21, in AFC championship. San Francisco 49ers defeat Los Angeles Rams, 30–3, in NFC championship.

1895 Teams from Latrobe, Pennsylvania and Jeanette, Pennsylvania play first professional football game, in Latrobe.

1902 First Tournament of Roses Association game (Rose Bowl) is held in Pasadena, California; Michigan defeats Stanford, 49–0. In 1923 games come to be called the Rose Bowl.

1905 Public outcry against increasing roughness of game follows season in which there are 18 deaths and 159 injuries in college football. President Theodore Roosevelt asks Harvard, Princeton, and Yale to find ways of saving game.

1905 Forerunner of National Collegiate Athletic Association (NCAA) and football rules committee are founded.

1906 Forward pass is introduced. Mass formations, hurdling, and other hazardous tactics are forbidden. Game also is shortened from 70 to 60 minutes. Neutral zone is established between offensive and defensive lines.

1907 Carlisle Indian School team, led by Jim Thorpe (1888–1953), one of football's greatest players, achieves fame.

1911 Jim Thorpe leads Carlisle Indian School to 18–15 victory over nationally rated Harvard team.

1912 Jim Thorpe scores 22 points in Carlisle Indian School's upset 27–6 victory over Army.

▶

Super Bowl Games

1967 I. Green Bay Packers (NFL) defeat Kansas City Chiefs (AFL), 35–10, at Los Angeles Coliseum (January 15).

1968 II. Green Bay Packers (NFL) defeat Oakland Raiders (AFL), 33–14, at Orange Bowl in Miami (January 14).

1969 III. New York Jets (AFL) defeat Baltimore Colts (NFL), 16–7, at Orange Bowl in Miami (January 12).

1970 IV. Kansas City Chiefs (AFL) defeat Minnesota Vikings (NFL), 23–7, at Tulane Stadium in New Orleans (January 11).

1971 V. Baltimore Colts (AFC) defeat Dallas Cowboys (NFC), 16–13, at Orange Bowl in Miami (January 17).

1972 VI. Dallas Cowboys (NFC) defeat Miami Dolphins (AFC), 24–3, at Tulane Stadium in New Orleans (January 16).

1973 VII. Miami Dolphins (AFC) defeat Washington Redskins (NFC), 14–7, at Los Angeles Coliseum (January 14).

1974 VIII. Miami Dolphins (AFC) defeat Minnesota Vikings (NFC), 24–7, at Rice Stadium in Houston (January 13).

1975 IX. Pittsburgh Steelers (AFC) defeat Minnesota Vikings (NFC), 16–6, at Tulane Stadium in New Orleans (January 12).

1976 X. Pittsburgh Steelers (AFC) defeat Dallas Cowboys (NFC), 21–17, at Orange Bowl in Miami (January 18).

1977 XI. Oakland Raiders (AFC) defeat Minnesota Vikings (NFC), 32–14, at Rose Bowl in Pasadena (January 9).

1978 XII. Dallas Cowboys (NFC) defeat Denver Broncos (AFC), 27–10, at Superdome in New Orleans (January 15).

1979 XIII. Pittsburgh Steelers (AFC) defeat Dallas Cowboys (NFC), 35–31, at Orange Bowl in Miami (January 21).

1980 XIV. Pittsburgh Steelers (AFC) defeat Los Angeles Rams (NFC), 31–19, at Rose Bowl in Pasadena (January 20).

1981 XV. Oakland Raiders (AFC) defeat Philadelphia Eagles (NFC), 27–10, at Superdome in New Orleans (January 25).

1982 XVI. San Francisco 49ers (NFC) defeat Cincinnati Bengals (AFC), 26–21, at Superdome in Pontiac, Michigan (January 24).

1983 XVII. Washington Redskins (NFC) defeat Miami Dolphins (AFC), 27–17, at Rose Bowl in Pasadena (January 30).

1984 XVIII. Los Angeles Raiders (AFC) defeat Washington Redskins (NFC), 38–9, at Tampa Stadium (January 22).

1985 XIX. San Francisco 49ers (NFC) defeat Miami Dolphins (AFC), 38–16, at Stanford Stadium in Palo Alto, California (January 20).

1986 XX. Chicago Bears (NFC) defeat New England Patriots (AFC), 46–10, at Superdome in New Orleans (January 26).

1987 XXI. New York Giants (NFC) defeat Denver Broncos (AFC), 39–20, at Rose Bowl in Pasadena (January 25).

1988 XXII. Washington Redskins (NFC) defeat Denver Broncos (AFC), 42–10, at San Diego Stadium (January 31).

1989 XXIII. San Francisco 49ers (NFC) defeat Denver Broncos (AFC), 55–10, at Miami's Joe Robbie Stadium (January 28).

1912 Four downs are allowed for advancing 10 yards. Score for touchdown is raised to six points from five.

1913 Notre Dame defeats Army in upset, 35–13. Knute Rockne (1883–1931) and Gus Dorais first popularize forward pass.

1918 Rockne becomes Notre Dame coach. His legendary teams between 1918 and 1931 win 105 games with only 12 losses.

1919 Green Bay Packers football team is founded.

1920 Decatur Staleys team is organized. It is renamed Chicago Bears in 1922.

1920 WTAW radio of College Station, Texas broadcasts Texas–Texas A&M game. This is first radio broadcast of a college football contest.

▶

1920s Football achieves enormous popularity in United States. Notre Dame games attract 500,000 fans each year.

1920 American Professional Football Association (later named National Football League) is established, with 11 team franchises. Jim Thorpe is first president.

1924 Harold (Red) Grange (b. 1903), star halfback for Illinois, leads his team to 39–14 victory over Michigan, rushing for total of 402 yards and five touchdowns and passing for sixth touchdown.

1924 Notre Dame fields famous backfield named (by sportswriter Grantland Rice) "Four Horsemen of the Apocalypse": quarterback Harry Stuhldreher, halfbacks Don Miller and Jim Crowley, and fullback Elmer Layden.

1925 Illinois star Red Grange joins Chicago Bears, greatly increasing interest in professional football.

1925 New York Giants team is organized.

1929 Scandal erupts in college football with revelations of dishonest recruiting tactics and payments to star players.

1930s Chicago Bears' coach George Halas (1895–1983) develops T formation; it becomes widely used.

1932 Formation of Washington Redskins team (as Boston Braves). Team is renamed Redskins in 1933 and moves to Washington, D.C., in 1937.

1933 Chicago Bears win first NFL football championship. (*See NFL and AFL Championships.*)

1933 Pittsburgh Steelers team, known as Pittsburgh Pirates until 1941, is founded by Art Rooney with $2,500 winnings from weekend at Saratoga Racetrack.

1934 Forward passing is allowed from any spot behind scrimmage line. After fumble, either team is allowed to advance ball.

1934 Size and shape of ball are standardized after various adjustments in 1912, 1929, and 1931.

1935 Heisman Trophy is awarded for first time. (*See Heisman Trophy Winners.*)

1935 NFL introduces player draft. It allows weaker teams earlier draft choice of eligible players.

c. 1935 Cantilevered shoulder pads are first used.

1937 Cleveland Rams football team is founded; it moves to Los Angeles in 1946.

1939 Helmet is first required in college football. (NFL adopts this requirement in 1943.)

1943 Free substitution is permitted in NFL.

1945 All American Football Conference (AAFC) is organized to rival NFL.

1945 Chicago Cardinals break 29-game losing streak—NFL's longest—by defeating Chicago Bears, 16–7.

1950 AAFC is absorbed into NFL by merger. Following year, NFL is reorganized into 12 teams.

1950s Professional football begins to rival college football in popularity for first time, with NFL attendance rising to 3 million a year by 1958.

1953 University of Oklahoma begins 47-game winning streak.

1959 American Football League (AFL) is established as rival to NFL. New teams include Dallas Cowboys, Denver Broncos, Houston Oilers, and Oakland Raiders. The two leagues struggle to outbid each other for star players throughout 1960s.

1960s Green Bay Packers, under head coach Vince Lombardi (1913–70), dominate pro football, winning six divisional titles, five NFL championships, and first two Super Bowls between 1960 and 1967.

1963 Fullback Jim Brown (b. 1936) of Cleveland Browns rushes for total of 1,863 yards in season, setting NFL rushing record.

1965 University of Alabama quarterback Joe Namath (b. 1943) signs with AFL's New York Jets for $400,000.

1965 Jim Brown sets NFL record for most lifetime touchdowns (126).

1965 Miami Dolphins is established as first AFL expansion franchise.

▶

Pro Football Hall of Fame Members

1963 Sammy Baugh, quarterback from Texas Christian

1963 Bert Bell, NFL commissioner and founder of Philadelphia Eagles

1963 Joe Carr, president of National Football League (1921–39) and co-organizer of American Professional Football Association

1963 Earl (Dutch) Clark, quarterback from Colorado State

1963 Harold (Red) Grange, halfback from Illinois

1963 George Halas, founder of Decatur Staleys (later Chicago Bears) and co-organizer of American Professional Football Association

1963 Mel Hein, center from Washington State

1963 Wilbur (Pete) Henry, tackle from Washington & Jefferson

1963 Robert (Cal) Hubbard, tackle from Centenary & Geneva

1963 Don Hutson, end from Alabama

1963 Earl (Curly) Lambeau, founder, head coach, and general manager of Green Bay Packers

1963 Tim Mara, founder of New York Giants

1963 George Preston Marshall, founder of Washington Redskins

1963 John (Blood) McNally, halfback from St. John's of Minnesota

1963 Bronko Nagurski, fullback from Minnesota

1963 Ernie Nevers, fullback from Stanford and head coach of Duluth Eskimos and Chicago Cardinals

1963 Jim Thorpe, halfback from Carlisle and president of American Professional Football Association

1964 Jimmy Conzelman, quarterback from Washington of St. Louis

1964 Ed Healey, tackle from Dartmouth

1964 Clarke Hinkle, fullback from Bucknell

1964 August (Mike) Michalske, guard from Penn State

1964 Art Rooney, founder of Pittsburgh Steelers

1964 George Trafton, center from Notre Dame

1965 Guy Chamberlin, end from Nebraska

1965 John (Paddy) Driscoll, quarterback from Northwestern

1965 Daniel J. Fortmann, M.D., guard from Colgate

1965 Otto Graham, quarterback from Northwestern

1965 Sid Luckman, quarterback from Columbia

1965 William Roy (Link) Lyman, tackle from Nebraska

1965 Steve Van Buren, halfback from Louisiana State

1965 Bob Waterfield, quarterback from UCLA

1966 Bill Dudley, halfback from Virginia

1966 Joe Guyon, halfback from Carlisle and Georgia Tech

1966 Arnie Herber, quarterback from Wisconsin and Regis

1966 Walt Kiesling, guard from St. Thomas of Minnesota and head coach of Pittsburgh Pirates and Pittsburgh Steelers

1966 George McAfee, halfback from Duke

1966 Steve Owen, tackle from Phillips and head coach of New York Giants

1966 Hugh (Shorty) Ray, technical adviser and supervisor of officials

1966 Clyde (Bulldog) Turner, center and linebacker from Hardin-Simmons

1967 Chuck Bednarik, center and linebacker from Pennsylvania

1967 Charles W. Bidwill, Sr., owner of Chicago Cardinals

1967 Paul E. Brown, head coach and general manager of Cleveland Browns and later of Cincinnati Bengals

1967 Bobby Layne, quarterback from Texas

1967 Dan Reeves, owner of Los Angeles Rams

1967 Ken Strong, halfback from NYU

1967 Joe Stydahar, tackle from West Virginia

1967 Emlen Tunnell, defensive back from Toledo and Iowa

1968 Cliff Battles, halfback and quarterback from West Virginia Wesleyan

1968 Art Donovan, defensive tackle from Boston College

1968 Elroy (Crazy Legs) Hirsch, halfback and end from Wisconsin and Michigan

1968 Wayne Millner, end from Notre Dame

1968 Marion Motley, fullback from South Carolina State and Nevada

1968 Charley Trippi, halfback and quarterback from Georgia

1968 Alex Wojciechowicz, center and linebacker from Fordham

1969 Albert Glen (Turk) Edwards, tackle from Washington State

1969 Earle (Greasy) Neale, head coach of West Virginia Wesleyan

Pro Football Hall of Fame Members

1969 Leo Nomellini, defensive tackle from Minnesota

1969 Fletcher (Joe) Perry, fullback from Compton Junior College

1969 Ernie Stautner, defensive tackle from Boston College

1970 Jack Christiansen, safety from Colorado State

1970 Tom Fears, end from Santa Clara and UCLA

1970 Hugh McElhenny, halfback from Washington

1970 Pete Pihos, end from Indiana

1971 Jim Brown, fullback from Syracuse

1971 Norm Van Brocklin, quarterback from Oregon

1971 Bill Hewitt, end from Michigan

1971 Frank (Bruiser) Kinard, tackle from Mississippi

1971 Vince Lombardi, head coach and general manager of Green Bay Packers and Washington Redskins

1971 Andy Robustelli, defensive end from Arnold

1971 Y. A. Tittle, quarterback from Louisiana State

1972 Lamar Hunt, founder of American Football League and owner of Kansas City Chiefs

1972 Gino Marchetti, defensive end from San Francisco

1972 Ollie Matson, halfback from San Francisco

1972 Clarence (Ace) Parker, quarterback from Duke

1973 Raymond Berry, end from SMU

1973 Jim Parker, guard and tackle from Ohio State

1973 Joe Schmidt, linebacker from Pittsburgh and head coach of Detroit Lions

1974 Tony Canadeo, halfback from Gonzaga

1974 Bill George, linebacker from Wake Forest

1974 Lou Groza, offensive tackle and place-kicker from Ohio State

1974 Richard (Night Train) Lane, cornerback from Scottsbluff, Nebraska Junior College

1975 George Connor, offensive tackle, defensive tackle, and linebacker from Holy Cross and Notre Dame

1975 Roosevelt Brown, offensive tackle from Morgan State

1975 Dante Lavelli, end from Ohio State

1975 Lenny Moore, flanker and running back from Penn State

1976 Ray Flaherty, end from Gonzaga and head coach of Washington Redskins

1976 Len Ford, defensive end from Morgan State and Michigan

1976 Jim Taylor, fullback from Louisiana State

1977 Frank Gifford, halfbacker and flanker from USC

1977 Forrest Gregg, tackle and guard from SMU

1977 Gale Sayers, halfback from Kansas

1977 Bart Starr, quarterback from Alabama

1977 Bill Willis, guard and middle guard from Ohio State

1978 Lance Alworth, wide receiver from Arkansas

1978 Weeb Ewbank, head coach of Baltimore Colts and New York Jets

1978 Alphonse (Tuffy) Leemans, running back from Oregon and George Washington

1978 Ray Nitschke, linebacker from Illinois

1978 Larry Wilson, safety from Utah

1979 Dick Butkus, linebacker from Illinois

1979 Yale Lary, linebacker from Texas A&M

1979 Ron Mix, offensive tackle from USC

1979 Johnny Unitas, quarterback from Louisville

1980 Herb Adderley, cornerback from Michigan State

1980 David (Deacon) Jones, defensive end from South Carolina State

1980 Bob Lilly, defensive tackle from TCU

1980 Jim Otto, center from Miami of Florida

1981 Morris (Red) Badgro, end from USC

1981 George Blanda, quarterback and place-kicker from Kentucky

1981 Willie Davis, defensive end from Grambling

1981 Jim Ringo, center from Syracuse

1982 Doug Atkins, defensive end from Tennessee

1982 Sam Huff, linebacker from West Virginia

1982 George Musso, guard and tackle from Millikin

1982 Merlin Olson, defensive tackle from Utah State

1983 Bobby Bell, linebacker and defensive end from Minnesota

1983 Sid Gillman, head coach of Los Angeles Rams and head coach and general manager of San Diego Chargers

1983 Sonny Jurgensen, quarterback from Duke

1983 Bobby Mitchell, wide receiver and halfback from Illinois

1983 Paul Warfield, wide receiver from Ohio State

1984 Willie Brown, cornerback from Grambling

Pro Football Hall of Fame Members

1984	Mike McCormack, tackle from Kansas and president of Seattle Seahawks
1984	Arnie Weinmeister, defensive tackle from Washington
1985	Frank Gatski, center from Marshall and Auburn
1985	Joe Namath, quarterback from Alabama
1985	Pete Rozelle, NFL commissioner
1985	O. J. Simpson, running back from USC
1985	Roger Staubach, quarterback from Navy
1986	Paul Hornung, quarterback from Notre Dame
1986	Ken Houston, safety from Prairie View A&M
1986	Willie Lanier, linebacker from Morgan State
1986	Fran Tarkenton, quarterback from Georgia
1986	Doak Walker, halfback from SMU
1987	Larry Csonka, fullback from Syracuse
1987	Len Dawson, quarterback from Purdue
1987	Joe Greene, defensive tackle from North Texas State
1987	John Henry Johnson, fullback from Arizona State
1987	Jim Langer, center from South Dakota State
1987	Don Maynard, wide receiver from Texas Western
1987	Charley Taylor, wide receiver and running back from Arizona State
1987	Gene Upshaw, guard from Texas A&I
1988	Fred Biletnikoff, wide receiver from Florida State
1988	Mike Ditka, tight end from Pittsburgh and head coach of Chicago Bears
1988	Jack Ham, linebacker from Penn State
1988	Alan Page, defensive tackle from Notre Dame
1989	Mel Blount, cornerback from Southern
1989	Terry Bradshaw, quarterback from Louisiana Tech
1989	Art Shell, tackle from Maryland State—Eastern Shore
1989	Willie Wood, safety from USC

1966 Washington Redskins defeat New York Giants, 72–41, on November 27, setting record for total number of points scored in game and establishing second-highest total of points scored by one team.

1967 Green Bay Packers (NFL) defeat Kansas City Chiefs (AFL) in first Super Bowl, in Los Angeles. Game is attended by 63,000. (*See Super Bowl Games.*)

1967 St. Louis Cardinals kicker Jim Bakken sets record by scoring seven field goals in game.

1968 NBC cuts off final minutes of critical New York Jets–Oakland Raiders game to broadcast production of children's classic *Heidi*. Oakland scores twice after broadcast cutoff and comes from behind to win, 43–32.

1969 Chester Marcol of Hillsdale College, Michigan, kicks 62-yard field goal, longest kick to date in college or professional football.

1969 Fourteen black student football players at University of Wyoming are dropped from team after defying coach's order to cease protest demonstrations on behalf of the Black Students' Alliance.

1970 New Orleans Saints' Tom Dempsey breaks Chester Marcol's field-goal record (1969) by kicking 63-yard goal.

1970 Merger of NFL and AFL becomes effective. NFL is divided into American and National conferences, with 13 teams each.

1971 Cornell University running back Ed Marinaro (b. 1950) sets new college football rushing record of 4,132 yards in three years.

1972 NCAA ruling allows freshmen to play on varsity teams.

1972 Robert Irsay swaps his Los Angeles Rams for Carroll Rosenbloom's Baltimore Colts in unusual no-cash trade.

1972 Inbounds lines are moved to 70 feet, 9 inches from the sidelines.

1973 World Football League (WFL) is formed, consisting of 10 teams; it goes out of business in 1975.

1973 Buffalo Bills running back O. J. Simpson breaks Jim Brown's season rushing record

▶

(1963), setting new pro record of 2,003 yards in season.

1974 NFL rules change. Goalposts are moved back 10 yards from goal line to end line to lower number of field goals; kickoff position is changed from 40-yard line to the 35; and 15-minute sudden-death period is introduced to help prevent tie games.

1975 Pass-interference penalty is lowered from 15 to 10 yards.

1980 Oklahoma defeats Colorado, 82–42, in highest-scoring game in college football.

1982 Half of NFL scheduled games are canceled due to first pro football strike held during regular season.

1982 NFL signs $2 billion, five-year television contract.

1983 Federal court awards $49 million to Los Angeles Coliseum and Oakland Raiders, claiming NFL, in attempting to prevent the Oakland franchise from moving to Los Angeles, had violated antitrust laws.

1983 United States Football League (USFL) kicks off its first season with games scheduled throughout spring and summer. Michigan Panthers defeat Philadelphia Stars, 24–22, in first USFL championship, on July 17.

1984 U.S. Supreme Court rules that NCAA can no longer enjoy exclusive control over televising of collegiate football, as such control violates Sherman Antitrust Act.

1984 U.S. Supreme Court rules that NFL cannot restrict teams from changing cities of affiliation, concluding that doing so violates antitrust laws.

1984 Philadelphia Stars defeat Arizona Wranglers, 23–3, in USFL championship.

1984 Los Angeles Rams running back Eric Dickerson (b. 1960) breaks O. J. Simpson's single-season rushing record (1973), running for season total of 2,105 yards.

1985 Baltimore Stars defeat Oakland Invaders, 28–24, in USFL championship.

1985 Grambling State University coach Eddie Robinson tallies his 324th victory, making him winningest college football coach of all time.

1986 Chicago Bears running back Walter Payton (b. 1954) sets lifetime pro football rushing record of 14,860 yards.

1986 USFL fails in $1.7 billion lawsuit against NFL and cancels its 1986 season; USFL receives only $3 in damages.

1986 NFL adopts video instant-replay rule and assigns new official to monitor instant replays. Replay official can override field official's call based on videotape review.

1987 NFL Players' Association goes on strike from September 22 to October 14.

1987 NFL outlaws dangerous blocking techniques. Tighter controls are instituted on roughing passers.

1990 NFL ends ban on drafting college players who are juniors.

Basketball

1500s Ollamalitzli, forerunner of basketball, is played by Aztecs in Mexico, using a stone hoop and rubber ball. Winning player gets as reward clothing of all those in audience; loser is beheaded.

1891 American physical education teacher James Naismith (1861–1939) invents basketball for his students in Springfield, Massachusetts YMCA training school. It is designed as an indoor sport to fill time between football and baseball seasons. Naismith uses soccer ball and closed-bottom

▶

NCAA Championships

1939	Oregon defeats Ohio State, 46–33	**1965**	UCLA defeats Michigan, 91–80
1940	Indiana defeats Kansas, 60–42	**1966**	Texas Western defeats Kentucky, 72–65
1941	Wisconsin defeats Washington State, 39–34	**1967**	UCLA defeats Dayton, 79–64
1942	Stanford defeats Dartmouth, 53–38	**1968**	UCLA defeats North Carolina, 78–55
1943	Wyoming defeats Georgetown, 46–34	**1969**	UCLA defeats Purdue, 92–72
1944	Utah defeats Dartmouth, 42–40	**1970**	UCLA defeats Jacksonville, 80–69
1945	Oklahoma A&M defeats NYU, 49–45	**1971**	UCLA defeats Villanova, 68–62
1946	Oklahoma A&M defeats North Carolina, 43–40	**1972**	UCLA defeats Florida State, 81–76
1947	Holy Cross defeats Oklahoma, 58–47	**1973**	UCLA defeats Memphis State, 87–66
1948	Kentucky defeats Baylor, 58–42	**1974**	North Carolina State defeats Marquette, 76–64
1949	Kentucky defeats Oklahoma State, 46–36	**1975**	UCLA defeats Kentucky, 92–85
1950	CCNY defeats Bradley, 71–68	**1976**	Indiana defeats Michigan, 86–68
1951	Kentucky defeats Kansas State, 68–58	**1977**	Marquette defeats North Carolina, 67–59
1952	Kansas defeats St. John's, 80–63	**1978**	Kentucky defeats Duke, 94–88
1953	Indiana defeats Kansas, 69–68	**1979**	Michigan defeats Indiana State, 75–64
1954	LaSalle defeats Bradley, 92–76	**1980**	Louisville defeats UCLA, 59–54
1955	San Francisco defeats LaSalle, 77–63	**1981**	Indiana defeats North Carolina, 63–50
1956	San Francisco defeats Iowa, 83–71	**1982**	North Carolina defeats Georgetown, 63–62
1957	North Carolina defeats Kansas, 54–53	**1983**	North Carolina State defeats Houston, 54–52
1958	Kentucky defeats Seattle, 84–72	**1984**	Georgetown defeats Houston, 84–75
1959	California defeats West Virginia, 71–70	**1985**	Villanova defeats Georgetown, 66–64
1960	Ohio State defeats California, 75–55	**1986**	Louisville defeats Duke, 72–69
1961	Cincinnati defeats Ohio State, 70–65	**1987**	Indiana defeats Syracuse, 74–73
1962	Cincinnati defeats Ohio State, 71–59	**1988**	Kansas defeats Oklahoma, 83–79
1963	Loyola (Ill.) defeats Cincinnati, 60–58	**1989**	UNLV defeats Duke, 103–73
1964	UCLA defeats Duke, 98–83		

peach basket hung from top of gymnasium.

1892 Naismith publishes rules of basketball in school newspaper.

1890s Basketball's popularity spreads rapidly through YMCAs in northeastern United States. Balls are originally emptied from baskets by climbing a ladder, a method replaced by jabbing ball out with poles, then by attaching strings to netting.

1893 Wire baskets replace wooden ones. Baskets have wire netting with hole in bottom.

1894 Backboards are first used. Soccer ball is replaced by larger ball.

1895 Number of points for field goals is lowered from three to two, and for free throws from three to one.

1896 First women's intercollegiate championship game is held between Stanford and University of California.

1897 First men's intercollegiate game is played between Yale and University of Pennsylvania.

1897–98 Five-player teams become norm.

▶

1898 First two professional basketball leagues are formed, National and New England. Each lasts only two years.

1899 Women's basketball, more popular than men's, is played according to modified rules.

1900 Dribbling is introduced.

1901 Amateur Athletic Union (AAU) is established to administer sport, as its popularity grows beyond the bounds of YMCA.

1904 Basketball is played as demonstration sport at Olympics in St. Louis.

1908 Glass backboards are introduced.

1908 NCAA takes charge in regulating college basketball rules.

1908–09 Limit of five personal fouls is introduced to prevent rough play.

1910–11 Personal foul limit is reduced to four.

1912 Open nets are introduced, eliminating need for retrieving balls.

1915 NCAA and AAU form joint rules committee.

1927 All-black Harlem Globetrotters team is founded. In addition to their ball-playing skills, Globetrotters perform entertaining highjinks.

1932 Ten-second rule is adopted to limit stalling in backcourt.

1934 Harlem Renaissance Five has 88-game winning streak.

1934 Total of 16,000 fans attend double-header college games at New York's Madison Square Garden, marking growing enthusiasm for college basketball.

1935 In new rule, offensive players are prohibited from standing on free-throw line during play for more than three seconds.

1935–38 Hank Luisetti of Stanford University develops innovative one-hand shot and breaks many records.

1936 Basketball becomes Olympic sport.

1937 Center jump after field goal is eliminated.

1937 National Basketball League (NBL) is founded.

1938 National Invitational Tournament for college basketball is established at Madison Square Garden.

1939 Position of backboards is changed from two to four feet in from end lines.

1940 Indiana University defeats University of Kansas, 60–42, in first NCAA championship.

1944 NCAA/AAU Joint Rules Committee issues new rules prohibiting goal-tending and increasing allowable number of personal fouls from four to five.

1946 Basketball Association of America (BAA) is organized.

1947 Philadelphia Warriors defeat Chicago Stags, 4–1, in first BAA championship.

1949 National Basketball Association (NBA) is formed by merger of BAA and NBL.

1949 Joe Fulks of Philadelphia Warriors develops jump shot and scores record 63 points in one game.

1950 Minneapolis Lakers defeat Syracuse Nationals, 4–2, in first NBA championship. (*See NBA Championships.*)

1950 NBA color barrier is broken when black player Chuck Cooper joins Boston Celtics.

1951 Scandal erupts in college basketball with revelations of game-fixing for gambling purposes.

1954 Frank Selvy of Furman College sets 100–point best individual scoring record for player from major college.

1954–55 Twenty-four-second rule, requiring attempted shot within 24 seconds of possession of ball, is introduced. Rule greatly increases pace of the game. Scores and attendance sharply increase.

1958 Syracuse Nationals' star Adolph Schayes breaks the lifetime pro scoring record, besting 11,764 points set by George Mihan. Schayes totals 19,209 points during career.

1959 Boston Celtics defeat Minneapolis Lakers, 173–139, in highest-scoring pro contest to date.

▶

NBA Championships

1950 Minneapolis Lakers defeat Syracuse Nationals, 4 games to 2.

1951 Rochester Royals defeat New York Knickerbockers, 4 games to 3.

1952 Minneapolis Lakers defeat New York Knickerbockers, 4 games to 3.

1953 Minneapolis Lakers defeat New York Knickerbockers, 4 games to 1.

1954 Minneapolis Lakers defeat Syracuse Nationals, 4 games to 3.

1955 Syracuse Nationals defeat Fort Wayne Pistons, 4 games to 3.

1956 Philadelphia Warriors defeat Fort Wayne Pistons, 4 games to 1.

1957 Boston Celtics defeat St. Louis Hawks, 4 games to 3.

1958 St. Louis Hawks defeat Boston Celtics, 4 games to 2.

1959 Boston Celtics defeat Minneapolis Lakers, 4 games to 0.

1960 Boston Celtics defeat St. Louis Hawks, 4 games to 3.

1961 Boston Celtics defeat St. Louis Hawks, 4 games to 1.

1962 Boston Celtics defeat Los Angeles Lakers, 4 games to 3.

1963 Boston Celtics defeat Los Angeles Lakers, 4 games to 2.

1964 Boston Celtics defeat San Francisco Warriors, 4 games to 1.

1965 Boston Celtics defeat Los Angeles Lakers, 4 games to 1.

1966 Boston Celtics defeat Los Angeles Lakers, 4 games to 3.

1967 Philadelphia 76ers defeat San Francisco Warriors, 4 games to 2.

1968 Boston Celtics defeat Los Angeles Lakers, 4 games to 2.

1969 Boston Celtics defeat Los Angeles Lakers, 4 games to 3.

1970 New York Knickerbockers defeat Los Angeles Lakers, 4 games to 3.

1971 Milwaukee Bucks defeat Baltimore Bullets, 4 games to 0.

1972 Los Angeles Lakers defeat New York Knickerbockers, 4 games to 1.

1973 New York Knickerbockers defeat Los Angeles Lakers, 4 games to 1.

1974 Boston Celtics defeat Milwaukee Bucks, 4 games to 3.

1975 Golden State Warriors defeat Washington Bullets, 4 games to 0.

1976 Boston Celtics defeat Phoenix Suns, 4 games to 2.

1977 Portland Trail Blazers defeat Philadelphia 76ers, 4 games to 2.

1978 Washington Bullets defeat Seattle SuperSonics, 4 games to 3.

1979 Seattle SuperSonics defeat Washington Bullets, 4 games to 1.

1980 Los Angeles Lakers defeat Philadelphia 76ers, 4 games to 2.

1981 Boston Celtics defeat Houston Rockets, 4 games to 2.

1982 Los Angeles Lakers defeat Philadelphia 76ers, 4 games to 2.

1983 Philadelphia 76ers defeat Los Angeles Lakers, 4 games to 0.

1984 Boston Celtics defeat Los Angeles Lakers, 4 games to 3.

1985 Los Angeles Lakers defeat Boston Celtics, 4 games to 2.

1986 Boston Celtics defeat Houston Rockets, 4 games to 2.

1987 Los Angeles Lakers defeat Boston Celtics, 4 games to 2.

1988 Los Angeles Lakers defeat Detroit Pistons, 4 games to 3.

1989 Detroit Pistons defeat Los Angeles Lakers, 4 games to 0.

1990 Detroit Pistons defeat Portland Trailblazers, 4 games to 0.

1960 Bill Russell (b. 1934), center for Boston Celtics, retrieves record 51 rebounds in a game.

1961 Collegiate players are again implicated in point-shaving scandal, following that of 1951.

▶

1962 Philadelphia Warriors defeat New York Knickerbockers, 169–147, breaking record for total points per game set in 1959. Wilt Chamberlain (b. 1936) scores 100 of Warriors' points.

1962 Wilt Chamberlain of Philadelphia Warriors establishes record for most points for season and for game, and for highest scoring average for season. He is also recognized as highest-scoring player in college basketball history.

1966 Bill Russell (b. 1934), becomes first black head coach in any professional sport when he is named head coach of Boston Celtics.

1966 Wilt Chamberlain leads NBA in scoring for seventh consecutive year.

1966 Boston Celtics win record eighth consecutive NBA championship.

1967 American Basketball Association (ABA) is formed.

1968 Pittsburgh Pipers defeat New Orleans Buccaneers, 4–3, in the first ABA championship.

1972 Immaculate College wins the first AIAW (women's) collegiate basketball championship.

1972 United States loses to Soviet Union at Munich Olympics, ending streak of 63 U.S. Olympic basketball victories.

1974 Moses Malone (b. 1955), a 6-foot, 11-inch 19-year-old, signs contract with Utah Stars, making him first player to join a major-league pro team directly from high school.

1975 UCLA Bruins wins its 10th college basketball championship in 12 years.

1976 NBA and American Basketball Association merge. NBA now has 22 teams.

1982 Louisiana Tech defeats Cheyney State, 76–62, in first NCAA women's basketball championship. NCAA women's contest virtually replaces AIAW game.

1983 Detroit Pistons and Denver Nuggets establish NBA single-game total scoring record with combined 370 points. Detroit wins, 186–184.

1985 Average height of NBA players is 6 feet, 7 inches. Their average salary is $300,000 per year. Total of 11 million fans attend NBA games each year by this date.

1986 45-second shot clock is introduced in collegiate basketball.

1987 Kareem Abdul-Jabbar (b. 1947) ends 10-year streak of double-digit scoring in 787 consecutive games.

1987 Record crowd of 52,000 attends NBA game between Detroit and Philadelphia.

1989 Kareem Abdul-Jabbar retires, holding NBA career record for most points scored, with 38,387.

1989 Three free throws are to be given to shooter fouled while attempting 3-point shot in college basketball.

Baseball

1833 Philadelphia's Olympic Ball Club plays rudimentary form of baseball, probably derived from English game called "rounders."

1845 New York baseball player Alexander Cartwright (1820–92) develops set of game rules for his club and draws field diagram including 90-foot baselines and home plate, where batter stands. Hit caught on first bounce constitutes an out; pitches are thrown underhand; first team to score 21 runs wins.

1846 First baseball game on record is played between New Yorks and Knickerbockers, at Elysian Fields in Hoboken, New Jersey. New Yorks win, 23–1.

▶

1848 New first-base rule is introduced, allowing fielder to tag base before runner reaches bag and so make an out.

c. 1850 Players may no longer throw ball at runner to put him out.

1853 *New York Mercury* publishes first news story on baseball.

1854 The baseball is standardized (until 1872) to weigh 5½ to 6 ounces and to measure 2¾ to 3½ inches in diameter.

1857 Baseball-club convention is held in New York to iron out differences in rules.

1858 National Association of Baseball Players is organized in New York City.

1859 Amherst defeats Williams, 60–32, in the first intercollegiate baseball game.

1860 Seventh-inning stretch is common feature at baseball games.

1863 New rules are established that allow for calling balls and strikes.

c. 1863 Rule stipulating that hit caught on first bounce constitutes an out is eliminated.

1864 William A. "Candy" Cummings (1848–1924) throws first curveball.

1864 Alfred Reach (1841–1928) proclaims himself first baseball "professional."

1865 Member level in National Association of Baseball Players reaches 91 clubs.

1866 Brooklyn Atlantics defeat Philadelphia Athletics, 27–10, in unofficial U.S. baseball championship.

1869 First edition of annual *Spalding's Official Baseball Guide* by sportswriter Henry Chadwick (1824–1908) is published.

1869 Term *battery*, denoting pitcher and catcher, comes into baseball parlance.

1869 Red Stockings of Cincinnati, Ohio, become first all-professional baseball team this year, enjoying undefeated season: 56 wins and 1 tie. Star shortstop George Wright earns top pay of $1,400.

c. 1870 Sliding into base is becoming common.

c. 1870 The bunt has been invented by this date.

1871 First switch hitter is major-league player Bob Ferguson (1845–94).

1871 National Association of Professional Baseball Players is organized, first professional baseball league. League lasts five years but is notable for gambling, liquor sales, and rowdiness at games.

1872 Regulations limiting ball size, still observed today, are set: weight of 5 to 5½ ounces, circumference of 9 to 9¼ inches.

1875 Fielding gloves, without padding, are introduced.

▶

World Series

1903 Boston Red Sox (American League) over Pittsburgh Pirates (National League), 5–1.

1904 No World Series.

1905 New York Giants (National League) over Philadelphia Athletics (American League), 4–1.

1906 Chicago White Sox (American League) over Chicago Cubs (National League), 4–2.

1907 Chicago Cubs (National League) over Detroit Tigers (American League), 4–1; 1 tie.

1908 Chicago Cubs (National League) over Detroit Tigers (American League), 4–1.

1909 Pittsburgh Pirates (National League) over Detroit Tigers (American League), 4–3.

1910 Philadelphia Athletics (American League) over Chicago Cubs (National League), 4–1.

1911 Philadelphia Athletics (American League) over New York Giants (National League), 4–2.

1912 Boston Red Sox (American League) over New York Giants (National League), 4–3; 1 tie.

1913 Philadelphia Athletics (American League) over New York Giants (National League), 4–1.

1914 Boston Braves (National League) over Philadelphia Athletics (American League), 4–0.

1915 Boston Red Sox (American League) over Philadelphia Phillies (National League), 4–1.

1916 Boston Red Sox (American League) over Brooklyn Dodgers (National League), 4–1.

World Series

1917 Chicago White Sox (American League) over New York Giants (National League), 4–2.

1918 Boston Red Sox (American League) over Chicago Cubs (National League), 4–2.

1919 Cincinnati Reds (National League) over Chicago White Sox (American League), 5–3.

1920 Cleveland Indians (American League) over Brooklyn Dodgers (National League), 5–2.

1921 New York Giants (National League) over New York Yankees (American League), 5–3.

1922 New York Giants (National League) over New York Yankees (American League), 4–0; 1 tie.

1923 New York Yankees (American League) over New York Giants (National League), 4–2.

1924 Washington Senators (American League) over New York Giants (National League), 4–3.

1925 Pittsburgh Pirates (National League) over Washington Senators (American League), 4–3.

1926 St. Louis Cardinals (National League) over New York Yankees (American League), 4–3.

1927 New York Yankees (American League) over Pittsburgh Pirates (National League), 4–0.

1928 New York Yankees (American League) over St. Louis Cardinals (National League), 4–0.

1929 Philadelphia Athletics (American League) over Chicago Cubs (National League), 4–1.

1930 Philadelphia Athletics (American League) over St. Louis Cardinals (National League), 4–2.

1931 St. Louis Cardinals (National League) over Philadelphia Athletics (American League), 4–3.

1932 New York Yankees (American League) over Chicago Cubs (National League), 4–0.

1933 New York Giants (National League) over Washington Senators (American League), 4–1.

1934 St. Louis Cardinals (National League) over Detroit Tigers (American League), 4–3.

1935 Detroit Tigers (American League) over Chicago Cubs (National League), 4–2.

1936 New York Yankees (American League) over New York Giants (National League), 4–2.

1937 New York Yankees (American League) over New York Giants (National League), 4–1.

1938 New York Yankees (American League) over Chicago Cubs (National League), 4–0.

1939 New York Yankees (American League) over Cincinnati Reds (National League), 4–0.

1940 Cincinnati Reds (National League) over Detroit Tigers (American League), 4–3.

1941 New York Yankees (American League) over Brooklyn Dodgers (National League), 4–1.

1942 St. Louis Cardinals (National League) over New York Yankees (American League), 4–1.

1943 New York Yankees (American League) over St. Louis Cardinals (National League), 4–1.

1944 St. Louis Cardinals (National League) over St. Louis Browns (American League), 4–2.

1945 Detroit Tigers (American League) over Chicago Cubs (National League), 4–3.

1946 St. Louis Cardinals (National League) over Boston Red Sox (American League), 4–3.

1947 New York Yankees (American League) over Brooklyn Dodgers (National League), 4–3.

1948 Cleveland Indians (American League) over Boston Braves (National League), 4–2.

1949 New York Yankees (American League) over Brooklyn Dodgers (National League), 4–1.

1950 New York Yankees (American League) over Philadelphia Phillies (National League), 4–0.

1951 New York Yankees (American League) over New York Giants (National League), 4–2.

1952 New York Yankees (American League) over Brooklyn Dodgers (National League), 4–3.

1953 New York Yankees (American League) over Brooklyn Dodgers (National League), 4–2.

1954 New York Giants (National League) over Cleveland Indians (American League), 4–0.

1955 Brooklyn Dodgers (National League) over New York Yankees (American League), 4–3; Johnny Podres (Brooklyn) MVP.

1956 New York Yankees (American League) over Brooklyn Dodgers (National League), 4–3; Don Larsen (New York) MVP.

1957 Milwaukee Braves (National League) over New York Yankees (American League), 4–3; Lew Burdette (Milwaukee) MVP.

1958 New York Yankees (American League) over Milwaukee Braves (National League), 4–3; Bob Turley (New York) MVP.

1959 Los Angeles Dodgers (National League) over Chicago White Sox (American League), 4–2; Larry Sherry (Los Angeles) MVP.

World Series

1960 Pittsburgh Pirates (National League) over New York Yankees (American League), 4–3; Bobby Richardson (New York) MVP.

1961 New York Yankees (American League) over Cincinnati Reds (National League), 4–1; Whitey Ford (New York) MVP.

1962 New York Yankees (American League) over San Francisco Giants (National League), 4–3; Ralph Terry (New York) MVP.

1963 Los Angeles Dodgers (National League) over New York Yankees (American League), 4–0; Sandy Koufax (Los Angeles) MVP.

1964 St. Louis Cardinals (National League) over New York Yankees (American League), 4–3; Bob Gibson (St. Louis) MVP.

1965 Los Angeles Dodgers (National League) over Minnesota Twins (American League), 4–3; Sandy Koufax (Los Angeles) MVP.

1966 Baltimore Orioles (American League) over Los Angeles Dodgers (National League), 4–0; Frank Robinson (Baltimore) MVP.

1967 St. Louis Cardinals (National League) over Boston Red Sox (American League), 4–3; Bob Gibson (St. Louis) MVP.

1968 Detroit Tigers (American League) over St. Louis Cardinals (National League), 4–3; Mickey Lolich (Detroit) MVP.

1969 New York Mets (National League) over Baltimore Orioles (American League), 4–1; Donn Clendenon (New York) MVP.

1970 Baltimore Orioles (American League) over Cincinnati Reds (National League), 4–1; Brooks Robinson (Baltimore) MVP.

1971 Pittsburgh Pirates (National League) over Baltimore Orioles (American League), 4–3; Roberto Clemente (Pittsburgh) MVP.

1972 Oakland A's (American League) over Cincinnati Reds (National League), 4–3; Gene Tenace (Oakland) MVP.

1973 Oakland A's (American League) over New York Mets (National League), 4–3; Reggie Jackson (Oakland) MVP.

1974 Oakland A's (American League) over Los Angeles Dodgers (National League), 4–1; Rollie Fingers (Oakland) MVP.

1975 Cincinnati Reds (National League) over Boston Red Sox (American League), 4–3; Pete Rose (Cincinnati) MVP.

1976 Cincinnati Reds (National League) over New York Yankees (American League), 4–0; Johnny Bench (Cincinnati) MVP.

1977 New York Yankees (American League) over Los Angeles Dodgers (National League), 4–2; Reggie Jackson (New York) MVP.

1978 New York Yankees (American League) over Los Angeles Dodgers (National League), 4–2; Bucky Dent (New York) MVP.

1979 Pittsburgh Pirates (National League) over Baltimore Orioles (American League), 4–3; Willie Stargell (Pittsburgh) MVP.

1980 Philadelphia Phillies (National League) over Kansas City Royals (American League), 4–2; Mike Schmidt (Philadelphia) MVP.

1981 Los Angeles Dodgers (National League) over New York Yankees (American League), 4–2; Ron Cey, Pedro Guerrero, Steve Yeager (Los Angeles) MVPs.

1982 St. Louis Cardinals (National League) over Milwaukee Brewers (American League), 4–3; Darrell Porter (St. Louis) MVP.

1983 Baltimore Orioles (American League) over Philadelphia Phillies (National League), 4–1; Rick Dempsey (Baltimore) MVP.

1984 Detroit Tigers (American League) over San Diego Padres (National League), 4–1; Alan Trammell (Detroit) MVP.

1985 Kansas City Royals (American League) over St. Louis Cardinals (National League), 4–3; Bret Saberhagen (Kansas City) MVP.

1986 New York Mets (National League) over Boston Red Sox (American League), 4–3; Ray Knight (New York) MVP.

1987 Minnesota Twins (American League) over St. Louis Cardinals (National League), 4–3; Frank Viola (Minnesota) MVP.

1988 Los Angeles Dodgers (National League) over Oakland A's (American League), 4–1; Orel Hershiser (Los Angeles) MVP.

1989 Oakland A's (American League) over San Francisco Giants (National League), 4–0; Dave Stewart (Oakland) MVP.

1876 National League of Professional Baseball Clubs is created to "clean up" the game. Constitution forbids teams from playing Sunday games or selling liquor.

1876 Boston's Joe Borden pitches first National League no-hitter.

1876 Chicago wins the first pennant with 52–14 season record. (*See National League Pennant Winners.*)

1879 By this date most teams have a "spare" pitcher.

1880 Number of balls required for a walk is reduced from nine to eight. The number is gradually reduced to four to speed up game.

1882 American Association is created to compete with National League. Rules covering players, teams, and minor leagues are drawn up.

1883 First baseball game is played under electric lights in Fort Wayne, Indiana; Fort Wayne Nine defeat Quincy, 19–11, in seven innings.

1884 Moses Fleetwood Walker, first black major-league player, signs with American Association.

1885 Cuban Giants, first black professional team, is formed on Long Island, New York.

1887 Standard strike zone is established. Also, batter may no longer call for high or low pitch.

1887 Stolen bases are allowed.

1888 Chicago White Stockings' owner Albert G. Spalding (1850–1915) organizes the first international postseason baseball tour.

1890 Women's baseball team is organized, but more interest is expressed in their costume—short-sleeved shirts and short pants—than their play.

c. 1890 Pitcher is allowed to throw overhand; short run before throwing is allowed.

1894 Boston Nationals outfielder Hugh Duffy slugs season batting average of .438, highest yet on record in major leagues.

1900 Five-sided home plate is introduced.

1901 Chicago wins first American League championship.

1901 Foul ball now is considered strike in National League. (American League adopts rule in 1903.)

1903 Boston wins first World Series. (*See World Series.*)

1904 Boston Red Sox pitcher Denton T. "Cy" Young (1867–1955) pitches the first perfect game.

1907 Special committee of baseball executives meets to confirm Cooperstown, New York as birthplace of baseball and American Abner Doubleday (1819–93) as its founder in 1839 (a fact later discredited).

1908 Cleveland pitcher Addie Joss pitches perfect game.

1909 First "modern" ballpark, permanent three-deck structure of concrete and steel, is built in Philadelphia.

1911 Cork-center baseball, invented in 1909, is marketed and adopted by both major leagues.

1911 Pitching legend Cy Young (1867–1955) retires.

1912 Boston's Fenway Park opens.

1912 Earned run average (ERA) becomes official National League statistic. (It is adopted by American League in 1913.)

1913 Washington Senators pitcher Walter Johnson (1887–1946) sets consecutive-inning shutout record, pitching 56 innings with no runs scored.

1914 Chicago businessman James A. Gilmore founds a third major league, the Federal League.

1917 Earned runs include those scored with the help of stolen bases.

1917 Boston Red Sox pitcher Ernie Shore throws perfect game against Washington Senators.

1918 Secretary of War Newton D. Baker closes major-league season on September 1 as patriotic response to military manpower shortage caused by World War I.

1919 World Series is allegedly thrown by members of Chicago White Sox in infamous Black Sox scandal.

▶

National Baseball Hall of Fame Members

1936 Tyrus R. Cobb
Walter P. Johnson
Christopher Mathewson
George H. "Babe" Ruth
John P. "Honus" Wagner

1937 Morgan G. Bulkeley
Bryon B. "Ban" Johnson
Napoleon "Larry" Lajoie
Connie Mack
John J. McGraw
Tristram E. Speaker
George Wright
Denton T. "Cy" Young

1938 Grover C. Alexander
Alexander J. Cartwright, Jr.
Henry Chadwick

1939 Adrian C. "Cap" Anson
Edward T. Collins
Charles A. Comiskey
William A. "Candy" Cummings
William B. "Buck" Ewing
H. Louis Gehrig
William H. "Willie" Keeler
Charles G. Radbourne
George H. Sisler
Albert G. Spalding

1942 Rogers Hornsby

1944 Kenesaw M. Landis

1945 Roger P. Bresnahan
Dennis "Dan" Brouthers
Frederick C. Clarke
James J. Collins
Edward J. Delahanty
Hugh Duffy
Hugh A. Jennings
Michael J. "King" Kelly
James H. O'Rourke
Wilbert Robinson

1946 Jesse C. Burkett
Frank L. Chance
John D. Chesbro
John J. Evers
Clark C. Griffith
Thomas F. McCarthy
Joseph J. McGinnity
Edward S. Plank
Joseph B. Tinker
George E. "Rube" Waddell
Edward A. Walsh

1947 Gordon S. "Mickey" Cochrane
Frank F. Frisch
Robert M. "Lefty" Grove
Carl O. Hubbell

1948 Herbert J. Pennock
Harold J. "Pie" Taylor

1949 Mordecai P. "Three Finger" Brown
Charles L. Gehringer
Charles A. "Kid" Nichols

1951 James E. Foxx
Melvin T. Ott

1952 Harry E. Heilmann
Paul G. Waner

1953 Edward G. Barrow
Charles A. "Chief" Bender
Thomas H. Connolly
Jay H. "Dizzy" Dean
William L. Klem
Aloysius H. Simmons
Roderick J. "Bobby" Wallace
William H. "Harry" Wright

1954 William M. Dickey
Walter J. "Rabbit" Maranville
William H. Terry

1955 J. Franklin Baker
Joseph P. DiMaggio
Charles L. "Gabby" Hartnett
Theodore A. Lyons
Raymond W. Schalk
Arthur C. "Dazzy" Vance

1956 Joseph H. Cronin
Henry B. Greenberg

1957 Samuel E. Crawford
Joseph V. McCarthy

1959 Zachariah D. Wheat

1961 Max G. Carey
William R. Hamilton

1962 Robert W. A. Feller
William B. McKechnie
Jack R. Robinson
Edd J. Roush

1963 John G. Clarkson
Elmer H. Flick
Edgar C. "Sam" Rice
Eppa Rixey

1964 Lucius B. "Luke" Appling
Urban C. "Red" Faber
Burleigh A. Grimes
Miller J. Huggins
Timothy J. Keefe
Henry E. "Heinie" Manush
John M. Ward

1965 James F. "Pud" Galvin

1966 Charles D. "Casey" Stengel
Theodore S. Williams

1967 W. Branch Rickey
Charles H. "Red" Ruffing
Lloyd J. Waner

1968 Hazen S. "Kiki" Cuyler
Leon A. "Goose" Goslin
Joseph M. Medwick

1969 Roy Campanella
Stanley A. Coveleski
Waite C. Hoyt
Stanley F. Musial

1970 Louis Boudreau
Earle B. Combs
Ford C. Frick
Jesse J. "Pop" Haines

1971 David J. Bancroft
Jacob P. Beckley
Charles J. "Chick" Hafey
Harry B. Hooper
Joseph J. Kelley
Richard W. "Rube" Marquard
Leroy R. "Satchel" Paige
George M. Weiss

1972 Lawrence P. "Yogi" Berra
Joshua Gibson
Vernon L. "Lefty" Gomez
William Harridge
Sanford Koufax
Walter F. "Buck" Leonard
Early Wynn
Ross M. Youngs

1973 Roberto W. Clemente
William G. Evans
Monford "Monte" Irvin
George L. Kelly

National Baseball Hall of Fame Members

Warren E. Spahn
Michael F. Welch

1974 James T. "Cool Papa" Bell
James L. Bottomley
John B. "Jocko" Conlan
Edward C. "Whitey" Ford
Mickey C. Mantle
Samuel L. Thompson

1975 H. Earl Averill
Stanley R. "Bucky" Harris
William J. Herman
William J. "Judy" Johnson
Ralph M. Kiner

1976 Oscar M. Charleston
Roger Connor
R. Cal Hubbard
Robert G. Lemon
Frederick C. Lindstrom
Robin E. Roberts

1977 Ernest Banks
Martin Dihigo
John H. Lloyd
Alfonso R. Lopez
Amos W. Rusie
Joseph W. Sewell

1978 Adrian Joss
Leland S. "Larry" MacPhail
Edwin L. Mathews

1979 Warren C. Giles
Willie H. Mays
Lewis R. "Hack" Wilson

1980 Albert W. Kaline
Charles H. Klein
Edwin D. "Duke" Snider
Thomas A. Yawkey

1981 Andrew "Rube" Foster
Robert Gibson
John R. Mize

1982 Henry L. Aaron
Albert B. "Happy" Chandler
Travis C. "Stonewall" Jackson
Frank Robinson

1983 Walter E. Alston
George C. Kell
Juan A. Marichal
Brooks C. Robinson, Jr.

1984 Luis E. Aparicio
Donald S. Drysdale
Richard B. Ferrell
Harmon C. Killebrew
Harold H. "Pee Wee" Reese

1985 Louis C. Brock
Enos B. "Country" Slaughter
Joseph F. "Arky" Vaughan
James Hoyt Wilhelm

1986 Robert P. Doerr
Ernest Lombardi
Willie L. "Stretch" McCovey

1987 Raymond E. Dandridge
James A. "Catfish" Hunter
Billy L. Williams

1988 Wilver D. "Willie" Stargell

1989 Albert J. Barlick
Johnny L. Bench
Albert F. "Red" Schoendienst
Carl M. "Yaz" Yastrzemski

1919 Farm system for training players is originated.

1920 Negro National League is formed.

1920 Only recorded fatality during major-league game occurs when Cleveland Indians' Roy Chapman (1891–1920) is killed by pitch. Spitball is afterward permanently outlawed (1921).

1921 First radio coverage of World Series is broadcast by WJZ radio, Newark, New Jersey.

1924 St. Louis Cardinal Rogers Hornsby sets major-league batting record for season with .424 average, highest ever in modern baseball.

1925 By this date, many teams have full-time relief pitchers.

1927 New York Yankees' Babe Ruth (1895–1948) hits 60th homer in 154 games, a season record that stands until 1961.

1928 Philadelphia Athletics player Ty Cobb (1886–1961) retires from baseball with lifetime batting average of .367. (He later becomes first player elected to National Baseball Hall of Fame.)

1933 First All-Star game is played, in Chicago. American League defeats National League, 4–2.

1935 Bullpen is in use for warming up relief pitchers.

1935 Cincinnati Reds become first major-league team to play night baseball.

1936 National Baseball Hall of Fame is established in Cooperstown, New York (*see Na-*

▶

tional Baseball Hall of Fame Members). Dedicated in 1939.

1939 Little League baseball is founded in Williamsport, Pennsylvania.

1939 Baseball game is telecast for first time, from Ebbets Field in Brooklyn.

1941 Brooklyn Dodgers experiment with first batting helmet, made of fiberglass. In 1955 batting helmets are required in National League; in 1956 American League requires helmets for all but those players who did not previously wear them.

1944 Fifteen-year-old Cincinnati Reds pitcher Joe Nuxhall is youngest player ever in major leagues.

1947 Maynard Little League team from near Williamsport, Pennsylvania wins the first Little League World Series.

1947 Jackie Robinson (1919–72) becomes first black to play in major leagues when he joins Brooklyn Dodgers. Robinson becomes first black player in National Baseball Hall of Fame, in 1962.

1953 First franchise shift occurs when Boston Braves move to Milwaukee, Wisconsin.

1957 Milwaukee Braves set National League season attendance record, drawing more than 2.2 million fans to home games.

1958 Brooklyn Dodgers move to Los Angeles and New York Giants move to San Francisco.

1960 Ebbets Field, once home to Brooklyn Dodgers, is demolished.

1961 New York Yankees' Roger Maris (1934–85) surpasses season home-run record of 60 set by Babe Ruth in 1927 by hitting 61st home run in last—162nd—game of season.

1961 American and National leagues each add two teams in first major-league expansion.

1964 Philadelphia Phillies pitcher Jim Bunning throws perfect game against New York Mets at Shea Stadium.

1965 Los Angeles Dodgers pitcher Sandy Koufax (b. 1935) hurls perfect game against Chicago Cubs. It was Koufax's fourth no-hitter, a major-league record.

1965 Houston Astrodome, first indoor professional baseball field, is completed.

1966 Milwaukee Braves' franchise moves to Atlanta.

1967 Kansas City Athletics move to Oakland.

1968 Jim "Catfish" Hunter (b. 1946) of Oakland A's pitches perfect game against Minnesota Twins.

1968 Los Angeles Dodgers' Don Drysdale pitches 58 ⅔ consecutive scoreless innings, surpassing old record of 56 innings set in 1913 by Walter Johnson.

1969 American and National leagues again expand, each adding two more teams and each dividing into Eastern and Western divisions. League playoff championships are introduced to decide World Series teams.

1970 Seattle Pilots' franchise moves to Milwaukee, where team is renamed Brewers.

1972 Washington Senators' franchise moves to Texas, where team is renamed Rangers.

1972 Baseball players strike. First 13 games of season are canceled. (Another players' strike, in 1981, shortens season by nearly two months.)

1973 American League adopts designated-hitter rule.

1974 California Angels' Nolan Ryan (b. 1947) sets pitching speed record of 100.9 mph, surpassing 98.6 mph record set in 1946 by Cleveland Indians' Bob Feller (b. 1918).

1974 St. Louis Cardinals' Lou Brock (b. 1939) steals 105 bases, defeating Maury Wills's major-league record of 104. Brock stole 938 bases in his major-league career.

1974 Atlanta Braves' Hank Aaron (b. 1934) breaks Babe Ruth's lifetime home-run record when he hits 715th home run.

1974 Cleveland Indians hire Frank Robinson (b. 1935) as first black manager of a major-league team.

1976 Philadelphia Phillies' Mike Schmidt (b. 1949) hits four consecutive home runs in one game, setting National League record.

1976 Major-league owners and players agree to reserve clause, allowing players to become free agents after five years' play; first free-agent draft is held.

▶

1981 First midseason baseball strike begins on June 12; nearly half of season's games are canceled.

1983 Oakland Athletics' Rickey Henderson (b. 1958) steals 130 bases, surpassing Lou Brock's (b. 1939) major-league record of 105 set in 1974.

1985 Cincinnati Reds' Pete Rose (b. 1941) reaches 4,192 hits, breaking major-league lifetime hits record set by Ty Cobb in 1928.

1985 Houston Astros' Nolan Ryan (b. 1947) sets major-league career record by striking out his 4,000th batter.

1986 Boston Red Sox pitcher Roger Clemens (b. 1963) sets major-league record when he strikes out 20 batters in a game.

1987 New York Yankees' Don Mattingly (b. 1961) sets major-league record of six season grand-slam home runs.

1988 Los Angeles Dodgers pitcher Orel Hershiser pitches 59 consecutive scoreless innings, breaking Don Drysdale's 1968 record.

1989 Bill White becomes National League president, first black named to the post.

1989 Cincinnati Reds manager Pete Rose (b. 1941), accused of betting on sports in violation of major-league rules, is banned from baseball for life.

1990 Season opens one week late after last-minute settlement of months-long dispute between players and owner.

Ice Hockey

c. 1850 British soldiers in Canada introduce hockey-type game and primitive ice skates. They play on frozen lakes and ponds around their camps, with teams composed of 15 to 20 players. Game probably was first played by American Indians in Canada.

1855 First organized game of hockey is played, in Kingston, Ontario.

1860 First recorded use of puck, in Kingston, Ontario.

1875 Students at Canada's McGill University propose "McGill Rules" of conduct during hockey games. They form first recognized team, McGill University Hockey Club, in 1880.

1876 First artificial ice is used for skating, in London.

c. 1880 Wooden, foot-high barriers around playing ice now are called *rink*, from Scottish term used in game of curling.

c. 1880 Montreal manufacturers produce hockey sticks and skates by or before this date as game's popularity increases.

1885 Amateur Hockey Association (AHA) is formed in Canada and includes teams from Montreal, Quebec, and Ottawa. They reduce number of players on a side to seven. (All hockey is considered amateur at this time, although many players are paid.)

1893 Ice hockey is introduced into the United States from Canada by college students. Several universities, including Yale and Johns Hopkins, sponsor games.

1893 Governor-general of Canada, Lord Stanley of Preston (1841–1908), donates cup to be presented annually to Canada's amateur hockey champions. Award becomes known as Stanley Cup.

1896 Amateur Hockey League, first amateur ice hockey league in United States, is organized by athletic clubs in the New York City area. First championship is won by New York Athletic Club. This league dissolved during World War I.

1899 U.S. and Canadian teams compete for first time.

1900 First season of Intercollegiate Hockey League, composed of Columbia, Dartmouth, Harvard, Princeton, and Yale.

▶

Stanley Cup Champions

Emblematic of World professional championship; NHL championship after 1967.

1894	Montreal A.A.A.	**1923**	Ottawa Senators	**1950**	Detroit Red Wings
1895	Montreal Victorias	**1924**	Montreal Canadiens	**1951**	Toronto Maple Leafs
1896	Winnipeg Victorias	**1925**	Victoria Cougars	**1952**	Detroit Red Wings
1897–99	Montreal Victorias	**1926**	Montreal Maroons	**1953**	Montreal Canadiens
1900	Montreal Shamrocks	**1927**	Ottawa Senators	**1954–55**	Detroit Red Wings
1901	Winnipeg Victorias	**1928**	New York Rangers	**1956–60**	Montreal Canadiens
1902	Montreal A.A.A.	**1929**	Boston Bruins	**1961**	Chicago Black Hawks
1903–05	Ottawa Silver Seven	**1930–31**	Montreal Canadiens	**1962–64**	Toronto Maple Leafs
1906	Montreal Wanderers	**1932**	Toronto Maple Leafs	**1965–66**	Montreal Canadiens
1907	Kenora Thistles	**1933**	New York Rangers	**1967**	Toronto Maple Leafs
1907	Montreal Wanderers	**1934**	Chicago Black Hawks	**1968–69**	Montreal Canadiens
1908	Montreal Wanderers	**1935**	Montreal Maroons	**1970**	Boston Bruins
1909	Ottawa Senators	**1936–37**	Detroit Red Wings	**1971**	Montreal Canadiens
1910	Montreal Wanderers	**1938**	Chicago Black Hawks	**1972**	Boston Bruins
1911	Ottawa Senators	**1939**	Boston Bruins	**1973**	Montreal Canadiens
1912–13	Quebec Bulldogs	**1940**	New York Rangers	**1974–75**	Philadelphia Flyers
1914	Toronto Arenas	**1941**	Boston Bruins	**1976–79**	Montreal Canadiens
1915	Vancouver Millionaires	**1942**	Toronto Maple Leafs	**1980–83**	New York Islanders
1916	Montreal Canadiens	**1943**	Detroit Red Wings	**1984–85**	Edmonton Oilers
1917	Seattle Metropolitans	**1944**	Montreal Canadiens	**1986**	Montreal Canadiens
1918	Toronto Arenas	**1945**	Toronto Maple Leafs	**1987–88**	Edmonton Oilers
1919	No champion	**1946**	Montreal Canadiens	**1989**	Calgary Flames
1920–21	Ottawa Senators	**1947–49**	Toronto Maple Leafs	**1990**	Edmonton Oilers
1922	Toronto St. Patricks				

1900 Modern face-off (practice of simply dropping puck at beginning of play) is introduced by referee Fred Waghorne in game in Ontario.

1903 First acknowledged professional team, Portage Lakes, is formed in Michigan by dentist who imports Canadian players. He organizes first professional league in 1904, International Pro Hockey League.

1904 Six-man teams are introduced.

1908 Canada's first professional league is formed, Ontario Professional Hockey League.

1910 Despite growing interest in hockey, there are only eight artificial rinks in world; none is in Canada.

1910 Three periods of 20 minutes each become standard play.

1910 National Hockey Association, a professional league, is formed in eastern Canada; includes teams from Toronto, Montreal, and Ottawa. Before this date professionals and amateurs played together on same teams.

1911 Pacific Coast Hockey League (PCHL) develops in western Canada, including Ameri-

▶

can and Canadian teams. Two enclosed artificial rinks are built for PCHL use in British Columbia. PCHL rule changes allow goalies to leave their feet to stop a puck and modernize offside rule; also divide ice into three areas, adding blue lines, and introduce assist in scoring.

c. 1911 First numbered player uniforms appear in PCHL games.

1912 NHA and PCHL participate in playoff, at which time Stanley Cup becomes award to best professional, rather than amateur, team.

1913 Teams in Michigan and Minnesota organize new hockey league, called American Amateur Hockey Association.

1917 Stanley Cup comes to represent championship of Canadian professional hockey.

1917 Seattle Metropolitans become first U.S. team to win Stanley Cup, defeating Montreal Canadiens.

1917 National Hockey League (NHL) replaces National Hockey Association and is composed entirely of teams representing Canadian cities.

1920 Ice hockey becomes Olympic sport, with Canada winning first Olympic title.

1924 Boston is first American city to be admitted into NHL.

1925 NHL accepts New York and Pittsburgh teams, making league truly international.

1925 Hockey game between Montreal Canadiens and New Yorks (later New York Americans) inaugurates new Madison Square Garden. (This is third Madison Square Garden, since replaced by current Garden.) It draws a crowd of 17,000, who watch Canadiens defeat New Yorks, 3–1.

1926 New York Rangers and teams in Detroit and Chicago enter NHL. League then splits into Canadian Division, with four Canadian teams plus New York Americans, and American Division, composed of five American teams.

1926 Most PCHL teams have joined NHL by this date. Stanley cup becomes exclusive NHL award.

1927 First player to be banished from NHL for life is Billy Couture (b. 1893), following brutal attack on referee.

1929 Rules of game are altered when forward passing within any of the zones becomes acceptable in professional hockey.

1931 Hockey writers and broadcasters begin selecting players for first and second All-Star teams.

1933 Riot in Boston Garden ensues when Eddie Shore of Boston Bruins seriously injures Ace Bailey of Toronto Maple Leafs.

1934 Penalty shot is introduced.

1936 Detroit Red Wings and Montreal Marroons set record for longest game in NHL history, which lasts 1 hour, 56 minutes, and 30 seconds beyond regulation period of 1 hour. Red Wings finally defeat Maroons, 1–0.

1938 Amateur Hockey Association of the United States (AHAUS) institutes championships that compete with Amateur Athletic Union (AAU) championships.

1942 Overtime periods, except during Stanley Cup playoffs, are dropped.

1942 NHL is reduced to six teams by dissolution of Brooklyn Americans, formerly New York Americans. NHL remains at six teams for next 25 years.

1943 Red center line, which divides neutral zone, is introduced by Frank Boucher.

1947 NHL begins pension plan for its players. Plan operates on contributions from players and a portion of proceeds from playoffs and All-Star games.

1947 Possibly hockey's biggest brawl occurs in Madison Square Garden, between Rangers and Montreal Canadiens. As many as 15 separate fights continue for over 20 minutes before eight security policemen finally restore order.

1947 First NHL All-Star game is held, in Toronto. All-Stars defeat Toronto Maple Leafs, 4–3.

1948 Two Boston Bruin players are banished from hockey for consorting with known gamblers.

▶

1948 NCAA institutes collegiate ice hockey playoffs. Michigan defeats Dartmouth for first championship.

1949 Invention of Zamboni, wheeled vehicle used for resurfacing ice between periods.

c. 1955 Introduction of the goalie's mask and curved stick (which can slap puck at speeds of 120 mph).

1956 Saturday NHL games are first televised.

1957 First NHL Players' Association is organized, in New York, to protect players' rights. Owners quickly disband union, which reorganizes in 1967.

1959 Jacques Plante of Montreal Canadiens becomes first goalie to wear padded plastic mask.

1960 Willie O'Ree breaks color barrier in hockey. He plays for Boston Bruins for one season.

1960 U.S. team wins International Ice Hockey Federation's amateur world championship, which has since been dominated by USSR.

1961 International Hockey Hall of Fame opens in Toronto.

1967 NHL expands from six to 12 teams.

1967 Average NHL player's salary is $18,200.

1970 Buffalo Sabres and Vancouver Canucks are admitted to NHL.

1972 World Hockey Association (WHA), new 12-team league, is formed. Rivalry with NHL causes sharp rise in player salaries. WHA fails in 1979.

1972 Soviet team plays eight games against Canadian professional players. Canada wins four and ties one.

1973 U.S. Hockey Hall of Fame opens in Eveleth, Minnesota.

1973 First WHA championships are held. New England Whalers defeat Winnipeg Jets, 4 games to 1.

1975 Dave Forbes of Boston Bruins becomes first professional athlete indicted for crime committed during play. He is tried for criminal assault, allegedly using excessive force against an opponent, but trial ends in hung jury.

1976 Canada wins first Canada Cup Tournament, which includes teams from Canada, United States, Czechoslovakia, Finland, Soviet Union, and Sweden.

1978 WHA shrinks to six teams when Indianapolis Racers become insolvent.

1978 Two hockey greats, Bobby Orr (b. 1948) and Bobby Hull (b. 1939), retire.

1979 Four WHA franchises are merged into NHL. The four teams admitted are New England (later Hartford) Whalers, Quebec Nordiques, Winnipeg Jets, and Edmonton Oilers.

1980 NHL enacts 5-minute overtime rule to break tie games.

1980 Hockey great Gordie Howe (b. 1928) retires, setting NHL career records for most points (1,850), most goals (801), and most assists (1,049).

1982 Colorado Rockies move to New Jersey. Team is renamed New Jersey Devils.

1982 Edmonton Oilers' Wayne Gretzky (b. 1961) sets NHL season records for most points (212), most goals (92), and most assists (120).

1986 Joe Murphy, Canadian student at Michigan State, is number-one draft choice of NHL team. It is first time a U.S. college student earns that honor.

1988 Minnesota North Stars' Dino Ciccarelli is jailed for day in what probably is first incarceration of hockey player for hitting opponent while on ice.

1989 NHL sets plans to expand from current 21 teams to 28 in early 1990s.

1990 Average NHL player's salary rises above $200,000.

Soccer

217 Legend has it that early Britons played a kicking game with Roman skulls at about this time.

1000 Danish chieftain is defeated by British, and his head is used as football.

1100s In Great Britain, a form of football is played by one whole village against another.

1314 English king Edward II (1284–1327) issues proclamation forbidding play of football.

1349 English king Edward III (1312–1377) also tries to ban football. Disruption of military duties and violence are reasons cited, but game is too popular to stop.

c. 1390 English king Richard II (1367–1400) issues ban on football and stresses need for archery practice.

1401 English king Henry IV (1367–1413) and Scottish kings try again to limit play of football.

c. 1670s English King Charles II (1630–85) finally approves game and encourages its play.

1801 Soccer football is standardized. Rules specify equal number of team members on each side and number of players; size of field is limited to 80–100 yards. Two sticks with tape between them are used for goals.

1820 In United States, Princeton students play kicking game called "ballown."

1820s Soccer-style games are popular on American college campuses.

1823 Rugby is distinguished from soccer as a separate "carrying" football game.

1827 Form of soccer is played at Harvard.

1857 First soccer club is organized in Sheffield, England.

1858 Soccer is banned at Yale by city of New Haven. Game is considered too violent.

1860 Harvard bans soccer, again because it is too violent.

1860 High-school teams play a gentler form of soccer on Boston Common.

1862 Oneida Football Club of Boston, first official soccer club in United States, is formed.

1863 Football Association is formed in England and becomes model for other national associations.

1869 Rutgers plays Princeton in first U.S. intercollegiate soccer game.

1870 Eleven-player teams are made standard. Teams usually are made up of nine forwards and two defenders.

1871 First English Football Association cup open soccer competition is held.

1871 Rugby Union is founded in England, formally establishing rugby as separate game with its own set of rules.

1873 Princeton, Yale, Columbia, and Rutgers agree to adopt English Football Association rules for intercollegiate play. Harvard refuses to comply and plays rugby-style football.

1875 Football Association makes goalpost bar mandatory.

▶

World Cup Soccer Championships

1930	Uruguay defeats Argentina, 4–2.
1934	Italy defeats Czechoslovakia, 2–1.
1938	Italy defeats Hungary, 4–2.
1950	Uruguay defeat Brazil, 2–1.
1954	West Germany defeats Hungary, 3–2.
1958	Brazil defeats Sweden, 5–2.
1962	Brazil defeats Czechoslovakia, 3–1.
1966	England defeats West Germany, 4–2.
1970	Brazil defeats Italy, 4–1.
1974	West Germany defeats Netherlands, 2–1.
1978	Argentina defeats Netherlands, 3–1.
1982	Italy defeats West Germany, 3–1.
1986	Argentina defeats West Germany, 3–2.
1990	West Germany defeats Argentina, 1–0.

NASL Championships	
1968	Atlantic Chiefs defeat San Diego Toros, 3–0, in first NASL two-game, total-point championship.
1969	Kansas City Spurs take NASL title over Atlantic Chiefs; no playoffs are held.
1970	Rochester Lancers defeat Washington Darts, 4–3, in two-game, total-point championship.
1971	Dallas Tornado defeat Atlantic Chiefs, 2–0, in three-game NASL championship.
1972	New York Cosmos defeat St. Louis Stars, 2–1.
1973	Philadelphia Atoms defeat Dallas Tornado, 2–0.
1974	Los Angeles Aztecs defeat Miami Toros, 4–3.
1975	Tampa Bay Rowdies defeat Portland Timbers, 2–0.
1976	Toronto Metros defeat Minnesota Kicks, 3–0.
1977	New York Cosmos defeat Seattle Sounders, 2–1.
1978	New York Cosmos defeat Tampa Bay Rowdies, 3–1.
1979	Vancouver Whitecaps defeat Tampa Bay Rowdies, 2–1.
1980	New York Cosmos defeat Fort Lauderdale Strikers, 3–0.
1981	Chicago Sting defeat New York Cosmos, 1–0.
1982	New York Cosmos defeat Seattle Sounders, 1–0.
1983	Tulsa Roughnecks defeat Toronto Blizzard, 2–0.
1984	Chicago Sting defeat Toronto Blizzard, 3–2.

1876 Intercollegiate Football Association is formed. However, Harvard convinces other members to adopt Rugby Union rules, effectively redirecting course of American football.

1880s Goalkeeper is distinguished as separate player position. He becomes only player allowed to carry ball.

1883 Pullman Railroad Car Company of Chicago encourages soccer play by setting up its own playing field. Chicago Soccer League forms later.

1884 American Football Association is founded.

1886 St. Louis Football Association is founded. St. Louis becomes major source of American soccer talent.

1886 First international soccer match for a U.S. team is played in Canada between a New Jersey eleven and a Canadian all-star team.

1887 New England Football Association is formed.

1888 Football League is organized in England due to success of FA cup.

1904 Federation of International Football Association (FIFA) is founded to coordinate different national associations. It enforces 17 rules that establish universal standard of play.

1905 Intercollegiate Association Football League is formed by Columbia, Cornell, Harvard, Haverford, and Pennsylvania.

1905 English Pilgrim Association defeats all-New York team, 7–1, at Polo Grounds in New York City before crowd of about 2,000.

1906 London's Corinthian Football Club tours United States, tallying 13–1–2 game record.

1911 Corinthian Football Club returns to United States, tallying 18–1–1 tour record.

1912 National Collegiate Athletic Association (NCAA) approves soccer for intercollegiate play.

1912 Sir Thomas Dewar, English football enthusiast, donates trophy to American Amateur Football Association for national championship soccer teams. Trophy is later known as National Challenge Cup (1915).

1913 American Football Association (professional) merges with American Amateur Football Association to form U.S. Football Association.

1915 Competition for Dewar Cup (later called National Challenge Cup) begins. At first open only to amateurs, it becomes forum

▶

for play for both amateur and professional teams.

1916 Team sponsored by Bethlehem Steel Company wins both American Challenge Cup and National Challenge Cup.

1921 Eight professional soccer clubs form American Soccer League.

1923 Riot breaks out at London soccer match; 1,000 are hurt.

1924 Competition for National Amateur Cup is started.

1926 Intercollegiate Association Football League is succeeded by Intercollegiate Soccer Football Association of America. By 1952 a total of 70 colleges are members; over 300 are members by 1990.

1930 First year of competition for World Cup, most prestigious prize in international soccer. (*See World Cup Soccer Championships.*)

1935 National Junior Cup is started for registered amateur teams with players 18 years or under.

1945 U.S. Football Association is officially renamed U.S. Soccer Football Association.

1956 European Cup for Champions holds first competition.

1959 St. Louis University defeats Bridgeport University, 5–2, in the first NCAA soccer playoff.

1960 International Soccer League is formed, providing summer tournament play between pro teams in Europe and South America. It encourages development of North American league.

1964 Stampeding fans at Peruvian soccer match kill 300.

1966–67 United Soccer Association (USA) and National Professional Soccer League (NPSL), two professional leagues, are started in United States.

1968 USA and NPSL merge to form 17-team North American Soccer League (NASL). (*See NASL Championships.*)

1969 Financial problems force 12 of 17 NASL teams to fold.

1970 Brazil becomes first country to win World Cup three times; with three wins they keep trophy.

c. 1970 Indoor soccer, played with six-player teams on artificial turf, begins to gain popularity. Season runs from November to May.

1971 First Women's World Cup match is played in Mexico City.

1971 NCAA holds collegiate soccer championship for the first time, in Orange Bowl.

1971 New York Cosmos team is awarded franchise in NASL.

1972 Southern Illinois University wins first NCAA Division II soccer championship.

1975 Pelé, Brazilian soccer superstar, joins New York Cosmos. He receives $2.8 million, three-year contract.

1977 Attendance at U.S. soccer games reaches high point. Cosmos playoff game draws record 77,691 fans.

1977 Cosmos' Pelé plays his farewell game. Attendance exceeds 75,600.

1978 The Major Indoor Soccer League is formed.

1978 NASL expands to high of 24 franchises just as American interest in soccer begins to decline. Only 14 franchises remain by 1982.

1982 Italy becomes second country to win World Cup three times.

1985 NASL breaks up as American interest in professional soccer continues to decline.

1985 Forty fans at London soccer match die when wooden stands suddenly go up in flames.

1989 Stampeding British soccer fans crush 95 of their number to death at Sheffield, South Yorkshire Stadium.

Golf

1100 Roman game paganica is popular in Scotland. Probably forerunner of golf, it is played with bent stick and leather ball stuffed with feathers.

1457 Earliest written record of golf. Scottish king bans "futeball and golfe" because they interfere with archery practice, needed for national defense. Ban ends when England and Scotland sign a peace treaty in 1502.

1744 First known golf club, Honourable Company of Edinburgh Golfers, is established in Scotland. Club now settles rules disputes, formerly resolved by senior player on course.

1754 World's oldest golf club still in existence, Royal and Ancient Golf Club of St. Andrews in Scotland (RAGC), opens. RAGC makes first written record of rules by a group and becomes a leader in establishing regulations and traditions worldwide.

United States Open Champions

Men

1900	Harry Vardon	**1925**	Willie MacFarlane	**1952**	Julius Boros
1901	Willie Anderson	**1926**	Bobby Jones	**1953**	Ben Hogan
1902	L. Auchterlonie	**1927**	Tommy Armour	**1954**	Ed Furgol
1903	Willie Anderson	**1928**	John Farrell	**1955**	Jack Fleck
1904	Willie Anderson	**1929**	Bobby Jones	**1956**	Cary Middlecoff
1905	Willie Anderson	**1930**	Bobby Jones	**1957**	Dick Mayer
1906	Alex Smith	**1931**	William Burke	**1958**	Tommy Bolt
1907	Alex Ross	**1932**	Gene Sarazen	**1959**	Billy Casper
1908	Fred McLeod	**1933**	John Goodman	**1960**	Arnold Palmer
1909	George Sargent	**1934**	Olin Dutra	**1961**	Gene Littler
1910	Alex Smith	**1935**	Sam Parks, Jr.	**1962**	Jack Nicklaus
1911	John McDermott	**1936**	Tony Manero	**1963**	Julius Boros
1912	John McDermott	**1937**	Ralph Guldahl	**1964**	Ken Venturi
1913	Francis Ouimet	**1938**	Ralph Guldahl	**1965**	Gary Player
1914	Walter Hagen	**1939**	Byron Nelson	**1966**	Billy Casper
1915	Jerome Travers	**1940**	Lawson Little	**1967**	Jack Nicklaus
1916	Chick Evans	**1941**	Craig Wood	**1968**	Lee Trevino
1917–18	No tournament	**1942–45**	No tournament	**1969**	Orville Moody
1919	Walter Hagen	**1946**	Lloyd Mangrum	**1970**	Tony Jacklin
1920	Edward Ray	**1947**	Lew Worsham	**1971**	Lee Trevino
1921	Jim Barnes	**1948**	Ben Hogan	**1972**	Jack Nicklaus
1922	Gene Sarazen	**1949**	Cary Middlecoff	**1973**	Johnny Miller
1923	Bobby Jones	**1950**	Ben Hogan	**1974**	Hale Irwin
1924	Cyril Walker	**1951**	Ben Hogan	**1975**	Lou Graham

United States Open Champions

Year	Champion	Year	Champion	Year	Champion
1976	Jerry Pate	1981	David Graham	1986	Raymond Floyd
1977	Hubert Green	1982	Tom Watson	1987	Scott Simpson
1978	Andy North	1983	Larry Nelson	1988	Curtis Strange
1979	Hale Irwin	1984	Fuzzy Zoeller	1989	Curtis Strange
1980	Jack Nicklaus	1985	Andy North	1990	

Women

Year	Champion	Year	Champion	Year	Champion
1946	Betty Berg	1963	Mary Mills	1976	JoAnne Carner
1947	Betty Jamieson	1964	Mickey Wright	1977	Hollis Stacy
1948	Babe Zaharias	1965	Carol Mann	1978	Hollis Stacy
1949	Louise Suggs	1966	Sandra Spuzich	1979	Jerilyn Britz
1950	Babe Zaharias	1967	Catherine Lacoste	1980	Amy Alcott
1951	Betsy Rawls	1968	Susie Maxwell Berning	1981	Pat Bradley
1952	Louise Suggs	1969	Donna Caponi	1982	Janet Alex
1953	Betsy Rawls	1970	Donna Caponi	1983	Jan Stephenson
1954	Babe Zaharias	1971	JoAnne Carner	1984	Hollis Stacy
1955	Fay Crocker	1972	Susie Maxwell Berning	1985	Kathy Baker
1956	Kathy Cornelius	1973	Susie Maxwell Berning	1986	Jane Geddes
1957	Betsy Rawls	1974	Sandra Haynie	1987	Laura Davies
1958	Mickey Wright	1975	Sandra Palmer	1988	Liselotte Neumann
1959	Mickey Wright			1989	Betsy King
1960	Betsy Rawls			1990	
1961	Mickey Wright				
1962	Murle Lindstrom				

1764 RAGC sets standard round of golf at 18 holes.

1786 Earliest U.S. golf course, established by British planters, opens. Game is played over public park.

1848 Feather-stuffed leather ball (called "feathery") is replaced with solid, gutta-percha ball called "gutty." Developers soon add indentations to ball's surface to make it fly properly.

1860 British Open Tournament is established. (*See British Open Winners.*)

1870 Golf bag is introduced for carrying clubs, replacing practice by wealthier players who employ caddies to carry loose clubs.

1888 First permanent U.S. golf course opens, set up by Royal and Ancient Golf Club (RAGC) member. Called "St. Andrews," it is located in Yonkers, New York.

c. 1890 First ball "pickups," forerunners of shag bags, are patented.

1891 British term *bogey*, named for fictitious Colonel Bogey, who is steady but not brilliant, comes to mean one shot over par.

1891 Muirfield Golf Course, considered one of world's finest, is designed in Scotland.

1892 Representative of U.S. sporting goods manufacturer A. G. Spalding and Brothers

▶

Masters Tournament					
1934	Horton Smith	**1955**	Cary Middlecoff	**1973**	Tommy Aaron
1935	Gene Sarazen	**1956**	Jack Burke, Jr.	**1974**	Gary Player
1936	Horton Smith	**1957**	Doug Ford	**1975**	Jack Nicklaus
1937	Byron Nelson	**1958**	Arnold Palmer	**1976**	Ray Floyd
1938	Henry Picard	**1959**	Art Wall, Jr.	**1977**	Tom Watson
1939	Ralph Guldahl	**1960**	Arnold Palmer	**1978**	Gary Player
1940	Jimmy Demaret	**1961**	Gary Player	**1979**	Fuzzy Zoeller
1941	Craig Wood	**1962**	Arnold Palmer	**1980**	Seve Ballesteros
1942	Byron Nelson	**1963**	Jack Nicklaus	**1981**	Tom Watson
1943–45	No tournament	**1964**	Arnold Palmer	**1982**	Craig Stadler
1946	Herman Keiser	**1965**	Jack Nicklaus	**1983**	Seve Ballesteros
1947	Jimmy Demaret	**1966**	Jack Nicklaus	**1984**	Ben Crenshaw
1948	Claude Harmon	**1967**	Gay Brewer, Jr.	**1985**	Bernhard Langer
1949	Sam Snead	**1968**	Bob Goalby	**1986**	Jack Nicklaus
1950	Jimmy Demaret	**1969**	George Archer	**1987**	Larry Mize
1951	Ben Hogan	**1970**	Billy Casper	**1988**	Sandy Lyle
1952	Sam Snead	**1971**	Charles Coody	**1989**	Nick Faldo
1953	Ben Hogan	**1972**	Jack Nicklaus	**1990**	Nick Faldo
1954	Sam Snead				

brings back $500 worth of golf balls and clubs from England. By 1894 Spalding sells its first club made in United States.

1892 First golf clubhouse, designed by U.S. architect Stanford White (1853–1906), is built at Shinnecock Hills, Southampton, New York.

1893 Shinnecock Hills creates separate nine-hole course for wives and daughters of members.

1894 First U.S. Open is held.

1894 Amateur Golf Association of the United States is established to govern the sport. Later renamed U.S. Golf Association (USGA).

1895 USGA holds Women's Amateur Championship.

1895 First U.S. 18-hole course opens, in Chicago.

1899 Liquid-center golf ball is invented in United States.

c. 1900 U.S. term *par* is used as base for figuring handicaps.

1905 Englishman patents dimpled-cover golf ball.

1906 Practice putting mat is developed in United States.

1910 Irons acquire markings—grooves or dashes—by about this time.

1910 Patent is issued for steel golf shaft despite the fact that steel shafts are illegal in United States and Great Britain. (United States legalizes steel shafts in 1926, Great Britain in 1929.)

1913 Pine Valley Golf Course, considered by many to be the best course built to date, is designed in New Jersey.

1913 U.S. amateur Francis Ouimet (1883–1967) wins U.S. Open. He helps popularize the game.

▶

Professional Golfer's Association Champions

Year	Champion	Year	Champion	Year	Champion
1919	James M. Barnes	1943	No tournament	1967	Don January
1920	Jock Hutchinson	1944	Bob Hamilton	1968	Julius Boros
1921	Walter Hagen	1945	Byron Nelson	1969	Ray Floyd
1922	Gene Sarazen	1946	Ben Hogan	1970	Dave Stockton
1923	Gene Sarazen	1947	Jim Ferrier	1971	Jack Nicklaus
1924	Walter Hagen	1948	Ben Hogan	1972	Gary Player
1925	Walter Hagen	1949	Sam Snead	1973	Jack Nicklaus
1926	Walter Hagen	1950	Chandler Harper	1974	Lee Trevino
1927	Walter Hagen	1951	Sam Snead	1975	Jack Nicklaus
1928	Leo Diegel	1952	James Turnesa	1976	Dave Stockton
1929	Leo Diegel	1953	Walter Burkemo	1977	Lanny Wadkins
1930	Tommy Armour	1954	Melvin Harbert	1978	John Mahaffey
1931	Tom Creavy	1955	Doug Ford	1979	David Graham
1932	Olin Dutra	1956	Jack Burke, Jr.	1980	Jack Nicklaus
1933	Gene Sarazen	1957	Lionel Hebert	1981	Larry Nelson
1934	Paul Runyan	1958	Dow Finsterwald	1982	Raymond Floyd
1935	Johnny Revolta	1959	Bob Rosburg	1983	Hal Sutton
1936	Denny Shute	1960	Jay Hebert	1984	Lee Trevino
1937	Denny Shute	1961	Jerry Barber	1985	Hubert Green
1938	Paul Runyan	1962	Gary Player	1986	Bob Tway
1939	Henry Picard	1963	Jack Nicklaus	1987	Larry Nelson
1940	Byron Nelson	1964	Bob Nichols	1988	Jeff Sluman
1941	Victor Ghezzi	1965	Dave Marr	1989	Payne Stewart
1942	Sam Snead	1966	Al Geiberger	1990	Wayne Grady

1914–29 U.S. professional Walter Hagen (1892–1969) wins 11 national championships.

1916 Professional U.S. golfers establish PGA. (*See PGA Winners.*)

1920 American dentist invents golf tee.

1920–30 Period often is called Golden Age in golf-course design. Courses become more elaborate and are more extravagantly financed.

1921 USGA and RAGC agree to standard 1.62-inch diameter and weight of 1.62 ounces.

1923–30 U.S. lawyer and amateur golfer Robert T. (Bobby) Jones, Jr. (1902–1971), wins 13 of 22 men's tournaments sponsored by USGA and RAGC. In 1930 he wins Open and amateur titles of both countries and retires from golf at age 28.

1925 First total fairway irrigation system is installed, in United States.

1925 USGA and RAGC ban golf clubs with deep grooves.

1926 British course designer and architect Alister Mackenzie (1870–1934) designs and builds Augusta National Golf Course in Georgia with golfer Bobby Jones. It becomes location for annual Masters tournament.

1928 Mackenzie creates Cypress Point in California. It becomes a classic U.S. course.

▶

British Open

Year	Winner	Year	Winner	Year	Winner
1860	Willie Park	**1901**	James Braid	**1951**	Max Faulkner
1861	Tom Morris, Sr.	**1902**	A. Herd	**1952**	Bobby Locke
1862	Tom Morris, Sr.	**1903**	Harry Vardon	**1953**	Ben Hogan
1863	Willie Park	**1904**	J. White	**1954**	Peter W. Thomason
1864	Tom Morris, Sr.	**1905**	James Braid	**1955**	Peter W. Thomason
1865	Andrew Strath	**1906**	James Braid	**1956**	Peter W. Thomason
1866	Willie Park	**1907**	Arnaud Massey	**1957**	Bobby Locke
1867	Tom Morris, Sr.	**1908**	James Braid	**1958**	Peter W. Thomason
1868	Tom Morris, Jr.	**1909**	John H. Taylor	**1959**	Gary Player
1869	Tom Morris, Jr.	**1910**	James Braid	**1960**	Kel Nagle
1870	Tom Morris, Jr.	**1911**	Harry Vardon	**1961**	Arnold Palmer
1871	No competition	**1912**	E. Ray	**1962**	Arnold Palmer
1872	Tom Morris, Jr.	**1913**	John H. Taylor	**1963**	Bob Charles
1873	T. Kidd	**1914**	Harry Vardon	**1964**	Tony Lema
1874	M. Park	**1915–19**	No tournament	**1965**	Peter W. Thomason
1875	Willie Park	**1920**	G. Duncan	**1966**	Jack Nicklaus
1876	B. Martin	**1921**	J. Hutchinson	**1967**	Roberto de Vicenzo
1877	Jamie Anderson	**1922**	Walter Hagen	**1968**	Gary Player
1878	Jamie Anderson	**1923**	A. G. Havers	**1969**	Anthony Jacklin
1879	Jamie Anderson	**1924**	Walter Hagen	**1970**	Jack Nicklaus
1880	Bob Ferguson	**1925**	J. Barnes	**1971**	Lee Trevino
1881	Bob Ferguson	**1926**	Bobby Jones	**1972**	Lee Trevino
1882	Bob Ferguson	**1927**	Bobby Jones	**1973**	Tom Weiskopf
1883	W. Fernie	**1928**	Walter Hagen	**1974**	Gary Player
1884	J. Simpson	**1929**	Walter Hagen	**1975**	Tom Watson
1885	B. Martin	**1930**	Bobby Jones	**1976**	Johnny Miller
1886	D. Brown	**1931**	Tommy D. Armour	**1977**	Tom Watson
1887	Willie Park, Jr.	**1932**	Gene Sarazen	**1978**	Jack Nicklaus
1888	J. Burns	**1933**	Denny Shute	**1979**	Seve Ballesteros
1889	Willie Park, Jr.	**1934**	T. Henry Cotton	**1980**	Tom Watson
1890	J. Ball	**1935**	Alf Perry	**1981**	Bill Rogers
1891	H. Kirkaldy	**1936**	Alf H. Padgham	**1982**	Tom Watson
1892	H. H. Hilton	**1937**	T. Henry Cotton	**1983**	Tom Watson
1893	W. Auchterlonie	**1938**	R. A. Whitecombe	**1984**	Seve Ballesteros
1894	John H. Taylor	**1939**	Richard Burton	**1985**	Sandy Lyle
1895	John H. Taylor	**1940–45**	No tournament	**1986**	Greg Norman
1896	Harry Vardon	**1946**	Sam Snead	**1987**	Nick Faldo
1897	H. H. Hilton	**1947**	Fred Daly	**1988**	Seve Ballesteros
1898	Harry Vardon	**1948**	T. Henry Cotton	**1989**	Mark Calcavecchia
1899	Harry Vardon	**1949**	Bobby Locke	**1990**	Nick Faldo
1900	John H. Taylor	**1950**	Bobby Locke		

c. 1930 Steel shafts begin to replace hickory as players look for power from clubs.

1932 USGA approves new 1.68-inch diameter and 1.62-ounce weight for golf balls.

1932 Owner of U.S. rubber company devises new method for manufacturing wound-rubber balls with round centers, using X-rays to check center of all balls made.

1934 U.S. pro Walter Hagen founds Masters Tournament, played annually at Augusta, Georgia. (*See Masters Tournament Champions.*)

1935 U.S. player Sam Snead (b. 1912) enters tournament play. He goes on to win more than 120 tournaments in some 40 years of golf.

1938 USGA limits number of clubs golfer may use in one round to 14.

1945 Average prize money for PGA tournament is less than $10,000.

1946 USGA introduces open championship for women players.

1950 LPGA (Ladies' Professional Golf Association) is incorporated in United States.

1953 U.S. golfer Ben Hogan (b. 1912) wins British and U.S. Open titles and Masters.

1958 World Amateur Golf Council is organized. It plays at St. Andrews, Scotland.

1958 U.S. golfer Arnold Palmer (b. 1929) wins Masters Tournament. He goes on to win it three more times; also wins some 60 other major tournaments.

1963 Palmer becomes first player to win over $100,000 in one year. He is also first to exceed $1 million in prizes during career.

1968 Spalding introduces two-piece ball with solid center and synthetic cover.

1975 U.S. golfer Jack Nicklaus (b. 1940) wins Masters for fifth time, a golfing record. Often considered greatest golfer of all time, he goes on to win record number of major championships.

1976 USGA adopts overall distance standard, limiting distance ball may be hit.

1976 American golf-course architect Robert Trent Jones (b. 1906) receives award from American Society of Golf Course Architects for outstanding contributions. He has built or redesigned over 450 courses.

1978 LPGA rookie Nancy Lopez (b. 1957) wins nine tournaments, five of them in a row.

1980 USGA adds Mid-Amateur Championship for golfers over 24 years of age. It adds Women's Mid-Amateur Championship in 1987.

1981 Kathy Whitworth (b. 1939) leads LPGA in earnings for eight years. She becomes first woman to win over $1 million in prize money.

1983 USGA adopts symmetry standard, which prohibits balls that self-correct in flight.

c. 1985 PGA players compete for prizes exceeding $25 million annually. LPGA prizes exceed $10 million annually.

1989 Estimated 25 million golfers, an increase of 1.5 million since 1988, use American golf courses, according to National Golf Foundation.

Tennis

1200s *Jeu de paume,* forerunner of tennis, is popular in France.

1245 Archbishop of Rouen, France bans *jeu de paume* in monasteries, because it distracts priests from their ecclesiastical duties.

▶

1352 English King Edward III (1312–77) has tennis court constructed in his palace.

1700s Court tennis (also known as royal tennis), played on an indoor court, is popular among aristocracy of Europe.

1873 British Army major Walter Wingfield (1833–1912) develops lawn tennis at garden party in Wales. It is played with rubber ball and oval racket on hourglass-shaped court with a 7-foot-high net.

1874 Wingfield patents his game, which he calls "Sphairistike." It soon becomes known as lawn tennis.

1874 Mary Outerbridge introduces tennis to United States at Staten Island Cricket and Baseball Club after encountering the new game in Bermuda. Its popularity spreads rapidly across country.

1875 Rules of new game are standardized by Marylebone Cricket Club in London. They include use of rectangular court and rubber ball with flannel coat.

1876 First recorded lawn-tennis tournament in United States is held in Nahant, Massachusetts.

1877 Wimbledon tournament is initiated at All-England Lawn Tennis and Croquet Club in Wimbledon, England. (*See Wimbledon Champions.*)

1879 Wimbledon tournament adds men's doubles competition.

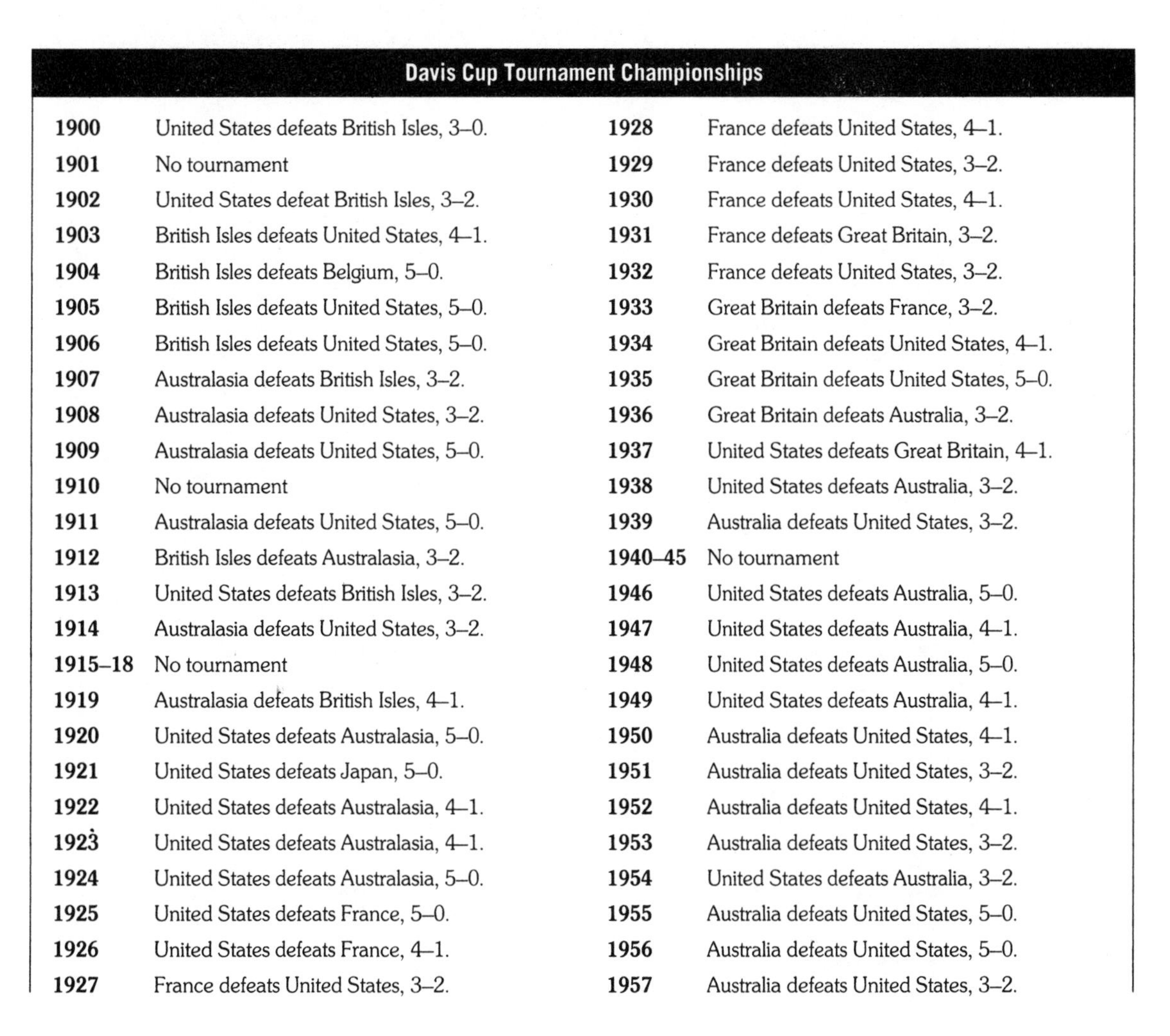

Davis Cup Tournament Championships

1900	United States defeats British Isles, 3–0.
1901	No tournament
1902	United States defeat British Isles, 3–2.
1903	British Isles defeats United States, 4–1.
1904	British Isles defeats Belgium, 5–0.
1905	British Isles defeats United States, 5–0.
1906	British Isles defeats United States, 5–0.
1907	Australasia defeats British Isles, 3–2.
1908	Australasia defeats United States, 3–2.
1909	Australasia defeats United States, 5–0.
1910	No tournament
1911	Australasia defeats United States, 5–0.
1912	British Isles defeats Australasia, 3–2.
1913	United States defeats British Isles, 3–2.
1914	Australasia defeats United States, 3–2.
1915–18	No tournament
1919	Australasia defeats British Isles, 4–1.
1920	United States defeats Australasia, 5–0.
1921	United States defeats Japan, 5–0.
1922	United States defeats Australasia, 4–1.
1923	United States defeats Australasia, 4–1.
1924	United States defeats Australasia, 5–0.
1925	United States defeats France, 5–0.
1926	United States defeats France, 4–1.
1927	France defeats United States, 3–2.
1928	France defeats United States, 4–1.
1929	France defeats United States, 3–2.
1930	France defeats United States, 4–1.
1931	France defeats Great Britain, 3–2.
1932	France defeats United States, 3–2.
1933	Great Britain defeats France, 3–2.
1934	Great Britain defeats United States, 4–1.
1935	Great Britain defeats United States, 5–0.
1936	Great Britain defeats Australia, 3–2.
1937	United States defeats Great Britain, 4–1.
1938	United States defeats Australia, 3–2.
1939	Australia defeats United States, 3–2.
1940–45	No tournament
1946	United States defeats Australia, 5–0.
1947	United States defeats Australia, 4–1.
1948	United States defeats Australia, 5–0.
1949	United States defeats Australia, 4–1.
1950	Australia defeats United States, 4–1.
1951	Australia defeats United States, 3–2.
1952	Australia defeats United States, 4–1.
1953	Australia defeats United States, 3–2.
1954	United States defeats Australia, 3–2.
1955	Australia defeats United States, 5–0.
1956	Australia defeats United States, 5–0.
1957	Australia defeats United States, 3–2.

Davis Cup Tournament Championships

1958	United States defeats Australia, 3–2.	**1975**	Sweden defeats Czechoslovakia, 3–2.
1959	Australia defeats United States, 3–2.	**1976**	Italy defeats Chile, 4–1.
1960	Australia defeats Italy, 4–1.	**1977**	Australia defeats Italy, 3–1.
1961	Australia defeats Italy, 5–0.	**1978**	United States defeats Great Britain, 4–1.
1962	Australia defeats Mexico, 5–0.	**1979**	United States defeats Italy, 5–0.
1963	United States defeats Australia, 3–2.	**1980**	Czechoslovakia defeats Italy, 4–1.
1964	Australia defeats United States, 3–2.	**1981**	United States defeats Argentina, 3–1.
1965	Australia defeats Spain, 4–1.	**1982**	United States defeats France, 3–0.
1966	Australia defeats India, 4–1.	**1983**	Australia defeats Sweden, 3–2.
1967	Australia defeats Spain, 4–1.	**1984**	Sweden defeats United States, 4–1.
1968	United States defeats Australia, 4–1.	**1985**	Sweden defeats West Germany, 3–2.
1969	United States defeats Romania, 5–0.	**1986**	Australia defeats Sweden, 3–2.
1970	United States defeats West Germany, 5–0.	**1987**	Sweden defeats Austria, 5–0.
1971	United States defeats Romania, 3–2.	**1988**	West Germany defeats Sweden, 4–1.
1972	United States defeats Romania, 3–2.	**1989**	West Germany defeats Sweden, 3–2.
1973	Australia defeats United States, 5–0.	**1990**	Argentina defeats West Germany, 3–2.
1974	South Africa defeats India by default.		

1880s Twin brothers William (1861–1904) and Ernest (1861–1899) Renshaw, British players, emerge as early stars of the game. William wins seven Wimbledon championships, and together the brothers win British doubles championship seven times.

1881 U.S. National Lawn Tennis Association is formed.

1881 First U.S. championship in men's singles is held at Newport, Rhode Island, in what would become known as U.S. Open tournament. (*See U.S. Open Champions.*)

1884 Standard three-foot height of net is established. Scoring, 26-yard rectangular court, and one fault for serves are standard by now.

1884 Wimbledon tournament adds women's singles competition.

1900 First international competition, Davis Cup, is initiated by Harvard student Dwight F. Davis in Boston. United States defeats British Isles, 3–0. (*See Davis Cup Championships.*)

1909 Clay court is first developed in England by Claude Brown.

1915 U.S. championship is moved to Forest Hills, New York.

1920 American player Bill Tilden (1893–1953), often considered greatest player ever in tennis, becomes first American to win men's singles title at Wimbledon. He also wins first of his seven U.S. singles championships.

1926 First professional tennis tour is organized by promoter C. C. Pyle, featuring six-time Wimbledon champion Suzanne Lenglen of France (1899–1938) and other star players.

1931 Bill Tilden becomes professional player. Move gives greatly increased prestige to professional tennis.

1935 British player Fred Perry (b. 1909) becomes first to win Grand Slam by holding Wimbledon, U.S., French, and Australian singles titles at same time.

1938 U.S. player Don Budge (b. 1915) becomes first to win Grand Slam in single year.

1950 Australia begins international dominance of game, winning Davis Cup 15 times between 1950 and 1967.

▶

Wimbledon Champions

Men's Singles

1877	Spencer Gore
1878	P. Frank Hodow
1879	J. T. Hartley
1880	J. T. Hartley
1881	William Renshaw
1882	William Renshaw
1883	William Renshaw
1884	William Renshaw
1885	William Renshaw
1886	William Renshaw
1887	Herbert Lawford
1888	Ernest Renshaw
1889	William Renshaw
1890	Willoughby J.Hamilton
1891	Wilfred Boddeley
1892	Wilfred Boddeley
1893	Joshua Pim
1894	Joshua Pim
1895	Wilfred Boddeley
1896	Harold Mahony
1897	Reginald Doherty
1898	Reginald Doherty
1899	Reginald Doherty
1900	Reginald Doherty
1901	Arthur Gore
1902	Hugh Doherty
1903	Hugh Doherty
1904	Hugh Doherty
1905	Hugh Doherty
1906	Hugh Doherty
1907	Norman Brookes
1908	Arthur Gore
1909	Arthur Gore
1910	Anthony Wilding
1911	Anthony Wilding
1912	Anthony Wilding
1913	Anthony Wilding
1914	Norman Brookes
1915–18	No tournament
1919	Gerald Patterson
1920	Bill Tilden
1921	Bill Tilden
1922	Gerald Patterson
1923	William Johnston
1924	Jean Borotra
1925	René Lacoste
1926	Jean Borotra
1927	Henri Cochet
1928	René Lacoste
1929	Henri Cochet
1930	Bill Tilden
1931	Sidney Wood
1932	Ellsworth Vines
1933	Jack Crawford
1934	Fred Perry
1935	Fred Perry
1936	Fred Perry
1937	Don Budge
1938	Don Budge
1939	Bobby Riggs
1940–45	No tournament
1946	Yvon Petra
1947	Jack Kramer
1948	Bob Falkenburg
1949	Fred Schroeder
1950	Budge Patty
1951	Dick Savitt
1952	Frank Sedgman
1953	Vic Seixas
1954	Jaroslav Drobny
1955	Tony Trabert
1956	Lew Hoad
1957	Lew Hoad
1958	Ashley Cooper
1959	Alex Olmedo
1960	Neale Fraser
1961	Rod Laver
1962	Rod Laver
1963	Chuck McKinley
1964	Roy Emerson
1965	Roy Emerson
1966	Manuel Santana
1967	John Newcombe
1968	Rod Laver
1969	Rod Laver
1970	John Newcombe
1971	John Newcombe
1972	Stan Smith
1973	Jan Kodes
1974	Jimmy Connors
1975	Arthur Ashe
1976	Björn Borg
1977	Björn Borg
1978	Björn Borg
1979	Björn Borg
1980	Björn Borg
1981	John McEnroe
1982	Jimmy Connors
1983	John McEnroe
1984	John McEnroe
1985	Boris Becker
1986	Boris Becker
1987	Pat Cash
1988	Stefan Edberg
1989	Boris Becker
1990	Stefan Edberg

Wimbledon Champions

Women's Singles

Year	Champion	Year	Champion	Year	Champion
1884	Maud Watson	1921	Suzanne Lenglen	1959	Maria Bueno
1885	Maud Watson	1922	Suzanne Lenglen	1960	Maria Bueno
1886	Blanche Bingley	1923	Suzanne Lenglen	1961	Angela Mortimer
1887	Lottie Dod	1924	Kathleen McKane	1962	Karen Hantze-Susman
1888	Lottie Dod	1925	Suzanne Lenglen	1963	Margaret Smith
1889	Blanche Hillyard	1926	Kathleen Godfree	1964	Maria Bueno
1890	L. Rice	1927	Helen Wills	1965	Margaret Smith
1891	Lottie Dod	1928	Helen Wills	1966	Billie Jean King
1892	Lottie Dod	1929	Helen Wills	1967	Billie Jean King
1893	Lottie Dod	1930	Helen Wills Moody	1968	Billie Jean King
1894	Blanche Hillyard	1931	Cecile Aussem	1969	Ann Haydon-Jones
1895	Charlotte Cooper	1932	Helen Wills Moody	1970	Margaret Smith Court
1896	Charlotte Cooper	1933	Helen Wills Moody	1971	Evonne Goolagong
1897	Blanche Hillyard	1934	Dorothy Round	1972	Billie Jean King
1898	Charlotte Cooper	1935	Helen Wills Moody	1973	Billie Jean King
1899	Blanche Hillyard	1936	Helen Jacobs	1974	Chris Evert
1900	Blanche Hillyard	1937	Dorothy Round	1975	Billie Jean King
1901	Charlotte Sterry	1938	Helen Wills Moody	1976	Chris Evert
1902	M. E. Robb	1939	Alice Marble	1977	Virginia Wade
1903	Dorothy Douglass	1940–45	No tournament	1978	Martina Navratilova
1904	Dorothy Douglass	1946	Pauline Betz	1979	Martina Navratilova
1905	May Sutton	1947	Margaret Osborne	1980	Evonne Goolagong
1906	Dorothy Douglass	1948	Louise Brough	1981	Chris Evert Lloyd
1907	May Sutton	1949	Louise Brough	1982	Martina Navratilova
1908	Charlotte Sterry	1950	Louise Brough	1983	Martina Navratilova
1909	Dorothea Boothby	1951	Doris Hart	1984	Martina Navratilova
1910	Dorothy Chambers	1952	Maureen Connolly	1985	Martina Navratilova
1911	Dorothy Chambers	1953	Maureen Connolly	1986	Martina Navratilova
1912	Ethel Larcombe	1954	Maureen Connolly	1987	Martina Navratilova
1913	Dorothy Chambers	1955	Louise Brough	1988	Steffi Graf
1914	Dorothy Chambers	1956	Shirley Fry	1989	Steffi Graf
1915–18	No tournament	1957	Althea Gibson	1990	Martina Navratilova
1919	Suzanne Lenglen	1958	Althea Gibson		
1920	Suzanne Lenglen				

1968 Wimbledon tournament is opened for first time to professional as well as amateur players.

1970s Popularity of tennis booms worldwide.

▶

United States Open Champions

Men's Singles

1881	Richard Sears
1882	Richard Sears
1883	Richard Sears
1884	Richard Sears
1885	Richard Sears
1886	Richard Sears
1887	Richard Sears
1888	Henry Slocum
1889	Henry Slocum
1890	Oliver Campbell
1891	Oliver Campbell
1892	Oliver Campbell
1893	Robert Wrenn
1894	Robert Wrenn
1895	Fred Hovey
1896	Robert Wrenn
1897	Robert Wrenn
1898	Malcolm Whitman
1899	Malcolm Whitman
1900	Malcolm Whitman
1901	William Larned
1902	William Larned
1903	Hugh Doherty
1904	Holcombe Ward
1905	Beals Wright
1906	William Clothier
1907	William Larned
1908	William Larned
1909	William Larned
1910	William Larned
1911	William Larned
1912	Maurice McLoughlin
1913	Maurice McLoughlin
1914	R. Norris Williams
1915	William Johnston
1916	R. Norris Williams
1917	Lindley Murray
1918	Lindley Murray
1919	William Johnston
1920	Bill Tilden
1921	Bill Tilden
1922	Bill Tilden
1923	Bill Tilden
1924	Bill Tilden
1925	Bill Tilden
1926	René Lacoste
1927	René Lacoste
1928	Henry Cochet
1929	Bill Tilden
1930	John Doeg
1931	H. Ellsworth Vines
1932	H. Ellsworth Vines
1933	Fred Perry
1934	Fred Perry
1935	Wilmer Allison
1936	Fred Perry
1937	Don Budge
1938	Don Budge
1939	Bobby Riggs
1940	Don McNeill
1941	Bobby Riggs
1942	Fred Schroeder
1943	Joseph Hunt
1944	Frank Parker
1945	Frank Parker
1946	Jack Kramer
1947	Jack Kramer
1948	Pancho Gonzales
1949	Pancho Gonzales
1950	Arthur Larsen
1951	Frank Sedgman
1952	Frank Sedgman
1953	Tony Trabert
1954	Vic Seixas
1955	Tony Trabert
1956	Ken Rosewall
1957	Malcolm Anderson
1958	Ashley Cooper
1959	Neale Fraser
1960	Neale Fraser
1961	Roy Emerson
1962	Rod Laver
1963	Rafael Osuna
1964	Roy Emerson
1965	Manuel Santana
1966	Fred Stolle
1967	John Newcombe
1968	Arthur Ashe
1969	Rod Laver
1970	Ken Rosewall
1971	Stan Smith
1972	Ilie Nastase
1973	John Newcombe
1974	Jimmy Connors
1975	Manuel Orantes
1976	Jimmy Connors
1977	Guillermo Vilas
1978	Jimmy Connors
1979	John McEnroe
1980	John McEnroe
1981	John McEnroe
1982	Jimmy Connors
1983	Jimmy Connors
1984	John McEnroe
1985	Ivan Lendl
1986	Ivan Lendl
1987	Ivan Lendl
1988	Mats Wilander
1989	Boris Becker
1990	Pete Sampras

United States Open Champions

Women's Singles

1887	Ellen Hansell
1888	Bertha Townsend
1889	Bertha Townsend
1890	Ellen Roosevelt
1891	Mabel E. Cahill
1892	Mabel E. Cahill
1893	Aline Terry
1894	Helen Helwig
1895	Juliette Atkinson
1896	Elisabeth Moore
1897	Juliette Atkinson
1898	Juliette Atkinson
1899	Marion Jones
1900	Myrtle McAteer
1901	Elisabeth Moore
1902	Marion Jones
1903	Elisabeth Moore
1904	May Sutton
1905	Elisabeth Moore
1906	Helen Homans
1907	Evelyn Sears
1908	Maud Bargar-Wallach
1909	Hazel Hotchkiss
1910	Hazel Hotchkiss
1911	Hazel Hotchkiss
1912	Mary Browne
1913	Mary Browne
1914	Mary Browne
1915	Molla Bjurstedt
1916	Molla Bjurstedt
1917	Molla Bjurstedt
1918	Molla Bjurstedt
1919	Hazel Wightman
1920	Molla Mallory
1921	Molla Mallory
1922	Molla Mallory
1923	Helen Wills
1924	Helen Wills
1925	Helen Wills
1926	Molla Mallory
1927	Helen Wills
1928	Helen Wills
1929	Helen Wills
1930	Betty Nuthall
1931	Helen Wills Moody
1932	Helen Jacobs
1933	Helen Jacobs
1934	Helen Jacobs
1935	Helen Jacobs
1936	Alice Marble
1937	Anita Lizana
1938	Alice Marble
1939	Alice Marble
1940	Alice Marble
1941	Sarah Palfrey Cooke
1942	Pauline Betz
1943	Pauline Betz
1944	Pauline Betz
1945	Sarah Palfrey Cooke
1946	Pauline Betz
1947	Louise Brough
1948	Margaret Osborne duPont
1949	Margaret Osborne duPont
1950	Margaret Osborne duPont
1951	Maureen Connolly
1952	Maureen Connolly
1953	Maureen Connolly
1954	Doris Hart
1955	Doris Hart
1956	Shirley Fry
1957	Althea Gibson
1958	Althea Gibson
1959	Maria Bueno
1960	Darlene Hard
1961	Darlene Hard
1962	Margaret Smith
1963	Maria Bueno
1964	Maria Bueno
1965	Margaret Smith
1966	Maria Bueno
1967	Billie Jean King
1968	Virginia Wade
1969	Margaret Smith Court
1970	Margaret Smith Court
1971	Billie Jean King
1972	Billie Jean King
1973	Margaret Smith Court
1974	Billie Jean King
1975	Chris Evert
1976	Chris Evert
1977	Chris Evert
1978	Chris Evert
1979	Tracy Austin
1980	Chris Evert Lloyd
1981	Tracy Austin
1982	Chris Evert Lloyd
1983	Martina Navratilova
1984	Martina Navratilova
1985	Hana Mandlikova
1986	Martina Navratilova
1987	Martina Navratilova
1988	Steffi Graf
1989	Steffi Graf
1990	Gabriela Sabatini

French Open Champions

Men's Singles

1891	Briggs
1892	J. Schopfer
1893	L. Riboulet
1894	A. Vacherot
1895	A. Vacherot
1896	A. Vacherot
1897	P. Aymé
1898	P. Aymé
1899	P. Aymé
1900	P. Aymé
1901	A. Vacherot
1902	A. Vacherot
1903	Max Decugis
1904	Max Decugis
1905	Maurice Germot
1906	Maurice Germot
1907	Max Decugis
1908	Max Decugis
1909	Max Decugis
1910	Maurice Germot
1911	André Gobert
1912	Max Decugis
1913	Max Decugis
1914	Max Decugis
1915–19	No competition
1920	André Gobert
1921	Jean Samazeuilh
1922	Henri Cochet
1923	Paul Blanchy
1924	Jean Borotra
1925	René Lacoste
1926	Henri Cochet
1927	René Lacoste
1928	Henri Cochet
1929	René Lacoste
1930	Henri Cochet
1931	Jean Borotra
1932	Henri Cochet
1933	John Crawford
1934	Gottfried von Cramm
1935	Fred Perry
1936	Gottfried von Cramm
1937	Henner Henkel
1938	Don Budge
1939	Don McNeill
1940–45	No competition
1946	Marcel Bernard
1947	Josef Asboth
1948	Frank A. Parker
1949	Frank A. Parker
1950	Budge Patty
1951	Jaroslav Drobny
1952	Jaroslav Drobny
1953	Ken Rosewell
1954	Tony Trabert
1955	Tony Trabert
1956	Lew Hoad
1957	Sven Davidson
1958	Marvyn Rose
1959	Nicola Pietrangeli
1960	Nicola Pietrangeli
1961	Manuel Santana
1962	Rod Laver
1963	Roy Emerson
1964	Manuel Santana
1965	Fred Stolle
1966	Tony Roche
1967	Roy Emerson
1968	Ken Rosewall
1969	Rod Laver
1970	Jan Kodeš
1971	Jan Kodeš
1972	Andres Gimeno
1973	Ilie Nastase
1974	Björn Borg
1975	Björn Borg
1976	Adriano Panatta
1977	Guillermo Vilas
1978	Björn Borg
1979	Björn Borg
1980	Björn Borg
1981	Björn Borg
1982	Mats Wilander
1983	Yannick Noah
1984	Ivan Lendl
1985	Mats Wilander
1986	Ivan Lendl
1987	Ivan Lendl
1988	Mats Wilander
1989	Michael Chang
1990	Andres Gomez

Women's Singles

1897	Cecilla Masson
1898	Cecilla Masson
1899	Cecilla Masson
1900	Cecilla Prévost
1901	P. Girod
1902	Cecilla Masson
1903	Cecilla Masson
1904	Katie Gillou
1905	Katie Gillou
1906	Katie Fenwick
1907	Mme. de Kermel
1908	Katie Fenwick
1909	Jeanne Mattey
1910	Jeanne Mattey
1911	Jeanne Mattey

French Open Champions

Year	Champion	Year	Champion	Year	Champion
1912	Jeanne Mattey	1946	Margaret Osborne	1968	Nancy Richey
1913	Marguerite Broquedis	1947	Patricia Canning Todd	1969	Margaret Smith Court
1914	Marguerite Broquedis	1948	Nelly Landry	1970	Margaret Smith Court
1915–19	No competition	1949	Margaret Osborne duPont	1971	Evonne Goolagong
1920	Suzanne Lenglen	1950	Doris Hart	1972	Billie Jean King
1921	Suzanne Lenglen	1951	Shirley Fry	1973	Margaret Smith Court
1922	Suzanne Lenglen	1952	Doris Hart	1974	Chris Evert
1923	Suzanne Lenglen	1953	Maureen Connolly	1975	Chris Evert
1924	Suzanne Lenglen	1954	Maureen Connolly	1976	Sue Barker
1925	Suzanne Lenglen	1955	Angela Mortimer	1977	Mima Jausovec
1926	Suzanne Lenglen	1956	Althea Gibson	1978	Virginia Ruzici
1927	Kea Bouman	1957	Shirley Bloomer	1979	Chris Evert Lloyd
1928	Helen Wills	1958	Susi Kormoczi	1980	Chris Evert Lloyd
1929	Helen Wills	1959	Christine Truman	1981	Hana Mandlikova
1930	Helen Wills Moody	1960	Darlene Hard	1982	Martina Navratilova
1931	Cilly Aussem	1961	Ann Haydon	1983	Chris Evert Lloyd
1932	Helen Wills Moody	1962	Margaret Smith	1984	Martina Navratilova
1933	Margaret C. Scriven	1963	Lesley Turner	1985	Chris Evert Lloyd
1934	Margaret C. Scriven	1964	Margaret Smith	1986	Chris Evert Lloyd
1935	Hilda Sperling	1965	Lesley Turner	1987	Steffi Graf
1936	Hilda Sperling	1966	Ann Haydon Jones	1988	Steffi Graf
1937	Hilda Sperling	1967	Françoise Durr	1989	A. Sanchez
1938	Simone Mathieu			1990	Monica Seles
1939	Simone Mathieu				
1940–45	No competition				

1975 Australian player Margaret Smith Court (b. 1942) wins her record 64th title in Big Four tournaments since 1960.

1975 American player Jimmy Connors (b. 1952) wins record $500,000 for single match when he defeats John Newcombe in Las Vegas.

1976 Controversy erupts when transsexual Dr. Renée Richards is barred from U.S. Open after refusing to take chromosome test designed to determine her sex. A 1977 court order allows her to play.

1978 U.S. Open is moved to Flushing Meadow, New York.

1979 Twenty-year-old John McEnroe (b. 1959) and 16-year-old Tracy Austin (b. 1963) win U.S. Open singles championships.

1979 American player Billie Jean King (b. 1943) holds record total of 20 Wimbledon wins, including six singles titles, 10 women's doubles, and four mixed doubles since 1961.

1980 Swedish player Björn Borg (b. 1956) wins fifth consecutive singles title at Wimbledon, a record.

1986 Record attendance is set at Wimbledon, with nearly 40,000 fans at single match.

1987 U.S. player Martina Navratilova (b. 1956) establishes record for highest career earnings, with $12.7 million.

1989 U.S. player Chris Evert retires from women's professional tennis circuit with lifetime tournament record of 157 wins.

Heavyweight Boxing Champions

1882–92	John L. Sullivan
1892–97	James J. Corbett
1897–99	Robert Fitzsimmons
1899–1905	James J. Jeffries
1905–06	Marvin Hart
1906–08	Tommy Burns
1908–15	Jack Johnson
1915–19	Jess Willard
1919–26	Jack Dempsey
1926–28	Gene Tunney
1928–30	No titleholder
1930–32	Max Schmeling
1932	Jack Sharkey
1933	Primo Carnera
1934	Max Baer
1935–37	James J. Braddock
1937–49	Joe Louis
1949–51	Ezzard Charles
1951–52	Joe Walcott
1952–56	Rocky Marciano
1956–59	Floyd Patterson
1959	Ingemar Johansson
1960–62	Floyd Patterson
1962–64	Sonny Liston
1964–67	Cassius Clay
1970–73	Joe Frazier
1973–74	George Foreman
1974–78	Muhammad Ali
1978–79	Leon Spinks, Muhammad Ali (WBA)
1978	Ken Norton (WBC), Larry Holmes (WBC)
1979	John Tate (WBA)
1980	Mike Weaver (WBA)
1982	Michael Dokes (WBA)
1983	Gerrie Coetzee (WBA)
1984	Tim Witherspoon (WBC); Pinklon Thomas (WBC); Greg Page (WBA)
1985	Tony Tubbs (WBA); Michael Spinks (IBF)
1986	Tim Witherspoon (WBA); Trevor Berbick (WBC); James Smith (WBA)
1986–90	Mike Tyson (WBC)
1987–90	Mike Tyson (WBA)
1990	James "Buster" Douglas (WBC, WBA)

America's Cup Races

1870 *Magic* beats British challenger *Cambria* in 1 race.

1871 *Columbia* beats British challenger *Livonia*, 4 races to 1.

1876 *Madeleine* beats Canadian challenger *Countess of Dufferin*, 2 races to 0.

1881 *Mischief* beats Canadian challenger *Atalanta*, 2 races to 0.

1885 *Puritan* beats British challenger *Genesta*, 2 races to 0.

1886 *Mayflower* beats British challenger *Galatea*, 2 races to 0.

1887 *Volunteer* beats British challenger *Thistle*, 2 races to 0.

▶

1893 *Vigilant* beats British challenger *Valkyrie II*, 3 races to 0.

1895 *Defender* beats British challenger *Valkyrie III*, 3 races to 0.

1899 *Columbia* beats British challenger *Shamrock*, 3 races to 0.

1901 *Columbia* beats British challenger *Shamrock II*, 3 races to 0.

1903 *Reliance* beats British challenger *Shamrock III*, 3 races to 0.

1920 *Resolute* beats British challenger *Shamrock IV*, 3 races to 2.

1930 *Enterprise* beats British challenger *Shamrock V*, 4 races to 0.

1934 *Rainbow* beats British challenger *Endeavour*, 4 races to 2.

1937 *Ranger* beats British challenger *Endeavour II*, 4 races to 0.

1958 *Columbia* beats British challenger *Sceptre*, 4 races to 0.

1962 *Weatherly* beats Australian challenger *Gretel*, 4 races to 1.

1964 *Constellation* beats British challenger *Sovereign*, 4 races to 0.

1967 *Intrepid* beats Australian challenger *Dame Pattie*, 4 races to 1.

1970 *Intrepid* beats Australian challenger *Gretel II*, 4 races to 1.

1974 *Courageous* beats Australian challenger *Southern Cross*, 4 races to 0.

1977 *Courageous* beats Australian challenger *Australia*, 4 races to 0.

1980 *Freedom* beats Australian challenger *Australia* in Australia, 4 races to 1.

1983 Australian challenger *Australia II* beats U.S. cup defender *Liberty*, 4 races to 3.

1987 U.S. challenger *Stars & Stripes* beats Australian cup defender *Kookaburra III* in Australia, 4 races to 0.

1988 *Stars & Stripes* beats New Zealand challenger *New Zealand* in New Zealand, 2 races to 0. In 1989 a New York court rules *Stars & Stripes'* design is illegal and awards cup to New Zealand. Cup is returned to United States following appeal to higher court (1989).

Olympic Games

Dates and locations of Olympic Games in modern times.

1896 Athens, Greece

1900 Paris, France

1904 St. Louis, Missouri

1908 London, England

1912 Stockholm, Sweden

1920 Antwerp, Belgium

1924 Paris, France; winter games in Chamonix, France

1928 Amsterdam, Netherlands; winter games in St. Moritz, Switzerland

1932 Los Angeles, California; winter games in Lake Placid, New York

1936 Berlin, Germany; winter games in Garmisch-Partenkirchen, Germany

1948 London, England; winter games in St. Moritz, Switzerland

1952 Helsinki, Finland; winter games in Oslo, Norway

1956 Melbourne, Australia; winter games in Cortina, Italy

1960 Rome, Italy; winter games in Squaw Valley, California

1964 Tokyo, Japan; winter games in Innsbruck, Austria

1968 Mexico City, Mexico; winter games in Grenoble, France

▶

1972 Munich, Germany; winter games in Sapporo, Japan

1976 Montreal, Canada; winter games in Innsbruck, Austria

1980 Moscow, USSR; winter games in Lake Placid, New York

1984 Los Angeles, California; winter games in Sarajevo, Yugoslavia

1988 Seoul, Korea; winter games in Calgary, Canada

Selected Bibliography

American Heritage History of Flight. American Heritage, 1962.

Annual Register: A Record of World Events. Longman.

Asimov, Isaac. Asimov's Chronicles of Science and Discovery. Harper & Row, 1989.

Asimov, Isaac, ed. *Biographical Encyclopedia of Science and Technology*, 2nd edition, Doubleday, 1982.

Banham, Martin. *The Cambridge Guide to World Theatre.* Cambridge University Press, 1988.

Benet's Reader's Encyclopedia. Harper & Row, 1987.

Boorstein, Daniel J. *The Discoverers.* Random House, 1985.

Boucher, François. *20,000 Years of Fashion.* Harry Abrams, 1987.

Brodie, Bernard, and Fawn Brodie. *From Crossbow to H-Bomb.* Indiana University Press, 1973.

Bronowski, Jacob. *The Ascent of Man.* Little, Brown, 1974.

Brooks, Tim, and Earle Marsh. *The Complete Directory to Prime Time Network TV Shows, 1946–Present*, 4th edition. Ballantine, 1988.

Brownstone, David, and Irene Franck. *Dictionary of 20th Century History.* Prentice Hall, 1990.

Bryant, Keith L., Jr., ed. *Railroads in the Age of Regulation, 1900–1980.* Facts on File, 1987.

Burbank, Richard. *Twentieth Century Music.* Harry Abrams, 1987.

Calvert, Michael, Brigadier General, and Peter Young, Brigadier. *Dictionary of Battles.* Mayflower, 1979.

Canby, Courtlandt. *Encyclopedia of Historic Places.* Facts on File, 1984.

Canby, Courtlandt. *The Past Displayed.* Phaidon, 1980.

Carruth, Gorton. *The Encyclopedia of American Facts and Dates.* Harper & Row, 1987.

Carruth, Gorton, and Eugene Ehrlich. *Facts and Dates of American Sports.* Harper & Row, 1988.

Cassin-Scott, Jack. *The Illustrated Encyclopedia of Fashion, 1550–1920.* Dorset, 1986.

Clark, Ronald. *Works of Man.* American Society of Civil Engineers, 1985.

Considine, Douglas M., P.E., ed. *Van Nostrand's Scientific Encyclopedia*, 6th edition. Van Nostrand Reinhold, 1983.

Cooke, David Coxe. *Bomber Planes That Made History.* Putnam, 1959.

Cooke, David Coxe. *Inventions That Made History.* Putnam, 1968.

The Crown Guide to the World's Great Plays. Crown, 1984.

Current Biography Yearbook. H.W. Wilson.

Delpar, Helen, ed. *The Discoverers.* McGraw-Hill, 1980.

Desmond, Kevin. *A Timetable of Inventions and Discoveries from Pre-History to the Present Day.* M. Evans, 1986.

Diagram Group. *Weapons.* St. Martin's, 1980.

Dickey, Glenn. *The History of the World Series Since 1903.* Stein & Day, 1984.

Dickson, Paul. *Baseball Dictionary.* Facts on File, 1988.

Discoverers of the New World. American Heritage, 1960.

Dummer, G. W. A. *et al. The Timetable of Technology.* Hearst, 1982.

Ecam Publications. *Chronicle of the Twentieth Century.* Prentice Hall, 1988.

Ecam Publications. *Chronicle of the World.* Prentice Hall, 1990.

Eliade, Mircea. *The Encyclopedia of Religion.* 4 vols. Macmillan, 1987.

Evans, Harold. *Front Page History.* Salem House, 1984.

Facts on File World News Digest. Facts on File, 1980–90.

Fage, J. D., and Roland Oliver. *A Short History of Africa*. Facts on File, 1989.

Ferrier, Jean-Louis. *Art of Our Century*. Prentice Hall, 1989.

Ferrill, Arther. *The Origins of War*. Thames & Hudson, 1985.

Ford, Boris. *The New Pelican Guide to English Literature, American Literature*, vol. 9. Penguin, 1988.

Frey, Robert L., ed. *Railroads in the Nineteenth Century*. Facts on File, 1988.

Giscard d'Estaing, Valerie-Anne. *The Second World Almanac Book of Inventions*. Pharos Books, 1986.

Glynn, Prudence. *In Fashion*. Oxford University Press, 1978.

Goetzmann, William. *New Lands, New Men*. Viking, 1986.

Grout, Donald. *A History of Western Music*, 4th edition. Norton, 1988.

Grun, Bernard. *The Timetables of History*. Simon & Schuster, 1982.

Guinness Book of Movies Facts and Feats. Bantam, 1988.

Guinness Book of World Records, 1990. Bantam, 1990.

Hastings, Max. *The Korean War*. Simon & Schuster, 1987.

Haylett, John. *The Illustrated Encyclopedia of World Tennis*. Exeter, 1989.

Hellemans, Alexander, and Bryan Bunch. *The Timetables of Science*. Simon & Schuster, 1988.

Hindley, Geoffrey, ed. *Larousse Encyclopedia of Music*. Hamlyn, 1974.

Illustrated Reference Book of Ancient History. Windward, 1982.

Information Please Almanac. Houghton Mifflin.

James, Bill. *The Bill James Historical Baseball Abstract*. Random House, 1988.

Janson, H. W. *History of Art*. Prentice Hall, 1986.

Katz, Ephraim. *The Film Encyclopedia*. Putnam, 1982.

Keegan, John. *Atlas of World War II*. Harper & Row, 1990.

Keegan, John. *The Second World War*. Viking, 1990.

Kemp, Peter. *The History of Ships*. Orbis, 1978.

Klein, Leonard S., ed. *Encyclopedia of World Literature in the Twentieth Century*. 4 vols. Frederick Ungar, 1981.

Kohn, George C. *Dictionary of Wars*. Facts on File, 1989.

Kohn, George C. *Encyclopedia of American Scandal*. Facts on File, 1979.

Krantz, Les, ed. *The New York Art Review*. American References, 1988.

Kurian, George. *Encyclopedia of the Third World*, 3rd edition. 3 vols. Facts on File, 1987.

Kurian, George. *World Press Encyclopedia*. Facts on File, 1982.

Langer, William L. *An Encyclopedia of World History*. Houghton Mifflin, 1980.

Levy, Leonard W. *Encyclopedia of the American Constitution*. 4 vols. Macmillan, 1986.

Livesey, Anthony. *Great Battles of World War I*. Macmillan, 1989.

Locher, Frances, C., ed. *Contemporary Authors*. Gale Research, 1982.

Low, W. Augustus, and Virgil A. Clift. *Encyclopedia of Black America*. Da Capo, 1984.

Magill, Frank. *Magill Survey of Cinema: First Series*. Salem, 1980.

Magnificent Voyages. Smithsonian Institution, 1985.

Maltin, Leonard, ed. *TV Movies*. New American Library, 1984.

Marshall Cavendish Illustrated Encyclopedia of World War I. Marshall Cavendish, 1986.

Mart, Gerald. *A Short History of the Movies*, 4th edition. Macmillan, 1986.

May, George S. *The Automobile Industry, 1920–1980*. Facts on File, 1989.

McGraw-Hill Encyclopedia of Science and Technology, 6th edition. 20 vols. McGraw-Hill, 1987.

McGrew, Roderick. *Encyclopedia of Medical History*. McGraw-Hill, 1985.
Messerole, Mike. *1990 Information Please Sports Almanac*. Houghton Mifflin, 1990.
Metropolitan Opera Encyclopedia. Simon & Schuster, 1987.
Milbank, Caroline Rennolds. *Couture: The Great Designers*. Stewart, Tabori and Chang, 1985.
Monaco, James. *Connoisseur's Guide to the Movies*. Facts on File, 1986.
Monaco, James. *How to Read a Film*, revised edition. Oxford University Press, 1981.
Morris, Richard B. *Encyclopedia of American History*. Harper & Row, 1982.
Morris, Richard, and Graham Irwin. *Harper Encyclopedia of the Modern World*. Harper & Row, 1970.
Mount, Ellis, and Barbara A. List. *Milestones in Science and Technology: The Ready Reference Guide to Discoveries, Inventions and Facts*. Oryx, 1987.
Nash, Jay Robert. *Darkest Hours*. Nelson-Hall, 1976.
Newby, Eric. *The World Atlas of Exploration*. Crescent, 1985.
New Cambridge Modern History. 4 vols. Cambridge University Press, 1957–79.
Olivora, Vera. *Sports and Games in the Ancient World*. St. Martin's, 1985.
Orr, Frank. *The Story of Hockey*. Random House, 1971.
Panati, Charles. *Browser's Book of Beginnings*. Houghton Mifflin, 1984.
Panati, Charles. *Extraordinary Origins of Everyday Things*. Harper & Row, 1987.
Paskoff, Paul. *The Iron and Steel Industry in the Nineteenth Century*. Facts on File, 1988.
Pennington, Piers. *The Great Explorers*. Bloomsbury, 1979.
Piper, David. *The History of Art*. Random House, 1981.
Plimpton, George. *Open Net*. Norton, 1985.
Porter, Glenn, ed. *Encyclopedia of American Economic History*. 3 vols. Scribner, 1980.
Ralston, Anthony, ed. *Encyclopedia of Computer Science and Engineering*, 2nd edition. Van Nostrand Reinhold, 1983.
The Right College. Arco, 1990.
Schuller, Gunther. *Early Jazz*. Oxford University Press, 1968.
Siegal, Scott, and Barbara Siegal. *The Encyclopedia of Hollywood*. Facts on File, 1990.
Siegman, Gita, ed. *Awards, Honors and Prizes*, 8th edition. 2 vols. Gale Research, 1989.
Steinberg, Cobbett. *Reel Facts*. Vintage, 1982.
Summers, Harry G., Jr., Col. *Vietnam War Almanac*. Facts on File, 1985.
Tapsul, R. F. *Monarchs, Rulers, Dynasties and Kingdoms of the World*. Facts on File, 1983.
Tobler, John. *30 Years of Rock*. Exeter, 1985.
Vecsey, George, ed. *The Way It Was: Great Sports Events from the Past*. McGraw-Hill, 1974.
Vernott, Edward, and Rima Shori. *The International Dictionary of 20th Century Biography*. New American Library, 1987.
The Vietnam War. Military Press, 1988.
Weis, Frank W. *Lifelines*. Facts on File, 1982.
Wetterau, Bruce. *Macmillan Concise Dictionary of World History*. Macmillan, 1983.
Who Did What. Galley, 1985.
Williams, Trevor. *History of Invention: From Stone Axes to Silicon Chips*. Facts on File, 1987–90.
Worldmark Encyclopedia of the Nations, 7th edition. Worldmark, 1988.
Wynne-Davies, Marion, ed. *Prentice Hall Guide to English Literature*. Prentice Hall, 1990.
Young, Peter, Brigadier. *Decisive Battles of World War II*. Galley, 1989.

Index

Aaron, Hank, 597
Abacus, 179
Abbey of St. Denis, 352
ABC network, 537
Abdul-Jabbar, Kareem, 590
Abelard, Peter, 337
Abensberg-Eckmuhl (battle), 458, 482
Abernathy, Rev. Ralph, 132
Abortion, 40, 68, 134
Aboukir (battle), 482
Abraham, 325
Abrams, Creighton (Gen.), 474
Abscam, 140
Academy Awards, 553–54, 556
Achebe, Chinua, 261
Acheson, Edward, 162
Acid rain, 422, 424
Acre (battle), 478, 479
Actium (battle), 477
Act of Disestablishment, 319; of Settlement, 318, 321; of Supremacy, 115, 318; of Toleration, 317, 318, 322; of Union, 33, 57
Acupuncture, 363
Adams, Henry, 248
Adams, John Quincy, 33, 133
Adding machine, 184, 412
Addison, Joseph, 238
Adhesive tape, 437
Adrianople (battles), 477
Adua (battle), 484
Aegates (battle), 476
Aegospotami (battle), 476
Aerosol cans, 438
Aesop (fabulist), 233
Africa, 13–15, 72–76, 505, 506
Agee, James, 259
Agincourt (battle), 479
Agnew, Spiro, 140
Agreement of the People, 454
Agriculture, advances in, 192–94
AIDS, 375, 529
Aiken, Conrad, 255
Air Commerce Act, 171
Air conditioner, 185
Airline industry (U.S), 171–73
Airplane crashes, 505–7
Aisne (battle), 464, 485
Akron (dirigible), 505
Alamo (siege), 483
Alaska, pipeline, 169; Anchorage, 518
Albee, Edward, 288
Albeniz, Isaac, 299
Albingensian Crusade, 307
Albrecht, archbishop of Mainz, 315
Alcoholism, 371
Alcott, Louisa May, 248
Aldrin, Edwin, 23
Alexander III, 62
Alexander the Great, 3, 19, 77, 95, 97, 476
Alexandria (battle), 478
Alfred the Great, 49
Alger, Horatio, 248
Alien and Sedition Acts, 533
Alighieri, Dante, 235
Aljubarrota (battle), 479
Allen, Joseph, 24
Allende, Salvador, 43
Allston, Washington, 270
Alma (battle), 483
Al-Mutamar al-Alam al-Islami, 334
Aluminum, 184
American Anthropological Association, 393
American Chemical Society, 403
American Colonization Society, 130
American Federation of Labor (AFL), 159
American Medical Association, 372
American Professional Football Association, 582
American Revolution, *see* War.
American Society of Composers, Authors, and Publishers, 568
Amin, Idi, 75
Amin, Qasim, 334
Amis, Kingsley, 260
Ammara (battle), 489
Ammonia, process for mass-producing, 206
Amniocentesis, 374
Amoco, 149
Ampère, André-Marie, 211, 408
Amphibpolis (battle), 476
Amundsen, Roald, 13, 16, 18
Anabaptist movement, 316
Anaxagoras, 336
Anaximander of Miletus, 335
Anaximenes, 335
Andersen, Hans Christian, 245
Anderson, Maxwell, 286
Anderson, Sherwood, 252
Andrea Doria (ship), 514
Angelico, Fra, 263
Angelou, Maya, 261
Anglicanism, 317–20
Anglo-Soviet Agreement, 118
Angola, 75, 453
Animal crackers, 428
Ankara (battle), 479
Annenberg, Walter Hubert, 153
Anouilh, Jean, 287
Antabuse, 374
Antarctic exploration, 17–19
Anthony, Susan B., 133
Anthropology, 392–96
Antiballistic missile technology, 189, 494
Anticollision devices, 223
Antietam (battle), 460, 484
Antimatter, 410
Antioch (battle), 478
Antony, Marc, 97
Anzio landings, 487
Ap Bac (battle), 488
Apollonius (poet), 233
Apollo missions, 21, 22, 24, 188, 217, 225, 229
Appel, Karel, 277
Aquae Sextiae (battle), 477
Aqua-lung, 186
Aquino, Benigno, 90
Arafat, Yasir, 71, 129
Arbuckle, Roscoe "Fatty," 556
Archaeology, 392–96
Archer-Daniels-Midland Co., 151
Archimedes, 403
Archipenko, Alexander, 275
Architecture 348–58
Arginusae (battle), 476
Argonaut (submarine), 198
Argonne Forest (battle), 464, 486
Aristotle, 336, 376, 403
Armani, Giorgio, 432
Armor, 489, 490
Armour, Philip D., 152
Arms race, 494
Armstrong, Edwin, 537
Armstrong, Neil, 21, 23
Arnhem (battle), 470, 487
Arnold, Benedict, 455
Arnold, Matthew, 247
Arp, Jean, 275
Arras (battle), 463
Arsenic, 396
Art, 264–77
Artois (battle), 463
Asbury, Francis, 324
Asch, Sholem, 253
Asculum (battle), 476
Ashbery, John, 261
Asia, 13, 76–91
Asoka, 363
Aspern-Essling (battle), 458
Aspirin, 373, 437
Assassinations, 34, 37, 89, 90, 141–42
Assyria, rulers of, 93
Astor, John J., 152
Astrolabe, 179, 376
Astronauts, 21–25
Astronomy, 375–85
Ataturk, Kemal, 70
Atlanta (battle), 198, 461
Atlantic Charter, 116
Atom, 226
Atomic bomb, 88, 186, 226, 471, 493
Attila the Hun, 101, 448
Atwood, Margaret, 262
Auden, W.H., 259, 287
Augsburg Confession, 316
Augustine, Saint, 29, 235, 307, 317, 337
Augustus, Caesar, 5
Augustus, Romulus, 100, 101
Aurelius, Marcus, 100, 101, 235, 336
Austen, Jane, 242
Austerlitz (battle), 458, 482
Austin, Tracy, 618
Australopithecus afarensis, 395
Automobile(s), 170–71, 184–85, 218–21, 230, 436, 507–8
Autumn Harvest Uprising, 144, 467
Averroës, 333, 337
Avogadro's Law, 400
Ayachucho (battle), 42, 483
Ayckbourne, Alan, 289
Aztecs, 29
Babbage, Charles, 227
Babeuf, François, 144
Babylon (battle), 475, 476
Bach, Johann Sebastian, 293
Bacon, Francis (painter), 277
Bacon, Sir Francis (essayist), 237, 337
Badr (battle), 332, 477
Baekeland, Leo, 162
Baird, J.L., 536
Bakelite, 162, 185, 207, 402
Baker, Newton D., 594
Baker, Robert, 140
Bakken, Jim, 585
Bakker, Jim and Tammy, 323
Balaklava (battle), 483
Baldwin, James A., 261
Balenciaga, Cristobal, 431
Ball bearings, 181
Balloon, hot-air, 181
Balzac, Honoré de, 243
Band-Aids, 437
Bannockburn (battle), 51
Baptist Church, 316, 322–23
Barbados, 110
Barber, Samuel, 301
Barbie doll, 436
Barcelona (battle), 486
Barium, 399
Barnard, Christiaan, 374
Barnard, Henry, 340
Barometer, 180
Barrow, Clyde, 441
Barth, Heinrich, 14
Barth, John, 261
Barthelme, Donald, 262
Bartók, Béla, 300
Barton, Otis, 20
Baseball, 435, 590–98
BASIC, 228
Basketball, 586–90
Bastille, 435, 456
Bataan (battle), 486
Bathysphere, 185, 421
Battery, 183, 212
Battle of Britain, 468, 486
Battle of the Bulge, 470, 487
Battle of the Trench, 332
Baudelaire, Charles Pierre, 247
Baudot, Emile, 214
Baum, L. Frank, 249
Baylen (battle), 458
Bay of Pigs, 39, 488
Bean, Alan, 24
Beatlemania, 436
Beaumarchais, Pierre-Augustin Caron de, 282
Beaumont, Francis, 279
Beauregard, Pierre (Gen.), 460
Beckett, Samuel, 287
Beckmann, Max, 275
Beds, 439
Beebe, William, 20
Beene, Geoffrey, 432
Beethoven, Ludwig van, 294
Behan, Brendan, 288
Behrens, Peter, 357
Belgrade (battle), 481
Bell, Alexander Graham, 212, 214, 567
Bell, Sir Charles, 372
Bellini, Giovanni, 265
Bellini, Vincenzo, 295
Bell Laboratories, 569
Belloc, Hilaire, 251
Bellow, Saul, 260
Bemis Heights (battle), 481
Benavente, Jacinto, 286
Benchley, Robert Charles, 255
Benedict, Ruth, 393
Benedict, Saint, of Nursia, 45
Benedict (Italian monk), 307
Benét, Stephen Vincent, 257
Bennington (battle), 481
Bentham, Jeremy, 338
Benz, Karl, 218
Beowulf, 235
Berezovoy, Anatoliy N., 24
Berg, Alban, 300
Berge Stahl (ship), 197
Beriberi, 369
Bering, Vitus, 11, 15
Berkeley, George, 337
Berkowitz, David, 442
Berle, Milton, 537
Berlin (battle), 471, 488
Berliner, Emile, 567
Berlin Wall, 69
Berlioz, Hector, 295
Bernard, Claude, 389
Bernini, Giovanni Lorenzo, 268
Bernoulli principle, 407
Bernstein, Leonard, 301
Berryman, John, 260
Berzelius, Jons Jacob, 399
Bessemer, 165, 166, 167, 183, 206
Beverages, 427–29
Beyle, Marie Henri. *See* Stendhal
Bhagavadgita, 331
Bhopal (India), 90, 424
Bhutto, Benazir, 91
Bible, The, 305–6
Bierce, Ambrose G., 249
"Big-bang theory," 383, 384
Big Ben, 200
Bigelow, Erastus, 440
Bikini (bathing suit), 431, 434, 436
Bill of Rights, 32, 115, 119, 134, 318, 533
Bingham, George Caleb, 271
Bingham, Hiram, 393
Biochemistry, 390
Biology, 385–96
Biometry, 390
Bionetics, 188
Birth control, 134, 374
Bishop, Elizabeth, 259
Bismarck (ship), 469, 486
Bizet, Georges, 296
Björnson, Björnstjerne, 248
Blackboards, 340
"Black Friday," 155, 158
Black holes, 382
"Black Maria" (film studio), 552
Blackmore, Richard Doddridge, 247
Black Panther Party, 132
Black Power Movement, 132
Black Sox scandal, 594
Blake, William, 241, 270
Blass, Bill, 432
Blenheim (battle), 481
Blizzard(s), 522–24
Block and tackle, 178
Bloomer, Amelia, 133, 430
Boardman, Richard, 324
Boccaccio, Giovanni, 235
Boccherini, Luigi, 294
"Body painting," 433
Boeing aircraft, 151, 172, 505–7
Boer War, 62, 73, 393, 451
Boethius (philosopher), 235, 337
Bollard, Wilfred, 20
Bombay, 512
Bonheur, Rosa, 271
Bonhomme Richard, 455
Bonnard, Pierre, 273
Books, 201, 545–52
Borg, Björn, 618
Borges, Jorge Luis, 257
Boring mill, 206
Borman, Frank, 22
Borodin, Aleksandr, 296
Borodino (battle), 459, 483
Bosch, Hieronymous, 265
Boston, 481, 511, 577, 602
Boston Braves, 582
Boston Tea Party, 32
Boswell, James, 241
Bosworth Field (battle), 479
Botticelli, Sandro, 265
Boucher, François, 268
Boucher, Frank, 600
Boudin, Eugène, 271
Bow and arrow, 191, 489
Bowen, Elizabeth D.C., 257
Boxer Rebellion, 144
Boxing, heavyweight champions, 619
Boyaca (battle), 42, 483
Boyle, Robert, 369, 396, 399, 404
Boyne (battle), 480
Bradstreet, Anne, 238
Brady, James D., 152
Bragg, Braxton (Gen.), 461
Brahms, Johannes, 296
Brakes, 164, 183, 205, 219, 220
Brancusi, Constantin, 273
Brandt, Willy, 67, 140
Brandywine Creek (battle), 455, 481
Bransfield, Edward, 17
Braque, Georges, 274
Braun, Ferdinand, 216
Brautigan, Richard, 262
Bread, 427
Brecht, Bertolt, 287
Breitenfeld (battle), 480
Brezhnev, Leonid, 146
"Brezhnev Doctrine," 67
Bricks, 177
Bridge(s), 181, 183, 190, 356–57
Brillo pads, 437

Britannica (ship), 196
British Empire, 110–11
British North America Act, 34
Britten, Benjamin, 301
Broadcasting, radio and television, 536–45
Brock, Lou, 597, 598
Brome, Richard, 280
Bromine, 162, 372
Bronfman, Samuel, 153
Brontë, Anne, 247
Brontë, Charlotte, 246
Brontë, Emily J., 246
Brown, Charles Brockden, 241
Brown, Claude, 612
Brown, Walter F., 172
Brownie camera, 210
Browning, Elizabeth Barrett, 245
Browning, Robert, 246
Brown v. Board of Education of Topeka, Kansas, 38, 131
Bruch, Max, 296
Bruckner, Anton, 296
Bruegel, Jan, 267
Bruegel, Pieter (the elder), 267
Bruegel, Pieter (the younger), 267
Brussels (Belgium), 131, 505
Bryant, William Cullen, 243
Buchner, Georg, 283
Buck, Pearl S., 256
Buckingham Palace, 54
Buddha, 77
Buddhism, 80, 330
Budge, Don, 612
Budweiser Rocket, 230
Buena Vista (battle), 483
Buffet, Bernard, 277
Bull Run (battle), 460, 484
Bulwark (battleship), 513
Bulwer-Lytton, Edward George Earle, 245
Bundy, Theodore, 443
Bundy Manufacturing Co., 150
Bunker Hill (battle), 454, 481, 491
Bunning, Jim, 597
Bunsen burner, 400
Bunyan, John, 238, 322
Buonarroti, Michelangelo, 266
Burgess, Anthony, 260
Burke, Edmund, 241
Burnett, Frances Hodgson, 249
Burney, Fanny, 241
Burns, Robert, 241
Burroughs, William S., 260
Burton, Richard, 14
Busch, August Anheuser, Jr., 153
Bush, George, 40
Businesses (U.S.), 149–52
Bustle, 430
Butler, L. Samuel (poet), 238
Butler, Samuel (novelist), 248
Buttons, 191, 430
Byrd, Richard (Admiral), 16, 18
Byrd, William, 291
Byzantine Emperors, 104–5
Byzantine Empire, 103–5
Cabala, 327
Cabbage Patch Kids, 436
Cable networks, 556
Cabral, João, 11
Cacella, Estêvão, 11
Caesar, Julius, 45, 97, 233, 376, 533
Caesarean section, 369
Cage, John, 301
Cagni, Umberto, 16
Caillié, René, 12, 14
Calcium, 399
Calculator, 180, 189, 227, 438
Calder, Alexander, 276
Caldwell, Erskine, 258
Caligula, 100
Calvin, John, 316, 320, 322
Cambrai (battle), 464, 485
Camden (battle), 482
Camera, 184, 209–10
Camera Notes, 210
Camera obscura, 179, 180, 182, 209
Camera tube, 217
Cameron, Verney, 14
Camian (battle), 477
Camorta (steamer), 513
Camp, Walter, 578
Campus Vogladensis (battle), 477
Camus, Albert, 260
Canadian Civil Service, 36
Canals, 353–54
Cancer, 373, 374
Canetti, Elias, 259
Cannae (battle), 476, 490
Cannon, 179, 491, 492
Canterbury, Archbishops of, 319–20
Canton (China), 511
Cão, Diego, 13
Capek, Karel, 287
Cape Matapan (battle), 469
Cape Saint Vincent (battle), 482
Capet, Hugh, 49
Cape Town (South Africa), 13
Capillary, 369, 385, 404
Capone, Al "Scarface," 441
Capote, Truman, 261
Carabobo (battle), 42, 483
Caravaggio, Michelangelo, 267
Carbon black, 161
Carbon dating technique, 186
Carbon paper, 437
Carborundum, 162
Carchernish (battle), 475
Carey, William, 323
Carlos, Juan, 67
Carlyle, Thomas, 243
Carmichael, Stokely, 132
Carnegie, Andrew, 152, 166
Carpaccio, Vittore, 266
Carpet industry, 439
Carr, Gerald, 24
Carrhae (battle), 477
Carroll, Lewis, 248
Carter, Harward, 393
Carthage (battle), 477, 478
Cartier, Jacques, 10, 29
Cartier-Bresson, H., 210
Cartwright, Alexander, 590
Caruso, Enrico, 568
Carver, George Washington, 194
Cassatt, Mary, 272
Castiglione, Baldassare, 237
Castles, 351–52
Castro, Fidel, 43, 146
Cather, Willa, 252
Catholicism, Roman, 306–13, 318
Cato (the elder), 233
Catullus (poet), 233
Cavelier, Robert, 108
Caxton, William, 545
Celaya (battle), 483
Cell, 370, 372, 386–87
Cellini, Benvenutto, 267
Cellophane, 207, 402, 437
Celluloid, 210, 400, 552
Celsius scale, 407
Celsus, Aulus Cornelius, 363
Cement, 182, 206
Censorship, 533, 555, 556, 557
Central America, 42–44
Central Intelligence Agency (U.S.), 38
Centrifugal force, 416
Cerf, Bennett, 551
Cervantes, Miguel de Saavedra, 237
Cézanne, Paul, 272
Chagall, Marc, 275
Chain letters, 436
Chairs, 438–40
Chaldea, kings of, 92
Challenger (space shuttle), 24, 25, 225
Chalons (battle), 477
Chamberlain, Arthur Neville, 65
Chamberlain, Wilt, 590
Champion v. Ames, 36
Champlain, Samuel de, 11, 108
Chancellorsville (battle), 461, 484
Chanel, Coco, 431, 434
Chaouia (steamer), 514
Chaplin, Charlie, 555
Chapman, Roy, 596
Chappaquiddick incident, 140
Chardin, Jean-Baptiste-Siméon, 268
Charlemagne, 45, 307, 340, 448
Charles I (king), 52
Charles II (king), 602
Charles's law (gases), 399
Charleston (battle), 481
Charleston (siege), 484
Charles X (king), 59
Chateaubriand, François René de, 241
Chattanooga (battle), 461, 484
Chatterton, Thomas, 241
Chaucer, Geoffrey, 237
Chautauqua movement, 340
Chavez, Carlos, 301
Chavez, Cesar, 160, 161
"Checkers speech," 139
Cheever, John, 259
Chekhov, Anton, 284
Chemicals industry (U.S.), 161–63
Chemistry, 396–403
Cherenkov effect, 410
Chernobyl (USSR), 424
Cherubini, Luigi, 294
Chiang Kai-shek, 87, 466
Chickamauga Creek (battle), 461, 484
Chickens, 192
Chi'in dynasty, 334
Chi'ing dynasty (Manchu), 83
Chioggia (battle), 479
Chippendale, Thomas, 439
Chirico, Giorgio de, 275
Chi-tsang, 329
Chlorination, 372
Chloroform, 372
Cholesterol, 372
Chopin, Frederic, 295
Chopin, Kate, 249
Christianity, 307
Chromatography, 402
Chromosome, 189, 373, 391
Chronometer, 181, 200
Chrysler, Walter P., 170
Chrysler's Farm (battle), 483
Chu Hsi, 80
Churchill, Sir Winston, 252, 468
Church of England, 308
Cicero, Marcus Tullius, 99, 233
Cimabue, 263
Circus Maximus, 511
City of God, The, 307, 337
Civil Aeronautics Board, 172
Civilian Conservation Corps (U.S.), 422
Civil rights, 39, 131–33, 134, 160
Civil war(s), Afghan, 453; American, 155, 451, 460–61; Angolan, 453; Chinese, 452, 466–68; English, 317, 449, 453–54, 480; Greek, 452; in Kinshasa, 452; Lebanese, 453; Nigerian, 452; Pakistani, 453; Roman, 448; Russian, 452, 465–66; Spanish, 65, 452, 486
Clapperton, Hugh, 12
Clare, John, 243
Clark, Mark (Gen.), 472
Clark, William, 12, 32
Claudian (poet), 235
Claudius I, 100
Clean Air Act (U.S.), 171, 422
Clean Water Bill (U.S.), 424
Cleland, John, 240
Clemens, Roger, 598
Clementi, Muzio, 294
Cleopatra, 97
Clermont (steamboat), 203
Clinton, Henry (Gen.), 455
Clock(s), 177, 179, 185, 186, 187, 199–201
Clontarf (battle), 478
Clothing, 429–34
Cloud chamber, 226
Coal mine, 509–10
Cobalt, 399
Cobb, Ty, 596, 598
COBOL (computer language), 228
Coca-Cola, 429
Cocaine, 372
Cocoa, 427
Cocteau, Jean, 255, 287
Codeine, 372
Code of Canon Law, 309
Code of Hammurabi, 115, 393
Coffee, 427
Coffeemate, 429
Coke, Thomas, 324
Coke (fuel), smelting, 206
Cold-fusion experiment, 411
Cold Harbor (battle), 484
Cold War, 145
Cole, Thomas, 270
Coleridge, Samuel Taylor, 242
Colette, Sidonie-Gabrielle, 252
Colgate, William, 152
Colgate-Palmolive Co., 152
Colleges, U.S. and Canadian, 341–44
Collins, Wilkie, 247
Colombo, Joseph, 442
Colosseum (Rome, Italy), 347
Columbia (space shuttle), 24, 40, 189, 225
Columbus, Christopher, 4–5, 8, 29, 52
Combine, 193
Comenius, John, 340
Comet, 375
Committee on Atmosphere and Biosphere (U.S.), 424
Commune of Paris, 62
Communism, 144–46
Communist First Army, 466
Compact disc(s), 189, 438, 569
Compass, 178, 185, 195
Complex numbers, 412
Composers, 289–301, 568
Compromise of 1850, 34, 131
Computer, 185–90, 222–29
Computerized Axial Tomography Scan, 189
Comte, Auguste, 338
Conan Doyle, Sir Arthur, 249
Concord (battle), 481
Concorde, 172, 229, 230
Concrete, 183
Confession of 1967, 321
Confucianism, 334–35
Confucius, 77, 334
Congresses (U.S.), composition of, 123–26
Congressional Black Caucus, 133
Congressional Transportation Act, 165
Congress of Industrial Organizations, 160
Congress of Racial Equality (CORE), 131
Congress of Vienna, 309
Congreve, William, 224, 280
Connors, Jimmy, 618
Conrad, Charles, 24
Conrad, Joseph, 249
Constable, John, 270
Constantine I, 101, 103, 307, 313
Constantinople (battle), 478, 479
Constellation (aircraft carrier), 513
Constitutional Act (U.S.), 32
Constitutional Congress (U.S.), 32
Constitution (U.S.), 115, 119–20, 138, 456, 457
Contact lenses, 188, 374, 437
Control Data, 188
Cook, James, 11, 15, 17
Cookies, 427
Coolidge, Calvin, 37
Cooper, Chuck, 588
Cooper, James Fenimore, 242
Cooper, L.Gordon, 21
Cooper, Peter, 152, 163
Cooperstown (New York), 594, 596
Copenhagen (Denmark), 511
Copland, Aaron, 301
Copley, John Singleton, 268
Copper ore, 177
Copying, 201–202
Copyright, 546, 552, 557
Coral Sea (battle), 469, 486
Corday, Charlotte, 456
Corelli, Arcangelo, 291
Corneille, Pierre, 280
Corneille, Thomas, 280
Corning Glass Works, 162
Coronary bypass operation, 374
Coronel (battle), 462, 485
Correggio, Antonio da, 266
Cortes, Hernando, 10, 29
Cortisone, 374
Corupedion (battle), 476
Cosmic rays, 382
Cosmonauts, 21–25
Cotton, 181, 192, 193, 203, 206
Council of Anuradhapura, 329; of Constance, 308; of Constantinople, 103; of Ephesus, 307; of Federated Organizations, 132; of Trent, 308
Couperin, Françoise, 291
Coupler, 205
Courbet, Gustave, 271
Courrèges, André, 431
Court, Margaret Smith, 618
Court tennis, 611
Cousteau, Jacques-Yves, 20
Couture, Billy, 600
Coward, Sir Noel, 287
Cowpens (battle), 455
Cowper, William, 241
Coxey, Jacob S., 159
Crabbe, George, 241
Cracker Jack, 428
Cranach, Lucas (the elder), 266
Crane, Hart, 257
Crane, Stephen, 251
Crecy (battle), 479
Criminal Justice System (U.S.), 138–39
Crinoline, 430, 436
Crippen, Robert, 24
Crivelli, Carlo, 265
Crocker, Charles, 152
Cromwell, Oliver, 322
Crosby, Bing, 569
Crossbow, 179, 490
Cruise missiles, 494
Crusades, 448
Crystal Palace (England), 355
Cuban missile crisis, 39, 41, 43
Cullen, Countee, 258

Culpeper's Rebellion, 29, 143
Cultural Revolution (China), 146
cummings, e.e., 256
Cummings, William A. "Candy," 591
Cunaxa (battle), 476
Cuneiform script, 392
Curaçao (cruiser), 514
Customs, 440–41
Cutty Sark (ship), 196
Cybernetics, 186, 228
Cyclops (warship), 514
Cyclotron, 410
Czerny, Karl, 294
da Gama, Vasco, 8, 13
Dali, Salvador, 276
Dallapiccola, Luigi, 301
Dalton, John, 399
Daly, Marcus, 152
Dams, 178, 353–54
Danbury Hatters Case, 36
Dandarah (boat), 514
Daniel, Book of, 305
D'Annunzio, Gabriele, 248
Danton, Georges, 457
Dara (ocean liner), 514
Darwin, Charles, 12
Daudet, Alphonse, 248
Daumier, Honoré, 271
David, Gerard, 266
David, Jacques Louis, 270
da Vinci, Leonardo, 179, 221, 266, 368
Davis, Dwight F., 612
Davis, Marvin Harold, 154
Davis, Sir Robert, 20
Davis, Stuart, 276
Davy, Humphry, 407
DC-aircraft, 505–7
DDT, 162, 186, 194, 403, 424
Dead Sea Scrolls, 328, 395
de Alarcon, Pedro A., 248
de Balboa, Vasco Nuñez, 8
de Beauvoir, Simone, 259
de Brazza, Savorgnan, 109
Debs, Eugene, 145, 160, 164
Debussy, Claude, 299
de Champlain, Samuel, 29
Declaration of Independence (U.S.), 115
Declaration of the Rights of Man and of the Citizen, 115, 456
de Coronado, Francisco, 10
Deere, John, 182, 193
de Falla, Manuel, 300
Defoe, Daniel, 238
De Forest, Lee, 213, 216, 536, 556
de Gamboa, Pedro Sarmiento, 10
Degas, Edgar, 271
de Gaulle, General Charles, 468
de Goes, Bento, 11
de Heredia, José Maria, 243
de Kerguelen-Tremarec, Yves-Joseph, 17
Dekker, Thomas, 278
de Kooning, Willem, 276
de la Barca, Pedro Calderón, 280
de la Condamine, Charles Marie, 11
Delacroix, Eugène, 270
De la Mare, Walter, 252
de Lamartine, Alphonse Marie Louis, 242
de la Renta, Oscar, 432
de la Verendrye, Sieur, 11
de Leon, Moses, 327
Delius, Frederick, 299
della Francesco, Piero, 265
Dello Joio, Norman, 301
De Long, George W., 15
de Lozier, Bouvet, J.B.C., 17
del Verrocchio, Andrea, 265
de Machaut, Guillaume, 289
de Mendoza, Antonio, 29
Democritus, 336, 376, 403
de Molina, Tirso, 278
de Montaigne, Michel E., 237
Demosthenes (orator), 233
Dempsey, Tom, 536, 585
de Musset, Alfred, 283
Denikin, Anton (Gen.), 466
Dentistry, 366
de Orellana, Francisco, 10
de Palestrina, Giovanni Pierluigi, 289
Department of Agriculture, 194
Department of Commerce and Labor, 155
Department of Energy, 424
Depth charges, 493
Derain, André, 274
DeSalvo, Albert, 442
Descartes, René, 337
de Soto, Hernando, 10
des Pres, Josquin, 289
de Staël, Nicholas, 277
de Staël-Holstein, Baronne Anne Louise, 241
de Troyes, Chrétien, 235
de Vega, Lope, 278
de Vigny, Comte Alfred Victor, 243
Dewar, Sir Thomas, 603
Dewey, John, 339, 340
Diamonds, 162
Dias, Bartolomeu, 13
di Buoninsegna, Duccio, 263
Dickens, Charles, 246
Dickerson, Eric, 586
Dickey, James, 261
Dickinson, Emily, 247
Dickson, William, 552
Diderot, Denis, 240, 282
Didion, Joan, 262
Dien Bien Phu (battle), 488
Dieppe Raid (battle), 487
Diesel, Rudolf, 207
Diesel engine, 184
Diet of Augsburg, 316; of Speyer, 316; of Worms, 307, 316
Diet-Rite, 429
Digitalization, 215
di Lasso, Orlando, 289
Dillinger, John, 441
Dime novels, 546
Dinesen, Isak, 253
Dinosaurs, 389
Diocletian, 101, 307
Diogenes, 5, 336
Dior, Christian, 431
Dirigible, 183
Disco, 436
Discovery (space shuttle), 225
Diseases, 364
Disraeli, Benjamin, 245
District of Columbia, 32
DNA, 187, 391
Dock, Christoph, 340
Doctorow, E.L., 262
Dolby, 188, 557
Dome of the Rock, 333
Domitian, 100
Doña Paz (ship), 515
Donatello, 263
Donizetti, Gaetano, 295
Donleavy, J. P., 261
Donne, John, 238
Doolittle, Hilda, 255
Doolittle, James, 171, 222
Doolittle's raid, 486
Doppler, Christian, 408
d'Orleans, Charles duc, 237
Dorylaeum (battle), 478
Dos Passos, John R., 257
Dostoyevsky, Fyodor, 247
Doubleday, Frank Nelson, 551
Doughnut, 427
Doughty, Charles Montague, 13
Downs (battle), 480
Dr. Pepper, 428
Drabble, Margaret, 262
"Draconian" laws, 94
Draft Riots, 143
Drake, Francis, 10
Drayton, Michael, 237
Dreadnought, 492
Dreiser, Theodore, 251
Dresden (battle), 459, 483
Drexel, Burnham, Lambert, Inc., 159
Dreyfus Affair, 62, 139
Drill, 182, 185
Drive-in theaters, 556
Droughts, 515–16
Dryden, John, 238, 280
Drysdale, Don, 597
Dubcek, Alexander, 146
Dubuffet, Jean, 276
Duchamp, Marcel, 275
Duffy, Hugh, 594
Dufy, Raoul, 273
Dukas, Paul, 299
Duke, James B., 153
Dumas, Alexandre (*fils*), 283; (*père*), 243
Du Maurier, Daphne, 259
Dunbar, Paul Laurence, 251
Dunbar, William, 237
Dunkirk (battle), 486
Dunlap, William, 282
Dunsinane (battle), 478
Du Pont, Eleuthère L., 152
Durant, William C., 153, 170
Dürer, Albrecht, 266
Durkheim, Emile, 393
Durrenmatt, Friedrich, 288
D'Urville, Jules, 18
Duryea, Charles and J. Frank, 170, 218
Dvorak, Antonin, 296
Dynamite, 183
Early, Jubal (Gen.), 461
Ear-piercing, 433
Earthquakes, 414, 516–19
Earth sciences, 414–21
Eastern Orthodoxy, 313–15
Easter Rebellion, 63
Eastland (steamer), 513
Eastman, George, 149, 152, 162
Ebers Papyrus, 363
Ebert, Friedrich, 63
Ebisu Maru (ship), 514
Ebro River (battle), 486
Eck, Johann, 315
Ecology, 422
Economy (U.S.), 154–57
Ecuador, 1985 revolution, 43
Ecumenical Councils, 314
Eddy, Mary Baker, 534
Edgehill (battle), 453
Edgeworth, Maria, 241
Edict of Emancipation, 115; of Milan, 101, 115, 307; of Nantes, 53, 316, 321; of Worms, 316
Edington (battle), 478
Edison, Thomas, 158, 162, 183, 212, 567
Edmund Fitzgerald (ship), 515
Education, 339–44
Edward III (king), 602, 611
Edward II (king), 602
Edwards, Jonathan, 240
Egg Watchers, 428
Egypt, Ancient, 95–97, 97–98
Egyptian Delta (battle), 475
Egyptian mummies, 393; pyramids, 347; temples, 347
Eielson, C.B., 16
Eight-hour workday, 166
Einstein, Albert, 382
Eisenhower, Dwight D., 166, 470
Eisenhower Doctrine, 38
Eisenstein, S., 556
El Alamein (battle), 470, 487
Electric chair, 138, 212
Electricity, 207, 211–14
Electric light bulb, 437
Electrolysis, 212
Electromagnetism, 211, 408
Electroplating process, 182
Electroscopes, 211
Electroweak theory, 410
Elements, 397–98, 410
Elevator, 183
Elgar, Sir Edward, 299
El Greco, 267
Eliot, George, 247
Eliot, T.S., 255, 286
Elizabeth I, 53, 316
Ellis, Perry, 432
Ellison, Ralph, 260
Ellsworth, Lincoln, 18
Emancipation Proclamation, 34, 131, 461
Embury, Philip, 324
Emerson, Ralph Waldo, 245
Empedocles, 336
Empire State Building, 505
Empress of Ireland (steamer), 513
Endangered Species, 422, 423
Engine(s), 184, 186, 187, 205–7, 218, 220–21, 222
English Reformation, 316
Ennius, Quintus, 233
Ensor, James Baron, 272
Environmentalism, 422–24
Environmental Protection Agency (EPA), 424
Enzymes, 188
Epidemics, 527–29
Episcopal Church, 318
Episcopalianism, 317–20
Epoxy glue, 186, 208
Equal Credit Opportunity Act (U.S.), 134
Equal Employment Opportunity Act, 133, 161
Equal Opportunity Commission, 160
Equal-pay-for-equal-work law, 160
Equal Rights Amendment, 134
Erasmus, Desiderius, 237
Ericson, Leif, 8, 29
Eric the Red, 8
Ernst, Max, 275
Escalator, 185
Eshghabad (steamer), 514
Ether, 372
Eurymedon River (battle), 476
Eustachio, Bartolommeo, 369
Evans, Oliver, 203, 218
Evans (destroyer), 515
Evers, Medgar, 132
Evert, Chris, 618
Explorer I, 421
Explosions, 509–10
Explosion seismology, 168
Exupéry, Antoine de Saint, 257
Eye(s), 374
Eyeglasses, 181, 207, 371, 433
Eylau (battle), 458, 482
Eyre, Edward John, 12
Fabian, John, 24
Fabian Society of Great Britain, 144
Fabriano, Gentile da, 263
Fads, 435–36
Fahrenheit temperature scale, 407
Fairbanks, Douglas, 555
Fair Oaks (battle), 460
Falkland Islands, 44, 68, 453, 462, 485
Fallout shelters, 436
Falvani, Luigi, 407
Falwell, Jerry, 323
Farnsworth, P.T., 537
Farrell, James T., 258
Faughart (battle), 479
Faulkner, William, 257
Fauré, Gabriel, 296
Fax machines, 202
Federal Aviation Agency, 172
Federal Bureau of Investigation, 138
Federal Bureau of Prisons, 138
Federal prison system, 138
Federal Reserve system, 155
Federal Trade Commission, 155
Federation of Organized Trades and Labor Unions, 159
Feininger, Lyonel, 273
Ferber, Edna, 255
Ferdinand, Archduke Francis, 462
Ferguson, Bob, 591
Fermi National Accelerator Laboratory, 227
Fermi, Enrico, 138
Fertilization, test-tube, 189
Fessenden, Reginald Aubrey, 536
Festivals, 434–35
Feydeau, Georges, 284
Feynman, Richard, 410
Fiction bestsellers (1945–89), 547–50
Field, Marshall, 152
Fielding, Henry, 240
Fielding gloves, 591
Filippo Lippi, Fra, 263
Firaudous, Jean, 286
Fires, 511–13
"Fireside chats," 537
First Antarctic Conference, 19
First Buddhist Council, 328
First Council of Nicaea, 101
First International Polar Year, 15
Fischer, Emil, 400
Fishing, 191
Fitzgerald, F. Scott, 256
Five Forks (battle), 461
Five-Power Naval Limitation Treaty, 64, 117
Flamethrowers, 492
Flash bulb, 210; flashcube, 438
Flaubert, Gustave, 247
Fletcher, John, 279
Fleurus (battle), 457, 480
Flight, 222
Flintlock gun, 491
Flint tools, 392
Flodden (battle), 479
Floods, 519–20
Floppy disks, 189, 228
Floyd, Charles "Pretty Boy," 441
Fluorescent lighting, 186, 213
Fluorine, 162, 400
Flying shuttle, 205
Foam rubber, 185
Foch, Ferdinand (Gen.), 464
Fokker, Anthony, 492
Foods, 427–29
Football, 577–86. *See also* Soccer
Foot fins, rubber, 20
Forbes, Malcolm, 154
Ford, Ford Madox, 252
Ford, Henry, 153, 170, 185, 218
Ford, John, 279
Forceps, 369
Forest Reserve Act (U.S.), 422
Formigny (battle), 480
Fornovo (battle), 479
Forrestal (aircraft carrier), 515

Forster, E.M., 253
Fort McHenry (battle), 483
FORTRAN (computer language), 228
Fort Stanwix (battle), 481
Fort Sumter, 484
Fort Ticonderoga, 454, 455
Fossils, 387, 414
Fotosetter, 186, 202
Fourier, Charles, 144
Four Noble Truths, 328
Four-Power Pacific Treaty, 37, 64, 87, 117
Fourteen Points, 115
Fourth Crusade, 104
Fowles, John, 261
Fox, George, 317
Fragonard, Jean, 268
France, Anatole, 249
Francis I (king), 53
Francis of Assisi, 307
Franck, Cesar, 295
Franco, Francisco (Gen.), 65
Frankfurters, 428
Franklin, Benjamin, 211, 240
Frazer, Sir James George, 393
Frederick I, 50
Fredericksburg (battle), 460, 484
Freedom 7 (space capsule), 21
Freeman's Farm (battle), 481
Freeze-drying, 185, 428
Fremont, John Charles, 12, 33
French Revolution, 450, 456–57, 482
Freon, 186, 402
Frequency modulation broadcasting, 537
Fresnel, Augustin, 408
Freyre, Emmanuel, 11
Frick, Henry C., 152
Friedan, Betty, 134
Friedland (battle), 458, 482
Frisbee, 436
Frisch, Max, 287
Frobisher, Sir Martin, 10, 15
Frontiers (battle), 462
Frost, Robert L., 252
Fry, Christopher, 287
Fulbright scholarships, 340
Fuller, Buckminster, 358
Fulton (warship), 203
Furman v. Georgia, 39
Furnace(s), 165, 179, 181
Furniture, 438–40
Gabo, Naum, 275
Gagarin, Yuri, 21
Gainsborough, Thomas, 268
Galaxies, 382, 383
Galen, 364
Galilei, Galileo, 377, 404, 412
Gallipoli (battle), 479
Gallo, Ernest, 153
Gallos, Joseph, 442
Gallus, Aelius, 5
Galsworthy, John, 251, 286
Galvanometer, 182
Gandhi, Indira, 89
Gandhi, Mohandas, 87, 88, 89, 331
Gang of Four, 90
García Lorca, Federico, 257, 287
García Márquez, Gabriel, 261
Garfield, James A., 36
Garland, Hamlin, 249
Garrett, Thomas, 130
Garrison, William Lloyd, 130
Gaskell, Elizabeth Cleghorn, 246
Gas lighting, 206
Gasoline, 168, 169, 185
Gas stove, 437
Gates, William Henry, III, 154
Gatling gun, 492
Gaugamela (battle), 476
Gauguin, Paul, 272
Gautama, Siddhartha, 328, 330
Gautier, Théophile, 246
Gay, John, 238, 280
Geiger, Abraham, 327
Geiger counter, 185
Gemini (space capsule), 21
Gene, 188, 375
General Belgrano Station (Antarctica), 19
General Slocum (steamer), 513
Generator, 183
Genet, Jean, 287
Genetic engineering, 189
Genetics, 390
Genovese, Kitty, 442
Geoffrey of Monmouth, 235
Geology, 416
Geophysical Service, Inc., 151
Geostationary Operational Environmental Satellites, 189
Géricault, Jean Louis André Théodore, 270
Germantown (battle), 481
Gernreich, Rudi, 431
Gershwin, George, 301
Getty, John Paul, 153
Gettysburg (battle), 461, 484
Ghazali, Abu Hamid Mohammed al-, 333
G.I. Bill (U.S.), 38, 340
Giacometti, Alberto, 276
Giannini, Amadeo, 153
Gibbons, Orlando, 291
Gibran, Kahlil, 253
Gide, André, 251
Gideon v. Wainwright, 139
Gilmore, Gary, 442
Gilmore, James A., 594
Gin, 427
Ginsberg, Allen, 261
Giorgione, Giorgio, 266
Giotto (painter), 263
Gjellerup, Karl A., 249
Glasnost, 68
Glass, 163, 177, 178, 208
Glazunov, Alexander, 299
Glenn, John, 21
Gliders, 221
Glinka, Mikhail, 295
Glomar Challenger (research ship), 20
Glorious Revolution, 54
Gluckhauf (ship), 184, 197
Gnomon, 199, 375
Goddard, Robert, 224
Goethe, Johann Wolfgang von, 241, 282
Goetz, Bernhard, 443
Gogol, Nikolai Vasilyevich, 246
Golden Bull of 1356, 51
Golden Spurs (battle), 479
Golding, William, 259
Goldoni, Carlo, 280
Gold rush, 34, 36, 44
Goldsmith, Oliver, 241, 282
Gold standard, 155
Golf, 605–10
Gompers, Samuel, 159
Goodyear, Charles, 161, 182
Gorbachev, Mikhail, 68, 69, 146
Gordimer, Nadine, 261
Gordon, George, 242
Gordon, Robert Jacob (Col.), 14
Gordon, Sir Charles, 108
Gorky, Arshile, 276
Gorky, Maxim, 251, 286
Gossamer Condor, 223
Gould, Jay, 152
Gounod, Charles, 295
Gower, John, 235
Goya y Lucientes, Francisco de, 270
Graham, Katharine, 154
Graham crackers, 428
Grahame, Kenneth, 249
Grainger, Percy Aldridge, 300
Grammy Awards, 568, 569
Gramophone, 184, 567
Granada (battle), 479
Grandcamp (ship), 509, 510, 512, 514
Grandma Moses, 272
Granite Railway, 163
Graphophone, 567
Grass, Günter, 261
Gravelines (battle), 480
Graves, Robert, 256
Great Awakening, 317, 322
Great Britain (steamship), 183
Great Chicago Fire, 35, 355
Great Depression, 37, 155, 158, 160, 170
Great Eastern (steamship), 196
Great Fire of London, 54, 511
Great Interregnum, 50
Great Mogul Empire, 107
Great Mosque of Kairouan, 333
Great Schism, 308
Great Wall of China, 347
Greece, Ancient, 94–95
Greek and Roman dramatists, 279
Green, Hetty, 152
Greene, Graham, 258
"Greenhouse effect," 421
Gregory, Lady Augusta, 283
Grenades, 491
Grey, Zane, 252
Grieg, Edvard, 296
Griffin, Merv, 154
Griffith, D.W., 555
Grissom, Virgil, 21
Grosz, George, 275
Groundhog Day, 434
Grünewald, Matthias, 266
Gryphius, Andreas, 280
Guadalcanal (battle), 487
Guano, 193
Guardi, Francesco, 268
Guarini, Battista, 278
Guggenheim, Meyer, 152
Guilford Court House (battle), 455
Gulf of Tonkin, 473, 488
Gunn, Thom, 261
Gunnbjörn, 8
Gunpowder, 179
Gutenberg, Johannes, 340, 545
Hadrian, 100
Hagen, Walter, 608, 610
Hagia Sophia, 314
Hairstyles, 431, 433
Halas, George, 582
Halberd, 490
Hall, Charles, 162
Hallaj, 333
Halleck, Henry (Gen.), 460
Halley, Edmund, 19, 381
Halley's comet, 376
Hals, Franz, 267
Hamburger Hill (battle), 474, 488
Hamburger, 428
Hammarskjöld, Dag, 128
Hammett, Dashiell, 256
Hammurabi, 363
Hamsun, Knut, 249
Handel, George Friedrich, 293
Handkerchiefs, 437
Han dynasty, 335
Hanging Gardens (Babylon), 347
Hanno, 3
Hanukkah, 434
HanYu, 80
Harcourt, Alfred, 551
Hardy, Thomas, 249
Hare, David, 289
Hari, Mata, 463
Harlem Globetrotters, 588
Harper's Ferry Raid, 34, 483
Harpoons, 191
Harris, Jean, 442
Harris, Joel Chandler, 249
Hart, Gary, 140
Hart, Moss, 287
Harte, Bret, 248
Harvey, William, 369, 385
Hastings (battle), 478
Hat(s), 430, 432, 433, 436
Hauck, Frederick, 24
Hauptmann, Gerhart, 284
Hawthorne, Nathaniel, 245
Hay-Bunau-Varilla Treaty, 117
Haydn, Franz Joseph, 294
Hayes, Rutherford B., 36
Haymarket Square Riot, 36, 159
Hays code for self-censorship, 556
Hazlitt, William, 242
Head, Edith, 431
Head Start, 340
Heaney, Seamus Justin, 262
Hearing aid, 184
Hearst, Patricia, 442
Hearst, William R., 153, 534
Heartbreak Ridge, 472
Heart-lung machine, 187, 374
Heat, 407
Hebbel, Friedrich, 283
Hedin, Sven, 13
Hegel, G.W.F., 338
Heidegger, Martin, 339
Heimlich maneuver, 375
Heine, Heinrich, 243
Helios B, 189, 229
Helium, 408, 410
Heller, Joseph, 261
Hellman, Lillian, 287
Helmsley, Harry Brakman, 153
Hemingway, Ernest M., 257
Henderson, Ricky, 598
Henley, Beth, 289
Henry, Joseph, 212, 214
Henry, Patrick, 454
Henry II (king), 317
Henry IV (king), 602
Henry VII (king), 318
Henry VIII (king), 316
Henson, William, 16
Hepworth, Cecil, 555
Heraclitus of Epheseus, 336
Herbert, George, 238
Herjulfsson, Bjarni, 8
Herodotus (historian), 233
Herrick, Robert, 238
Hersey, John, 260
Hershey bar, 428
Hershiser, Orel, 598
Hertz, Heinrich, 216
Herzl, Theodore, 327
Hess, Rudolf, 469
Hesse, Hermann, 252
Hewlett, William Redington, 154
Heyse, Paul J., 248
Heywood, Thomas, 278
Hieroglyphics, 95, 392
High-definition television (HDTV), 539
Hijacking, 172
Hilton, Conrad N., 153
Himalayas, 11, 13
Himara (steamer), 514
Himera (battle), 475
Hindemith, Paul, 300
Hindenburg (dirigible), 64, 505, 537
Hinduism, 330–32
Hippocrates, 363, 385
Hiroshima, 471
Hirsch, Samson Raphael, 327
Hitler, Adolf, 65, 145, 219, 468, 470, 471
HMCS Labrador (vessel), 17
Hobbema, Meyndert, 268
Hobbes, Thomas, 337
Hobson (vessel), 514
Ho Chi Minh, 474
Hockney, David, 277
Hoffa, James "Jimmy," 160, 161, 442
Hoffman, E.T.A., 242
Hogarth, William, 268
Hohenlinden (battle), 482
Holbein, Hans (the elder), 266
Holbein, Hans (the younger), 267
Holdheim, Samuel, 327
Holidays, 434–35
Holland V (submarine), 20, 184, 198
Hollerith, Herman, 227
Holley, A.L., 165
Hollick-Kenyon, Herbert, 18
Hollywood (California), 555
Holmes, Oliver Wendall, 246
Holocaust, 328
Holst, Gustav, 300
Holy Bible, The, 305–6
Holy Roman Emperors, 46
Holy Roman Empire, 57
Holz, Arno, 249
Home Insurance Building, 355
Homer (writer), 233
Homer, Winslow, 271
Homestead strike, 159
Homo erectus, 393
Homo habilis, 395
Homo sapiens, 396
Hong Kong (steamer), 514
Hopkins, Gerard Manley, 249
Hopper, Edward, 274
Horace (poet), 233
Hornemann, Frederick, 14
Hormone, 390
Horses, 192
Horseshoes, 177
Hospital(s), 363, 509–10, 513
Hot dogs, 428
Household items, 437–38
Houses of Parliament (London, England), 355
Housman, A.E., 249
Hovercraft, 187, 223
Howard, John, 230
Howe, Gordie, 601
Howe, Julia Ward, 247
Howells, William Dean, 248
Hsin Hsu-Tung (steamer), 514
Hsin Wah (steamer), 514
Hsin Yu (steamer), 513
Hsuan Tsang, 329
Hsuan-t'ung, 86
Hsun-tzu, 334
Hubble (space telescope), 25, 190, 383
Hudson, Henry, 11, 15, 29
Hughes, Howard, 153, 222
Hughes, Langston, 258
Hughes, Ted, 262
Hugo, Victor, 245, 283
Hula hoop, 436
Hume, David, 338
Humperdinck, Engelbert, 299
Hundertwasser, Friedrich, 277
Hunt, H. L., 153
Hunter, Jim "Catfish," 597
Hurricane(s), 521–22
Hussein, King, 71
Hussein, Saddam, 72
Husserl, Edmund, 339

Huxley, Aldous L., 256
Hyatt, John, 161
Hydrogen bomb, 187, 226, 410, 494
Hydrogen peroxide, 372
Iacocca, Lee, 171
Ibáñez, Vicente Blasco, 251
Ibsen, Henrik, 283
Icahn, Carl, 154
Ice cream, 427, 428
Ice hockey, 598–602
Ice-making machine, 183
Ikhanton, King, 96
Imhotep, 363
Immigration Act, 160
Immunology, 373
Imo (steamer), 513
Ince, T.H., 555
Inchon landing, 488
Indian Mutiny, 143
Indigirka (steamer), 514
Industrial Revolution, 205–7
Industrial Workers of the World, 160
Infrared Astronomy Satellite, 189
Infrared detection devices, 165
Inge, William, 288
Ingemann, Bernard Severin, 242
Ingres, Jean-Auguste Dominique, 270
Inoculation, 387
Insecticide, 192
Instant-replay, 217
Insulin, 373
Integrated circuit, 187, 213
Intercontinental ballistic missile, 494
Interferon, 374
Intermediate Nuclear Forces, 40, 68, 118
Intermediate-range ballistic missile, 187, 494
International Geophysical Year, 19, 421
International Whaling Commission, 424
Interstate Commerce Commission, 164
Iodine, 372
Ionosphere, 419
Ipsus (battle), 476
Iraklin (ferryboat), 515
Iran hostages, 39
Iranian Revolution, 144
Irenaeus, 307
Iron lung, 373
Iroquois Theater (Chicago), 512
Irving, Washington, 242
Isherwood, Christopher, 258
Islam, 332–34
Isonzo (battle), 463
Issus (battle), 476
Isotopes, 408
Italian Fascist Party, 145
Italy, 60
Ituzaingo (battle), 42
Ivan III the Great, 51
Ivan IV, 53, 314
Ives, Charles, 300
Iwo Jima (battle), 470, 487
Jackets, 430, 433
Jackson, Michael, 569
Jackson, Shirley, 260
Jacobs, Henry, 322
Jahl, Abu, 332
James, Henry, 249
James, William, 338
James I (king), 306
James River Corp., 151
Jamestown, Virginia, 29, 511
Janáček, Leoš, 299
January Insurrection, 144
Japan, 78–80, 144, 462, 467
Jarratt, Jim, 20
Jarvis, Gregory, 25
Java Sea (battle), 487
Jazz, 435, 572–73
Jeans, 430, 432, 434
Jeffers, Robinson, 255
Jefferson, Thomas, 32, 340
Jellicoe, Ann, 288
Jemappes (battle), 456
Jena-Auerstadt (battle), 458, 482
Jerome, Jerome, 284
Jerome, Saint, 306
Jerusalem (battle), 475, 477, 478
Jesuits (Society of Jesus), 308
Jesus, 100, 305, 306
"Jeu de paume," forerunner of tennis, 610
Jewelry, 440
Jewish Revolt(s), 142–43, 326
Jigsaw puzzles, 435
Joan of Arc, 51
Joffre, Joseph (Gen.), 463
Johns, Jasper, 277
Johnson, Andrew, 34
Johnson, Lyndon, 132, 473
Johnson, Samuel, 240
Johnson, Walter, 594
Johnston, Joseph (Gen.), 461
Jolliet, Louis, 11
Jones, John Paul, 455
Jones, Rev. Jim, 43, 442
Jones, Robert "Bobby," 608
Jones, Robert Trent, 610
Jonson, Ben, 278
Jorn, Asger, 277
Joss, Addie, 594
Joule's law, 408
Journalist(s),-ism, 533, 535
Joyce, James, 253
Judaism, 325–28
Jukeboxes, coin-operated, 569
July Revolution, 143
Justinian (emperor), 313
Justinian I, 103
Justin Martyr, Saint, 306
Jutland (battle), 463, 485
Juvenal (poet), 235
Kadesh (battle), 475
Kafka, Franz, 253
Kaiser, Henry J., 153
Kalka River (battle), 478
Kamali, Norma, 432
Kandinsky, Wassily, 273
Kant, Immanuel, 338
Kaplan, Mordechai, 327
Karlfeldt, Erik A., 251
"Katzenjammer Kids," 534
Kaufman, George, 287
Kawachi (battleship), 514
Kazantzakis, Nikos, 253
Keats, John, 243
Keds, 431
Kellogg, Will K., 153
Kellogg-Briand Pact, 64, 118
Kelly, William, 165
Kenesaw Mountain (battle), 461
Kennedy, John F., 21, 39, 167, 473
Kennedy, Robert F., 39
Kenzo, 432
Kepler, Johannes, 404
Kerouac, Jack, 260
Kesey, Ken, 262
Ketchup, 427
Kewpie dolls, 435
Khachaturian, Aram, 301
Khan, Kublai, 81
Khartoum (battle), 484
Khesanh (siege), 488
Khomeini, Ayatollah, 72
Khrushchev, Nikita, 66, 129, 315
Kiang-Kwan (steamer), 514
Kiangya (steamer), 514
KicheMaru (steamer), 513
Kidney machine, 186, 374
Kierkegaard, Søren, 338
Kinetoscope, 184
King, Billie Jean, 618
King, Martin Luther, 39, 132
Kingsley, Charles, 247
Kings Mountain (battle), 455, 482
Kipling, Rudyard, 251
Kirchner, Ernst Ludwig, 274
Kissinger, Henry, 474
"Kitchen debate," 66
Kizim, Leonid, 24
Klee, Paul, 274
Klein, Yves, 277
Klimt, Gustav, 273
Klopfer, Donald, 551
Knox, John, 316
Kodachrome, 210
Kodály, Zoltan, 300
Koestler, Arthur, 259
Kohl, Helmut, 69
Kokoschka, Oskar, 275
Kolchak (Admiral), 466
Koldewy, Robert, 393
Kolin (battle), 481
Komarov, Vladimir, 21
Koniggratz (battle), 484
Koran, 115, 333
Koryo Dynasty, 329
Koufax, Sandy, 597
Kress, Samuel H., 153
Kroc, Ray A., 153
Kulikovo (battle), 51, 479
Kunersdorf (battle), 481
Kyd, Thomas, 278
Labor Movement (U.S.), 159–61
Lafayette, Marquis de, 455
Lageos (satellite), 421
Lagerlof, Selma, 249
La Hogue (battle), 480
Laing, Alexander Gordon, 12
Lake Champlain (battle), 483
Lake Erie (battle), 483
Laker, Freddie, 173
Lamaism, 330
Lamb, Charles, 242
Lambeth Conference, 318
Lamps, 164
Land, Edwin, 186, 210
Lander, John, 14
Lander, Richard, 14
Landini, Francesco, 289
Landru, Henry Desire, 441
Landsats, 189, 194
Lantern, 180
Laon (battle), 459
Lardner, Ring, 253
Largs (battle), 478
La Salle, Ferdinand, 144
La Salle, Robert, 11, 29
Laser(s), 188, 189, 227, 229, 410
Las Navas de Tolosa (battle), 478
Lathe(s), 178, 180, 206
Lauren, Ralph, 154, 432
Lavoisier, Anton, 371
Lawrence, D.H., 253
Lawrence, T.E. (Col.), 463
Lazarus, Emma, 249
League of Nations, 131
League of Woman Voters, 134
Leakey, Louis, 393, 395
Lear, Edward, 246
Lebanon, 71, 453, 489
Lebedev, Velentin, 24
Lechfeld (battle), 478
Lee, Robert E., 460
Le Figaro, 533
Léger, Fernand, 274
Legionnaires' disease, 375
Legnano (battle), 478
Leibniz, Gottfried Wilhelm, 337
Leica camera, 185, 210
Leipzig (battle), 459, 483
Leland, Henry M., 170
Lend-Lease Act, 469
Lenglen, Suzanne, 612
Lenin, Vladimir Ilyich, 63, 145, 466
Leningrad (siege), 469, 486
Leo III, 313
Leopardi, Count Giacomo, 243
Lepanto (battle), 107, 480
Le Plongeur (submarine), 20
Leprosy, 365
Lermontov, Mikhail Yuryevich, 246
Lessing, Doris, 260
Lethbridge, John, 19
Leuctra (battle), 476
Leviathan, 337
Lewes (battle), 479
Lewis, C.S., 257
Lewis, John L., 160
Lewis, Meriwether, 12, 32
Lewis, Sinclair, 253
Lewis and Clark Expedition, 9
Lexington (battle), 481
Leyden jar, 181, 211, 407
Leyte Gulf (battle), 487
Liberty Bell 7 (space capsule), 21
Libya, 70
Li Chih, 81, 335
Lichtenstein, Roy, 277
Light bulb, 185, 212, 213
Lighthouse, 180, 212
Lightning, 407, 509
Lightning Bolt (motorcycle), 230
Lightning rod, 211
Lincoln, Abraham, 34, 131, 460, 461
Lindbergh, Charles, 37, 222
Lindbergh kidnapping, 441
Lindsay, Vachel, 253
Linnaeus, Carolus, 387
Linotype, 184, 202, 534
Lipchitz, Jacques, 275
Lipstick, 437
Liszt, Franz, 295
Lithography, 181, 202
Little Bighorn (battle), 484
Live Aid, 539
Livingstone, David, 12, 14, 72
Livy (historian), 233
Locarno Pact, 117
Locke, John, 337
Locomotive, 163, 164, 183, 203, 204, 205, 229
Lodi (battle), 457, 482
Loincloths, 429
London, Jack, 252
London Bridge, 45, 511
London (England), 355, 511
London Naval Treaty, 37, 64, 118
Longbow, 490
Longfellow, Henry Wadsworth, 245
Long Island (battle), 455, 481
Loom(s), 181, 182, 206
Lopez, Nancy, 610
Lorenzo, Francisco A., 154
Lorrain, Claude, 268
Loti, Pierre, 249
Louisiana Purchase, 32, 109
Love Canal, 424
Lovejoy, Rev. Elijah, 130
Lovelace, Richard, 238
Lovell, James A., 22
Lowell, Amy, 252
Lowell, James Russell, 247
Lowell, Robert, 260
Lowry, Malcolm, 259
Lucan (poet), 235
Lucas, Henry Lee, 442
Luciano, Charles "Lucky," 442
Lucite, 208
Lucretius (poet), 233, 336
Ludwig II (king), 62
Luisetti, Hank, 588
Lully, Jean-Baptiste, 291
Lundy, Benjamin, 130
Lundy's Lane (battle), 483
Lurçat, Jean, 276
Lusitania (steamship), 197, 463, 513
Luther, Martin, 115, 308, 315, 316
Lutheranism, 316
Lutzen (battle), 480
Luzon (battle), 487
Lyly, John, 278
Lyons (battle), 477
MacArthur, Douglas (Gen.), 469, 471
Macaulay, Thomas Babington, 243
MacDiarmuid, Hugh, 256
MacDowell, Edward, 299
Machiavelli, Niccolò, 237, 278
Machine gun, 183
Mack, Ted, 537
Mackenzie, Allister, 608
Mackenzie, Sir Alexander, 12
Macklin, Charles, 280
MacLeish, Archibald, 256
MacNeice, Louis, 259
Madison Square Garden, 600
Madrid (battle), 486
Maeterlinck, Maurice, 284
Magazines, 533–36
Magellan, Ferdinand, 6, 10
Maginot Line, 493
Magna Carta, 50, 115
Magnesium, 399
Magnetic chart, 416
Magnetic recording, 184, 568
Magnetic resonance imager, 189, 374
Magritte, René, 276
Magsat (satellite), 189
Maharishi, Ramana, 331
Mahaviranhara, 329
Mah-jongg, 435
Mahler, Gustav, 299
Mailer, Norman, 261
Maillol, Aristide Joseph, 272
Maimonides, 326
Maipu (battle), 42, 483
Majnoon (battle), 489
Malamud, Bernard, 260
Malevich, Kazimir, 273
Malone, Moses, 590
Malory, Sir Thomas, 237
Malplaquet (battle), 481
Malraux, André, 257
Mame (battle), 462, 464
Mamet, David, 289
Mammography, 373
Manarov, Musa, 25
Mandela, Nelson, 76
Manet, Edouard, 271
Manganese, 191
Manhattan (tanker), 17
Manhatten Project, 162
Manila Bay (battle), 484
Mann, Thomas, 252
Mansfield, Katherine, 255
Manson, Charles, 442
Mansur, Abu Jaffar al-, 106
Mantegna, Andrea, 265
Mantua (siege), 482
Manziker (battle), 478
Mao Tse-tung, 467
Marat, Jean Paul, 456
Marathon (battle), 475
Marathons, 435
Marbury v. Madison, 32
Marc, Franz, 274

Marconi, Guglielmo, 213, 215, 216, 536
Marcos, Ferdinand, 90
Mardi Gras, 435
Marengo (battle), 482
Marinaro, Ed, 585
Maris, Roger, 597
Maritsa River (battle), 479
Marlowe, Christopher, 278
Marne (battle), 485
Marquette, Jacques, 11
Marquis de Sade, 241
Marriot, John Willard, 153
Marshall, George (Gen.), 467
Marshall Plan, 38
Marston Moor (battle), 453, 480
Martí, José, 249
Martineau, Harriet, 243
Martini, 428
Martini, Simone, 263
Marvell, Andrew, 238
Marx, Karl, 61, 144, 338
Mary Rose (ship), 395
Masaccio (painter), 263
Masefield, John, 252, 286
Massacre of St. Bartholomew's Day, 316
Massiaen, Oliver, 301
Massinger, Philip, 279
Masters, Edgar Lee, 251
Matches, 182
Maternus, Julius, 8
Mathematics, 411–14
Mathieu, Georges, 277
Matisse, Henri, 273
Mattingly, Don, 598
Maugham, W. Somerset, 252, 286
Mau Mau, 74
Maupassant, Guy de, 249
Maxell, James, 216–18
Maxfield, J.P., 569
Mayan civilization, 29
Mayonnaise, 428
Mazurian Lakes (battle), 462, 485
McAuliffe, Christa, 25
McCarthy, Mary, 259
McCarthy hearings, 38
McClellan, George (Gen.), 460
McClure, Robert, 15
McCormick, Cyrus, 182, 193
McCullers, Carson S., 260
McDonald's Restaurants, 153, 428
McDonnell Douglas, 151
McEnroe, John, 618
McFadden, Mary, 432
McKay, Claude, 256
McKinley, William, 36
McKissick, Floyd, 132
McNair, Ronald, 25
Mead, Margaret, 393
Meany, George, 160
Medicine, 363–75
Megaliths, 348
Meggido (battle), 475
Melba toast, 428
Melies, George, 555
Mellon, Andrew, 152
Melville, Herman, 247
Memlinc, Hans, 265
Mencius, 334
Mencken, H.L., 253
Mendelssohn, Felix, 295
Mendelssohn, Moses, 327
Mendoza, Juan Ruiz de Alarcón y, 279
Menotti, Gian Carlo, 301
Mercator projection, 180
Merchants' Exchange, 157
Mercury (flight), 21
Meredith, George, 247
Meredith, James, 132
Mérimée, Prosper, 243
Merrill, Charles E., 153
Merrimac (warship), 460
Mesopotamia, Ancient, 91–93
Metaurus (battle), 477
Meteorology, 417–18
Methodist Church, 323–25
Metro-Goldwyn-Mayer (MGM), 556
Meyerbeer, Giacomo, 294
Miami Dolphins, 582
Michener, James A., 259
Michinaga, 80
Mickey Mouse watches, 433
Mickiewicz, Adam, 243
Microcomputer, 189, 340, 438
Microphone, 215
Microscope(s), 180, 182, 186, 187, 188, 189, 190, 419
Microwave cooking, 438
Middle East, 70–72
Middleton, Thomas, 279
Midori Maru (ferryboat), 514
Midway (battle), 469, 487
Mihan, George, 588
Mikasa (battleship), 512, 513
Milan (battle), 477
Milhaud, Darius, 300
Military history and technology, 447–502
Milken, Michael, 159
Milky Way, 376, 383
Mill, John Stuart, 133, 338
Millais, Sir John Everett, 271
Millay, Edna St. Vincent, 256
Miller, Arthur, 288
Miller, Henry, 256
Millet, Jean Françoise, 271
Milne, A.A., 253
Milne-Edwards, Henri, 20
Milton, John, 238
Milvian (battle), 307
Mimimum-wage laws, 160
Minh, Ho Chi, 89, 109
Miniature golf, 436
Minoxidil, 375
Minuteman II missile, 494
Miranda v. Arizona, 39
Miranda warning, 139
Miró, Joan, 276
Mirror system, 190
Mir (space station), 25
Mishima, Yukio, 261
Mishna, 326
Missouri Compromise, 33, 130
Mistral, Frédéric, 248
Mistral, Gabriela, 255
Mitchell, Margaret, 257
Mitterrand, François, 68, 146
M & Ms, 428
Mobile Bay (battle), 461
Mo Ching, 363
Model T, 170, 185, 219
Modigliani, Amedeo, 275
Mohacs (battle), 480
Moholy-Nagy, László, 276
Molière, Jean-Baptiste Poquelin, 280
Mollwitz (battle), 481
Molly Maguires, 159
Molnar, Ferenc, 286
Moltke, Helmuth von, 462
Mondrian, Piet, 273
Monet, Claude, 272
Mongol invasions, 449
Monitor (warship), 460
Monmouth (battle), 455
Monopoly, 436
Mons (battle), 462, 485
Montemaggiore (battle), 478
Montes Claros and Villa Viciosa (battle), 480
Monteverdi, Claudio, 291
Montezuma (emperor), 29
Montgomery, L.M., 252
Monuments, 349–50
Moore, George, 249
Moore, Henry, 276
Moore, Marianne C., 255
Moore's Creek Bridge, 454
Mop, 438
Moral, Don Jose Zorrilla y, 283
Moral Majority, 323
More, Hannah, 241
More, Sir Thomas, 237
Morgan, John P., 152, 155
Morley, Thomas, 291
Moro, Aldo, 68, 442
Morphine, 372
Morris, William, 248
Morrison, Toni, 262
Morro Castle (luxury liner), 512, 514
Morse, Samuel, 34, 182, 214
Moscow (battle), 486
Moscow (USSR), 465, 511
Moses, 305, 325
Motherwell, Robert, 277
Motion pictures, 185, 207, 552, 553–67
Mott, Lucretia, 133
Mounds (burial chambers), 348
Mountbatten, Lord Louis, 68, 88
Mounted cavalry soldiers, 489, 490
Mozart, Wolfgang Amadeus, 294
Muckrakers, era of, 36
Muhammad, 106, 332
Muhammad ibn Abd al-Wahhab, 333
Muhammad ibn Isma'il al-Bukhari, 333
Muir, Edwin, 255
Mukden (battle), 485
Munch, Edvard, 273
Munich Beer Hall Putsch, 64
Munich Pact, 65, 118
Munzer, Thomas, 316
Murdoch, Iris, 260
Murdoch, Rupert, 534
Murfreesboro (battle), 460, 484
Murillo, Bartolomé Estêban, 268
Murphy, Joe, 601
Murrow, Edward R., 537
Mursa (battle), 477
Musical instruments, western, 292–93
Muslims, 448
Mussolini, Benito, 64, 145, 470
Mussorgsky, Modest, 296
Mutsuhito, Emperor, 84
Mycale (battle), 476
Mylae (battle), 476
My Lai massacre, 474
Myriocephalon (battle), 478
Nabokov, Vladimir, 257
Nagasaki, 471
Naipaul, V.S., 262
Naismith, James, 586
Namhong-Ho (ferryboat), 515
Napoleon, 57, 109, 309, 457, 491
Napoleonic code, 115
Narses, 103
Narva (battle), 480
Naseby (battle), 454, 480
Nash, Ogden, 258
Nash, Thomas, 237, 278
National Aeronautics and Space Administration, 225
National American Women's Suffrage Association, 134
National Association of Book Publishers, 551
National Association of Security Dealers, 158
National Audubon Society, 422
National Banking System, 155
National Board of Censorship, 555
National Congress of Mothers, 340
National Environmental Policy Act, 423
National Industrial Recovery Act, 160
Nationalist CCP alliance, 467
Nationalist Constitution, 467
National Labor Union, 159
National Liberation Front, 472
National Organization of Women, 134
Nautilus (submarine), 187, 198
Navarino (battle), 483
Navratilova, Martina, 618
Nazism, 64, 70
Nebuchadnezzar, 92
Neerwinden (battle), 456
Nehru, Jawaharlal, 88
Nei Ching, 363
Nemea (battle), 476
Nemerov, Howard, 260
Neon, 185, 213, 400
Neoprene, 402
Nero, 8, 100, 306
Neruda, Jan, 248
Neruda, Pablo, 258
Neutrons, 410
Nevelson, Louise, 276
Newark (New Jersey), riot, 132
Newbury (battle), 453, 480
New Castle, 204; locomotive, 182
New Deal, 160, 164
Newhouse, Samuel Irving, 153
New Jersey (battleship), 474
New Orleans (battle), 483
Newspaper(s), 178 533–36
New Testament, 115, 305
Newton, Huey, 132
Newton, Isaac, 224, 377, 404
New York Stock Exchange, 155, 157, 157–59
Ngo Dinh Diem, 473
Nguyen Cao Ky, 473
Nickel, 399
"Nickelodeon" (theater), 555
Nicklaus, Jack, 610
Nicopolis (battle), 479
Nietzsche, Friedrich, 338
Nightingale, Florence, 372
"Nightstalker," 443
Nikon, 315
Nile, battle, 482
Nile flooding, 515
Nin, Anaïs, 258
Nina (ship), 8
Nineteen Propositions, 453
Nineveh (battle), 478
Ninja Turtles, 436
Nivelle, Robert (Gen.), 463
Nixon, Richard, 39, 67
Noble, Richard, 230
Nobopolassar (king of Chaldea), 92
Nordingen (battle), 480
Norell, Norman, 431
Norge (steamer), 513
Noriega, Manuel (Gen.), 44
Norman Conquest, 49, 448
Normandy invasion, 470, 487
Noronic (vessel), 514
Norris, Frank, 251
North Atlantic Treaty Organization, 38, 66
Northeast Passage, 15
North Pole, 16, 17
Notium (battle), 476
Novo Amapa (steamer), 515
Novocaine, 373
Novorossiisk (warship), 514
Novum Organum, 337
Nuclear arms race, 65
Nuclear power plant, 187, 213
Nuclear technology, 226–27
Nuclear Test-Ban, 39, 67, 118, 422
NutraSweet, 428
Nuxhall, Joe, 597
Nyerere, Julius, 75
Nylon, 162, 186, 207, 208, 403
Nylon stockings, 431
O. Henry, 249
Oates, Joyce Carol, 262
O'Casey, Sean, 286
Oceanic (luxury liner), 197
Ocean liners, 183
Ochs, Adolph, 534
O'Connor, Flannery, 261
Octavian, 100
Odets, Clifford, 287
Oersted, Hans Christian, 408
O'Faoláin, Sean, 257
Offenbach, Jacques, 295
O'Flaherty, Liam, 257
O'Hara, John, 258
Ohm, George Simon, 211
Ohm's law, 408
Oil industry (U.S.), 168–69, 171
O'Keefe, Georgia, 275
Okinawa, 90, 471, 487
Oldenberg, Claes, 277
Olds, Ransom, 218
Olives, 192, 427
Olympic Games, 94, 620–21
Omar Khayyám, 235
Omduman (battle), 485
O'Neill, Eugene, 286
Onizuka, Ellison, 25
Opera, 297–98
Optical fibers, 187
Optics, 406
Orff, Carl, 300
Organization of American States Charter, 118
Organization of Petroleum Exporting Countries, 169
Orleans (siege), 479
Ortega, President Daniel, 44
Ortelius, Abraham, 10
Orwell, George, 258
Osborne, John, 289
Oswald, Lee Harvey, 39
Ouija boards, 435
Ouimet, Francis, 607
Outer Space Treaty, 67, 118, 129
Oven, 213, 437
Overmyer, Robert, 24
Ovid (poet), 233
Owen, Robert, 144
Oxygen, 399
Ozone layer, 424
Pacemaker, 188, 374
Packard, David, 153
Pact of Steel, 118
Paddleboat, 195
Paderewski, Ignace, 299
Paganini, Niccolò, 294
Paine, Thomas, 133, 241
Painpat (battle), 480
Painters, 263–77
Pajamas, 430
Palaces, 349–50
Paley, William S., 153, 536
Palmer, Arnold, 610
Palmer, Nathaniel, 17
Panama Canal, 36, 39, 43, 118
Pantheon (Rome, Italy), 347
Papandreou, Andreas, 68
Paper, 178
Paper clip, 437
Papyrus, 177, 411
Paracelsus, 368
Paré, Ambroise, 369

Paricutin (volcano), 527
Paris (France) Opera House, 355
Park, Mungo, 14
Parker, Bonnie, 441
Parker, Dorothy R., 256
Parkinson's disease, 375
Parks, Rosa, 131
Parry, Sir William, 15
Parthenon (Athens, Greece), 347
Pasta, 427
Pasternak, Boris, 256
Pasteur, Louis, 372; Institute, 373
Patay (battle), 479
Pathé film company, 552
Paton, Alan, 258
Paul, Saint, 100
Pavia (battle), 477, 478, 479
Payola scandal, 569
Peace of Augsburg, 309, 316; of Prague, 117; of Vereeningning, 117; of Westphalia, 317, 321
Peale, Charles William, 270
Peanuts, 192, 194, 427
Pearl Harbor, 469, 486
Pearson, C. Arthur, 534
Peary, Robert, 13, 16
Peasant's Revolt, 51, 52, 143
Peele, George, 278
Peirce, Charles, 338
Peking (siege), 488
Pelagius, 307
Pelé, 604
Penicillin, 374, 390
Penology, 138
Pens, 186, 187, 438
Pentateuch, 305
Pepin the Short, 45
Pepsi-Cola, 428
Pepsin, 372
Pepys, Samuel, 238
Percy, Walker, 260
Perdue, Franklin Parsons, 154
Perelman, Ronald Owen, 154
Perelman, S.J., 258
Pergolesi, Giovanni Battista, 294
Pericles, 94
Periodic law of elements, 408
Periscope, 198
Perón, Juan, 43
Perot, H. Ross, 154
Perry, Fred, 612
Perryville (battle), 460
Pershing, John J., 463
Pesticides, 194
Petersburg (siege), 484
Petrarch (poet), 235
Petrified wood, 415
Petrochemicals industry, 162
Pet Rocks, 436
Pevsner, Antoine, 275
Pharsalus (battle), 477
Philadelphia Times, 533
Philby, Harry St. John, 13
Philippi (battle), 477
Philippine Sea (battle), 470, 487
Philosophy, 335–39
Phipps, Constantine, 15
Phnom Penh, 488, 489
Phoenician explorers, 3
Phoenix (steamboat), 196
Phonograph, 184, 569
Phosphorus, 399
Photoelectric cell, 213
Photoengraving, 183
Photography, 209–10
Photovoltaic cell, 187, 213
Phu, Dien Bien, 109
Phyfe, Duncan, 439
Physics, 403–411
Picabia, Francis, 274
Picasso, Pablo, 274
Piccard, Jacques, 20
Pichincha (battle), 42
Pickford, Mary, 555
Picturephone, 188, 215, 217
Pike, Zebulon, 12
Pilgrims, 317
"Piltdown man," 393
pi (mathematical notation), 411, 413
Pindar (poet), 233
Pinero, Sir Arthur, 283
Ping-Pong, 435
Pinochet, Augusto (Gen.), 44
Pinta (ship), 8
Pinter, Harold, 289
Pioneer (spacecraft), 189
Pipeline, 183
Pirandello, Luigi, 286
Pisano, Nicola, 263
Pissarro, Camille, 271
Piston, Walter, 300
Pitts, Hiram, 193
Pitts, John, 193
Pittsburgh Plate Glass Co., 149
Pizarro, Francisco, 10
Pizarro, Gonzalo, 10
Pizza Hut, 428
Plague(s), 5, 527–29
Plante, Jacques, 601
Plastic(s), 162, 190, 207–8
Plastic wrap, 438
Plataea (battle), 476
Plath, Sylvia, 262
Platinum, 399
Plato, 336, 340, 403
Plautus, 278
Playwrights, 278–89
Plessy v. Ferguson, 36
Plexiglas, 208
Pliny (the elder), 235
Plotinus, 337
Plow, 179, 182, 192
Plumbicon, 217
Plumbing, 177
Plutarch (biographer), 235
Pluto, 383
Pockets, 430
Poe, Edgar Allan, 245
Poets, 233–62
Pogroms, 327
Pogue, William R., 24
Poitiers (battle), 478, 479
Polaroid camera, 186, 210, 438
Polio vaccine, 374
Political scandals, 139–41
Pollen, 387
Pollock, Jackson, 277
Pollution, 220, 422, 424
Polo, Maffeo, 8; Marco, 8; Nicolò, 8
Poltava (battle), 481
Polybius (historian), 3, 233
Polyesters, 186, 208
Polyethylene, 186, 208
Polypropylene, 208, 403
Polystyrene, 207, 208
Polyurethane, 208, 403
Polyvinyl acetate, 208
Polyvinyl chloride, 186, 207, 402
Ponce de León, 9
Pontiac's Rebellion, 31, 143
Pope, Alexander, 240
Pope(s), 310–13
Popped corn, 427
Popper, Karl, 339
Popsicles, 428
Pop-top can, 428
Porcelain, 178
Port Arthur (battle), 485
Porter, Edwin, 555
Porter, Gene Stratton, 249
Porter, Katherine Anne, 256
Postage stamp, 437
Postal service, 31
Postmodernism, 359
Pot, Pol, 90
Potassium, 399
Potato chips, 428
Pottery, 177, 191
Poulenc, Francis, 301
Poulsen, Valdemar, 568
Pound, Ezra L., 253
Poussin, Nicholas, 267
Powys, John Cowper, 252
"Prague Spring," 67, 69
Prehistoric tools, 190–92
Presbyterian Church, 317, 318, 320–22
Presidents (U.S.), 30–31, 120–23
Presley, Elvis, 569
Pressure cooker, 180, 437
Preston (battle), 454, 480
Pretzels, 427
Priestley, J.B., 256
Priestley, Joseph, 369
Prince Edward Island, 111
Princess Sophia (steamer), 514
Printing, 179, 181, 183, 184, 201–3, 214, 340, 533
Probability theory, 412
Project Mercury, 21
Prokofiev, Sergei, 300
Protestantism, 315–17
Proton, 408
Protoplasm, 387
Proust, Marcel, 251
Provence (cruiser), 513
Psychology, 367–68
PTL Club, 323
Ptolemy, 8, 403
Publishing, 201, 533–36, 545–52
Puccini, Giacomo, 299
Pulitzer, Joseph, 533
Pullman sleeper cars, 204
Pullman strike, 160, 164
Pump(s), 178, 180, 181, 203, 205
Puranas, 330
Purcell, Henry, 291
Pushkin, Aleksandr, 243, 283
Pyle, C.C., 612
Pynchon, Thomas, 262
Pyroscaphe (vessel), 203
Pythagoras of Samos, 336
Pythagorean school, 411
Pytheas, 3, 15, 95
QE2 (ocean liner), 197
Quakers, 130, 317
Quang Tri City (battle), 488
Quant, Mary, 431
Quantum mechanics, 408
Quantum theory, 410
Quark theory, 410
Quasimodo, Salvatore, 257
Quebec City (Canada), 511
Queen Elizabeth (ocean liner), 197
Queen Elizabeth II (ocean liner), 197
Queen Mary (ocean liner), 514
Quiberon Bay (battle), 481
Rabe, David, 289
Rabelais, François, 237
Rachmaninoff, Sergei, 299
Racine, Jean, 280
Radar, 172, 186, 213, 222, 493
Radiation, 226, 384, 401, 408
Radio, 184–85, 216–18, 437, 438, 536–45, 581, 596
Radiocarbon dating, 395
Radiotelephony, 185
Radium, 401, 408
Raeburn, Sir Henry, 270
Raggedy Ann dolls, 435
Rahman, Abd al-, 106
Railroad(s), 163–65, 182, 204–5
Ramakrishna, 331
Ramananda, 331
Rameau, Jean-Philippe, 293
Ramillies (battle), 481
Ramirez, Richard, 443
Ramses II, 97
Ramses III, 97
Ransom, John Crowe, 255
Rasputin, 63
Rauschenberg, Robert, 277
Ravel, Maurice, 300
Ray, Man, 275
Razor, 184, 437
Reach, Alfred, 591
Reactor, 227
Reagan, Ronald, 40, 161
Reagan–Gorbachev Summit, 68
Reaper, 182, 193
Recording industry, 567–69
Red Baron (German air ace), 464
"Red scare," 160
Refrigerator, 183, 213
Reger, Max, 299
Regnard, Jean, 280
Reign of Terror, 57, 456
Reinhardt, Ad, 277
Remarque, Erich Maria, 257
Rembrandt van Rijn, Paul, 268
Remek, Vladimir, 24
Renault, Louis, 219
Renoir, Pierre-Auguste, 272
Renshaw, Ernest, 612
Renshaw, William, 612
Reproductive system, 390
Reproductive technology, 194
Resnik, Judith, 25
Respighi, Ottorino, 300
Revere, Paul, 454
Revolts, 142–44
Reymont, Wladyslaw, 251
Reynolds, Joshua, 268
Rice, Elmer, 287
Rich, Adrienne, 261
Richard II (king), 602
Richards, Renée, Dr., 618
Richardson, Henry, 355
Richter scale, 421, 516
Ride, Sally, 24
Ridgway, Matthew (Gen.), 472
Riley, James Whitcomb, 249
Rilke, Rainer Maria, 252
Rimbaud, Arthur, 249
Rimsky-Korsakov, Nikolai, 296
Ring Theater, 511
Ritz crackers, 428
Rivera, Diego, 275
Robes, 429
Robinson, Eddie, 586
Robinson, Edwin Arlington, 251
Robinson, Frank, 597
Robinson, Jackie, 597
Robots, 194
Rockefeller, John D., 152, 153, 168
Rockefeller Center, 358
Rocket(s), 185, 186, 224–25
Rock music, **1955–89**, 570–71, 573–74
Rockne, Knute, 581
Rocroi (battle), 480
Rodin, Auguste, 272
Roethke, Theodore, 259
Roe v. Wade, 39, 40, 134
Rogers, Will, 253
Rolleiflex camera, 210
Roller-skate dancing, 436
Rolling Thunder, 473
Rolvaag, Ole E., 252
Roman baths, 395
Roman Colosseum, 347
Roman Empire, 99–102, 101–3, 313
Roman numeral system, 411
Rome (Italy), 99, 100, 347–48, 477
Romer, Olaf, 406
Romnanenko, Yuri, 25
Roncesvalles (battle), 478
Roosevelt, Franklin D., 126, 131, 160, 469, 471, 537
Rose, Pete, 598
Rosenberg conviction, 140
Ross, James C., 17
Rossetti, Christina Georgina, 248
Rossetti, Dante Gabriel, 247, 271
Rossini, Gioacchino, 294
Rostand, Edmond, 286
Roth, Philip, 262
Rothko, Mark, 276
Rouault, Georges, 273
"Rounders," 590
Rousseau, Henri, 272
Rousseau, Jean-Jacques, 240, 338
Rowe, Nicholas, 280
Rowley, William, 279
Rubber, 182, 186
Rubens, Peter Paul, 267
Ruckert, Friedrich, 242
Rudolf I (king), 50
Rugby, 602
Runeberg, Johan Ludvig, 245
Rush-Bagot Agreement, 33, 116
Rushdie, Salman, 262
Russell, Bertrand, 252, 339
Russell, Bill, 590
Russian Social Democratic Labor Party, 144
Ruth, Babe, 596
Ryan, Nolan, 597, 598
Saarinen, Eero, 359
Sabbath, 325
Sacco, Nicola, 441
Sachs, Nelly, 256
Sadat, Anwar, 71
Saint Albans (battle), 479
Saint-Germain-en-Laye Convention, 131
Saint-Saëns, Camille, 296
St. Mihiel (battle), 464, 486
St. Peter's Basilica, 308
St. Philibert (ship), 514
Ste. Geneviève (library), 355
Sakarya (battle), 486
Salamis (battle), 475
Salang Tunnel, 510
Salerno (battle), 487
Salinger, J.D., 260
Salk, Jonas, 374
Salt March, 87
Salyut (space station), 24
Samarkand (battle), 478
Samarkand observatory, 377
Sand, George, 245
Sandals, 429
Sandburg, Carl, 252
Sandwich, 428
Sanger, Margaret, 134
San Jacinto (battle), 483
Sanka, 428
San Simeon castle, 153
Santa Maria (ship), 8
Santayana, George, 249, 339
Santiago (battle), 485
Sargent, John Singer, 272
Sargon, 91
Sargon II, 92
Sarnoff, David, 216, 536
Saroyan, William, 259, 287
Sartre, Jean-Paul, 258, 287, 339
Sasbach and Altenheim (battle), 480
Satie, Erik, 299

Saul of Tarsus, 306
Sausages, 427
Savannah (nuclear cargo ship), 187, 197
Savannah (paddle wheel), 203, 226
Save (steamer), 514
Savings and loan bailout, 157
Saw, 179
Sayers, Dorothy L., 256
Scandals, 37, 39, 139-68, 534, 569
Scarlatti, Alessandro, 291
Scarlatti, Domenico, 293
Schayes, Adolph, 588
Schickard, Wilhelm, 227
Schiele, Egon, 275
Schiller, Friedrich von, 282
Schleiermacher, Friedrich, 317
Schmalkaldic War, 53, 316
Schmidt, Mike, 597
Schoenberg, Arnold, 300
Schools, 340
Schopenhauer, Arthur, 338
Schubert, Franz, 294
Schuman, William, 301
Schumann, Robert, 295
Schuster, Max, 551
Schwitters, Kurt, 275
Scobee, Francis, 25
Scorpion, 515
Scott, David R., 21
Scott, Robert F., 18
Scott, Sir Walter, 241
Scott Paper Co., 149
Scotus, Duns, 337
Scrabble, 436
Screw propeller, 182, 196
Scriabin, Alexandr, 299
Scribe, Augustin, 283
Sculptors, 263–77
Scurvy, 372
Sea Devil (submarine), 198
Seagram Building, 359
Sealab 1, 20
Sea-Land Commerce (containership), 229
Seale, Bobby, 132
Sears, Richard W., 153
Seat-belt law, 171
Sea Witch (ship), 183
Second Strategic Arms Limitation Treaty, 39
Securities and Exchange (U.S.), 158
Securities trading, 155
Sedan (battle), 484
Sedgemoore (battle), 480
Segal, George, 277
Seiden, George, 218
Seismograph, 178
Selassie, Haile, 75
Selden, George, 170
Selective Service Act, 469
Selenium, 216
Selfridge, Thomas E., 505
Selim III (Sultan), 107–8
Selvy, Frank, 588
Semaphore, 181, 204
Seneca (philosopher), 278
Sennett, Mack, 555
Serao, Matilde, 249
Sessions, Roger, 300
Seurat, Georges, 272
Sevastpol, German siege of, 469
Seven Days Battle, 460, 484
Sewall, Samuel, 130
Sewing machine, 183, 206, 437
Sewing needles, 191
Sexagesimal, 411
Sex-change operation, 374
Sextant, 181, 416
Sexton, Anne, 261
Shackleton, Ernest Henry, 18
Shaffer, Peter, 288
Shahn, Benjamin, 276
Shakespeare, William, 278, 284
Shaw, George Bernard, 283
Shelley, Mary Wollstonecraft, 133, 243
Shelley, Percy Bysshe, 243
Shensi, 82
Shepard, Alan, 21, 39
Shepard, Sam, 289
Sheridan, Richard Brinsley, 282
Sherman Antitrust Act, 36, 149, 155, 159
Sherman's March to Sea, 461
Sherwood, Robert, 287
Shiloh (battle), 460, 484
Shirley, James, 279
Shoes, 429, 430, 432
Sholokhov, Mikhail A., 259
Shore, Eddie, 600
Shore, Ernie, 594
Shostakovich, Dmitri, 301
Shoulder pads, 434
Siam, kingdom of, 84
Sibelius, Jean, 299
Sibir (icebreaker), 17
Sicilian mafia, 443
Sicilian Vespers, 50
Sickles, 192
Sidney, Sir Philip, 237
Siebe, Augustus, 19
Sienkiewicz, Henryk, 249
Sikorsky, Igor, 222
Silicone, 185, 207, 208, 401
Silkworms, 192
Silly Putty, 436
Simon, Neil, 288
Simon, Richard, 551
Simpson, O.J., 585, 586
Sinclair, Upton, 253
Singer, Isaac Bashevis, 258
Sisley, Alfred, 271
Sitwell, Dame Edith, 255
Sivapithecus, 395
Six Articles, 318
Skateboarding, 436
Skate (submarine), 17
Skelton, John, 237
Skirts, 430–33
"Skull Famine," 83, 515
Skyjacking, 172
Skyscrapers, 358
Slavery, 130, 130–131
Slide rule, 412
Sloan, John, 273
Sluis (battle), 479
Slumber parties, 436
Smallpox, 528
Smetana, Bedrich, 295
Smith, David, 276
Smith, Jedediah Strong, 12
Smith, Michael, 25
Smyth, John, 322
Snow, C.P., 258
Snowstorms, 522–24
Soap, 161
Sobral Santos (riverboat), 515
Soccer, 602–4
Socialism, 144–46
Social Security Act, 160
Society for Applied Anthropology, 394
Society of Authors, 551
Society of Automobile Engineers, 219
Society of Friends, 130, 317
Socket bayonet, 491
Socrates, 95, 336
Sodium, 399
Soil Conservation Service, 422
Soil erosion, 193, 422
Solar cell, 189; eclipse, 376; energy conversion, 162
Solar system, 377–81
Solar wind, 384
Solferino (battle), 483
Solidarity (organization), 144
Solon, 94
Solovyov, Vladimir, 24
Solzhenitsyn, Aleksandr, 67, 260
Somme (battles), 463, 464, 485
Sonar, 185, 198, 421, 493
"Son of Sam" killer, 442
Sontag, Susan, 262
Sony Walkman, 438
Soranus of Ephesus, 364
South America, 42–44
Southeast Asia Treaty, 118
Southey, Robert, 242
South Ford Dam, 519–20
Soutine, Chaim, 275
Sovereign of the Seas (ship), 197
Soyuz (spacecrafts), 22, 24
Space exploration, 21–25
Spain, 52, 57
Spalding, Albert G., 594
Spam, 428
Spanish Inquisition, 51, 308, 309
Spark, Muriel, 260
Spears, 190, 191, 489
Speck, Richard, 442
Spectrometer, 227, 421
Spectrum analysis, 400, 408
"Speech of the 100 Flowers," 89
Speke, John, 14
Spenser, Edmund, 237
Spinning jenny, 206
Spinning wheel, 179, 180
Spinoza, Baruch, 337
Spirit of Australia (hydroplane), 230
Spirit of St. Louis (monoplane), 222
Spitball, 596
Splinting, 363
Spontaneous generation, 390
Spruce Goose (airplane), 223
Sputnik I (satellite), 21, 66, 187, 224
Squash, 427
Stahl, George, 396
Stainless steel, 185
Stalin, Joseph, 64, 145, 315
Stalingrad (battle), 470, 487
Stamp Act, 31
Standards (battle), 478
Stanley, Henry, 14
Stanley Steamer (automobile), 184
Stanton, Elizabeth Cady, 133
Stark (Navy frigate), 72
Starkweather, Charles, 442
Stars, 382
"Star Wars," 494
Statue of Liberty, 36
Statute of Westminster, 37, 64, 116
Steam, 181, 197, 203–4, 206
Steamboat, 195, 203
Steamship, 197
Stearns, Shubael, 322
Steel, 175–67, 194
Steele, Sir Richard, 238
Steen, Jan, 268
Stein, Gertrude, 252
Steinbeck, John, 258
Stella, Frank, 277
Stendhal, 242
Sterne, Laurence, 240
Stevens, John, 163
Stevens, Wallace, 253
Stevenson, Robert Louis, 249
Stirrup, 490
Stockholm, Thomas, 569
Stockholm (ocean liner), 514
Stock-market crash, 37, 158
Stoker, Bram, 249
Stonehenge (England), 375
Stoppard, Tom, 289
Stowe, Harriet Beecher, 131, 246
Stowe-on-the-Wold (battle), 454, 480
Strabo (geographer), 233
Strand theater, 555
Strategic Air Command, 494
Strategic Arms Limitation Talks, 39, 67, 118
Stratosphere, 419
Strauss, Johann, 295, 296
Strauss, Richard, 299
Stravinsky, Igor, 300
Streaking, 436
Streptomycin, 374
Strindberg, Johan, 283
Strontium, 399
Structuralism, 395
Stuart, Gilbert, 270
Stuart, John McDouall, 12
Styron, William, 261
Submarine(s), 20, 180, 184, 198–99, 226, 230, 493
Suckling, Sir John, 238
Suetonius (biographer), 235
Suez Canal, 108
Suez Crisis, 71
Suffrage, 134
Sufism, 333
Sugar, 161, 193, 427
Sui dynasty, 80
Sulfa drugs, 374
Sulfur, 161, 489
Sullivan, Ed, 537
Sunday, Billy, 317
Sundials, 94, 177, 178, 199, 200
Sunspot cycle, 382
Sun Yat-sen, 87
Superconductors, 190, 408, 411
Supermarkets, 428
Supertankers, 197
Supreme Court (U.S.), 135, 136–37
Suspenders, 430
Sutton, Willie, 442
"Sweating sickness," 528–29
Swift, Jonathan, 238
Swinburne, Algernon Charles, 248
Swiss Reformation, 315
Sydenham, Thomas, 369
Sykes-Picot Agreement, 63, 117
Synge, John, 286
Synthetics, 207–8
Syphilis, 369, 528–29
Syracuse (battle), 476
Tacitus (historian), 235
Taft-Hartley Act, 160, 166
T'ai-hsu, 330
Tai Ping (steamer), 514
Taiping Rebellion, 84
Taj Mahal, 83
Tallis, Thomas, 289
Talmud, 326
Tammany Hall, 139
Tamponas II (ship), 515
T'ang dynasty, 80
Tanks, 492, 496–97
Tannenberg (battle), 485
Tàpies, Antoni, 277
Tarkington, Booth, 251
Tarnower, Herman, 442
Tasman, Abel, 11
Tasso, Torquato, 237, 278
Tate, Sharon, 442
Tattoo, 433, 436
Taxonomy, 387
Taylor, Sir Edward, 393
Tchaikovsky, Peter Ilyich, 296
Tea, 427
Technicolor, 556
Technology, 177–90
Teddy bear, 435
Teflon, 208, 403
Tegner, Esaias, 242
Telegraph, 178, 181, 182, 183, 214–15
Telephone, 158, 214–15, 437
Telescope, 180, 184, 185, 186, 188, 190, 383
Television, 187, 188, 190, 217, 428, 536–45
Telex system, 215
Telstar (satellite), 188, 200, 217
Temple(s), 325, 347, 349–50
Temple Scroll, 328
Tenneco, 151
Tennessee Valley Authority, 422
Tennis, 610–18
Tennyson, Alfred Lord, 246
10th of Ramadan (riverboat), 515
Terence (Roman dramatist), 278
Tereshkova, Valentina V., 21
Terrorism, 129
Tesla, Nikola, 183, 212
"Test-tube" baby, 375, 392
Teutoburg Forest (battle), 477
Thackeray, William Makepeace, 246
Thales of Miletus, 335, 376, 403
Thalidomide, 374
Thames (battle), 483
Thapsus (battle), 477
Thatcher, Margaret, 68
Thaw, Harry K., 441
Theater(s), 281–82, 511, 512, 513, 555, 556
Theocritus (poet), 233
Theodolite, 180
Theodosius, 101, 103
Theory(ies): analytic probability, 403; atomic structure of matter, 407; cell, 390; of functionalism, 393; heliocentric system, 404; probability, 412; relativity, 408, 413; of structural functionalism, 395; of the Universe, 383; wave theory of light, 406, 407
Thermodynamics, 408
Thermometer, 180, 181, 416
Thermopylae (battle), 475, 477
Thetis (submarine), 514
Thomas, Dylan M., 260
Thomas, Joseph, 213
Thomas, Virgil, 301
Thomas Aquinas, Saint, 308, 337
Thompson, David, 12
Thomson, Joseph, 14
Thomson-Houston Electric, 150
Thoreau, Henry David, 246
Thorpe, Jim, 580
Thor Power Tool Company tax ruling, 552
Three Mile Island, 39
Thresher (submarine), 514
Threshing machine, 193
Thucydides (historian), 233
Thurber, James, 256
Ticonderoga (battle), 481
Tidal wave(s), 421, 519–20
Tientsin Massacre, 84
Tiepolo, Giovanni, 268
Tilden, Bill, 612
Time-telling devices, 199–201
Tin can, 437
Tinguely, Jean, 277
Tintoretto, 267
Tipitaka, 328, 329
Tippecanoe (battle), 483
Tire(s), 170, 184, 218
Tiros 1 (satellite), 188, 421
Titanic (luxury liner), 20, 396, 513
Titanium, 400

Titan (rocket), 225, 510
Titian, 266
Tito, Josip, 65, 68, 145
Toaster, 437
Tobacco, 427
Tobruk (battle), 469, 486
Togas, 429
Toilet, 437
Toleration Act, 321
Tolkien, J.R.R., 256
Toller, Ernst, 287
Tolstoy, Count Leo, 247
Tom Thumb, 204
Tools, prehistoric, 190–92
Toothbrush, 437, 438
Topmast, 180, 195
Topology, 413
Torgau (battle), 481
Tornadoes, 524–25
Torpedoes, 183, 198
Torrey, Rev. Charles, 130
Toulon (siege), 482
Toulouse (battle), 459
Toulouse-Lautrec, Henri de, 273
Tourneur, Cyril, 279
Tourniquet, 369
Tov, Baal Shem, 327
Townshend Acts, 32
Towton (battle), 479
Toya Maru (ferryboat), 514
Tractor(s), 184, 194
Trafalgar (battle), 458, 482
Traffic lights, 185
Traherne, Thomas, 238
Train(s), 186, 507–08
Trajan, 100
Transatlantic cable, 196
Transformer, 184
Transistor, 228
Transportation Act (U.S.), 164
Trans-Siberian Railroad, 63, 205
Tran Van Huoung, 473
Treaties, 116–18
Trekking, 13
Trenton (battle), 481
Trevithick, Richard, 218
Trilling, Lionel, 258
Trojan War, 447
Trollope, Anthony, 246
Trotsky, Leon, 38, 64
Trousers, 430, 431, 432
Troy (battle), 475
Troy (city), 393
Truck(s), 219, 507–8, 510
Truman, Harry S, 131, 164, 166, 471
Truman Doctrine, 116
Trump, Donald, 154, 173
Tsunamis, 519–20
Tsushima Strait (battle), 485
Tuberculosis, 373, 529
Tuileries Palace, 456
Tupolev Tu-144 (supersonic airplane), 172, 189
Tupperware, 438
Turbine, 206, 207
Turbinia (steamship), 204
Turgenev, Ivan Sergeyevich, 246
Turner, Joseph Mallord William, 270
Turner, Nat, 130
Turner, Robert Edward II, 154
Tutankhamen (king), 96, 435
TV dinners, 428
Twain, Mark, 248
"Twist" dance, 436
Tylenol capsule contamination, 442
Tyler, Anne, 262
Typesetting, 184, 202
Typewriter, 202
Typhoon(s), 521–22
Typhus epidemic, 528–29
Tyre (battle), 476
Uccello, Paolo, 263
Udall, Nicholas, 278
Uhland, Ludwig, 242
Uhuru (satellite observatory), 188
Ultrasound, 187, 374
Unamuno y Jugo, Miguel de, 251
Underground railroad, 130
Underwater exploration, 19–20
Undset, Sigrid, 253
Union Carbide, 90, 150, 162, 424
United Nations, 116, 118, 126–29, 153, 424
United States (ocean liner), 197
United States v. Weber, 133
Universal Copyright Convention, 552
Universal Declaration of Human Rights, 116, 128
Universal Exposition of 1889, 62
Universal joint, 180
Updike, John, 262
Uprisings, 142–44
Uranium, 399, 410
Uranus, 381
U.S. Liberty (ship), 509, 510
U.S. National League of Cities, 424
U.S. National Research Council, 421
U.S. National Wildlife Federation, 424
U.S. Skylab (orbiting work station), 24
U.S.S. Forrestal, 510
U.S.S. Great Republic, 183
Uthman, 106
Utrillo, Maurice, 274
Vaccination, vaccines, 372, 390
Vacuum tube, 184, 185, 213, 216
Valâzques, Diego Rodriguez de Silva y, 268
Valbanera (steamer), 514
Valdez (tanker), 169
Valentine's Day Massacre, 441
Valēry, Paul, 251
Valmy (battle), 456, 482
Vanderbilt, Cornelius, 152
Vanderbilt University, 152
van der Weyden, Rogier, 263
Van Diemen's Land, 11
van Dyck, Anthony, 268
van Eyck, Jan, 263
van Gogh, Vincent, 272
Vanguard 2 (satellite), 22
Vanguard (warship), 514
van Leeuwenhoek, Anton, 369
Vanzetti, Bartolomeo, 441
Varaville (battle), 478
Vardhamana, Nataputta, 330
Varna (battle), 479
Vaughan, Henry, 238
Vaughn Williams, Ralph, 299
V-E Day, 471
Velcro fastener, 438
Venerable Bede (historian), 235
Venice, 511
Venus de Milo statue, 392
Verdi, Giuseppe, 295
Verdun (battle), 463, 485
Verhaeren, Émile, 249
Verlaine, Paul, 249
Vermeer, Jan, 268
Verne, Jules, 247
Veronese, Paolo, 267
Versailles Peace Conference, 87, 464
Vesalius, Andreas, 369
Vesco, Donald, 230
Vice Presidents (U.S.), 30–31
Vicksburg (siege), 461, 484
Victor Emmanuel II (king), 61
Vidal, Gore, 261
Videocassette recorders, 438, 556
Video recorder, 438
Vienna (battle), 480
Vienna Secession, 357
Villa, Pancho, 37
Village Voice, 534
Villains, 441–43
Villa-Lobos, Heitor, 300
Virgil (poet), 233
Virgin Atlantic Challenger II (motorboat), 230
Virus, 188, 373
VisiCalc, 228
Vitamin(s), 373, 390
Vittoria (battle), 459
Vittorio Veneto (battle), 464
Vivaldi, Antonio, 291
V-J Day, 471
Vladimir I, 314
Vlaminck, Maurice de, 273
Voice of America, 537
Voiturette, 219
Volta, Alessandro, 211, 399, 407
Voltaire (satirist), 240
von Eschenbach, Wolfram, 235
von Falkenhayn, Erich (Gen.), 462
von Furstenberg, Diane, 432
von Gluck, Christoph Willibald, 294
von Heidenstam, Verner, 250
von Kleist, Heinrich, 282
Vonnegut, Kurt, 261
von Weber, Carl Maria, 294
von Webern, Anton, 300
von Wrangel, Baron Ferdinand, 15
Voskhod 1 & 2 (space capsules), 21
Vostok I (capsule), 188
Voting Rights Act, 132
Voyager (airplane), 223
Voyager 1 & 2 (space probes), 189–190
Vuillard, Édouard, 273
Vulcan (ship), 182
W. & R. Chambers Ltd., 546
Wagner, Otto, 357
Wagner, Richard, 295
Wagner Act, 160
Wagram (battle), 459, 482
Waldheim, Kurt, 129, 140
Walker, Alice, 262
Walker, Moses Fleetwood, 594
Waller, Edmund, 238
Wallpaper, 439
Wall Street, 157
Wal-Mart Stores, 154
Walpole, Horace, 240
Walsh, Don (Lieutenant), 20
Walt Disney (motion picture company), 556, 569
Walton, Sam Moore, 154
Wandiwash (battle), 481
Warburton, Peter Egerton, 13
Warby, Kenneth, 229–30
War crimes, 139
Warhol, Andy, 277
Warner Bros. Pictures, 556
War of the Worlds, The (radio broadcast), 537
Warplanes, 500–02
Warren, Robert Penn, 259
War(s) of 1812, 33, 57, 450; American Civil War, 143, 155, 451, 460–61; American Revolution, 32, 56, 154, 450, 454–55, 481–82; from ancient Greece to the modern world, 447–53; Arab–Israeli, 70, 71, 452–53; with Austria, 457; of Austrian Succession, 56, 108; Chinese Civil War, 466–68; civil, *see* Civil war(s); of Conquest (Alexander the Great), 476; of Devolution, 449; *Crimean*, 61, 451; of the Diadochi, 95, 447; *Dutch*, 54, 449; *1812*, 450, 483; ending in Bulgaria, 464; English Civil War, 317, 449, 453–54, 480; famous battles, 475–89; French and Indian, 31, 56, 450; Franco-Prussian, 62; French Revolution and Revolutionary Wars, 450, 456–57, 482; *Gallic*, 99, 448; of the Grand Alliance, 54, 449; of Greek Independence, 451; Hundred Years, 51, 449, 479, 480; of independence in Mozambique, 74; for Independence of Angola, 74; Iran–Iraq, 71, 453, 489; Italo-Ethiopian, 452; Italo-Turkish, 451; King George's, 31, 450; King William's, 31; Korean, 452, 471–72; Latin American, 50; Macedonian, 95, 99, 448; of Marcomanni and Quadi, 448; Messenian, 94, 447; Mexican, 34, 451; Napoleonic, 42, 57, 458–59, 482–83; Opium, 84; of the Pacific, 43, 451; Peloponnesian, 94, 447, 476; Peninsulas, 57; Persian, battles of, 475–76; of the Polish Succession, 450; Punic, 45, 99, 448; of Religion, 316, 449; between Rome and Tarentum, 447; of the Roses, 449, 479; Russian Revolution and Civil War, 465–66; Russo-Japanese, 62, 86, 451; Russo-Polish, 63, 452; Russo-Turkish, 54, 107, 449; Sacred, 95, 447; Samnite, 99, 447; Seven Years, 56, 83, 110, 450; Sikh, 84, 451; Sino-Japanese, 85, 88, 451, 452; Spanish-American, 36, 451; Spanish Civil War, 65, 452, 486; of Spanish Succession, 54, 450; Syrian, 447; Texas (of independence), 33; Thirty Years, 53, 316, 321, 449, 480; of the Triple Alliance, 42, 451; Tripolitan, 32, 108; Trojan, 447; of the Vendee, 143, 456; Vietnam, 39, 156, 452, 472–75, 488; Vietnamese-Cambodian, 489. Worldwide. *See* World War I; World War II
Warsaw Treaty, 118
Warsaw Uprising, 144, 470
Warship, 198, 203, 497–500
Washington, George, 454–455
Washington Post Co., 154
Wasp (vessel), 514
Wasserstein, Wendy, 289
Watches, 185, 200, 433
Water hoses, 180
Waterloo (battle), 459, 483
Water Quality Act (U.S.), 423
Water-supply system, 177
Waterwheel, 178, 179
Watteau, Jean-Antoine, 268
Wattignies (battle), 457, 482
Waugh, Evelyn, 258
Wave theory of light, 406, 407
Weapons, 224
Weaving loom, 177
Weber, Max, 274
Weber, Wilhelm, 214
Webster-Ashburton Treaty, 33, 116
Wedekind, Frank, 286
Weirton Steel Works, 167
Weiss, Peter, 288
Welding technique, 187
Weldon, Fay, 262
Wells, H.G., 251
Welsh Calvinistic Methodist Church, 321
Welty, Eudora, 259
Werfel, Franz, 255, 287
Werner, Zacharias, 282
Wesley, John, 323
West, Benjamin, 270
West, Dame Rebecca, 256
West, Nathaniel, 258
Western Union, 214
Westinghouse, George, 183, 212
Westinghouse Electric Co., 149
Westminster Confession, 321
Westmoreland, William (Gen.), 473
"Wet look" raincoats, 432
Weyand, Frederick C., 475
Weyerhaeuser Co., 150
Wharton, Edith, 249
Wheatstone, Charles, 214
Wheelbarrow, 193
Wheels, 177
Whiskey Rebellion, 32
Whistler, James Abbott McNeill, 271
White, Bill, 598
White, E.B., 257
White, Edward H.,II, 21
White, Patrick, 259
White, T.H., 259
Whitefield, George, 324
White Hill (battle), 480
White Lotus Rebellion, 143
White Plains (battle), 455, 481
White Russians, 466
Whitman, Charles, 442
Whitman, Walt, 247
Whitney, Eli, 181, 206, 491
Whittier, John Greenleaf, 245
Whittle, Frank, 222
Whitworth, Kathy, 610
Whole Booke of Psalms, 546
Wiene, Robert, 555
Wigs, 430
Wilde, Oscar, 249, 283
Wilder, Thornton N., 257, 287
Wilderness Act (U.S.), 422
Wilderness (battle), 461, 484
Wiles Group Ltd., 151
Wilhelm Gustoff (ship), 514
Wilhelm II, 464
Wilkes, Charles, 17
Wilkins, G. H., 16, 18
William I (king), 62
Williams, Roger, 322
Williams, Tennessee, 287
Williams, Wayne, 442
Williams, William Carlos, 253
Wills, William John, 12
Will to Power, The, 338
Wilson, August, 289
Wilson, Lanford, 289
Wilson, Woodrow, 164
peace plan, 464
Wilson cloud chamber, 408
Wincroft Hotel, 512
Windmill, 178, 179
Windshield washer, wipers, 219
Wine, 427
Wingfield, Walter, 611
Wittgenstein, Ludwig, 339
Wittstock (battle), 480
"Wobblies," 160
Wodehouse, P.G., 253
Wohler, Friedrich, 400
Wolfe, Thomas, 257
Women's Rights, 133–34
Women's World Cup (soccer), 604
Wood, Grant, 275
Woodblock printing method, 201
Woodcuts, 545
Woodstock Music Festival, 569

Woolf, Virginia, 253
Woolworth, Frank, W., 152
Woolworth & Co., 152
Worcestershire sauce, 428
Wordsworth, William, 241
Workingmen's Party, 144, 159
World Amateur Golf Council, 610
World as Will and Idea, The, 338
World Bank, 128
World Council of Churches, 317
World Football League (WFL), 585
World Health Organization (WHO), 128
World Hockey Association (WHA), 601
World News Roundup **(CBS network),** 537
World of Islam festival, 334
World Series (baseball), 591–93; **Little League,** 597
World Trade Center, 359
World War I, 63, 70, 108, 155, 158, 161, 168, 452, 462–64, 485–86
World War II, 65, 155, 168, 452, 468–71, 486–88
Woven paper, 202
Wright, Frank Lloyd, 357
Wright, Orville, 184, 221
Wright, Richard, 259
Wright, Wilbur, 184, 221
Wristwatches. See Watches
Wu Ching, 335
Wu Ti, 78, 334
Wycherley, William, 280
Wycliffe, John, 306
Wyeth, Andrew, 277
Wyeth Chemical Co., 151
Xenon, 400
Xenophon, 336
Xenophon (historian), 233
Xeres de la Frontera (battle), 478
Xerography, 186, 202
Xerox Corp., 150
X-ray, 184, 373, 408
Yachts, America's Cup Races, 619–20
Yat-sen, Sun, 466
Yazid, 106
Yeager, Charles, 223
Yeats, William Butler, 251, 286
Yellow Turban Rebellion, 143
Yi dynasty, 81
Yogurt, 428
York, Alvin, 464
Yorktown (siege), 455, 482
You Are There, 538
Young, Denton T. "Cy," 594
Young, John, 24
Yo-yo, 436
Ypres (battle), 462, 485, 492
Yuan Shi-kai, 87
Zaire, 75
Zama (battle), 477
Zamboni, 601
Zebra serial murders, 442
Zeno of Citium, 336
Zenta (battle), 480
Zevi, Sabbatai, 327
Ziggurat of Ur, 91
Zionism, 327
Zipper, 430, 437
Zirconium, 399
Zodiaç circle, 376
Zola, Émile, 249
Zoopraxiscope, 209–10
"Zoot suit," 436
Zurich (battle), 479
Zwingli, Huldreich, 315
Zworykin, Vladimir, 537